Isotopes of the Earth's Hydrosphere

V. I. Ferronsky • V. A. Polyakov

Isotopes of the Earth's Hydrosphere

Springer

V. I. Ferronsky
Water Problems Institute of the
Russian Academy of Sciences
Moscow
Russia

V. A. Polyakov
Isotope Laboratory
Institute of Hydrogeology and
Engineering Geology
Moscow
Russia

Every effort has been made to contact the copyright holders of the figures and tables which have been reproduced from other sources. Anyone who has not been properly credited is requested to contact the publishers, so that due acknowledgement may be made in subsequent editions.

The Work is a translation of the book in Russian "Isotopes of the Earth's Hydrosphere" (Isotopiya Gidrosfery Zemli), World Scientific (Nauchniy Mir), 2009, 632 p., ISBN 978-5-91522-139-9 by V. I. Ferronsky and V. A. Polyakov.

ISBN 978-94-007-2855-4 e-ISBN 978-94-007-2856-1
DOI 10.1007/978-94-007-2856-1
Springer Dordrecht Heidelberg London New York

Library of Congress Control Number: 2011946279

Printed on acid-free paper

Springer is part of Springer Science+Business Media (www.springer.com)

Preface

The results of experimental and theoretical studies of isotopic composition of natural waters of the Earth's hydrosphere are analyzed and summarized in this book. These studies represent a modern direction of science of the natural waters actively developed in hydrology, hydrogeology, oceanography, and climatology starting from the middle of last century. The book is a further development and improvement of the earlier published author's works: Cosmogenic Isotopes of the Hydrosphere, Nauka, Moscow, 1984; Isotopy of the Hydrosphere, Nauka, Moscow, 1983; Environmental Isotopes in the Hydrosphere, Nedra, Moscow, 1975 and Environmental Isotopes in the Hydrosphere, Wiley, Chichester–New York, 1982.

More than 25 years have passed since the above books were published. During that period, together with everyday application of isotopic content of water for solution of different problems in hydrology, new original fields of use of isotope tracers for scientific and practical purposes have appeared. Many of such applications found a place in this work, but the fundamentals of methods and interpretation of isotope studies, which were developed mainly in the second part of last century, were preserved in the book together with references of their authors.

The successful development of the isotope methods and practical application for natural water studies was in many respects obliged to the international cooperation of scientists from different countries. Its start was initiated by UNESCO in implementation of the International Hydrological Decade (1965–I974) Program which is continued until now. The other important form of cooperation was the International program on collection and publication of experimental data about concentration of the tritium, deuterium, and oxygen isotopes in precipitation, sampling of which was organized on the global network of the IAEA/WMO and on the national hydrometeorological stations. This program was initiated and headed by Isotope Hydrology Section of the IAEA which up to now continues this work including processing and publication of the data and organizing regular symposia and advisory group meetings and publication of proceedings.

Active role in research and cooperation on isotope hydrology in Russia belongs to the Water Problems Institute and Institute of Ecology (St. Petersburg branch) of the Russian Academy of Sciences, All Russian Research Institute in Hydrogeology and Engineering Geology and All Russian Geological Research Institute of the

Ministry of Natural Resources, Institute of Experimental Meteorology of the Russian Hydrometeorological Service, which organized and implemented the research, international cooperation, and national symposia.

In this book, the authors summarized large experimental material and data on isotopic content of atmospheric moisture, oceanic water, surface and subsurface continental waters. The theoretical fundamentals on the fractionation of isotopes in water and of interacting water and rocks are presented. The theoretical and experimental data on dating of natural waters using the cosmogenic and radiogenic isotopes are discussed. Based on isotope data and theoretical solution, the problem of the origin of the Earth and its hydrosphere is also discussed.

In preparation of a number of the book's chapters the following colleagues have taken part: Yu. B. Seletsky put significant contribution into the preparation of Chaps. 6–8; Chap. 12 was written together with V. V. Romanov, Chaps. 15 and 16 were prepared together with V. M. Kuptsov, Chap. 19 was written together with S. V. Ferronsky.

The authors gratefully acknowledge their co-workers A. F. Bobkov, V. S. Brezgunov, A. L. Cheshko, V. T. Dubinchuk, N. V. Isaev, B. V. Karasev, Yu. A. Karpychev, L. N. Kolesnikova, I. K. Morkovkina, V. V. Nechaev, N. V. Piatnitsky, L. V. Sal'nova, A. E. Tkachenko, L. S. Vlasova, T. V. Yakimenko B. C. for their fruitful cooperation while carrying out the isotope studies of natural water.

The book was published in Russian by financial support of the Russian Foundation for the Fundamental Research (Project No: 09-05-07 107).

Permission Acknowledgements

The authors are indebted to the Publishing House "Nauka" for the permission to use figures and tables from their publications "Isotopy of the Hydrosphere" (Ferronsky and Polyakov, Nauka, 1983) and "Cosmogenic Isotopes in the Hydrosphere (Ferronsky, Polyakov and Romanov, Nauka, 1984), and to the International Atomic Energy Agency for the permission to use figures and tables from their Proceedings of International Symposia on Isotope Hydrology problems".

Contents

Part II Cosmogenic Radioisotopes

Chapter 1
Introduction

The hydrogen and oxygen isotope ratios are unique characteristics of the matter of which natural waters are composed of and therefore are directly highlight their history. It is for this reason that the stable isotopes of hydrogen, oxygen, and tritium (the radioisotope of hydrogen) are in common use now for studying the dynamics of natural waters, their genesis, and hydrochemical effects resulting from their interaction with rocks.

Natural waters are complicated chemical solutions. They always contain a number of dissolved and suspended minerals and organic substances. The stable and radioactive isotopes of these substances provide information about the processes occurring in the hydrosphere. These isotopes, in common use for studying natural waters, are primarily those of carbon, sulfur, silica, and isotopes of the heavy elements of the uranium-thorium series.

According to their origin natural radioisotopes are usually divided into two groups: (1) the isotopes produced by nuclear reactions between cosmic rays and the elements of the atmosphere and rocks are called cosmogenic, and (2) the radioisotopes of the heavy elements of the uranium-thorium series, created at the time of nucleosynthesis, are called radiogenic.

The natural stable and radioactive isotopes most widely used in the investigation of natural waters are the stable isotopes of oxygen and hydrogen, the cosmogenic radioisotopes such as tritium, carbon-14, and some radiogenic isotopes of the uranium-thorium series. Besides these, the stable and radioactive isotopes of helium, carbon, sulfur, sodium, chlorine, argon, strontium, krypton, and others are in use.

Isotope studies of natural waters are concerned with the principles governing the distribution of the stable and radioactive isotopes in the hydrosphere. Such studies aim to estimate the factors which determine these principles and interpret the hydrological, hydrogeological, hydrochemical, oceanographycal, and meteorological processes involved on the basis of data on isotopic composition of the various elements in the waters.

The idea that isotopes of an element exist, was suggested by Soddy in 1910. In 1913, he proposed the term 'isotopes' for identifying the atoms of the same element but having different masses. The famous English physicians Thomson and Aston played an important role in the discovery and study of stable isotopes. In 1911, Thompson developed the method of parabolas for determination of the ratio

V. I. Ferronsky, *Isotopes of the Earth's Hydrosphere*,

DOI 10.1007/978-94-007-2856-1_1, © Springer Science+Business Media B.V. 2012

of the particle charge to its mass by mass spectra recorded on film. Based on this, using the apparatus designed in the Cavendish Laboratory, Thomson discovered the neon isotopes of mass number 22, which was reported by him in January 1913. In 1919, Aston proved the existence of the neon isotopes and soon found the isotopes of chlorine and mercury. Several years later, more than 200 isotopes of various elements were discovered, with the exception of those of hydrogen and oxygen which were then considered 'simple' elements.

In 1929, Giauqie and Johnston, applying the new techniques of research by absorption air spectra, discovered the isotopes of oxygen of mass number 17 and 18.

In 1931, Johnston, on the basis of the free electrons and protons rule developed from analysis of established isotopes, concluded that hydrogen isotopes with masses 2 and 3 should exist. Soon after this in 1932, Urey, Brickwedde, and Murphy, using the mechanism of isotopic fractionation, detected the Balmer lines of deuterium in the enriched spectrum of the gaseous hydrogen using the spectral method. A detailed history of isotope discovery and isotope studies was reported by Aston in his monograph (Aston 1942) and also by Trifonov with his coauthors (Trifonov et al. 1974).

Soon after the discovery of hydrogen and oxygen isotopes, studies were undertaken for the distribution of these isotopes in natural waters. The first works by Bleckney, Gould, Smythe, Manian, Urey, Brodsky, Vernadsky, Vinogradov, Tays, and Florensky were carried out in 1932–1942. They investigated river, lake, sea, rain, and ground waters. Thus, the first comparative data concerning hydrogen and oxygen isotope variations in natural waters were obtained. But many of these measurements were based upon density techniques, leading to an inadequate accuracy.

Only in the 1940s, when more perfect mass spectrometers were designed by Dempster and Nir, and especially after improvements by McKinny in 1950, did the studies of the isotopic composition of natural waters become more common. Numerous data on the first isotope studies of natural waters and the results of study of physical and chemical properties of the heavy water and methods of its investigation were presented in the work of Kirshenbaum (1951). The early mass spectrometric studies, carried out by Kirshenbaum, Graff, and Forstat in 1945, Silverman in 1951, Friedman in 1953, and Epstein and Mayeda in 1953, dealt with the distribution of deuterium and oxygen-18 in natural waters and other objects. These studies initiated systematic research of the natural principles governing the distribution of the stable isotopes of hydrogen and oxygen in the hydrosphere and their application to the solution of scientific and practical problems. Great credit for the organization and supervision of numerous studies on the hydrogeochemistry of hydrogen and oxygen in different countries must be given to such researchers of the old school as A. P. Vinogradov, K. Rankama, and H. Urey, who made important contributions to the development of the geochemistry of isotopes as a whole.

The theoretical and experimental principles of fractionation of the stable isotopes of hydrogen during the phase transitions of water in the liquid–vapor–solid body system and also during interaction of water with gases and rocks in the liquid–gaseous and liquid–rock systems were discovered later. This has led to many studies of great applied and scientific importance in recent years.

Among the general natural principles, it has been found that the water of the ocean, being the main reservoir of the hydrosphere, has a fairly homogeneous hydrogen and oxygen isotope composition. Thus Craig (1961b) proposed it as a standard, relative to which all other measurements of deuterium and oxygen-18 content have been carried out. It was found that the average isotopic composition of the ocean water remains constant over the geologic time. Slight regional deviations in the average isotope composition of oceanic masses of water are the result of the continuous evaporation of water from the oceanic surface, which continually enriches the upper layers with heavy isotopes. The isotopic composition of the ocean as a whole is maintained to be constant by the balance between evaporated water and precipitation (including continental runoff) and the persistent mixing of all oceanic waters. The experimental data on the distribution of hydrogen and oxygen isotope ratios in the ocean reinforce those of salinity distribution and provide a more reliable base for the study of the dynamics of water in the ocean on the global scale.

One problem related to the ocean, and being solved with the help of isotope data, is that of paleotemperature reconstruction. Considerable progress has been made both on the basis of the analysis of oxygen isotopes in carbonates of fossil marine fauna (deposited during the past 730 million years) and chert paleothermometry, developed in recent years by Epstein and his coworkers. Progress has also been made on the longer time perspective with the work on ancient geological epochs, including the Precambrian.

It should be pointed out that the isotope composition of the world ocean water and that of the pore water of oceanic sediments and basement rocks in particular, has not been studied sufficiently and there is plenty of scope for further research.

The isotopic composition of atmospheric moisture and surface continental waters indicates a great variation resulting from considerable latitudinal and altitudinal temperature variations in time. The principles of the temporal and spatial distribution of the isotopic composition in atmospheric moisture and surface waters, related to these factors, provide a basis for the interpretation of isotopic data and the elucidation of the condition of global and regional water circulation in nature. At present, comprehensive factual material concerning the isotope composition of atmospheric precipitation has been obtained for the whole world. These data were obtained in the course of systematic measurements of the deuterium and oxygen-18 concentrations in precipitation, sampled at the global International Atomic Energy Agency/World Meteorological Organization (IAEA/WMO) network of stations. They are published by IAEA and are available for use by specialists.

The picture of the distribution of the isotopic composition of groundwater is most indicative due to the complicated processes involved in the formation of groundwater in the past. The interpretation of the data on isotopic composition from the point of view of isotope fractionation provides the key to explaining the processes involved in the formation of groundwater over geologic time. The most important fact obtained from an examination of the isotopic composition of groundwater is that its formation in artesian basins has involved admixing, in different proportions, of the more recent meteoric waters with the ancient waters of marine origin. There is no conclusive evidence indicating the presence of juvenile water in groundwater. This is clearly demonstrated by investigations carried out in the regions of recent volcanism where, in all areas of the world, hydrothermal waters have been found to be similar to the waters of local atmospheric precipitation. However, it has been shown no less convincingly that in most of the investigated ore deposits of hydrothermal origin the majority of the original ore-bearing fluids are meteoric origin.

The work which is undertaken by the Isotope Hydrology Section of the IAEA in inter-comparison of the main isotope standards plays an important role in unification of procedure for presentation of the isotope content of hydrogen and oxygen and some other elements of natural objects. In this work the leading mass spectrometry laboratories from countries take parts. The numerical characteristics of the main standards, expressed in terms of per mile (‰), of present use are as follows:

		δD (‰)	$\delta^{18}O$ (‰)
V-SMOW	Vienna standard of mean ocean water	0	0
SLAP	Standard of light Antarctic precipitation	−428	−55.5
GISP	Greenland precipitation ice standard	−189.8	−24.85
NBS-1	Distilled water of the Potomac River	−47.6	−7.94
NBS-1A	Water of snow from the Yellowstone Park	−183.3	−24.33
Standard of groundwater from the near Moscow upper carboniferous sediments used by the authors		−94.5	−13.08

In this book all the data of isotopic composition of hydrogen and oxygen presented are relative to the SMOW standard. The isotope results of earlier works of other authors cited in the book were recalculated with respect to the SMOW standard.

The story of cosmogenic isotopes dates back to 1912, when the Austrian physician Victor Hess discovered cosmic-rays by the effect of ionization of the air molecules. Later on it was found that the initial cosmic radiation of the solar, galactic, and meta-galactic origin of higher energy particles (mainly protons), as a result of interaction with atomic nuclei of the air, produce secondary and less-energetic radiation. This secondary radiation in nuclear reactions with nuclei of air atoms produces great number of cosmogenic radioactive isotopes. The tritium and radiocarbon are cosmogenic radioisotopes which have the most wide practical application in the study of natural water.

The search for tritium was started in 1931, when Johnston predicted its existence among the hydrogen isotopes. But the attempts of a number of scientists (Aston among them), undertaken over a number of years and aimed at its discovery in enriched hydrogen and natural water spectrum, failed. Rutherford, analyzing these studies, came to the conclusion that either the amount of this isotope was too small or it does not exist at all.

While the search for natural tritium was unsuccessful, Olefant, Harteck, and Rutherford using irradiation of the deuterium-bearing compounds with deuterons, detected the tracks of particles with mass number 3, which could have been related to both tritium and helium. In the same year, this discovery was confirmed by Dees. At first, it was assumed that both the isotopes discovered were stable. But Bonner found in 1938, that the mass of ^3He is less than that of ^3H. The assumption of radioactive transformation of the tritium to helium was confirmed by Alvarez and Cornog in 1939. They made the first estimates of the half-life of tritium to be 150 ± 40 days. Later, they found it to be several years. In 1940, O'Neil and Goldhaber, studying the production of tritium using the reaction $^6Li(n, \alpha)^3H$, determined the half-life of tritium to be 12.1 year.

The search for radioisotope ^{14}C was carried out in parallel with the search for tritium. In 1934, Kurie (and later, in 1936, Bonner and Brubaker, and Burcham and Goldhaber) irradiated air with neutrons in a chamber and discovered that radiocarbon is formed from nitrogen by the reaction $^{14}N(n, p)^{14}C$. The search for the latter radioisotope was undertaken only after the discovery of the secondary neutron flux produced in the atmosphere by cosmic-rays, which was made by Korff and Libby in 1937–1939. For this purpose Libby designed a special neutron counter, being elevated by a balloon up to a height of 50,000 ft. After these investigations, Korff in 1940, reported that the production of ^{14}C in the atmosphere is likely to proceed according to the (n, p) reaction on nitrogen under irradiation by the secondary neutrons. In 1941, Cornog and Libby showed that in the (n, p) reactions on nitrogen the fast neutrons produce tritium and the slow ones ^{14}C. In 1946, Libby detected radiocarbon in the atmosphere and came to the conclusion that the biosphere could be traced with it. Thus, radiocarbon dating was discovered. The half-life period of ^{14}C was not determined at once. At first it was assumed, by analogy with ^{35}S, to be equal to four months. Then it was considered to be 25,000 year and later was found to be equal to 5,730 year.

The search for natural tritium was continued simultaneously. The increasing sensitivity of the law background radiometric apparatus seemed to be advantageous. Tritium was detected almost simultaneously in 1950 by Faltings and Harteck in atmospheric hydrogen and in 1951 by Grosse, Johnston, Wolfgang, and Libby in lake water. They found concentrations of tritium in atmospheric hydrogen $^3H/H \approx 3.5 \times 10^{-15}$ and in lake water $^{3H}/H \approx 3.5 \times 10^{-18}$ which appeared to be in agreement, to order of magnitude, with those established later. A detailed history of the tritium and radiocarbon techniques may be found in the works by Libby (1967, 1973).

Tritium, being a perfect indicator of water motion, was immediately used by a number of researchers in the investigation of natural waters. But in the pre-nuclear test era only a few individual measurements of its content in natural waters were carried out. A large amount of tritium released into the atmosphere in the course of thermonuclear weapons tests increased its natural content by two orders of magnitude. Therefore, the work aimed at re-estimating the equilibrium content and natural concentrations in natural waters is presently being carried out.

As for radiocarbon, its content has been increased by the bomb, although by lesser amounts. The work by Münnich (1957) reported the first application of ^{14}C, to the dating of natural waters.

From the investigations carried out on the basis of data on tritium concentrations in atmospheric precipitation, gathered over a number of years at the IAEA/WMO network of stations, it was concluded that atmospheric air masses mix rather thoroughly. Valuable evidence highlighting the distribution of tritium in atmospheric moisture has been provided by observing the decline of bomb-tritium ejected before 1963 in atmospheric precipitation. These observations have thrown light upon the general principles of the temporal and spatial distribution on the global scale of tritium in atmospheric moisture, oceans, and surface waters. Similar studies for obtaining the radiocarbon distribution in natural waters were undertaken. Thus, a good experimental and theoretical basis involving the tritium and ^{14}C, isotope tools has been developed for the solution of a wide range of scientific and practical problems related to the investigation of dynamics of water in the hydrosphere.

The radiometric methods are used for measurement of the cosmogenic isotope radioactivity by means of the low-level beta-spectrometers with scintillation detectors or by a proportional counters with internal gas filling. Some other methods and techniques are developed for the measurement of natural concentrations of the cosmogenic isotopes.

Tritium concentration in natural objects is accepted to be expressed in the tritium units (TU). One TU corresponds to the content of 1 atom of tritium per 10^{18} atoms of protium. Up to recent times, it was assumed that the half-life period of the tritium decay is equal to 12.26 year. In September 1979, a group of experts gathered in the IAEA, Vienna has discussed new experimental data of the National Bureau of Standards (NBS), USA and accepted new value of the half-life period of the tritium decay equal to 12.430 year. The new decay constant $\lambda = 0.55764$ year^{-1} corresponds to the above value of the half-life period of tritium decay (standard NBS SRM-4526 C) and also 1 TU $= 7.088$ counts/min $\times$ 1 kg^{-1} water or 1 TU $= 3.193$ pCi $\times$ кг$^{-1}$ of water. In 2000, as a result of the work over the IAEA program TRIC-2000, the half-life period of tritium decay was corrected to 12.32 year and the decay constant $\lambda = 0.05625$ year^{-1} (1 TU $= 0.119$ Bq per 1 kg of water).

Radioactivity of the other cosmogenic radionuclides is measured in the units of decay per minute on 1 kg or 1 t of water. The total activity of radionuclides is measured in curie (Ci) or in becquerel (Bq); 1 Ci is equal to 3.7×10^{10} counts/s or 3.7×10^{10} Bq.

Specific activity of radiocarbon in natural objects is measured in counts/min on 1 g of carbon. But this is not convenient for practice because it needs to know precise efficiency of the counting techniques. This form of expression of the ^{14}C content gives a rise to difficulty in obtaining of consistent data for different laboratories having non-identical measurements. In order to unify measurements, it was proposed to express radiocarbon contents in percents (or per miles) with respect to a standard of the modern carbon (Olsson 1979; Stuiver and Suess 1966). Because the modern natural objects have variable ^{14}C, specific activity, a standard of the radiocarbon is used for unification of the measurements. For this purpose the standard of NBS, USA is used which is the oxalic acid having radioactivity of 14.3 count/min on 1 g of carbon ($1.176 \times 10^{-10}\% \times {}^{14}C$). The standard of the modern carbon is defined as specific activity of ^{14}C, in wood in 1950, after correction on isotope fractionation and presence in the atmosphere of the 'industrial' carbon dioxide (Stuiver and Suess 1966). Such a definition of standard for the modern carbon corresponds to 0.950 (95%) of the oxalic acid activity of the NBS standard, i.e., 13.6 count/min on 1 g of carbon.

The radiocarbon activity in the studied sample expressed in (%, ‰) with respect to the standard of modern carbon is equal to

$$A = [(A_{sp} - A_b)/(A_{st} - A_b)] \times 100\% \; (1{,}000‰),$$

где A is the radioactivity of ^{14}C, in the sample (%, ‰); A_{sp}, A_b, A_{st} are the radioactivity in the sample, the background and the standard detected accordingly by a radiometric device.

In the literature devoted to radiocarbon studies, by analogy with mass spectrometric measurements, the data of radiocarbon in a sample relative to the standard of the modern carbon are reported. These measurements are presented in percent or in per mile and calculated by the formula

$$\delta^{14}C = [(A_{sp} - A_{st})/A_{st}] \times 100\% \; (1{,}000‰).$$

Fixation of the carbon oxide in natural objects is accompanied by the isotope separation which can lead to some isotope dilution or enrichment of samples (especially biogenic) by the ^{14}C isotope. The effect of isotopic fractionation in the studied samples is determined by isotopic content of stable carbon ($^{13}C/^{12}C$), which is measured by mass spectrometric techniques. As a rule, normalization of the radiocarbon content to the mean value of isotopic composition of wood is provided ($\delta^{13}C = -25‰$ relative to the Pee Dee Belemnite (PDB) standard). In this case the corrected value of the radiocarbon content $\Delta^{14}C$ is equal to

$$\Delta^{14}C = \delta^{14}C - 2(\delta^{13}C + 25)\left(1 + \frac{\delta^{14}C}{1000}\right) ‰.$$

One more expression for the radiocarbon in a sample, presented through $\nabla^{14}C$, is found in the literature. This value $\Delta^{14}C$ is corrected with respect to the atmospheric 'industrial' carbon dioxide without radiocarbon. It is assumed at correction of $\Delta^{14}C$ with respect to the industrial carbon dioxide that before 1890 Г correlation factor was equal to 0 and at 1950 Г it has reached about 10%. If one takes a linear dependence of the CO_2 content in the atmosphere since 1890 through 1950, then the error of $\Delta^{14}C$ approximation will be negligible. As an example, of such correlation of the radiocarbon content, registered in the fragments of arctic mammals, we present the results obtained by Tauber (1979).

Year of mammal death	$\delta^{13}C$, (‰) PDB	$\delta^{14}C$, with age correction (‰)	$\Delta^{14}C$, (‰)	$\nabla^{14}C$, (‰)
1886	−16,1	−41 ± 6	−58	−58 ± 6
1915	−10,5	−41 ± 6	−73	−71 ± 6
1932	−14,5	−36 ± 6	−56	−52 ± 6
1931	−17,2	−44 ± 6	−59	−55 ± 6
1750 ± 30	−13,0	−39 ± 8	−62	−62 ± 6
1750 ± 50	−12,6	−37 ± 6	−61	−61 ± 6

Note that the values $\delta^{14}C$, and $\Delta^{14}C$, are defined in ‰ or % units, depending on the scale of the effect.

Some information related to other cosmogenic isotopes in the hydrosphere like ^{7}Be, ^{10}Be, ^{22}Na, ^{26}Al, ^{32}Si, ^{32}P, ^{33}P, ^{36}Cl, ^{37}Ar, ^{39}Ar is presented in the book. Investigation of the regularities on distribution of these isotopes in natural waters is mainly limited by technical difficulties of sampling, concentration and measurement of the corresponding samples.

The radiogenic isotopes of the uranium-thorium series, the lifetime of which is comparable to the age of the hydrosphere and the Earth as a whole, have, up to recent times, been used mainly for the purposes of geochronology. But the studies of their distribution in natural waters have shown them to be valuable indicators of the processes involved in the motion of natural waters, both on global and regional scales, for a large period of geologic time. The most fruitful in this respect was the trend, related to application of the effect of the disequilibrium ratio of $^{234}U/^{238}U$ in natural waters, which was discovered in 1953 by Cherdyntsev and Chalov (Cherdyntsev 1969; Chalov 1975).

The application of the disequilibrium of uranium and other heavy radioactive elements being rapidly delivered seems to be advantageous for discovering the main periods of evolution of the hydrosphere, and in the solution of many practical problems.

In this book an attempt has been made to generalize the published works available, including the author's own works, related to the examination and application of the environmental isotopes in the hydrosphere. The authors decided primarily to show the general principles of the distribution of stable and radioactive isotopes in the hydrosphere. Based on this, the possibility of solving the most important and complicated problems associated with the study of natural waters is illustrated. The applicability of the isotope techniques for the solution of practical problems in hydrogeology and hydrology, posed in the course of examination and management of water resources, is shown. Based on the quantitative data of the isotope composition of natural waters together with those of meteorites and lunar rocks, the most complicated problem that of the formation of the Earth and its hydrosphere, is discussed. Proceeding from this analysis and also using recently obtained theoretical solutions in the field of Jacobi's dynamics and dynamics of the Earth, the authors have shown that the Earth's hydrosphere is of condensational meteoric origin. The problem of the origin of the hydrosphere is naturally considered together with the problem of the origin of the Earth. In view of this, it has been found that the envelope structure of the Earth results from its chemical differentiation, which took place at its protoplanetary stage of evolution. This conclusion may lead to new discussion and further study, aimed at the solution of this fundamental problem.

Part I
Stable Isotopes

Chapter 2
Isotope Geochemistry of Natural Waters

2.1 Some Properties of Waters and Solutions

From the physical point of view, all natural waters are solutions of various inorganic and organic compounds. They also contain gases, colloidal and suspended particles of organic and inorganic origin, and many species of microorganisms. Thus, when studying the dynamics and phase transition of natural waters which result in their isotopic fractionation, one should not use the physicochemical constants for distilled water but rather those for solutions of given composition and concentration, which are features of the given type of natural water.

In a general sense, phase transitions in a one-component system are represented by the Clapeiron–Clausius equation

$$\frac{\partial p}{\partial T} = \frac{\Delta H}{T \Delta V} \tag{2.1}$$

where T and p are temperature and pressure of the phase transition; ΔV is the molal change in volume for this; ΔH is the molal heat of this transition (i.e., melting, vaporization, etc.).

Equation 2.1 is valid only for the reversible processes. Plotted on the (p, T)-diagram, the equation describes the two-phase equilibrium (Dreving and Kalashnikov 1964). For most of the phase transitions, we have $\partial p/\partial T > 0$. The only exceptions are the phase transition for the ice–water system and also those for a small number of other substances (such as bismuth, gallium, and germanium). The schematic thermodynamic diagram of water in terms of the p and T variables is given in Fig. 2.1.

The curves TA and TC corresponds to the pressure of vapor which is in equilibrium with water and ice respectively and the curve TB shows the dependence of ice melting temperature on pressure. The curve TB has a negative gradient, which means that the melting temperature decreases with an increase in pressure ($\partial p/\partial T < 0$). The point T is called the triple point of the thermodynamic equilibrium. It corresponds to the pressure and temperature at which coexistence of the three phases (solid–water–vapor) is possible. For water, the triple point pressure is equal to 609 Pa and the temperature is +0.0076°C; the ice melting temperature under pressure 1.013×10^5 Pa

V. I. Ferronsky, V. A. Polyakov, *Isotopes of the Earth's Hydrosphere*,
DOI 10.1007/978-94-007-2856-1_2, © Springer Science+Business Media B.V. 2012

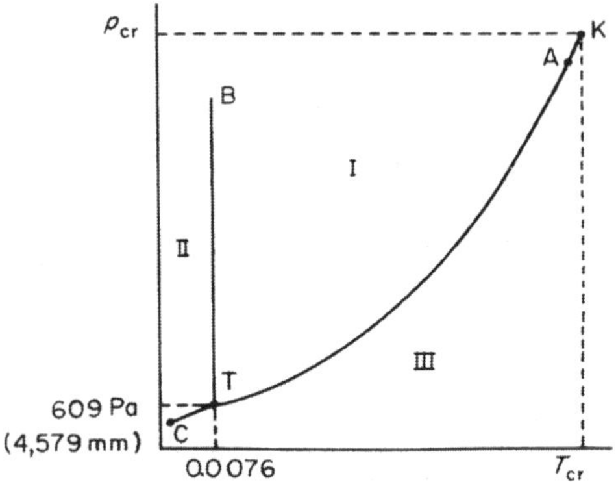

Fig. 2.1 Schematic thermodynamic diagram of water (p and T-diagram): ***I*** is the domain of water existence in the liquid phase; ***II*** is the domain of ice existence; ***III*** is the domain of water existence in the gaseous phase

is 0°C. The dash represents the thermodynamic equilibrium between vapor and super-cooled water since ultra-distilled water permits super-cooling down to −50°C.

The point K in the diagram is called the critical point. It corresponds to the critical values of temperature and pressure. For water, the critical value of pressure is $p_{cr} = 2.21 \times 10^7$ Pa and the critical temperature is $T_{cr} = 374.2°C$. At these values, the water has density of 0.324 g/ml (the critical density). The critical point on the graph TK corresponds to the temperature and pressure at which the difference between the liquid and the vapor completely disappears. The critical point for liquids was discovered by Mendeleev in 1860. At the critical point, the graph TK pictured in the p and T coordinates breaks down. Indeed, putting the values $T = T_{cr}$, $H_{gas} = H_{liquid}$, and $V_{gas} = V_{liquid}$ into Eq. 2.1, we see that the left-hand side of the equation becomes indeterminate, which indicates that there is a discontinuity of the function $p = f(T)$. All the properties of the coexisting phases are the same at the critical point. Moreover, for temperatures $T > T_{cr}$, water cannot exist in a liquid and vapor state simultaneously at any pressure. In contrast to the liquid phase, the solid phase (ice) can theoretically exist at temperature $T > T_{cr}$, but the phase transition of vapor into solid for these temperatures can occur only at much distilled water, that is for a one-component system.

Natural waters from the thermodynamic point of view are complicated multi-component solutions. Therefore, the coexistence of water in the liquid phase in equilibrium with vapor (more precisely with the abovementioned critical phase) is possible in groundwater at depths of scores of kilometers. Figure 2.2 is a thermodynamic diagram of water at high pressures.

Fractionation, migration, and accumulation of the stable and radioactive isotopes which are components of water molecules—and also those of substances—contained in water solutions are dependent on the properties of the water solutions. These properties are in turn functions of the chemical composition of the given solution, its temperature and pressure.

A solution is a homogeneous system (in solid, liquid, or gaseous state) which consists of two or more substances (Karapetyan 1953). From the thermodynamic viewpoint, all the components of a solution are equivalent and therefore the division of the components into solvent and solute is relative.

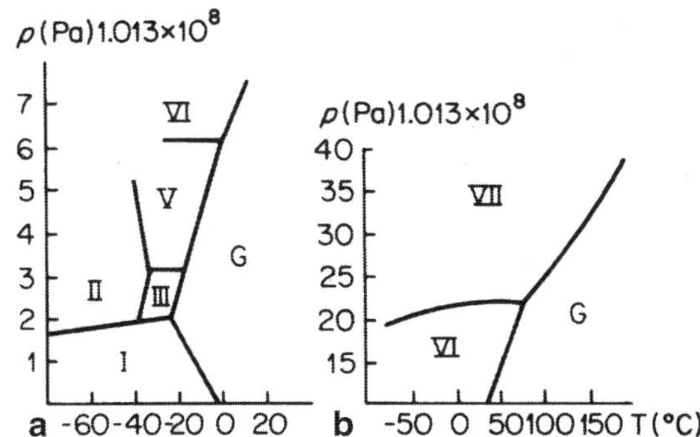

Fig. 2.2 Water state diagram at high (**a**) and superhigh (**b**) pressures for ice modification densities (in g/ml): **I** corresponds to 0.92; **II** corresponds to 1.12; **III** corresponds to 1.03; **V** corresponds to 1.09; **VI** corresponds to 1.13; **VII** corresponds to 1.46; *G* is the domain of liquid water existence

2.1.1 The Notion of Activity and the Activity Coefficient

The notion of activity in thermodynamics of real solutions was introduced by Lewis in 1905–1908. When calculating the thermodynamic state of real systems, the analytical concentration of a given component is replaced by its activity equal to the concentration of this component in an ideal system. The introduction of the value of activity makes it possible to describe the behavior of the real solutions using the thermodynamic equations obtained for ideal system. In this case, the functional dependence between the parameters remains unchanged after the replacement of the substance concentration in the solution by its activity. This made it possible to apply the ideal solution relationships to the real systems.

The activity is a function of the concentration c and differs from it by some factor γ which is called the activity coefficient:

$$a = \gamma c. \tag{2.2}$$

It is known from the theory of solutions that the chemical potential of a substance in an ideal solution can be given by the following equation:

$$\mu_{\text{ideal}} = \mu_0 + RT \ln c, \tag{2.3}$$

where μ_{ideal} is the chemical potential of the ideal solution; μ_o is the standard chemical potential; T is the absolute temperature; R is the gas constant.

For the real system, Eq. 2.3 must be rewritten so that the value of the concentration is replaced by the activity

$$\mu_{\text{sol}} = \mu_0 + RT \ln \gamma c = \mu_0 + RT \ln a, \tag{2.4}$$

where, μ_p is the chemical potential of the real solution.

It is evident from Eqs. 2.3 and 2.4 that the activity can be considered as a function of the thermodynamic properties of the solution:

$$\ln a = \frac{\mu_{\text{sol}} - \mu_0}{RT} \tag{2.5}$$

Here the activity coefficient can be defined as a function of the difference between the chemical potentials of the substance in real and ideal systems.

In fact, from Eq. 2.4, it is evident that

$$\mu_{sol} = \mu_0 + RT \ln \gamma c = \mu_0 + RT \ln \gamma + RT \ln c = \mu_{ideal} + RT \ln \gamma. \quad (2.6)$$

Then

$$RT \ln \gamma = \mu_{sol} - \mu_{ideal}, \quad (2.7)$$

$$\ln \gamma = \frac{\mu_{sol} - \mu_{ideal}}{RT} \quad (2.8)$$

or

$$\gamma = \exp(\Delta\mu/RT). \quad (2.9)$$

Since the activity coefficient is a function of the thermodynamic properties of a system, the arbitrary standard thermodynamic state can be used for their estimation. One should keep in mind that during the calculations, all the values must be referred to the one standard thermodynamic state both for the solvent and for the solute.

2.1.2 *The Relationship between Solvent and Solute Activity*

It is known (Izmaylov 1966) that the total differential of the isobaric–isothermal potential G (Gibbs' free energy) is equal to

$$dG = -Sdp + VdT + \sum \mu_i dn_i, \quad (2.10)$$

where, S is the entropy of the system; T, V, p are the temperature, volume, and pressure respectively; μ_i is the chemical potential of the ith component of the system; n_i is the molar quantity of the ith component.

At constant p and T, the equation has the form

$$dG = \sum \mu_i dn_i.$$

For the system in equilibrium, $dG = 0$, so $\sum \mu_i \, dn_i = 0$. The total differential of the function G for constant p and T and variable μ_i and n_i is

$$dG = \sum n_i d\mu_i + \sum \mu_i dn_i. \quad (2.11)$$

When the system is in equilibrium, the right-hand equation is equal to zero and since

$$\sum \mu_i dn_i = 0,$$

then

$$\sum n_i d\mu_i = 0. \tag{2.12}$$

Equation 2.12 is called the Gibbs' equation and for binary systems it can be written as

$$n_1 d\mu_1 + n_2 d\mu_2 = 0. \tag{2.13}$$

Subscripts 1 and 2 denote the solvent and the solute, respectively. Dividing Eq. 2.13 by $n_1 + n_2$, we obtain an expression related to the molar fractions of substance

$$N_1 d\mu_1 + N_2 d\mu_2 = 0. \tag{2.14}$$

Substituting in Eq. 2.14, the expression for the chemical potential $\mu = \mu_0 + RT$ $\ln c$, we get the equation for the ideal solutions, which is called the Duhem–Margules

$$N_1 d \ \ln c_1 + N_2 d \ \ln c_2 = 0.$$

Substituting the activity $a = \gamma c$ for c, we finally derive the Gibbs–Margules–Lewis equation

$$N_1 d \ \ln a_1 + N_2 d \ \ln a_2 = 0, \tag{2.15}$$

where a_1 is the activity of the solvent; solutions; a_2 is the activity of the solute.

Equation 2.15 describes the relationship between the activities of the solvent and the solute. It is commonly used in physical chemistry for determining the activity coefficients of components of real solutions by indirect methods.

2.2 Water Vapor Pressure over the Water and Solution

In practice, the decisive role in the processes of fractionation of the isotopic species of water (molecules with different masses) is played by the phase transitions in the ice–water–vapor system, but for the quantitative analysis of the process of vaporization one should take into account that the vapor pressure of water (the solvent) over solutions is unequal to the pressure of the pure aqueous vapor at the identical temperature value T. For ideal solutions, at constant temperature, the vapor pressure of the solvent for all concentrations follows Raule's law which is:

$$p_1 = p_1^0 N_1, \tag{2.16}$$

where p_1 is the vapor pressure over the solution; p_1^0 is the vapor pressure over the distilled water; N_1 is the concentration of solvent (the molar fraction).

For real solutions, one should replace N_1 in Eq. 2.16 by the value of the solvent activity $a = \gamma N_1$. Then Raule's law for real solutions is represented by the equation

$$p_1 = p_1^0 a_1. \tag{2.17}$$

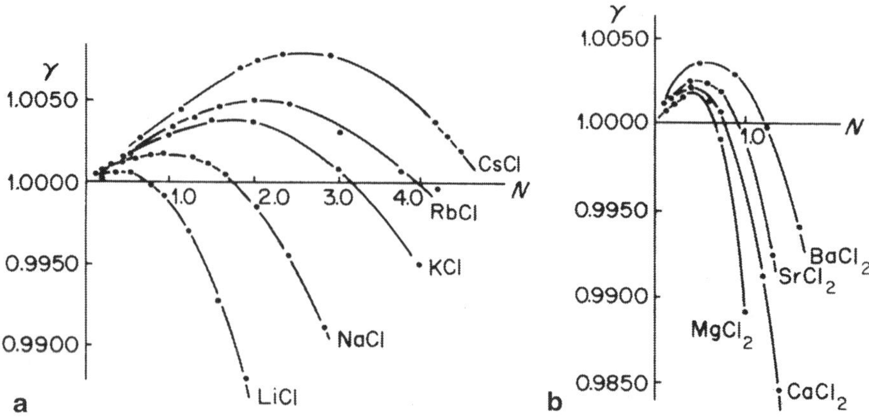

Fig. 2.3 The dependence of the activity coefficient of water γ in some aqueous electrolytes on the molality concentration of the solution N for cations Li–Cs (**a**) and Mg–Ba (**b**)

From Eq. 2.17, it is evident that the activity of water in the solution can be expressed by the ratio of water vapor pressure over the solution p_1 and the vapor pressure over distilled water p_1^0 (at the same temperature T):

$$a_1 = \frac{p_1}{p_1^0},$$

or,

$$\gamma N_1 = \frac{p_1}{p_1^0}. \tag{2.18}$$

Here γ is said to have a negative deviation from Raule's law if, in the real solution, $\gamma < 1$, and it has positive deviation if $\gamma > 1$.

The dependence of the activity coefficient of water in some aqueous electrolytes on the molal concentration of the solution[1] is shown in Fig. 2.3.

It can be seen from the diagrams that the aqueous electrolytes exhibit both positive and negative deviations from Raule's law depending on concentration of the solutions. It is typical that in the series of cations Li→Cs and Mg→Ba, there is a tendency towards positive deviations from Raule's law (the curves' maxima are shifted to the domain of greater values of concentration). Samoilov (1957) explains this phenomenon in terms of the difference in hydration of the ions in water solutions. According to his theory, ions of electrolytes affect the translational motion of the surrounding molecules of water in the solution. Ions with great heat of hydration (Li, Na, Mg, Ca) weaken the translational motion in the hydrate shells, which results in the decrease of activity coefficient. In contrast, the ions with negative hydration

[1] The molality of the solution is the quantity of gram-molecules of the substance dissolved in 1,000 g (55.55 moles) of water

Table 2.1 Vapor pressure decrease (in Pa) over water solutions of some salts

Salts	Concentrations (in moles/l of water)						
	0.5	1.0	2.0	3.0	4.0	5.0	6.0
$CaCl_2$	2,261	5,293	12,635	22,158	32,120	42,495	–
K_2CO_3	1,915	4,123	9,044	14,032	20,216	27,787	34,384
KCl	1,523	3,245	6,384	9,855	13,420	17,091	20,243
$MgCl_2$	2,234	5,187	13,367	24,379	36,841	50,141	–
Na_2CO_3	1,902	3,631	7,116	10,667	14,763	–	–
NaCl	1,636	3,352	6,929	10,640	14,763	19,019	23,475
Na_2SO_4	1,676	3,325	6,504	9,863	–	–	–

(with loss heat hydration), disturbs the water structure in the hydration shells, which intensifies the translational motion of water molecules and results in an increase in the activity coefficient (such ions are K, Rb, Cs, Ba) .

It is noteworthy that the process of cation hydration in water solution leads to the isotopic fractionation of hydrogen and oxygen in their hydrated shells (Sofer and Gat 1972, 1975). For instance, ^{18}O is concentrated in the hydrated shells of ions with great heat of hydration (fractionation coefficient $\alpha \geq 1$) and those with negative hydration (using Samoylov's notation) with, as a rule, $\alpha \leq 1$ (Sofer and Gat 1972; Taube 1954).

The isotopic fractionation of hydrogen and oxygen upon the hydration of ions is of some importance to the isotopic fractionation of natural waters.

Table 2.1 shows the data on decreases in vapor pressure over water solutions of some salts contained in natural waters. The amount of vapor pressure decrease over solutions of given concentrations at 100°C is given and can be compared with the pressure of pure water vapor (1.01×10^5 Pa) at the same temperature.

In studying the fractionation of hydrogen and oxygen isotopes in the process of the vaporization of water of different salinity, it is necessary to take into account the decrease of pressure of the saturated vapor over the salt solution. In fact, the change of the salinity in surface reservoirs during the vaporization process will lead to decreases in the saturated vapor pressure over their surface. However, the total pressure of water vapor, as a rule, depends not only on the total area of vaporization in the given region but also on the atmospheric moisture which is transported from the other regions. In this case, the isotopic equilibrium in the closed basin will depend on the salinity of its waters, temperature of evaporation, the humidity, and the isotopic composition of the water in the basin and of the atmospheric vapor.

2.3 Physicochemical Foundations of Isotope Separation

In considering the environmental phenomena and processes of geochemistry and cosmochemistry based on studies of isotopic composition of chemical elements, one makes the assumption that the initial isotopic composition of any chemical element is a natural constant (Rankama 1963). Then, as a result of natural physicochemical processes, the isotopic composition of any element changes through geological history.

If the chemical properties of individual elements depend mainly on their electronic configurations, then the differences observed in the physicochemical processes, where the isotopes and the isotopic varieties of molecules take part, are due to the differences in the masses of the isotopes and their nuclear properties. As an illustration, consider the typical reaction of the isotopic exchange which can be written in the form (Godnev 1956):

$$aC_1 + bD_2 \leftrightarrow aC_2 + bD_1, \tag{2.19}$$

where C_1, C_2, D_1, D_2 are molecules of different isotopes of the same element.

The following exchange processes serve as an example of such reaction:

$$^{16}O_2 + 2H_2{}^{18}O \leftrightarrow {}^{18}O_2 + 2H_2{}^{16}O, \tag{2.20}$$

$$^{16}CO_2 + 2H_2{}^{18}O \leftrightarrow {}^{18}CO_2 + 2H_2{}^{16}O, \tag{2.21}$$

The thermodynamic expression for the equilibrium constant of these reactions is

$$K = \frac{Q^a_{C_2} Q^b_{D_1}}{Q^a_{C_1} Q^b_{D_2}} = \frac{(Q_{C_2}/Q_{C_1})^a}{(Q_{D_2}/Q_{D_1})^b}, \tag{2.22}$$

here Q_{c1}, Q_{c2}, Q_{D1}, Q_{D2} are partition functions of isotopic molecules.

According to quantum, statistics for thermodynamic functions, the expression for the partition function of the individual molecules is, in slightly simplified form:

$$Q = Q_1 Q_r Q_v \exp\frac{E_0}{kT} \tag{2.23}$$

where Q_1 is the translational partition function of the molecule; Q_r is the rotational partition function of the molecule; Q_v is the oscillation partition function of the molecule; $\exp(-E_0/kT)$ is a factor of the zero-point energy of the molecule.

The partition function of the molecule's translational motion during the isomolar reactions is given by

$$Q_1 = \frac{(2\pi MkT)^{3/2}}{h^3}, \tag{2.24}$$

where M is the mass of the molecule; K is the Boltzmann's constant; h is the Plank's constant; T is the Kelvin's temperature.

A simplified expression for the partition function of the rotational motion of molecules is

$$Q_K = \frac{8\pi^2 IkT}{\sigma h^2}, \tag{2.25}$$

where I is the moment of inertia, that is the sum of the products of the individual atom masses constituting the molecule and its squared distances up to the axis of rotation of the molecule; σ is the symmetry number; other notations are as in Eq. 2.24.

The partition function of the oscillation motions of the molecule can be written in the following simplified form:

$$Q_v = \left[1 - \exp\left(-\frac{hv}{kT}\right)\right]^{-1} \tag{2.26}$$

where v is the frequency of the atom oscillation in the molecule.

It follows from Eq. 2.22 that the equilibrium constant of the reaction expressed by Eq. 2.19 is determined by the ratio of the partition function of the isotopic molecules C_2C_1 and $D_2\,D_1$. If we denote the partition function for each isotopic molecule pair by Q_1 and Q_2 then (at average temperatures) we obtain

$$\begin{aligned}\frac{Q_2}{Q_1} &= \exp\left(-\frac{\Delta E_0}{kT}\right)\frac{(Q_1)_2(Q_r)_2(Q_v)_2}{(Q_1)_1(Q_r)_1(Q_v)_1} \\ &= \exp\left(-\frac{\Delta E_0}{kT}\right)\frac{\sigma_1}{\sigma_2}\frac{I_2}{I_1}\left(\frac{M_2}{M_1}\right)^{3/2}\left(\frac{1-\exp\left(-hv_1/kT\right)}{1-\exp\left(-hv_2/kT\right)}\right),\end{aligned} \tag{2.27}$$

where σ_1, σ_2, I_1, I_2, M_1, M_2, v_1, v_2 are the symmetry numbers, moments of inertia, masses (molecular weights), and frequencies of the internal oscillations of the isotopic molecules, respectively.

The factor $\exp\,(-\Delta E_o/kT)$ takes into account the difference in the zero-point energies of the two isotopic molecules C_2C_1 and D_2D_1.

The difference in the zero-point energies of oscillations may be given

$$\Delta E_0 = \frac{hv_2}{2} - \frac{hv_1}{2}. \tag{2.28}$$

If we introduce the new oscillations

$$u_1 = \frac{hv_1}{kT} \tag{2.29}$$

and

$$u_2 = \frac{hv_2}{kT}, \tag{2.30}$$

then Eq. 2.27 can be rewritten in the form

$$\frac{Q_2}{Q_1} = \frac{\sigma_1}{\sigma_2}\frac{I_2}{I_1}\left(\frac{M_2}{M_1}\right)^{3/2}\left(\frac{\exp\left(-u_2/2\right)}{1-\exp\left(-u_2\right)}\right)\left(\frac{1-\exp\left(-u\right)}{\exp\left(-u_1/2\right)}\right). \tag{2.31}$$

For polyatomic molecules, the expressions for the partition functions have the same form as of Eqs. 2.24 and 2.25, but all the moments of inertia and frequencies of the molecules' oscillations should be taken into account. In this case, Eq. 2.27 can be written as

$$\frac{Q_2}{Q_1} = \exp\left(-\frac{\Delta E_0}{kT}\right)\frac{\sigma_1}{\sigma_2}\prod_n\left(\frac{I_{2n}}{I_{1n}}\right)^{1/2}\left(\frac{M_2}{M_1}\right)^{3/2}\prod_i\frac{1-\exp(-hv_1/kT\ )}{1-\exp(-hv_2/kT\ )}, \tag{2.32}$$

where $I_{11}, I_{12}, \ldots, I_{1n}$ and $I_{21}, I_{22}, \ldots, I_{2n}$; $\nu_{11}, \nu_{12} \ldots \nu_{1i}$ and $\nu_{21}, \nu_{22} \ldots \nu_{2i}$ are, respectively, the moments of inertia and the oscillation frequencies of the first and second isotopic molecules.

The difference in zero-point oscillation energies can be introduced in the form

$$\Delta E_0 = \sum_i \frac{h\nu_{2i}}{2} - \sum_i \frac{h\nu_{1i}}{2}. \tag{2.33}$$

Substituting Eq. 2.33 into Eq. 2.32, one obtains

$$\frac{Q_2}{Q_1} = \frac{\sigma_1}{\sigma_2} \prod_n \left(\frac{I_{2n}}{I_{1n}}\right)^{1/2} \left(\frac{M_2}{M_1}\right)^{3/2} \prod_i \left(\frac{\exp(-h\nu_{2i}/kT)}{1-\exp(-h\nu_{2i}/kT)}\right) \left(\frac{1-\exp(-h\nu_{1i}/kT)}{\exp(-h\nu_{1i}/kT)}\right) \tag{2.34}$$

Introducing the variables in Eq. 2.29 and 2.30, with substituting of $h\nu_{1i}$ and $h\nu_{2i}$ for $h\nu_1$ and $h\nu_2$, and putting them into Eq. 2.34, an analogy with Eq. 2.31 is obtained:

$$\frac{Q_2}{Q_1} = \frac{\sigma_1}{\sigma_2} \prod_n \left(\frac{I_{2n}}{I_{1n}}\right)^{1/2} \left(\frac{M_2}{M_1}\right)^{3/2} \prod_i \left(\frac{\exp(-u_{2i}/2)}{1-\exp(-u_{2i})}\right) \left(\frac{1-\exp(-u_{1i})}{\exp(-u_{1i}/2)}\right) \tag{2.35}$$

Using the Teller–Redlich rule (Redlich 1935), which relates the ratio of the products of the frequencies for each symmetry type of the two isotopes to the ratio of their masses and the moments of inertia, Eq. 2.35 can be rewritten as

$$\frac{Q_2}{Q_1} = \frac{\sigma_1}{\sigma_2} \left(\frac{M_2}{M_1}\right)^{3/2} \prod_i \frac{u_{2i}}{u_{1i}} \left(\frac{\exp(-u_{2i}/2)}{1-\exp(-u_{2i})}\right) \left(\frac{1-\exp(-u_{1i})}{\exp(-u_{1i}/2)}\right), \tag{2.36}$$

which, in the case of the diatomic isotopic molecule, may be simplified to the form

$$\frac{Q_2}{Q_1} = \frac{\sigma_1}{\sigma_2} \left(\frac{M_2}{M_1}\right)^{3n/2} \frac{u_2}{u_1} \left(\frac{\exp(-u_2/2)}{1-\exp(-u_2)}\right) \left(\frac{1-\exp(-u_1)}{\exp(-u_1/2)}\right), \tag{2.37}$$

where n is the number of isotopic atoms exchanged; M is their mass.

Equations 2.36 and 2.37 are used to calculate equilibrium constants for the isotopic exchange reactions of isotopic molecules. In order to carry out the calculations, the experimental data on the oscillation spectra frequencies, the molecules' masses, and their symmetry numbers are required. The theory for approximating equilibrium constants of isotopic exchange reactions was developed by Bigeleisen and Mayer (1947), Urey (1947), Tatevsky (1951), and other authors.

In calculating the equilibrium constants for such isotopic exchange reactions as in Eq. 2.19, the following function is introduced:

$$f = \frac{Q_2}{Q_1} \left(\frac{M_1}{M_2}\right)^{3n/2}. \tag{2.38}$$

Using Eq. 2.38, the expression for the equilibrium constant, Eq. 2.22 can be written in the form

$$K = \frac{(Q_{C_2}/Q_{C_1})^a}{(Q_{D_2}/Q_{D_1})^b} = \frac{[f(M_2/M_1)^{3n_1/2}]^a}{[f(M_2/M_1)^{3n_2/2}]^b}. \tag{2.39}$$

Let the number of exchangeable isotopic atoms in both molecules be equal, that is $n_1 = n_2$. Then

$$K = f_C^a / f_D^b. \tag{2.40}$$

Using the Bigeleisen–Mayer function and corresponding experimental data, an approximate value of the function is obtained

$$\ln f = \ln\frac{\sigma_1}{\sigma_2} + \sum_i \left(\frac{1}{2} - \frac{1}{u_i} + \frac{1}{e^{u_i} - 1}\right)\Delta u_i = \ln\frac{\sigma_1}{\sigma_2} + \sum_i G(u_i)\Delta u_i, \tag{2.41}$$

where $G(u) = 1/2 - 1/u_i + 1/(e^{u_i} - 1)$ is the Bigeleisen–Mayer function.

The value of the function $G(u)$ have been given in a number of studies (Bigeleisen and Mayer 1947; Godnev 1956).

The following is an example of a calculation of the equilibrium constant for the isotopic exchange reaction according to the method described by Godnev (1956).

Let us determine the equilibrium constant at $T = 600\ K$ for the following reaction of isotopic exchange,

$$C^{16}O_2 + 2H_2{}^{18}O \leftrightarrow C^{18}O_2 + 2H_2{}^{16}O.$$

Using Eq. 2.39, we can write

$$K = \frac{(Q_{C^{18}O_2}/Q_{C^{16}O_2})}{(Q_{H_2{}^{18}O}/Q_{H_2{}^{18}O})^2} = \frac{f_{CO_2}}{(f_{H_2{}^{16}O})^2} \tag{2.42}$$

With the help of the data on the function $G(u)$ (Godnev 1956) and for the corresponding values of the oscillating frequency of the molecules $C^{16}O_2$, $C^{18}O_2$, $H_2{}^{16}O$ and $H_2{}^{18}O$ for symmetry numbers of all the molecules $\sigma_1 = \sigma_2 = 2$, we may determine the parameters u_i and Δu_i (Table 2.2).

Putting the data given in Table 2.2 in Eq. 2.41, one obtains

$$f_{CO_2} = 1.076;\ f_{H_2O} = 1.026.$$

Finally, making use of the Eq. 2.42, one obtains the unknown isotopic exchange equilibrium constant for the system under investigation at $T = 600\ K$

$$K_{(600K)} = 1.023.$$

Similar methods to those developed for calculating equilibrium constants, based on the spectroscopic data, and are widely used for calculating the fractionation factors of the stable isotopes of light elements in reactions between the gaseous molecules

Table 2.2 Parameters of the isotopic exchange reactions of CO_2 and H_2O

ν_i, cm^{-1}	$\nu_i + \Delta u_i$	ν_i, cm^{-1}	u_i	Δ u$_i$	G_i
$C^{16}O_2$		$C^{18}O_2$		$C^{18}O_2$	
1,342	3.219	1,266	3.037	0.182	0.22128
667	1.600	657	1.576	0.024	0.12649
667	1.600	657	1.576	0.024	0.12649
2,355	5.650	2,320	5.566	0.084	0.32435
$H_2{}^{16}O$		$H_2{}^{18}O$		$H_2{}^{18}O$	
3,650	8.756	3,642	8.737	0.019	0.38574
1,595	3.826	1,590	3.814	0.012	0.26018
3,756	9.011	3,740	8.972	0.039	0.38864

(Richet et al. 1977) and in heterogenic systems (Urey 1947; Bigeleisen and Mayer 1947).

When u ranges from 0 to ∞, the value $G(u)$ ranges from 0 to 0.5. At high temperatures Δu tends to zero. Therefore, in Eqs. 2.38–2.41, we have $Q_2/Q_1 \approx \sigma_1/\sigma_2$ and

$$K{=}K_0 = \frac{(\sigma_1/\sigma_2)^a_C}{(\sigma_1/\sigma_2)^b_D}. \tag{2.43}$$

The limiting value of the equilibrium constant of the isotopic exchange reaction $K = K_0$ corresponds to the equally probable distribution of isotopes between the molecules of the system and is determined by their symmetry numbers. Further, in this case, the separation of the isotopes in the exchange reactions does not occur. In some studies (Bowen 1966; Galimov 1968a) concerned with the application of isotope techniques in geology, it is stated that the equilibrium constants for individual isotopic exchange reactions tend to unity with increasing temperature. This is not true if we consider general case of isotopic exchange, which may be represented by such a reaction as Eq. 2.19.

Thus, the limiting value of the equilibrium constant is equal to 2 for the isotopic exchange reaction $DCl + H_2O \leftrightarrow HCl + HD$. In fact, the symmetry numbers of the molecules DCl, HCl, HD are equal to unity and $\sigma_{H_2} = 2$, and hence $K_0 = 1 \times 2/1 \times 1 = 2$. For this reaction the equilibrium constant K increases with temperature and tends to the limiting value $K_0 = 2$ (Brodsky 1957). But the isotopic fractionation factor between the molecules of hydrochloric and hydrogen actually tends to unity with an increase in temperature.

As a second example, let us consider the reaction of isotopic exchange between ammonia and hydrogen:

$$NH_3 + HD \leftrightarrow NH_2D + H_2.$$

The limiting value of the equilibrium constant for this reaction is $K_0 = 1.5$. In fact,

$$K_0 = \frac{(\sigma_{NH_3}/\sigma_{NH_2D})}{(\sigma_{H_2}/\sigma_{HD})},$$

$\sigma_{NH_3} = 6, \sigma_{NH_2D} = 2, \sigma_{HD} = 1$, and hence $K_0 = (6 \times 1)/(2 \times 2) = 1.5$.

For the isotopic exchange reaction between carbonate-ion and water (Bowen 1966) written in a general form as

$$1/3[C^{16}O_3]^{2-} + H_2{}^{18}O \leftrightarrow [C^{18}O_3]^{2-} + H_2{}^{16}O],$$

where the value of K_0 is equal to unity since, in this case, the symmetry numbers of the molecules do not change. The process of equilibrium isotopic fractionation occurring due to the energy state difference of the isotopic molecules is sometimes called the thermodynamical isotopic effect.

In a series of physicochemical processes occurring at non-equilibrium conditions, the isotopic fractionation can be the result of kinetic effects.

In the case when the initial concentrations of the isotopic molecules are equal, the reaction rate constant and the true rate of the process according to the theory of the absolute rates of reactions can be given by the equation (Glasstone et al. 1941):

$$K_v = k\frac{kT}{h}\frac{Q^*}{Q_1 Q_2}\exp\left(-\Delta E_0/RT\right), \tag{2.44}$$

where K_v is the reaction rate constant; k is the Boltzmann's constant; T is the absolute temperature; h is the Plank's constant; Q^* is the oscillation-rotational partition function of the activated complex of reacting substances; R is the gaseous constant; ΔE_0 is difference between zero of energy activated complex and reacting matter (energy of activation of the reaction at 0°K); κ is the constant for a given reaction (coefficient of the transmission coefficient by Glesstone).

It follows from the Eq. 2.44 that the differences in the rates of processes involving reacting substances, containing both heavy and light isotopes of an element, are due to the differences in the molecular weights of the isotopic species of molecules (as was shown earlier, the partition function of each molecule includes the mass and the moment of inertia of the molecule) and also to the change in activation energy ΔE_0.

Hence, for the light isotopes (such as deuterium and protium) the difference in the rates of reaction on which the isotopic molecules are participating is influenced both by the difference in the mass of the molecules and by the difference in the activation energy. If both molecules are heavy, then the isotopic composition has little effect on their masses. In this case the difference in kinetics is practically completely dependent on the change in the activation energy of the activated complex formation.

The isotope separation, carried out in non-equilibrium conditions and provided by difference in the rate of reactions of the isotopic molecules, is called isotopic kinetic effect.

It has been shown by Varshavsky and Veisberg (1955) that the total distribution of deuterium (D) between the hydrogen-bearing substances should depend on individual distribution, characterizing each of the reactions involved. Under such circumstances, the total α can be found by calculating the partial values of α_i for all the reactions and taken proportional to the mass of the D concentration. Varshavsky and Veisberg assumed that the values of α_i for all the exchange reactions between given substances are equal to each other and to the total α. This assumption is equivalent to the assumption that there is a statistical distribution between the different D-bearing molecules of the same substance. As Varshavsky and Veisberg have shown, this assumption is true with an accuracy of 5% (at 25°C) for water and ammonia. The exception is molecular hydrogen, for which this assumption is true with an accuracy of about 20%. Taking this assumption into account, one can estimate α of any reaction of the hydrogen isotope exchange. In practice, it is more convenient to use that expression for the total α which contains all the oscillation frequencies of the molecules containing D atoms and those which do not contain them:

$$\alpha = \frac{\left[\prod (\mathrm{v_n}/\mathrm{v_0}) \frac{1-\exp(-h\nu_0/kT)}{1-\exp(-h\nu_n/kT)}\right]^{1/n} \exp\left\{(-h/2kTn)\left[\sum \mathrm{v_n} + \mathrm{Z_n} - \sum \mathrm{v_0} - \mathrm{Z_0}\right]\right\}}{\left[\prod (\mathrm{v_m}/\mathrm{v'_0}) \frac{1-\exp(-h\nu_0/kT)}{1-\exp(-h\nu_m/kT)}\right]^{1/m} \exp\left\{(-h/2kTm)\left[\sum \mathrm{v_m} + \mathrm{Z_m} - \sum \mathrm{v'_0} - \mathrm{Z'_0}\right]\right\}},$$

where v_n, v_0, v_m, v'_n are the fundamental frequencies of the corresponding molecules; m, n are numbers of hydrogen atoms in the molecules of the corresponding substances; Z_n, Z_0, Z_m, Z'_0 are the total corrections, accounting for the effect of anharmonicity of oscillations.

In the given expression, the numerator contains the oscillation frequencies of one substance only and the denominator contains those of the other one. Therefore, α may be expressed by any of the two quantities β, each one representing an individual substance and being independent of the other substance in the isotopic exchange reaction. Varshavsky and Veisberg called these quantities the β-factors:

$$\beta = \left[\prod (\mathrm{v_n}/\mathrm{v_0}) \frac{1-\exp(-h\nu_0/kT)}{1-\exp(-h\nu_n/kT)}\right]^{1/n} \exp\left\{(-h/2kTn)\left[\sum \nu_n + Z_n - \sum \nu_0 - Z_0\right]\right\}.$$

It follows from this expression that the β-factor depends on temperature. For small temperature ranges, the dependence of the factor appearing before the exponent can be neglected. Then the expression for the β-factor becomes

$$\beta = a \exp(-b/T),$$

where

$$\mathrm{a} = \left[\prod (\mathrm{v_m}/\mathrm{v_0}) \frac{1-\exp(-h\nu_0/kT)}{1-\exp(-h\nu_m/kT)}\right]^{1/m} \approx \text{const},$$

$$b = (-h/2kTn)\left[\sum \nu_n + Z_n - \sum \nu_0 - Z_0\right] \approx \text{const}.$$

Varshavsky and Veisberg have calculated and tabulated the values of a and b and also of the β-factor at 20, 50, and 75°C, for some of the hydrogen-bearing substances (Table 2.3). The anharmonicity of the oscillation has not been taken into account and the observed frequencies were substituted for the zero-point ones.

It has been shown that β-factors of hydrides (compounds and radicals) vary with increase in the atomic number of an element bounding hydrogen (Fig. 2.4). Here

Table 2.3 *a*, *b* and β-factors for some hydrogen-bearing substances. (After Varshavsky and Veisberg 1955)

Compounds	a	b	β_{20}	β_{50}	β_{75}
H_2	0.8484	418.7	3.54	3.10	2.83
CH_4	0.5971	894.5	10.74	8.09	6.63
C_2H_6	0.4640	932.8	11.20	8.33	6.77
C_2H_4	0.4632	936.8	11.33	8.42	6.84
C_6H_6	0.4128	984.7	11.89	8.70	6.99
NH_3	0.5611	889.4	11.68	8.81	7.23
H_2O	0.6344	852.1	11.62	8.87	7.34
HF	0.7340	757.3	9.73	7.66	6.47
H_2S	0.6291	633.1	5.46	4.47	3.88
HCl	0.7245	572.1	5.11	4.26	3.75
HBr	0.7192	517.3	4.20	3.57	3.18
HI	0.7178	453.7	3.38	2.93	2.64

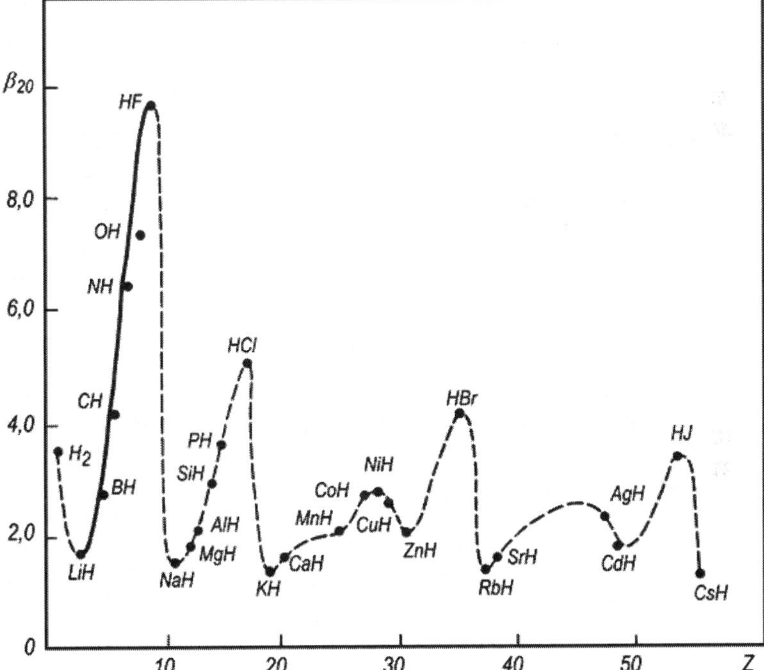

Fig. 2.4 Dependence of β-factors of two-atomic hydrides on the atomic number of element bounding hydrogen. (After Varshavsky and Veisberg 1955)

values of the β-factors of the diatomic hydrides, in contrast to the polyatomic ones, increase sharply within each period. The authors explained this in terms of the number of electrons in the outer shell of an atom linked to the hydrogen one. An examination of the resultant data shows that the maximum possible factor of D distribution

($\alpha = 8.0 - 8.5$ at 20°C) should correspond to the isotopic exchange reaction between hydrogen and a hydride of the most heavy alkaline metal and also that of the compound containing non-metal of the first period. The utilization of the β-factors makes it possible to estimate α-factors of the D distribution between the hydrogen-bearing compounds without complicated calculations.

The fractionation of isotopes, under non-equilibrium conditions and caused by differences in the rates of the exchange reactions, is referred to as the kinetic isotope effect.

It is obvious that in general the same physicochemical process (e.g., phase transitions, isotopic exchange reactions, biochemical processes, etc.) may occur in natural conditions both exchange under equilibrium and non-equilibrium conditions.

On one hand, this complicates the clear interpretation of the observed facts. On the other hand, it enables an estimation of the degree of equilibrium of this—or the other natural process—to be made.

For a quantitative estimation of the isotopic fractionation effect in a physicochemical process the fractionation factor α is used (Brodsky 1957):

$$\alpha = \frac{N/(1-N)}{n/(1-n)} = \frac{R_1}{R_2}, \tag{2.45}$$

where $N/1-N = R_1$ is the atomic part of the isotope in the enriched fraction (component); $n/(1-n) = R_2$ is the atomic part of the isotope in the depleted fraction (compound).

In the case of a minimal content of the component being enriched which is the case, for example, during the fractionation of a natural mixture of isotopic species of water, one might assume that $1 - N \approx (1-n) = 1$ and hence,

$$\alpha = \frac{N}{n}. \tag{2.46}$$

If the process of fractionation is successively repeated k times, then the total isotopic fractionation factor will be α^k.

2.4 Hydrogen and Oxygen Isotope Separation at Phase Transition of Water

During the fractionation of isotopes in a liquid–vapor system, the content of the isotopic components in liquid is N and $1-N$ and their vapor pressures in a pure form will be p_1^0 and p_2^0. According to Raule's law for ideal liquid mixtures (Rabinovich 1968), the vapor pressure of each component above the mixture can be expressed as

$$\rho_1 = N\rho_1^0; \rho_2 = (1-N)\rho_2^0. \tag{2.47}$$

Then

$$\frac{p_1}{p_2} = \frac{p_1^0}{p_2^0}\frac{N}{1-N} \tag{2.48}$$

The partial pressures of the isotopic species of molecules of liquid in the vapor phase are proportional to the ratio of their molar ratios n and $(1-n)$. Hence, Eq. 2.45 can be rewritten in the form

$$\frac{n}{1-n} = \frac{N}{1-N}\frac{p_1^0}{p_2^0}.$$

or

$$\alpha = \frac{p_2^0}{p_1^0} \tag{2.49}$$

If subscript 2 corresponds to the light (more volatile) isotopic component, then $\alpha > 1$.

The vapor pressure is a function of temperature and is described by Eq. 2.1. If one uses Mendeleev–Klapeiron's equation $pV = RT$ for the saturated vapor, then, neglecting the volume of the condensed phase for one mole of the vapor, it follows that

$$\frac{dp}{dT} = \frac{\Delta Hp}{RT^2} \quad \text{or} \quad \frac{d(\ln p)}{dT} = -\frac{\Delta H}{RT^2}. \tag{2.50}$$

With sufficient accuracy one can assume that the value of ΔH is a constant over a narrow range of temperatures. Then integration of Eq. 2.50 yields

$$\ln p = -\frac{\Delta H}{4.576T} + \text{const}, \tag{2.51}$$

that is a linear relationship between the logarithm of vapor pressure and the inverse temperature holds

$$\ln p = A - \frac{B}{T}. \tag{2.52}$$

For the ideal liquid mixture of the two isotopic species of molecules (e.g., H_2O and HDO), the logarithm of the vapor pressure ratio, corresponding to more and less volatile components, follows the equation

$$\ln\frac{p_2}{p_1} = A' - \frac{B'}{T} = \ln\alpha, \tag{2.53}$$

where p_2 is the pressure of the saturated vapor of H_2O; p_1 is the pressure of the saturated vapor of HDO; A', B' are constants.

It is easy to show that $B' = \Delta H' R$, where $\Delta H'$ is the change in heat of vaporization for the liquids H_2O and HDO.

It follows from Eq. 2.53 that below some fixed temperature, one has $\ln \alpha > 0$ and $\alpha > 1$. At a certain temperature the value of $B'/T = A'$ and $\alpha = 1$, and above this temperature $\alpha < 1$. In this case the enrichment of the vapor phase with respect to the heavy isotope occurs. The temperature at which $\alpha = 1$, is called the temperature (point) of intersection (Rabinovich 1968).

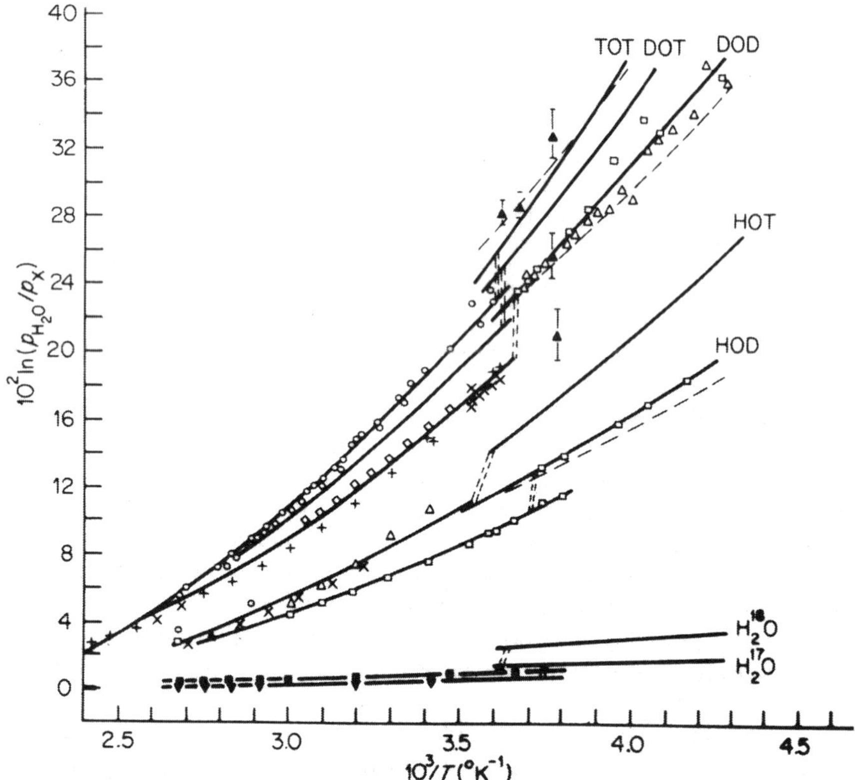

Fig. 2.5 Experimental relations between vapor pressure of the heavy and light water molecules and the temperature. Solid lines are obtained by calculations. (After Van Hook 1968)

The Eq. 2.53 is shown in Fig. 2.5, which illustrates the influence of isotopic substitutions in the water molecule upon the pressure of the saturated vapor.

Equation 2.53 satisfactorily describes the dependence of the fractionation factor on temperature over a comparatively narrow temperature range. Over a wide temperature range, the experimental data, as one can see from Fig. 2.5, does not conform to the straight line expressed by Eq. 2.53, since the enthalpy of the reaction is, in turn, a function of temperature.

Sometimes the experimental data can be linearized successfully with the help of the polynomial

$$\ln \alpha = A + \frac{B}{T^2}. \tag{2.54}$$

More exactly, the experimental dependence $\ln \alpha = f(T)$ may be described using more complicated polynomial function of the form:

$$\alpha = A + \frac{B}{T} + \frac{C}{T^2} C/T^2. \tag{2.55}$$

Table 2.4 Values of the fractionation factors α for the water hydrogen and oxygen isotopes in liquid–vapor systems. (From Brodsky 1957; Rabinovich 1968)

t, °C	$\alpha_{D_2O} = \frac{p_{H_2{}^{16}O}}{p_{D_2O}}$	$\alpha_{HDO} = \frac{p_{H_2{}^{16}O}}{p_{HDO}}$	$\alpha_{H_2{}^{18}O} = \frac{p_{H_2{}^{16}O}}{p_{H_2{}^{18}O}}$	$\alpha_{T_2O} = \frac{p_{H_2{}^{16}O}}{p_{T_2O}}$	$\alpha_{HTO} = \frac{p_{H_2{}^{16}O}}{p_{HTO}}$
1.0	–	1.104	–	–	–
3.815	1.215	1.000	–	–	–
10	1.190	1.091	–	1.265	1.125
20	1.162	1.080	1.0092	1.230	1.110
30	1.139	1.070	1.0084	1.198	1.096
40	1.116	1.060	1.0071	1.182	1.076
50	1.100	1.051	1.0068	1.145	1.065
60	1.090	1.046	1.0061	1.105	1.056
70	1.079	–	1.0054	1.098	1.049
80	1.066	1.032	1.0050	1.083	1.041
100	1.050	1.027	1.0040	1.059	1.030
140	1.029	1.012	1.0021	–	–
180	1.012	1.006	1.0000	–	–
220	1.000	1.000	–	–	–
240	0.997	0.998	–	–	–
260	0.994	0.996	–	–	–

Thus, according to the data of Bottinga (McKenzie and Truezdell 1977), the relationship $\ln \alpha = f(T)$ for the liquid–vapor system and isotopic species of water molecules $H_2{}^{18}O$ and $H_2{}^{16}O$ over the range of temperatures from 3 to 360°C can be expressed by

$$10^3 \ln \alpha = -3.494 + 1.205(10^3/T) + 0.7664(10^6/T^2).$$

The values of the factor α for the binary liquid–vapor system consisting of various isotopic species of water are given in Table 2.4, taken from the study of Rabinovich (1968).

The equations obtained by Bottinga and Craig (1968) expressing the dependence of the fractionation factor on temperature in the liquid–vapor system gives slightly different values of α_D and α_{18_O} from those shown in Table 2.4, in the range of temperatures 100–300°C. Since these values are applied by different researchers while studying the conditions of hydrogen and oxygen isotopic fractionation in hydrothermal systems and are used for calculating base temperatures, we introduce them in Table 2.5 based on calculations of Arnason (1977a).

The values of the coefficients α_D and α_{18_O} are given for the range of temperatures from −20 to +100°C, which were obtained by Merlivat and by Zhavoronkov, respectively, and summarized by Dinçer (1968).

As can be seen from a comparison of the values of α_D and α_{18_O}, given in Tables 2.4 and 2.5, the accuracy of experimental determinations of the above is likely to be ~0.005 for α_D and ~0.0005 for α_{18_O}.

Table 2.5 Values of the fractionation factors α_D and $\alpha_{^{18}O}$

t, °C	$\alpha_D = \frac{p_{H_2{}^{16}O}}{p_{HDO}}$	$\alpha_{^{18}O} = \frac{p_{H_2{}^{16}O}}{p_{H_2{}^{18}O}}$
Arnason (1977a)		
100	1.0256	1.0053
120	1.0212	1.0045
140	1.0166	1.0039
160	1.0118	1.0034
180	1.0074	1.003
200	1.0035	1.0025
220	1.0004	1.0022
240	0.9981	1.0018
260	0.9966	1.0015
280	0.9958	1.0012
300	0.9957	1.0009
Dinçer (1968)		
−20	1.1469	1.0135
−10	1.1239	1.0123
0	1.106	1.01119
20	1.0791	1.00915
40	1.06	1.00746
60	1.046	1.00587
80	1.037	1.00452
100	1.029	1.0033

2.5 Relationship Between the Isotope Reaction Change Constant and the Fractionation Factor

As demonstrated above, the isotopic exchange in the system under equilibrium conditions may be described both by the equilibrium constant K and the fractionation factor α.

For such a reaction as $Ax + Bx^* = Ax^* + Bx$, where there is only one atom of exchangeable isotopes x and x^* in each molecule of the reacting substances, the equilibrium constant is equal to the fractionation factor

$$\alpha = \left(\frac{x^*}{x}\right)_{Ax} : \left(\frac{x^*}{x}\right)_{Bx} = \frac{[Ax^*]}{[Ax]} : \frac{[Bx]^*}{[Bx]} = K. \tag{2.56}$$

In more complicated reactions, where several atoms of exchangeable isotopes in each molecule are involved, $\alpha \neq K$.

For example, let us consider the isotopic exchange of oxygen in the carbon dioxide–water system

$$C^{16}O_2 + H_2{}^{18}O \leftrightarrow C^{16}O^{18}O + H_2{}^{16}O.$$

The equilibrium constant of this reaction is

$$k = \frac{[C^{16}O^{18}O][H_2{}^{16}O]}{[C^{16}O_2][H_2{}^{18}O]} \tag{2.57}$$

and the isotopic fractionation factor is equal to

$$\alpha = \left(\frac{^{18}O}{^{16}O}\right)_{CO_2} : \left(\frac{^{18}O}{^{16}O}\right)_{H_2O}. \tag{2.58}$$

Symbols ^{18}O and ^{16}O denote here the atomic ratios of oxygen-18 (^{18}O) and oxygen-16 (^{16}O). For water, the molecule of which contains one atom of oxygen, it follows that

$$\left(\frac{^{18}O}{^{16}O}\right)_{H_2O} = \frac{[H_2{}^{18}O]}{[H_2{}^{16}O]}. \tag{2.59}$$

For carbon dioxide, one should take into account that the molecule $C^{16}O_2$ contains two atoms of ^{16}O. Besides, the molecule $C^{16}O^{18}O$ also contains an atom of ^{16}O. Then

$$\left(\frac{^{18}O}{^{16}O}\right)_{CO_2} = \frac{[C^{16}O^{18}O]}{[C^{16}O^{18}O] + 2[C^{16}O_2]}. \tag{2.60}$$

Putting the values of $^{18}O/^{16}O$ from Eqs. 2.58 and 2.59 into Eq. 2.60 and multiplying both numerator and denominator by $C^{16}O_2$ one obtains

$$\alpha = \frac{[C^{16}O^{18}O][H_2{}^{16}O]}{[C^{16}O_2][H_2{}^{16}O]} \frac{[C^{16}O_2]}{[C^{16}O^{18}O] + 2[C^{16}O_2]}, \tag{2.61}$$

or

$$K = 2\alpha \frac{[C^{16}O^{18}O] + 2[C^{16}O_2]}{2[C^{16}O_2]}. \tag{2.62}$$

If the atomic ratio of ^{18}O in the carbon dioxide after equilibrium is reached is denoted by R_2, then

$$2[C^{16}O_2] = 1 - 2R_2, \tag{2.63}$$

$$[C^{18}O^{16}O] + 2[C^{16}O_2] = 1 - R_2. \tag{2.64}$$

Substituting Eq. 2.63 and 2.64 into Eq. 2.62, we obtain expressions for K and α

$$K = \frac{2\alpha(1 - R_2)}{(1 - 2R_2)}, \tag{2.65}$$

$$\alpha = \frac{K(1 - 2R_2)}{2(1 - R_2)}. \tag{2.66}$$

If $R_2 \ll 1$, which remains the same for natural concentrations of ^{18}O, then $K \approx 2\alpha$, or $\alpha \approx K/2$.

The relationship between α and K for isotopic exchange reactions was studied by Brodsky (1957), who demonstrated theoretically that, in general

$$a = (K/K_0)^{1/ab},$$

where a and b are the stoichiometric coefficients of the reaction.

Thus, for the isotopic exchange reaction between carbon dioxide and water $a = b = 1$ the symmetry numbers; σ_{H_2O} for both isotopic molecules of water are equal to σ_b (the substitution of ^{16}O for ^{18}O does not change the symmetry of the molecules); the symmetry number $\sigma_{C^{16}O_2}$, and $\sigma_{C^{16}O^{18}O} = 1$.

Hence,

$$K_0 = \frac{\sigma_b x_2}{\sigma_b x_1} = 2.$$

From this, the importance of the 2 in the denominator of Eq. 2.66 becomes apparent. It is the limiting value of the factor K of the isotopic exchange reaction. One can easily find from Eq. 2.66 that the factor α depends on the concentration (atomic factor) of the isotope in the exchange system. The value of α does not depend on concentration only when $R_2 \to 0$. This condition is satisfied in particular while considering the reactions of exchange in the system with natural concentrations of D and ^{18}O.

The above-mentioned oxygen isotopic exchange reaction between water and carbon dioxide, which is in common use for the mass spectrometric measurement of isotopic ratios of oxygen in water applying the isotopic equilibration method at 25°C (Epstein and Mayeda 1953; Craig 1957), has been proposed in the following form by Urey (1947):

$$1/2C^{16}O_2 + H_2{}^{18}O \leftrightarrow 1/2C^{18}O_2 + H_2{}^{16}O.$$

In this case, the equilibrium constant K is equal to the isotopic fractionation factor α. In fact,

$$K = \frac{[C^{18}O_2]^{1/2}}{[C^{16}O_2]^{1/2}} : \frac{[H_2{}^{18}O]}{[H_2{}^{16}O]}. \qquad (2.67)$$

The expression for the coefficient α can be written as

$$\alpha = \left(\frac{^{18}O}{^{16}O}\right)_{CO_2} : \left(\frac{^{18}O}{^{16}O}\right)_{H_2O} = \frac{2C^{18}O_2 + C^{16}O^{18}O}{2C^{16}O_2 + C^{16}O^{18}O} : \frac{[H_2{}^{18}O]}{[H_2{}^{16}O]}.$$

The equilibrium constant of the reaction

$$C^{18}O_2 + C^{16}O_2 = 2C^{16}O^{18}O \qquad (2.68)$$

can be given in the form

$$\sigma_1\sigma_2 : \sigma_{12^2},$$

where σ_1, σ_2, σ_{12} are the symmetry numbers of the isotopic species of the carbon dioxide molecules $C^{18}O_2$, $C^{16}O_2$, $C^{16}O^{18}O$. It is obvious that $\sigma_1 = \sigma_2 = 2$ and $\sigma_{12} = 1$. Therefore, the limiting value of the equilibrium constant will be equal to 4.

Equation 2.68 can be written now as

$$\frac{[C^{16}O_2]}{[C^{16}O_2]^{1/2}[C^{18}O_2]^{1/2}} = 2.$$

After simple transformations, we get

$$\alpha = \frac{[C^{16}O_2]^{1/2}}{[C^{16}O_2]^{1/2}} : \frac{[H_2{}^{18}O]}{[H_2{}^{18}O]}.$$

One can see that the right-hand side of last expression and the right-hand side of Eq. 2.67 are same.

On the basis of the theoretical calculation of Urey (1947), carried out for the CO_{2gas}–H_2O_{liq} system, taking into account the isotopic fractionation factor of oxygen between vapor and liquid water ($\alpha = 1.009$), the value $\alpha_{H_2O\text{–}CO_2} = 1.038$ has been obtained at 25°C. Later on, this value was determined experimentally with a higher degree of precision by Compston and Epstein (1958) and it was found that $\alpha_{H_2O\text{–}CO_2} = 1.0407$. Since 1975, most of the studies on isotopes of oxygen use the value $\alpha_{H_2O\text{–}CO_2} = 1.0412 \pm 0.0001$, obtained by O'Neil et al. (1975).

2.6 Hydrogen and Oxygen Isotope Fractionation at Interaction Between Water and Gases and Rocks

In the hydrological cycle, the process of interaction in rock-water and water-gas systems are important for the isotopic fractionation of oxygen and hydrogen. We shall now consider the experimental results of determining the isotopic fractionation factors for various natural systems, obtained by different researchers.

The isotopic exchange reactions in the water–carbonate or silicate rocks may be written (Urey 1947) in a general form as

$$1/3C^{16}O_3^{2-} + H_2{}^{18}O \leftrightarrow 1/3C^{18}O_3^{2-} + H_2{}^{16}O. \tag{2.69}$$

$$1/2Si^{16}O_2 + H_2{}^{18}O \leftrightarrow 1/2Si^{18}O_2 + H_2{}^{16}O. \tag{2.70}$$

For such a presentation of isotopic exchange reactions, the symmetry numbers of the molecules with exchangeable isotopes do not change. Therefore, according to Eq. 2.56, the fractionation factor will be equal to the equilibrium constant, that is

$$\alpha \equiv K. \tag{2.71}$$

In this case, the dependence of the equilibrium constant K on temperature in the water–mineral system over a wide range of temperatures can be described with a sufficient degree of accuracy by Eq. 2.54, that is

$$\ln \alpha = A + B/T^2. \tag{2.72}$$

At high temperature $\alpha \to 0$ and $\ln\alpha \to 0$, therefore, the constant $A \to 0$. Thus, the dependence of the equilibrium constant on temperature plotted with $\ln \alpha - 1/T^2$ coordinates can be approximated by a straight line passing through the origin.

On the other hand, according to Eq. 2.45, the isotopic fractionation factor between the two phases a and b can be written in general form as

$$\alpha = \frac{R_a}{R_b}, \tag{2.73}$$

where R_a and R_b are the isotopic ratios of each isotope in a corresponding phase.

Let us use the generally accepted expression for reporting isotope data in a given phase versus some standard:

$$\delta = \frac{R_{\text{sample}} - R_{std}}{R_{std}} \times 1000‰. \tag{2.74}$$

where R_{sample} and R_{std} are isotopic ratios for sample and standard, respectively.

Using Eq. 2.74, the Eq. 2.73 can be written as:

$$\alpha = \frac{1 + \delta_\alpha/1000}{1 + \delta_b/1000}. \tag{2.75}$$

Taking logarithms of Eq. 2.75, we obtain

$$\ln \alpha = \ln\left(1 + \delta_\alpha/1000\right) - \ln\left(1 + \delta_b/1000\right). \tag{2.76}$$

As $\ln(1+x) \approx x$, if $x \ll 1$, we obtain the approximate relationship

$$10^3 \ln \alpha \approx \delta_\alpha - \delta_b = \Delta. \tag{2.77}$$

With the help of Eq. 2.77, the isotopic fractionation factor can be expressed by the difference of the δ-values for corresponding phases or compounds if their isotopic compositions have been measured relative to some standard.

For graphical presentation of the $\ln \alpha$ dependence on temperature it is convenient to use Eq. 2.72 in the form

$$10^3 \ln \alpha = \mathrm{A} + \mathrm{B}\left(10^6/\mathrm{T}^2\right), \tag{2.78}$$

or with the help of the more complicated polynomial

$$10^3 \ln \alpha = \mathrm{A} + \mathrm{B}\left(10^3/\mathrm{T} + \mathrm{C}(10^6/\mathrm{T}^2\right). \tag{2.79}$$

For the solution of practical problems, it is important to have reliable values of the fractionation factors of the oxygen and hydrogen in water—principal rock formation minerals. The experiment consists of mixing the finely crushed sample of the mineral (approximately 0.2 g) together with water, in an ampoule at a known temperature, before soldering it. After equilibrium is attained and mass spectrometric measurements are taken, the fractionation factor can be calculated by the formula

$$\alpha_{^{18}O} = \frac{[^{18}O/^{16}O]_M}{[^{18}O/^{16}O]_B} = \frac{(1000 + \delta^{18}O)_M}{(1000 + \delta^{18}O)_B},$$

$$\alpha_D = \frac{[D/H]_M}{[D/H]_B} = \frac{(1000 + \delta D)_M}{(1000 + \delta H)_B}.$$

However, as experimental study has shown (Nortrop and Clayton 1966; Suzuoki and Epstein 1976), the isotopic equilibrium between the exchangeable phases for silicate, carbonate, and clayey minerals, even at high temperatures (700–800°C), requires 300–400 hours to settle. To avoid continuous laboratory experiments, Nortrop and Clayton (1966) used the so-called method of partial isotopic exchange. This method is based on the kinetics of the isotopic exchange reactions in heterogeneous system and permits determination of the fractionation factor by interpolating experimental data. This method was employed for the first time by Nortrop and Clayton while studying oxygen isotope exchange in the dolomites-water system and later on the method was modified by Suzuoki and Epstein (1976) to study the hydrogen isotopic exchange between water and hydroxyl-bearing minerals. The method consists of placing pre-weighed amounts of crushed mineral into gold soldered ampoules together with water of varying isotopic composition. After the corresponding mass spectrometric measurements have been taken the value of the fractionation factor α_e is calculated by the graphical method of least squares from the equation

$$10^3(\alpha_i - 1) = 10^3(\alpha_e - 1) + A(\alpha_f - \alpha_i),$$

where α_i is quotient of division of isotopic ratios in the sample of the mineral and in water before the experiment starts; α_f is the same after a period of time t; α_e is equilibrium fractionation factor at $T°\kappa$.

The slope A in the equation is dependent on the time of exchange, that is a function of the isotopic exchange completeness. The equation is given in the form convenient for calculations, since, as a first approximation,

$$10^3(\alpha - 1) \approx \delta_{\text{mineral}} - \delta_{\text{water}} \approx 10^3 \ln \alpha.$$

2.7 Isotope Geothermometry

Equations 2.78 and 2.79 are in common use for calculating modern and paleotemperatures of the formation of minerals and also for determining the base temperatures of the hydrothermal systems.

Thus, Clayton (1961) has determined experimentally the oxygen isotope fractionation factor for a calcite-water system at high temperatures. He employed the parameters of the paleotemperature scale drawn up by Epstein et al. (1951) and plotted the curve describing the equilibrium in the calcite-water system, in the range of temperatures of approximately 1,000°C. This dependence in the range of temperatures 0–750°C can be described by the following empirical equation

$$10^3 \ln \alpha_{^{18}O} = 2.73(10^6/T^2) - 2.56.$$

Clayton et al. (1972) determined experimentally the dependence of the fractionation factor on temperature in the quartz-water system in the range of temperatures. However, this dependence, as Knauth and Epstein (1976) have pointed out, requires correction for real geological temperatures in the low-temperature range. Interpolating Clayton's data in the low-temperature range and using the value $\alpha = 1$ at 0°C they reported the following relationship for the quartz-water system, suitable for temperatures 0 – 100°C:

$$10^3 \ln \alpha_{^{18}O} = 3.09(10^6/T^2) - 3.29.$$

In the case of the water-quartz system, Kawabe (1978) obtained by thermodynamic calculations the following values of the parameters in Eqs. 2.78 and 2.79 for temperatures below 100°C:

$$10^3 \ln \alpha_{^{18}O} = -5.533 + 3.2763(10^6/T^2),$$

$$10^3 \ln \alpha_{^{18}O} = -18.977 + 8.582(10^3/T) + 1.9189(10^6/T^2).$$

Numerical values of the fractionation factor in Eqs. 2.78 and 2.79 for the quartz-water system have been determined by several researchers. Some results are shown in the range of temperatures from 0 to 100°C in Fig. 2.6.

In the narrow ranges of temperatures used for the paleotemperature analysis of the ancient ocean, the obtained empirical relationships differ from those expressed by Eqs. 2.78 and 2.79.

Urey et al. (1951) and Epstein et al. (1951, 1953) found experimentally the following dependence of temperature upon the isotopic composition of oxygen in the sea carbonates:

$$t = 16.5 - 4.3\delta^{18}O + 0.14(\delta^{18}O)^2,$$

where the $\delta^{18}O$ values in this case are expressed versus the PDB-I standard, which is 30.6‰ heavier than the standard of the ocean water (SMOW).

Later on, Epstein and Mayeda (1953) proposed a more general form for a paleotemperature scale:

$$t = 16.5 - 4.3\,(\delta - A) + 0.14(\delta - A)^2,$$

where A is parameter describing the isotopic composition of water in which the shell of a mollusk grew up. The value of A is equal to the $\delta^{18}O$ value of the sea water, obtained on the basis of the isotopic analysis of carbon dioxide equilibrated with the examined water at 25°C and measured relative to the PDB-I standard. However, as Bowen (1966) has pointed out, it is impossible to determine the value of A for

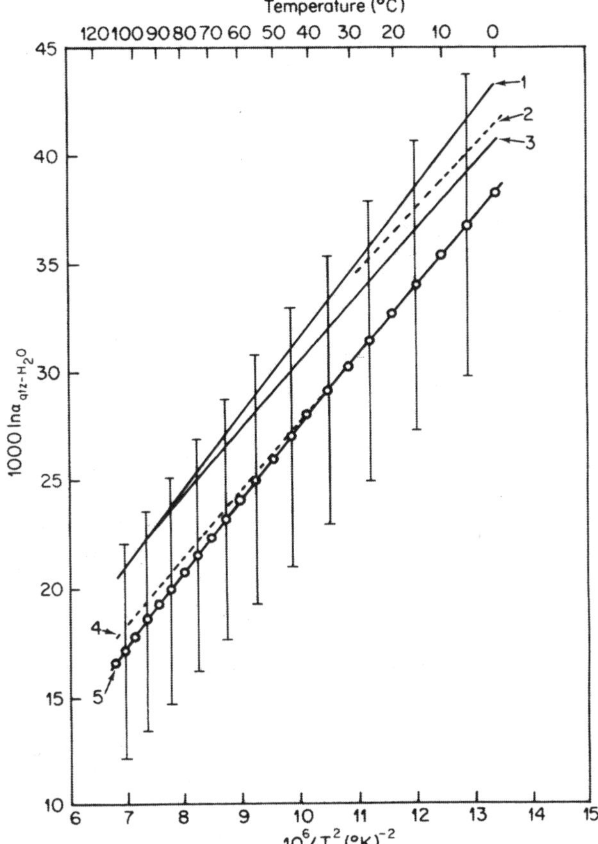

Fig. 2.6 Temperature dependence of the oxygen isotopes for quartz–liquid water system: (*1*) Shiro and Sakai (1972); (2) Labeyrie (1974); (*3*) Backer and Clayton (1976); (*4*) Knauth and Epstein (1976); (*5*) Kawabe (1978). (After Kawabe 1978)

paleooceans using only the carbonate thermometer. This means that all measurements made for the oxygen isotopes in the carbonate shells of fossil mollusks will only give relative temperatures. The absolute temperatures would be imprecise. More precise results seem to be obtained by eliminating the parameter *A*, for example, simultaneously using data for phosphate paleothermometry (Longinelly and Nuty 1965). Thus, one obtains

$$t = -90 - 4.8\left(\delta_p - \delta_w\right),$$

where δ_p is the oxygen isotopic composition of the phosphates; δ_w is the isotopic composition of water from which sedimentation of the phosphates occurred.

Analyzing the formula of Epstein, Mayeda and Craig (1965) have shown that several corrections should be involved, taking into account the systematic errors of the oxygen isotopic composition analysis on the mass spectrometer when the ionic source is being filled in with the carbon dioxide. In this case, the formula of Epstein and Mayeda should be rewritten in the corrected form

$$t = 16.5 - 4.3\left(\delta_c^o - \delta_w^o\right) + 0.14\left(\delta_c^o - \delta_w^o\right)^2,$$

where δ_w^o is the oxygen isotopic composition in the investigated carbonates, measured relative to the PDB-I standard in the carbon dioxide which was extracted from the carbonate sample by reaction with 100% phosphoric acid; δ_w^o is the oxygen isotopic composition of water measured in relation to the PDB-I standard of the carbon dioxide, equilibrated with water from which the precipitation of the carbonates had occurred.

The corrected δ_c' and δ'_w values, according to Craig, can be obtained from the expressions

$$\delta_c' = f_c\delta_c^o + 0.009\delta^{13}C_c,$$
$$\delta'_w = f_w\delta_w^o + z,$$
$$z = 0.009\delta^{13}C_t - (\alpha/\rho)\,\delta^{18}O_t,$$

where $\delta^{13}C_c$ is the isotopic composition of the carbon in a sample of precipitated sea limestone; f_c is the correction coefficient of experimentally measured δ_w^o value of the carbon dioxide extracted from the sample being investigated while taking into account apparatus errors (for apparatus of mean resolution $f_c = (1.028 - 1.029)$; f_w is the correction coefficient of experimentally measured δ_w^o value of the carbon dioxide being in equilibrium with water while taking into account the systematic and apparatus errors (f_c); $\delta^{13}C_t$ is the carbon isotopic composition of the carbon dioxide used for equilibration purposes; $\delta^{18}O_t$ is the isotopic composition of oxygen of the same gas; α is the isotopic fractionation factor of oxygen in the CO_2–H_2O system, which can be assumed to be 1.04; ρ is the ratio of atoms (gram-atoms) of oxygen in water to atoms (gram-atoms) of oxygen in carbon dioxide in the equilibrium flask. All the values are taken in ‰.

As Craig (1965) has shown $f_w \approx 1.04$, and $z \approx -0.09$‰.

The difference between the measured values $(\delta_c^o - \delta_w^o)$ is related to those corrected by

$$\delta_c^o - \delta_w^o = \frac{\delta_c' - \delta'_w}{f_c} + \left(\frac{f_w}{f_c} - 1\right)\delta_w^o + \frac{z}{f_c}.$$

The sum of the last two terms on the right-hand side of this equation is approximately equal to -0.092‰. Substituting $(\delta_c - \delta_w)$ from the difference $(\delta_c' - \delta_w')$ Craig obtained from the last relationship, the corrected carbonate-water isotopic temperature scale equation is in the form

$$t = 16.9 - 4.2\left(\delta_c' - \delta'_w\right) + 0.13\left(\delta_c' + \delta'_w\right)^2.$$

Here δ' is the corrected $\delta^{18}O$ value, which is measured in the carbon dioxide extracted from the carbonates by reaction with 100% H_3PO_4 (δ_c^o) and is measured in CO_2 equilibrated with water at 25°C (δ_c).

The equation is applicable for any laboratory standard on condition that measurements of the oxygen isotopes of CO_2 extracted from the carbonates and equilibrated with water are made relative to the same standard.

Note that the correction factor f_c and f_w for the modern mass spectrometers can be different from those given above for the old instruments such as M-86 (Atlas-Werke) or MI-1305 (SCB APS Academy of Sciences of the USSR).

For the carbonate paleotemperature measurements, the PDB-1 standard is obtained by crushing and grinding the belemnite (mollusk of the Belemnitela Americana type) taken from Peedee formation of South Carolina (Upper Creteaceous, Maastrichtian). At present, this standard is completely exhausted but in 1957 Craig compared the PDB-1 standard with the etalon isotopic specimen No. 20 from the National Bureau of Standards, USA. The NBS-20 standard is a $CaCO_3$ specimen of Solenhofen limestone, for which the corrected value of $\delta^{18}O = 4.14‰$ has been found (Bowen 1966).

The isotopic composition of sample X in relation to the PDB standard $\delta_{(x-PDB)}$ can be calculated if one takes some other standard B whose $\delta_{(x-B)}$ value is (Craig 1957). Then

$$\delta_{(x-PDB)} = \delta_{(x-B)} + \delta_{B+PDB} + 10^{-3}\delta_{(x-B)}\delta_{(x-PDB)},$$

where all δ values are expressed in terms of per mile.

Craig (1965) has calculated the $\delta^{18}O$ value of the PDB-1 standard on the SMOW scale. A comparison of oxygen isotopic composition of the carbon dioxide, obtained from the PDB-1 standard and from the analogous La Jolla standard, with the isotopic composition of CO_2 equilibrated with the ocean water at 25°C, has shown that on the SMOW scale the oxygen isotopic composition of the carbon dioxide extracted from the PDB-1 standard and from the La Jolla standard are 0.22‰ and 0.18‰ heavier than the oxygen isotopic composition of the carbon dioxide equilibrated with SMOW at 25°C. In this case, as Craig has demonstrated, the following relation exists:

$$R_{(PDB-CO_2)}/R_{(SMOW-CO_2)} = 1.00020,$$

where $R_{(PDB-CO_2)}$ and $R_{(SMOW-CO_2))}$ are the $\delta^{18}O/\delta^{16}O$ isotopic ratios in the carbon dioxide obtained from the PDB-1 carbonate by the acid extraction with 100% H_3PO_4 and another sample of CO_2 equilibrated with the ocean water (SMOW) at 25°C.

It is also known that during the CO_2 extraction from $CaCO_3$ with 100% H_3PO_4 isotopic fractionation between liquid and gaseous phases occurs. The value of $\alpha_{CaCO_3-CO_2} = 1.01000$ (Clayton 1961) has been accepted, that is the extracted carbon dioxide is enriched by approximately 10‰ with ^{18}O in relation to the $CaCO_3$ sample. According to Craig, the value of fractionation factor is $\alpha_{H_2O-CO_2} = 1.0407$ (25°C). If both fractionation factors are known, one can write

$$\frac{R_{(H_2O-CO_2)}}{R_{(CaCO_3-CO_2)}} = \frac{\alpha_{(H_2O-CO_2)}R_{H_2O}}{\alpha_{(CaCO_3-CO_2)}R_{CaCO_3}} = 1.03040\frac{R_{H_2O}}{R_{CaCO_3}}.$$

Combining the last equation with the previous one, we obtain

$$\delta_{(PDB\text{-}SMOW)} = (1.00020 \times 1.03040) \times 10^3 = +30.6‰,$$

that is the oxygen isotopic composition in the PDB-1 standard is 30.6‰ heavier than the SMOW one.

Using the new value of $\alpha_{H_2O-CO_2} = 1.0412$ (O'Neil et al. 1975), one obtains

$$\delta_{(PDB-SMOW)} = 31.1‰,$$

which is likely to be more precise value than which was obtained earlier (+30.6‰). It follows that the oxygen isotopic composition of any carbonate sample on the SMOW scale will be related to the PDB-1 scale by the equation

$$\delta_{SMOW} = 1.031\delta_{PDB} + 31.1‰.$$

If the values of the fractionation factors $\alpha_{CaCO_3-CO_2}$ and $\alpha_{H_2O-CO_2}$ are corrected in the future, then the constant factor in the last expression will be different from those obtained.

The above paleotemperature scales are in common use in paleotemperature analysis (Bowen 1966; Teys and Naydin 1973).

Labeyrie (1974) derived an equation applicable to the determination of surface water temperature using the oxygen isotopic composition of water and that of silica of the diatomic silt which has the form

$$t = 5 - 4.1(\delta_{SO_2} - \delta_{H_2O} - 40).$$

In the range of low temperatures, this equation gives the value of temperatures which are close to those for the fractionation factor related to the equation obtained by Knauth and Epstein (1976).

For the study of the base temperatures of the hydrothermal systems, the scale which is dependent on the fractionation factor of oxygen isotopes in the sulfate-water system is often used. According to Lloyd's (1968) evidence, this dependence, in the case of the dissolved sulfate-water system, is

$$10^3 \ln \alpha_{^{18}O} = 3.251(10^6/T^2) - 5.6.$$

Mizutani and Rafter (1969) and Mizutani (1972) have reported the same relationship for slightly different initial conditions:

$$10^3 \ln \alpha = 2.88(10^6/T^2) - 4.1.$$

McKenzie and Truesdell (1977) have employed these relationships for determining base temperature of hydrothermal systems in the range of 100 to 400°C. They noted that the equations of Lloyd and Mizutani and Rafter must be corrected for the following considerations. For the experimental determination of the corresponding $\delta^{18}O$ values, the isotopic composition of water was analyzed by the method of water equilibration with CO_2. The fractionation factor in the carbon dioxide-water system being used was equal to $\alpha_{25^\circ C} = 1.0407$. However, the new experimental evidence, obtained by O'Neil et al. (1975), has shown that a more precise value of the factor is $\alpha_{25^\circ C} = 1.0412$. If we take this into account then the numerical values obtained for the base temperatures at corresponding conditions deviate slightly from those given above:

$$10^3 \ln \alpha = 3.251(10^6/T^2) - 5.1.$$

$$10^3 \ln \alpha = 2.88(10^6/T^2) - 3.6.$$

However, the obtained corrections do not considerably affect the temperatures being calculated. Sakai (1977) reported the equation for the barytes-water system which was obtained earlier by Kusakabe and Robinson in the range of temperatures 100–500°C:

$$10^3 \ln \alpha = 3,01(10^6/T^2) - 7.3.$$

In the case where notable differences occur in the fractionation factor of oxygen isotopes in coexisting phases on the basis of $\delta^{18}O$ values of mineral pairs, the temperatures of the mineral formation are calculated as well as the isotopic composition of the coexisting water phase. As Degens has pointed out, studies of systems such as calcite–apatite, syenite–magnetite, calcite–dolomite, calcite–quartz, or other oxygen-bearing minerals are essential (Becker and Clayton 1976).

In a similar way to oxygen isotopic fractionation in rock-water systems, the hydrogen isotopic fractionation occurs during the formation of cristallohydrates, clay minerals, and in hydrothermal processes during serpentinization and chloritization of silicate minerals of primary magmatic rocks. All these processes are directed so the OH-hydrogen in slightly depleted in D in relation to the equilibrated water phase. By the opinion of Hamza and Epstein (1980), in perspective, the direction of the isotope geothermometry is the study of oxygen isotope distribution between different oxygen-bearing groups of a mineral and its OH-groups.

For example, according to data obtained by Fontes and Gonfiantiny (1967) for the water-gypsum system, the fractionation factors are $\alpha_{^{18}O} = 1.004$ and $\alpha_D = 0.98$. Therefore, during crystallization of gypsum, its bounded water becomes enriched in ^{18}O by 4‰ and depleted in D by 20‰ compared to the solution.

In the process of the formation of clay minerals, the isotopic fractionation of hydrogen and oxygen occurs between water and hydroxide ions of minerals, for example, those of kaolinite $Al_4(OH)_8[Si_4O_{10}]$ or montmorillonite $Mg_3(OH)_4[Si_4O_8(OH)_2]$. nH_2O. In the process of clay mineral formation from mountain rocks weathered with the help of precipitation, when the conditions of the reaction in the water-rock system are close to those at equilibrium, these minerals, as a rule, are enriched in ^{18}O by approximately 27‰ and depleted in D by 30‰ compared with the water participating in their formation (Savin and Epstein 1970a; Sheppard et al. 1971; Lawrence and Taylor 1971; Suzuoki and Epstein 1976; Sofer 1978). In Table 2.6, the corresponding fractionation factors are presented according to various authors.

It is most likely that the hydrogen fractionation factor in the mineral-water system undergoes insignificant change in the range of temperatures from normal up to 400°C. This conclusion can be drawn from the summarized data reported by Taylor (1974) and also from the results of Suzuoki and Epstein (1976) who have studied the isotopic fractionation of hydrogen between hydroxide-bearing minerals and water. These authors determined that in the kaolinite-water system at 400°C, the fractionation factor $\alpha_D = 0.977$. This α_D value is close to the value of the fractionation factor given in Table 2.6 for normal temperatures of the hypergenic zone (Savin and Epstein 1970a; Taylor 1974). Hence, when the temperature increases to more than 300°C kaolinite, being in isotopic equilibrium with water, becomes depleted in D by $\sim$ 10‰

Table 2.6 Hydrogen and oxygen isotope fractionation factors for water-clay minerals and water-gypsum systems at $t = 10–20°C$

Mineral	$\alpha_D = \frac{(D/H)_{mineral}}{(D/H)_{water}}$	$\alpha_{^{18}O} = \frac{(^{18}O/^{16}O)_{mineral}}{(^{18}O/^{16}O)_{water}}$	Reference
Montmorillonite	0.94	1.027	Savin and Epstein (1970a)
Kaolinite	0.97	1.027	Savin and Epstein (1970a)
Glauconite	0.93	1.026	Savin and Epstein (1970a)
Quartz (Si–OH), 0°C	0.92	1.039	Kolodny and Epstein (1976)
Gypsum	0.98	1.004	Sofer (1978); Fontes and Gonfiantini (1967)

Table 2.7 Hydrogen and oxygen isotope fractionation factors for water–hydroxide-bearing mineral systems (After Suzuoki and Epstein 1976)

Mineral	t, °C	$\alpha_D = \frac{(D/H)_{mineral}}{(D/H)_{water}}$
Muscovite	750	1.000
	650	0.992
	550	0.988
	450	0.976
Biotite	850	0.981
	650	0.972
	550	0.965
	450	0.957
	400	0.955
Phlogopite	650	0.990
	575	0.985
Hornblende	670	0.982
	570	0.974
	450	0.963
	400	0.961
Lepidolite	650	1.015
Serpentinite	400	0.980
Kaolinite	400	0.977
Boehmite	400	0.936

compared with kaolinite at equilibrium with water of the same isotopic composition at normal temperature (at 20°C). In Table 2.7, referring the data of Suzuoki and Epstein (1976), the hydrogen isotopic fractionation factors are given in the system of water-hydroxide-bearing minerals for temperatures above 400°C (Suzuoki and Epstein 1976).

In the range of temperatures 450–800°C, Suzuoki and Epstein (1976) obtained the following dependence of the fractionation factors on temperature for oxygen-bearing minerals:

$$10^3 \ln \alpha_{\text{muscovite-water}} = -22.1(10^6/T^2) + 19.1,$$

$$10^3 \ln \alpha_{\text{biotite-water}} = -21.3(10^6/T^2) + 2.8,$$

$$10^3 \ln \alpha_{\text{hornblende-water5}} = -23.9(10^6/T^2) + 7.9.$$

Suzuoki and Epstein also obtained the general equation for amphiboles and micas taking into account the cation substitution on the minerals' crystal grating:

$$10^3 \ln \alpha_{\text{mineral-water}} = -22.4(10^6/T^2)) + 28.2(2X_{Al} + X_{Mg} + X_{Fe} = 1).$$

where X_{Al}, X_{Mg}, X_{Fe} are the molal ratios of the corresponding $(X_{Al} + X_{Mg} + X_{Fe} + 1)$.

It is obvious that this equation can be applied to other oxygen-bearing minerals such as serpentine and kaolinite.

The data on the temperature dependence of the fractionation factor for the serpentine-water system obtained by Suzuoki and Epstein are in accordance with those obtained by Sakai and Tsutsumi (1978). In the range of temperatures 100–500°C and at water pressure of 2 *Kbar*, they have obtained the following relationship:

$$10^3 \ln \alpha_{\text{serpentine-water}} = 2.75(10^7/T^2) - 7.69(10^4/T) + 40.8.$$

This relationship contradicts the semi-empirical evidence of Wenner and Taylor (Taylor 1974). According to these, serpentines become progressively depleted in D, compared to water being in isotopic equilibrium with them, if the temperature decreases from 500 to 100°C. According to the experimental evidence of Suzuoki and Epstein (1976) and also that of Sakai and Tsutsumi (1978), the opposite process takes place (Fig. 2.7).

Figure 2.8 show the graphical dependence of the equilibrium fractionation factors of oxygen isotopes for natural mineral-water systems based on data obtained by various authors (Taylor 1974). The figure shows that practically all the oxygen-bearing rocks are enriched with ^{18}O compared with the equilibrated water. With increase in temperature, the difference in ^{18}O values for water and rock decreases. At high temperatures, under natural conditions, this effect results in ^{18}O enrichment of the groundwaters compared with ^{18}O content in precipitations. This effect is often called 'the oxygen shift'.

For isotopic investigations of natural waters, some isotopic fractionation curves for liquid and gaseous reacting components are of great hydrologic interest. Figure 2.9 and 2.10 are graphs of experimental and calculated dependences of oxygen and hydrogen isotopic fractionation factors at equilibrium for such systems. These graphs are based on data from various authors and are taken from the work of Taylor (1974).

It is evident from Fig. 2.9 that in the CO_2–H_2O gas system, the carbon dioxide is enriched with ^{18}O by about 50‰ compared to the water vapor with which it is isotopically equilibrated at 20°C. In the range of temperatures 100–300°C, this equilibrium can be calculated with the help of an equation obtained by Bottinga (1969):

$$10^3 \ln \alpha = -10.55 + 9.284(10^3/T) + 2.561(10^6/T^2). \qquad (2.80)$$

Interaction of carbon dioxide with groundwaters takes place in nature, for example, in regions of modern volcanic activity where it leads to the depletion of ^{18}O in the groundwaters. An analogous phenomenon has been observed by Ferrara et al. (1965) in thermal waters of the Larderello region.

Equation 2.80 can be used in the calculation of isotopic base temperatures in hydrothermal systems based on the oxygen isotopic composition data in water vapour and carbon dioxide. This study was done in the Larderello region (Panichi et al. 1977, 1979).

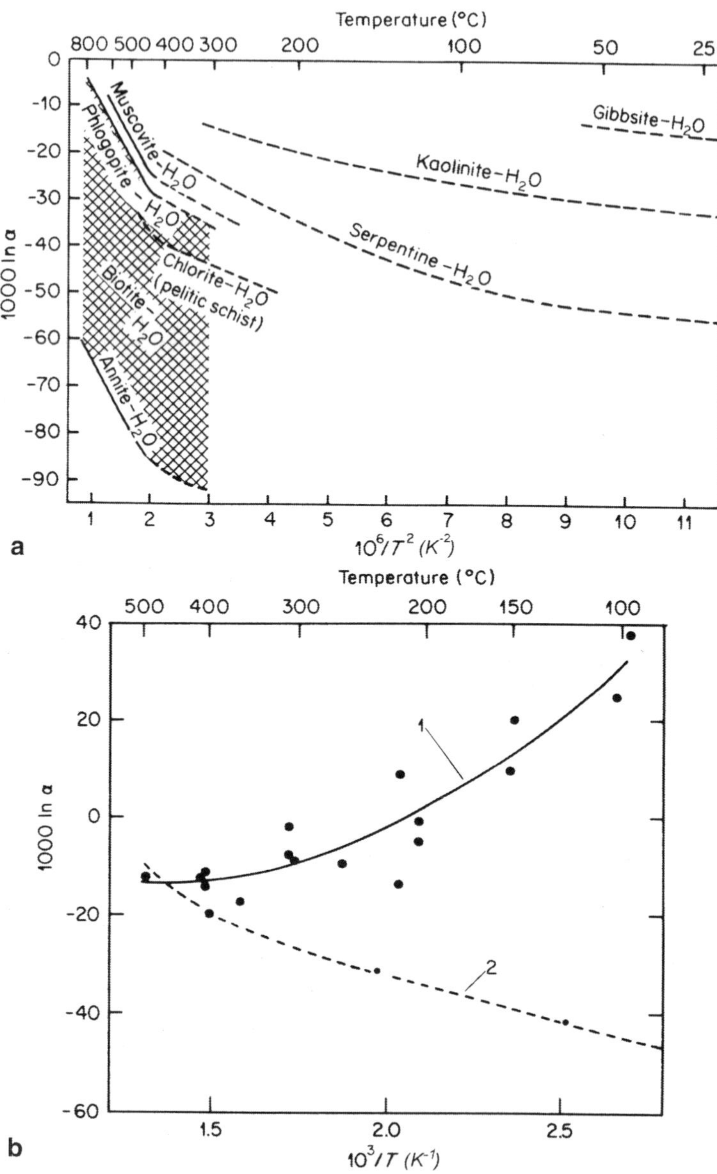

Fig. 2.7 Temperature dependences of the hydrogen isotope fractionation factor for a mineral-water system (**a**) are obtained by various authors (from Taylor 1974) and for a serpentine-water system (**b**). (*1*) experimental results of Suzuoki and Epstein (1976) and by Sakai and Tsutsumi (1978); (*2*) empirical curve of Wenner and Taylor. (From Taylor 1974)

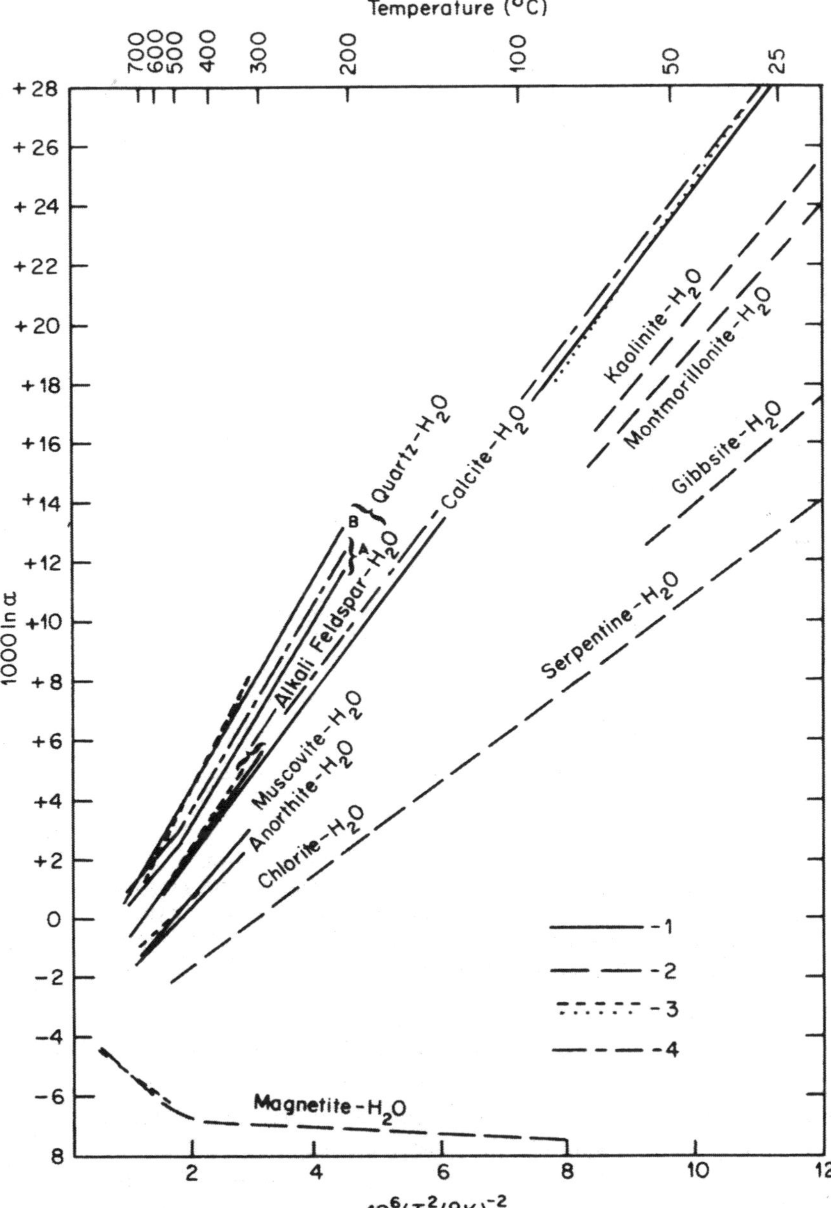

Fig. 2.8 Equilibrium oxygen isotope fractionation curves for natural mineral-water systems are obtained by various authors. (*1*) experimental curves (A—'complete' isotopic exchange; B—'patrial' isotopic exchange); (*2*) empirical curves; (*3*) calculated curves; (*4*) experimental curves based on changing the CO_2–H_2O liquid fractionation factor at 25°C from 1.0407 to 1.0412. (From Taylor 1978. © IAEA, reproduced with permission of IAEA)

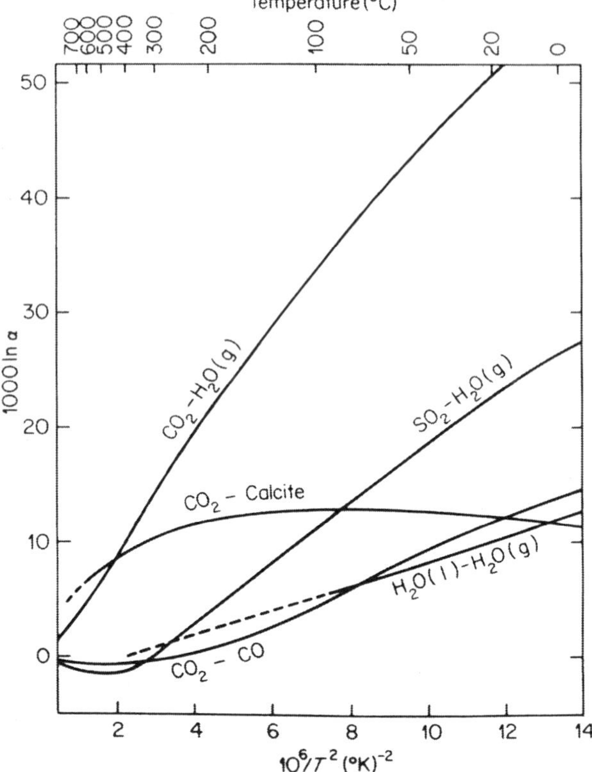

Fig. 2.9 Calculated equilibrium oxygen isotope fractionation factors as a function of temperature for CO_2–H_2O and CO_2–calcite and experimentally determined curve for liquid water–vapor water. (From Taylor 1974. © IAEA, reproduced with permission of IAEA)

Figure 2.10 shows several curves calculated for the isotopic fractionation of hydrogen, which are of interest from the geological viewpoint. From the plotted curves, one can see that under equilibrium at 400°C water will be enriched in D by 400‰ compared to the gaseous hydrogen.

Hence, if some water is reduced to gaseous hydrogen in the hydrothermal system, then, knowing the initial isotopic composition of water and the hydrogen fractionation factor in the system H_2O–H_2, the isotopic composition of the remaining water can be calculated. Let the initial value of deviation $\delta D_{H_2O} = -70‰$. Suppose 5% of the gaseous hydrogen is released after heating the water above 400°C, then for hydrogen, $\delta D_{H_2} = -470‰$. One can find the $\delta D^*_{H_2O}$ value of the heated water from the condition of normalization of the total quantity of molecules (or molal fractions) of H_2O and H_2 per unit as

$$\delta D^*_{H_2O} = \frac{\delta D_{H_2O} - 0.05\delta D_{H_2O}}{0.95} = \frac{-70 - 0.05(-470)}{0.95} \approx -50‰.$$

The approximate calculations show that exchange processes in the water-hydrogen system in principle may lead to the enrichment of water with D, but it is unlikely that the processes considered play a decisive role in the hydrogen isotopic change of

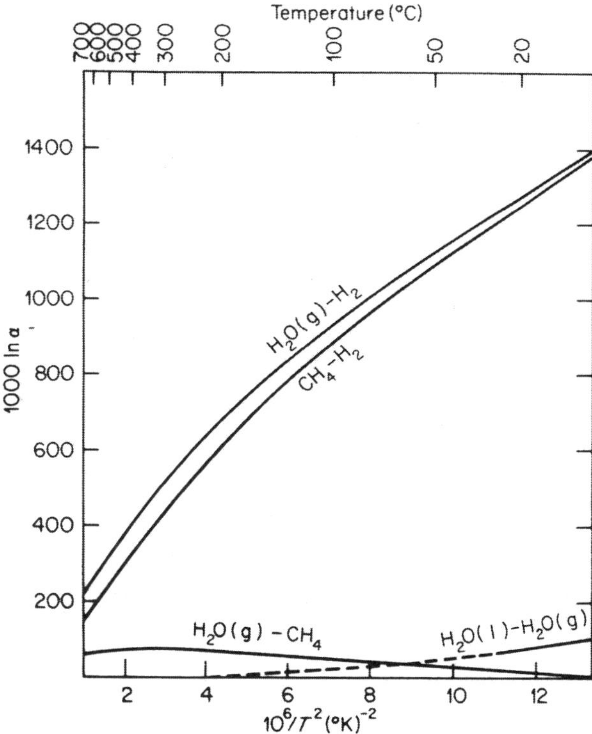

Fig. 2.10 Calculated equilibrium hydrogen isotope fractionation factors as a function of temperature for H_2O–H_2, CH_4–H_2 (experimental data) and H_2O–CH_4 (calculated values). (From Taylor 1974. © IAEA, reproduced with permission of IAEA)

groundwater since, under the changed conditions methane will be formed. However at real temperatures, as shown by Taylor (1974), the hydrogen fractionation in the H_2O–CH_4 system does not exceed 70‰ (see Fig. 2.9). In order to change the isotopic composition of water by 20‰, as in the previous case, 30% of the hydrothermal solution must react with carbon-bearing components to produce methane. As Taylor pointed out, this reaction might only occur in systems which contain carbon in large amounts (graphite, coal, and oil). It is not likely to occur under natural conditions.

The reactions of this isotopic equilibrium in such systems as water-hydrogen and water-methane can be used for determining water temperature in hydrothermal systems. For example, Arnason (1977a) has used the data on hydrogen isotopes of water and the dissolved hydrogen in water for determining base temperatures in the hydrotherms of Iceland.

The formation of the isotopic composition of oxygen and hydrogen in natural waters and the influence of different natural factors on isotopic separation processes are considered below.

Chapter 3
Isotopic Composition of Ocean Water

On an average, there are 320 molecules of HDO, 420 molecules of $H_2{}^{17}O$, and about 2,000 molecules of $H_2{}^{18}O$ for each 10^6 molecules of $H_2{}^{16}O$ in natural waters of the Earth. The ratio of isotopic abundances of deuterium (D) to protium (H) is D/H = 0.000155 (0.0150 atom.%) and that of oxygen is $^{18}O/^{16}O = 0.002$ (0.2 atom.%).

3.1 Distribution of Hydrogen and Oxygen Isotopes and Water Dynamics

The maximum limits of the variation in D concentrations, which have been found, are about 70% and those of oxygen-18 (^{18}O) are about 9%. As one may expect, precipitation over the Southern polar region is most depleted in heavy water isotopes: D/H = 0.0109 atom.% and $^{18}O/^{16}O = 0.1887$ atom.% (Schotterer et al. 1996). The most enriched waters are in the closed reservoirs of the arid zone: D/H = 0.0178 atom.% and $^{18}O/^{16}O = 0.2055$ atom.% (Fontes and Gonfiantini 1967). For other natural objects, the fixed limits of the D content variations are about 150% (in atmospheric molecular hydrogen D/H = 0.0079 atom.% (Begemann and Friedman 1968) and in the chondrites and lunar rocks D/H = 0.0195 atom.% (Briggs 1963; Friedman et al. 1971). For oxygen, these limits are about 10 atom.% (in the polar regions, the precipitation $^{18}O/^{16}O = 0.189$ atom.% (Gonfiantini 1978), and for the oxygen dissolved in ocean waters $^{18}O/^{16}O = 0.2083$ atom.% (Craig and Gordon 1965)).

It follows that the differences in the limits of D variation are one or more orders of magnitude higher than those of heavy oxygen. This can be explained by the more effective separation of D and H in natural processes than that of any other pair of stable isotopes due to the large difference in their atomic weights (Table 3.1).

For this reason, D can be considered as one of the most interesting isotopes from the geochemical viewpoint. Since the end of the 1950s, researchers studying the isotopic composition of natural waters have preferred to determine both isotopes of water hydrogen and oxygen simultaneously.

Ocean waters account for 90% of the total amount of water in the hydrosphere (Poldervaart 1955; Ronov and Yaroshevsky 1967; L'vovich 1969) (see Table 3.2).

V. I. Ferronsky, V. A. Polyakov, *Isotopes of the Earth's Hydrosphere*,
DOI 10.1007/978-94-007-2856-1_3, © Springer Science+Business Media B.V. 2012

Table 3.1 Limiting values of stable isotope variations for some elements in natural objects. (From Larionov 1963; Begemann and Friedman 1968; Briggs 1963; Craig 1963; Hoefs 1973; Gonfiantini 1978)

Element	Isotope ratio	Variation limits (absolute values)	Relative variation of limiting values
H	D/H	0.000079–0.000195	147
Li	$^{8}Li/^{7}Li$	0.079–0.084	6
B	$^{11}B/^{10}B$	0.226–0.234	3.5
C	$^{13}C/^{12}C$	0.01079–0.01225	13
O	$^{18}O/^{16}O$	0.001893–0.002083	10
Si	$^{30}Si/^{28}Si$	0.0332–0.0342	0.5
S	$^{34}S/^{32}S$	0.0427–0.0491	13

Table 3.2 Total amount of water in the hydrosphere reservoirs. (From Poldervaart 1955 and L'vovich 1969)

Reservoir	Amount of water (million km^3)		Time of water exchange, L'vovich 1969
	Poldervaart 1955	L'vovich 1969	
Oceans	1370	1370.323	2,600 years
Lakes and rivers	0.5	0.231	From 12 days to 3.3 years
Ice sheets	22	24	~11,000 years
Atmospheric waters	0.013	0.014	10 days
Water in marine and continental sediments	196	–	–
Groundwaters	–	60	~5,000 years
Soil waters	–	0.075	0.9 year
Bounded waters in the Earth's crust (Ronov and Yaroshevsky 1967)	180	380	–
Total amount in the hydrosphere	1968.513	1834.643	–

According to Ronov and Yaroshevsky (1967), Poldervaart (1955), and Vinogradov (1967a), the total amount of water in the Earth's crust and oceans is estimated to be 1,800–2,700 million km^3 (1.82.7) $\times$ 10^{24} g.

Of all natural waters, the ocean, which is a unique reservoir, remains most constant regarding its isotopic and other physicochemical properties. Several authors (Craig and Gordon 1965; Craig 1961b; Epstein and Mayeda 1953; Friedman et al. 1964) have shown that ocean water at depth of more than 500 m is homogeneous in isotopic composition. This allowed Craig (1961b) to propose it as a standard for reporting concentrations of D and ^{18}O content in natural water. The standard of the ocean water (SMOW) has the following values of hydrogen and oxygen isotope ratios: D/H = (155.76 $\pm$ 0.08) $\times$ 10^{-6} (Hagemann et al. 1970) and $^{18}O/^{16}O$ = (2005.20 $\pm$ 0.45) $\times$ 10^{-6} (Baertschi 1976). Fluctuations in the ratios of the absolute values for hydrogen and oxygen isotopes arise from the difficulties incurred in the precise preparation of the synthetic isotope mixtures for mass spectrometer calibration.

The variation in the relative content of D in deep ocean layer is about 4‰ and that of ^{18}O is of 0.3‰. In the surface ocean layer, the regional variations, depending upon water temperature, are 35‰ for D and about 3‰ for ^{18}O. The lowered content of D in the surface ocean layer occurs in those regions where ice-melting water

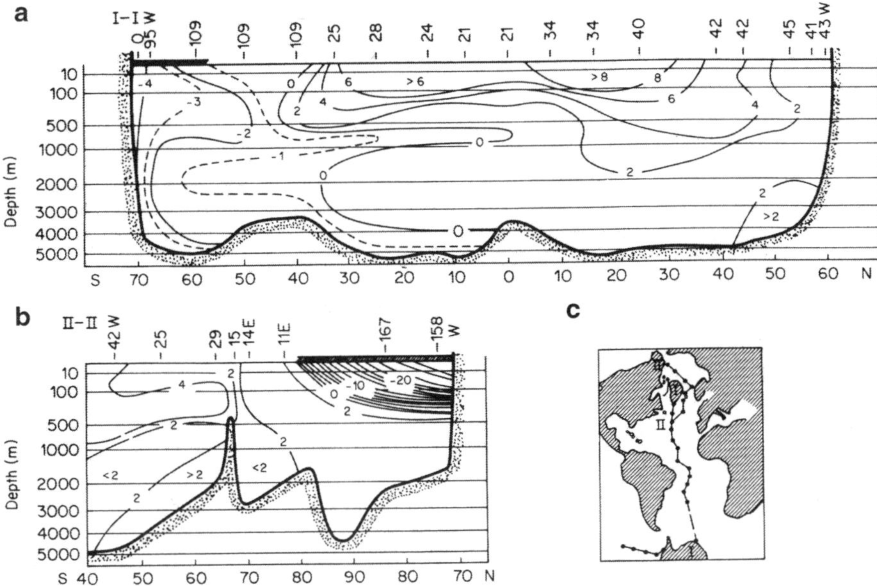

Fig. 3.1 Distribution of D concentrations (in δD‰) in a vertical section of Atlantic Ocean and Arctic basin waters. **a** section from Antarctic through the Atlantic Ocean to Greenland. **b** section from North Atlantic across the Norwegian Sea and Arctic Ocean to Alaska. **c** schematic plan of sections. (After Redfield and Friedman 1964)

affects isotopic composition. In high latitudes, where the surface layer of the ocean is freezing, isotopic fractionation factor is ~ 1.0180 for D and ~ 1.0030 for ^{18}O. In the equatorial region of the ocean where intensive evaporation of water takes place, there is an enrichment of the surface layer with heavy isotopes.

Despite the considerable homogeneity of the isotopic composition of ocean water, the ranges of D and ^{18}O variations with latitude and depth are wide enough to be reliable indicators of processes occurring in the ocean. These ranges, for deep ocean waters, are about ten times greater than the accuracy of mass spectrometer measurements with which modern techniques and apparatus permit determination of D (±0.2‰) and ^{18}O (±0,02‰).

On the basis of investigations carried out by Epstein, Friedman, and Craig with co-authors (Epstein and Mayeda 1953; Friedman et al. 1964; Redfield and Friedman 1964; Craig and Gordon 1965), we can compose a sufficiently detailed picture of D and ^{18}O distribution in ocean waters. Numerous samples of ocean water obtained during expeditions in the main oceans have been studied. The number of water samples is finite due to the ocean size; however the choice of samples, which have been taken from the most characteristic points of individual basins according to oceanographic data in common use, combined with the high precision of measurement and the coincidence of experimental data with principal conclusions which have been made by different authors, suggests that the results obtained are reliable.

Figure 3.1 and 3.2 show isolines of D distribution in the profile and in the surface layers of the Atlantic and Arctic Oceans plotted by Redfield and Friedman (1964).

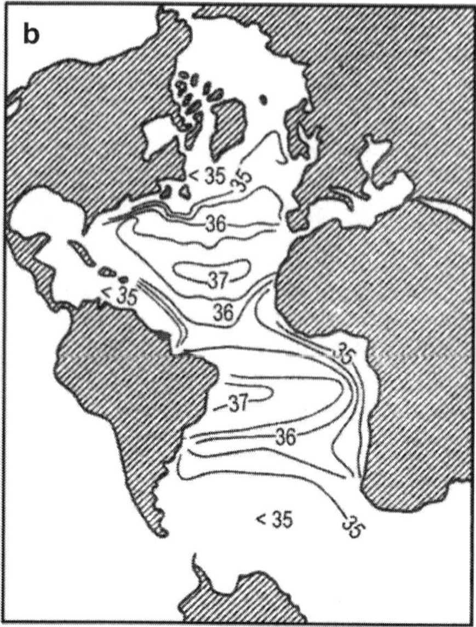

Fig. 3.2 Distribution of **a** D concentrations (in δD‰) and **b** salinity (in‰) on the surface waters of the Atlantic Ocean. (After Redfield and Friedman 1964)

Table 3.3 Regional distribution of average D concentrations in deep oceanic-water masses. (From Redfield and Friedman 1964)

Ocean	δD, ‰	Number of stations	Number of samples
Arctic Ocean	+2.2 ± 1,0	6	22
Norwegian Sea	+2.2 ± 0,7	6	22
North Atlantic	+1.2 ± 0,8	18	44
South Atlantic	−1.3 ± 0,6	7	11
Pacific Ocean	−1.4 ± 0,4	5	8
Antarctic Intermediate	−0.9 ± 0,8	10	19
Antarctic Circumpolar	−1.7 ± 0,8	6	16

Three layers are commonly distinguished according to the conditions of the distribution of individual isotope concentrations and the observed picture of the ocean water mixing at different depths. The surface layer (down to ~500 m) is characterized by the greatest ranges of local and regional variations of isotopic composition. The deep layer (below ~1,000 m) is distinguished by thorough mixing of water and uniform isotopic composition in the entire ocean. The intermediate—or mixing—layer (from ~500 to ~1,000 m) is characterized by some intermediate parameters. Sometimes in the oceanic water profile, a bottom layer can be distinguished which has characteristic regional peculiarities in some basins.

The deep ocean layer, including approximately three-quarters of its total mass, can be practically considered representative of all the Earth's hydrosphere. On the basis of experimental data, Redfield and Friedman give the following average D δ-values (in relation to the SMOW standard) for the major deep water masses in individual oceans (see Table 3.3).

Table 3.4 Regional distribution of average ^{18}O concentrations in deep oceanic-water masses. (From Craig and Gordon 1965)

Ocean	$\delta^{18}O$, ‰	Salinity, ‰
North Atlantic	+0.12	34.93
Antarctic (bottom water)	−0.45	34.65
Indian Ocean	−0.18	34.71
Pacific Ocean:		
Antarctic basin (55–65°S)	−0.21	34.700
Shousern region (22–40°S)	−0.17	34.707
Equatorial basin (6°S–30 N)	−0.17	34.692
Northern region (44–54°N)	−0.17	34.700
Antarctic Circumpolar	−0.3 to 0.2	34.69

The picture of ^{18}O δ-values for deep ocean waters according to data of Craig and Gordon (1965) is given in Table 3.4.

From the analysis of these data, the following principles of D and ^{18}O variations in deep ocean waters were obtained (Craig and Gordon 1965).

1. The ocean waters in the northern hemisphere are heavier in D and ^{18}O than those of the southern hemisphere.
2. The waters of the Arctic Ocean are highly uniform in isotopic composition and differ a little from those of the Norwegian Sea but differ significantly from the North Atlantic deep waters.
3. The waters of the Atlantic Ocean are mixing preferentially northward and have considerable variations even in the subtropical latitudes of the northern hemisphere.
4. The deep waters of the Indian and pacific Oceans are highly homogeneous in isotopic composition but have some variations in high southern latitudes. The isotopic composition of water becomes heavier in the direction of low latitudes.
5. The Pacific and Atlantic ocean waters and circumpolar Antarctic waters are approximately similar in isotopic composition.
6. In deep Pacific trenches, the water is enriched with ^{18}O by about 0.2‰ in relation to the bulk of deep waters in this ocean.
7. For deep ocean waters as a whole there is a relationship between D and ^{18}O content which is sufficiently close to linear: $\delta D = n\delta^{18}O$. The value of n is approximately equal to 10, with some regional deviations.

It appears from the information on heavy isotope distribution that waters of the ocean display a natural tendency towards constant isotopic composition. The main factors governing the observed D and ^{18}O variations are the processes which occur in surface ocean waters and have a regional character. In surface waters, the constant enrichment or depletion of water with heavy isotopes takes place due to evaporation of water and exchange with the atmosphere. Further redistribution of isotopes in the ocean proceeds as a result of mixing of surface and deep waters on global and regional scales.

The data on ^{18}O variations in the North Atlantic surface waters, obtained by Epstein and Mayeda (1953)—and recalculated by Craig in relation to the SMOW standard—are given in Table 3.5, which supplements Fig. 3.2.

Table 3.5 Variation of ^{18}O concentration in North Atlantic surface waters

Location of sampling place	$\delta^{18}O$, ‰	Salinity, ‰
43°04′N, 19°40′W	+0.68	35.,8
Off Bermuda	+1.11	36.1
Off Bermuda	+1.00	36.4
Off Bermuda	+1.30	36.8
28°05′N, 60°49′W	+1.32	36.78
Golfstream, off Norway	+0.26	35.2
44°09′ N, 68°18′W	−0.95	33.0
Off Greenland	−11.33	16.2
East cost, Greenland	−3.34	29.3

The isotopic composition of the surface ocean waters undergoes considerable variations with latitude; however, for equatorial and temperate latitudes the linear relationship between D and ^{18}O contents in the form $\delta D = n\delta^{18}O$ remains the same. The variation of n for a given region depends on the ratio of the amounts of evaporation and precipitation and is equal to 7.5 for the North Pacific, 6.5 for the North Atlantic, and 6 for the Red Sea.

3.2 Effect of Evaporation and Vertical Water Exchange

The factors governing regional isotopic variations in the surface ocean waters are: the ratio between evaporation and precipitation; the continental run-off and in a number of cases the submarine discharge of groundwater run-off; the freezing of sea water; the contribution of water from Polar and Circumpolar glaciers and also ice-melt water; the mixing of surface and deep ocean waters. Among all the enumerated factors, the most important is evaporation from the ocean surface and its ratio with precipitation in a given region.

According to Craig and Gordon (1965), the values of the ratio of evaporation to precipitation is 0.67 for the equatorial belt, 2.0 (maximum value) for the subtropical trade winds region, and 0.5 for high latitudes (above 40°).

The isotopic composition of precipitation varies with latitude in accordance with principles given in Chap. 4. The primary precipitation, condensing from vapor, has the maximum content of heavy isotopes, since the standard vapor pressure for heavy molecules of water is lower than that for the light ones and, thus, they are condensing preferentially before the light molecules. Practically, the first precipitation has an isotopic composition close to that of the sea water in a given region. Consequently, during the movement of vapor away from the evaporation region they are depleted in heavy isotopes. Therefore, in those latitudes where the balance of evaporating and precipitating moisture is observed, the isotopic composition of surface waters remains constant and is close to the mean ocean isotopic composition. In those regions, where evaporation exceeds precipitation, enrichment D and ^{18}O of the surface waters occur. In the Polar Regions where precipitation exceeds evaporation and the incoming water vapor has the lowest D and ^{18}O composition, the surface waters are considerably depleted. Craig and Gordon (1965) have given the following regional limits for

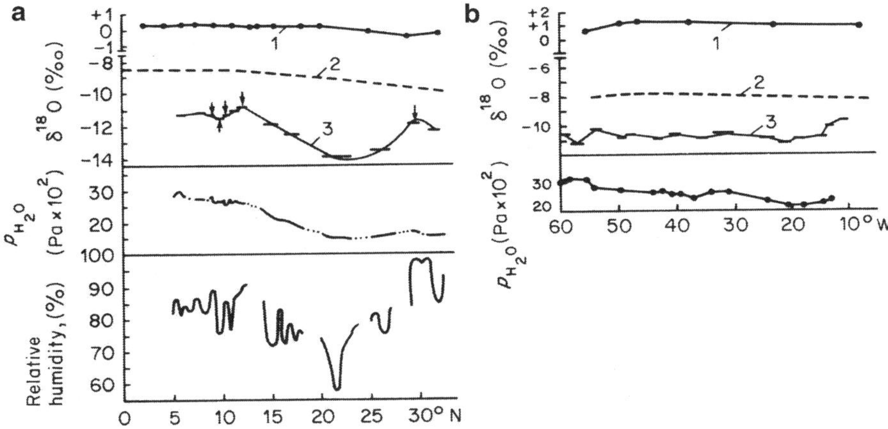

Fig. 3.3 ^{18}O variations in vapor over sea surface for **a** North Pacific latitudes and **b** North Atlantic longitudes: (1) sea surface water; (2) equilibrium vapor; (3) observed vapor data. (After Craig and Gordon 1965)

^{18}O variations in precipitation: from 0 to −5‰ for tropical and subtropical regions, from −5 to −15‰ for temperate latitudes, and below −15‰ for the polar regions.

The process of vapor isotopic composition formation during evaporation and molecular exchange above the ocean surface has a complicated and non-equilibrium character. Figure 3.3 shows the results of Craig and Gordon's work, for determining ^{18}O variations in water vapor, which had been carried about in the Pacific Ocean. The samples of vapor were taken at the height of the ship's mast. The figure also shows the curve of ^{18}O variation in surface waters in the regions under investigation and the curve of ^{18}O content in equilibrium vapor in relation to surface waters at the observational temperatures.

While comparing the experimental data with those calculated for equilibrium vapor, one can see that the observed difference in ^{18}O content is 4.5‰ and occurs in the northern part of the trade winds region (18–26° N), that is in the region of maximum evaporation. The analogous investigations carried out by Craig and Gordon across the North Atlantic along 20°N latitude have shown that, within small regional deviations, the isotopic composition of vapor is sufficiently homogeneous and is about 3‰ lighter than that at equilibrium. Thus, the following conclusion has been drawn: the latitudinal isotopic variations of vapor are more easily observed than longitudinal variations. The physicochemical effects provided by non-equilibrium conditions of condensation and water evaporation from the surface are considered in Chap. 4.

Further fractionation of oxygen and hydrogen isotopes proceeds with the freezing of water in the Polar Regions. The process goes further at equilibrium conditions and results in enrichment of ice in relation to the surface water by approximately 2‰ in ^{18}O and 20‰ in D (Craig and Gordon 1965; Redfield and Friedman 1964). In general, this effect is not huge. The main influence upon the formation of isotopic

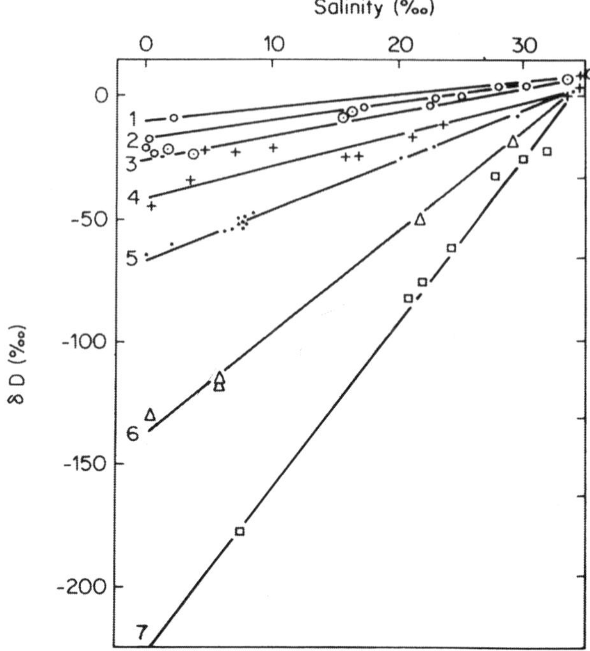

Fig. 3.4 Relationship between D content and salinity for various coastal waters: (1) Gulf of Venezuela; (2) Georgia Coast; (3) Albemarle, NC; (4) Chesapeake Bay, VA.; (5) Baltic Sea; (6) Labrador, Hamilton Inlet; (7) Greenland Fjords. (After Redfield and Friedman 1964)

composition of surface water in polar basins is expected by melt-waters from glaciers and snow formed by the accumulation of atmospheric precipitation. The analogous effect arises from continental run-off and the submarine discharge of groundwaters in the continental shelf area. The formation of the isotopic composition of the ocean waters finally depends on the process of surface and deep water mixing. The observed picture of D and oxygen isotope distribution, together with other classical experimental data, may be applied to investigate ocean water dynamics.

It is commonly known that the classical approach for experimental oceanographic studies is based on the relationship between the temperature and salinity of ocean water. The logical question arises whether or not data on isotopic composition of sea water contains any new information. It appears they do for the following reasons. Salinity characterizes sea water in terms of a solution, whereas D and ^{18}O composition characterizes just water, which is a solvent. During changes of salinity and heavy isotope composition in surface ocean waters, due to evaporation and precipitation, the salinity depends on the isotopic ratios of water in precipitation and evaporation. Therefore, the data of isotopic composition and salinity of sea water complement each other, characterizing the solution from different sides: from the side of components dissolved in it and from the side of water, the solvent.

Experimental data obtained from a number of coastal regions (Redfield and Friedman 1964) representing the relationship between D content and salinity are shown in Fig. 3.4. From this figure, it is evident that the experimental points fall on lines representing different proportions of fresh and salt water with gradients representative of given regions. All the lines meet at a point, which is characteristic of the relationship between D content and salinity for North Atlantic deep waters or their modifications at lower latitudes and shallower depth.

The D content corresponding to zero salinity is characteristics of fresh water, it varies with latitude over a wide range, attaining extreme values in the fjords of Greenland. These data are close to those for the D content in continental surface waters of corresponding regions.

3.3 Dynamics of the Ocean Water

The relationship between the isotopic composition and salinity for subsurface and deep ocean waters is illustrated by the experimental evidence provided by Craig and Gordon (1965) which is shown in Fig. 3.5 for the Pacific and the Atlantic Oceans. Craig and Gordon gives the following interpretation of the data on the relationship and motion of subsurface and deep ocean waters.

1. The position of deep waters on the isotopic composition-salinity plot reflects the average chemical and isotopic composition of ocean water. The starting point for near-surface waters on this plot is their geographical position, corresponding to zero balance of water in the precipitation–evaporation process. The position of this point is determined by the intersection of equatorial, trade winds, and polar relationships. On a geological time scale, the composition of near-surface and deep ocean waters should vary over a certain range. The variations of oxygen isotopic composition with time are limited by products of ocean activity and provide the base for oxygen paleothermometry.
2. In Pacific Ocean deep waters (see Fig. 3.5a), the identical relationship between isotopic composition and salinity found for near-surface waters of this **basic** are not observed, which suggest the absence of convective transfer water masses between near-surface and deep waters in this ocean. The South Pacific waters have a higher salinity than waters of the North Pacific although they have similar ^{18}O content. The near-surface isotopic composition of water in high southern latitudes of the Pacific and Atlantic Oceans is approximately similar and is governed by circumpolar conditions of the Antarctic region. The slope of the straight line which is characteristic for the isotopic composition-salinity relationship for near-surface waters of the southern hemisphere is greater than that for the northern hemisphere, which suggest different latitudinal variations of water composition in the two hemispheres, probably reflecting the continental effect of the northern hemisphere. In addition, high latitudes in both hemispheres generally have a greater isotopic composition-salinity ratio than low ones, reflecting the latitudinal dependency of isotopic ratios in precipitation. In the southern hemisphere in the Pacific Ocean from 30°S to the equator, an isotopic loop is observed, whereas in waters with similar ^{18}O contents the salinity varied from 35 to 34.5‰. The nature of this loop has not yet been explained.
3. From the data given in Fig. 3.5a, it is evident that the Pacific Ocean surface waters do not exert a significant influence on the formation of its deep waters.

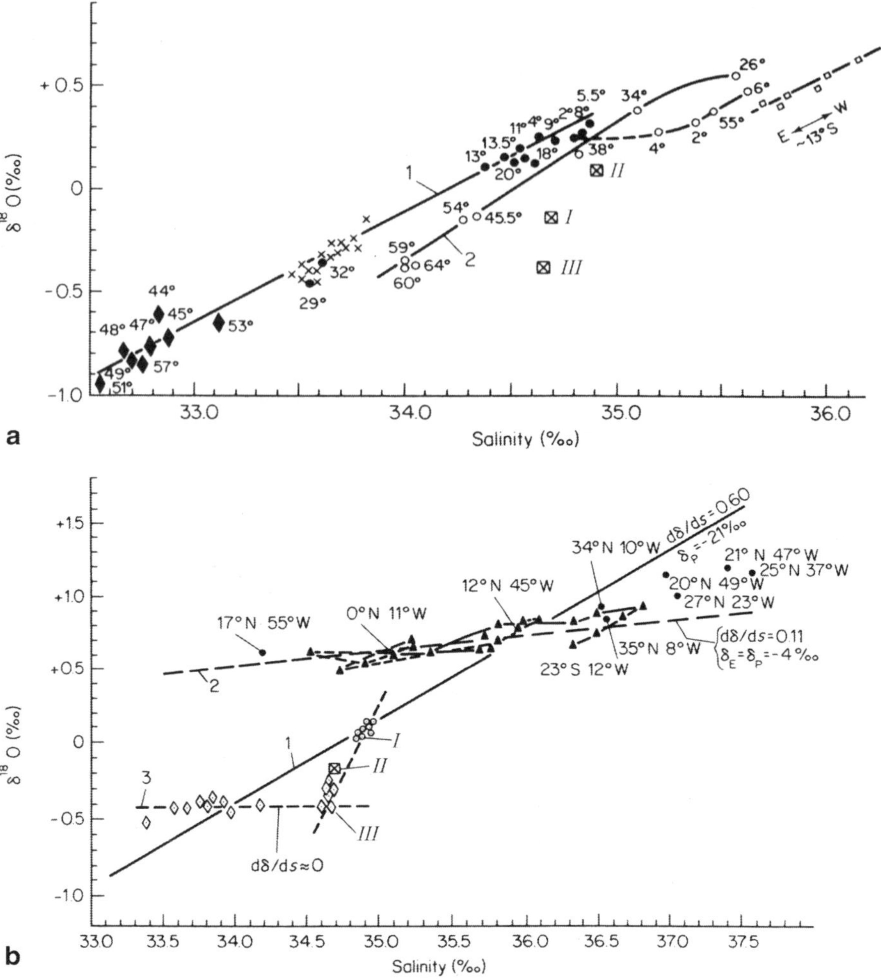

Fig. 3.5 Relationship between ^{18}O content and salinity. **a** for northern (1) and southern (2) latitudes of the surface Pacific Ocean: (I) deep and bottom Pacific; (II) North Atlantic; and (III) Antarctic waters. (The samples are from different expeditions). **b** for northern (1) and equatorial (2) regions of the Atlantic and Weddell Sea (3) waters and for deep North Atlantic (I), Indian and Pacific (II) and Antarctic waters (III). Here δE and δp refer to the isotopic composition of vapor and precipitation respectively. (After Craig and Gordon 1965)

An analogous picture is observed in the Indian Ocean. Waters of the Pacific and Indian oceans being replenished from the Atlantic and Antarctic deep and bottom waters are not simply formed by the mixing of these two components, because they lie off the range of the line connecting these two source waters. To obtain the water of the required composition, a third component is needed which has less water but the same isotopic composition. Investigations by other authors based on a temperature–salinity relationship are leading to the same conclusion. As a third possible component, the Pacific and Indian intermediate waters are considered. Besides the possibility of these waters taking part in the formation of Pacific and Indian deep waters, Craig and Gordon have also considered circumpolar deep waters having the necessary isotopic composition and salinity and being found in the South Atlantic in the Weddell Sea waters. To answer this question, more involved investigations are required.

The circumpolar waters in the North Pacific (see Fig. 3.5a) are characterized by a mean slope of the isotopic composition-salinity line of about 0.54. In contrast to Epstein and Mayeda (1953), who considered the continental run-off of fresh waters to be of great importance, Craig and Gordon (1965) found that in the formation of water composition in the sub-arctic region, the decisive role is played by the typical prevalence of sea precipitation over evaporation.

4. The North Atlantic deep waters, in contrast to the Pacific waters, are formed due to mixing with subsurface.This is observed from their position on the line of subsurface waters on the isotopic composition-salinity diagram (see Fig. 3.5b). This conclusion is in agreement with certain classic data on the convective mixing of subsurface and deep waters in this region. The data on the isotopic composition of near-surface waters are also in agreement with those of precipitation in high latitudes.

A sharp difference in the average values of the isotopic composition-salinity ratio between the equatorial region and the trade winds belt on the one hand and the high latitudinal region on the other is typical for the Atlantic subsurface waters. In equatorial regions of this ocean, in contrast to the Pacific Ocean at low salinity, rather high concentrations of ^{18}O content are observed in subsurface waters.

According to Eremeev (1972), who has studied the isotopic composition of ^{18}O in Atlantic surface waters, there is an empirical relationship between ^{18}O content and salinity of water. For high latitudes from 60°N and 40°N this relationship is

$$\delta_{34.8}{}^{18}O = 0.64S - 21.1‰,$$

where $\delta_{34.8}{}^{18}O$ is ^{18}O content in a sample relative to that in intermediate ocean waters with salinity 34.8‰; that is the observed salinity of the surface waters per mile.

For middle latitudes between 40°N and 20°N the relationship is

$$\delta_{34.8}{}^{18}O = 0.26S - 6.7‰.$$

For the equatorial region, 0–15°N.)

$$\delta_{34.8}{}^{18}O = 0.17S - 5.4‰.$$

For southern latitudes

$$\delta_{34.8}{}^{18}O = 0.34S - 11.8‰.$$

These data were used by Eremeev for calculating the evaporation–precipitation ratio in corresponding latitudes.

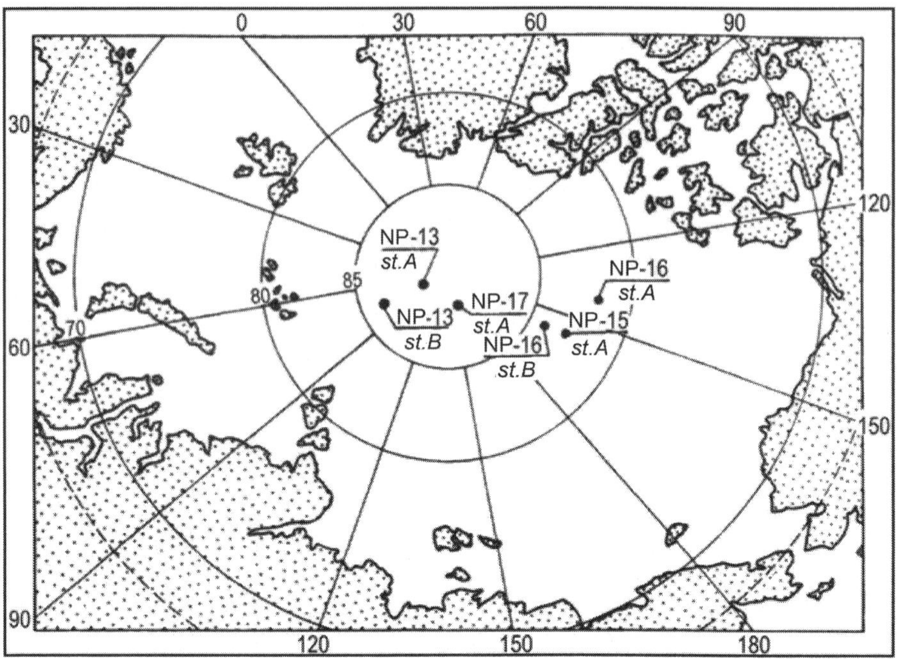

Fig. 3.6 Location of water sampling places for study of ^{18}O distribution in waters of Arctic basin. (After Vetshtein et al. 1974)

In the South Atlantic, the surface and near-surface waters in the Weddell Sea are characterized by a relatively constant ^{18}O content within a considerable range of the salinity variations ($^{18}O = -0.45‰$, when salinity ranges from 33.4‰ to 34.6‰). The salinity of these waters is not much lower than that of Pacific surface waters. The decrease of salinity in the circumpolar waters of this region reflects the dilution effect provided by ice melting.

The Antarctic bottom waters studied in two locations (64°55′S and 52°10′W at depth 2,935 m; 64°00′S and 40°50′W at depth 4,609 m) have the same values of salinity, 34.66‰ and $^{18}O = -0.45‰$, which corresponds to the minimum value of salinity required for sea water at freezing temperature to sink to the bottom. The isotopic composition of this water is the same as that of the surface water. Therefore, the process of Antarctic bottom water formation due to sinking of the surface waters agrees with the data obtained by both ^{18}O and D isotope studies (see Fig. 3.1).

In Fig. 3.5b, one can see that the Antarctic intermediate water may possibly contribute to the Pacific and Indian deep waters as the elusive 'third component'.

Isotopic composition of oxygen in the Antarctic basin during drifting expedition stations, 'North Pole–13, 15, 16 and 17' has been studied.

Figure 3.6 shows location of water sampling places during these investigations and Table 3.6 demonstrates the analysis results. On the basis of interpretation of the obtained data the authors derived, in the Arctic basin, the surface water layer up to

Table 3.6 Experimental data of $\delta^{18}O$ values obtained in the central part of Arctic basin. (From Vetshtein et al. 1974)

Sampling depth, M	Temperature, °C	Salinity S‰	$\delta^{18}O$, ‰
Station NP-13 (winter 1967)			
15	−1.71/−1.80	31.13/32.79	−2.3/−1.0
75	−1.82/−1.79	33.22/33.33	−2.0/−1.5
400	+0.88/+0.82	34.85/34.88	+0.2/+0.1
3500	−0.80/−0.80	34.92/34.90	−0.4/−0.8
Station NP-15 (winter 1967)			
5	−1.71/−1.67	31.04/30.97	−3.6/−3.8
35	−1.68/−1.66	31.00/31.00	−2.8/−2.6
130	−1.54/−1.50	33.75/33.68	−1.4/−1.6
400	+0.76/+0.76	34.87/34.88	+0.1/+0.5
Station NP-15 (winter 1969)			
10	−1.8/−1.78	29.65/32.14	−4.5/−2.5
50	−1.32/−1.74	31.73/32.14	−2.7/−2.1
100	−1.56/−1.67	32.68/33.13	−2.0/−1.4
250	+0.04/+0.38	34.63/34.72	−0.6/−0.4
400	+0.71/+0.88	34.87/34.79	−0.2/0,00
750	+0.16/−0.01	34.88/34.87	−0.2/+0.2
1500	−0.49/−0.56	34.92/34.90	−0.6/−0.3
2000	−/−0.77	−/35.92	−/+0.3
3000	−/−0.83	−/34.92	−/−0.3
4000	−/−0.70	−/34.92	−/−0.4

/Variation limits of values measured by a number of samples

the depth of 250 m, which is characterized by −4.5 to −0.8‰ value of $\delta^{18}O$ with the temperature of water from 1.65 to 1.75°C and salinity up to 32‰ (Vetshtein et al. 1974). The water layer of the depth from 250 to 750 m is characterized by higher value of $\delta^{18}O$ equal to -0.4 ± 0.3‰, temperature from 0.04 to 0.88°C, and salinity up to 34.88‰. The bottom waters have small fluctuations in $\delta^{18}O$ values in the range of 0 to −0.7‰, temperature about −0.8°C, and salinity close to 34.92‰.

By the above investigation, the surface waters located to the west from Lomonosov Ridge are formed with the Atlantic waters having high salinity and $\delta^{18}O$ about 1.0‰. The waters of the east part of the basin are formed by the ice-melting waters, with the influence of river run-off and also with the Pacific waters coming through the Bering's Strait. The following experimental relationship between ^{18}O and salinity for the Arctic basin is found by the study of Vetshtein et al. (1974): $\delta^{18}O = 0.7S - 24.6$‰. This dependence gives general idea about the degree of dilution of the regional waters by fresh waters, which has slope of the straight line as $\Delta\delta^{18}O/\Delta S = -0.7$. At zero value of salinity, one has $\delta^{18}O = -25$‰, which relates with the mean value of $\delta^{18}O$ for the fresh waters of the Arctic latitudes. The mean value of content of the fresh waters in the Arctic basin is estimated to b eabout 0.2–0.3%.

Only the initial steps in investigation of the conditions and principles of oxygen and D isotopic distribution in oceanic waters have been undertaken, mainly to study their transfer on global and regional scales. The studies have shown the considerable potential of stable isotope techniques which together with the classical parameters of the ocean water, salinity, and temperature, greatly expand our experimental ability to study the ocean dynamics.

3.4 Isotopic Composition of Ocean Water in the Past

Paleotemperature studies based on oxygen isotopic analysis of the ancient ocean sediments and glaciers (Emiliani- 1970, 1978) show that the temperature variations of the ocean surface in the equatorial region during the last 730,000 years have not exceeded 5–6°C. With regards to the variations of $\delta^{18}O$ values of the sea water, they have not deviated by more than 0.5‰ from the modern level. Similar results were obtained based on the analysis of the D content in water of the clay minerals of ocean origin and of different ages.

By Imbrie's (1985) data, temperature of the surface oceanic layer during the ice epoch has felt down not by 5–6°C, as it follows from works of Emiliani, but only by 2°C. Imbrie notes that Emiliani 's curves express not the temperature change of the ocean surface waters, but changes in isotopic and chemical composition due to pumping-over the oceanic part of water to the continental access. In this case a change of A parameter in the Epstein and Mayeda isotope paleothermometry formula (see Chap. 1) takes place. The parameter A is characterized oxygen isotopic composition of water, in which the foraminifera's shells are growing in the surface layer of water and after their death falling down to the bottom and forming the bottom sediments.

The isotopic and chemical composition of the ocean water has more than likely remained the same at least during the last 250–300 million years. This is confirmed by many facts summarized during paleotemperature studies and based on the analysis of the isotopic composition of oxygen in the shells of modern and ancient mollusks (Bowen 1966) and also of oxygen and hydrogen isotopes in the charts of different ages (Kolodny and Epstein 1976).

Arguments in favor of the stability of isotopic and chemical composition of the ocean water were considered by Lowenstam (Bowen 1966). They are:

1. The crystalline form of carbonate of the castle brachiopodas remained the same from the Mississippian time. This supports the theory that the crystallochemical process of Mg and Sr accumulation has remained stable during the last 250 million years, the period which corresponds to the time interval of the studied samples.
2. Modern organisms, living in ocean waters of variable chemical composition, have different ratios of Sr/Ca and Mg/Ca. This fact demonstrates the absence of any homeostatical mechanism due to which the definite Sr/Ca and Mg/Ca ratios might remain constant.
3. Differences in the Sr and Mg content found in fossil species of mollusks are analogous to those observed in modern ones. The interdependence between Sr and Mg content and oxygen isotopic ratios in both the fossil and modern mollusks is very close.

Bowen notes that since the isotopic composition of oxygen, the concentrations of $SrCO_3$ and Mg CO_3, and the values of the Sr/Ca and Mg/Ca ratios in those shells of the fossil mollusks studied by Lowestam always varied within the limits which are characteristic for modern samples, it seems probable that the concentration of ^{18}O in ocean waters has remained constant, at least during the last 250 million years.

One may suppose that isotopic composition of hydrogen in ocean waters has also remained unchanged during geological time. This conclusion does not contradict theoretical and experimental results related to the value of isotopic balance of hydrogen on the Earth during its geological history (Soyfer et al. 1967).

The total accumulation of D on the Earth might occur as a result of the interaction of initial cosmic rays with the atmosphere and the Earth's surface, but according to Korff's data, the output of D in nuclear reactions of neutron capture (the secondary component of cosmic rays) by H nuclei, assuming that the intensity of the cosmic rays was a constant during 3.2×10^9 years, would make up only $1/10^{10}$ part of all the D in the hydrosphere at present.

In the literature, the question of the accumulation of D in the Earth's hydrosphere caused by the preferential loss of H—by the escape of the free hydrogen from the gravitational field of the Earth—has been discussed.

It should be noted that estimations of the isotopic composition of ocean water in the Precambrian era remain uncertain. The study of the isotopic ratios of oxygen in sea carbonates and siliceous sediments shows that the ^{18}O content decreases progressively from modern time to ancient geological epochs (Degens and Epstein 1962, 1964; Perry 1967; Chase and Perry 1972, 1973; Knauth and Epstein 1975, 1976; Kolodny and Epstein 1976; Veizer and Hoefs 1976; Becker and Clayton 1976; Bowen 1991).

At present, there are three possible reasons for this phenomenon: (1) the change in the isotopic composition of rocks during their metamorphism as a result of the isotopic exchange with 'light' meteoric waters; (2) the lighter isotopic composition of waters of ancient oceans; (3) higher average temperatures of the ocean waters in ancient geological epochs. A reliable solution to this problem has not yet been obtained. Degens and Epstein (1962, 1964) explain the observed tendency in the siliceous and carbonate sediments in terms of isotopic exchange process with meteoric waters. Perry (1967), Perry and Tan (1972), and Chase and Perry (1972, 1973) have assumed on the basis of the oxygen isotopic composition analysis of the charts that the possible depletion of the ocean in Precambrian era had attained the maximum value of 20‰. Knauth and Epstein (1975, 1976) and Kolodny and Epstein (1976) have come to a conclusion based on hydrogen and oxygen isotopic composition analysis that the isotopic composition of ancient oceans was already close to the modern one in the late Precambrian era and the change with time of the oxygen isotopic composition of the charts should be completely explained by higher temperatures of the Earth's surface (up to 70°C) in the Precambrian era. Becker, and Clayton (1976), analyzing the isotopic composition of oxygen of Precambrian sediments in Western Australia of the age of 2.2×10^9 years, assume that the oxygen isotopic ratios in the Precambrian oceans could have been 10–11‰ lower than those of modern times, assuming that modern temperatures are valid for the past. If one assumes that formation of the investigated sediments occurred at $t = 60°C$ (the temperature determined by the isotopic composition of the co-existing minerals), then the $\delta^{18}O$ value of the ancient ocean waters had an upper limit equal to −3.5‰.

The process of the isotopic exchange in the system water–oxygen-bearing sediments should play an important role during the evolution of the isotopic composition of the ocean water in geological history. If one assumes that all the oceans were formed due to degassing of the mantle, then the initial ocean waters should have been enriched with ^{18}O relative to the recent waters by 6–8‰, and depleted in D by 50–80‰. This assumption follows from the supposition that the water degassing occurred at high temperatures, of which the equilibrium constant of the isotopic exchange in the water–rock system tends to unity. The hypothesis of the cold accretion of the Earth followed by the processes of warming up and mantle degassing has led several researchers (Taylor 1974) to search for the mechanism which could provide the ocean with zero values of $\delta^{18}O$ and δD during geological history. The ^{18}O in charts ($\delta^{18}O = +30 - +38$‰), carbonates ($\delta^{18}O = +25 - +30$‰), clays ($-18 - +25$‰), and other authigenic minerals may provide an example of such processes. By assuming that the ocean has had a constant volume during geological history, Savin and Epstein (1970c) and earlier Silverman (1951) have shown that, due to sedimentation, the oxygen isotopic composition of the ocean water could undergo variations of about 1‰ during 5×10^8 years. The process of weathering and sedimentation during the whole Earth's history, according to data of Savin and Epstein (1970c), might result in the enrichment of its hydrosphere with D by 0.3‰ and depletion in oxygen by 3‰.

The accumulation of D in ocean waters resulting in an increase in its mean value from about $\delta D = -65$‰ (characteristic for the upper mantle minerals) to the recent 'zero' level can be explained, as mentioned above, by the processes of H dissipation. However, we do not possess evidence in favor of this process taking place.

The authors of the present book are developing the hypothesis of the condensational formation of the Earth and its hydrosphere (see Chap. 19) based on the solutions obtained in Jacobi dynamics. This hypothesis allows one to assume that the isotopic composition of the ocean water has remained constant during the whole history of the Earth by conditions of the hydrosphere formation.

At present, an attempt can be made to estimate the hydrogen and oxygen isotopic balance from the viewpoint of plate tectonics (Le Pichon et al. 1973). If we assume that the residence time of substances in the ocean plates does not exceed 1.5×10^8 years(for continents 1.5×10^9 years), then we can assume that during geological history the ocean plates have undergone about 30 exchange cycles and the continental crust about 3 cycles. During each cycle, there were large masses of hydroxide and oxygen-bearing minerals with $\delta^{18}D$ from −50 to −80‰ and $\delta^{18}O$ from +18 to +38‰ (Taylor 1974). These processes should result in enrichment of the ocean waters with D and depletion in ^{18}O. On the contrary, however, in the region of the destructive boundaries of the plates at the middle-ocean ridges, some enrichment of waters in oxygen-18 and depletion in D may occur as a result of the exchange reactions of ocean water with basalts and serpentinization and chloritization processes at high temperatures ($t = 280$–$380°C$; Taylor 1974; Hoernes and Friedrichsen 1979). These two processes have continued further during geological history in such a way that the isotopic composition of the ocean waters, as Taylor (1974) suggested, may be 'buffered' in time near a value of zero, but this conclusion is not indisputable and requires more detailed study. At the same time, the approach to the problem of the circulation of waters in the Earth's crust and mantle from the viewpoint of plate tectonics permits one to assume that the isotopic composition of hydrogen and oxygen in igneous rocks, and therefore magmatogenic and juvenile waters, reflects the isotopic composition of the sea hydroxide-bearing minerals cycling through the mantle (Taylor 1974).

Glacial epochs may have played an important role in varying the isotopic composition of ocean waters. The total amount of water in the ocean in non-glacial time, for example during the Mesozoic, was 5% greater than that in glacial periods according to Fairbridge (1964). If we accept that all excess water was provided by ice melting, then the amount of D in the ocean in non-glacial time would be 10‰ less than during glaciation. Fairbridge estimated the amount of continental ice during glaciation to be 80 million km^3, which exceeds by factor 2 the data given by other authors. According to Bowen (1966), the excess of ice during the Pleistocene was 40.2×10^6 km^3, which corresponds to an ocean volume increase of 4% in non-glacial times. According to Emilliani (1970), the increase in the ^{18}O content of the ocean in glacial time was equal to 0.5‰. In order to carry out these calculations, the increase in the volume of ice during Pleistocene was assumed to be equal to 40.106 km^3 and the average isotopic composition of ice to be $\delta^{18}O = -15$‰.

Emilliani assumed that during the Pleistocene glaciation, the average ^{18}O content in the North American ice sheet was −9‰. At the same time, Yapp and Epstein (1977) found from the analysis of the hydrogen isotopic composition in the cellulose of plants of glacial age that, for the Pleistocene glaciers of North America, the characteristic values are $\delta^{18}O$ from −12 to −15‰. Assuming during the Wisconsin maximum glaciation, the value of $\delta^{18}O = -15$‰ for North American glaciers, $\delta^{18}O = -30$‰ and $\delta^{18}O = -40$‰ for all the others, Yapp and Epstein calculated the average enrichment of ocean waters with ^{18}O as −0.8‰. This value is in agreement with the results obtained by Craig and Gordon (1965) and Craig (1966). A greater value, $\delta^{18}O = -1.1$‰, of the average enrichment of the ocean in that period was given by Shackleton and Opdyke (1973). This result was obtained on the basis of the oxygen isotope analysis of the shells of foraminifera of Pleistocene and recent ages in equatorial parts of the Pacific ocean. However, as Yapp and Epstein (1977) have pointed out, the deep ocean waters could have been depleted in ^{18}O by 0.2‰.

Therefore, the variations in concentrations of D and ^{18}O in the oceans during glacial and non-glacial times do not exceed ±10‰ in D and ± 1‰ in ^{18}O. The isotopic composition of the oceans has remained practically unchanged at least during the last 250 million years. There is some evidence to suppose that the D and ^{18}O contents in the oceans in more ancient period (e.g., during Cambrian era) also did not differ considerably from those in the recent ocean waters.

Chapter 4
Isotopic Composition of Atmospheric Moisture

4.1 Hydrogen and Oxygen Isotope Fractionation in the Hydrological Cycle

The main factor controlling the fractionation of isotopic species in surface waters is the difference in the saturated vapor between various water molecules: $P_{\text{H}_2{}^{16}\text{O}} > P_{\text{H}_2{}^{18}\text{O}} > P_{\text{HDO}}$. The fractionation factor in isotopic species of water molecules under equilibrium conditions is determined by the ratio of the saturated vapor pressure of light (p) and heavy (p') components (Eq. 2.49): $\alpha = p/p'$. The fractionation factor α at 20°C is 1.08 for HDO and 1.009 for H_2^{18}O. In this case, the vapor in equilibrium with water will be depleted in deuterium (D) by 80‰ and in oxygen-18 (^{18}O) by 9‰. The isotopic composition of the vapor is $R_v = R_w/\alpha$. If water is taken as a standard, then

$$\delta_v = \frac{(R_v - R_w)}{R_w} = \frac{1}{\alpha - 1}, \tag{4.1}$$

and the isotopic composition of the water relative to the vapor in equilibrium (vapor is taken as a standard) will be, by analogy, equal to

$$\delta_v = \alpha - 1. \tag{4.2}$$

The fractionation factors of D and ^{18}O increases with decrease in temperature (see Table 2.4). The dependence of the factor α on temperature in general is expressed by the Eq. 4.3

$$\alpha = a \exp\left(\frac{b}{RT}\right). \tag{4.3}$$

For oxygen, this dependence in the temperature range from −20 to +100°C can be described by Zhavoronkov's equation:

$$\alpha_{^{18}\text{O}} = 0.982 \exp\left(\frac{15.788}{RT}\right).$$

V. I. Ferronsky, V. A. Polyakov, *Isotopes of the Earth's Hydrosphere*,
DOI 10.1007/978-94-007-2856-1_4, © Springer Science+Business Media B.V. 2012

The most important process involved in the formation of the isotopic composition of atmospheric precipitation is the condensation of water vapor. In a closed system, a small amount of initial condensate obviously has the same isotopic composition as that of liquid in equilibrium with vapor. With the further condensation of moisture, the isotopic composition of the condensate changes in accordance with the Eq. 4.4 (Dansgaard 1964):

$$\delta_k = \left(\frac{1}{\alpha_0}\right)\left(\frac{1}{\varepsilon F_v + 1}\right) - 1, \tag{4.4}$$

where $\varepsilon = (1/\alpha) - 1$; α_0 is the fractionation factor at the beginning of fractionation; F_v is the remaining part of the vapor.

If the temperature remains constant, then $\alpha_0 = \alpha$. The isotopic composition of the remaining vapor phase δ_v can be described by the equation

$$\delta_v = \left(\frac{1}{\alpha_0 \alpha}\right)\left[\frac{1}{(\varepsilon F_v + 1)} - 1\right]. \tag{4.5}$$

It follows from Eqs. 4.4 and 4.5 that with a decrease in temperature of the system the vapor, and therefore the liquid phase being formed, becomes depleted in heavy isotopes of hydrogen and oxygen to a greater extent.

Equations 4.4 and 4.5 describe the closed system in a state of equilibrium. If the condensate precipitates from the vapor phase directly into the solid one, the isotope exchange does not take place between the sublimate and the vapor. In this case, the isotopic composition of the condensate and the vapor phase can be given by the Rayleigh's equation (Dansgaard 1964).

$$\delta_c = \left(\frac{\alpha}{\alpha_0}\right) F_v^{\alpha-1} - 1, \tag{4.6}$$

$$\delta_v = \left(\frac{1}{\alpha_0}\right) F_v^{\alpha-1} - 1. \tag{4.7}$$

It is evident from Eqs. 4.6 and 4.7 that, in this case, the vapor phase becomes depleted in heavy isotopes to a greater extent during the cooling of the system. Figure 4.1 shows the dependence of isotopic fractionation of vapor and condensate as a function of the remaining amount of vapor F_v.

If Eq. 4.7 is expressed in logarithms and differentiated with respect to temperature t, the equation describing the dependence of the rate of isotopic fractionation $d\delta/dt$ on temperature changes is obtained. In other words one obtains a relationship describing the dependence of the isotopic composition of precipitation on temperature of condensation. The dependence is experimentally determined between the concentrations of ^{18}O and D isotopes in precipitation and the surface average temperature t_a of air (Dansgaard 1964). It is expressed for a broad range of temperatures by the equations

$$\delta^{18}O = 0.695 t_a - 13.6‰, \tag{4.8}$$

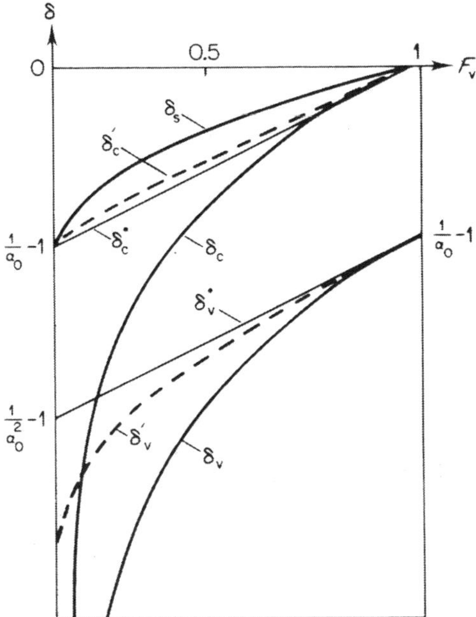

Fig. 4.1 Isotope separation in vapor–condensate system as a function of the vapor portion F_v remained: $\delta_v^*(\delta_c^*)$ is the equilibrium curve between the liquid and vapor phases for isothermal condensation; $\delta_v'(\delta)_c'$ is the same for condensation at cooling; $\delta_v^*(\delta_s)$ is the equilibrium curve for the sublimation process; δ_s is the averaged isotopic composition of the solid phase; $\delta_v(\delta_c)$ is the Rayleigh's curve for sublimation or condensation at cooling on the newly formed condensate. (After Dansgaard 1964)

$$\delta D = 5.6t_a - 100‰. \tag{4.9}$$

The dependence of isotopic composition on annual average temperature expressed by Eqs. 4.8 and 4.9 is only true for coastal regions with mild and cold climate. With the increase in distance from the sea, the continental effect appears. In this case, the calculated and experimental data do not often coincide. For example, for Vienna the annual average is $\delta^{18}O = -9‰$ whereas Eq. 4.9, using the annual average temperature (+9.5°C), produces $\delta^{18}O = -7‰$ (Drost et al. 1972). Mook (1970) has observed the continental effect even in the Netherlands.

The dependence of the oxygen and hydrogen isotopic composition upon the monthly average temperature t_m has been expressed in various ways. For example, for the Moscow region, the dependence is expressed by the following empirical equations (Polyakov and Kolesnikova 1978):

$$\delta^{18}O = (0.34 \pm 0.03)t_m - (12.6 \pm 0.3)‰, r = 0.82;$$

$$\delta D = (2.4 \pm 0.2)t_m - (101 \pm 2)‰, r = 0.89.$$

For Vienna, the corresponding equations are (Polyakov and Kolesnikova 1978; Hübner et al. 1979b):

$$\delta^{18}\mathrm{O} = (0.40 \pm 0.04)t_{\mathrm{m}} - (13.2 \pm 0.5)‰, r = 0.74;$$

$$\delta\mathrm{D} = (2.8 \pm 0.3)t_{\mathrm{m}} - (96 \pm 4)‰, r = 0.78;$$

$$\delta\mathrm{D} = (2.94 \pm 0.19)t_{\mathrm{m}} - (99.3 \pm 16.6)‰, r = 0.805.$$

For England, we have (Evans et al. 1979):

$$\delta^{18}\mathrm{O} = 2.93t_{\mathrm{m}} - 8.62‰, r = 0.77.$$

For the stations Thule, Groenedal, Nord, and Vienna, we have (Gat and Gonfiantini 1981; Yurtsever and Gat 1981):

$$\delta^{18}\mathrm{O} = (0.521 \pm 0.014)t_{\mathrm{m}} - (14.959 \pm 0.208)‰, r = 0.893, n = 363.$$

It is evident from Eqs. 4.8 and 4.9 that

$$\frac{\mathrm{d}\delta^{18}\mathrm{O}}{\mathrm{d}t} \approx \frac{0.7‰}{1°\mathrm{C}}, \tag{4.10}$$

$$\frac{\mathrm{d}\delta\mathrm{D}}{\mathrm{d}t} \approx \frac{5.6‰}{1°\mathrm{C}}, \tag{4.11}$$

The annual average value of δ^{18}O in precipitation from different regions is shown in Fig. 4.2 as a function of surface annual average temperature. (Dansgaard 1964)

The relationship between the content of D and ^{18}O in precipitation based on Eqs. 4.8 and 4.9 can be written in a general form as:

$$\delta\mathrm{D} = a\delta^{18}\mathrm{O} + \mathrm{b}. \tag{4.12}$$

The value of the coefficient a can be determined from the relation

$$a = \frac{\mathrm{d}\delta\mathrm{D}}{\mathrm{d}\delta^{18}\mathrm{O}} = \frac{\dfrac{\mathrm{d}\delta\mathrm{D}}{\mathrm{d}t}}{\dfrac{\mathrm{d}\delta^{18}\mathrm{O}}{\mathrm{d}t}} = \frac{5.6}{0.69} \approx 8.0.$$

Dansgaard's theoretical calculations (Dansgaard 1964) and the experimental evidence of several researchers have shown that for such precipitation as rain and snow in most regions ranges from 8.0 ± 0.2 to 8.1 ± 0.4. Some exceptions are the data obtained at island and ship stations located mainly in the tropical and subtropical zones. In this case, the slope of the line obtained for 15 stations is equal to 4.6 ± 0.4. This fact has not yet been explained satisfactorily on theoretical grounds since the minimal theoretical value of the coefficient a in equilibrium during condensation (following Rayleigh's law in the range of temperatures ±20°C) is equal to 7.5 (Dansgaard 1964).

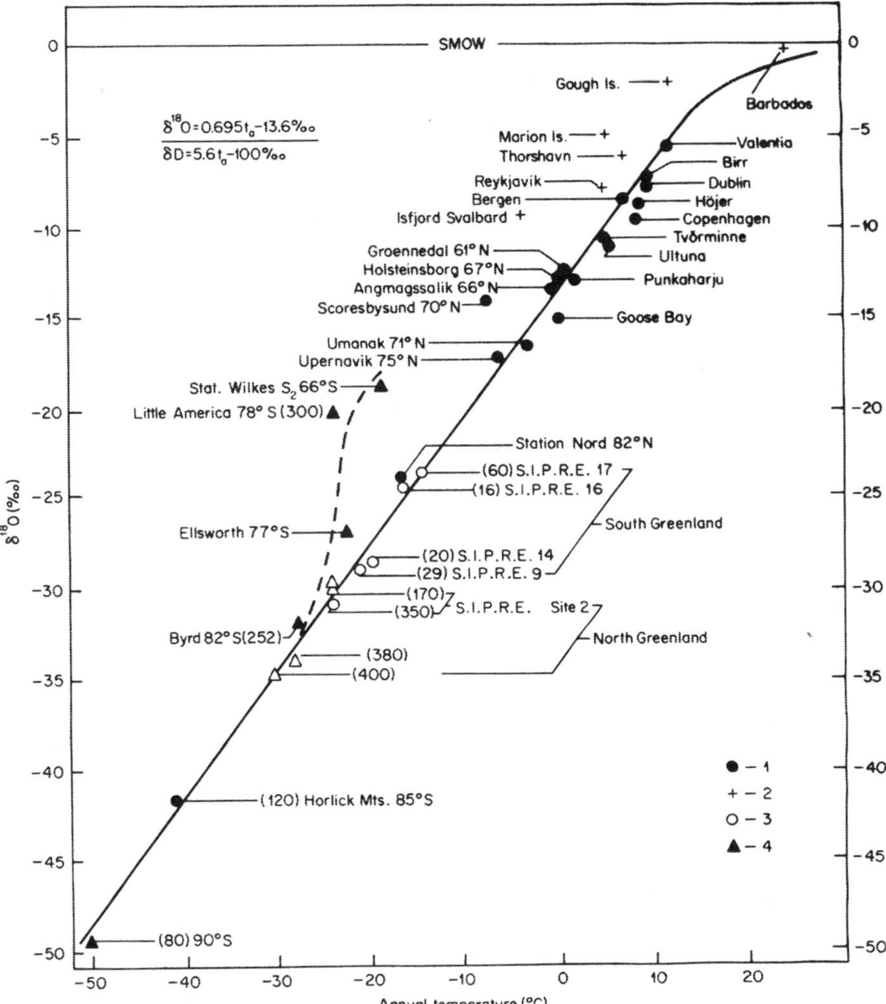

Fig. 4.2 Annual mean values of $\delta^{18}O$ in precipitation as a function of air temperature of the Earth's surface. (*1*) Continental stations of the North Atlantic; (*2*) Island stations; (*3* and *4*) Greenland and Antarctic stations. Figures in the brackets mean the snow layer thicknesses investigated. (After Dansgaard 1964)

Constant b in Eq. 4.12 characterizes the degree of disequilibrium in the system caused by the evaporation of initial masses of the ocean water, that is the rate of its evaporation. If the evaporation of the ocean water occurs under equilibrium conditions, b will be equal to zero; however, due to kinetic effects provided by the evaporation of water under non-equilibrium conditions the vapor formed becomes depleted in ^{18}O to a slightly greater extent. Therefore $b \neq 0$; for most stations on the

Earth, where experimental determination of D and ^{18}O in precipitation were carried out, $b = 10‰$. The values of b obtained on some island and ship stations are lower and sometimes below zero. Parameter b in Eq. 4.12 was defined by Dansgaard (1964) as $d = \delta D - \delta^{18}O$ (where d is the 'excess parameter' determining the D excess in the atmospheric precipitation relative to its amount in the equilibrium process, when $d = 0$).

For some regions, the value d is >10‰. For example, on the East Mediterranean coast $d = 22‰$ (Gat and Carmi 1970), for Japanese islands $d = 17.5‰$ (Sakai and Matsubaya 1977), for Alexandria $d = 15.9$, for Karizimir $d = 23.35‰$, and for Invercargill $d = -0.3‰$ (Yurtsever and Gat 1981). The parameter d is affected by the process of condensation of the precipitation and its value is decreased with distance from the region of vapor formation and the place of precipitation (continental effect). For Valentia $d \approx 10$, for Vienna $d \approx 5‰$, for Moscow $d \approx 0$.

As per Yurtsever and Gat (1981), in general case the mean weighted value of $\overline{\delta D}$ and $\overline{\delta^{18}O}$ is described by many-parametric dependence of form

$$\delta^{18}O = a_0 + a_1 T + a_2 P + a_3 L + a_4 A,$$

where T is the mean monthly temperature in °C; P is the mean monthly amount of precipitation in mm; L is the geographic latitude in grad.; A is the altitude over the sea level in m; a_0, a_1, a_2, a_3, a_4 are the regressive coefficients.

Calculation by the least square method for the 91 IAEA network stations has given the following values of the coefficients: $a_1 = 0.815$; $a_2 = 0.303$; $a_3 = 0.722$; $a_4 = 0.007$.

The relationship between δD and $\delta^{18}O$ for precipitation, plotted from experimental data, is given in Fig. 4.3. This dependence was first reported by Craig (1961a) based on a great number of experiments. Therefore, the regression line

$$\delta D = 8\delta^{18}O + 10‰ \tag{4.13}$$

is often called the Craig line.

Correlation dependence between the mean weighted (annual) values of $\overline{\delta D}$ and $\overline{\delta^{18}O}$ for many stations ($n = 74$) calculated by Yurtsever and Gat (1981) have the form

$$\overline{\delta D} = (8.17 \pm 0.08)\overline{\delta^{18}O} + (6.03 \pm 3.08), ‰; r = 0.997.$$

The equation is close to the Craig's dependence. For stations of the North American continent, the dependence obtained by Yurtsever and Gat is

$$\overline{\delta D} = (7.95 \pm 0.22)\overline{\delta^{18}O} + (10.56 \pm 0.64), ‰; r = 0.997.$$

It should be noted that Eq. 4.13 is only true for the annual average hydrogen and oxygen isotopic data with respect to different regions of the Earth and needs to be corrected for the analysis in the other time ranges (Gat and Dansgaard 1972).

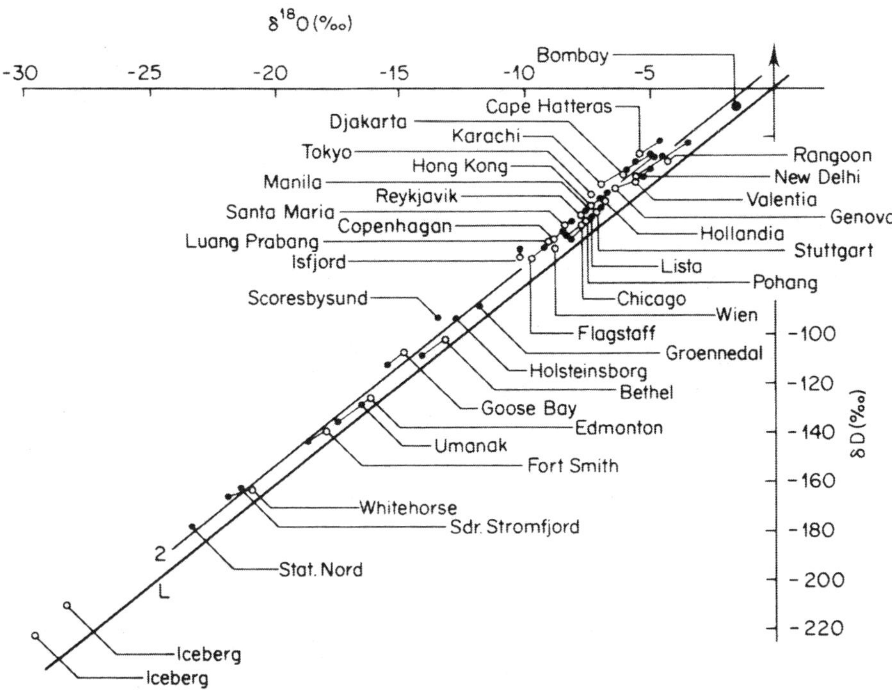

Fig. 4.3 D and ^{18}O relationship in precipitation and continental meteoric water based on the global network data of the Northern hemisphere and expressed by $\delta D = 8\delta^{18}O + 10‰$. (After Dansgaard 1964)

Regression analysis of many years' observations of D and ^{18}O concentrations in precipitation at individual stations on monthly average basis have shown that for the equation $\delta D = a\delta^{18}O + b$, the regional coefficients given in Table 4.1 should be taken into account.

For the United Kingdom, the following relationship has been obtained on a monthly average basis (Bath et al. 1979; Evans et al. 1979):

$$\delta D = 6.6\delta^{18}O + 1.3‰; r = 0.83;$$

$$\delta D = 6.6\delta^{18}O + 1.4‰; r = 0.85;$$

For the Europe, as a whole (Evans et al. 1979):

$$\delta D = 6.9\delta^{18}O - 0.3‰; r = 0.95.$$

For the northeast of Brazil, this dependence has the form (Salati et al. 1980):

$$\delta D = 6.4\delta^{18}O + 5.5.$$

Table 4.1 Calculated numerical factors ($\delta D - \delta^{18}O$ relationship) for monthly average precipitation data of some observational stations. (From Polyakov and Kolesnikova 1978)

Type of Station	Station	Annual Average Data			Number of Month's Observations	Monthly Average Factors		Correlation Factors
		Precipitation (mm)	Vapour Pressure (mb)	Temperature (°C)		*a*	*b*	*r*
Island	Havaii	3470	21.3	23	92	4.0 ± 0.5	3.0 ± 1.1	0.66
–	Azores	1200	15.4	17	48	5.5 ± 0.6	4.0 ± 1.0	0.92
–	Valentia	1400	10.7	11	72	6.1 ± 0.6	−3.7 ± 0.5	0.80
–	Adak	1400	6.0	5	66	6.1 ± 0.6	−8.9 ± 5.4	0.80
–	Isfiord	350	3.7	−4	42	6.1 ± 0.4	−8.5 ± 3.7	0.93
–	Nord	150	1.1	−18	78	7.2 ± 0.2	−8.2 ± 3.4	0.99
Coastal	Gibraltar	815	15.4	18	48	5.6 ± 0.4	−0.6 ± 1.4	0.91
–	Bethel	490	4.3	−2	48	5.9 ± 0.2	−39.7 ± 2.8	0.97
–	Barrow	110	1.7	−12	66	5.5 ± 0.5	−39.7 ± 9.5	0.84
Continental	Vienna	660	8.5	10	108	6.1 ± 0.3	−11.5 ± 3.5	0.92
–	Moscow	575	6.5	4	60	6.0 ± 0.5	−26.2 ± 6.7	0.89

As Table 4.1 shows, the coefficient a for all stations appeared to be less than 8. There is tendency of the coefficient b to decrease with a decrease in the annual decreases to approximately −8.3 and remains constant within the wide range of temperature from 5 to 18°C. The dependence of the coefficient b upon the annual average temperature t (°C) can be approximated by:

$$b = -(24 \pm 5) + 1,2\bar{t}.$$

As the atmospheric moisture travels further from the Atlantic across the Eurasian continent, b decreases by approximately 0.7‰ per 100 km and the decrease of the relative values of D and ^{18}O concentrations is 2.5 ± 0.5‰ and 0.35 ± 0.05‰ per 100 km, respectively. Similar relationship for Europe and Africa has been reported by Sonntag et al. (1979).

The dependence of b upon temperature is probably the result of the processes involved during the non-equilibrium condensation of the precipitation and during the partial evaporation of the liquid-droplet part of the moisture cloud whilst the primary source of moisture vapor is continental.

As pointed out by Polyakov and Kolesnikova (1978), for all regions with an annual average temperature $\bar{t} < 20$°C, the relationship between δD and δ^{18}O in the total annual precipitation follows the empirical relationship

$$\delta\mathrm{D} = 8\delta^{18}\mathrm{O} + 10 - 0.7l^2‰,$$

where l is the distance from the ocean (the source of water vapor) in thousands of kilometers. For the coastal and island stations, $l = 0$.

Therefore, Craig's equation (Eq. 4.13) branches out on sets of parallel lines with varying slopes which are dependent on distance from the ocean.

The relationship between the isotopic composition of the monthly average precipitation for the stations with annual temperature in the range of $-15°\mathrm{C} < t < 20°\mathrm{C}$ can be approximated by the empirical equation

$$\delta\mathrm{D} = 6\delta^{18}\mathrm{O} - 0.7l^2 + 0.7\bar{t} - 7‰.$$

The relationship between δD and δ^{18}O for different regions is often used for the interpretation of data for hydrogeological reconstructions. Craig's equation, which determines the global relationship between the D content and the ^{18}O on an annual average basis is often employed. As seen from Table 4.1, the regional effects are not taken into account, which might lead to appreciable errors.

The description of the formation processes of isotopic composition of precipitation based on equilibrium, the Rayleigh fractionation is the approximate natural model of moisture condensation and precipitation. Nevertheless, in a series of cases, the use of Rayleigh's formula for natural processes coincides satisfactorily with the experimental data. Thus, Eriksson (1965b), suggested that atmospheric moisture be considered as the uniform vapor phase in which the liquid phase forms under isothermal conditions in accordance with the Rayleigh mechanism. With such an assumption, the average isotopic composition of atmospheric precipitation R_p is a function of the atmospheric moisture W and the water content in the vertical atmospheric column, that is

$$\frac{dR_p}{Rp} = (\alpha-1)\left(\frac{dW}{W}\right). \tag{4.14}$$

Under isothermal conditions, $\alpha = \text{const}$. Integrating Eq. 4.14 and substituting $\ln R_p = \ln(1+\delta_p) \approx \delta_p$, one obtains

$$\delta_p = 2.3(\alpha - 1)lgW + B. \tag{4.15}$$

Equation (4.15) describes the process of advective transition of moisture in the atmosphere. For turbulent process, Eriksson (1965b) suggests the substitution of $\sqrt{\alpha - 1}$ for $\alpha - 1$; then

$$\delta_p = 2.3\sqrt{(\alpha - 1)}lgW + R. \tag{4.16}$$

The experimental justification of Eqs. 4.15 and 4.16 carried out by Eriksson by comparing the annual average isotopic composition of atmospheric precipitation for a series of island stations showed that, despite the considerable scattering of the experimental evidence, the points in the diagram δ^{18}O–lgW are predicted by the above equations for $\delta^{18}\text{O} = 1.009$.

Brezgunov (1978), on the basis of the analysis of a large amount of experimental data, obtained empirical relationship between δD and lgW for regions of meridional water vapor transition above the ocean and inland transition above Europe and North America. For winter, this relationship in the case of the sea profile is

$$\delta\text{D} = 106lg\ p - 138‰, \tag{4.17}$$

and for the inland profile, it is

$$\delta\text{D} = 193lg\ p - 215‰. \tag{4.18}$$

For summer, in the case of sea profile, the relationship is

$$\delta\text{D} = 174lg\ p - 219‰, \tag{4.19}$$

and for the North American continent, it is

$$\delta\text{D} = 313lg\ p - 456‰. \tag{4.20}$$

Brezgunov noted that the greater rate of δD decrease in summer precipitation with the decrease of lg p (dδD/d$lg\ p$), for the stations of the ocean and North American inland profiles, in comparison with winter precipitation, may be explained by the additional isotopic fractionation provided by raindrop evaporation below the level of the clouds.

The dependence of the content of heavy hydrogen and oxygen isotopes on the vapor condensation temperature leads to seasonal deviations of D and ^{18}O in precipitation reaching a maximum in summer and minimum in winter. For the same

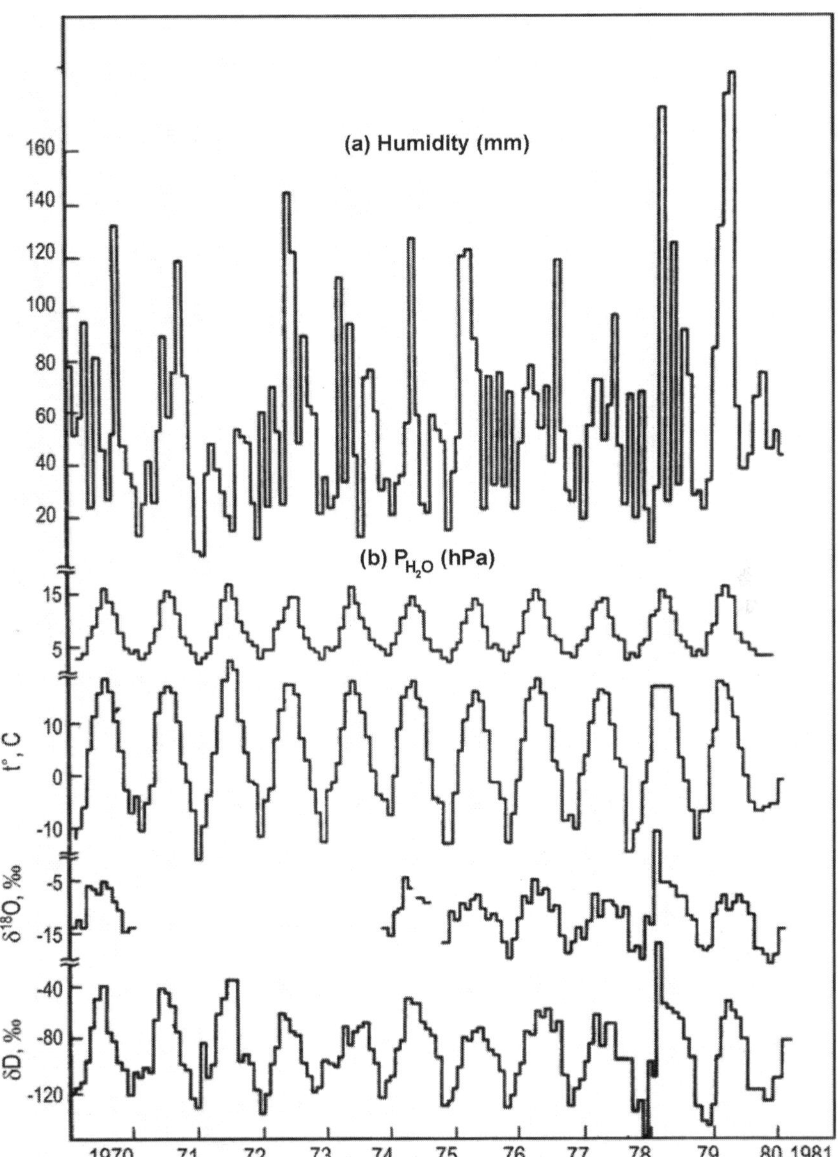

Fig. 4.4 Seasonal variation of δD and $\delta^{18}O$ in precipitation, its amount, vapor pressure and temperature for Moscow station in 1970–1980

reason, the oxygen and D concentrations in meteoric waters decrease in higher latitudes and altitudes. Figure 4.4 demonstrates seasonal variation in contents of the heavy isotopes of hydrogen and oxygen for Moscow station (Zeleny settlement).

It was found that the values of δD and δ^{18}O depend on the amount of precipitation. This phenomenon is called 'amount effect' (Dansgaard 1964). The calculations done by Yurtsever and Gat (1981) for tropical stations with close mean annual temperatures give the following relationship between mean annual values of $\sqrt{\delta^{18}\mathrm{O}}$ and precipitation P:

$$\sqrt{\delta^{18}\mathrm{O}} = (-0.015 \pm 0.0024)P - (0.047 \pm 0.419); r = 0.874.$$

It is seen from this equation that precipitation is decreased in ^{18}O by about 1.5‰ in each 100 mm. The average mean monthly values for two stations (Apia and Madang) gives mean correlated relationship between the mean monthly values of δ^{18}O and P (Yurtsever and Gat 1981):

$$\text{for Apia } \delta^{18}\mathrm{O} = (-0.010 \pm 0.003)P - (1.56 \pm 0.42);$$

$n = 52$
$r = 0.67$

$$\text{for Madang } \delta^{18}\mathrm{O} = (-0.011 \pm 0.002)P - (3.95 \pm 0.69);$$

$n = 48$
$r = 0.60$

For both these stations, the average decrease of precipitation in ^{18}O is equal to about 1‰ in 100 mm.

It is evident from Fig. 4.1 and from Eqs. 4.4 and 4.6 that development of the amount effect seems to relate to the degree of condensation of the initial vapor cloud at the formation of precipitation. Levin et al. (1980) observed depletion in ^{18}O of the subsequent portion of precipitation in the mountain region Negev, Israel. The last portions of precipitation (on 1 mm) were collected automatically by special hermetic device. The results obtained were as follows:

Precipitation (mm)	δ^{18}O
1st	−3.27‰
2nd	−4.79‰
3rd	−5.56‰
4th	−7.27‰
Mean value	−5.22‰

The δ^{18}O value of precipitation in total sample taken by standard sampler was −5.45‰.

Because of dependence of isotopic composition of precipitation on temperature, D and ^{18}O concentrations in meteoric waters decrease at increase of latitude and absolute mark of the land. The highness effect is developer different in regions with different climatic conditions. The gradient values $\Delta\delta\mathrm{D}/\Delta h$ and $\Delta\delta^{18}\mathrm{O}/\Delta h$, as a rule, are varying in ranges of 1.5–4‰ (per 100 m for δD) and 0.15–0.5‰ (per 100 m for δ^{18}O) (Yurtsever and Gat 1981).

4.2 Isotopic Balance in the Global Hydrologic Cycle at Evaporation and Condensation of Water

The unity of natural waters on the Earth consists of their genetic relationship which has continued during the whole history of the hydrosphere. The principal reservoir of the hydrosphere—the ocean—contains about 97% of the total amount of water, taking no account the water being bonded in rocks. The other 3% is represented by the polar and continental glaciers, which have accumulated about 2.2% of water in the solid phase, the underground waters (0.7%), and the surface continental waters, which make up less than 0.05% of the total amount of water. The water vapor in the atmosphere represents a small part of the water (about 0.0015%) but is the most dynamic element of the hydrosphere, responsible for the genetic relationship between the ocean and the continental surface and groundwaters. While the total mass of the atmosphere is about 5.2×10^{21} g the mass of water vapor equals to about 1.3×10^{19} g, which amounts to about 0.25%. The mean synoptic rate of motion of the air in the atmosphere is 10^3 cm/s and the time of the average water cycle is equal to 11 days. Thus the process of hydrosphere mixing proceeds rather intensively. In the previous section, it was pointed out that out of $\sim 4 \times 10^{20}$ g of annual average precipitation a quarter falls over the continents and three-quarters over the oceans. At the same time evaporation from the ocean surface exceeds that from the continents, being 85% and 15%, respectively. The difference between precipitation and evaporation over the continents is equal to 10% or $\sim 3.7 \times 10^{19}$ g, which corresponds to that mass of water which is being annually exchanged between the ocean and continents through the atmosphere. The continents return to the ocean the same amount of water in the form of surface and groundwater run-off.

The most important problem in hydrology is the determination of natural principles governing the process of the water cycle on the Earth and the variation of the ratio of its elements with time. The existence of the evaporation-condensation-atmospheric precipitation process is absolutely indisputable from the moment when water appeared on the Earth. However, now the volume of the hydrosphere and the water balance within it has changed with time, and it is not yet understood if its origin is a result of the continuous degassing of water from the Earth's interior. If the bulk of the Earth's ocean has not changed during the greater part of its history, then the question arises as to the timing and mechanism of its appearance. The temperature history of the Earth has played a decisive role in the origin and evolution of the hydrosphere and also in the evolution of water cycle elements. The geological and paleontological facts give evidence on secular variations of temperature in the range of 6–8°C or more on the Earth's surface. The causes of these variations remain unknown. In the meantime, they play an important role in the evolution of the surface 'shells' of the Earth since the atmosphere, the hydrosphere, and the biosphere are very sensitive to these small temperature variations. In extreme times of temperature variation, the catastrophic phenomena have taken place. Temperature variations in the surface shells of the Earth were observed up to Archean times, that is approximately 2.7×10^9 year ago. The age of the most ancient rocks discovered on the Earth is about 4×10^9 years. Any other evidence of temperature processes which took place earlier have not been observed as they happened in the times of high-temperature evolution of the Earth.

The study of oxygen and hydrogen isotopic ratios in water and other objects (rocks, gases, organic and mineral remains of buried organisms, etc.), which have interacted with water in the past, throws light upon the origin and evolution of the hydrosphere. In the present section, we consider some principal facts concerning the formation of the isotopic composition of water which is constantly being removed from the ocean into the atmosphere, transferred in the form of water vapor and precipitation, and finally rained out over the ocean and the continents. This will help us to better understand the mechanism of water movement in nature and, on the basis of data of isotopic composition of water molecules observed in natural conditions, to produce more reliable accounts of the global and regional water balance. Besides, the understanding of the principles of the distribution of stable isotopes of water during its circulation broaches the question of the hydrosphere's evolution and origin.

The problem is to find out what values of isotopic ratios are characteristic for water vapor coming from the ocean surface into the atmosphere and how the isotopic composition of various forms of vapor, from which precipitation is formed, is distributed when falling over continents and oceans. Finally, the problem is reduced to the composition of water balance on genetic basis.

The isotopic composition of hydrogen and oxygen in precipitation in global scale was studied for last 50 years, on the stations of the International Atomic Energy Agency/World Meteorological Organization (IAEA/MWO) Isotopes-in-Precipitation Network which are located in different climatic zones. This international project started working in 1961 and included more than 100 stations (see Fig. 4.5). The network covered continental, coastal, island, and ship stations. About half of the samples and the meteorological data from all network stations were sent to the IAEA Laboratory in Vienna for analysis. The remainder samples were forwarded to cooperating laboratories in the member states. The IAEA has acted as the collection agency for data on the distribution of the stable isotopes and tritium in hydrological cycle. All the information is available in special report series publications and on Internet site of the IAEA.

The picture of the ^{18}O global distribution in atmospheric precipitation prepared by Yurtsever and Gat (1981) and complemented by the authors with the data of the former USSR area is presented on Fig. 4.6. It gives general view about distribution of ^{18}O concentration in precipitation before 1981 and does not take into account possible regional changes because of climatic variations.

The global estimation of the average isotopic ratios of hydrogen and oxygen for ocean vapor may be given from the condition of mass balance of the evaporating and precipitating water. From the material balance of evaporation–precipitation ($E = P$) it follows that $\delta E = \delta P$, that is when the water comes back into the ocean its isotopic composition should be restored. When considering the isotopic balance of the ocean during the evaporation–precipitation cycle, the continents play an insignificant role since they obtain, in the form of precipitation, only 10% of the water evaporated from the oceanic surface. Thus, Craig and Gordon (1965) proposed to consider the ocean

Fig. 4.5 Schematic map of the IAEA/MWO Isotopes-in-Precipitation Network Stations at 1975

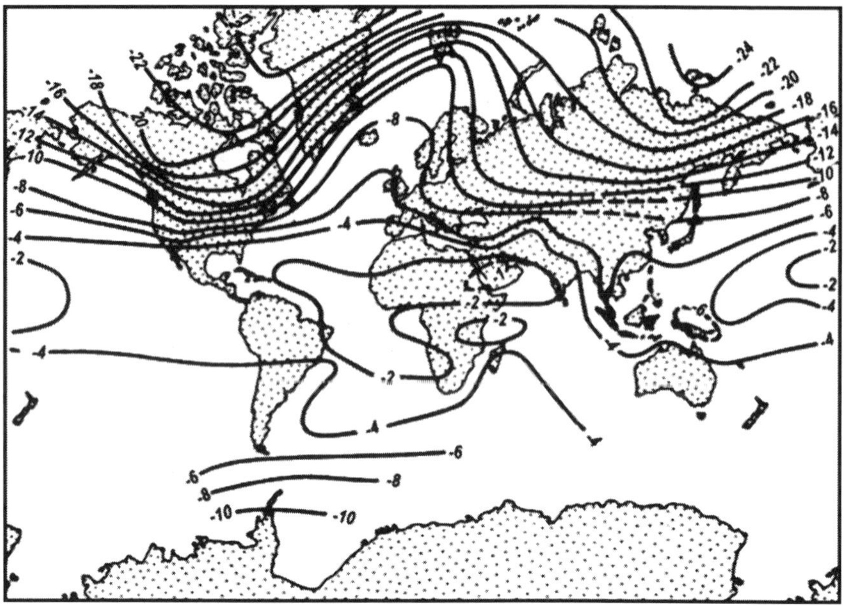

Fig. 4.6 Global distribution of $\delta^{18}O$ in atmospheric precipitation based on data having at least two series of observation. (From Yurtsever and Gat 1981; Ferronsky and Polyakov 1983)

as a closed system. It is logical that for separate regional oceanic conditions, the influence of continental run-off and contribution of melt-water from polar glaciers should be considered.

On the basis of experimental data on precipitation distribution, the amounts of which are in accordance with latitudinal belts, we can expect the following approximate values of latitudinal ^{18}O abundances:

Latitudes	Ratio of Precipitation	$\delta^{18}O$ (‰) (mean value)
0°—20°	0.5	–2
20°—40°	0.4	–5
40°—90°	0.1	–15

Using these data, Craig has derived the following theoretical values of isotopic ratios for oceanic vapor and precipitation, which are equal to $\delta^{18}O = -4‰$ and $\delta D = -22‰$. These magnitudes are close to the average ones in continental precipitation, in contrast to high latitudes where the above-mentioned values are significantly lower, but the amount of precipitation for these regions of the isotopic balance for the North American continent (Craig and Gordon 1965) the mean values are $\delta^{18}O = -5‰$ and $\delta D = -30‰$, which agree with the accepted data above.

In the case of equilibrium evaporation from the ocean surface without the influence of atmospheric moisture and kinetic effects, the mean values of isotopic ratios are $\delta^{18}O = -9‰$ and $\delta D = -70‰$, as reported by Epstein and Mayeda (1953) and Dansgaard (1964). The experimentally observed data on the isotopic composition of vapor over the ocean (see Fig. 4.3) is evidence in favor of the isotopic ratios being considerably lower than those suggested by the material balance and even lower than the values corresponding to equilibrium conditions ($\delta^{18}O$ is varying from -11 to $-14‰$). According to Craig and Gordon (1965), the given facts can be explained by the character of steady global circulations of atmospheric air and also those of condensation of atmospheric precipitation. This process can be outlined as follows. Dry atmospheric air descends to the ocean surface in the trade wind regions where it picks up the water vapor and moves it to the equatorial belt, where the majority of the vapor rises to high atmospheric layers and precipitates its moisture by cooling and condensation. The remaining vapor moves away from the trade winds to the high latitudes where it undergoes an analogous process. This general study system of atmospheric circulation, in which the isotopic variations and the humidity of air are reflected in transport and mixing conditions, is restricted by the local vertical convective flows, rich in water vapor, which form areas of cumulus clouds saturated with moisture. Therefore, we never observe in nature the full saturation of any atmospheric layer with moisture but only local domains of condensation in a general flow of circulating masses of air and water vapor. In these conditions, the vapor above the sea will never be in equilibrium with water even if there is no kinetic effect.

The isotopic variations in vapor and precipitation are explained by Craig and Gordon with the help of a simplified single-stage precipitation model. They assumed that there is a homogeneous layer of atmospheric moisture and in the precipitation region, there is a large ascending flux of moisture being evaporated from the ocean surface. Thus, compared with this flux, insignificant amount of moisture precipitates. In such case, the isotopic composition of precipitation will be formed by a single-stage equilibrium fractionation between the condensing moisture and atmospheric vapor. From the isotopic balance equation, one obtains the following values of δD and $\delta^{18}O$, characteristic for elements of the considered system. The slope of the straight line precipitation ($\delta^{18}O = -4‰$, $\delta D = -22‰$)—atmospheric vapor ($\delta^{18}O = -13‰$, $\delta D = -94‰$) is equal to 6 and for the line atmospheric vapor-equilibrium vapor ($\delta^{18}O = -8‰$, $\delta D = -67‰$) the slope is equal to 5.5. The slope of the line atmospheric vapor-surface sea water ($\delta^{18}O = +1‰$, $\delta D = +5‰$) is about 6.5. For the isotropic ratios in the equilibrium vapor–evaporating vapor system, ascending from the ocean surface, both kinetic effects and atmospheric humidity are important. If kinetic effect is low enough and humidity is large enough, the relationship can be reversed. At normal conditions for the observed parameters in the ocean–atmosphere system, the equilibrium vapor will be lighter than the ascending flux of evaporating moisture. In nature, the reverse may occur only in high latitudes.

The difficulties have arisen from the instability of evaporation and moisture exchange for different oceanic regions as one tries to develop the model, describing formation of isotopic composition of moisture transferred in the atmosphere. The most important oceanic region from the viewpoint of moisture exchange is the trade

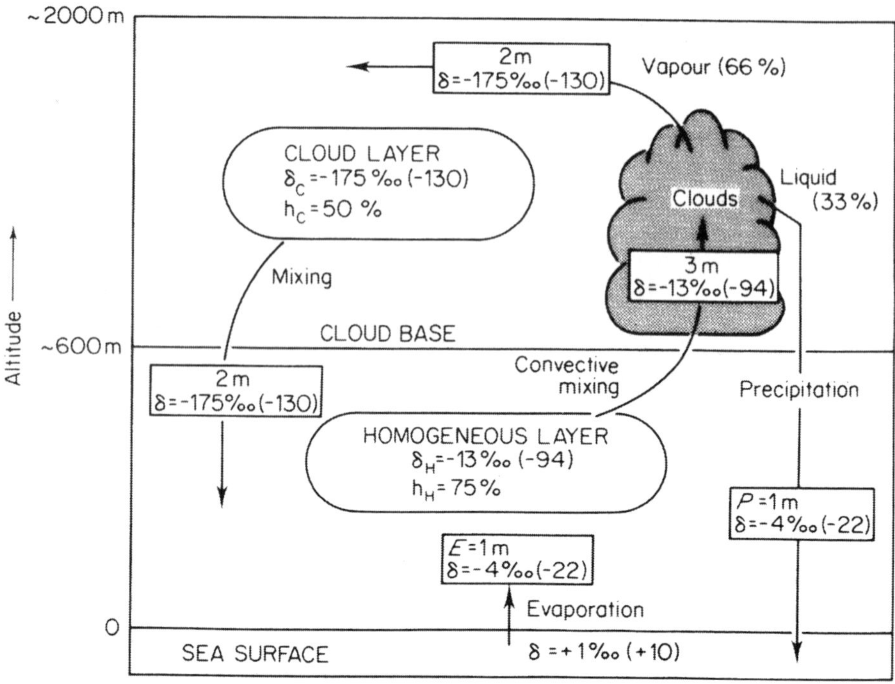

Fig. 4.7 Two-layer model for vapor and precipitation isotopic composition formation and their motion in the ocean–atmosphere–ocean system for trade winds regions. (After Craig and Gordon 1965)

wind region between 30°N and 30°S, which incorporates about 50% of the total area of the Earth and receives three-quarters of annual amount of precipitation and evaporation. The structure of the atmosphere in this region, derived from numerous experimental investigations, and the characteristic average values of isotopic ratios of individual atmospheric elements, according to Craig and Gordon (1965), are shown in Fig. 4.7.

The model consists of two atmospheric layers. Up to 600 m above sea level, a lower homogeneous layer can be distinguished which is well mixed by turbulent stirring. The specific humidity of this layer changes sharply in the first 10–20 m above the sea and then remains almost constant up to 600 m with humidity equal to 75%. The second, cloud layer is characterized by the active process of convective mixing with large-scale descent of dry air and local jets of rising air rich in water vapor around which cumulus clouds are formed. The thickness of the second layer is restricted by the extent of the moist, preferentially convective mixed layer of the outer trades. About this layer is the trade-wind inversion section, separating the second layer from the dry upper troposphere above. The location of the inversion section varies from an altitude of 2 km in the outer trade wind belt to 4 km near the equator. The two principal layers are separated by an intermediate layer with a thickness of about

100–300 m, which is characterized by a specific humidity which, approximately, decreases linearly towards the cloud layer. The cloud layer displays a mostly uniform humidity of about 50%. If the problem of the isotope balance of the exchanging moisture is considered from the principal viewpoint of a system in a steady state, then this model may, to some extent, represent the relationship between the ocean and the atmosphere as a whole. Thus, both the material and isotopic balances in the system during evaporation–precipitation process should be taken into account. The loss of moisture evaporated from the ocean surface (10%) and precipitating over the continents is not considered. Assuming the lower homogeneous and upper cloud layers have common values of humidity equal to 75% and 50%, respectively, evaporation–precipitation amounts to about 1 m per year, and the isotopic ratios, according to the observed data, equal $\delta^{18}O = -13‰$ and $\delta D = -94‰$ for the vapor in the homogeneous layer, and $\delta^{18}O = +1‰$ and $\delta D = +10‰$ for the surface ocean water, the model may be described by the following equations.

The first equation is just the mass balance equation of moisture in the system

$$P = I(h_1 - h_2), \tag{4.21}$$

where P is the amount of precipitation; h_1 and h_2 are humidity of the upper and lower cloud layers; I is the exchange constant between the upper and lower layers, equal to the flux of moisture between them when the air is saturated.

As shown in Fig. 4.7, the flux of moisture characterized by Eq. 4.21 is equal to a water layer 3 m thick being transported from the lower layer to the upper one and a layer 2 m thick from the upper layer to the lower one.

The second equation relates the isotopic composition and humidity of these layers:

$$h_1(\delta_P - \delta_1) = (\delta_P - \delta_2), \tag{4.22}$$

where δ_p is the isotopic ratio for precipitation; δ_1 and δ_2 are isotopic ratios of water vapor in the upper and lower layers.

From the last equation for the upper cloud layer, other parameters being known, we obtain $\delta_2^{18}O = -17.5‰$ and $\delta_2 D = 130‰$.

The third equation, relating the isotopic composition in precipitation and water vapor, can be written on the basis of the model chosen for the formation of precipitation. One may assume that condensation of precipitation takes place in the upper layer from the water vapor ascending from the lower layer for certain values of the parameters describing the cloud layer, but in this case the isotopic composition of vapor of the cloud layer can be described by Eq. 4.22. To obtain the fixed isotopic ratios for precipitation δ_p and vapor of the cloud layer δ_2, at known vapor pressures, the fractionation factors of ^{18}O and D (1.0135 and 1.124) are required. The initial precipitation temperature corresponding to these values must range between the limits from $-10°C$ to $-20°C$, which is significantly lower than the observed temperature in nature. The isotopic data show independently that precipitation is formed from the ascending flux of moisture flowing from the lower layer, being a component of the exchange process between the two layers. At the same time the observed temperature criteria lead to a conclusion that condensation of vapor rising from the lower layer is not just a process of equilibrium formation of precipitation but is governed by the more complicated mechanism of inversion distillation similar to those observed in a rectification column.

The atmospheric model, made up of two layers, which has been proposed by Craig and Gordon, principally for estimating the isotopic balance of moisture resulting from the interaction between ocean and atmosphere, can be considered as a basis for the development of more detailed models of water movement through the atmosphere based on isotopic investigations both in oceanic regions and in the ocean—continent system, but in contrast to the oceanic conditions in the evaporation–precipitation cycle, where a 10% loss of evaporating moisture (being precipitated over the continents) may be neglected, a gain of 10% of ocean water represents 40% of the continental moisture as a whole. The other 60% is secondary evaporating moisture from the continental surface, which was previously precipitated. Thus, assuming the previous considerations (see Fig. 4.7), about 40% of water vapor in a continental region would be contributed by lower and upper atmospheric layers with fixed isotopic composition, depending upon proportions of the vapor mixing. The isotopic ratios of moisture evaporating from the continental surface will have lower values compared with the moisture evaporating from the ocean surface. At the same time the net water run-off from continents into the ocean will be enriched in D and ^{18}O compared with their content in precipitation and will correspond (in accordance with the isotopic balance) to the isotopic ratios of water vapor moving from the ocean surface to the continents.

4.3 Isotopic Composition of Atmospheric Water in the Past

While considering conditions of formation of isotopic composition of precipitation, it is interesting to give data on the evolution of isotopic ratios with time, using as an example the Antarctic ice sheet. Epstein et al. (1970) studied ice samples from a borehole drilled in 1968 at the Antarctic Byrd Station (80°01′S, 119°31′W), located on 1,530 m above sea level. Isotopic studies of the ice were carried out for depths ranging from 99 to 2,162 m. The total depth of the borehole was 2,164 m. The geomorphologic surface conditions in this region are such that the danger of any catastrophic contribution of ice from other regions is improbable. In addition, there is no evidence that it has been found more than 600 m below sea level.

In Fig. 4.8, the results mentioned above are given, each point representing the mean value of a water sample obtained from the ice core ranging in size from 30 to 151 cm. The exceptions are the last points representing samples taken from the borehole bottom. The relationship between δD and $\delta^{18}O$ for the investigated profile follows the equation $\delta D = 7.9\ \delta^{18}O$. The variation in the age of the ice with depth was calculated from the rate of snow deposition using recent values of accumulation of snow equal to 12 g/cm^2 of water per year. On the left side of the figure the vertical scales of the borehole depth and age are not linear, since the age scale increases disproportionately with the depth scale. On the right-hand side of the figure pictured

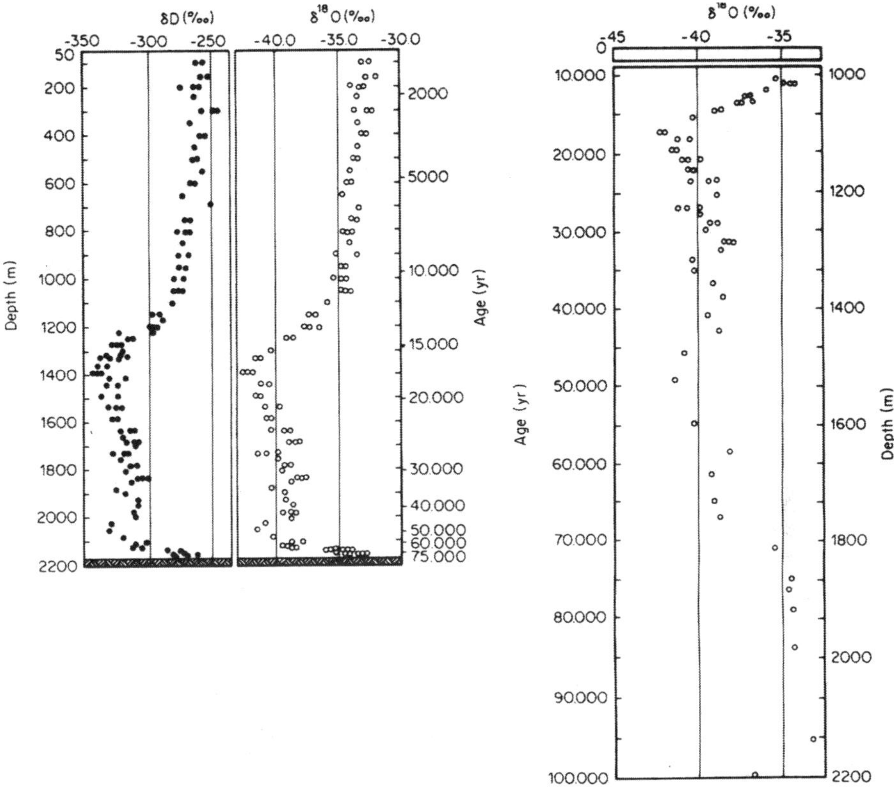

Fig. 4.8 Relationship between D and ^{18}O content in the ice sheet profile for the Antarctic Byrd Station. (After Epstein et al. 1970)

for $\delta^{18}O$, the age of the layers and their thickness, starting from 1,000 m, are given in a linear scale.

From the analysis of the magnitudes and character of isotopic variations in the ice sheet with time, the authors of the study provided the following interpretation. If the observed isotopic variations in the upper 300 m during the last 2,500 years could be related to some random factors resulting in temperatures 2–3°C lower than the modern ones, then, in the range of depths from 1,050 to 2,160 m, they certainly show considerable climatic variations in the past. The corresponding time interval 11,000–75,000 years ago is commonly related to the last period of cooling on the Earth during Pleistocene. Using the variations of isotopic ratios with time, it is possible to obtain the colder periods, corresponding to 17,000, 27,000, 34,000, and 40,000 years ago and warmer ones, corresponding to 25,000, 31,000, and 39,000 years ago. It is rather interesting that the last phase of maximum cooling (being 17,000–11,000 years ago) corresponds with the end of the Wisconsin period of cooling for the northern hemisphere. An analogous coincidence in the two hemispheres is also observed for the other cold and warm phases during the entire time period. The data characterizing the bottom part of the ice sheet then modern ones for the Byrd station. This warmer period should ended 75,000 years ago. Emiliani's data (Emiliani 1970, 1978), based on oxygen isotope studies of fossil pelagic foraminifera, and the data provided by study of the Greenland ice sheet estimated for the time period of Wisconsin glaciation, are close to those given above. The temperature varies by 7–8°C between glacial and non-glacial periods (see Fig. 4.8).

Similar results were obtained by Dansgaard et al. (1969) for 1,300 m ice column from a borehole at Camp Century Station in Greenland and also by Gordienko and Kotlyakov (1976) while studying an ice column 1,000 m in height obtained at the Antarctic Vostok Station, but according to Gordienko and Kotlyakov, the temperature variations between the Holocene and Pleistocene glaciation in the Antarctic have only been of the order of 5°C. Thus, it is possible that Epstein et al. (1970) did not take into account the decrease of heavy hydrogen isotopic composition resulting from difference in altitude marks in the ice sheet provided by the ice melting, which could be 500 m for the Byrd station.

According to Emiliani and other authors the temperatures varied in these periods by ∼5–6°C for the equatorial region of the Atlantic ocean, ∼7–8°C for the Caribbean basin and ∼3–4°C for the equatorial part of the Pacific Ocean. As for variations of isotopic composition in glacial-nonglacial times, according to Epstein et al. (1970) and Dansgaard et al. (1971), the Greenland and Antarctic ice sheets were depleted in oxygen by average of 10‰ and the ocean waters were enriched in oxygen by an average of 0.5‰ during glacial periods.

Isotopic composition change in atmospheric precipitation in the past because of climate change on the basis of continental ice sheets study was also considered by Vasilchuk and Kotlyakov (2000). During the period of 1970–1998 at the Russian Vostok Station in Antarctica, a borehole up to 3,623 m in depth was done and covered the time interval from 0 to 420,000 years. The age of several thousand years for the upper part of ice sheet was studied by the annual layer counting. The deeper accumulations determined by modeling of the ice flowing. The acceleration mass spectrometry techniques based on the cosmogenic radioactive isotopes ^{14}C, ^{10}Be, ^{36}Cl, ^{26}Al was also used.

Hydrogen isotopic composition of the ice core from Vostok Station varies in the range from −420 to −480‰. Its higher values (from −420 to −460‰) reflect relatively short periods of interglacial warming and more lower values (from −460 to −480‰) characterize long periods of the glacial cooling. Until now, this is the only borehole covering the last glacial–interglacial cycles: the Würm glacial epoch (10,000–120,000 years); Riβ–Würm interglacial (120,000–140,000 years); Riβ glacial (140,000–220,000 years); Great interglacial (220,000–320,000 years); and Mindel glacial (320,000–420,000 years). It was found that the last glacial epoch is characterized by three temperature minimums which happened 20,000, 60,000, and 110,000 years ago. The interglacial peak came130,000 years ago. 1.5–2.0°C. The warmer were XII, XVI and XX centuries, and cooler were XII–XV and XVII–XIX centuries. The last interval is called as small glacial period. The paleoclimatic studies of the Earth were carried out on the basis of analysis of ace cores from Camp

Century, Dye-3 and Summit (Greenland). A good agreement between the above data of hydrogen and ^{18}O was found. It means that the paleoclimatic variations are characteristic for both hemispheres and very probably appear to be a consequence of the Milancovich's astronomical effects.

The dependence of heavy hydrogen content and oxygen isotopes in atmospheric precipitation upon temperature makes it possible to carry out paleotemperature studies, not only based on isotopic composition in atmospheric precipitation (e.g., accumulated in glaciers) but also using natural chemical compounds of organic and inorganic origin in the formation of which water was essential. Such compounds are, for example, the clay minerals, formed by weathering of silicate rocks, and also the polysaccharides (starch, cellulose) of plants.

The formation of kaolin during the process of weathering of feldspar rocks may be described by the equation

$$\underset{\text{(albite)}}{2NaAlSi_3O_8} + 2CO_2 + 3H_2O = \underset{\text{(solution)}}{2Na^+} + \underset{\text{(solution)}}{2HCO_3^-} + \underset{\text{(solution)}}{4SiO_2} + \underset{\text{(kaolinite)}}{Al_2Si_2O_5(OH)_4} \downarrow .$$

During the clay formation, hydroxides of silicon and aluminum, as described in the above reaction, undergo a stage of precipitation from the water solution, where they attain isotopic equilibrium with water.

Investigations of pure kaolinites taken from the area of weathering, carried out by Savin and Epstein (1970a), have shown that, during their formation, kaolinites, and montmorillonites come close to the condition of isotopic equilibrium with meteoric waters participating in the hypergeneous process of geological weathering. This was substantiated by Lawrence and Taylor (1972) and Taylor (1974). The values of the fractionation factors at temperature $\sim 20°C$ indicated that clay minerals were enriched in ^{18}O approximately by 27‰ and depleted in D approximately 30‰, compared with the water participating in their formation.

Figure 4.9 shows the relationship between δD and $\delta^{18}O$ for clay minerals and hydroxides from recent soils in the USA, formed on igneous parent rocks (Lawrence and Taylor 1971). Here $\delta^{18}O$ values have been calculated approximately as they were determined by $\delta^{18}O$ for gross samples of soils, reduced by relative ^{18}O values obtained by the analysis of parent rock minerals. In addition, the so-called kaolinite line ($\delta D = 7.6\delta^{18}O - 220‰$) is shown in the Fig. 4.9 (Savin and Epstein 1970a), as is data on the isotopic composition of typical meteoric waters from different regions in the USA which are plotted on the meteoric line.

It is evident from the data given in Fig. 4.9 that the points corresponding to soil samples enriched with hydroxides of metals (e.g., gibbsite $Al(OH)_3$), are plotted on the left of the points relating to clay samples. This is probably because the fractionation factors between water and gibbsite are different from those in the water–kaolinite and water–montmorillonite systems.

It has been shown by Taylor (1974), and by Lawrence and Taylor (1972) and Sheppard et al. (1969), that data on oxygen and hydrogen isotopic content in clay minerals, and soil of kaolinite weathering zones of the Tertiary, provide an opportunity to reconstruct the content of these isotopes in the tertiary meteoric waters.

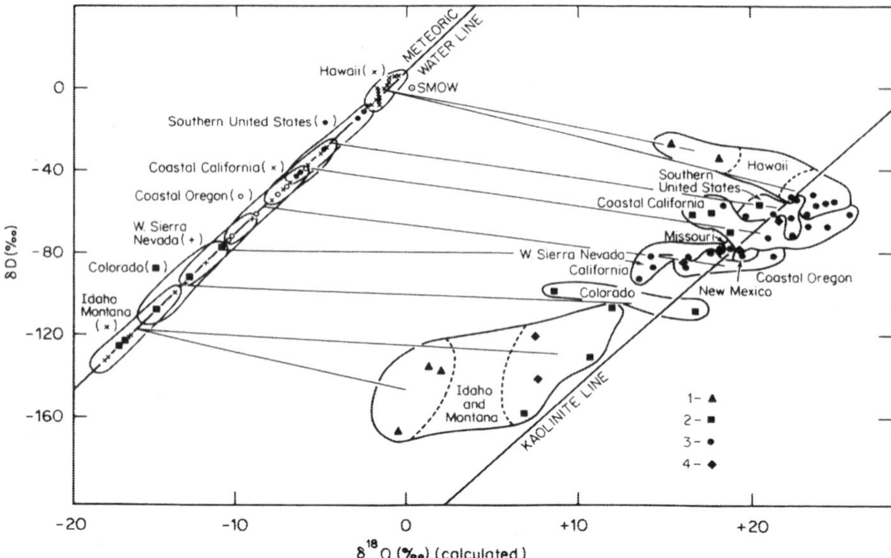

Fig. 4.9 Relationship between δD and δ^{18}O values of clay minerals and hydroxides from modern soils formed on igneous parent rocks in the United States. (*1*) gibbsite and amorphous Al-Si-Fe hydroxides; (*2*) montmorillonite; (*3*) kaolinite; (*4*) mixed kaolinite and montmorillonite. Also shown is the kaolinite line of Savin and Epstein (1970a) and isotopic composition of typical meteoric waters from various US regions. (After Lawrence and Taylor 1971)

In Fig. 4.10, the map of North America is shown with isolines δD values for Tertiary meteoric waters calculated by Sheppard et al. (1969) and Taylor (1974) based on the isotopic studies of various Tertiary meteoric–hydrothermal systems and also upper zone clay and weathering deposits of Tertiary age. Comparing this map with recent δD values for meteoric waters of the North American continent, which are shown in Fig. 4.11 and based on data collected by Dansgaard (1964), Friedman et al. (1964), and Hitchon and Krouse (1972), one finds that for the Tertiary isolines of δD there is a tendency towards enrichment about 1–2‰. As Taylor (1974) has pointed out, these data are in accord with the theory that the climate was warmer in the Tertiary age not only in the North America but also in the other regions of the Earth and particularly in former USSR (Petrov 1975).

It should be noted that, although the conclusions of Taylor (1974) on the milder climate in the Tertiary (between the Paleogene and Neogene, i.e., about 26 million years ago) have been confirmed by several geological arguments, the differences in the δD values between the recent meteoric waters and atmospheric precipitation during Tertiary, are not likely to be as great as depicted in Figs. 4.10 and 4.11. Yapp and Epstein (1977), citing a private communications with Friedman, considered that all the δD values in Fig. 4.11 should be divided by factor of 1.04, correcting laboratory error. Besides, the question concerning the period during which the information is covered and enclosed in the hydrogen isotopic composition of hydroxide groups of

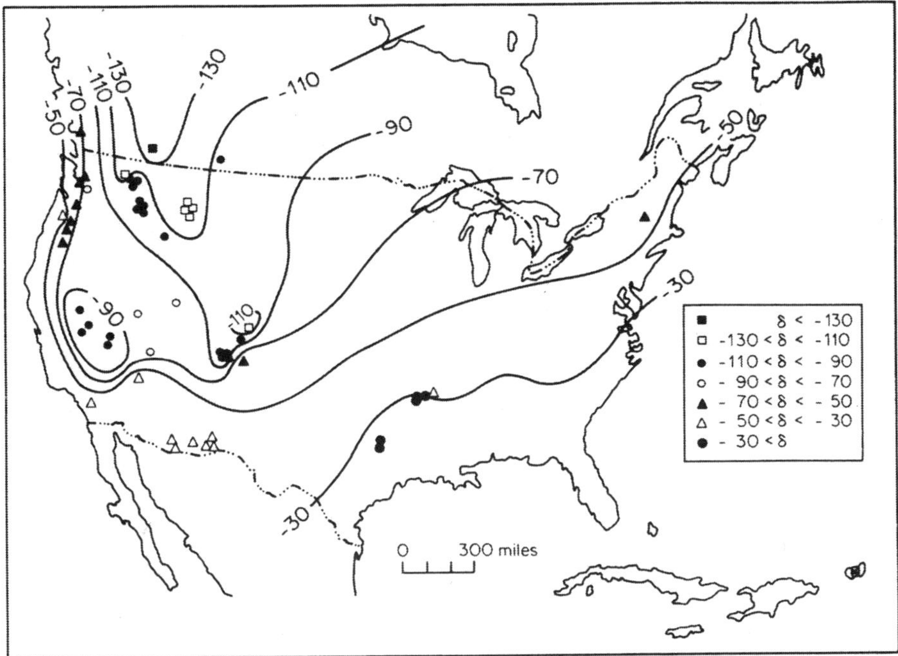

Fig. 4.10 Schematic map of the North American continent showing calculated δD values of Tertiary meteoric waters. (After Taylor 1974. © IAEA, reproduced with permission of IAEA)

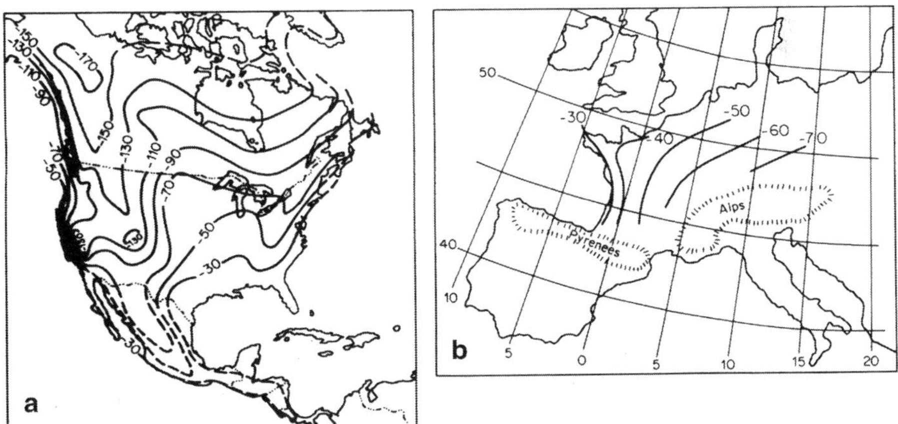

Fig. 4.11 Map of contours of the average δD values of recent meteoric surface waters; (**a**) for North America (from Taylor 1974); (**b**) for Western Europe. (From Sonntag et al. 1979. © IAEA, reproduced with permission of IAEA)

Fig. 4.12 Map of δD values of recent meteoric waters obtained for various locations of the United States by means of hydrogen isotopes in plant cellulose (points) and by direct measurements. (From Yapp and Epstein 1977. © IAEA, reproduced with permission of IAEA)

clay minerals has not yet been solved. There is evidence (James and Baker 1976) to suggest that isotopic exchange between the hydroxide groups and interlayer water takes place at room temperatures. Therefore, caution is needed when making the paleotemperature reconstruction of individual geological periods using hydrogen isotope analysis of hydroxide groups of clay minerals.

In recent years, Epstein et al. (1976), Epstein and Yapp (1976), and Yapp and Epstein (1977) have developed techniques of paleotemperature studies based on hydrogen isotopic content in the cellulose of wood plants. Cellulose is a polysaccharide of the form $(C_6H_{10}O_5)_n$, formed in plants from monosaccharides, for example glucose $C_6H_{12}O_6$, which is formed in turn by photosynthesis involving water and carbon dioxide

$$6CO_2 + 6H_2O \rightarrow C_6H_{12}O_6 + 6O_2 \uparrow .$$

It is obvious that, during photosynthesis the hydrogen of glucose and cellulose—being formed it—is in isotopic equilibrium with meteogenic waters of the region of the plants' growth. Experiments show that cellulose extracted from plants by nitration and extraction with alcohol–benzol mixture is depleted by about 20‰ in D compared with water (Epstein et al. 1976). Therefore, the corresponding average isotopic composition of atmospheric precipitation can be reconstructed on the hydrogen isotopes content in plant's cellulose. Yapp and Epstein (1977), taking modern plants as an example, found a good relationship between δD values obtained indirectly from the hydrogen isotopic content in the cellulose of plants and direct measurements of δD in precipitation for various regions of the United States (Fig. 4.12). They showed experimentally that hydrogen isotopic exchange in C-H groups is extremely limited. These circumstance makes it possible to use variations in the hydrogen isotopic content in the fossil cellulose of plants for paleotemperature reconstructions. In particular, the climatic change on the North American continent during the last Pleistocene (Wisconsin) glaciation between 9,500 and 22,000 years before present has been estimated. The age of fossil plants was determined to be 22,000 years ago and the relative D content in atmospheric precipitation was 19‰ higher than the recent one. Yapp and Epstein explained this in terms of a gentle temperature gradient between the ocean and inland surface; cooling of the oceans; higher δD values for the surface oceanic waters; variation of the ratio of summer and winter precipitation; and a positive shift in the δD value in the oceanic vapor provided by the depleted influence of kinetic factors during evaporation of water at lower temperatures.

The reconstruction of the isotopic composition of glacial lakes—Whittlesey and Aquassiz—had made it possible to estimate the average hydrogen isotopic content (and the average oxygen isotopic content calculated by Craig's equation) of the North American ice sheet. It has been found that δD values for Whittlesey Lake vary from -89 to -113‰ and δ^{18}O values vary from -12 to -15‰. From data obtained while studying the woods grew on the shore of Lake Aquassiz, the average hydrogen and oxygen isotopic content for this lake was found to be (δD from -103 to -129‰, δ^{18}O from -14 to -17‰). The average δ^{18}O value of -15‰ for the North American ice sheet during the Wisconsin glaciation was estimated by Yapp and Epstein.

The application of hydrogen isotopic composition in the C-H group of cellulose as a paleothermometer needs to be carefully tested under natural and laboratory conditions despite the coincidence of independent geological estimations presented by Epstein and Yapp (1976, 1977). In particular, as pointed out by Ferhi et al. (1977) and Long and Lerman (1978), the formation of isotopic composition of hydrogen, oxygen, and carbon in cellulose is influenced not only by the isotopic composition of atmospheric precipitation but also by the temperature at the time of cellulose formation, the type of plant, and the environmental humidity. At low air humidity (Gonfiantini 1965), an intensive evaporation of water from the leaves of plants results in considerable enrichment of the leaves' moisture and, therefore, in the heavy isotopes of hydrogen and oxygen which are the products of photosynthesis.

Epstein and Yapp (1976) were criticized by Wilson and Grinsted (1975), who had previously proposed a biochemical thermometer based on the temperature dependence of the hydrogen fractionation factor between water and plant cellulose of the form C-3 with photosynthesis in the same way as described by Kalvin. They determined the temperature coefficient for a biochemical thermometer to be equal to -5‰/°C (Wilson and Grinsted 1975). This effective dependence of the hydrogen isotope composition of cellulose of plants upon temperature disagrees with the data obtained by Epstein and Yapp (1976). The latter found that, for 25 tests on the cellulose of various plant types, growing in different climatic conditions with δD values in atmospheric precipitation varying over the range of 180‰, the δD values in the C-H group of cellulose differ from those in atmospheric precipitation by -22‰ with an average standard deviation of ± 11‰ (1σ).

Answering the critical remarks of Wilson and Grinsted, Epstein (1978) explained the marked deviations of the experimental data by the presence of OH-groups which readily exchanged isotopes of hydrogen during Wilson and Grinsted's isotopic analyses of cellulose. As an example of δD determination, Epstein made 19 cuts across the three annual rings of a New Zealand pine *Pinus Radiata*, which had grown between May 1915 and April 1917, and produced a satisfactory qualitative coincidence of the δD values in plant cellulose with seasonal variation of δD in modern atmospheric precipitation (1963–1967) for the northern region of New Zealand. Epstein (1978) made the following conclusions:

1. Variations of δD in C-H groups of the plant's cellulose reflect hydrogen isotopic variations in atmospheric precipitation.
2. Wilson and Grinsted's statement that a biological thermometer with a temperature coefficient $-5‰/°C$ permits estimation of variations of limiting temperatures with an error better than 0.1°C is not supported by the available experimental data.
3. The method of analysis of hydrogen isotopic composition of cellulose, suggested by Wilson and Grinsted (1975, 1977), requires further verification.
4. Despite the fact that hydrogen isotopic content in plant cellulose gives agreeable qualitative coincidence with variations in atmospheric precipitation, a number of factors which are related to the kind of plants and the intensity of evapotranspiration may influence the hydrogen isotopic fractionation, depending on the climatic conditions of the region where the plant is growing.

All these circumstances require further experimental studies which should account for the influence of the above-mentioned factors upon the isotopic composition of the plant's cellulose.

Chapter 5
Isotopic Composition of Surface Continental Waters

5.1 Isotopic Balance of the Continental Waters

The isotopic composition of natural and artificial surface reservoirs is determined by several factors, the most important of which are (Gat et al. 1968; Merlivat 1970; Fontes 1976):

1. The isotopic composition and the amount of precipitation directly feeding the reservoir.
2. The isotopic composition of the surface and underground sources of water recharge as well as the rate of water inflow.
3. The isotopic composition of atmospheric vapor moisture and also the air humidity.
4. The rate of water outflow and also the process of evaporation.

The mass balance of reservoir during a period of time Δt may be given by the equation:

$$\Delta V = \left(\sum I - \sum Q - E\right)\Delta t + P, \tag{5.1}$$

where ΔV is the volume change of the reservoir; I is the component of the surface and subsurface water recharge; Q is the component of the surface and subsurface discharge; E is the mean rate of evaporation; P is the amount of precipitation during the time Δt.

The isotopic balance equation in this case is

$$R_L \Delta V + V \Delta R_L = \left(\sum R_I I - \sum R_Q Q - R_E E\right)\Delta t + R_P P, \tag{5.2}$$

where R is the value of mean isotopic ratio of the components of recharge and discharge.

The value R in Eq. (5.2) can be changed by value δ. Then, the equation acquires the form

$$\Delta(\delta_L V_L)\Delta t = \sum \delta_I I - \sum \delta_Q Q - \delta_E E + \delta_P P.$$

V. I. Ferronsky, V. A. Polyakov, *Isotopes of the Earth's Hydrosphere*,
DOI 10.1007/978-94-007-2856-1_5,

It is known (Craig and Gordon 1965; Merlivat 1970) that a liquid's evaporation rate in the atmosphere can be described by the equation

$$E = \kappa \frac{dC}{dz},$$

where E is the rate of evaporation (vapor flux from the unit of liquid surface); κ is the coefficient of turbulent diffusion; dC/dz is the vertical gradient of humidity.

For isotopic species of water ($H_2{}^{18}O$ or HDO), one has

$$E_i = \kappa_i \frac{dC_i}{dz}.$$

The values of the isotopic fractions in vapor R_E in this case can be given by

$$R_E = \frac{E_i}{E} = \frac{\kappa_i}{\kappa} \frac{dC_i/dz}{dC/dz}.$$

This equation in finite differences, after the coordinate dz is eliminated, obtains the form

$$R_E = \frac{\kappa_i}{\kappa} \frac{R_L/\alpha - hR_\alpha}{I - h}, \tag{5.3}$$

where h is the relative air humidity at altitude z above the water surface at a given temperature; R_L and R_a are isotopic ratios for liquid and atmospheric vapor, respectively; α is the equilibrium fractionation factor.

In the Eq. (5.3), it is accepted

$$\Delta C = 1 - h, \; \Delta C_i = R_i/\alpha - hR_\alpha.$$

Designating $\kappa/\kappa_i = k$, Eq. (5.3) can be written as (Gat et al. 1968):

$$R_E = (R_L/\alpha - hR_\alpha)/k(1 - h). \tag{5.4}$$

Equations 5.3 and 5.4 give an idea about the parameters which determine the isotopic composition of the vapor flux. As Brezgunov (1978) has pointed out, these formulae are realistic in the physical sense only when both fluxes have the same sign that is when either evaporation or condensation of water and isotopic species occurs. If the fluxes have different signs, they should be considered separately while estimating the isotopic and mass balance of water in the reservoir.

An additional atmospheric resistance effect of fractionation of isotopes during the water evaporation in atmosphere, being defined by humidity and the parameters of the turbulent diffusion of vapor, Craig and Gordon (1965) is evaluated by

$$\Delta\varepsilon = (\kappa/\kappa_i - 1)(1 - h).$$

The isotopic composition of evaporating water expressed in δ values after such a transformation as described in Eq. 5.3 will be as follows:

$$\delta_E = \frac{\alpha^* \delta_L - h\delta_\alpha - \varepsilon^* - \Delta\varepsilon}{(1 - h) + \Delta\varepsilon}, \tag{5.5}$$

where $\alpha^* = 1/\alpha$, $\varepsilon^* = 1 - 1/\alpha$; δ_L, δ_a are the isotopic compositions of liquid and vapor.

Equation 5.5 can be conveniently used to obtain an equation describing the isotopic balance of reservoirs. To obtain a precise value of δ_E, it is necessary to measure the parameters δ_L, δ_α and h directly and know the values of ε^* and $\Delta\varepsilon$. The value ε^* can be calculated easily with the help of equilibrium factor at given temperatures. The $\Delta\varepsilon$ value has been estimated in a number of experimental and theoretical works which were estimated in detail in Brezgunov's review paper (1978). According to experimental data, $\Delta\varepsilon_{^{18}\mathrm{O}}$ is dependent on humidity:

$$\Delta\varepsilon_{^{18}\mathrm{O}} = (1 - h)16‰.$$

At constant humidity, the ratio $\Delta\varepsilon_D/\Delta\varepsilon_{^{18}\mathrm{O}}$, as shown by the experimental and theoretical estimations of Craig and Gordon (1965) and Merlivat (1970) ranges from 0.5 to 0.85, but in a number of experiments (Craig and Gordon 1965; Gat 1970) $\Delta\varepsilon_D$ exceeds $\Delta\varepsilon_{^{18}\mathrm{O}}$ by 2–4 times. Such deviations between experimental and theoretical values were explained by Craig and Gordon in terms of extra isotopic effects at the boundary of the liquid–vapor system, which might be provided by differences in the condensation (vaporization) coefficients for isotopic species of the molecules. The values of the parameter $k = \kappa/\kappa_i$ in Eq. 5.4 have been found, experimentally, to range between 1.016 and 1.020 for oxygen-18 (^{18}O) and 1.009 and 1.0136 for deuterium (D) (Merlivat 1970). Gat et al. (1968) have considered particular values on the basis of work by Craig and Gordon (1965) calculated to be $k_D = 1.009$ and $k_{^{18}\mathrm{O}} = 1.016$. However, such an assumption, as was noted above, is based on the fact that the condensation coefficient for isotopic molecules does not undergo visible variations.

It follows from Eq. 5.3 that the isotopic composition of water vapor above the reservoir is determined both by the isotopic composition of evaporated water and by the isotopic content of atmospheric moisture. If under natural conditions the evaporation occur in accordance with the Rayleigh's distillation law, then water and vapor should be in isotopic equilibrium at any moment. In this case, the isotopic composition of vapor over the reservoir is equal to $R_E = R_L/a$ (R_L is the isotopic composition of the reservoir water at time t) and the isotopic composition of water during evaporation would be given by the Rayleigh's formula (Brodsky 1957)

$$R_L = R_0(V_0/V)^{(\alpha-1)/\alpha} = R_0(V/V_0)^{(\alpha-1)/\alpha}, \tag{5.6}$$

where R_0 and R_L are the ratios of the isotopic species of water initially and at time t; V_0 and V are the initial and final volumes of water.

Taking logarithms of Eq. 5.6, we obtain

$$\ln R_L = \ln R_0 + \varepsilon \ln F,$$

where $\varepsilon = 1/\alpha - 1$; $F = V/V_0$.

Substituting δ values for R and taking into account that $\ln(1+\kappa) \approx \kappa$ at $\kappa \ll 1$, one obtains

$$\delta_L - \delta_0 = \varepsilon \ln F.$$

In other words, during the Rayleigh's evaporation, the expression for $(\delta_L - \delta_0)$ becomes a linear function of ln F. This is convenient for analyzing experimental data while studying the change in a liquid's isotopic composition during evaporation (Figs. 5.2 and 5.5).

If under conditions of Rayleigh's evaporation, the volume of the reservoir V_0 remains unchanged due to recharge, then the isotopic composition of the water at any time may be found from the equation (Brodsky 1957).

$$\alpha \ln [(\alpha - 1)/(\alpha - R_{\mathrm{L}}/R_0) = V/V_0. \tag{5.7}$$

The maximum enrichment, which might be attained in this case at $V \to \infty$, is $R_0\alpha$. Under real conditions, the changes in the isotopic composition of water in the drying reservoir do not follow the Rayleigh's law because the thermodynamic and isotopic equilibrium between water and vapor does not occur here.

As shown in the previous chapter, in case of thermodynamic disequilibrium in a system the rate of transition of molecules HDO and $H_2{}^{18}O$ from the liquid into the gaseous phase will be determined by difference in the diffusion coefficient of the molecules H_2O, HDO, and $H_2{}^{18}O$. In contrast to the processes of equilibrium fractionation when the fractionation factors in the water–vapor system are 1.08 for H_2O–HDO and 1.009 for H_2O–$H_2{}^{18}O$, the difference in the coefficients in the case of diffuse separation is not so great. For the above-mentioned pairs of molecules, the coefficients are 1.009 and 1.016, respectively.

The observed relationship between D and ^{18}O content in precipitation, expressed by the equation $\delta D = 8\delta^{18}O + 10‰$, is evidence that the moisture condensation occurs under conditions which are sufficiently close to those at equilibrium. Under these conditions, the D and ^{18}O fractionation factors might be assumed to be equal to those for systems in equilibrium such as 1.08 for D and 1.009 for ^{18}O at 20°C.

During evaporation, under nonequilibrium conditions, the difference between the D and ^{18}O content is limited not only by the value of the fractionation factors but also by the ratio of the diffusion rates of the molecules such as the coefficient $k = \kappa_i/\kappa$ in Eq. 5.3. This leads to a more gentle sloping curve of δD against $\delta^{18}O$ for water in open reservoirs.

As Dansgaard (1964) has pointed out, in this case the fractionation factor can be given by the expression $\alpha^k = \alpha k$. Hence,

$$\alpha^k_{\mathrm{D}} = 1.08 \times 1.009 = 1.09; \quad \alpha^k_{^{18}\mathrm{O}} = 1.009 \times 1.016 = 1.025.$$

The value of the ratio dδD/dδ^{18}O can be obtained by differentiating Eqs. 4.6 and 4.7, which are similar to the relationship described by Eq. 5.6, since $V/V_0 = F$, and $\delta = R_L - 1$ since $R_0 = 1$:

$$\mathrm{d}\delta\mathrm{D}/\mathrm{d}\delta^{18}\mathrm{O} = (\varepsilon_{\mathrm{D}}/\varepsilon_{^{18}\mathrm{O}})F^{\varepsilon_{\mathrm{D}}-\varepsilon_{^{18}\mathrm{O}}}, \tag{5.8}$$

where $\varepsilon = (1 - \alpha)/\alpha$.

Equation 5.8 can be approximated by ^{18}O

$$\mathrm{d}\delta\mathrm{D}/\mathrm{d}\delta^{18}\mathrm{O} \approx \varepsilon_{\mathrm{D}}/\varepsilon_{^{18}\mathrm{O}}, \tag{5.9}$$

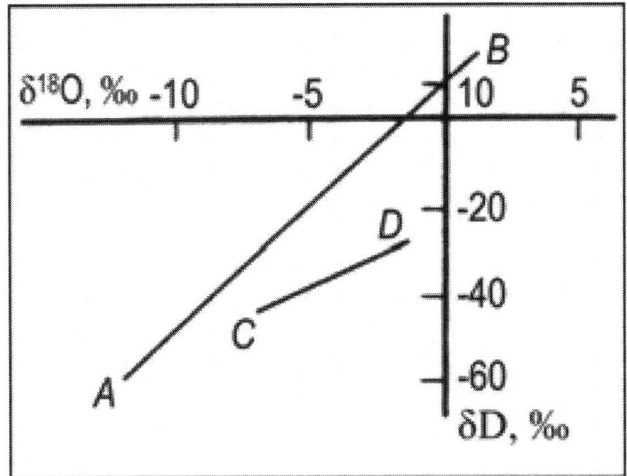

Fig. 5.1 Schematic relationship between δD and δ^{18}O values for precipitation (*AB*) and for closed continental basins (*CD*)

since the index $\varepsilon_{\mathrm{D}} - \varepsilon_{^{18}\mathrm{O}}$ differs only slightly from zero and at a given temperature is a constant.

The slope of the line describing changes in D and ^{18}O concentrations in a solution during Rayleigh's evaporation (at $t = 20\,°\mathrm{C}$ $\alpha_{\mathrm{D}} = 1.08$ and $\alpha^k_{^{18}\mathrm{O}} = 1.009$) will be

$$\mathrm{d}\delta\mathrm{D}/\mathrm{d}\delta^{18}\mathrm{O} \approx (\alpha_{\mathrm{D}} - 1)\alpha_{^{18}\mathrm{O}}/(\alpha_{^{18}\mathrm{O}} - 1)\,\alpha_{\mathrm{D}} \approx 8. \tag{5.10}$$

Under nonequilibrium conditions of evaporation, when kinetic factors are important (at 20°C $\alpha^k_{\mathrm{D}} = 1.09$; $\alpha^k_{^{18}\mathrm{O}} = 1.025$), one has

$$\mathrm{d}\delta\mathrm{D}/\mathrm{d}\delta^{18}\mathrm{O} \approx (\alpha^k{}_{\mathrm{D}} - 1)\alpha^k_{^{18}\mathrm{O}}/(\alpha^k_{^{18}\mathrm{O}} - 1)\alpha^k{}_{\mathrm{D}} < 8. \tag{5.11}$$

It follows from Eqs. 5.10 and 5.11 that for nonequilibrium water vaporization from closed reservoir the values of δD and δ^{18}O lie below Craig's line plotted for atmospheric precipitation on a line with a smaller gradient (Fig. 5.1).

At a steady state, when no significant changes in the volume of the reservoir occur, the isotopic composition of well-mixed water tends toward a constant value

$$R_L = \frac{(1-h)R_I s + hR_a/k}{1/\alpha k + (1-h)(1-s)}, \tag{5.12}$$

where s is ratio of the mean rates of recharge and evaporation.

Equation 5.12 is only true if short time variations of the parameters α, R_I, R_a, h permit one to consider their mean values to be constant in time.

Assuming that water loss from the reservoir takes place only due to evaporation that is $s = I/E = 1$, then Eq. 5.12 becomes

$$R_L = \alpha k[(1-h)R_{\mathrm{I}} + h(R_a/k)]. \tag{5.13}$$

Equation 5.13 gives the value of maximum isotopic enrichment of reservoirs with constant volume. It is obvious that at $h = 0$ (for evaporation when air humidity equals zero) $R_L = \alpha_k R_I$. This expression corresponds to the maximum isotopic enrichment for evaporation of a liquid at constant level during Rayleigh's distillation where α is substituted by $\alpha k = \alpha^k$.

If the level of the reservoir is reduced only by evaporation and recharge does not occur ($I = 0$), then the integration of Eqs. 5.1 and 5.2 under these conditions yields the following expressions for the isotopic composition of the reservoir as a function of V/V_0:

$$R_L = \frac{[R_0(1/\alpha k - 1 + h) - hR_\alpha/k](V/V_0)^{(1/\alpha k - 1 + h)/1 - h} + hR_\alpha/k}{1/\alpha k - 1 + h}. \quad (5.14)$$

The Eq. 5.14 is similar to the formula for Rayleigh's distillation (Eq. 5.6), but in this case the kinetic factors, the isotopic composition of the atmospheric vapor, and the air humidity should be taken into account.

In drying reservoirs, that is when $(V/V_0) \to 0$, the final isotopic composition of water can be found from the expression

$$R_L = \frac{hR_\alpha/k}{1/\alpha k - 1 + h}. \quad (5.15)$$

It follows from the analysis of Eqs. 5.14 and 5.15 that under natural conditions of an existing reservoir, the isotopic composition of its water does not increase to infinity—as the Rayleigh's equation (Eq. 5.6) suggests—but tends toward some stationary condition determined by parameters α, k, h and R_α (see Figs. 5.5 and 5.7).

For $h = 1$, when the reservoir is in thermodynamic equilibrium with the vapor, the isotopic composition R_α which is independent of the isotopic composition of the reservoir, one has

$$R_L = R_\alpha \alpha. \quad (5.16)$$

Under such conditions, the water in the reservoir attains isotopic equilibrium with the atmospheric moisture. It follows from Eq. 5.12 that the isotopic composition of water is governed both by evaporation and by recharge of water having an isotopic composition R_I. Therefore, for seas where rivers play an important role in the formation of water bodies, D and ^{18}O content is lower than that in the ocean, so for the Black Sea $\delta D_{SMOW} = -18.8‰$ and $\delta^{18}O = -3.15‰$; for the Caspian Sea, the values are even lower.

As it was earlier noted, by experimental results and natural observations, it was found that the isotopic composition δ_L of an evaporated reservoir in natural conditions tends to have some stationary state δ_L^S (in the δ-values). It was shown in the works of Craig and Gordon (1965), Gonfiantini (1965), Fontes et al. (1979) that the value δ_L^S for a basin with stationary level is determined by equation

$$\delta_L^S = \frac{\lfloor (E/I)h\delta_\alpha + (\alpha = h)\delta_I(\varepsilon_V + 1) \rfloor + (E/I)\alpha\varepsilon_V}{(\alpha - h)(I - E/I)(1 + \varepsilon_V) + \alpha(E/I)}.$$

For the drying lakes, one has

$$\delta_L^S = \frac{\alpha\varepsilon_V + h\delta\alpha(1 + \varepsilon_V)}{h(1 + \varepsilon_V) - \alpha\varepsilon_V},$$

where $\varepsilon_V = \varepsilon^* + \Delta\varepsilon$ is the total effect of enrichment of the evaporated water and a is activity of water in the solution.

Kinetic term $\Delta\varepsilon$ depends on the air relative humidity and the water activity and δ^{18}O is described by the following dependency (Fontes et al. 1979a): $\Delta\varepsilon = [14.4 \times (a - h)/a] \times 10^{-3}$. The δ-value, substituted into above formula, is expressed in parts of the unit ($\delta = \delta_{\text{measured}} \times 10^{-3}$).

If the lake discharges water Q, then in stationary state for full mixing lakes $\delta_{\text{Q}} = \delta_{\text{L}}^{\text{S}}$. In this case, the total water recharge and the evaporating flux have the following relationship: $I = E(\delta_{\text{L}}^{\text{S}} - \delta_{\text{E}})/(\delta_{\text{L}}^{\text{S}} - \delta_{\text{I}})$ (Zimmermann 1979).

If the lake volume and also the hydrological and climatic parameters can be accepted constant, then isotopic composition of the lake water in time comes to the stationary state in accordance with the equation (Zimmermann 1979)

$$\ln(\delta_L^S - \delta_L) = -(1/\tau)t + \ln(\delta_L^S - \delta_Q),$$

where τ is the mean value of the residence time of water in the lake which is determined by relation V/I; V is the reservoir volume; I is the total water recharge which includes the surface and ground inflow and atmospheric precipitation over the lake surface.

For the reservoir which has not reached steady state in isotopic content, the value τ can be determined by the time dependence relation $\ln(\delta_{\text{L}}^{\text{S}} - \delta_{\text{L}}) = -(1/\tau)t + \ln(\delta_{\text{L}}^{\text{S}} - \delta_{\text{Q}})$ (Zimmermann 1979).

If one plots dependence $\ln(\delta_L^S - \delta_L)$ with reference to time, then the point should be put on the straight line with the slope $1/\tau$.

The combination of water (Eq. 5.1) and isotope (Eq. 5.2) balance assuming that $\delta_I = \delta_P$, gives the following relationship between the rate of the water evaporation E and the total discharge from the lake (Hübner et al. 1979a):

$$E = Q(\delta_L - \delta_Q)/(\delta_E - \delta_I).$$

This equation makes it possible to calculate the values E and Q by one known parameter and experimental values δ_L and δ_Q. In this case, the value δ_E can be calculated by Eq. 5.5. The isotope methods have wide application in study of the lake water balance. The methods are used for indirect assessment of the underground recharge and discharge of lake. As a rule, determination of these parameters by traditional hydrological means have difficulties. The examples of such a solution can be found in the IAEA proceedings on Isotopes in Lake Studies (IAEA 1979).

5.2 Isotopic Composition of the River and Lake Water

Let us now consider some peculiarities of the formation of isotopic composition of surface continental waters. The most complete regional survey of characteristic D contents in surface continental waters is given by Friedman et al. (1964) and for Canada by Brown (1970). The isotopic composition of water in rivers, lakes, and other reservoirs is dependent on the isotopic composition of atmospheric precipitation

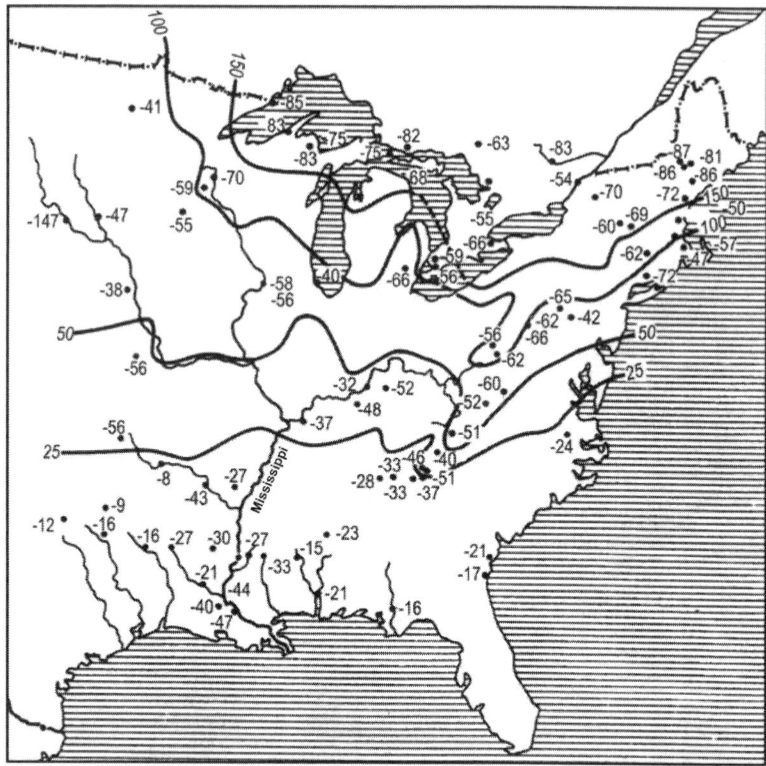

Fig. 5.2 Map of δD values for the Missouri–Mississippi river basin and East Atlantic coast of the United States plotted on the data of 1956–1959. Contourr show thickness of snow cover in centimeter. (After Friedman et al. 1964)

feeding these reservoirs. The surface continental waters are less subject to isotopic variations with time compared with atmospheric precipitation, and the springs and wells which do not dry up are the most reliable sources of water for determination of isotopic ratios characteristic of a given region.

Figure 5.2 shows the regional characteristic of D concentrations in Mississippi river water and the East Atlantic coast of the United States (Friedman et al. 1964). The basin of the Mississippi river covers a considerable area of the eastern region of USA, characterized by a comparatively low height above the sea level and wide ranges of temperature, which determine the initial isotopic composition of atmospheric precipitation. Figure 5.2 shows the isolines of annual snow cover well correlated with relative D content in the surface water. As expected, an even closer correlation is observed between the relative D content and the ratio of the amounts of snow and rain falling down as atmospheric precipitation. In this case, an increase in the proportion of snowfall results in a decrease of the D content and vice versa.

Figure 5.2 may serve as an example of regional D distribution. Equation 4.13 can be used as a first approximation in obtaining a picture of oxygen isotope distribution.

Table 5.1 Estimates of the apparent hydrogen isotopic ratio of vapor pressures and water from the saline lakes and their recharge. (Friedman et al. 1964)

Lakes	δD, ‰		Estimated ratio of vapor pressure	Origin of source water
	Lake water	Source water		
Piranid Lake	– 30	– 87	0.942	Truckee river
Mono Lake	– 62	– 130	0.927	Sierra Nevada near crest
Soda Lake	– 51	– 130	0.918	
Owens Lake	– 97	– 135	0.958	Los Angeles aqueduct
Salton Sea	– 41	– 136	0.901	Colorado and Alamo rivers
Great Salt Lake	– 89	– 130	0.934	Yellowstone plateau
Harney Lake	– 40	– 114	0.923	Silvies river
Mean value			0.923	
Corresponding temperature			18°C	

Evaporation from the basin as well as from the surface of its catchment area is an important factor governing the formation of the isotopic composition of surface waters. As a result of this process, water becomes enriched in heavy isotopes compared with the local precipitation. In this case, the final effect on basins from the regional point of view depends upon the ratio of precipitation to evaporation and conditions of supply. In arid regions, where evaporation exceeds precipitation, the total isotopic balance shows a deficit of the light isotopes. For the humid regions, where precipitation exceeds evaporation, a deficit of heavy isotopes should be observed. Finally, for regions having a balanced amount of evaporation and precipitation, the isotopic balance should also be held. On the continents, the role of balancing the isotopic composition is played by large rivers, the basins of which lie in different climatic areas from humid to arid.

The influence of evaporation on the isotopic composition of basins may be demonstrated by investigating closed and particularly saline lakes (see Table 5.1). According to theoretical data obtained for this group of lakes, the fractionation should occur at 18°C but as the authors themselves pointed out, their estimations are imprecise since the initial data on the isotopic composition of the inflow water from tributaries are not accurate enough, the water losses due to underground discharge have not been taken into account, and the water in the lakes does not attain complete mixing.

Another example of elucidating the relationship between lake water, sources feeding them, and atmospheric vapor is the investigation of hydrogen and oxygen isotope distribution in the basin of the Issyk-Kul Lake (Brezgunov et al. 1979, 1981). The Issyk-Kul Lake is located in a high mountain valley in Tien Shan at an altitude of 1,600 m above sea level. Its total area is 6,200 km^2 with an average depth of 280 m. More than 50 rivers flow into the lake from the slopes of the surrounding mountains and are fed mainly by ice-melt waters. The annual run-off of the rivers is about 4 km^3 and the mean annual amount of precipitation is about 250 mm ($\sim$1,5 km^3). The temperature of the lake's surface water in July–August is equal to 18–19°C and in January–February never drops below +2°C.

Due to natural conditions of its location, the Issyk-Kul basin, with a total area of about 22,000 km^2, a surface which never freezes, and being a source of intensive

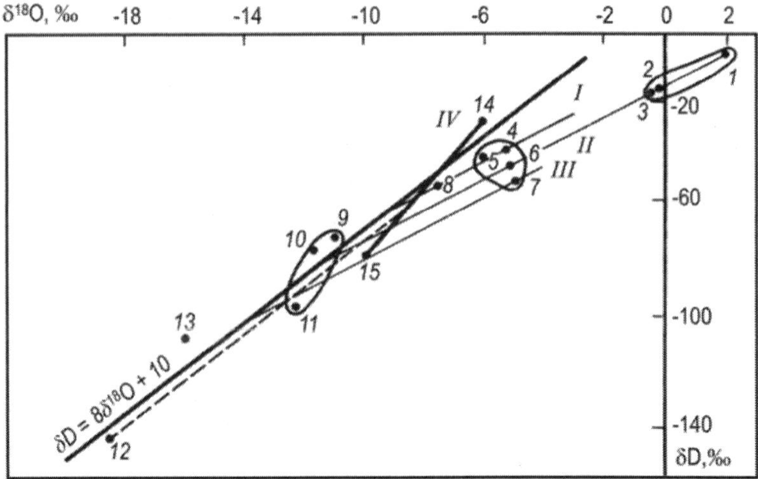

Fig. 5.3 Plot of δD versus $\delta^{18}O$ for natural waters sampled in June 1975 in the Issyk-Kul Valley. Kara Kel Lake water; *2–3* Issyk-Kul Lake water; *4* precipitation from Cholpon-Ata Station; *5* precipitation from Pokrovka Station; *6* precipitation from Przhevalsk Station; *7* precipitation from Rybachye Station; *8* precipitation from Bolshaya Kyzyl-Su Station; *9–10* Chon Kyzyl-Su river water; *11* Dzhergalan River water; *12* precipitation from Bolshaya Kyzyl-Su Station in December 1975; *13* calculated values of 'lake' vapor isotopic composition; *I, II, III* evaporation lines with slope of 5; *IV* condensate 'mixture' line reflecting the 'lake' and 'regional' vapor mixture degree. (After Brezgunov et al. 1979)

evaporation (condensation of atmospheric moisture is localized on the slopes of the mountain ridges surrounding the lake), can be considered as unique natural model of the global circulation of waters.

The authors (Brezgunov et al. 1979) carried out systematic investigations of the isotopic content in the lake and river waters, sources of groundwaters, atmospheric precipitation, glaciers, and water vapor in the valley from 1974.

The relationship of natural waters of the Issyk-Kul basin is given in Fig. 5.3, plotted in δD–$\delta^{18}O$ coordinates, drawn the data of June 1975. Here the experimental points may be subdivided into three groups. The first group (points 1, 2, 3) corresponds to lake water, characterized by the greatest values of δD and $\delta^{18}O$ and the minimum value of parameter b in Eq. 4.12. Close to the line for meteoric waters lie points 9, 10, 11, which correspond to the isotopic composition of water from rivers. Points 4, 5, 6, 7 belong to the third group and lie between those of the first two groups; they correspond to the isotopic composition in precipitation, measured at the coastal stations, and form a sufficiently independent set. Far from this set of points lies the point 8, which correspondsto the isotopic content in precipitation at high mountain station of Bolshaya Kyzyl-Su. Point 12, corresponding to the isotopic composition in atmospheric precipitation obtained in December 1975 for the same station, makes it possible to compare the range of seasonal variations of hydrogen and oxygen isotopic content for the same region. Note that the mean δD and $\delta^{18}O$

values calculated for samples obtained in June and December for Bolshaya Kyzyl-Su station lie in the domain corresponding to the river water in δD–$\delta^{18}O$ diagram.

The data given in Fig. 5.2 visibly reflect the process of hydrogen and oxygen isotopic fractionation. Here, the relative values of D and ^{18}O content in waters of the Issyk-Kul lake are considerably greater than the rivers which are the major sources of lake water. This is due to the evaporation. The straight line in the diagram, connecting data for the river and lake waters, has slope of about 5, which is characteristic for the line of evaporating water. Maximum values of δD and $\delta^{18}O$ (point 1) are found for the Kara Kel Lake water. The closed lake is small (area is equal to 0.5 km^2) and shallow (depth is ~10 m), and it is situated 500 m from the southern shore of the Issyk-Kul Lake. The excess of D and ^{18}O in the water of this small lake reflects its high rate of water exchange.

On the right of the meteoric water line in the diagram there are points which characterize the isotopic composition of precipitation in the coastal region, and for which the parameter b (Eq. 4.12) is negative. Brezgunov et al. (1979) have explained this phenomenon in terms of the evaporation of raindrops in the atmosphere below the cloud layer, resulting in the enrichment in heavy isotopes of atmospheric precipitation taken near the surface. The reduction of the parameter b to negative values in the summer precipitation is a feature of most high latitude stations, which was noted first by Dansgaard (1964). It was found that, other things being equal, the lower the relative humidity of air below the cloud layer, the more is the rainwater enriched in heavy isotopes and the value of parameter b becomes lower.

Using the experimental data on the isotopic composition of water in lakes and rivers and those of atmospheric vapor from the five stations situated around the Issyk-Kul Lake, Brezgunov et al. (1979) evaluated hydrogen and oxygen isotopic content in atmospheric moisture about the lake. In order to carry out the calculations, they used the model of isotopic fractionation by interaction of the reservoir with the atmosphere and the corresponding expressions (Eq. 5.5).

Expression (Eq. 5.5), representing the ratio of heavy isotope flux to the flux of moisture, can be written in terms of δ values as follows:

$$\delta_E = \frac{(\delta_L/\alpha) - \delta_A h - \varepsilon - \Delta\varepsilon}{1 - h + \Delta\varepsilon},$$

where δ_E, δ_L, and δ_A correspond to the isotopic composition in the evaporating flux of moisture, reservoir water, and atmospheric moisture; ε and $\Delta\varepsilon$ are values, expressed in δ units, related to equilibrium and kinetic fractionation factors.

For the stationary evaporation process and the model of the reservoir with constant level $\delta_E = \delta_I$ (where δ_I is the heavy isotopic content in the inflow sources).

For a reservoir with a constant level, such as the lakes in question, for known δ_L, δ_I values and air humidity above the reservoir and known values of the equilibrium kinetic fractionation factors from the previous expression, we found that:

$$\delta_A = \frac{\delta_L - \delta_I(1 - h) - \varepsilon - \Delta\varepsilon}{h}.$$

According to this expression, in which the values of ε and $\Delta\varepsilon$ were derived from Dansgaard (1964) and Merlivat (1970), the values $\delta D = -108‰$ and $\delta^{18}O = -16‰$

were found to be close to the corresponding values of the water vapor determined experimentally.

In addition, by analyzing the data on the isotopic composition of precipitation in the investigated valley and water vapor above the lakes, the atmospheric vapor mixing proportions were estimated. The position of the intersection points of the evaporating lines *I*, *II*, and *III* in Fig. 5.3, which has a slope equal and close to 5 and which passes through the points with isotopic content of precipitation with mixing line *IV*, made it possible to estimate the proportion of these two components in the atmospheric vapor (the regional component and that component removed from the lake) in the precipitation which fell in the June 1975 in different locations of the Issyk-Kul valley. It was found that in the western part of the valley (Rybachye station) the precipitation was almost completely formed of water vapor from outside the region. Precipitation in the eastern part of the valley (Przhevalsky station) was composed from 20% of the lake vapor and 80% of the regional vapor. The maximum contribution of lake vapor (up to 40%) was found in the central part of valley at stations Cholpon-Ata, Pokrovka, and Bolshaya Kyzyl-Su.

Variation of isotopic composition of the river water due to evaporation can be observed when examining, as an example, the Nile River (Friedman et al. 1964). During flood periods, the relative D content in river water near Khartoum is −23‰ and near Cairo it is −21‰. By the calculations of Friedman, the data correspond to R/R_0 value of 1.002 and a value of 0.97 for the fraction of remaining water. Thus, the Nile appears to have lost about 3% of its water by evaporation between Khartoum and Cairo (~3,000 km). At low water, these values are +15‰ and +16‰ respectively. Again, evaporation loss is about 3%. In Friedman's opinion, the difference in δ values reflects the fact that at high water the Blue Nile is the dominant water source and at low water the Wight Nile predominates.

Another example, illustrating conditions of isotopic composition formation and the influence of evaporation in a large continental water system, is the Great Lakes basin and that of the St. Lawrence River in North America.

Figure 5.4 shows the data obtained by Brown (1970) on D content in the St. Lawrence River and in the main tributaries of the system and also the data on D content in precipitation. As the author has pointed out, the D concentration in this water system is the highest compared to all the other reservoirs of water in Canada due to intensive evaporation of water and its enrichment with heavy isotopes. Within the range of these lakes, there is a slight (up to 10‰) enrichment in D when water flows from one lake to another. Down the river course, the D content drops due to the dilution of its water with higher waters from the tributaries and also due to the discharge of groundwaters of the valley. After Quebec (see Fig. 5.4, after point 21) the D content increases due to contributions of sea waters from the gulf mixing with the river water. Further (after point 18), some variations of isotopic composition are observed, being lower than those in the ocean water.

The following main principles of isotopic composition formation of continental surface waters in a regional scale should be pointed out. Since, surface continental waters are represented by a number of separate regional basins in the framework of which accumulation, mixing, and run-off of atmospheric precipitation in the ocean occur, the isotopic content of surface inland waters reflects some average

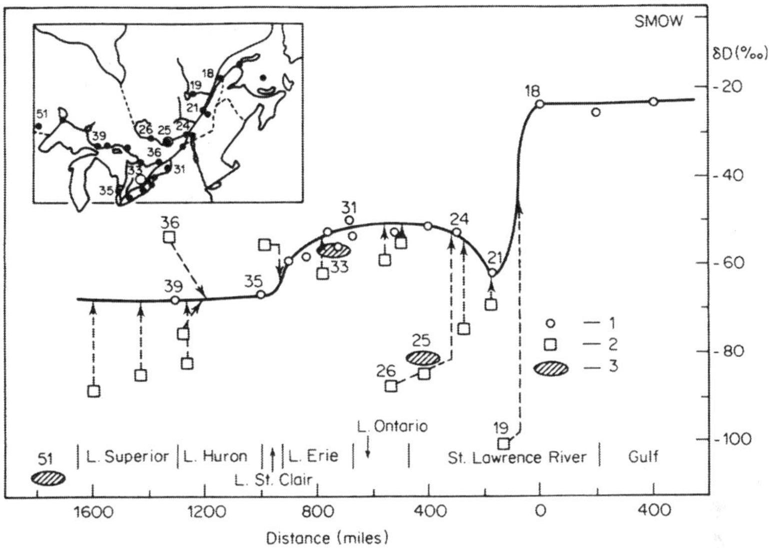

Fig. 5.4 D concentrations along the Great Lakes–St. Lawrence system in October 1967. Numbers refer to the points of investigation. *1* main stream; *2* tributaries; *3* precipitation. (After Brown 1970)

regional isotopic ratios in precipitation during a certain period of time. The isotopic composition of water of the basin may reflect both average annual isotopic ratios and seasonal variations in precipitation, which depend upon the climate zone, size of the basin, and the supply conditions. A rather important factor in the formation of the isotopic composition of water is the evaporation of water from the basin and the catchment area. This process has particular importance for conditions in lakes where heavy isotope content is always higher than in the source.

The latitudinal and altitudinal effects, which are a feature of isotopic variations in precipitation, are also valid for the surface waters. In high latitudes, the river and lake waters are depleted in heavy isotopes compared to those in low latitudes. Rivers and lakes of the arid zones, being fed by water from high mountain glaciers and snow melting, are distinguished by lower values of heavy isotopes. In contrast, the drying rivers and lakes of the arid zone have excessive heavy isotopic contents.

It should be noted that when considering the difficulty of the formation of isotopic composition of surface continental waters , one of the main problems is to establish a genetic relationship between the isotopic composition of the surface waters in a given region and that of precipitation. This problem, which should be based on the development of regional water exchange models, awaits solution.

5.3 Isotopic Composition of Water in Evaporating Basins

The problem of D accumulation in closed evaporating basins connected genetically with ocean waters is of great theoretical and practical interest. The study of D behavior during the evaporative concentration of sea water is important to answer

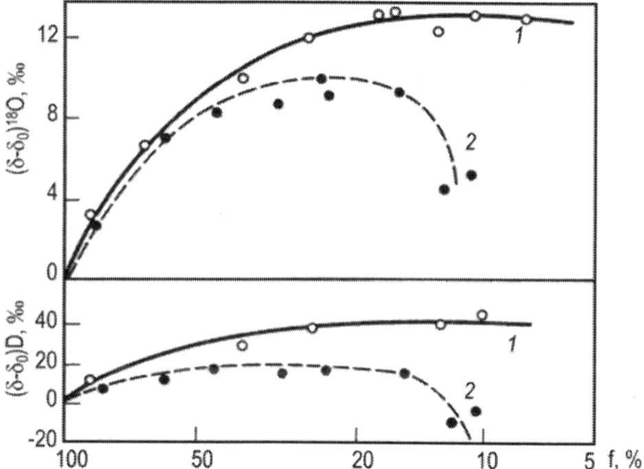

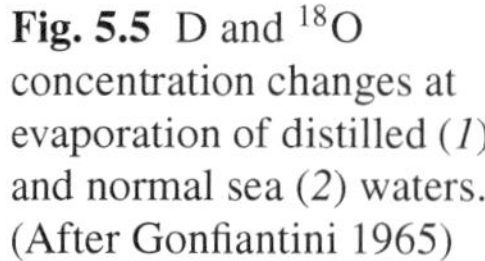
Fig. 5.5 D and ^{18}O concentration changes at evaporation of distilled (*1*) and normal sea (*2*) waters. (After Gonfiantini 1965)

paleogeological questions, since the formation of thalassogenic sedimentary waters might occur not only from basin waters with normal salinity but also from concentrated lagoon brines, shallow bays, etc. If the subsurface waters of connate marine basins can be adequately identified by their 'marine' D content, the history of sedimentary waters formed from waters of saline basins becomes ambiguous. There is an opinion (Polivanova 1970) that brines which are formed during continuous evaporation from reservoirs have the same isotopic content as infiltration waters. If this is true then, the D content serves as indicator of the marine origin of subsurface brines.

As shown above, the accumulation of D in nature does not follow Rayleigh's equation but is complicated by kinetic factors and by isotopic exchange between liquid and atmospheric moisture vapor. Whilst studying the behavior of heavy isotopes of hydrogen and oxygen during the evaporation of sea water Gonfiantini (1965) found an inversion of the isotopic content in the water–vapor system when the water body has decreased 10 times (Fig. 5.5). He correctly explained this phenomenon in terms of the increasing influence of the condensing vapor of atmospheric moisture when the rate of evaporation of the solution decreases due to decrease in the water activity provided by increases in the concentration of salts. Such an approach provides explanation of the observed phenomenon of the process studied, but the equation derived by Gonfiantini contains several assumptions which preclude its use to describe the evaporation of very small amounts of sea water.

Because of this, Polivanova (1970) concluded that the D content of water with high salinity may decrease sharply, attaining magnitudes characteristic for meteoric waters. Polivanova tried to confirm this hypothesis by considering theoretical variations in the activity coefficient of magnesium chlorides and also by the changes in water chemistry. She did not take into account the role of evaporation rates in formation of the isotopic composition of the liquid being evaporated, although Craig (1963), Lloyd (1966), and Gonfiantini (1965) showed this effect to be of great importance.

While studying the process of accumulation of heavy isotopes of oxygen during the evaporation of sea water, Lloyd (1966) came to the conclusion that changes in the

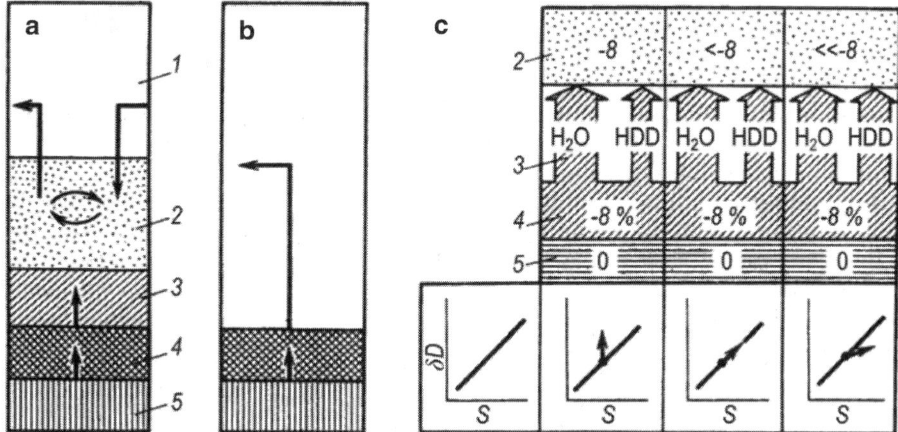

Fig. 5.6 Sverdrup (**a**), equilibrium (**b**), and deviating from equilibrium (**c**) models of water evaporation: *1* atmosphere; *2* mixed layer; *3* diffusion layer; *4* equilibrium layer; *5* liquid water

isotopic composition of liquids may be explained by Sverdrup's evaporation theory which, unlike the equilibrium evaporation theory, assumed the presence of diffuse and mixed layers (Fig. 5.6a, b). A somewhat modified scheme of Lloyd is pictured in Fig. 5.6c, illustrating the influence of the isotopic composition of vapor of mixed layer on the formation of isotopic composition of evaporating water. If the content of HDO in the mixed layer is greater than, or equal to, their equilibrium concentration with respect to the evaporating liquid (which, at 20°C, corresponds to a depletion in the D vapor by approximately 80‰ in comparison with the liquid), then the outflow of HDO vapor through the diffuse layer occurs less intensively than the outflow of H_2O vapor. In this case, the Raxleigh's process would predict.

The change in concentration of heavy isotopes in vapor of mixing layers (the degree of deviation of this concentration from the equilibrated value) provides for deviations in the accumulation of D in evaporating water on both sides of the line described by Rayleigh's equation (see Fig. 5.6c).

In natural conditions, the D concentration in mixed layer (in atmospheric moisture), as a rule, is appreciably lower than that in equilibrium, which results in a negative deviation from the Rayleigh's law. The influence of the isotopic composition of atmospheric moisture vapor upon the isotopic composition of evaporated water is clearly illustrated by Lloyd's scheme but according to this scheme it is practically impossible to explain inversions in the changes of isotopic composition being conditioned by the increase of saline concentration in the solution. Undoubtedly the rate of evaporation and the isotopic composition of atmospheric moisture are important factors in the formation of isotopic composition of evaporating liquids (Craig et al. 1963; Gonfiantini 1965; Sofer and Gat 1975; Polyakov and Seletsky 1972; Gutsalo 1980).

The isotopic composition of the atmospheric moisture can be considered in laboratory experiments to be independent of the isotopic composition of the evaporating liquid. This is also true within certain assumptions for natural evaporating basins (e.g., where the lagoon is isolated from the sea). In this case, the evaporating rate of the liquid will be proportional to the difference between the equilibrium pressure of vapor above the liquid and the vapor pressure of the atmosphere:

$$G = k(p_1 - p_a), \tag{5.17}$$

where G is the rate of liquid evaporation; p_1 is the equilibrium pressure of vapor above the liquid; p_a is the vapor pressure of the atmosphere; k is the coefficient of proportionality.

According to Raule's law, $p_1 = p_1^0 a$, (where p_1^0 is the activity of water in the solution). If we assume $p_a = p_1^0 h$ (where h is relative humidity), then

$$G = kp_1^0(a - h)k(p_{1/}p_1^0) - h. \tag{5.18}$$

The activity of water in the solution remains constant during the evaporation of pure water or saturated solutions ($a = 1$ for pure water and $a \approx 0.8$ for saturated solution of NaCl).

If the evaporation led to an accumulation of salts in the reservoir to such an extent that the vapor pressure above the liquid equals to the pressure of vapor of the atmospheric moisture ($a = h$), then the final isotopic composition of the reservoir water by Eq. 5.15 will be equal to $R_L = R_a a$, that is the reservoir will be in equilibrium with atmospheric moisture. If water loss from the reservoir is due only to evaporation and the water body is decreasing, then the isotopic composition of the remaining liquid can be calculated by the equation

$$R_L = \frac{[R_0(a/\alpha k - a + h) - hR_\alpha/k](V/V_0)^{(a/ak-a+h)/(a-h)} + hR_\alpha/k}{a/\alpha k - 1 + h}, \tag{5.19}$$

where a is the activity of water being changed by evaporation; V/V_0 is the remaining body fraction of the solution; V_0 is the initial volume; R_0 is the initial isotopic ratio of the liquid.

In the limiting case of $V/V_0 \rightarrow 0$, the final isotopic composition of water will be equal to:

$$R_L = \frac{hR_a/k}{a/\alpha k - a + h}. \tag{5.20}$$

This formula is similar to Eq. 5.15 but the former takes into account the activity of water in the solution, where the value of $a \neq 1$.

It follows from Eq. 5.19 that the limiting isotopic composition obtained by evaporating distilled water ($a = 1$) and saturated salt solution ($a < 1$) are not the same as those discovered by experimentation. If water activity in the solution during experiment (during the sea-water evaporation) is not a constant, then the rate of evaporation by Eq. 5.18 will be equal to zero.

Inserting the value $a = h$ in Eq. 5.20, one obtains

$$R_L = aR_a, \tag{5.21}$$

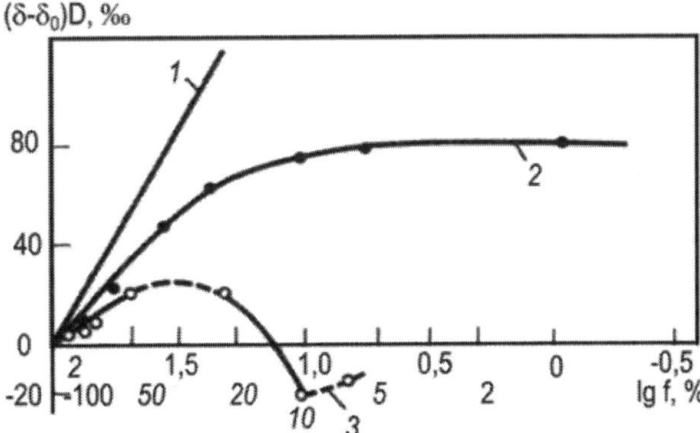

Fig. 5.7 Air moisture influence on the D accumulation in liquid at the evaporation of water with marine salinity (from Atlantic ocean): *1* Rayleigh distillation case; *2* the experimental data at air relative moisture equal to 40–50%; *3* the data at relative air moisture equal to 70–80%

that is the limiting isotopic composition of liquid with respect to vapors of atmospheric moisture will be at equilibrium. This result is similar to that obtained from Eq. 5.13 since in both cases it is assumed that the water reservoir is in dynamic equilibrium with atmospheric moisture vapor.

It follows from sea-water evaporation experiments (Fig. 5.7) that the isotopic composition of water being evaporated under certain conditions may undergo inversion. This is caused by a decrease in the evaporation rate due to decreasing activity of water in the liquid phase (Polyakov and Seletsky 1972). When $V \to 0$, the D content tends to a stationary value at which $R_L = aR_a$.

In natural evaporating basins with initially normal sea-water salinity, the consequent sedimentation of minerals and the increase in the salt concentration in the residue of the solution occur due to the evaporation of water. The changes in the composition of salts of sea-water at different stages of concentration are given in Table 5.2 (data obtained by Galakhovskaya 1967).

It is evident from Table 5.2 that at the stage of halite sedimentation, a gradual change in the water salt composition occurs (the sodium chloride water transforms into magnesium chloride water). If evaporation continues, the water activity also changes (from 0.8 at the stage of halite sedimentation to ~0.6 at the stage of bischofite sedimentation), resulting in a decrease in the evaporation rate of the brine at the final

Table 5.2 Sea-water composition changes at different stages of concentration

Stage of concentration	Liquid content relative initial volume (%)	Density (g/cm^3)	Ion composition (mass %)						
			SO_4^{2-}	Cl−	Ca^{2+}	Mg^{2+}	K^+	Na^+	∑ ions[a]
Calcium carbonate	25	1.10	1.00	7.31	0.12	0.50	0.16	4.05	13.1
Gypsum	15	1.16	1.28	11.36	0.07	0.75	0.25	6.36	19.90
Halite	13	1.21	1.67	15.44	0.03	1.10	0.36	8.52	27.31
Epsomite	1.0	1.30	6.33	14.37	0.11	4.80	1.86	2.16	29.83
Silvine	1.6	1.30	5.53	18.21	0.01	6.39	2.22	1.25	34.44
Bishofite	0.7	1.35	3.03	23.58	–	8.81	0.02	0.28	36.25
Carnollite	0.9	1.32	4.35	19.59	0.03	6.95	1.82	0.77	33.88

[a]Together with the other ions not shown in the table

stage of evaporation. Under conditions of high humidity, this process might cause a certain decrease in D concentration in the solution. The magnitude of this decrease depends on the relative humidity, isotopic composition of atmospheric moisture, and also on the evaporation rate of the solution. If relative air humidity during evaporation is 70–80%, then obviously sea water cannot be evaporated to the stage of bischofite sedimentation and the stationary isotopic composition of the solution $R_L = aR_a$ will be attained at earlier stages of evaporation.

In coastal regions (if the saline basin is marine lagoon), the isotopic composition of vapor R_a will be close to that in equilibrium with sea water, that is equal to R_s/a (where R_s is the isotopic composition of sea water).

During the condensation of vapor, the liquid phase is enriched with heavy isotopes of hydrogen by a factor α, that is the isotopic composition of the coastal evaporating basin, in any case, does not differ considerably from the isotopic composition of seawater and will be in isotopic equilibrium with the marine water vapor ($\delta D \approx -94‰$, $\delta^{18}O \approx -13‰$, see Chap. 3). For small inland saline basins (saline lakes), the isotopic composition can differ considerably from the marine one, being depleted in heavy isotopes for two reasons. Firstly, such basins could be fed by groundwaters with an isotopic composition lower than that of seawater. Secondly, the isotopic composition of the vapor of atmospheric moisture is much lower than that in equilibrium with ocean water. Besides, precipitation greatly influences the isotopic composition of some of these basins because of their small storage in comparison with large evaporating basins of marine origin. Precipitation is decreasing in response to a decrease in the vapor condensation temperature.

These conclusions were confirmed by model calculations reported by Vlasova and Btezgunov (1978) for brines with certain isotopic composition formed from standard seawater at different stages of evaporation. Two models have been considered by these authors such as the one assumes a constant storage (with replenishment) and the exhaustion model (the model of a drying basin). Calculations have been carried out using formulae similar to Eqs. 5.13 and 5.19 taking into account the changes in the water activity in the solution during evaporation. Substituting the isotopic ratios in the replenishing water R_I, in the basin R_L, in the vapor outflow from the basin R_E ($R_E = R_I$) and in the vapor of the atmospheric moisture R_a with the corresponding $1+\delta_i$ values, an expression is obtained which permits the calculation of the liquid's isotopic composition in relative units for given values of air humidity h, activity of water in the solution a, isotopic composition of the inflowing water δ_I and the isotopic composition of atmospheric moisture vapor δ_a:

$$\delta_L = \frac{\alpha[k(1+\delta_I)(a-h) + (1+\delta_a)h]}{a} - 1. \tag{5.22}$$

The α-values for a given temperature were taken from reference books and the values of k were taken to be $k_D = 1.009$ and $k_{^{18}O} = 1.016$.

In the case of a drying basin, calculations have been made using the formula

$$\delta_L - \delta_L^0 = \frac{h[\delta_a - k\delta_L^0 - (k-1)] - \Delta\varepsilon_a(1000 + \delta_L^0)}{hk + \Delta\varepsilon_a}[1 - f^{hk+\Delta\varepsilon_a/k(a-h)}], \tag{5.23}$$

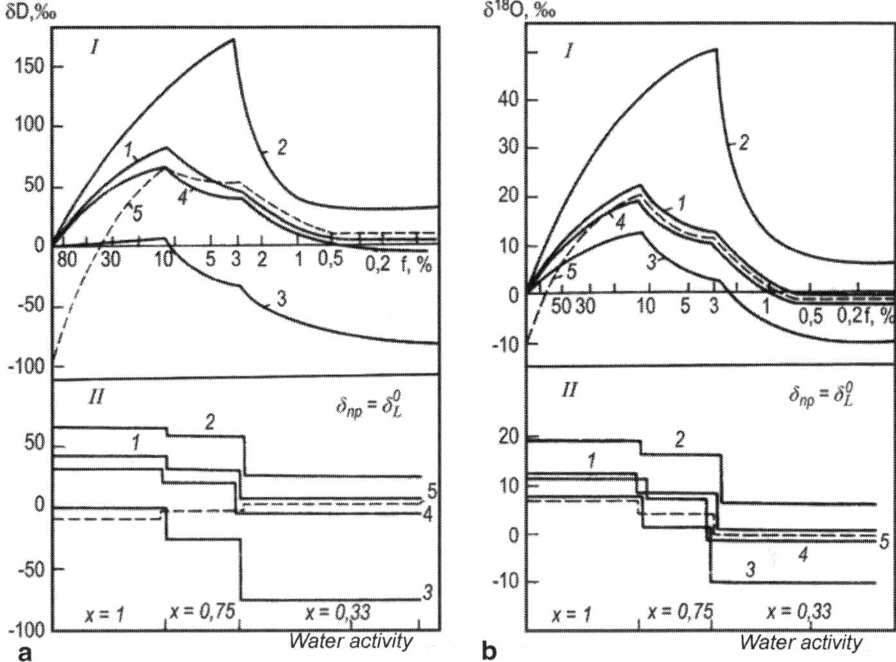

Fig. 5.8 Isotopic composition of **a** hydrogen and **b** oxygen of evaporated sea water for changed evaporation conditions (calculated curves): *I* exhaustion model; *II* constant level model

where δ_L^0, δ_L, δ_a are the relative isotopic contents of water at the beginning of evaporation ($f = 1$), at an arbitrary moment after evaporation has begun, the vapor of atmospheric moisture, respectively; f is the degree of water loss, equal to the ratio of volume of the remaining water to the initial volume; $\Delta\varepsilon = 1/\alpha - k$; α is the equilibrium fractionation factor of isotopes at a given temperature; k is the kinetic fractionation factor.

Three constant values of water activity were taken in order to carry out the calculations: $a = 1$, corresponding to evaporation without precipitation of salts when the water body decreases from 100 to 12%; $a = 0.75$, corresponding to precipitation of halite when the water body decreases from 12 to 3%; $a = 0.33$, corresponding to precipitation of magnesium salts when the water body decreases from 3 to 0.1%.

Equations (5.22) and (5.23) were solved for the following parameter values: humidity $h = 25\%$, 50%, 75%; temperature of evaporating water $t + 10°C$, +20°C, +30°C; isotopic composition of the atmospheric vapor δ_a for δD −70‰, –150‰, −200‰ and for $\delta^{18}O$ −10‰, −20‰, −40‰; isotopic composition of inflowing water for the model with a constant storage δ_I for δD 0‰, −50‰, −100‰ and for $\delta^{18}O$ 0‰, −5‰, −10‰.

The results of the calculations are given in Fig. 5.8 (Vlasova and Brezgunov 1978).

The following parameters were accepted to plot the graphs:

(a)

	h, %	δ_a, ‰	t, °C	δ^0_L, ‰
1	50	– 10	20	0
2	25	– 10	20	0
3	30	– 150	20	0
4	50	– 10	30	0
5	50	– 10	20	– 100

(b)

	h,%	δa, ‰	t, °C	δ^0_L, ‰
1	50	– 10	20	0
2	25	– 10	20	0
3	50	– 20	20	0
4	50	– 10	30	0
5	50	– 10	20	– 10

The results show that under natural conditions, the isotopic composition of evaporating basins may vary greatly depending on the conditions of evaporation. The climatic conditions of the formation of large saline basins from the sea restrict the range of δD and δ^{18}O variations. According to Vlasova and Brezgunov (1978), under less favorable conditions, the minimum D and ^{18}O content in brines of sea origin at the evaporation stage will not be less than $\delta D = -40‰$ and $\delta^{18}O = -3‰$.

These conclusions enable one to use isotopic data for estimating the genesis of connate brines. Theoretical and experimental data for modern saline basins have been reported to be in agreement with the author's conclusions. They are also in accordance with calculations made by Sofer and Gat (1975) and Gutsalo (1980).

Chapter 6
Isotopic Composition of Water in the Unsaturated and Saturated Zones

On the basis of experimental data, it has been pointed out that the isotopic composition of shallow groundwaters, which are replenished by infiltration of atmospheric precipitation through the unsaturated zone, is characterized on average by the heavy isotopes of typical precipitation of a given region. Under certain conditions, there are some differences in isotopic ratios for the two types of water mentioned above. This is explained by the fact that precipitation in the spring–summer season is partially and often completely re-evaporated from the Earth's surface and from the unsaturated zone both directly and due to transpiration of moisture by plants.

As Zimmermann et al. (1967) have shown, the water which infiltrates from the Earth's surface into the unsaturated zone is already enriched with the heavy isotopes compared with precipitation due to the evaporation process. Under identical soil conditions, water enriched in places which are covered with plants is, on average, 10‰ greater than in those areas which have plant cover. The above mentioned values, as Zimmermann et al. (1967) noted, are characteristic for Central Europe where the average relative deuterium (D) concentration is about −70‰. The above authors have also shown that variations in the isotopic composition of the soil moisture remain unaffected by the observed ratio between moisture movement velocity in the unsaturated zone, the rate of evaporation of moisture from the surface, and the process of diffusion and exchange through transpiration by plants. At the same time, Gonfiantini et al. (1965) reported that the process of water evaporation by leaves results in its isotopic fractionation. It should also be noted that precipitation water in the spring–summer season is enriched with heavy isotopes compared to groundwaters. The autumn–winter precipitation is more depleted in δD and $\delta^{18}O$ and also undergoes enrichment due to evaporation from the surface. Finally, this results in isotopic balance between the atmospheric precipitation and shallow groundwater.

Analogous investigations regarding formation of the isotopic composition in near-surface ground waters of the arid zone of the south–east Mediterranean coastal region were carried out by Gat and Tzur (1967). They found that groundwaters which are fed by local precipitation are enriched with oxygen-18 (^{18}O) by 1–3‰ relative to standard of the ocean water (SMOW). Studies carried out using lysimeter have shown that under experimental conditions, an excess of ^{18}O of up to 2‰ has also been observed in the infiltrating moisture due to surface evaporation. In general,

V. I. Ferronsky, V. A. Polyakov, *Isotopes of the Earth's Hydrosphere*,
DOI 10.1007/978-94-007-2856-1_6, © Springer Science+Business Media B.V. 2012

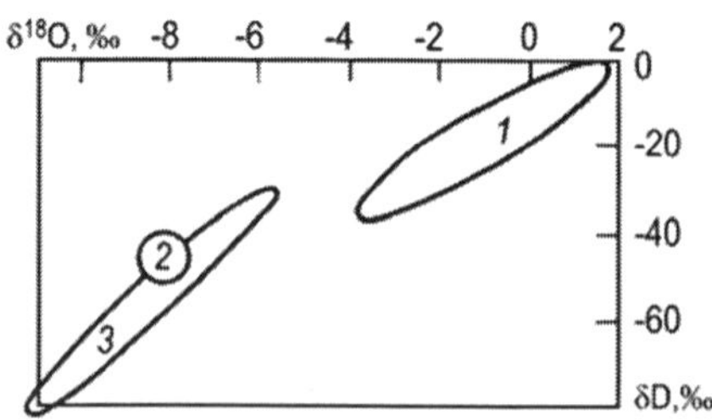

Fig. 6.1 δD–$\delta^{18}O$ relationship in waters of south–central Turkey: *1* lakes; *2* costal springs; *3* precipitation. (After Dinçer and Halevy 1968)

surface evaporation results in the visible enrichment of ground waters in D and ^{18}O in arid and semi-arid regions (Gat and Tzur 1967; Gonfiantini et al. 1974; Dinçer et al. 1974; Gonfiantini et al. 1976).

The knowledge of the formation of groundwater isotopic composition in active water exchange zones helps to solve a number of practical problems. Some characteristic examples of solutions to such problems are given below.

6.1 Relationship Between Surface and Ground Water

This problem is rather widespread in hydrological and hydrogeological investigations. When solving it on the basis of D and ^{18}O content of the waters being investigated, the following effects are used. Enrichment with heavy isotopes occurs during evaporation of the reservoir water. Direct recharge to the reservoirs by precipitation does not result in enrichment of water with D and ^{18}O. Thus, the existence or absence of the relationship between surface and groundwaters can be established by plotting and analyzing δD–$\delta^{18}O$ diagrams (see Fig. 6.3). In this diagram, for precipitation which has not undergone evaporation, the relationship $\delta D = 8\delta^{18}O + 10$ (or a similar characteristic for a given region) holds. While evaporation of the reservoir water—which is being fed by precipitation—takes place, the slope of the line in the diagram δD–$\delta^{18}O$ decreases to values between 3 and 6.

In a study in south–western Turkey with the method outlined above, it was possible to reject the hypothesis of interconnection of two large lakes, where the water is lost through fractured limestone formations with some large limestone springs (Fig. 6.1). In another case, a similar conclusion was reached regarding the connection of Lake Chala in Africa and some springs in the same region (Dinçer and Halevy 1968).

By means of isotopic and hydrochemical methods, Mazor (1976) checked the ancient groundwater tracing experiment. They say, 2,000 years ago, the ruler Philip established by his belief that Ram Lake is the feeder of the Banias spring, located 6 km to the west of the slopes of Mount Hermon. Philip's experiment consisted of throwing chaff into Ram Lake and observing it in the Banias spring. The modern studies do not prove the above version. The crater is surrounded by tuff, the highest point of which has an altitude of 1,000 m, whereas the altitude of the water table of the lake is about 940 m and its D and ^{18}O contents are close to the sea water. This

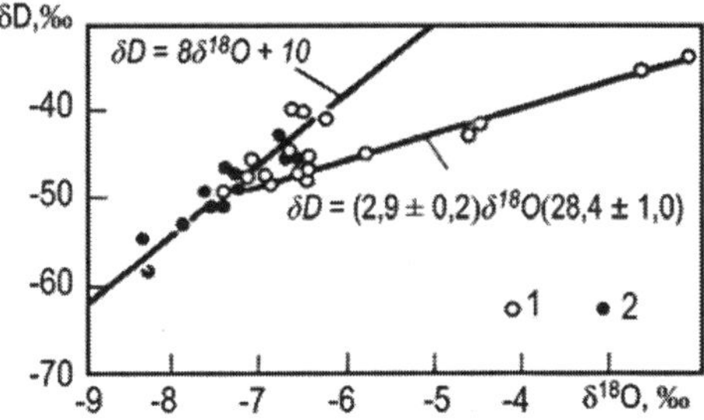

Fig. 6.2 δD–$\delta^{18}O$ relationship in groundwaters of salt plane Chott-el-Hodna, Algiers. Ground waters: *1 shallow*; *2 deep*. (After Gonfiantini et al. 1974)

enrichment with the heavy isotopes is a result of intensive evaporation because of the arid climate. In spring Banias, which is 600 m lower than the lake, $\delta D \approx -40‰$ and $\delta^{18}O \approx -8‰$. The D, ^{18}O, and Cl^- values in the well Masada, 500 m south–west of the lake, show that its water is a mixture of the regional groundwater with water infiltrated from the lake.

Enrichment of heavy isotopes in shallow waters due to evaporation with respect to the deep groundwaters was used for identification of local recharge areas in the salt plane of the Chott-el-Hodna, Algiers (Gonfiantini et al. 1974). The shallow groundwater was reported to be recharged by deep groundwater. Stable isotopes demonstrate that this does not occur. Some local low salinity shallow groundwater and the shape of the watertable were found to be recharged from wadis (dry river beds); however, the shallow waters in the alluvial Quaternary and continental Cretaceous formations have different origin. According to the isotope data, to the south of the Chott, the groundwaters evaporated more intensively compared to the deep waters (Fig. 6.2). This can be explained by discharge of the deep waters to the shallow horizons and intensive evaporation from the surface of the groundwater which locate close to land surface (less than 1 m). This assumption is also proved by the increase in water mineralization (100 g/l) and $\delta^{18}O$ content from $-7‰$ to $-2‰$.

In the areas north and east of the Chott-el-Hodna, the recharge of the deep aquifer appears to be limited. Isotope data demonstrate more active infiltration of atmospheric water from wadis. The experimental points of δD and $\delta^{18}O$ here lie along straight line $\delta D = 8\delta^{18}O + 10$ characteristic for atmospheric precipitation.

In another study (Dinçer 1968; Dinçer and Payne 1971), the hydraulic relationship of a group of karstic lakes was studied. The lakes are situated on the northern slope of the Taurus Mountains, at a distance of 100–150 km from the coast of the Mediterranean Sea, with several rivers and springs which start their course in the same region. The area of investigation was a typical karstic region of Mesozoic and Eocenic limestones. Based on results of isotopic measurements for water and precipitation at various points over the region and of the lake water, it was found that the conditions and sources of river recharge during different seasons of the year are in response to surface runoff and supply from the lakes through the karst-like rocks. Data on other environmental isotopes (tritium, carbon-13, carbon-14 contents) in precipitation, in surface reservoirs, and in groundwater were used in most studies to estimate the hydraulic relationship and the parameters of water exchange time.

6.2 Groundwater Recharge at Present Time

The seasonal variations of isotopes in precipitation, the feeding groundwater may be used for identification of the season of groundwater recharge (Winograd et al. 1998). As a rule, D and ^{18}O content in groundwater corresponds to the mean-weighted content of the isotopes in the inflow water for a large period of time. Thus, short periodic variations in D and ^{18}O content can be identified in a high-dynamic groundwater like, for example, in karstic regions.

Bowen and Williams (1973), while investigating groundwaters in the western region of Ireland, found that ^{18}O content in karstic waters of the Gort plateau was lower ($\delta^{18}O \leq -7.2‰$) than the average one in precipitation ($\delta^{18}O = -5.4‰$). The authors concluded that the seasonal water supply of this region is the result of depleted winter precipitation. The spring–summer precipitation enriched with heavy isotopes is lost by evapotranspiration.

Groundwater aquifers in the Sahara (Conrad and Fontes 1970) and the Kalahari Desert (Mazer et al. 1974a) were found to be depleted in heavy isotopes compared with the average δD and $\delta^{18}O$ values detected in precipitation. This evidence, combined with the tritium and radiocarbon studies, allowed the authors to determine that they are supplied with water of shower precipitation. With the help of known data about the duration of showers, the authors estimated the conditions of formation of the groundwater storage in this region.

6.3 Groundwater Recharge in the Past

The mean yearly values of temperature for all the Earth's regions in the last Pleistocene glacial epoch were lower than the modern (Emiliani 1970). Climatic conditions of the many arid regions were also different from the present days with respect to mean yearly temperature and humidity. Shower precipitation was found in the northern parts of the desert belts in glacial epochs (Bowen). It was formed because the western winds' belt in the Northern hemisphere was deviated to the south due to very wide glacial areas. Climatic changes in the northern latitudes of the modern arid regions were characterized by decreased concentration of D and ^{18}O in the covering glacial due to precipitation during the pluvial periods.

By the data of Emiliani (1970) and Epstein et al. (1970), the end of the Pleistocene glacial period (the Pleistocene–Holocene border) happened about 10,000 years ago. The influence of the pluvial epochs on groundwater in the arid regions' recharge is fixed by isotope data for the modern deserts in Africa, Arabian Peninsula, and Middle Asia. Vogel and Ehhalt (1963) have determined the age by radiocarbon method for fresh groundwater in western desert of Egypt which appeared to be 20,000–30,000 years, that is main part of the groundwater resources was recharged during the pluvial period of Pleistocene. This conclusion is well agreed with the groundwater data of D and ^{18}O, which are even lower than the modern precipitation for this region and even in modern precipitation of the Central Europe. Analogous data were obtained during groundwater study in Sahara (Conrad and Fontes 1970; Gonfiantini et al. 1974; Sonntag et al. 1979), Sinai Peninsula (Issar et al. 1972), Saudi Arabia (Champine et al. 1979; Yurtsever and Payne 1979), Potiguar Basin, Northern Brazil (Salati et al. 1974), and other regions.

6.4 Identification of Area of Groundwater Recharge

The solution of the problem of identification of recharge areas in mountain regions is based on the observed orographic effect on the distribution of the stable oxygen and hydrogen isotopes (see Eqs. 4.10 and 4.11).

Arnason and Sigurgeirsson (1967) were the first who attempted to identify drainage basins for a number of hot and cold springs in the eastern part of Iceland. Systematic sampling was carried out monthly (for 2–8 years) and the D content was measured in precipitation, springs, and ground waters. A map of δD distribution in precipitation and groundwater was plotted (Fig. 6.3). Later on a map like this (Arnason 1977b) was plotted for all the Iceland territory.

It is seen from the figure that the D content in precipitation decreases toward the northeast direction depending on the distance from the southern coastal area. This distribution is controlled by the southern winds which determine the air moisture transfer over the island. In average, D content decreases by 2‰ per km from the coast. This regularity is superimposed by the altitude effect and while the absolute marks of the land are dropped, the D concentration decreases. For example, maximum values of $\delta D = -50‰$ were registered in the southern coastal area and the lower were $\delta D = -94‰$ near glacier Hofsjökull on the northeast of the studied area. A large cold spring Keldur emerging from an extensive lava field produced by the volcano Hekla has local value of $\delta D = -58‰$, but the spring has $\delta D = -67‰$.

In the city of Reykjavik, there is a rich geothermal area where many boreholes have been drilled down to 700–800 m and one to a depth of 2,000 m. The base temperature is 140°C. The water in all these boreholes has δD from −65‰ to −66‰. This water could come from Lake Thingvallavatn, but a more likely recharge area would be a group of mountains north of the lake some 40–50 km away. At the same time in Krisuvik—a geothermal area on the Reykjanes peninsula, 25 km southwest from Reykjavik, 140 m above sea level—the water with a base temperature of 230°C was observed. The value of δD for this water is −51‰ (taking into account the evaporation effect on the way up through the borehole) which, in the author's opinion, says about the local recharge of the geothermal system located on the peninsula or little farther to the northeast where δD values are lower. In this case, say the authors, we can assume that δD values are increased while the water moves along the aquifer. Theodorsson (1967) continued the study in the same region, applying tritium indicator, and improved the results and conclusions.

Payne and Yurtsever (1974), while studying distribution of D and ^{18}O, determined relative importance of local atmospheric precipitation and mountain slope drainage in recharge of groundwater for the Chinandega Plain, the area of about 1,100 km^2

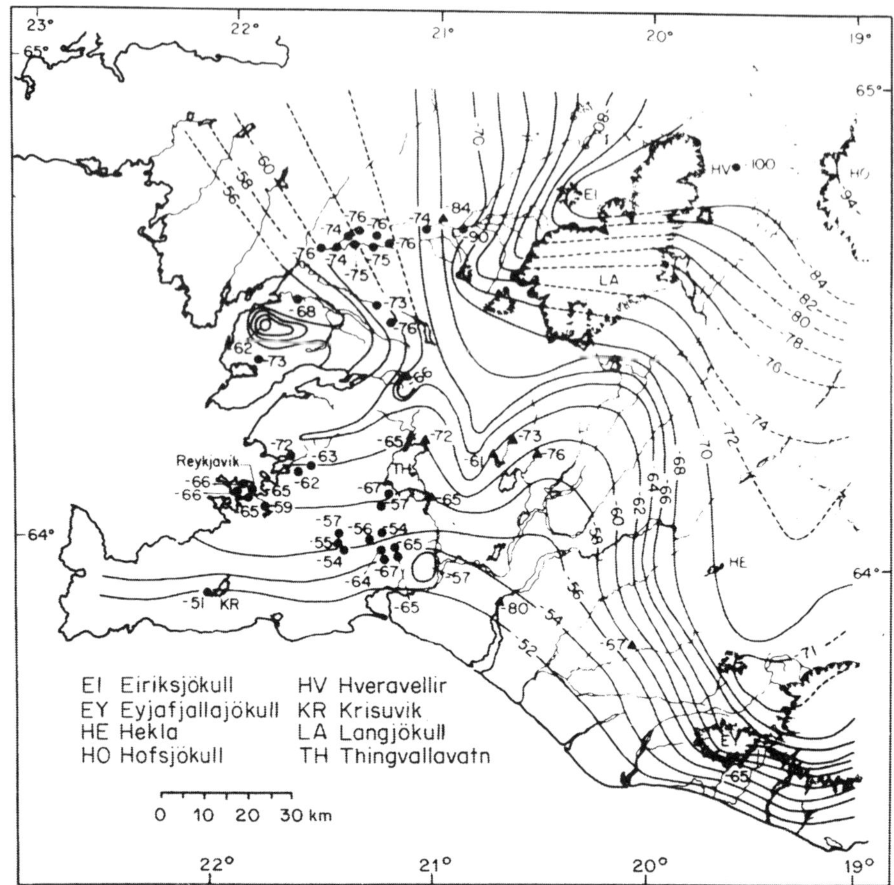

Fig. 6.3 Map of contours of δD values in precipitation and in geothermal springs (points) and in cold springs (three angles) of south-west Iceland. (After Arnason and Sigurgeirsson 1967)

located between the Pacific Ocean coast and the Cordillera de Marrabios in Nicaragua. It was found that δ^{18}O value of the direct plane precipitation partly recharged groundwater is equal to −5.9‰. On the basis of water and isotopic balance, the calculations show that ^{18}O in precipitation along the slope of Cordillera is decreased by equation $\delta^{18}O = -5.653 - 0.0026\,H$ (here H is the height of the sampling place above sea level, m). It is evident from the equation that the slope of the straight line is −0.26‰ per 100 m. In the borehole on the plane at an altitude of 280 m above sea level, the water has $\delta^{18}O = -7.18$‰, which means that the recharge area of this water is higher than the 600 m above-sea level.

Analogous investigation was done by Stahl et al. (1974) in Greece. It was found by use of the height gradients in the area of study ($\Delta\delta D = -1.2$‰/100 m, $\Delta\delta^{18}O = -0.16$‰/100 m) and the D concentrations that the potential areas of recharge for the artesian waters situate on different hypsometric levels differed more than 1,000 m. The same situation was found to be for the recharge areas of thermal waters.

The natural isotope tracing in precipitation depending on temperature of their condensation was used for identification of water recharge to mountain tunnels.

Rauert and Stichler (1974) applied isotope techniques to study seepage water in a 7-km-long tunnel in the Austrian Central Alps. It was found that the water comes to the tunnel from the top of the fractured rocks. The conclusion was made on the basis that the D content in recharge water along the tunnel practically reflected its high-level location ($\Delta\delta D = -3‰/100\,m$) for the given region. It was found using tritium that the minimum velocity of the infiltrated water was equal to about 20 m per year.

Fontes et al. (1979) have undertaken investigations close to above in Mont Blanc tunnel (11.6-km-long). The study allowed not only localizing the recharge area and determine the velocity of vertical movement of water but also to contribute to the study of hydrogeological and hydrochemical processes in the massif of Mont Blanc. In particular, it was found that the residence time of water in the massif does not exceed a number of years and the infiltration of the surface water through the fractured granite rocks is accompanied by its significant evaporation. Concentration of chlorides in the water having contact from the French side of the tunnel with the crystal slates has increased. The data on hydrothermal water obtained in a number of points along the tunnel have not solved the problem about the source of water recharge because the water temperature seems to be determined mainly by its vertical movement velocity.

After studying the distribution of hydrogen and oxygen isotopes in the possible sources of water recharge, Rodriges (1979) localized the area of water flow into large tunnel Palacio, which was constructed for water supply of Bogota (Columbia) in Andean Cordillera. After the main part of the tunnel was constructed, the works were cut off because of catastrophic water recharge (4 m^3/min) to the face. A number of versions were assumed based on general geological and hydrodynamic date. In particular, it was assumed that the water enters from a local lake. The isotopic and chemical content of water from the lake and also from water of the modern, paleogene, and cretaceous sediments, situated around the tunnel, were measured. The mean value of hydrogen $\delta D = -76‰$ was found to be close to that of water from the fractured carbonate rocks of the upper Guadalupe formation (K_2). The waters from lake and some other possible sources were enriched with heavy hydrogen (δD from -49 to $-71.1‰$) compared with water from the carbonate massif. The macro-component contents of groundwater from the Guadalupe sediments and from the tunnel were found to be close. Those data allowed the author to conclude that the tunnel water recharge comes mainly from the groundwater of the carbonate rocks massif.

The author of the same work reports about study on groundwater migration in a large landslide Quebradablanca composed by alluvial sediments be deposited on the slope of a massif on metamorphic rocks. Earlier, on the basis of traditional observation and hydrogeological calculation, it was concluded that a cause of the landslide motion is the infiltrating atmospheric precipitation of local origin. The water from

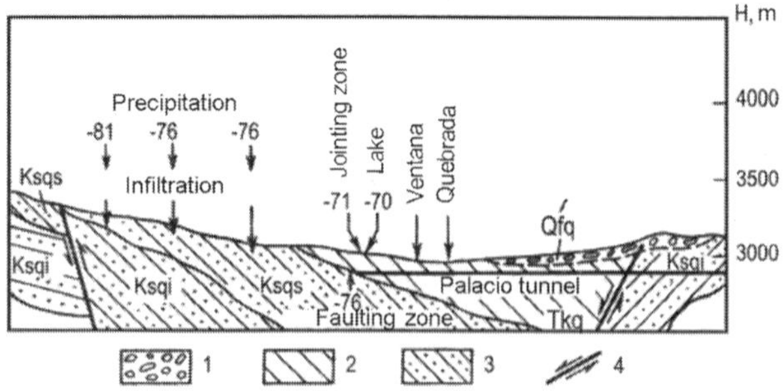

Fig. 6.4 Hydrogeological situation in construction of tunnel Palacio: *1* fluvial-glacial sediments; *2* Tertiary clays; *3* Cretaceous sandstones; *4* fractures. Figures show values of δD, ‰. (After Rodriges 1979)

the metamorphic rocks was not taken into account. In order to cut the infiltrating precipitation water access, a special drainage system was constructed which has not solved the problem. It was found by isotopic composition of hydrogen and ^{18}O that the groundwater of the alluvial deposits (δD from −57 to −58‰; $\delta^{18}O$ from −8.2 to −8.5‰) differ from those in drainage system (δD = −54‰). This water was formed by mixing water from the metamorphic rocks (δD = −64‰) and the infiltrating atmospheric precipitation. The water from the metamorphic rocks with lighter isotopic content (δD = −64.5‰; $\delta^{18}O = -9.37$‰) have a higher located recharge area. In principle, the new hydrogeologic conclusions related to the dynamics of groundwater in the landslide were made. A schematic picture of the isotope application for solution of hydrogeologic problems in Columbia is given in Figs. 6.4 and 6.5.

In the above considered examples, the localization of areas of groundwater recharge is based on isotope data where difference in isotopic composition of precipitation depends mainly on the process of condensation of the atmospheric moisture. At the same time, the kinetic factors of marine water evaporation in the regions with different climatic conditions change the general correlation dependence δD–$\delta^{18}O$‰. For instance, this dependence for the East Mediterranean has the form $\delta D = 8\delta^{18}O + 22$

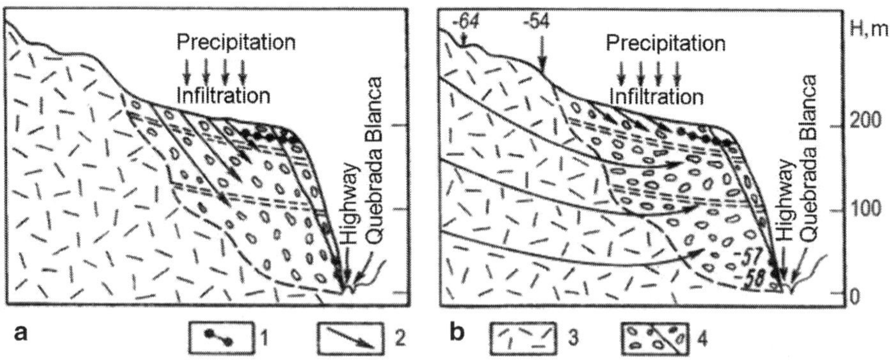

Fig. 6.5 Hydrogeological conditions of landslide Quebradablanca: (**a**) before studies; (**b**) after studies; *1* drainage row; *2* direction of flow; *3* hard rocks; *4* landslide cone. Figures show values of δD, ‰. (After Rodriges 1979)

(Gat and Tzur 1967) which differs from the Craig's correlation $\delta D = 8\delta^{18}O + 10$ for the regions obtaining the Atlantic Ocean's moisture.

Sentürk et al. (1970) carried out studies in central Anatolia for the semi-arid region of the Konya valley in the south–western part of the Mediterranean coast of Turkey. The Konya plain is surrounded by mountains and is situated in the semi-arid zone of the central part of the Anatolian Peninsula. In order to widen the usage of groundwater for irrigation, the investigations of their resources were undertaken. The isotopic composition of groundwater and local precipitation has been examined to identify the sources and recharge conditions of the two most important aquifers. It was found that in the first shallow aquifer, the isotopic ratios of water plot on a line is defined by the equation $\delta D = 8\delta^{18}O + 22$. This equation, relating the D and ^{18}O contents, is characteristic for atmospheric precipitation in the eastern part of the Mediterranean basin. This led to the conclusion that the shallow aquifer is recharged by precipitation of Mediterranean origin. The isotopic ratios of the deep aquifer fell on the lower line defined by the equation $\delta D = 8\delta^{18}O + 10$. The latter is a feature of precipitation originating in the northern continental regions. Thus, conclusions regarding the feeding of the deep aquifer by precipitation could be drawn. The isotopic analyses were also supported by conclusions which were made on the basis of geological and hydrogeological data concerning the principal direction of motion of groundwater in the region.

6.5 Relationship Between Aquifers

The study of water percolation from one aquifer into another through weakly permeable rocks is an important problem of groundwater dynamics. Isotope composition of the two aquifers is a good indicator of their degree of connection and in same cases the only method in solving the problem.

Deak (1974, 1979) has been solving this problem in Hungary using isotope techniques. He has measured the D content in two aquifers in order to find out if the storage of groundwater is replenished from the Upper Pleistocene artesian aquifer. It has been found that the difference in average δD values for ground ($-63‰$) and artesian pressure water ($-86‰$) is too great to play any significant role in groundwater supply due to percolation of the Pleistocene waters. Later on, this conclusion was proved by radiocarbon measurements of both aquifer waters (Deak 1979).

In the earlier cited work of Gonfiantini (1974) on distribution of hydrogen and oxygen isotopes in the salted area Shott-el-Hodna (Algiers), the conclusion was made about discharge of the pressurized deep groundwater into shallow horizon. In zones to the north and east from this area, a percolation of water from the shallow to the deep horizon is observed. This phenomenon takes place in a narrow zone inside the area but outside of the artesian basin. Within the northern part of the area $\delta^{18}O$ values are decreased with the depth from $-6.6‰$ to $-8.4‰$. It is found by radiocarbon dating that if ^{18}O values are increased then the age of the groundwater is increased. The author interprets this result as a proof of the fact that deep water here are recharged by the shallow one.

Dinçer et al. (1974) investigated the problem of replenishing of shallow and artesian groundwater in Saudi Arabia. The Neogene, Eocene, and Paleocene aquifers were studied by distribution of ^{18}O and chemical contents and the conclusions were made about interconnection of water horizons and the conditions of their recharge. The criteria for the conclusions are as follows.

The groundwaters with values of stable isotopes identical to the modern precipitation (^{18}O values for the studied region varied from −2.5‰ to −3.5‰) and mineralization (dry residue) less than 1 g/l are recharged mainly by means of infiltration from irrigation channels and vadis (dry river beds). The groundwaters with the moderate isotope enrichment (from about 1–2‰ in ^{18}O) compared with the modern precipitation and moderate mineralization (from 1–3 g/l) are possibly recharged from the irrigated channel and vadis. The groundwaters of such isotopic and chemistry content are observed in the irrigated areas. The groundwaters with considerable enrichment of heavy isotopes (especially ^{18}O) and relatively low mineralization have additional feeding from precipitation within the areas of sandy dunes. The groundwaters with considerable enrichment of heavy isotopes and high salinity in their origin have relation to marine intrusion of intensive evaporation in the zone of salt areas. In this case, the ratios of D and ^{18}O contents may indicate the source high mineralization. Concentration of D and ^{18}O lower than in the modern precipitation (^{18}O less than −3.5‰) indicates that groundwater recharged in colder epochs. Mineralization of such waters on Arabian Peninsula varies between 1–10 g/l. Based on the above criteria, Dinçer et al. (1974) presented their study of groundwater recharge in Saudi Arabia.

Three identified aquifers, in the northeast coast of New Zealand's south Island on the Kaikoura Plain, were investigated with the help of isotopic and hydrogeologic methods by Brown and Taylor (1974). Distribution of hydrogen and oxygen isotopes in precipitation was obtained. Concentration of ^{18}O values in groundwater water in different parts of island varies from −5.9 to −8.5‰. These data point out on a complicated character of the groundwater recharge which originate directly by infiltration of the atmospheric precipitation ($^{18}O = -5.9$‰ and −6.4‰), by surface run-off from the Mount Fyffe ($^{18}O = -8.0$‰), and also by near-shore filtration from the Kowhai River ($^{18}O = -8.5$‰). Water of the pressurized horizon is characterized by $^{18}O = -9.4$‰ and in weakly pressurized water $^{18}O = -8.9$‰. The tritium content in the weakly pressurized horizon is close to the tritium content in the river. The tritium content in the pressurized horizon is very low. It follows from those data that there is no close hydraulic relationship despite they both are recharged from the Kowhai River. The authors explain the lower content of ^{18}O in both horizons and in the Kowhai River at present by possible change in climatic conditions observed in the last decade and also by disappearance of vegetation on the higher landmarks. All the above effects together led increase in the surface run-off to the river from the lower landmarks.

Analogous investigations on study of relationship between aquifers carried out in Algeria, Tunis, and Brazil (Gonfiantini et al. 1974; Klitsch et al. 1976; Salaty et al. 1974), on the Siberian polygon for nuclear wastes disposal (Polyakov et al. 2008; Tokarev et al. 2009), and in the Asov-Kuban artesian basin (Sokolovsky et al. 2007).

6.6 Mixing Proportions of Groundwater of Different Genesis

The solution of this problem is based on the assumption that waters of different genesis differ in their isotopic composition. Note that one does not succeed to obtain two independent equations based on the data of D and ^{18}O value because, as a rule, for atmospheric precipitation there is close correlation between the two isotopes. The only possible case of the above isotope data use when one of the mixing components was subjected, for example, to evaporation and changed its isotope values (Dinçer and Halevy 1968). In general, the process of mixing can be written by two mass and tracer conservation equations:

$$Q_{mix} = \sum Q_i,$$

$$Q_{mix}\delta_{cmix} = \sum Q_i\delta_i,$$

where Q_{mix} and Q_i are quantitative mass values of mixture components; δ_{mix} and δ_i isotope values of the mixture components.

Evidently, for a multi-component mixture, the equation $n+1$ can be written using the results of measurement of n for independent isotopes, for instance, D, tritium, carbon-14 and so on. It was pointed out that using the data of D and ^{18}O values, it is possible to separate binary systems even if the chemical contents are identical. In the work of Vinograd and Friedman (1972), using the equation of mass balance, the proportions of waters from different regions feeding one group of springs was calculated. Earlier this problem, when solved by classical hydrogeological methods, has not found single-valid interpretation. The mean value of D content in the first discharging area was $\delta D = -113‰$, in the second are $\delta D = -103‰$, in the recharged area $\delta D = -106‰$, the mean value of the water recharge was 57,500 m^3/day. By solving these two equations, it is not difficult to find that from the first discharging area, about 20,000 m^3 per day or 35% of the total discharging water flow out. The authors pointed out that this method can be used if the following conditions are satisfied: the statistic values of D and ^{18}O should be significantly different differed in both components of the mixture; the isotope content should be constant in time; the isotopic contents of water should not be affected due to interaction with water-bearing rocks; and the other components of mixing waters should be ignored.

Payne and Schotterer (1979) have studied the process of water infiltration from the Chimbo River in Equador to groundwater. It was found that isotopic composition of the Chimbo River, which leaves in the Andes Mountains, does not vary too much during the year. Mean values of isotopic composition in the river water and accounted for $\delta D = -47.0 \pm 0.7‰$, $\delta^{18}O = -7.34 \pm 0.09‰$. Mean local precipitation on the plane during 7 months in the year ($\sim$2000 m) were determined as regular observation and characterized by $\delta D = -24.1 \pm 0.8‰$, $\delta^{18}O = -4.43 \pm 0.07‰$. The groundwater infiltrated from the river has mean values of $\delta D = -30‰$ and $\delta^{18}O = -4‰$. Based on the above data, the authors calculated the recharged portion to the groundwater from the local precipitation, the minimum limit of which appeared to be $73 \pm 5\%$. It was discovered that the deeper aquifers are also recharged from the river bed. Distribution of D in the depth of the aquifer is also studied: at 10 m depth $\delta D = -31.9$, at 40 m it is -39.2, and at 80 m it is -46.6‰. The calculated parts of water from precipitation in the studied aquifer were 66% at the depth 10 m, 23% at 40 m, and about 1% at 80 m.

Analogous investigations regarding the study of mixing proportions of water of different genesis and in mining recharge were done by Verhagen et al. (1979), Zuber et al. (1979) and Karasev et al. (1981b).

6.7 Groundwater Residence Time in an Aquifer

Radioactive isotopes, like tritium and radiocarbon, are often used for solution of this problem, but in several cases the time-dependent characteristics can be successfully obtained by analysis of the stable isotope content. The seasonal δD and $\delta^{18}O$ variations at the input of the hydrological system, and the regularity of their deviation with time in the water-bearing complex, may provide basis for solving this problem. In general case, the residence time τ of water and any dynamical hydrogeological system can be determined as $\tau = V/q$, where V is the system's volume and q is the water discharge.

Eichler (1965) has measured D contents in precipitation and the soil moisture along profile of the unsaturated zone. He found that δD values in the moisture profile are the repeated seasonal variation of D in precipitation, but these variations are equalized with the increase in infiltration time. On the basis of these measurements, Eichler proposed a model for determination of the residence time of precipitation water in the unsaturated zone and also in groundwater.

The method was used by Kusakabe et al. (1970) to study the residence time of thermal waters in the volcanic region of Nasudake (Japan). Isotope balance for a system can be described by a simple equation:

$$d\,(VR_v)/dt = q_i R_i - q_o R_o,$$

where V is the volume of a mixing reservoir; q_i, q_o are velocities of the recharge to and discharge from the system; R_v, R_i, R_o are the ratios of the isotopes and their corresponding components in the system.

In the δ values, for example for $\delta^{18}O$, the equation obtains the form

$$d[V(1 + \delta^{18}O \times 10^{-3})]/dt = q_i(1 + \delta^{18}O \times 10^{-3}) - q_0(1 + \delta^{18}O \times 10^{-3}).$$

At the stationary state, $V_i = \Phi$ and the volume of the reservoir does not depend on time. Seasonal variations of the oxygen isotopic content in precipitation can be presented in the form of equation of harmonic oscillations $\delta_i^{18}O = \overline{\delta^{18}O} + A\cos 2\pi t$, where $\delta_i^{18}O$ is the ^{18}O content at the system enter; $\overline{\delta^{18}O}$ is the mean value of ^{18}O in precipitation; A is the amplitude of the ^{18}O in precipitation relative to its mean value. Taking into account that $\tau = V/q$, the equation can be used for obtaining solution describing change of the ^{18}O ($\delta_v{}^{18}O$) at the exit of the system,

$$\begin{aligned}\delta_v{}^{18}O = \overline{\delta^{18}O} + [A(1 + 4\pi^2\tau^2)](2\pi\tau \sin 2\pi t + 2\cos 2\pi t)\\ + \delta_{vo}{}^{18}O - \overline{\delta^{18}O} - [A/(1 + 4\pi^2\tau^2)]\exp(-t/\tau),\end{aligned}$$

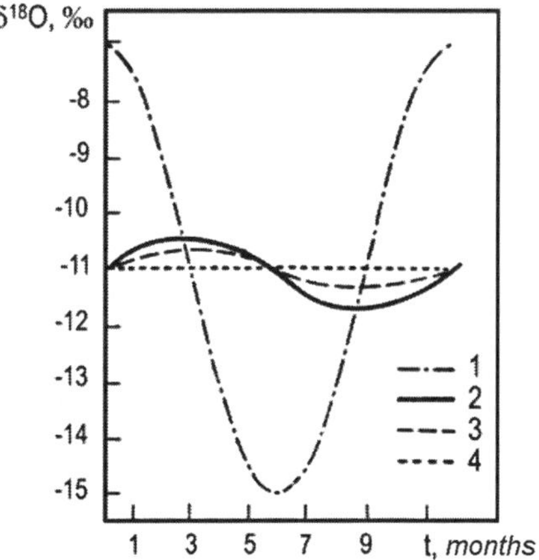

Fig. 6.6 Variation of δ^{18}O values in time τ versus residence time in the system: *1* entering function; *2* $\tau = 1$ year; 3–5 years; 4–10 years. (After Kusakabe et al. 1970)

where $\delta_v{}^{18}$O and $\delta_{v0}{}^{18}$O are the ^{18}O content of water in the system's outflow at time t and t_0 accordingly. Figure 6.6 shows variation of $\delta_v{}^{18}$O values in time (during 1 year) for values $\overline{\delta^{18}\mathrm{O}} = -11‰$, $A = 4‰$ and $\delta_{v0}{}^{18}\mathrm{O} = -11‰$ characteristic for the studied region. One can see that the system practically equalizes the entering oscillations at outflow for $\tau = 10$ years.

Analysis of the above equation for the thermal waters of the volcanic region Nasudake based on variation of ^{18}O in hot springs in time says that the residence time of thermal waters is accounted by several years or longer periods.

Analogous studies have been carried out by Dubinchuk et al. (1974) using the hydrogen and oxygen isotope variation in precipitation and mines.

6.8 Relationship of Waters in Conjugate Hydrologic Basins

The determination of the portion of groundwater in surface flow recharge is based on the analysis of seasonal variations of stable isotope concentration in surface and subsurface waters. An example of the solution of this problem is the investigation undertaken in the Alpine region for the Dishma basin, Switzerland, in the range 1668–3146 (Martinec et al. 1974). The experimental data on the ^{18}O content in precipitation, snowpack, and river water of the basin during the period 1969–1972 are shown in Fig. 6.7. One can see that variation in δ^{18}O in summer and winter precipitation range considerably (from $-3‰$ in summer to $-25‰$ in winter). The δ^{18}O values in snow cover were found to be close to those in winter precipitation. At the same time, seasonal variation of ^{18}O content in waters discharged from the basin was gradually diminishing. Martinec et al. (1974) explained this in terms of the main contribution to the surface run-off being derived from water flowing out of subsurface reserves. Simultaneous investigations with tritium as a tracer have shown that the average age (the residence time) of groundwater in the basin is about 4.5–4.8 years.

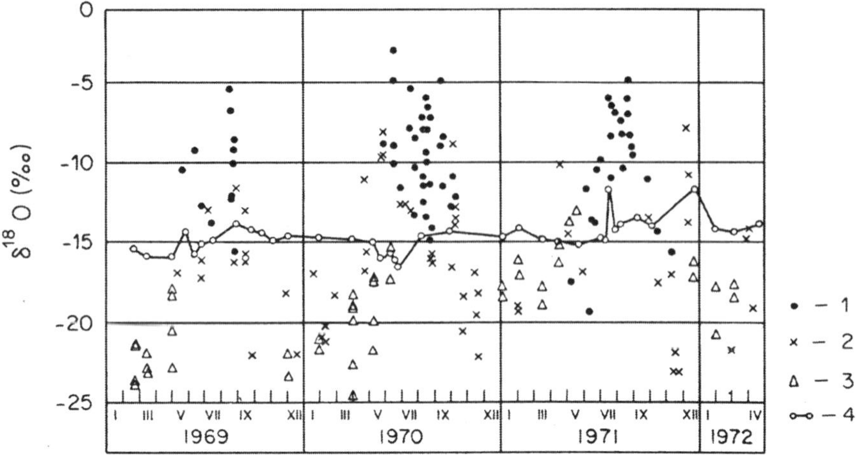

Fig. 6.7 ^{18}O content in *1* precipitation, *2* snow precipitation, *3* snow cover, and *4* discharge water of Dishma basin, Switzerland. (After Martinec et al. 1974)

Thus, in general, as it is seen from Fig. 6.7, isotopic composition of discharged water is close to its mean value in precipitation if the residence time of groundwater is equal to several years.

The other works on solution of the balance problems for the river basins, lakes, and reservoirs are presented by many authors (Brown 1970; Payne 1970; Dinçer et al. 1970; Merlivat 1970; Gat 1970; Behrens et al. 1979; Baonza et al. 1979; Brezgunov and Nechaev 1981).

In conclusion it is noteworthy that in addition to the above described problems, based on the use of natural concentration of D and ^{18}O, there are other hydrological problems that can be solved, which are: seasonal recharge of groundwater by precipitation and portion of recharge by precipitation in different seasons in karstic regions (Dinçer and Payne 1971); study of groundwater dynamics around boreholes and reservoirs for municipal water supply (Fritz et al. 1974); recharge of modern and ancient marine water to the fresh water aquifers (Salati et al. 1974; Cotecchia et al. 1974; Yurtsever and Payne 1979). Combination of different stable and radioactive isotopes with conventional hydrological and hydrochemical data gives more reliable results (IAEA 1976).

Chapter 7
Isotopic Composition of Formation Waters

7.1 Relationship Between Hydrogen and Oxygen Isotopes in Formation Waters

It is observed from the previous chapters that the processes of water evaporation and condensation are of great importance in the fractionation of isotopes of natural waters. At the same time, the evaporation is primarily an attribute of surface conditions. It might occur in shallow underground waters but it is generally agreed by hydrogeologists that groundwater evaporation does not occur on a regional scale (Zaitsev 1967; Stankevich 1968; Smirnov 1971); however, in local zones, the underground evaporation is likely to take place. An example of such phenomenon is the evaporation of groundwater accompanying oil and gaseous deposits (Sultanov 1961). As a rule, these processes of water evaporation occur at elevated temperatures (~80°C). In this case the isotopic fractionation factors are $\alpha_{\mathrm{D}} = 1.032$ and $\alpha_{^{18}\mathrm{O}} = 1.0042$. The vapor phase differs insignificantly in isotopic composition from layer waters to a deposit. The water vapor which has been formed migrates with oil gases. During the migration of the vapor–gaseous mixture through porous layers at lower temperatures, underground fresh water deposits with mineralization less than 1 g/l and δD and δ^{18}O values greater than, those which are characteristic of meteoric waters might form.

Thus, in the region of the Dnieper–Donets depression at depths exceeding 2,000 m waters with mineralization of up to 4 g/l and values of δD from −21‰ to −53‰ and ^{18}O from −2.5‰ to −4.6‰ were found. In this region for oil waters with mineralization ranging from 150 to 330 g/l, the deuterium (D) and oxygen-18 (^{18}O) content varies within the limits δD from −21 to −54‰ and ^{18}O from +2.0 to −4.6‰ (Yakubovsky et al. 1978).

Analogous waters were also found in the east part of the Terek–Sundzha oil and gas region at depth of 4–5.5 km (Nikanorov et al. 1980), but naturally underground evaporation cannot result in considerable changes of the isotopic composition of the layer waters since the amount of evaporated water is always negligible in comparison to the amount of native water.

The basic process determining the isotopic composition of water undergoing underground circulation is that of isotopic exchange in the water–rock system. The

V.I. Ferronsky, V. A. Polyakov, *Isotopes of the Earth's Hydrosphere*,
DOI 10.1007/978-94-007-2856-1_7, © Springer Science+Business Media B.V. 2012

isotopic exchange of water with gases (H_2S, H_2, CH_4, CO_2) and liquid hydrocarbons of oil also takes place, but these processes are considerably less effective than the exchange processes with water-bearing rocks. For example, according to Soyfer et al. (1967), the change in the isotopic composition of hydrogen in groundwaters due to exchange reactions with gaseous hydrogen and hydrogen sulfide is negligible.

The absence of the influence of isotopic exchange in the H_2O–H_2S system upon the isotopic composition of formation waters has been demonstrated by Clayton et al. (1966) and Hitchon and Friedman (1969). By indirect means, the possible scale of changes of hydrogen isotopic composition in groundwater due to exchange processes with liquid hydrocarbons of the oil series can be estimated. Thus, Mason (1966), with reference to Smith, has pointed out that the formation of oil deposits in the USA, in the Gulf of Mexico region, Louisiana, and Saint Croix in California, was accompanied by the release of about 3.4×10^5 t of liquid hydrocarbons from 1 km^3 of sediments ($\sim 2 \times 10^9$ t). Assuming that moisture made up 10% of the mass of sediments at the stage of diagenesis, that is there were $\sim 2 \times 10^8$ t of water in 1 km^3 of sediments, than the ratio of hydrogen atomic fractions in water and oil becomes $w/o \approx 300$. Here we have assumed that the exchangeable amount of hydrogen in oil per mass unit is twice as large as that of water. It is obvious that for such a low value of w/o, the isotopic composition of water remains even when the isotopic equilibrium between hydrocarbons $\delta D_n = -200‰$ and water is maintained such that $\alpha = 1$ ($\delta D_{oil} - \delta D_{water} = 0$). This was confirmed by Clayton et al. (1966) studies of isotopic composition of groundwater in the oil deposits mentioned above. The results of this research are presented in Figs. 7.3 and 7.5. Some observable changes in the hydrogen isotopic composition are likely to be expected in the region of water–oil contact, when $w/o \ll 300$.

The change in oxygen isotopic composition of water due to exchange with carbon dioxide may take place under certain conditions in hydrothermal systems of groundwater gas CO_2. Thus, Ferrara et al. (1965) observed the negative 'oxygen shift' in the thermal waters of Toscana which they explained in terms of the exchange of water with carbon dioxide. These results are presented later on.

Isotopic exchange with water-bearing rocks at high temperatures has the greatest influence upon the isotopic composition of the water. These processes mainly affect the oxygen isotopic composition of groundwater since, in sedimentary rocks, oxygen content amounts to 40% while hydrogen is only 0.3%. In granite rocks, these values amount to 48% and 0.12% and in basalt a 46% and 0.1%, respectively (Beus 1972).

The scales of changes of δD and $\delta^{18}O$ values of water provided by isotopic exchange reactions with water-bearing rocks depend on the following factors (Ohmoto and Rey 1974). (1) The initial isotopic composition of water δ_w and rocks δ_r, participating in isotopic exchange reactions. (2) The ratio of the amount of exchangeable oxygen or hydrogen atoms in water to the same in rocks (w/r). (3) The temperature which determines the equilibrium isotopic fractionation factor between water and rocks. As pointed out earlier, the effect of isotopic fractionation between the two phases, for example between rocks and water, can be approximated by the difference of the values $\Delta = \delta_r - \delta_w \approx 10^3 \ln \alpha$. (4) The degree of exchange in the rock-water system, depending on the average residence time of the water in the aquifer.

Assuming that water and rocks are in isotopic equilibrium at a given temperature, the final isotopic composition of water δ_w^f can be expressed as (Ohmoto and Rey 1974)

$$\delta_w^f = \frac{\delta_r - \Delta + (w/r)\delta_w}{1 + w/r}.$$

This formula can be used not only for calculating the isotopic composition of water in isotopic equilibrium with rocks but also for estimating the ratio (*w/r*) when the parameters are known (Sheppard et al. 1969, 1971; Taylor 1974, 1978).

Using the above formula, let us estimate the change in the isotopic composition of hydrogen and oxygen of normal sea water ($\delta D_w = 0$, $\delta^{18}O_w = 0$) syngenetic with the clay sediments. For example, this process may occur by wringing out the inner-layer water of the clay minerals during the formation of deep waters in sedimentary basins (Kartsev and Vagin 1973). Sergeev et al. (1963) have shown that the dominant clay minerals in the rocks of Mesocainozoic sediments are those of the more ancient ages are hydromicas. Next in abundance are minerals of the montmorillonite group and then of the kaolinite group. The clay minerals of hydromicas and montmorillonites are very similar in their crystalline structure and hydrogen (water) content of the hydroxide groups coming into the crystalline lattice of the complete dehydrated minerals (Lazarenko 1958; Kulchitsky 1975). For minerals of these groups, the values of hydrogen isotopic fractionation factors are also close to each other in the same way as the oxygen ones. They most readily start the oxygen exchange with water (Savin and Epstein 1970a, c; James and Baker 1976; O'Neil and Kharaka 1976).

These circumstances permit one to consider, in reliable details, isotopic exchange in the montmorillonite-water system as an illustration of the scale of changes in water of δD and $\delta^{18}O$ from initial values. The assumption is made that during clay diagenesis, isotopic exchange occurs under conditions of a closed system that is only the water which was bounded by rocks at the stage of sedimentation is involved in isotopic exchange reactions with clay minerals. Let us take into consideration the following parameters of the process: the humidity of the clay is 20% of the total mass; the hydrogen content in hydroxide groups per mass unit of the dry rock is 0.4% (which is equivalent to about 4 mass % of H_2O). Then $w/r \approx 5$ for hydrogen atoms and 0.3 for oxygen atoms. Let us consider further that prior to sedimentation, the clay minerals were formed under hypergeneous conditions in regions with a temperate climate. The isotopic composition of atmospheric precipitation may then be taken as $\delta D_w = -90‰$ and $\delta^{18}O_w = -12‰$ and the isotopic composition of the clay minerals of the montmorillonite group, equilibrated with meteoric waters, is $\delta D_r\Delta = -15‰$ and $\delta^{18}O_r = +15‰$ (Savin and Epstein 1970a). For the purpose of estimation, we consider the isotopic composition of pore water, at the initial stage of sedimentation, to be the same as that of normal sea water, and isotopic exchange in system to occur at 100°C. For these conditions, one has $\Delta D_{\text{mineral-water}} \approx -50‰$ and $\Delta^{18}O_{\text{mineral-water}} \approx +12‰$. The values Δ for oxygen isotopic exchange in the mineral-water system are estimated on the basis of data provided by Taylor (1974). For the isotopic exchange of hydrogen between montmorillonite and water at normal temperatures of hypergenic processes, the value Δ will be obtained using Savin and Epstein's (1970a) data. In addition, let us consider that fractionation factor of hydrogen isotopes between hydroxyl-bearing minerals and water does not doesn't undergo any significant changes up to ~400°C (Taylor 1974; Suzuoki and Epstein 1976).

Now putting δ_w, δ_r, Δ, and *w/r* into the Ohmoto-Rye formula, one obtains $\delta D_w^f = -17‰$ and $\delta^{18}O_w^f = +2.3‰$. Note that the conditions described above are the most favorable for the isotopic exchange (due to filtration of water through the clay). If such filtration takes place, the value of *w/r* calculated per unit of mass of rock should be increased by factor *n* (where *n* is the number of water exchange cycles). For $n = 2$, $\delta D_w^f \approx 9‰$ and $\delta^{18}O_w^f \approx +1.2‰$.

From these estimations, it is observed that the isotopic exchange of water with the clay minerals at layer temperatures below 100°C will not result in significant changes in the content of D and ^{18}O in groundwater. In natural conditions ($w/r = 1$), the interaction of water with basic (basalts, $\delta^{18}O \approx +7‰$) and acid (granites, $\delta^{18}O \approx +10‰$) volcanic rocks at temperatures above 400°C, when the value of $\Delta_{\text{rock-water}} \to 0$, may result in a maximum oxygen shift from 3 to 5‰ when $\delta^{18}O = 0$. The value of oxygen shift increases as $\delta^{18}O_w$ decreases; the maximum oxygen shift for the above conditions is 10‰ when $\delta^{18}O_w = -10‰$ and 12‰ when $\delta^{18}O \approx -15‰$. Other conditions being equal, the maximum oxygen shift will take place at isotopic equilibrium for groundwater in circulation in carbonate rocks composed of limestones and dolomites with $\delta^{18}O_r = +30‰$.

The hydrogen isotopic exchange with hydroxyl-bearing minerals of siliceous rocks of volcanic genesis does not, in practice, play any significant role in the change of the isotopic composition of underground waters since the hydrogen content in these rocks does not exceed an average of 0.1% (Beus 1972).

Many studies have shown how this process occurs on a very limited scale. For example, according to Arnason and Sigurgeirsson (1967), who studied Iceland's thermal waters, there is reason to believe that during the movement of groundwater, large distances from the infiltration zone to the place of discharge changes in D content do not occur.

By analyzing numerous experimental data in various hydrothermal regions of the world, Craig (1963, 1966) has shown that D content in hydrothermal waters corresponds to its content in meteoric waters in region where the former is fed by the latter. According to data of several researchers (Clayton 1961; Craig 1963, 1966; Clayton et al. 1966; Hitchon and Friedman 1969; Savin and Epstein 1970a; Bottinga and Javoy 1973; Taylor 1974; Becker and Clayton 1976; Kawabe 1978), at higher temperatures the water is enriched with heavy oxygen due to isotope exchange with oxygen-bearing rocks (limestones, silicates, and so on). This phenomenon of oxygen isotopic exchange has been called the oxygen shift. Figure 7.1 demonstrates the data obtained by investigation of geothermal brines brought out through boreholes in the Salton-Sea region of California, USA (Craig 1966)[1].

[1] The work was undertaken to refute that the above-mentioned brines are deep ore-bearing fluids.

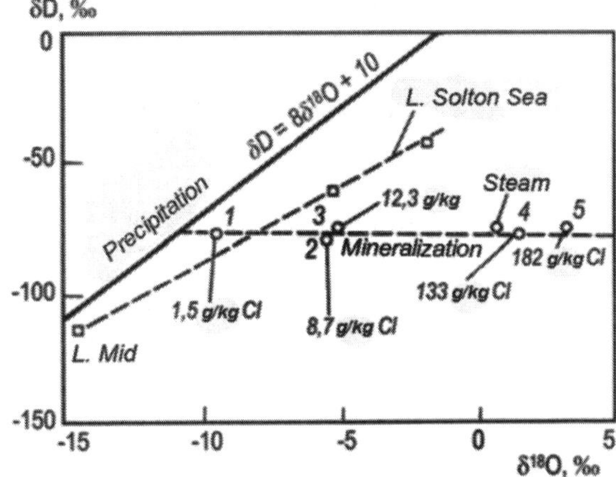

Fig. 7.1 Isotopic composition of waters in the Salton-Sea geothermal area: *1* a spring at the base of the Chocolate Mountains (17°C); *2* water from the 'mud volcanoes' on the east shore of the lake (39°C); *3* water from the old CO_2 production wells about 300 m deep (42°C); *4* steam and brine from the two geothermal wells WGS and IID-1; *5* brine from a deep geothermal well (>300°C). (After Craig 1966)

It follows from Fig. 7.1 that with increases in temperature of the waters being studied, the D concentration remains constant at the level corresponding to its content in atmospheric moisture in this region, while the ^{18}O content increases sharply for the same conditions. In contrast, on the surface (for the Salton-Sea and Mead Lakes) the content of these isotopes changes in parallel.

Thus, it should be noted that the statement made by several authors (Soyfer et al. 1967) on the parallel behavior of D and ^{18}O concentration in groundwater cannot be considered as universal. It holds only for waters of the surface cycle as Craig (1961a) has shown. For groundwater, the behavior of D and ^{18}O differs due to oxygen isotopic exchange between water and water-bearing rocks, especially at elevated temperatures, and this must be taken into consideration in the course of isotope studies. The variation of ^{18}O concentration in rocks and groundwater is illustrated by another example characteristic of the Larderello region of Italy (Ferrara et al. 1965). The sea limestones, for which values of $\delta^{18}O = +30‰$ are typical, are notably depleted in this isotope. At the same time, for water migrating through these limestones, range from $\delta^{18}O = -7.3‰$ in the region of supply -4.8 and even $-1.9‰$ in the areas of emergence at springs on the surface or in the boreholes. Nevertheless, the D concentration is practically a constant equal to that in meteoric waters in the supply region.

Coplen and Hanshow (1973) have studied the isotopic composition of oxygen and hydrogen in ultra-filter and sedimentary solution by infiltration of distilled water and NaCl solution through the montmorillonite membrane. The porosity of the disk under pressure has reached 35%. They have found that the liquid being infiltrated through the montmorillonite membrane becomes depleted in δD and $\delta^{18}O$ by 2.5‰ and 0.8‰, respectively. The concentration of NaCl up to 0.01 NH does not affect the process of isotopic fractionation. Higher saline concentrations are likely to exert influence upon the isotopic separation in the course of infiltration of a liquid through the clayey membranes. It should be noted that the experimental data obtained by Coplen and Hanshow concerning the δD and $\delta^{18}O$ deviations in the infiltrated liquid may be considered as insignificant and the scale of the experiment does not apply itself to natural conditions where, in the range of experimental error, this phenomenon has not been observed.

It is worth to note that in the study of Yeh and Epstein (1980) the authors, with reference on Coplen and Hanshow, by the process of ultra-filtration explains some increase (~20‰) of D content in pore waters in the clays profile opened by boreholes in the Gulf of Mexico, but this phenomenon has another explanation. For example, the observed fact can be related with decrease with the depth in the fractionation factor of hydrogen in the water-rock system because of increase in temperature; the effect can be also explained by the *w/r* ratio decrease due to decrease of the moisture in the clayey thickness and by the climate changes in the area of the North America in the previous epochs. The latter explanation must effect on isotopic composition of the hydroxide group of the clay minerals discharged by Mississippi.

On the other side, on the basis of study undertaken, it has been shown (Soyfer et al. 1967) that fractionation of hydrogen isotopes does not occur during the motion of groundwater in sandstones and argillites of the Hot Springs suite, North Caucasus, within a distance of hundreds of meters (photoneutron methods are accurate within ±3%). By using more accurate mass spectrometric measurements, Degens (1961) found that the oxygen isotopic composition remains essentially unchanged during filtration at a distance of 1,100 km through the sandstone aquifer of the Numibian suite in Northern Africa (deviations of only 9%). Forty samples of water were taken at intervals of 35–40 km along the profile. It is obvious that although in this case the chemical composition of groundwater has undergone considerable change, the conditions for oxygen isotopic exchange did not exist.

Other natural processes of groundwater isotopic fractionation are practically unimportant. In fact, the process of gravitational isotopic fractionation of hydrogen is buffered by the more powerful hydrodynamic process. The process of bacterial fractionation, which is rather intensively energetic under laboratory conditions, can be considered as insignificant under natural conditions since even at the maximum value of bacteria content (tens of millions per ml) their total weight does not exceed 1 mg/l. This conclusion is a natural one since isotopic exchange is not only a result of practical process but also indicates the scales of their activity. The latter effects of hydrogen separation in natural waters were classified as second-order effects (Soyfer et al. 1967).

7.2 Isotopic Composition of Formation Water in Sedimentary Basins

The formation of the deep groundwater of sedimentary basins of the Earth has had a long and complicated history. The deep waters of sedimentary basins were formed within the ancient seas and lagoons located at the edge of continents. A close relationship exists between sedimentary thickness and horizontal tectonic displacements of lithospheric plates (Le Pichon et al. 1973; Bullard 1978) resulting in the vertical tectonic displacements of the Earth's crust (Verhoogen et al. 1970).

The main sedimentary basins which originated in various geological times have been related to boundaries of plates, that is with those places on the Earth's crust where the maximum vertical tectonic displacements have occurred. It is no coincidence that the majority of giant petroleum fields discovered before 1970 are related to sedimentary basins and located in the shelf areas of modern continents (Moody 1978).

While studying the processes of groundwater formation in a sedimentary basin with the help of commonly used techniques, researchers are normally faced with considerable difficulties. These consist of a marked transformation in the chemical composition of primary marine water both during the stage of the surface exposure and as a result of interaction of the groundwater (primary or infiltrational) with water-bearing rocks. Studies of the isotopic composition of sedimentary basin groundwater help to overcome these difficulties to a great extent.

As mentioned earlier (see Chap. 4), during the exposure of a sedimentary basin where the precipitation of evaporites have occurred, the hydrogen and oxygen isotope composition of the water could have been a little different from that of the sea water recharging it, due to evaporation. During evaporation, there is an increase in the concentration of salts in such a basin resulting in a decrease of water activity and, also, of the evaporation rate. This, in turn, affects the water hydrogen and oxygen isotope composition at the final stages of evaporation. As Sofer and Gat (1975) have shown, the D content in such basins could decrease compared with the primary marine water ($\delta D = 0$), depending upon humidity. Thus, at a relative air humidity of 40–50% and a vapor isotope content $\delta D = -80‰$, when the sedimentation of bishofite might occur, the water hydrogen isotope composition is greater than the D content of the ocean water at practically all stages of evaporation. At humidity of 80%, when basin in natural conditions may be evaporated to the stage of halite sedimentation, the hydrogen isotope composition of the remaining water might decrease, reaching $\delta D = -20‰$.

In the underground conditions at the stage of diagenesis of the sedimentary strata, the isotope composition of the groundwater could also have been changing. As a rule, at elevated temperatures compared with the surface, oxygen isotope exchange between groundwater and water-bearing rocks occurs, resulting in an increase of the ^{18}O content in groundwater. The D content also changes but its variation, provided by the exchange processes with the hydroxide-bearing minerals such as clays, is more diminished than that for oxygen because the oxygen content exceeds the hydrogen content by more than one order of magnitude (Taylor 1974; Ohmoto and Rye 1974).

Therefore, the original groundwater of sedimentary rocks differ considerably, in terms of their D content, from the infiltration meteogenic water entering the basin during its period of exposure at the surface. All this leads to the conclusion that, in the groundwater of ancient sedimentary basins, it is possible to distinguish connate marine water from infiltration meteogenic ones, even when their chemical composition is the same, by virtue of the D content (Degens et al. 1964; Graf et al. 1965, 1966; Craig 1966; Clayton et al. 1966; Taylor 1974; White 1974; Hitchon and Friedman 1969; Kharaka et al. 1973; Gutsalo 1980).

Concentration of ^{18}O in groundwater is a less conservative parameter. Its high values are explained not only by the marine origin but also by the isotopic shift. This is why high concentration of ^{18}O cannot be an indicator of marine origin of the studied water. At the same time, the ratio of concentration of D and ^{18}O can be an index of the depth of sedimentary basins.

Many studies have been devoted to the principles of stable isotope distribution in deep sedimentary waters and rocks and on the determination of the isotopic content in water using its physical properties (the techniques based on density measurements and other). Later on, due to their accuracy and restricted applicability, these techniques were replaced at first by the photoneutron method and later by modern mass spectrometric methods which are presently in use.

Alekseev et al. (1966) and Tyminsky et al. (1966) used the photoneutron method measurements to carry out studies of the D distribution in the deep formation waters of the Tashkent artesian basin. With the help of experimental data they found that the infiltrating waters of this basin are depleted in D and the ancient waters are enriched in it.

While studying the absolute ages of groundwater in the chalky water-bearing complex in the Tashkent artesian basin using the helium–argon method, which was suggested by Ovchinnikov, it was found that the most ancient waters (up to 5 million years) are situated in the immersed parts of the depression and the side-wards regions and the northern part of the depression, with rocks of the Cretaceous age at the surface, have waters of more recent age (tens or hundreds of thousands of years). It was found that D content increases with increases of groundwater age. In the northern and north-eastern part of the basin, relatively young waters occur (up to 60,000 years) having a lower D content (from −160 to −120‰). The maximum D content (from −26 to −7‰) was found in waters at depths of about 2,000 m below the land surface, the age of which is about 4–5 million years.

For the paleohydrological consideration, the dilution proportions of infiltration and sedimentary waters in chalky sediments of a basin were calculated. The proportions were determined with the help of mixing formulae in common use. The D concentrations accepted were −160‰ for infiltration waters and +26‰ for sedimentary waters. The researchers distinguished zones of abundance of waters with proportions of infiltration waters in the mixture up to 0.25, from 0.25 to 0.5, and more than 0.5.

Using these studies, and also investigating the D content in waters of a number of oil and gas field regions (Brezgunov et al. 1966), the applicability of isotope techniques has been shown for the identification of formation and infiltration waters in artesian basins.

Seletskt et al. (1973), Ferronsky et al. (1974), and other researchers have carried out a considerable amount of mass spectrometric measurements for the investigation of D and ^{18}O distribution in deep groundwater in the former USSR. The results of these studies are generalized in Table 7.1.

The isotopic composition of groundwater in the North Caucasus region was studied for two hydrogeological levels Nikanorov et al. (1976). The waters of the neogen-quaternary upper floor are of infiltration origin (δD from −67 to −46‰,

Table 7.1 Distribution of D and ^{18}O in groundwater of the artesian basins within the former USSR

Artesian basin	δD, ‰	$\delta^{18}O$, ‰	Reference
Near Baltic	From −119 to −48	From −18.8 to −6.3	Pelmegov et al. (1978)
Pripyat Depression	From −95 to +2	From −12.7 to +4	Tkachuk et al. (1975)
Ukrainian Carpathy	From −108 to −7	From −13.1 to +3	Babinets et al. (1971) Vetshtein et al. (1972) Polyakov et al. (1974)
Near Carpathy Mineral Waters type 'Naftusya' and 'Shklo'	From −102 to −6	–	Vetshtein et al. (1973)
Dnepr-Donetsk	From −55 to +11	From −7.2 to +0.3	Vetshtein et al. (1973)
Near Caspian Depression (north-west)	From −107 to −33	From −10.1 to −2.4	Alekseev et al. (1975)
North Caucasus	From −67 to −15	From −12 to −5.8	Nikanorov et al. (1976)
Azov-Kuban	From −120 to −28	From −17.7 to +7	Sokolovsky et al. (2007)
Caucasus Mineral Waters	From −118 to −42	From −12.7 to +1.1	Seletsky et al. (1973)
Sochi-Adler	From −73 to −35	From −9.8 to −0.5	Gorbushina et al. (1972, 1974)
Middle Caspian	From −125 to −17	From −12.4 to −2.2	Seletsky et al. (1973)
West Turkmenian	From −79 to −4	From −13.4 to +4.4	Seletsky et al. (1973) Alekseev et al. (1975)
Yaskhan Lens of Fresh Water	From −89.2 to −54.4	From −12.4 to −2.5	Seletsky et al. (1973)
Mud Volcanoes of West Turkmenia	From −57 to −31	–	Seletsky et al. (1973)
North Caucasus3	From −115 to −28	From −14.2 to +5	Alekseev et al. (1975)
Moscow	From −114 to −22	From −14.5 to −5	Sokolovsky et al. (1999)
Ferganian	From −106 to −78	From −10.9 to ±1.7	Seletsky et al. (1973)
Amu-Darian	From −102 to −20	From −12.3 to −6.1	Alekseev et al. (1974, 1975)
Sibirian Platform	From −180 to −23	From −21.5 to 0.0	Pinneker et al. (1973, 1974, 1975)

$\delta^{18}O$ from −12.0 to −6.8‰). The waters of the lower floor are separated from the upper one by a thick layer of Maykop argillaseous suite. These waters are enriched in heavy isotopes (δD from −55 to −15‰, $\delta^{18}O$ from −4 to +5.8‰) and are similar to thalassogenic ones in terms of origin.

Investigations carried out in the Azov–Kuban artesian basin have shown that the D content in waters of oil and gasfield ranges from −68 to 58‰. These values corresponds to an intermediate, average D content in meteoric (−107.5) and sea waters, which is evidence in favor of the presence of connate sea sedimentary waters up to about 50%. About 50% of Yessentuki saline–alkali waters (Nagut aquifers) are also of marine origin, which refutes the existing opinion that these types of water originate from meteoric local recharge areas.

A greater portion of ancient sea waters are contained in mud volcanoes in this region, where δD ranges from −56.9 to −54.9‰. At the same time, a considerably lower D content has been detected (from −80.8 to −80.0‰) in the iodine waters of the Slaviano–Troitsk aquifers. These data indicate that these waters are diluted with meteoric water to a greater extent, which is in disagreement with the option that there is a genetic relationship between the former waters and those occurring in oil and gas field.

Within the region of the Caucasus mineral waters, it has been reported (Seletsky et al. 1974) that the waters of the Olkhovsk prospecting area have the lowest D content (δD = −118‰). These waters are typically fissured and are completely recharged by precipitation.

The 'Narzan' group of waters in Kislovodsk as a whole, are intermediate, in terms of D content (D = −108 to −76‰), between the waters of the Olkhovsk prospecting area and waters of the deep wells situated in the Yessentuki area.

The 'Yssentuki-20' spring is clearly distinguished by its D content (δD = −117‰) from other waters in the region of the Yssentuki aquifer. The low D content is the result of the spring being recharged by groundwater flowing from the sandy-pebble sediments of the Podkumok River, starting at a height of more than 2 km. The other waters in the Yessentuki area have greater values of δD (up to −42‰). This suggests that they also contain ancient waters of marine genesis from the more deeply lying complexes, in particular the Valazhinsky complex which exerts great influence on the formation of all main aquifers of the Caucasus mineral waters.

The relative D content in formation waters of the oil–gas fields in the South Mangyshlal Peninsula ranges from −82.3 to −16.5‰. The waters most depleted in D have been contaminated by fresh water pumped into the productive oil-bearing layers. In water of the West Turkmenian artesian basin δD ranges from −76.6 to −37.3‰. The elevated D concentrations in waters of the Jurassic complex in the Mangyshlal Peninsula is evidence that waters of marine genesis participate in their formation to a greater extent compared with waters of the Tertiary sedimentation in the Western Turkmenia.

The investigations of deep waters of Pliocene sediments in the Western Turkmenia artesian basin (Seletsky et al. 1973; Alekseev et al. 1975) have shown, by ^{18}O and D isotope analyses that brines in the red colored sediments contain a large portion of marine waters.

The hydrothermal ore sediments in the Cheleken area (Gutsalo et al. 1978; Esikov et al. 1979) have been reported to be colored sediments (δD from −41 to −31‰, δ^{18}O from −3.5 to −2.1‰). They are likely to be similar in origin to connate marine waters, being metamorphosed in salt composition. However, in order to elucidate the conditions of high salt saturation and enrichment with heavy metals, further prospecting of this area is required.

While studying the isotope composition of the Yaskhan Lens fresh water in West Turkmenia (Seletsky et al. 1973), considerably different conditions of the lens water formation from those of the underlying Kara Kum saline waters have been reported. The lower δD and δ^{18}O values in waters of the diffusion zone of the lens are evidence that waters of pluvial age exist in the underlying aquifer, whereas the waters of the lens are likely to have been formed in conditions of arid climate.

In the Fergana artesian basin, more careful studies of the sulfide mineral water fields of Chimion and Obi-Shiro (Gorbushina and Tyminsky 1974) have been carried out. The D content variations were reported to be insignificant (δD from −98 to −70‰), although there are large differences in mineral content (0.7–120 g/l), δD values (from −16.1 to −3.7‰) shows that processes resulting in oxygen isotope shift are involved. The low D content in the Fergana artesian basin waters probably reflect the importance of the mountains surrounding the depression is supplying infiltrating waters to recharge the basin complexes.

In the Amu-Darya artesian basin (Alekseev et al. 1974, 1975) it has been reported that δD and δ^{18}O variations are respectively −52–(−20‰) and −6.3–(+6.1‰) in waters of the Jurassic aquifer; −102–(−43‰) and −11.8–(−5.1‰) in the Lower Cretaceous aquifer; −93–(−57‰) and −11.7–(−6.8‰) in the Upper cretaceous aquifer; and −97–(−60‰) and −12.3–(−8.4‰) for the surface waters. On the basis of these data, the authors have reported that waters of the Jurassic sediments are preferentially of marine genesis and the Cretaceous complex waters originated due to the dilution of marine connate waters with modern meteoric ones which follows from the fact that the corresponding points fit the line plotted for δD–δ^{18}O coordinates connecting the plots of ocean and meteoric waters. Assuming $\delta D = -25‰$ for sea waters and $\delta D = -90‰$ for meteoric waters, Alekseev et al. have calculated that the proportion of infiltrational waters composing sedimentogenic waters may vary from 9 to 94%.

The relative D content in highly saline calcium chloride brines in the Angara–Lena artesian basin ranges from −70 to −23‰ and in this respect are similar to the waters of petroleum and gas provinces in the South Mangyshlak and West Turkmenia. The D content in NaCl brines of this basin is considerably lower. They vary from −168 to −75‰ and differ insignificantly from local meteoric waters. The D content of waters here is even lower than that sampled in the artesian basin.

The ^{18}O concentrations in the considered basins are characterized by the following ranges of values. In the Fergana artesian basin, they range between −10.9 to +1.2‰; in the Western Turkmenian basin from −7.6 to +3.4‰. The data obtained from these two basins support the idea that large portion of infiltrating waters dilute the Fergana basin waters and the process of oxygen isotope exchange with water-bearing rocks takes place in these two basins. There are no great differences in the range of ^{18}O values compared with to the amplitude of D content variations. By comparing the data from both isotope studies in the West Turkmenia artesian basin, the relatively high ^{18}O content can be explained both in terms of the inheritance of marine concentrations and isotopic exchange with rocks. In the Fergana basin, it can be explained as a sequence of oxygen isotopic exchange.

In the Angara–Lena artesian basin, the ^{18}O content ranges between −17.0 to −0.6‰, which agrees with the D content variations for this basin. Using the above-mentioned knowledge corresponding to the examined basins, it was concluded that the calcium chloride brines in the Angara–Lena artesian basin contain a considerable proportion of waters of marine genesis, thus being distinguished from all other waters of this region. The brines of sodium chloride composition do not differ from waters which are obviously of meteoric origin. Thus, the experimental data on water isotope composition disagree with the existing opinion that initial sedimentary water resources of this region have been completely renewed.

The isotope studies in the West Turkmenian artesian basin suggest that the origin of the red colored Pliocene aquifer was not completely sedimentary in origin and that the degassation process from the interior may have occurred.

The isotope data obtained for waters of the Alb-Senoman and Jurassic complexes in the South Mangyshlak Peninsula are in agreement with the hypothesis of the sedimentary origin of the latter and in conflict with the hypothesis suggesting a common hydrodynamic model for all the complexes.

The low D content in waters of the Fergana artesian basin reflects the important role of the mountain surrounding the depression in recharging all complexes in the basin from infiltrating waters.

The waters of mud volcanoes are distinguished from all other deep waters occurred in the Earth's sedimentary shell. The D content in them, according to mud volcano studies in Western Turkmeniya and Northern Caucasus, normally corresponds to the D content of waters of oil and gasfields of the artesian basins. As to the ^{18}O concentrations, they attain the highest values for all known groundwaters, not only in the former USSR area but on the Earth as a whole. The δ^{18}O values detected in them range from +3.4 to +5.0‰. More detailed discussion on the above results may be found in Seletsky et al. (1973), Alekseev et al. (1975), and Esikov et al. (1979).

While studying the formation conditions of the Matsesta mineral waters in the Sochi–Adler artesian basin (Gorbushina et al. 1972, 1974), it was found that δ^{18}O values in these waters range from −9.8 to −2.0‰, in the Black Sea waters from −4.0 to −2.3‰, and in local lakes and rivers from −14.5 to −10.9‰. On the basis of these data, it was concluded that recharge to mineral springs up to depths of 500 m occurs due to precipitation and melt waters flowing from the snow-packs, and deeper mineral waters (at depth of 1,200–2,000 m) are formed due to the dilution of sea water with surface water.

In the sulfide waters of the Jurassic sediments in the same region (Gorbushina et al. 1972), the δD values have been reported to vary from −51 to −35‰ and δ^{18}O from −5.8 to −0.5‰. In the sulfide waters of chalky sediments, these values for D and ^{18}O range from −73 to −52‰ and from −9.7 to −5.0‰, respectively. Comparing these data with those of the surface waters of the region, the authors concluded that the sulfide waters of the Jurassic sediments are connate marine waters and the waters in chalky sediments ate marine and meteoric waters mixed in some proportion.

For the conditions of the distant parts of the Baltic artesian basins the hypothesis of the glacial origin of the Strelninsky aquifer and the weathering core water has been confirmed, as was the assumption of the hydraulic isolation of the Strelninsky and Gdovsky aquifers from each other and from the overlying water-bearing complexes (Sobotovich et al. 1977). Moreover, isotopic data on the solutions shows that the formation waters of the Strelninsky aquifer are situated in the reduced water exchange zone and their recharge by the surface waters is about 5% per thousand years. Analogous waters with anomalous light isotopic composition (δD up to −170‰ and δ^{18}O up to −22.2‰) were also found in Cambrian–Vendian sediments near the shore of the Tallinn in Estonia Bay (Yezhova et al. 1996).

Within a complicated region of the southern slope of the Ukrainian crystalline shield dipping towards the Black Sea depression, the hydrogen isotope composition was investigated in a profile of the South Belozersk iron ore deposits area (Voytov et al. 1976). It was found that waters of the Buchakovsky and chalky aquifers and weathering core waters are depleted in heavy isotopes (δD = −107–(−90‰)) compared with the fissured waters of the iron–siliceous formation (δD = −92–(−70‰) and even −33‰). On the basis of these data the authors have assumed that the mine waters are a mixture of two water types. The first being a relict of the pluvial period, since they are considerably lighter than the recent waters of the river run-off (δD = −77–(−61‰)). The second type of water has the modern meteoric origin.

With the north-western region of the Caspian depression, D and ^{18}O abundances have been studied in a profile from surface waters to those of the deep Paleozoic aquifer. In this region, high δD values (up to −33‰) are observed in waters of the Devonian sediments. The reduction of the δD values (−64–(−47‰)) is a feature of the carboniferous sediments as a whole (for depths from 500 to 3,200 m). At depth of 50–300 m, the D content is even lower (δD = −107–(−74‰)). On the basis of the isotope analysis data (Alekseev et al. 1975), it was reported that within the Don-Medveditsk heights the water of the carboniferous sediments were formed under certain influence from infiltration processes involving the dilution of the marine sedimentary waters with those of the surface discharge. The influence of the infiltration processes is observed up to depths of 1,200–1,400 m. The waters of the Devonian sediments are the connate marine waters which have been insignificantly diluted by meteoric waters probably during the short periods of infiltration regime.

The δ^{18}O values have been found (Vetshtein et al. 1973) to range from −7.2 to +0.3‰ in waters of deep (1,900–4,000 m) Devonian, carboniferous, and Lower Permian brines (150–300 g/l) of the oil and gas in the Dniepr–Donetsk depression, whereas in the surface waters of this region, the δ^{18}O values range from −11.1 to −8.6‰. The relative D content ranges from −55 to +11‰. This suggests these waters are 'connate' waters of the sedimentary basin (metamorphosed by the salinity content). Using the isotope evidence while studying the genesis of groundwater in the Dniepr–Donetsk depression, the authors have found that the hypothesis of the juvenile of infiltration origin of the Paleozoic waters fails to explain the fact that the content of stable isotope of water (solvent) is close to that in recent sea waters.

The carbonic acid saline groundwater in the Ukrainian Carpathy region (Babinets et al. 1971; Vetshtein et al. 1972), mineralized to 4 from 13 g/l and taken from the internal flush zone have, in three cases, shown a high ^{18}O content (−3–(+6‰)) and in the fourth case it was −8.6‰. The authors related the first three samples with relicts of the ancient seas and the last one with water of meteoric origin. In aquifers of the same region used for balneological purpose (Vetshtein et al. 1972), the δD varies between −108 and −67‰ and δ^{18}O from −13.1 to −9.4‰. Therefore these

Table 7.2 D content in Japanese gasfield formation waters. (Kobayakawa and Horibe 1960)

Sampling area	Sampling depth, below surface (m)	δD (‰)
Niigata wells	180	−48
	460	−77
	800	−42
	1500	−47
Shinagawa wells	1100	−9
	1300	−7
Chiba wells	136	+15
	505	−5
	507	−8
	508	−9
Miyazaki mine	–	+2

waters are recharged from surface waters despite considerable differences in gas and salt composition.

D and ^{18}O content was studied in deep waters of the Pripyat depression oil field (Tkachuk et al. 1975). It was found that waters from the aquifer, situated above the salt-bearing sediments, have δD content between −95 and −75‰ and $\delta^{18}O$ content between −12.7 and −10.5‰ and are of meteoric origin. The brines situated between the salt-bearing sediments with δD values ranging from −65 to −10‰ and $\delta^{18}O$ from −9.7 to +4‰, and waters below the salt-bearing sediments with δD values ranging from −49 to −8‰ and $\delta^{18}O$ values ranging from −6.4 to +2‰, where attributed to ancient marine origin by the authors.

While studying the ^{18}O content in saline formation waters sampled in the Sibirian Platform, it was found (Pinneker et al. 1973, 1974, 1975) that in the sulfate hydrocarbonate fresh and saline waters $\delta^{18}O$ ranges from −20.3 to −17.8‰. In the sodium and calcium chloride saline waters and brines of the salt-bearing—and above the salt-bearing sediments—sulfate hydrocarbonate ranges from −20.1 to −15.5‰; for the calcium chloride brines from −7 to −1.3‰; and for the calcium chloride brines below the salt-bearing sediments from −8.7 to −7.3‰. The relative ^{18}O content in the surface waters in this region ranges from −19.9 to −15‰ (in the Baikal Lake $\delta^{18}O = -16.5$‰ is characteristic of meteoric waters in the south of Eastern Siberia). On this basis, the authors concluded that the sulfate and hydrocarbonate calcium-sodium waters have a completely meteoric origin and the deep brines originated from waters of marine genesis being markedly diluted with meteoric waters. In spite of the low temperatures (below 44°C) and almost complete absence of carbonic acid in brines, oxygen isotope exchange with rocks had obviously not markedly affected the formation of their isotopic composition.

The study of the D content in the concomitant waters of the Japanese gas fields was reported by Kobayaki and Horibe (1960) (Table 7.2: data recalculated relative to the SMOW standard). The D concentrations in the formation waters of Niigata, Shinagava, Chiba, and Miyazaki Prefectures are approximately equal to those of sea water. The authors have suggested that this should result from the penetration of sea water into the gas field waters. Unfortunately, Kobayaki and Horibe did not give

Table 7.3 The results of oxygen isotope measurements in the formation of some oil and gas fields In the United States. (From Degens et al. 1964)

Geological age of water-bearing rocks	Number of samples	Total dissolved solids (g/kg)		$\delta^{18}O$ (‰)	
		Variation range	Average	Variation range	Average
Marine sediments					
Cambrian	12	77–206	146	−9.98–(−1.39)	−4.1
Ordovician	4	231–262	256	−0.06–(−1.77)	+0.4
Devonian	1	132	132	−0.46	−0.5
Carboniferous	6	79–269	153	-2.63 ± 2.05	+0.7
Creteceous	9	2.9–5.4	4.0	-11.67 ± 10.29	−11.08
Tertiary	2	103–104	104	$+3.04 \pm 3.34$	+3.2
Freshwater sediments					
Tertiary	10	0.2–15.8	5.6	−16.86–(−3.03)	−12.7
Data for comparison					
Oceans	–	–	35	–	0.0
Surface waters (Utah)	–	–	0.5	–	−16.8
Great Salt Lake	–	–	220	–	−7.4

the geologic structural peculiarities of the profiles studied. It is obvious that such a conclusion, though probably true, cannot be drawn from the D content alone, since similar D concentrations could be exhibited by ancient sedimentary waters.

In the concomitant waters of the Niigata field, the D concentrations were reported to be close to the Tokyo standard (0.01489 ± 0.00005 at. %) which is the tap water of Tokyo University. The similarity of the isotope ratios for those two waters should be explained by the percolation or pumping of the local surface waters into the productive gas layers. The particularly low D content has been found in waters at depths of about 400 m. The different chemical composition of these waters and those lying above or below the mentioned level indicates differences in their origin.

Degens et al. (1964) studied the oxygen isotopic composition in 44 samples of different ages from water-bearing layers in the oil and gas-bearing fields of Oklahoma, Texas, Colorado, and Utah, USA. The data are shown in comparison with those for surface water in Table 7.3 and are recalculated relative to the SMOW standard.

Both in highly and partly mineralized waters there is a tendency for the ^{18}O content to increase with increasing mineralization (Fig. 7.2). The $\delta^{18}O$ values in highly mineralized waters are similar to those for modern ocean waters. Negative deviations from the ocean mean values of ^{18}O agree well with decreasing mineralization, resulting in the penetration of the recent meteoric waters or dilution with ancient infiltration waters, during the geological history at positive tectonic dislocations. The positive deviations of ^{18}O content in some samples were explained by the authors to be a result of continuous evaporation before the waters were connected, for example, during the isolation of small parts of the ocean from the whole basin, that is during lagoon formation.

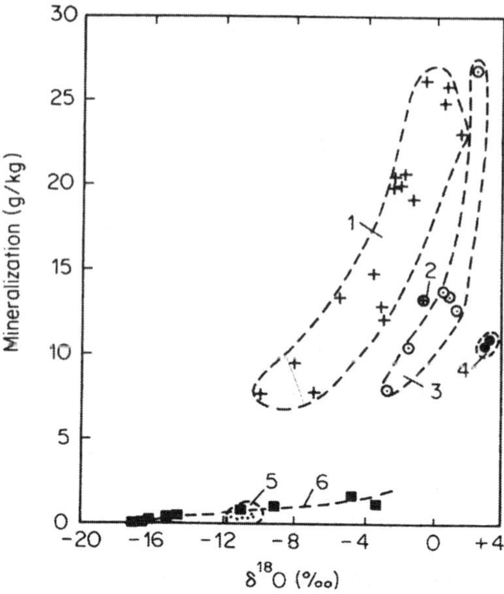

Fig. 7.2 Relationship between ^{18}O values and mineralization of the oil-field formation waters from marine and fresh water sediments: Cambrian–Odrovician (Oclahoma) *1* Devonian (Oklahoma) *2* Pennsylvanian (Oklahoma) *3* Tertiary (Texas) *4* Cretaceous (Colorado) *5* Tertiary (Utah) *6*. (From Degens et al. 1964)

From this data, the authors (Degens et al. 1964) concluded that the brines under study represent ancient marine water, assuming that the ratio in sea waters has remained more or less unchanged since Cambrian times. This suggestion is confirmed by the fact that calculations fossils of Paleozoic age, according to Compston, have the same ^{18}O content as analogous modern fossils. Further, Degens has pointed out that the similarity of the isotope characteristics of ground brines and less mineralized recent ocean waters suggests that the concentration of inorganic salts was not completed by evaporation. The completion of such a process might possibly occur during the process of compaction of sediments accompanied by ion infiltration through the clays. Degens has shown that isotopic data might serve as an indicator of the proportions of meteoric and 'connate' water in samples under study.

Wight (1965) also reported his results of isotopic composition of formation waters of oil-field sedimentary rocks. He carried out rather detailed investigations in the course of which both ^{18}O and D isotopes were determined (Table 7.4).

White, citing Friedman, has pointed out that the majority of highly saline oil-field brines have lower D content than those of the sea water, ranging from 0 to −50‰. As Table 7.4 shows, the D variations for the investigated waters from oil-bearing sediments fall within these limits and are typical. Brines with lower D content are not considered to be of sea water origin.

From Table 7.4 it is observed that, according to the results obtained, heavy oxygen is present in large amounts than in modern sea waters. The author has obtained this in terms of equilibrium exchange reactions of ^{18}O between water and water-bearing rocks at high temperatures which takes place in deep parts of pressurized water systems. Therefore, White stresses that the temperature of waters and the degree of

Table 7.4 D and ^{18}O content in some saline waters and brines of California, United States. (From White 1965)

Sampling location	Chemical composition	δD (‰)	$\delta^{18}O$ (‰)
Cympric oil-field, 900 m depth from Miocene sandstones	$M_{17.7}\frac{Cl}{Na}$	−17.1	+2.93
The same place, 1400 m depth	$M_{17.7}\frac{Cl}{Na}$	−16.1	+3.14
The same place, 1600 m depth from Eocene sandstones	$M_{25.09}\frac{Cl}{NaCa}$	−11.4	+5.93
El Dorado oil-field, 870 m depth from Mississippian limestones	$M_{146.4}\frac{Cl}{NaCa}$	−22.0	–
Wilbur oil-field, from Cretaceous rocks	–	−17.9	+3.07
Wilbur Springs, mixture of 'relict' and meteoric waters	–	−22.2	+5.58
Tuscany Springs, 'relict' waters	–	−13.8	+5.27
Searls Lake, nonmarine evaporitic brines	–	−26.0	+4.20
Sulfur Bank, metamorphic waters	–	−24.1	+5.62
Salton Sea, thermal waters	–	−75.3	+3.27

equilibrium between the brines and solid phases are factors which should always be taken into account during more detailed investigations.

Miller et al. (1966) pointed out that oil-field brines are similar in heavy oxygen isotope content to normal sea water and may exhibit mineralization to a great extent. The insignificant enrichment in ^{18}O content might be due to low evaporation occurring during sedimentation. They pointed out another process involved, that of isotopic exchange with water-bearing rocks, and also emphasized that ^{18}O decrease in oil brines is always accompanied by the process of dilution of the aquifer with meteoric waters.

The data on ^{18}O isotope ratios have been used by these authors to explain the origin of hot acid highly mineralized (310 g/l) brines, localized in the central part of the Red Sea at depth of about 2,000 m. Before the investigations, it was supposed that studies would confirm one of three possible hypotheses of their origin: (1) due to inflow of brines, concentrated by local surface evaporation; (2) due to evaporation from the whole surface of the Red Sea during its isolation stage; (3) due to submarine outflow of the brines.

The ^{18}O concentrations in brines and surface waters in the Red Sea and other sea waters insignificantly enriched by evaporation are similar to each other. Therefore, the observed values of oxygen isotope ratios do not support the idea that the investigated brines are evaporates of the normal Red Sea waters. In the last case, the ^{18}O content should be considerably higher. In relation to this, the authors Miller et al. (1966) concluded that the hypotheses of submarine discharge of deep groundwaters into the Red Sea are more convenient. Later Craig (1969) studied in detail the abundances of D, ^{18}O, and argon dissolved in water and the temperatures and salinities of the Red Sea waters. He concluded that Red Sea brines are forming in the near-surface layer of the southern part of the sea near the Bab-el-Mandeb Strait under conditions of the high temperature and salinity. An important role of the process which took place in the past, during glacial times when the sea level had markedly dropped, and in more distant times when the sea was dried up forming large amounts of evaporates, has been emphasized by the author. Craig's model uses a similar argument. White (1974) pointed out that, in accordance with this model, the existence of the groundwater flow 500–900 km long and passing along the Middle Rea Sea valley should be assumed. Moreover, the brines, in some incomprehensible manner, avoid some other basins which are deeper than those containing hot brines.

There are also some other models, explaining the formation of the chemical and isotope composition of the Red Sea brines, which are free of the hydrogeological contradictions present in Craig's model. For example, White (1974) has assumed the hot brines to be recharged from the nearby edges of the Red Sea which are represented by highly evaporated sea waters often diluted with meteoric waters due to discharge from inland Saudi Arabia. In any case, White noted that the hot brines are genetically connected with the Red Sea waters.

The most detailed studies of isotopic composition of the formation waters in the United States were carried out by Clayton and Graf with co-authors (Clayton et al. 1966; Graf et al. 1965, 1966). They studied formation waters in the Illinois, Michigan, Alberta, and Gulf Coast artesian basins which have a relatively simple geological structure, vast geologic documentation, and a considerable number of boreholes available for sampling. In the profile of the Michigan basin, there are thick deposits of salts and anhydrites and in the Illinois basin only anhydrites of restricted thickness were found. According to the data of isotope studies, it has been discovered that meteoric waters play a significant role in the formation of deep saline waters on the Alberta and Gulf Coast basins. Ninety-five water samples were taken in order to determine D and ^{18}O content. D content variations about ±20‰ were indicated in each basin. In the Michigan and Golf Coast basins, the total mineralization is, in general, insignificantly related to the D concentration. In the Illinois and Alberta basins, the D content increases proportionally with mineralization (Fig. 7.3).

In general, while comparing data obtained for individual basin, marked variations in D content were observed. For example, the Gulf Coast basin waters are similar in D content to sea water, while in the Alberta basin the D content of water is about 100% lower than that in the ocean.

More objective conclusions on the water's genesis, as emphasized by the authors, may provide a hypothesis that original marine waters were pressed-off during compaction from syngenetic rocks and were then mixed with meteoric waters formed under climatic conditions close to modern ones.

Clayton et al. (1966) reported that the main process affecting the observed D/H variations from basin to basin is climatic change with time which has altered the isotopic composition of meteoric waters. The effect of climatic variations can be demonstrated by a few extreme values indicating very low D and ^{18}O contents, which correspond to climatic conditions during the Pleistocene glaciation.

The authors have concluded the processes of isotope exchange between water and other hydrogen-bearing geological objects as a result of hydrogen dispersion to be of no importance since only slight variations in D content have been detected in each basin. These slight variations should more likely be due to isotope fractionation by filtration of water through micropores in clay minerals (Clayton et al. 1966).

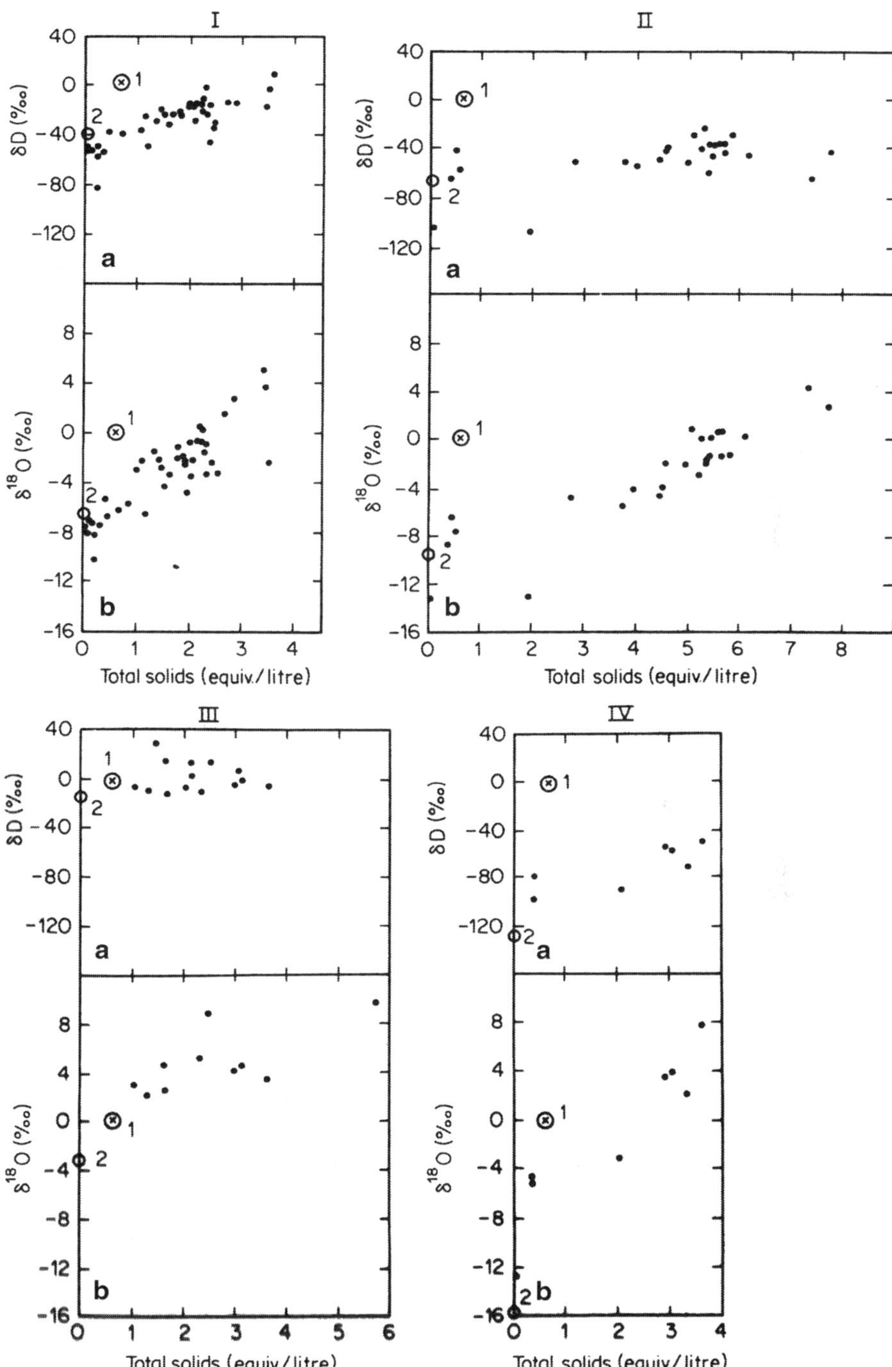

Fig. 7.3 Variation of brine isotope composition with salinity for Illinois (**I**), Michigan (**II**), Gulf Coast (**III**), and Alberta (**IV**) basins: *1* ocean water; *2* present-day meteoric water. (After Clayton et al. 1966)

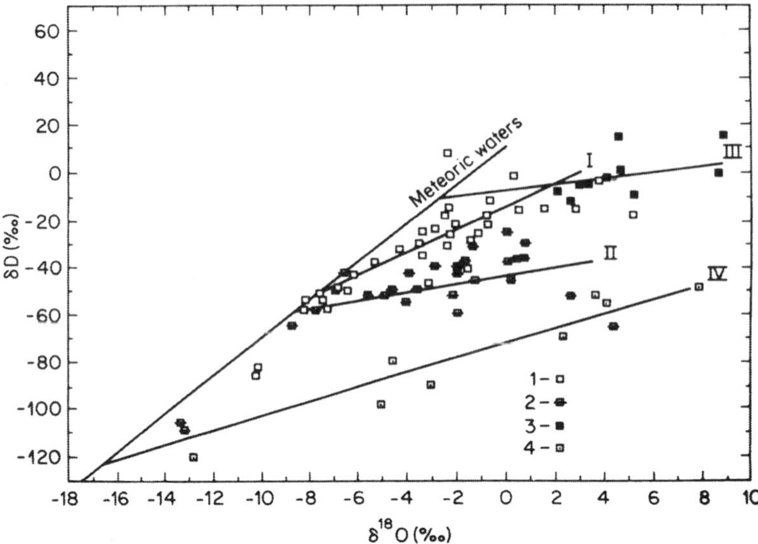

Fig. 7.4 Relationship between δ^{18}O and δD values for Illinois (**I**), Michgan (**II**),Gulf Coast (**III**), and Alberta (**IV**) basins. *1* Ocean water; *2* meteoric waters throughout the world. (After Clayton et al. 1966)

Marked ^{18}O content variations, strongly correlated with salinity variations, were detected in the investigated basins. Measurements of the ^{18}O content at zero mineralization coincide with its content of meteoric waters. At first sight, this seems to be due to the dilution of formation water with meteoric water. This argument was reported by Degens with his co-authors (1964), but Clayton et al. (1966) found another explanation of the observed data. From Fig. 7.4, it is observed that the least saline waters and the highly saline waters corresponding to D and ^{18}O content in meteoric waters and the highly saline waters fall markedly off this line. This picture resembles the case of hot springs such as the sets of points obtained within each basin lie aside from the meteoric water line so that a slight D enrichment corresponds to a great enrichment in ^{18}O. In fact, there is an oxygen shift in this case. The authors pointed out that a wider range of oxygen shift here is observed than in hydrotherms. This is likely to be due to a marked time of water exchange with rocks in the exchangeable system of the considered brines.

The main factor governing oxygen isotope composition in oil brines is water exchange between water and water-bearing rocks.

Figure 7.5 shows the theoretical curves corresponding to the isotope equilibrium between calcite and formation waters relative to the heavy oxygen content in the latter and dependent on temperature. The relative oxygen contents reported in the calcite of water-bearing limestones are +24.2, +25.3, +22.8, +22.3‰ in the Illinois, Michigan, Gulf Coast, and Alberta, respectively. It has been found that the majority of points lay within the 'band' of the theoretical line ranging about ±20‰ side of this line. Thus the following two conclusions have been drawn: (1) in all the basins

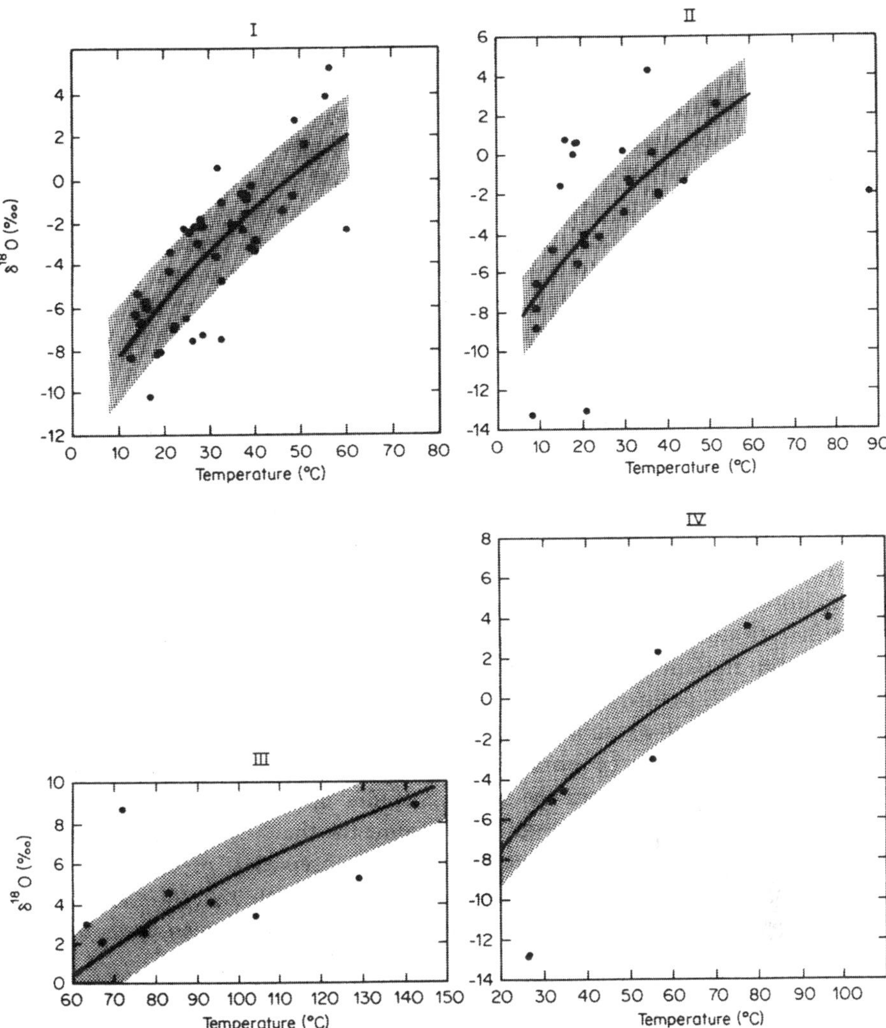

Fig. 7.5 Variation of brine $\delta^{18}O$ value with groundwater temperature for Illinois (**I**), Michgan (**II**), Gulf Coast (**III**), and Alberta (**IV**) basins. The equilibrium curve is drawn for calcite of $\delta^{18}O$ equal to +24.2, +25.3, +22.8, +22.3‰. (After Clayton et al. 1966)

and at all possible temperatures the $\delta^{18}O$ content corresponds to isotope equilibrium lines; (2) the ^{18}O abundances in calcite are practically the same for all the basins. The scatter of points around the equilibrium line may be result of: (a) the lack equilibrium between the waters and the carbonates (especially in the case of the low-temperature waters and calcite-deficient rocks); (b) temperature-measurement errors; (c) local deviations of the oxygen isotope content of carbonates from the mean value.

Graf et al. (1966) investigated the conditions of formation of saline waters and brines in the Illinois and Michigan basins. They found that during the formation of the chemical composition of these basins, the marine and fresh waters participated to different degree. Further, the processes of substitution of magnesium for calcium during the dolomite formation are of importance. These processes involve the transition of calcium into solution and filtration of the solution through clay sediments working as ultra-filters. It was found that brines from the Illinois basin originated mainly from marine waters, the volume of which should be five times as much as the pore volume in the modern basin. For more calcium-rich brines in the Michigan basin, the role of fresh waters was highly essential. According to the author's estimation, the formation of these brines required a mixture of marine and fresh waters in the proportion of 1:200.

Hitchon and Friedman (1969) reported their results of formation waters in the western Canada sedimentary basin from the Upper Cretaceous waters to the Upper Devonian. They took 20 samples of surface waters, 8 samples from shallow aquifers, and 79 samples from different regions of gas and oil fields. Analyzing the data on isotopic composition of waters the authors concluded that formation waters of the investigated artesian basin originated by the mixing of surface water and marine water modified during the process of diagenesis. This process has been accompanied by enrichment of waters in heavy oxygen. The material balance of D and dissolved substances in the basin water shows that the observed D content can be obtained by the mixing of diagenetically modified marine and fresh waters from the same latitudes in proportions not greater than 1:2.9. During its passage through the sediments, the fresh water redistributed dissolved substances up to the salinity variations observed at past. The oxygen shift results from the oxygen exchange of water with that in the carbonate minerals. The effect of the process depends on the ratio between water and rock masses.

On the basis of works of Clayton et al. (1966), and Hitchon and Friedman (1969), Taylor (1974) has come to the conclusion that, in the majority of the North American sedimentary basin, meteoric waters are the main component of the brines. Thus, Taylor suggested the use of the term 'metamorphic connate waters' or 'formation waters' instead of 'connate waters', which is the term commonly used. These formation waters, in Taylor's opinion, are present in all sedimentary basins. In a number of cases these waters may be important, ore-forming fluids.

Rozkowski and Przewlocki (1974) studied the hydrogeological conditions of two coal basins in Poland with the aid of isotope techniques. In the Lublin coal basin, located at the boundary of the pre-Cambrian and Paleozoic platforms, the infiltration waters contribute greatly to sediments of different ages up to a depth of about 900 m. This phenomenon has been identified by decreases of mineralization and D and ^{18}O concentrations and also by a charge in the chemical making up of water from the carbonate sediments (of the Cambrian times) below the reference 900 m (Fig. 7.6). In the lower branch of the curve, the mineralization increases from 2.4 g/l (sample 17) up to 25 g/l (sample 27). The results of the chemical analyses have not been reported by the authors for sample 12. In water of the Cambrian formation, the mineralization

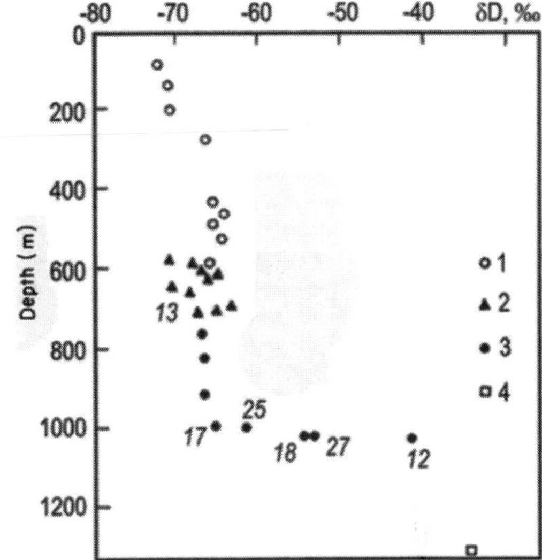

Fig. 7.6 $\delta^{18}O$ and δD values plotted against depth for Lublin Coal Basin waters from Cretaceous *1*, Jurassic *2*, Carboniferous *3*, and Cambrian *4* formations. (After Rozkowski and Przewlocki 1974)

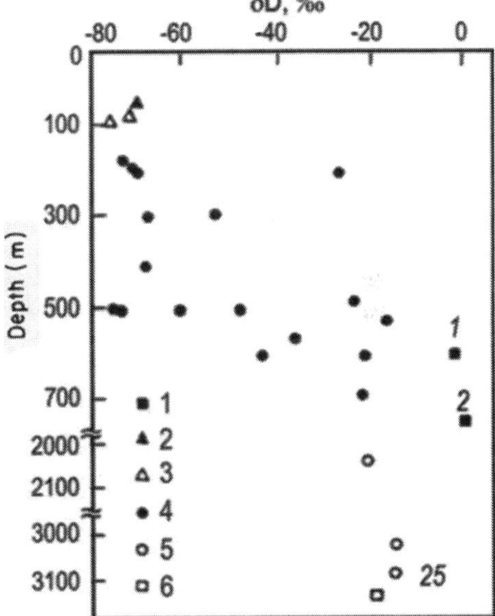

Fig. 7.7 ^{18}O and δD values plotted against depth for Upper Silesian Coal Basin waters from Tertiary *1*, Jurassic *2*, Triassic *3*, Carboniferous *4*, Devonian *5*, and Cambrian *6* formations. (After Rozkowski and Przewlocki 1974)

is 65 g/l ($\delta D = 33.2‰$). The type of water in a series (samples 17, 25, 18, 27) varies from the chloride–hydrocarbonate–sodium to the chloride–sodium one.

The Upper Silesian coal basin exhibits different characteristics. Here, the infiltration component is seen up to depth of about 500 m. Below this, the mineralization of water and the heavy Tertiary formation have the greatest hydrogen and oxygen isotope ratios, close to that of SMOB (Fig. 7.7). The mineralization of these chloride–sodium

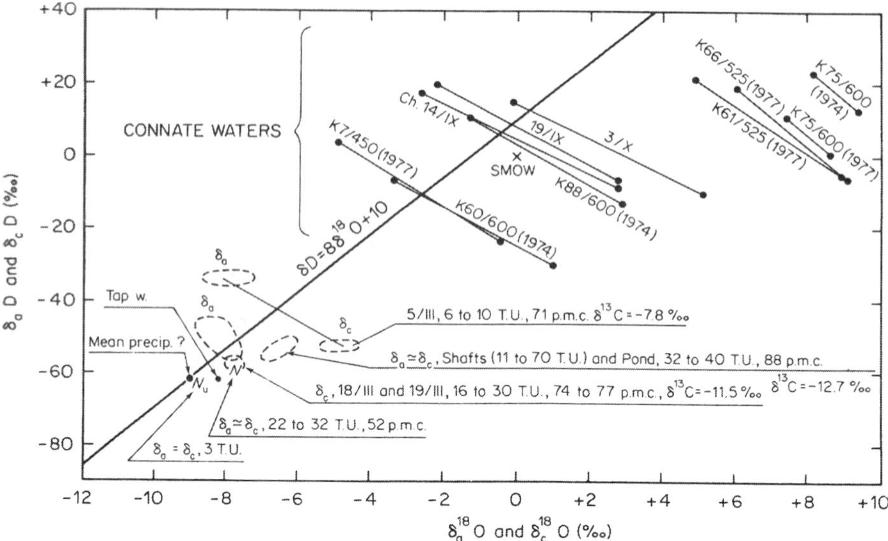

Fig. 7.8 Isotope data used for investigating brines of the Wapno and Klodava salt mines (Poland): δ_a is isotope activities; δ_b is isotope concentrations. (After Zuber et al. 1979)

waters is equal to 32 g/l (δD = −3.6‰, δ^{18}O = +0.07‰) and 52 g/l (δD = −1.2‰, δ^{18}O = +0.34‰). The highest values of mineralization (223 g/l, sample 25) are found in the Devonean. The authors have reported this to be a result of infiltration and relict waters mixing in the deep basin horizons. This process began at the end of the Tertiary period after regression of the sea. In the Lublin, there are more favorable conditions for the contribution of infiltration waters in the productive sandy–clay horizons of the carboniferous complex since the Cretaceous and Tertiary rocks are markedly thinner than in the Upper Silesian basin.

Zuber et al. (1979), with the help of a complex of isotopes (D, ^{18}O, T, ^{14}C), studied the genesis of water recharge in three salt mines in Poland. In particular, on the basis of the D and ^{18}O distribution in the aquifers at different mine horizons, it has been found that the amount of connate metamorphic waters, characterized by increased D and ^{18}O content, increases with depth. This is particularly evident for the Klodava salt mine located in the salt dome of the Zeichstein formation (Upper Permian). It is indicated also that in a number of cases, a considerable amount of modern infiltrational water ends up in the deep aquifers of the mines. Thus, at the third level of the Vapno mine values have been obtained for the tritium content (20–30 TU) and radiocarbon (74% relative to the modern standard), the total mineralization ranging from 360 to 422 g/l. The chlorine–bromine coefficient reached 80 in these typical alkaline brines with a chloride–magnesium type of water. Some of the results of the isotope studies at the Vapno and Klodava mines are shown in Fig. 7.8.

Moser et al. (1972) determined the D content in 110 water samples from the pressurized aquifers in the Styria basin, carried out paleoclimatic reconstructions, and identified several local zones related to different ages of groundwater formations.

At present, sufficiently comprehensive information on the isotopic composition of deep groundwater in sedimentary basins in the former USSR territory and other countries is available, making it possible to further develop isotope techniques and apply them in solving problems of deep groundwater origin. The most successful solution of problems concerning the conditions affecting the formation of sedimentary basin waters may be attained by the interpretation of both isotopic composition data and other hydrological and hydrochemical evidence (IAEA 1976; Bath et al. 1979; Sonntag et al. 1979; Gutsalo 1980; Dubinchuk et al. 1988; Cook and Herzeg 2000; Glynn and Plummer 2005).

Chapter 8
Hydrogen and Oxygen Isotopic Composition of Sedimentary Rocks of Marine Genesis and Implications for Paleothermometry

8.1 Isotopic Composition of Sediments and Pore Water

Important information regarding the evolution of the ocean and climatic changes of the Earth during geological history becomes available from investigating oxygen and hydrogen isotope abundances in pore waters and in the minerals of sedimentary rocks.

The deposits of marine genesis are divided into two groups which are detrital and homogenous (Verhoogen et al. 1970). Different water transported particles of various types (sand, gravel, shell fragments, and clay minerals) form the detrital component of sedimentary rocks. Homogeneous sediments are those which precipitate from water in the form of various minerals (the evaporite minerals, concretions of minerals) or are the products of living sea organisms (carbonates and silicates formed from the remains of corals, mollusks, foraminifera, diatomites, radiolarius, etc.). Authigeneous minerals and evaporates (carbonates, sulfates, chlorides, and other minerals) are deposited during the evaporative concentration of the sea water and also homogeneous sediments. The conditions of marine sedimentation are dependent upon several factors, the major ones are the depth, climate, distance from the shore, and water mass circulation. Under the bottom conditions of recent and ancient oceanic depressions, the major role in the accumulation of sea sediments has been played by pelagic sediments, being formed, as a rule, in the upper ocean layer and precipitated to the floor. The process of fine, dispersed particle accumulation is related to the sedimentation of mainly calcareous shells of foraminifera, coccolithes of single cell algaes, siliceous skeletons of diatomites and radiolaria, and also organic components. Initial sediments, mainly from the ooze layers, are compacted and changed during the low-temperature metamorphism. For instance, at deep bottom conditions, where the temperature is lower than in the upper ocean layers, the higher content of dissolved carbon dioxide favors the solution of carbonate sediments. Therefore, at depths of more than 4,000 m, the calcite in sediments is present only in insignificant amounts. Carbonate sediments in the ocean are known to be present along the narrow belt ranging from 10°N to 10°S (Verhoogen et al. 1970). Thus, as a rule, deep sediments are represented by various types of siliceous shales and cherts of different

V. I. Ferronsky, V. A. Polyakov, *Isotopes of the Earth's Hydrosphere*,

DOI 10.1007/978-94-007-2856-1_8, © Springer Science+Business Media B.V. 2012

ages from the Eocene to Jurassic. The other component of the fine-grained pelagic sediments is continental dust. Part of it is represented by a fine dispersive fraction of volcanic ash. The remainder is a dust which is the result of wind erosion of the continents. Cosmic dust is present in insignificant amounts. In open oceanic basins, the sediments accumulated due to the transport of suspensions of clay particles do not play a significant role in the formation of sediments. This material, as a rule, is accumulated in the coastal region several hundreds of kilometers in width.

The homogeneous sediments of abiogenic origin also precipitate to the sea bottom. In deep water sediments, the minerals such as phillipsite (potassium-calcium zeolite, containing hydroxide group), glaukonite, and even ortoklase (potassium feldspar) are found. At the conditions of the ocean bottom, iron-manganese concretions (the crusts of manganese acids enriched in iron and some other rare elements such as cobalt, nickel, copper) are also commonly dispersed. The accumulation rate at the deep ocean bottom is extremely slow. In the northern part of the Pacific Ocean, where dust transported by wind is a main component of sediments, this rate is about 1 mm per thousand years. In the southern part of the Pacific Ocean, where the major part of sediments are represented by authigenic material of biogenic and homogenic origin, the accumulation rate is about 0.5 mm per thousand years (Verhoogen et al. 1970). According to the data of Kolodny and Epstein (1976), the rate of sedimentation in general is equal to 10 mm per thousand years. According to Emiliany (1978), the sedimentation rate in the Caribbean Sea is about 25 mm per thousand years and in the Pacific Ocean on the Solomon Plateau, this value is about 17 mm per thousand years. For the Southern Atlantic, as Savin (1977) has reported, the rate is 3 mm per thousand years.

On the basis of the data obtained during deep sea drilling, which was carried out by the *Glomar Challenger* expedition ship, the modern oceanic sediments started to form only 150–200 million years ago. The basal layer was reached everywhere. While drilling, the oldest sediments (siliceous) of 140 million years of age were found in the North Atlantic. The age of siliceous sediments from the Southern Atlantic does not exceed 70 million years. The oldest of them was represented by a volcanic ash. The vertical and horizontal motion of the oceanic bottom seems to be the main factor in distribution of the ocean sediments. As it was earlier mentioned, calcium carbonates in the deep ocean are absent because they were dissolved at high pressure and corresponding temperature. The depth at which carbonate sediments are dissolved is called the level of carbonate compensation. As a rule, 'red' clays are accumulated lower of the above level. The horizontal bottom spreading is estimated by a number of effects including the age of the sediments and is characterized for the North Atlantic by 1.2 cm per year, for the Southern Atlantic by 2.0 cm per year, and for the Pacific Ocean by 12 cm per year.

In order to elucidate the conditions of interrelation between water and rocks—and to solve the genetic problems—a number of researchers have been studying the hydrogen and oxygen isotopic composition of pore waters of the deep ocean sediments, of calcareous sediments of marine genesis, clay minerals, and the quartz of cherts. There have been also isotope composition studies of minerals of evaporite basins. The results of some of these investigations are given below.

Table 8.1 D content in pore and residue water from ocean sediments. (From Friedman 1965)

Sediments	Sample depth below the floor (m)	δD (‰)	
		Pore water	Residue water
Calcareous	30.2	−56	−36
Silty, siliceous	86	−26	–
Siliceous, calcareous	95.8	−27	−12
Calcareous, siliceous	105.5	−9	–
Siliceous, volcanic	114.7	−10	–
Siliceous	125.6	−12	+16
Siliceous	131	−30	–

The deuterium (D) content of water has been determined in ocean sediments by Friedman (1965) in the course of drilling under the Mohole Project in the California region. The water has been extracted from silt sediment samples with the help of a vacuum system, in which the pressure, which is lower than the atmospheric pressure, is maintained for several seconds. Then the water is distilled at 60°C before being cooled by liquid nitrogen. Distillation continued until the sample was dry. The D content was also studied in the remainder of the sample subjected to temperature up to 1,400°C (the 'remaining' water). The results of the described analysis are shown in Table 8.1.

From Table 8.1, it is evident that the D content varies in more than 100 m of profile without any regularity being observed. The pore water is depleted in D both relative the standard of the ocean water (SMOW) and relative to the near-bottom water portion of water, depleted in D by 5% relative to the same standard. The analysis of the dried up remainder has shown that the residue bounded water remaining in the sample after heating up to 1,400°C is enriched in D compared with the pore waters, but the amount of remaining water is small by comparison, so that the total amount of D cannot be altered by more than several tenths of percent relative to its content in the pore water.

According to the authors' opinion, observed depletion in D of the pore water compared with the modern sea water cannot be satisfactorily explained and should be repeated by the analysis of extra samples with the obligatory condition of the elimination of possible contamination during sampling and analysis.

Babinets and Vetshtein (1967), who studied the oxygen isotope composition of water from silty sediments from the Atlantic Ocean, found a lower oxygen-18 (^{18}O) content (+5.8 to +6.4‰) than in the near-bottom water layers (+6.3 to +8.4‰) relative to the top water standard. The observed difference in the oxygen content between the silt and near bottom water was explained by the authors in terms of the exchanged reactions between the molecules of the water and silica. The relative depletion of the silt water in ^{18}O compared with the near-bottom waters, according to the evidence from the cited work, is about 2–3‰. These data are in agreement with the results obtained by Lawrence (1973) who has found that the pore water in the Caribbean Sea sediments is depleted by 3‰ in ^{18}O. Later on Lawrence et al. (1975) and Perry et al. (1976) explained this phenomenon by insertion of the basalt weathering products and volcanic matter into the phase enriched in ^{18}O . This assumption is likely to have some physical basis. For example, Kolodny and Epstein (1976) have observed in several cases, depletion in ^{18}O by 1–2‰ of deep sea cherts close to the basalt layer (~25 m) compared with cherts of the same age lying over them in the section. This depletion has obviously occurred as a result of the oxygen isotope exchange between quartz and water, the ^{18}O content of which is lower than in the near-bottom water.

The hydrogen isotope content studies of pore water in the Caribbean Sea sediments were carried out by Friedman and Hardcastle (1973). Their experimental data shows that, for pore water from one borehole, the δD values increase from +5.3‰ at a depth of 2.5 m below the sea floor to +11.2‰ at a depth of 51 m and then drops to +6.3‰ at a depth of 176 m. For another borehole δD values increase from −2.3‰ at a depth of 4 m below the sea floor to +5‰ at a depth of ~90 m and then decrease to −6.8‰ at a depth of 375 m. The profile in last borehole spans sediments from the Pleistocene to the Late Eocene age, from 250,000 years to 40 million years. Here D abundance in the pore waters is characterized by increasing concentrations from the Pleistocene to the Miocene and decreases during the Late Eocene ages. According to Friedman and Hardcastle (1973), the D content variations in the pore waters reflect the δD variations in the ocean due to removal of the light isotope in the Arctic and Antarctic ice sheets.

The authors have reported that D content results in deep sea water sediments are conservative up to at least one million years, but they have assumed that δD variations with time and place as a result of isotopic exchange reactions with clay minerals involved into the composition of sediments. It has also been pointed out that effect the mixing of waters with the different D content due to water seeping out from the underlying layers is negligible since the water seeping out and leaching along the fissures does not interfere with the pore water. Besides, it has been pointed out that the process of self-diffusion of water is too slow to cause any marked differences in D content of pore waters inside for 100,000 years for layers up to 0.5 m and inside for 40 million years for layers up to 5 m thick.

Interesting results were obtained by Savin and Epstein (1970b) on oxygen and hydrogen isotope studies of the ocean sediments. They took 27 samples of sediments from the Northern and Southern Atlantic, Northern and Southern Pacific, and Indian Oceans (between 30°N and 30°S) and investigated the carbonate-free fractions of the samples in order to estimate the scale of isotopic exchange between clay minerals of the initial composition and the sea water. The results of their analysis of core samples treated by the least square method are shown in Fig. 8.1a. For the clay minerals, the δD and $\delta^{18}O$ values obtained were presented in Table 8.2.

The δD and $\delta^{18}O$ values found for montmorillonite are evidence that it has not been subjected to isotope exchange with the sea water and has preserved the isotope ratios corresponding to the conditions of equilibrium with meteoric water at the temperature of sedimentation. Two authigenic montmorillonites from the sea floor had $\delta^{18}O$ values of +26.1‰ and +28.5‰ and δD values of −78‰ and −68‰. Analogous conclusions have been drawn for illite. The isotope data is evidence in favor of its detrial origin. Otherwise, the $\delta^{18}O$ values should range between +26 and +30‰, corresponding to isotope equilibrium with sea water at bottom temperatures.

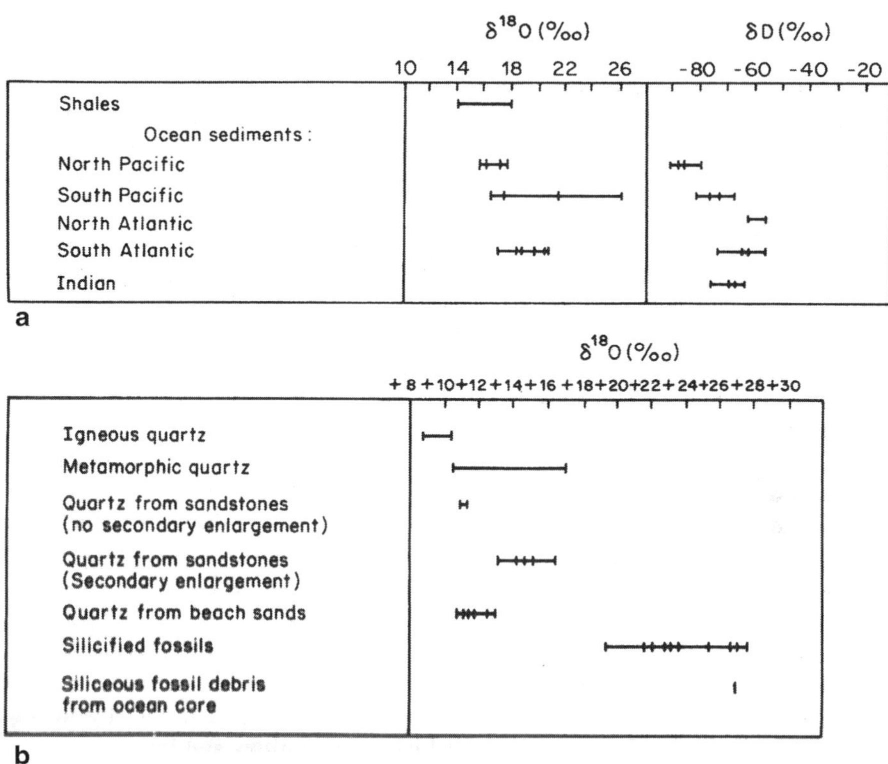

Fig. 8.1 **a** ^{18}O and D content in some non-carbonate ocean core samples (Savin and Epstein 1970b). **b** ^{18}O content of detrital and autogenic quartz from sedimentary rocks (Savin and Epstein 1970c)

Table 8.2 Isotopic composition of clay minerals in ocean sediments after least squares treatment. (From Savin and Epstein 1970c)

Mineral	δD (‰)	$\delta^{18}O$ (‰)
Montmorillonite	+17.2	−70
Illite	+15.4	−60
Chlorite	+14.9	−145
Kaolinite	+24.9	−32

Both in the first and second cases, the formation of montmorillonite and illite could occur on continents at higher temperatures with subsequent partial isotopic exchange at the conditions of the ocean floor.

The data obtained for chloride have shown the continental conditions of its origin as a product of weathering in the presence of light polar type water and an absence of the influence of isotope exchange with sea water. It was difficult for the authors to draw conclusions regarding the conditions of kaolin formation, since, for the tropical region from which the sample was obtained, the equilibrium conditions with sea water and precipitation give similar results.

During the analysis of the samples, the authors investigated three samples of phyllipsite, which is mineral of authigenic origin frequently found among silicates in marine sediments. The $\delta^{18}O$ values for them appeared to be +33.6‰ (for two samples) and +34.1‰ (for the third one). They are close to the ^{18}O content in the diatomous silt. The results are evidence that phyillipsite is formed under conditions of isotope equilibrium with sea water and does not inherit isotope ratios characteristic of its initial basalt material.

In some later studies (Yeh and Savin 1976; O'Neil and Kharaka 1976; Yeh and Epstein 1978), the effect of isotope exchange between clay minerals and sea water has been studied in relation to the size of mineral and its residence time in water. In studies of deep ocean sediments from the North Atlantic and the northern part of the Pacific, and also on the basis of the laboratory experiments it has been found that in clay particles greater than 0.1 μm in diameter the oxygen and hydrogen isotope composition has remained unchanged for several million years since their origin. When the diameter of particles is less than 0.1 μm, this time is reduced to about $n \times 10^4$ years.

On this basis, it has been concluded that at equilibrium conditions of clay minerals interaction with water, the isotope composition of minerals reflects the conditions at which the clay sediments are formed. The synchronous variations of hydrogen and oxygen isotope ratios for minerals suggest that the recrystallization of minerals has taken place.

Among the other noncarbonaceous minerals in ocean sediments, the quartz and feldspar are of importance. Savin and Epstein (1970c) reported data on the oxygen isotope ratios of detrital and authigenic quartz, separated from different sedimentary rocks. The most enriched quartz in terms of heavy oxygen was in the cherts from marine limestones. The $\delta^{18}O$ values of this authigenic quartz range from +33 to +34‰, that is as with those of phyllipsite they are close to $\delta^{18}O$ of dehydrated diatomous silt (+32 and 33.5‰). The maximum $\delta^{18}O$ value, detected by Mopper and Garlick (1968), in radiolarian silica was +38‰ and in dredged quartz from the Pacific Ocean was +36‰.

Figure 8.2b shows Savin and Epstein's results for oxygen isotope ratios detected in quartz of different origin and conditions of formation. The authigenic quartz is markedly enriched in heavy oxygen and thus may be distinguished by its isotope composition from quartz of detrital origin. The detrital quartz in the form of sand grains gave values of about +11.2‰ and quartz from igneous rocks +10‰. The quartz samples from a Devonian marine limestone, where it is often found as a result of replacement of fossil carbonaceous remains, shows $\delta^{18}O$ values ranging from +19.1 to +27.1‰. These data provide evidence that the formation temperature of the quartz has not exceeded 100°C. The detected variations resulted from the process of silification of fossils in the presence of water of varying composition and the subsequent process of dehydration and recrystallization of sediments.

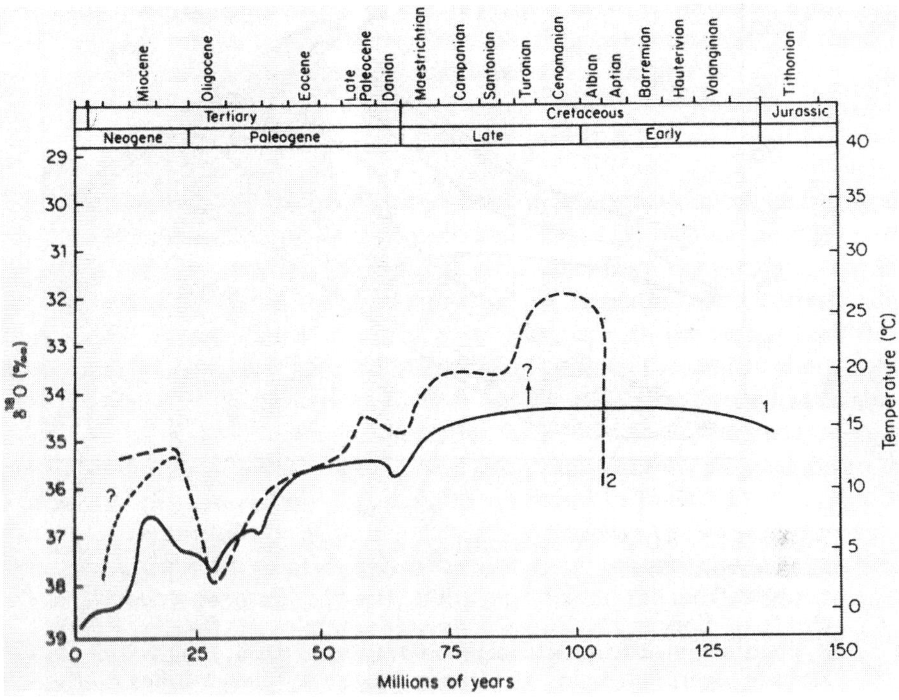

Fig. 8.2 Sea floor paleotemperature curves obtained by biogenic carbonates (*1*) and by cherts (*2*) when shifting the time scale by 40 million years. (From Kolodny and Epstein (1976))

The $\delta^{18}O$ values of silica, extracted from a specimen of siliceous organic remains, gives +26.6‰ and for the modern coastal sands, which contain more than 95% of pure quartz, the values ranges between +10 and +12.5‰ for seven separate determinations.

The studied specimens of feldspar represented a core with $\delta^{18}O$ of about +8.5‰, being characteristic of igneous formation samples, and the outer shell of authigenic origination with $\delta^{18}O = +18.8‰$. Assuming for water—where the authigenic feldspar was formed—the $\delta^{18}O$ to be +3‰, the authors reported the temperature of its formation to be 110°C.

8.2 Paleothermometry Based on the Isotopic Composition of Cherts

Among the sea sediments, the hydrogen and oxygen isotopic composition of cherts (siliceous shales) is of greater interest to researchers. Cherts represent the siliceous rocks practically completely composed of SO_2. As a rule, they are formed in a marine environment during the sedimentation of the siliceous skeletons of micro-organisms. In subareal conditions, the concentrations of siliceous rocks may have formed due to abiogenic process involving the solution and resedimentation of silica dioxide.

Cherts are composed of quartz crystals of 2–3 μm in size. After deposition, the diagenetic recrystallization of initially hydrated silicates (opals) takes place with the formation of the phase of α-crystobolite at the first stage and α-quartz later on. Kolodny and Epstein (1976) distinguished among siliceous quartz, the cherts—composed of granular microcrystalline α-quartz, and porcellanites—containing mainly α-crystobolite. In the cherts, bulks of well-faceted quartz and individual crystals of α SiO_2 are often found.

It is not known if the microcrystalline quartz from cherts is a hydrated mineral, but it was found that it is possible to extract from it about 0.3% of free water and between 0.5–1% of water bonded in hydroxide groups, being present at the surface of the lamellated crystals of quartz. This water does not exchange with D_2O at 101°C and, as shown by the authors, is depleted in D by 80‰ compared with the sea water in which the cherts have formed. For hydrogen isotope studies, the water from cherts is extracted at 1,000°C in a vacuum in amounts of 0.4–1.4 weight %.

Studies of isotope composition in cherts and porcellanites of marine origin, the samples of which were derived from the deep oceans and continental sediments, have led to the following conclusions (Kolodny and Epstein 1976).

Cherts formation is a complicated diagenetic process. At the primary stage of formation, the siliceous algae skeletons are in isotopic equilibrium with the sea water at an appropriate temperature. The hydrogen of the hydroxide group is depleted in D by 80‰ and the oxygen of SiO_2 is enriched in ^{18}O by 42‰, but during diagenesis of the sediments at high temperatures and because of exchange with pore waters depleted in ^{18}O, the $\delta^{18}O$ values of cherts decrease. In the porcellanite ($\delta^{18}O = 30 - 42‰$) – microcrystalline quartz ($\delta^{18}O = 27 - 39‰$) system the $\delta^{18}O$ value of coexisting minerals decreases which proves ones more that it is possible for the isotope composition of cherts to vary over time. The hydrogen isotopic composition of cherts of marine genesis remains practically unchanged. It ranges from −78 to −95‰ and is independent of hydroxide groups which form water by a process of thermal recombination. When comparing the paleotemperature records of deep ocean fossils from the Jurassic to resent times, obtained from the oxygen isotope content of carbonaceous sediments formed from skeletons of foraminifera (Douglas and Savin 1975), Kolodny and Epstein (1976) concluded that analogous changes in $\delta^{18}O$ (and, therefore, temperatures) also take place in cherts, with a shift in the curve, plotted on temperature–time coordinates on the basis of the cherts' isotopic composition, of 40 million year (the average time of chert formation after sedimentation) forward in time. In this case Kolodny and Epstein suggest that the paleotemperature curves plotted by oxygen isotopic composition of carbonates and cherts are satisfactorily similar (Fig. 8.2).

The absence of temperature maxima in the curve plotted by Duglas and Savin in Albian and Coniacian–Santomian, observed earlier by Lowenstam and Epstein (1954) and also by Coplen and Shlanger (1973), has been explained by Kolodny and Epstein (1976) in terms of a lack of experimental points for this time range which

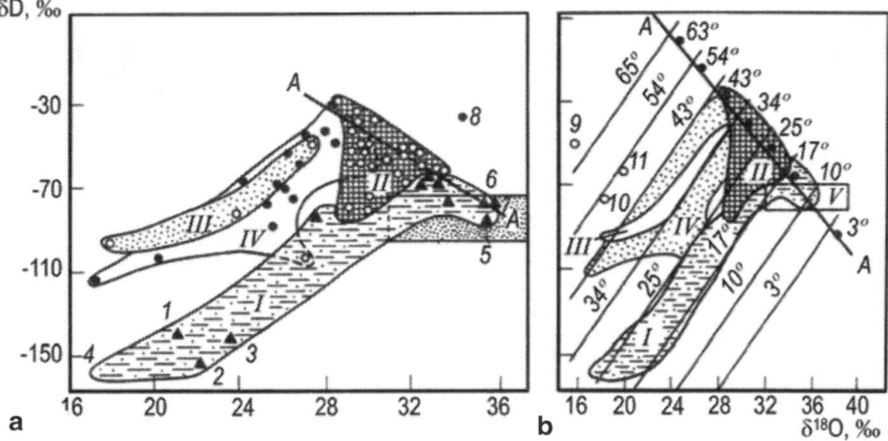

Fig. 8.3 3 δD $-$ δ^{18}O relationship for deep sea and continental Phanerozoic cherts **a** and calculated temperatures of their crystallisation based on a quartz-water isotope fractionation scale and given by the equation $10^3\alpha = 3.09 \times 10^6/T^2 - 29$. **b** Line A–A represents cherts in isotopic equilibrium with sea water (SMOW): nodular cherts of Miocene age from Humboldt formation, Nevada (fresh) water (*1–4*); cherts of Eocene and Cretaceous ages, sampled at JOIDES DSDP (*5–7*); chert sampled from Mississippian evaporitic basin, Illinois (*8*); Precambrian cherts (*9–11*): from S. African Fig Tree formation, 3,000 million years (*9*); from Death Valley Beck Spring Dolomite (California, US), 1.2 (?) thousand million years (*10*); from Belt Supergroup (Louisiana, US), 1.3 thousand million years (*11*); Post-Jurassic cherts domain (*I*); Upper Paleozoic cherts domain (*II*); Triassic cherts domain (*III*); Lower Paleozoic cherts domain (*IV*); deep sea water cherts domain. (From Knauth and Epstein (1976); Kolodny and Epstein (1976))

may easily result in maxima being missed. The details of the authors' interpretation of the 'chert paleotemperature' are not discussed here but it is worth mentioning that cherts as well as carbonaceous sediments reflect the δ^{18}O decrease in sedimentary oxygen-bearing minerals over geological time, if deep ocean cherts are also taken into account, for a range of about 150 million years.

On the basis of oxygen and hydrogen isotope variations in layer and nodular cherts of different geologic ages (from the Miocene to pre-Cambrian) Knauth and Epstein (1976) attempted estimations of climatic variations on the Earth during a period of 3,000 million years. Analyzing the δD and δ^{18}O data obtained for cherts sampled in the United States, England, and South Africa, and also in deep sea cores, Knauth and Epstein have drawn the following conclusions.

In δD–δ^{18}O coordinates the points corresponding to the cherts of different geological ages of the Phanerozoic time fall close to straight lines, being parallel to the meteoric water line (Fig. 8.3a). For cherts of the Lower Paleozoic (Cambrian-Silurian) and Triassic times all the δ-values cluster close to the line $\delta\text{D} = 8\delta^{18}\text{O} - 268$. For the Upper Paleozoic (Devonian-Permian) time the experimental points nay be approximated with the line $\delta\text{D} = 8\delta^{18}\text{O} - 298$, and for the Cretaceous and Tertiary cherts the points range near the line $\delta\text{D} = 8\delta^{18} - 328$.

Making a number of assumptions, Knauth and Epstein reported that the oxygen and hydrogen isotopic composition of cherts, equilibrated with sea water at various temperatures, plotted in $\delta D - \delta^{18}O$ coordinates may be described by the equation $\delta D = -6\delta^{18}O + 135$ (the line A–A in Fig. 8.3a). The authors further assumed that all the δ-values lying on or near the line A–A are characteristic for cherts, formed in sea water with a constant isotopic composition close to that of SMOW but at various temperatures. The cherts with different δD and $\delta^{18}O$ values but ranging around the line parallel to the line of meteoric waters were formed at the same temperatures with the participation of meteoric waters of different isotopic composition. According to the author's hypothesis, the cherts with isotopic composition characterized by the points lying on the right side and above the line A–A (point 8 in the Fig. 8.3a.) has been derived in evaporite basins, where water subjected to evaporation should have been enriched in D and ^{18}O . Then the authors attempted to show that after the formation of the cherts in the latter geological epochs, marked changes of hydrogen and oxygen isotopic composition by isotopic exchange with groundwater have not occurred. This statement has been illustrated by a number of samples from the Upper Cambrian cherts taken from different depths and from various locations in the United States. All the δD and $\delta^{18}O$ values for these samples are satisfactorily close to the line, being parallel to the meteoric water line.

From the isotopic composition of cherts formed in sea water under equilibrium conditions but at various temperatures, Knauth and Epstein calculated the temperatures of their formation using the modified formula (see Chap. 2):

$$10^3 \ln\alpha_{\text{chert-water}} = \frac{3.09 \cdot 10^6}{T^2} - 3.29.$$

Figure 8.3b shows the temperatures of the cherts' formation, calculated by the above equation, assuming that the hydrogen and oxygen isotope composition of the ocean water remains constant during geological time. Here the points on the line A–A denote the calculated temperatures of chert formation with corresponding oxygen isotope composition at equilibrium conditions with the mean ocean waters (SMOW). It follows from Fig. 8.3b that the majority of the studied cherts, when adopting the above assumptions of Knauth and Epstein, were formed at temperatures of 35°C or less. For cherts formed after Jurassic times the temperatures range between 10 and 25°C, which is in accord with carbonaceous paleotemperatures. The calculated chert formation temperatures are shown in Table 8.3 (Knauth and Epstein 1976).

The temperatures calculated by Knauth and Epstein on the basis of the oxygen isotopic composition of cherts are mostly in accord with the data and notions based on considerations of the evolution of the Earth's crust. But a number of assumptions which were made in the work require further investigation in this interesting and important avenue of research.

The work of Knauth and Epstein is a valuable contribution to paleotemperature studies of the early geologic epochs for which carbonate paleothermometry evidence is unreliable (Bowen 1966; 1991). The studies based on hydrogen and oxygen isotope investigations of cherts provided for the application of an extra isotope like D in order to solve the problem and, thus, to estimate the isotopic composition of water in which

Table 8.3 Calculated isotopic temperatures of chert formation for some geologic epochs

Geologic epoch and sample location	Age (billion years)	Temperature of formation (°C)
Paleogene (US)	–	17
Triassic (US)	–	35–40
Permian (US)	–	20–25
Carboniferous (US)	–	20–25
Cambrian (US)	–	35
Precambrian (US, Arizona)	1.2	20–33
Precambrian (US, California)	1.2–1.3	52
Precambrian (US, Ontario)	2	38
Precambrian (South Africa)	3	70

cherts have formed. But the work makes a number of assumptions which require further verification and justification. The basic assumption, which is made in order to justify their techniques, is that the ocean isotopic composition remains constant over geological time (3 billion years). In addition to that, considering the possibility of variation in the ocean isotopic composition due to contribution of marine water, Knauth and Epstein found that sea water could only have been enriched in ^{18}O by 5‰ if it were diluted with 42% of the modern amount of marine water. Such contributions are only one way in which the ocean becomes enriched in heavy oxygen.

From the viewpoint of ideas developed by the authors of this monograph (see Chap. 20), the primary water content in the mantle could have been affected by hydration process of silicate minerals during the formation of chemical compounds in the protoplanetary cloud, in the case its content could not exceed a tenth of one per cent, which is confirmed by experimental evidence of other researchers (Pugin ad Khitarov 1978). Ferronsky suggest that the most probable processes which resulted in the formation of the hydrosphere involve the chemical differentiation of the protoplanetary cloud due to the Coulomb forces, with a subsequent condensation of water at a certain stage of formation of the Earth as a planet (see Chap. 20). Hence, it follows, as Knauth and Epstein have assumed, that the ocean volume most likely remained the same over geological time. But increases in the ocean volume due to contributions of mantle water are not an obligatory condition of the ocean's change in isotopic composition. The oxygen isotopic composition of ocean waters could have changed with time as a result of interaction of the hydrosphere with the silicate shell of the Earth, the temperature of which has changed both during its global cooling and large-scale magmatic processes.

As Knauth and Epstein have pointed out, the majority of the Phanerozoic cherts were initially formed in sea basins where they have undergone a stage of diagenesis with the fixation of the isotopic composition of 'initially' marine waters. The elucidation of the role of meteoric water is also of importance when considering the further metamorphism of cherts after regression of the seas. It is also not clear why the cherts record the isotopic composition of precipitation of those geological epochs to which the sedimentary rocks are related. As Knauth and Epstein have pointed out, citing other researchers, the formation of cherts takes place over dozen of million years after deposition of amorphous SO_2 of pelagic diatomous sea algae. Therefore

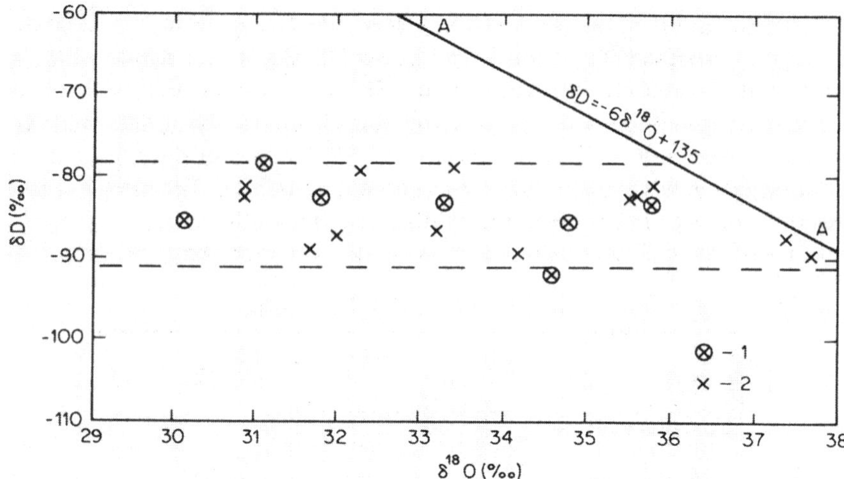

Fig. 8.4 Hydrogen and oxygen isotopic composition of deep sea cherts. The line A–A represents cherts in equilibrium with sea water at various temperatures. Marked points represent various sets of deep sea drilling: (*1*) is from set No167; (*2*) is from other sets. The calculated regression line is $\delta D = 0.54\delta^{18}O - 66.4$, $r = 0.3$. (From Kolodny and Epstein (1976))

the temperature recorded by cherts in their isotopic composition should be identified and related to that part of the hydrosphere or lithosphere concerned. From the data of Kolodny and Epstein (1976) corresponding to the deep ocean cherts, it follows (Fig. 8.4) that their hydrogen isotopic composition is independent of temperature.

The hydrogen isotopic composition of all the studied cherts, the deposition of which took place over 150 million years at equilibrium conditions with sea water and various temperatures (at least from 0–25°C), has not significantly changed, ranging from −80 to −90‰. At the same time, taking into account the data given in Fig. 8.4, from the status of the line A–A if follows that the hydrogen isotopic composition of cherts should change by about 1.3–1.4‰ per 1°C. In general, such dependence of hydrogen isotopic composition on temperature is not characteristic of the majority of hydroxide-bearing minerals, for which the $\Delta\delta D/\Delta T$ values, most probably exceed 0.1‰ per 1°C (see Chap. 2).

The temperatures calculated using the data on isotopic composition of the Precambrian cherts are obviously too high to reflect temperatures of the ocean floor or land surface. At present there is reason to consider that the Precambrian carbonaceous and siliceous sedimentary rocks are of biogenic origination (Lugovaya et al. 1978; Shidlowski et al. 1979). It is hard to assume that micro-organisms could live in water reservoirs at temperatures higher than 50°C. Lowenstam and Epstein (1959) have pointed out that the possible range of temperatures favorable for living organisms is 0–30°C. Bowen (1966) has considered as reliable only those determined temperatures which lie between the above limits with the upper bound of

30°C lowered to some extent. But the temperature of 50°C determined by the Devonian brachiopods he reported to be 'anomalious' and relate it to oxygen isotopic exchange after deposition.

It should be noted that Knauth and Epstein obtained more reliable temperatures (20–25°C) for four samples of Precambrian cherts ranging from 1.2–2 billion years. In the later work in which the isotopic composition of cherts of about 3.4 billion years in the Overwacht Group from Transvaal, Southern Africa, was investigated by Knauth and Lowe (1978), it was shown that in the Archean cherts tend to be depleted in ^{18}O by 8‰ compared with cherts of Precambrian age. In this work the oxygen isotopic composition determined from 87 samples range from +16 to +22‰. The $\delta^{18}O$ values in the fine-grained and druse quartz differ from each other by no more than 0.7‰. The samples, taken in the zone of contact metamorphism, exhibit practically the same oxygen isotopic values as those taken from unaltered parts of rocks. On the basis of these data the authors concluded that the isotopic composition of cherts has not undergone any marked changes due to metamorphism or continuous contact with meteoric waters and has remained the same from the moment of deposition of sediments, that is during 3.4 billion years. The authors have explained the causes of the Archean cherts depletion in ^{18}O but this tendency is observed for practically all sedimentary rocks of this age.

According to the data of Kolodny (1978) meteoric waters were effective in the second stage of the diagenesis of primary cherts of marine origin. This conclusion has been drawn on the basis of oxygen isotope studies for detritus or primary cherts and secondary siliceous cement from the Milash formation (Israel) of the Creteceous, where the $\delta^{18}O$ values +32.1‰ and +27.6‰, respectively, were fixed, and also on the basis of the higher boron content in the detritus. This conclusion is confirmed by a number of α-tracks fixed in a nuclear reaction (n, δ) in the ^{10}B nuclei.

As has been shown by Matthews and Kolodny (1978), variation of oxygen isotopic composition in carbonaceous-siliceous shales during their metamorphism may result from thermal decomposition of initial rocks and the release of significant amounts of carbon dioxide. The stage of high-temperature metamorphism of primary carbonaceous–siliceous shales to rocks enriched in calcium silicate (larnite) may be expressed by the following pertogenic equations:

$$3\mathrm{Cal} + \mathrm{Kaol} + \mathrm{Q} \rightarrow \mathrm{Gr} + (\mathrm{H_2O}) + (\mathrm{CO_2}),$$

$$\mathrm{Dol} + 2\mathrm{Q} \rightarrow \mathrm{Di} + 2(\mathrm{CO_2}),$$

$$\mathrm{Cal} + \mathrm{Q} \rightarrow \mathrm{Wo} + (\mathrm{CO_2}),$$

$$2\mathrm{Cal} + 2\mathrm{Wo} \rightarrow \mathrm{Sp} + 3(\mathrm{CO_2}),$$

$$\mathrm{Sp} + \mathrm{Wo} \rightarrow \mathrm{La} + (\mathrm{CO_2}) \text{ and so on.,}$$

where

Cal — calcite
Kaol — kaolinite

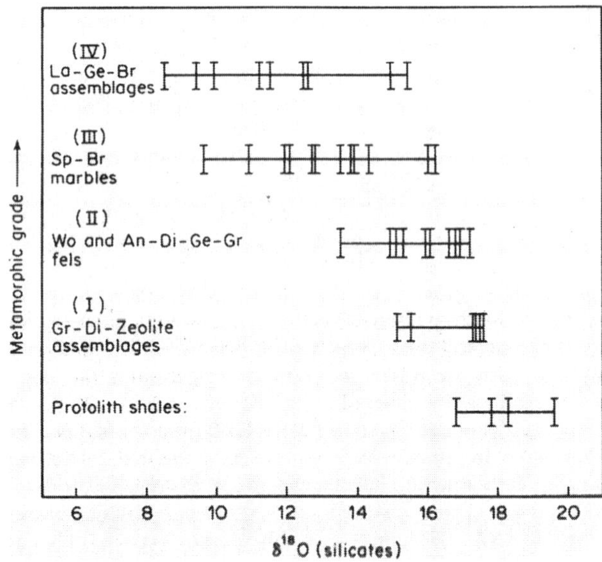

Fig. 8.5 Variation of $\delta^{18}O$ of Mottled Zone silicates with metamorphic grade. (From Matthews and Kolodny (1978))

Q — quartz
Gr — grossular
Dol — dolomite
Di — diopside
Wo — wollastanite
Sp — sperrite
La — larnite.

With increasing metamorphism of sedimentary rocks, the $\delta^{18}O$ values in silicates decreases progressively, from an average of +18.1‰ to an average of +11.7‰ in rocks metamorphosed to high degree and containing larnite (Fig. 8.5).

Matthews and Kolodny (1978) explained this decrease of ^{18}O content in silicates in terms of the thermal dissociation of primary calcites and the above-mentioned reactions of decarbonization which are accompanied by CO_2 release. Assuming carbon dioxide to be enriched by 16‰ at the temperature of metamorphism compared with the gross rock or by 8.5‰ relative to the parent carbonates, the variation of isotopic composition in carbonates may be theoretically considered with the help of the Rayleigh's distillation of ^{18}O isotopes in the rock–carbon dioxide system. Therefore the isotopic composition of the ancient cherts could vary over geologic time due to metamorphic process involving both water and carbon dioxide release during decarbonization of primary sediments.

Finally, it should be noted that the potential application of isotope paleothermometry involving cherts, which first applied in the work of Knauth and Epstein, is very attractive, since it widens the time scale of paleotemperature study; however, a number of new problems arise which have to be solved using data on the lithology and the mineralogy of cherts, their genesis, and metamorphism, in order to realize this potential.

8.3 Paleothermometry Based on the Isotopic Composition of Carbonate Rocks

Among the continental and marine rocks, the carbonaceous sediments of biogenic and non-biogenic origin have played an important role in the evolution of the isotopic composition of the hydrosphere. In connection with the solution of a number of important problems, the investigation of the isotopic composition of carbonate rocks is of interest. The paleotemperature studies of the ancient seas, where carbonate rocks have formed, have been carried out after the establishment of an isotopic temperature scale, obtained by Epstein and Urey with their co-workers (Epstein et al. 1951, 1953; Bowen 1966, 1991) and based on the oxygen isotopic equilibrium during the formation of carbonates in water. The equations obtained by these authors and modified by other researches, are given in Chap. 2. They are valid for paleotemperature determinations on the basis of oxygen isotope ratio measurements in carbonates.

Oxygen isotope ratios in modern carbonates have been shown by numerous studies (Clayton 1961; Craig 1963, 1965; Gross and Tracey 1966; Silverman 1951) to vary from +25 to +34‰. O'Neil (1979) has pointed out that the temperature fractionation factor of oxygen isotopes in the calcium carbonate-water system over the temperature range of interest in geology is equal to about 0.2‰ per 1°C. On account of the accuracy of the modern mass-spectrometric techniques developed for oxygen isotopic analysis (0.01% relative), it is theoretically possible to determine paleotemperatures of ancient oceans with accuracy better than ±0.5°C. In many studies devoted to paleotemperature analysis, following four main questions are discussed (Bowen 1969, 1991; Tays and Naydin 1973; Hoefs 1973; Savin 1977; O'Neil 1979):

1. Does sedimentation of calcium carbonates take place at isotopic equilibrium of oxygen with sea water?
2. Does the isotopic composition of the ocean water change over geologic time?
3. How long is the original isotope-paleotemperature record conserved in the sedimentary strata?
4. What relationship exists between isotope temperature and climate?

Many organisms deposit their lime skeletons at isotopic equilibrium with sea water but the isotopic composition of the shells of some marine organisms is governed by kinetic factors related to metabolism process (O'Neil 1979). In particular, there is substantial experimental evidence (Bowen 1969; Savin 1977) to suggest that a number of mollusks and foraminifera, the shells of which are used in paleotemperature analysis, build their carbonate shells at isotopic equilibrium with sea water. A more complicated problem which has not yet been solved is the problem of ocean water change over geologic time (Imbrie and Imbrie 1979). Thus, for example, a difference of 1‰ in the salinity of the ocean water may result in an error for the isotopic temperature of 1°C (Bowen 1991). As shown earlier, the variation of the oxygen isotopic composition of ocean water in glacial times could range from 0.5–1.5‰. Undoubtedly, this should be taken into account in paleotemperature studies. At first sight, it seems advantageous to use data on the isotopic composition

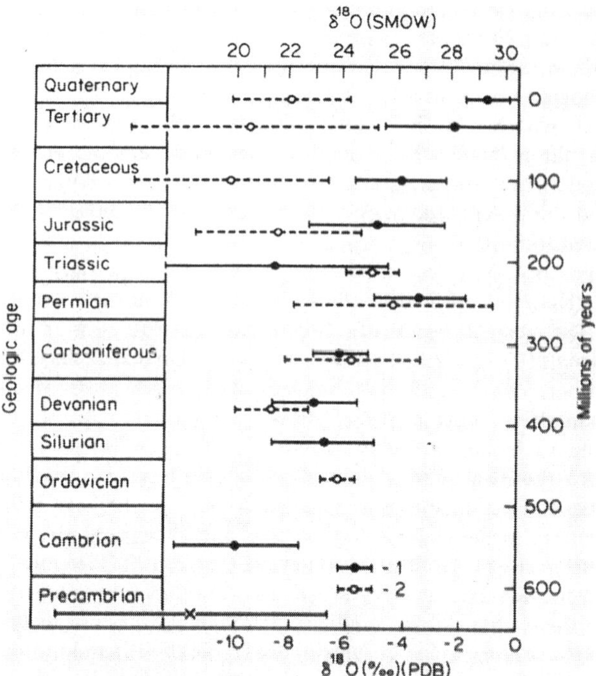

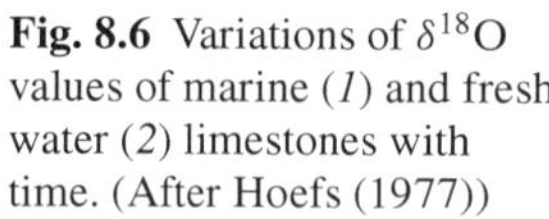
Fig. 8.6 Variations of $\delta^{18}O$ values of marine (*1*) and fresh water (*2*) limestones with time. (After Hoefs (1977))

of mollusk shells built out of phosphates, silicates, and sulfates in order to eliminate the uncertainties in the isotopic composition of water, but as O'Neil (1979) has shown, this problem is far from being solved.

It can be shown experimentally that the oxygen isotopic composition of shells of sea origin has progressively decreased with increases in the geological age of the carbonate formations. This is illustrated in Fig. 8.6, plotted using data obtained by different authors. One of the possible explanations of this phenomenon is based on the assumption that isotopic exchange between carbonate rocks and groundwater had taken place after deposition. However, as shown in Chap. 3, there are other explanations, based, for example, on the assumption that the ancient ocean was lighter in isotopic composition than the modern one.

Laboratory and natural experiments aimed at the investigation of paleotemperature record conservation and the role of the recrystallization and diffusion processes in its destruction have shown that at 20°C a calcium carbonate crystal 1 mm in diameter conserves, during 7×10^8 year, about 96.4% of the original concentration of ^{18}O (Bowen 1966). At 100°C, the record in the same crystal is conserved for only 6.4×10^4 year. As Bowen has pointed out, on the basis of Urey's estimations, even the record of daily temperature variations can be partly conserved in shells of ancient mollusks. However, the paleotemperature record can be completely lost during recrystallization of argonite skeletons of ancient mollusks into the calcite ones. Such a specimen should normally be excluded from paleotemperature analysis.

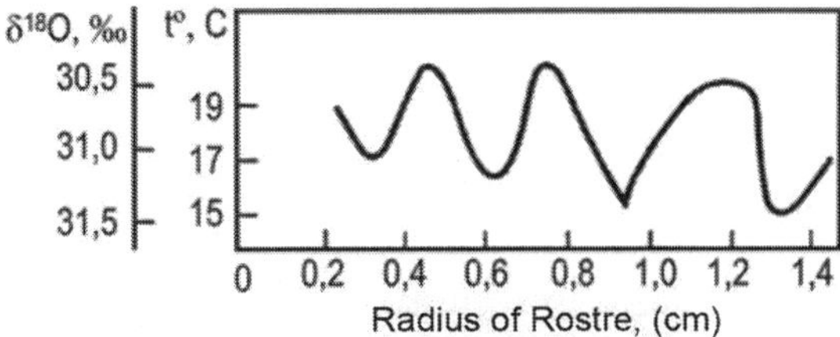

Fig. 8.7 $\delta^{18}O$ and temperature variations in the shells of belemnite from the Sky Isle. (After Urey (1951))

A criterion of the initial paleotemperature record conservation is the existence of oxygen isotope variation related to seasonal temperature variations. Such variations can be fixed by a precise analysis of yearly layers in the shells of belemnite mollusks. Figure 8.7 demonstrates this study (Urey et al. 1951).

In the recent years, as O'Neil (1977) has pointed out, studies of the climatic variations from the Late Mesozoic to the Cainozoic have attained considerable success and become common due to a great number of specimens available in the course of the Deep Sea Drilling Project, improvements in modern biostratigraphic techniques, and an increase in the number of research laboratories equipped with modern mass-spectrometric instruments available for the oxygen isotope analysis in paleotemperature studies.

In Fig. 8.8a, comparison is made by Savin (1977) between Cretaceous paleotemperatures obtained during the oxygen isotope studies of belemnites rostra sampled on land in different countries (Fig. 8.8a) and obtained from carbonate sediments from the central part of the Pacific Ocean (the Shatsky Rise) (Fig. 8.8b). From the temperature curves and the distribution of the experimental points in the figure, one can conclude that during the Albian time the most elevated temperatures existed in Northern Europe and in the central part of the Pacific Ocean whereas in the floor water layer (the specimens of the benthic foraminifera), the maximum is not sharp. The minimum temperatures are observed in the early Maastrichtian. It was also evident that the climatic changes were of a global character and the general cooling between the Albian time and the late Cretaceous time amounted to 8–10°C.

The oxygen isotope paleotemperature curves of the Tertiary age were shown by Savin (1977) (Fig. 8.9). The most authentic values were obtained on the basis of Pacific Ocean specimens. The general trend of temperature decrease during the Tertiary is observed.

Savin (1977), citing Emiliany, has pointed out that paleotemperatures of the ocean floor, obtained on the basis of the oxygen isotopic composition of benthic foraminifera, most probably reflect the surface ocean water temperature corresponding to higher latitudes, were the more dense, cooled water masses submerged into deep ocean layers forming cold bottom streams. Regional and local climatic changes in different geological times could have been affected by changes occurring in the major ocean streams and changes in air mass transference due to displacements of the lithospheric plates.

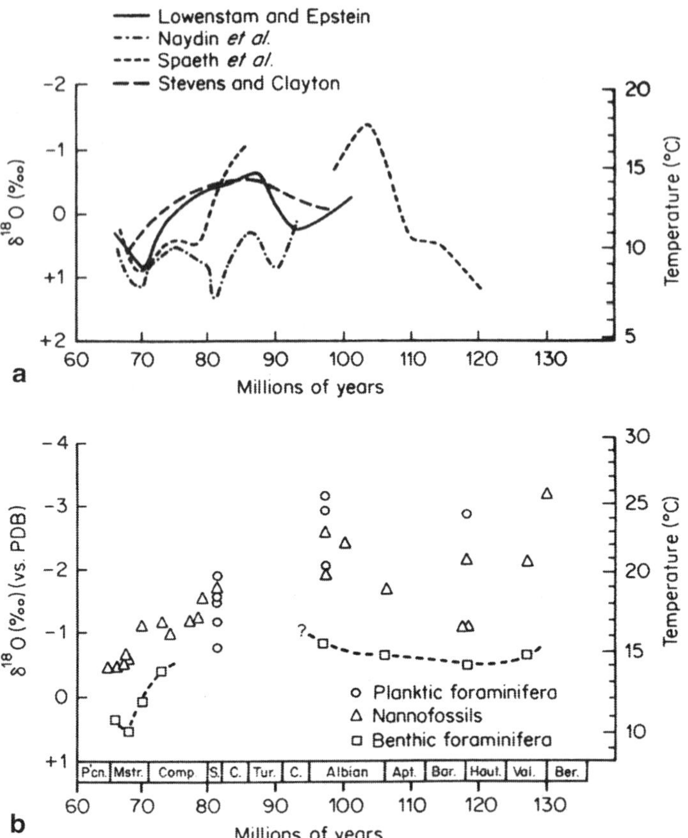

Fig. 8.8 Cretaceous paleotemperature curves based on **a** belemnite data and **b** carbonates of Pacific Ocean (Shatsky Rise) by the data of (*1*) *Lowenstam and Epstein*; (2) *Naydin et al.*; (*3*) *Spaeeth et al.*; (*4*) *Stevens and Clayton*; (*5*) *planktic foraminifera*; (*6*) *nanofossils*; (*7*) *benthic foraminifera*. (After Savin (1977))

Figure 8.10 shows data on oxygen isotopic variations for *Globigerinoides Sacculifera* mollusk shells obtained by Emiliany (1978) based on core analysis carried out during deep sea drilling in the Caribbean Sea (7 boreholes) and in the Pacific Ocean in the region of the Solomon plateau (1 borehole). The data shows sympathetic variations of ^{18}O content in shells with the Earth climatic changes during the Pleistocene. Here the minimum values of $\delta^{18}O$ correspond to warm periods and maximum ones to cold periods.

It is likely that the alteration of cold and hot periods did not only occur in the Tertiary. Savin (1977) has carried out detailed oxygen isotope studies of carbonates of biogenic origin from Site 15 in the framework of the Deep Sea Drilling Project. The results of their studies are shown in Fig. 8.11. It is seen from the character of the paleotemperature curve that during the early Miocene, the ^{18}O values

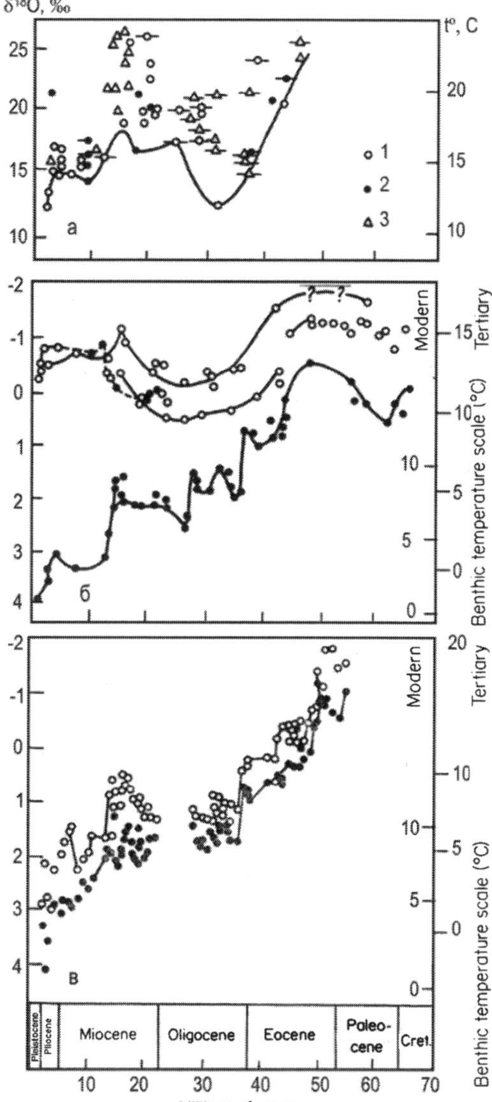

Fig. 8.9 Tertiary paleotemperature curves obtained by various authors for New Zealand (**a**), North Pacific Ocean (**b**), and Sub-Antarctic Pacific (**c**) based on the data of: *(1) planctic foraminifera*; (2) *benthic foraminifera*; (3) *meggafossils limestones*. (From Savin (1977))

fluctuate over a range of 0.2–0.5‰, which correspond to periodic temperature variations with an amplitude ranging 1–2.5°C. The average period of these fluctuations is 80,000–90,000 years. The duration of these cycles is similar to that observed in the Pleistocene glacial–nonglacial times (see Fig. 8.10).

The well preserved remains of ancient marine flora show that their isotopic composition differs little from the modern one, but this similarity of isotopic ratios is observed up to the Mesozoic times only. The more ancient carbonates were found

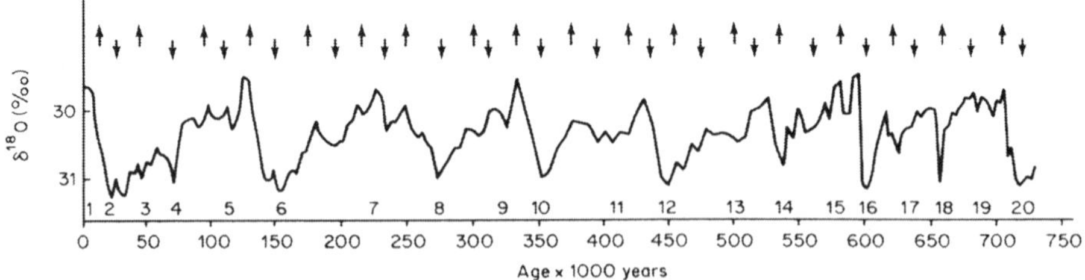

Fig. 8.10 Variation of the $\delta^{18}O$ values of *Globigerinoides Sacculifera* mollusk shells for time interval 0–730,000 years. (After Emiliany (1978))

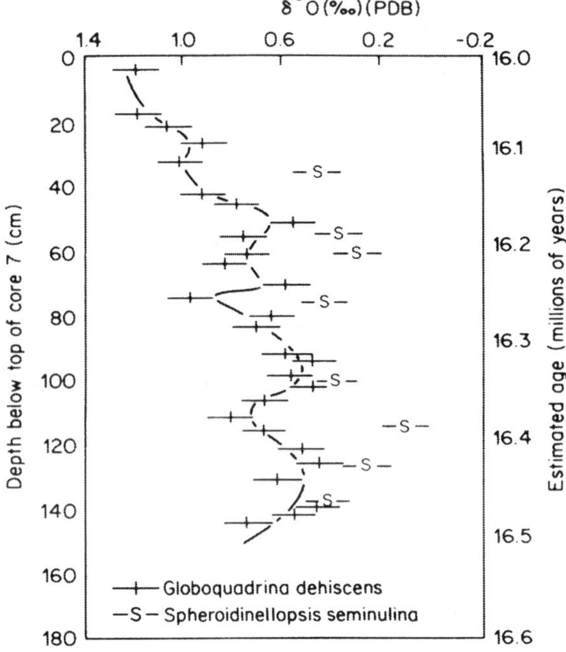

Fig. 8.11 Early Miocene section of paleotemperature curve based on high resolution investigation of core 7, DSDP Site 15. (*1*) *Globoquadriana dehiscens*; (2) *Spheroidinellopsis seminulina*. (After Savin (1977))

to be depleted in heavy oxygen everywhere and exhibit $\delta^{18}O$ values varying from +7 to +25‰ (see Fig. 8.6). using the above-mentioned Epstein–Urey temperature scale for these carbonates, the sea water temperatures were found to be too high to be tolerated by living organisms.

Weber (1965) has analyzed about 600 ancient calcite limestones formed throughout geologic time from modern to the Cambrian. Among them, there were several hundreds of freshwater origin and 300 of sea water limestones. All those specimens which exhibited signs of incipient metamorphism or any other evidence of natural alterations were excluded. As a result of these studies, a linear relationship was obtained which shows the continuous heavy oxygen increase in calcium carbonates with time up to the present. The studied dolomitic specimens have shown the same trend. Thus, the author puts forward the hypothesis that there is a gradual enrichment in ^{18}O of sea water with time, although other possible explanations of the observed facts are not excluded.

Craig (1963) has assumed that evidence for considerable depletion of ancient carbonates in heavy oxygen can be explained by their continuous isotope exchange with meteoric waters on the continents. During this exchange, some portion of heavy oxygen is again returned to the ocean. He has reported that the total depletion effect of carbonates is about 10‰. According to Craig, similar processes of isotopic exchange have taken place in siliceous sedimentary rocks.

It should be pointed out that because of another equation based on oxygen isotope composition in phosphates—and applicable to more ancient epochs—it is very important that analogous investigations of sedimentary thicknesses of modern oceans are undertaken.

Another problem involved in studies of isotopic composition of sedimentary rocks is the estimation of the sedimentation effect during which enrichment of sedimentary rocks in heavy oxygen occurs for the hydrosphere as a whole. Assume that during sedimentation enrichment of carbonaceous and siliceous sediments, at temperature close to the modern ones amounted to +30‰. Then 10‰ of heavy oxygen has returned to the ocean due to the exchange of sedimentary rocks with the water. The initial material forming sedimentary rocks should have $\delta^{18}O$ of about 10‰, since granites exhibit values between +7 and +12‰ and basalts of about +7‰. Finally, the ocean water during sedimentation should lose up to 10‰ of the heavy oxygen per unit of sedimentary rock and, therefore, the ocean should become depleted with time.

Savin and Epstein (1970b) estimated this effect on a global scale. The data of their calculations is shown in Table 8.4.

On the basis of the calculations, they have drawn the conclusion that over the whole Earth's history the hydrosphere should be depleted in heavy oxygen by 3‰ and enriched in D by 0.3‰ due to sedimentation. These values do not exceed the possible experimental errors and therefore it may be assumed that within the range of accuracy of initial data the sedimentation process has not resulted in considerable variations of the isotopic composition of the hydrosphere over the Earth's history. Because of the insufficient reliability of the data on amounts of precipitation, the calculations obviously do not take into account possible resedimentation effects and subcrustal moving of the sediments during the early stages of the Earth's history. The probability of this process, at least within the oceans, is very high if the age of the ocean sediments is borne in mind.

8.4 Isotopic Composition of Evaporates

Among sedimentary formations of marine genesis, those originated by evaporation of water from saline basins are of considerable interest. The geological value of isotopic investigations of the crystalline mater of salt minerals and initial brines of

Table 8.4 Oxygen and hydrogen material and isotope balance. (From Savin and Epstein (1970b))

Rock type	Amount on the Earth surface (kg/cm^2)	Isotope weight amount in rock type (kg/cm^2)	Isotope amount on the Earth surface (kg/cm^2)	$\delta^{18}O$, δD of rock type (‰)
Oxygen				
Pelagic sediments	23	45	10	+19
Carbonate	8.5	49	4	+25
Sandstone and shale	145	47	68	+15
Total sediments	176.5	–	82	+16
Weathered igneous rock	185	48	89	+8.5
Ocean water	278	89	249	0.0
Hydrogen				
Pelagic sediments	23	0.56	0.13	−60
Carbonate	8.5	0.18	0.015	−60
Sandstone and shale	145	0.4	0.69	−60
Total sediments	176.5	–	0.84	−60
Weathered igneous rock	185	0.13	0.24	−60
Ocean water	278	11.1	30.8	0

salt deposits, lies in the possibility of the solution of such problems as the degree of salinization of the basin, the evaporation rate and the temperature of evaporation, the character of connections between a saline basin and the ocean, the sequence of primary salt minerals precipitated, the role of secondary processes in the formation of salt deposits, the influence of groundwater on salt layers, etc.

In Chap. 5, some physicochemical aspects involving the formation of water isotopic composition in evaporite basins were considered in detail. It was shown that if a saline basin is a sea lagoon, then, finally, the isotopic composition of evaporite basins should not differ markedly from that of sea water in any event. The isotopic composition of small inland saline basins can differ considerably from that of sea water being depleted in heavy hydrogen and oxygen isotopes due to evaporation conditions and atmospheric water recharge.

Baertschi (1953) studied the oxygen isotope composition of gypsum crystalline water from various deposits and Maass (1962) reported results of D and ^{18}O measurements in the Strassfurt carnallite crystalline water. The ^{18}O content here was found to be considerably higher compared with that of the ocean water, while the D content equals to meteoric water. This data of isotopic composition cannot be yet explained from any theoretical point of view.

The low value of D content detected in the carnallite of saline fields of the Dead Sea, and the high ^{18}O content compared with the mean ocean water in all the investigated carnallites, is in contradiction with the assumption that carnallites have undergone recrystallization under the influence of surface waters.

It was found that D content in crystalline water of carnallite ($KCl \times MgCl_2 \times 6H_2O$) is equal to -36.9 ± 0.6‰ (80% of theoretically possible amount of water was obtained). Barrer and Denny (1964) found that the value of D content in the hydrated salts is always less of water solution equilibrated with the salt.

Fontes et al. (1965) studied the variation of D and ^{18}O content in water flowing from the sea to places where evaporational concentrations results in the halite precipitation. It was found that enrichment of water in heavy isotopes takes place up to a mineralization of 65 g/l. After attainment of this value of mineralization, the process is sharply reduced and even inversion occurs. It has been experimentally detected that under the investigated conditions ^{18}O content does not increase in parallel with the mineralization of water. This fact has been affirmed by calculations made by Borshchevsky and Khristianov (1965).

While studying the isotope composition of gypsum crystalline water, Fontes and Gonfiantini (1967) found that it is depleted in D by 15‰ relative to sea water whereas it is enriched in ^{18}O by 4‰. In accordance with those D and ^{18}O values, they proposed a relationship which is applicable for the reconstruction of the isotopic composition of the basin where the gypsum was formed: $\delta^{18}O_{sea} = \delta^{18}O_{gypsum} - 4$ and $\delta D_{sea} = D_{gypsum} + 15$. The authors did not consider the relationship to be typical since they did not investigate the dependence of the degree of water fractionation on gypsum formation temperature.

The authors of this book have investigated D content in gypsum specimens from Middle Asia. It was found that $\delta D = -70.9$‰. Using the above expression of Fontes and Gonfiantini, it may be concluded that the studied salt deposit was formed under conditions of contribution of large amount of meteoric water in the saline basin.

As Sofer and Gat (1975) and Sofer (1978) have pointed out, the isotopic composition of water in gypsum is a sensitive diagnostic, elucidating its formation from evaporite brines by anhydrite hydration or by sulfide oxidation. They have shown that values of fractionation factors in the gypsum-water system equal to $\alpha_{18_o} = 1.0040, \alpha_D = 0.980$ and independent of temperature over the range 20–40°C but depend on mineralization of the solution. Since gypsum is dehydrated and is converted into anhydrite at $t = 42$°C, that is at depth of 600 m (at normal geothermal gradient), Sofer and Gat concluded that the water isotopic composition of marine gypsum specimens cannot indicate the primary sea water isotopic composition.

It is of interest to note that gypsum and halite crystals, together with saponite and phillipsite, were found in basaltic veins at depth of 625 m below the sea floor during drilling in the region of the mid-Atlantic ridge (Drever et al. 1979). Using the gypsum sulfur isotopic composition ($\delta^{34}S = +19.4$‰), it was found that the sea water is a source of sulfate, and on the basis of oxygen isotopic composition of saponite ($\delta^{18}O = +9.9$‰) and phillipsite ($\delta^{18}O = +18$‰) Drever et al. have drawn the conclusion that their formation occurred in normal sea water at $t = 55$°C or in water depleted in ^{18}O at lower temperatures. They assumed that gypsum in this case has been formed during reactions between the Ca^{2+} of basalts and the SO_4^{2-} of circulating water. They considered the primary precipitation process of anhydrite at $t > 60$°C with the consequent transition of gypsum in the course of a drop in temperature.

Matsubaya and Sakai (1973) have found that gypsum of nonmarine origin in the Kuroko ore hydrothermal deposits, Japan, exhibit δD and $\delta^{18}O$ values lying between narrow limits: −75 to −90‰ and −6 to −8‰, respectively.

Table 8.5 D and ^{18}O content for macro- and micro-inclusions in salt deposits

Sample location	Salt	Inclusion form	δD (‰)	$\delta^{18}O$, ‰
Solotvino	Halite	Macro	−71 to −40	−1.5 to +2.1
Stebniki	Potassium chloride	Macro	−35 to −1.4	+0.9 to +2.7
Artemevsk	Halite	Micro	−35	−3.5

As the practice of salt deposit exploration shows, lenses of clay and breccia layers can be found in salt thicknesses from which small amounts of water may emerge. These waters are considered to be the remains of parental brines, syngenetic with the salt. Gutsalo et al. (1978b) have investigated the isotopic composition of such water with a mineralization of between 280–400 g/l and micro-inclusions of a number of the Ukrainian salt deposits. The results of these investigations are shown in Table 8.5.

From the data presented in the Table 8.5, it is evident that in the course of salt concentration from sea water, evaporates of the later stage of deposition, as should be expected, are enriched in heavy oxygen and hydrogen isotopes. Some interesting results on the question can be found in the studies of Martines and Kumar (1980) and Nikanorov et al. (1980).

An experimental study has been reported by Matsuo et al. (1972). The authors carried out a study of hydrogen isotopic fractionation during the synthesis from solution of borax ($Na_2B_4O_7 \times 10H_2O$), heilussite ($Na_2CO_3 \times CaCO_3 \times 5H_2O$), nahcolite ($NaHCO_3$), and trona ($Na_2CO_3 \times NaHCO_3 \times 2H_2O$). The synthesis of these minerals was carried out at −35°C. Except for borax, where only a slight enrichment of the solid phase is observed, all the other minerals are depleted in D relative to the solutions from which they were recrystallized, i.e., the value of the fractionation factor $\alpha_D < 1$.

The most reliable dependence of δD on temperature was discovered for the water–trona system which can be used as a geothermometer during salt deposition investigations. Matsuo et al. proposed a formula expressing the relation between temperature and isotope ratios in trona and water:

$$\text{In}\frac{(\text{D/H})_{\text{trona}}}{(\text{D/H})_{\text{water}}} = -1.420\left(\frac{10^4}{\text{T}^2}\right) + 23.56\left(\frac{1}{\text{T}}\right).$$

Matsuo et al. (1972) also obtained the graphical dependence of the fractionation factor on temperature and derived the limitation of the geothermometer based on measurements of the carbon isotopic ratios $^{13}C/^{12}C$ in minerals and carbonate ions of the solutions.

Chapter 9
Hydrogen and Oxygen Isotopic Composition of Groundwater in Volcanic Regions

9.1 Use of Isotopes in Studying the Origin of Thermal Water

Groundwaters of modern volcanic regions are of interest to specialists as are the deep groundwaters of sedimentary basins, using isotope techniques in their studies, in order to elucidate the problems of their formation and solve a number of applied problems of interest concerning their utilization. Let us consider the most important results obtained during these investigations.

Kirshenbaum, Graf, and Forstat (Kirshenbaum 1951) determined deuterium content in two specimens of vapor condensate from the Steamboat thermal springs in Nevada (USA) as early as 1945. The high precision of deuterium measurements, which were carried out by using a technique independent of oxygen isotope concentration, allowed them to account the obtained data as well as the subsequent mass-spectrometer measurements. The deuterium content in the samples was -120 and -107‰, respectively (in the mean ocean water standard (SMOW)). The low deuterium content was explained by the researchers as being a consequence of different HDO and H_2O volatilities during the evaporation of thermal waters and not a reflection of the isotopic content of water before its discharge to the surface where it could be subjected to evaporation.

Boato et al. (1952) studied the hydrogen isotope abundances in the Larderello region hydrotherms both in water vapor molecules and molecular hydrogen which amounted to 1/1,000 of the total vapor from the tube, collecting vapor from all the wells, have shown that the deuterium content varies within ± 0.1 of the working top water standard (water from the local water-supply system). The data for samples taken from an individual water-vapor borehole varied within the same limits. The deuterium content in hydrogen gas was reported to be lower than in water vapor, equal to 0.79 ± 0.05 of the standard (Rankama 1963). Boato et al. (1952) pointed out that at 600 ± 100°C (corresponding to above-critical temperatures at considerable depths) the reaction $H_2O + HD \rightarrow HDO + H_2$ takes place. As shown in Chap. 2, in such isotopic exchange reactions deuterium is concentrated in water whilst gaseous hydrogen is depleted in deuterium (see Fig. 2.10). However, the expected enrichment of vapor in deuterium has not been found due to the small amount of the molecular

V. I. Ferronsky, V. A. Polyakov, *Isotopes of the Earth's Hydrosphere*,
DOI 10.1007/978-94-007-2856-1_9,

Table 9.1 Isotopic composition of selected steam samples collected in the Larderello area wells. (The data recalculated to the SMOW standard). (From Ferrera et al. 1965; Panichi et al. 1974. © IAEA, reproduced with permission of IAEA)

Well number or name	Sampling data	δD (‰)	$\delta^{18}O$ (‰)
5	10.1964	−41	−3.1
6	08.1965	−43	−3.9
38	04.1968	−43	−4.7
59	08.1965	−40	−3.7
85	08.1965	−40	−2.8
89	08.1965	−41	−2.3
109	04.1968	−41	−1.2
112	03.1968	−40	−3.4
114	03.1965	−40	−4.2
141	03.1968	−41	−2.4
Pacciana 3	08.1965	−42	−3.4
Cabattino	08.1965	−41	−2.8
Acquabona	04.1968	−43	−4.0
S. Luigi	06.1969	−42	−5.2
Fornace	09.1964	−43	−5.5
Oliveta	03.1968	−44	−1.7
V.C 10	03.1968	−43	−0.7
V.C. 2	03.1968	−43	+0.1
La Prata	03.1968	−44	−2.4
Grotte 2	03.1968	−42	−4.0
Covo	03.1968	−39	−3.5
V. Madama	03.1968	−40	−2.4

hydrogen (0.001 for the vapor volume) whereas the low deuterium content in the gaseous hydrogen has been detected.

Later on, the isotope composition of the Larderello region hydrotherms was investigated by Ferrara et al. (1965) and Panichi et al. (1974). As some Italian hydrogeologists believe that Larderello therms are recharged by atmospheric precipitation, typical samples of atmospheric water were taken, in the region of their supply, several kilometers to the south of the Larderello. The deuterium and oxygen −18(^{18}O) values there range from −48 to −41‰ and from −7.3 to −8.1‰, respectively. The concentration of the same isotopic composition in vapor carefully sampled from several boreholes in the Larderello varied from −48 to −43 ‰ and from −4.8 to −2.3‰, respectively (Table. 9.1.)

Plotting the meteoric water line for the region $\delta D = 8\delta^{18}O + 10$ and the results of isotope analysis of atmospheric water and sampled vapor on the δD–$\delta^{18}O$ diagram, one obtains the picture shown in Fig. 9.1. It follows from this figure that deuterium content in vapor is analogous to that in atmospheric waters whereas the ^{18}O content is shifted toward deuterium enrichment of the vapor. This fact was accounted for by the authors in terms of the process of oxygen isotope exchange between thermal water and water-bearing carbonates. All the carbonate specimens taken during well drilling in the Larderello were depleted in ^{18}O compared with common marine limestone.

Final evidence connected with ^{18}O and deuterium abundances in thermal and meteoric water supports the hydrogeologists concept that water vapor from the

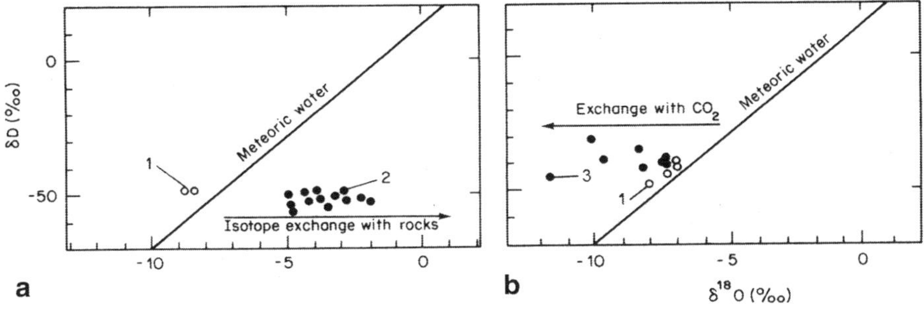

Fig. 9.1 Isotopic composition of selected steam and represented local meteoric water sampled from the Larderello area **(a)** and Tuscany **(b)**: (*1*) atmospheric water; (*2*) water vapor; (*3*) hydrogen-sulfide springs water. (After Ferrera et al. 1965; Panichi et al. 1974. © IAEA, reproduced with permission of IAEA)

Table 9.2 Isotopic composition of Tuscany springs water. (The data recalculated with the SMOW standard). (From Ferrara et al. 1965)

Spring location	Type of spring	Sampling data	δD (‰)	$\delta^{18}O$ (‰)
Momialla (Florence)	Hydrogen-sulfide $t = 3 - 18°C$	23.09.1965	−40	−7.3
Petriole (Siene)	Hydrogen-sulfide $t = 44°C$	07.04.1965	−41	−7.0
Bollore (Florence)	Hydrogen-sulfide $t = 45°C$	22.09.1965	−39	−9.3
Le Puzzolr (Siena)	Hydrogen-sulfide Cold	08.06.1965	−32	−9.9
Perdgine (Arezzo)	Mofette	26.03.1965	−46	−11.7
Doccio (Siena)	Hydrogen-sulfide Cold	07.04.1965	−38	−8.1
Torrike (Florence)	Mofette	22.09.1965	−42	−7.8

Larderello springs should be of meteoric origin and that its "juvenile" component (if it exists) should not exceed 5%. The deep groundwater circulation, assumed for interpretation of the data on ^{18}O concentrations, is also in accord with the geologists' assumption that there is a continuous convective water circulation due to the existence of a large deep heat source. In this case, deep groundwater may emerge at a depth of about 5,000 m. The authors have also discussed an interesting question regarding isotopic composition of hydrogen-sulfide springs and mofettes of the Tuscany (Table 9.2).

Large amount of gases, among them CO_2 is predominant, are a feature of waters from all the springs indicated in the table. From Table 9.2 and Fig. 9.1b it follows that deuterium content in spring waters is generally in accord with its content in meteoric waters. The only exception is the Le Puzzole hydrogen-sulfide spring which is insignificantly enriched in deuterium and situated not far from Pienzo (Siena). This may be accounted for by evaporation of water, since the sample was taken from the surface of the spring's catchment area.

Table 9.3 Deuterium content in Yellowstone National Park gas samples. (The data recalculated to the SMOW standard). (From Friedman 1953)

Sampling place	Type of sample	δD (‰)
Daisy Geyser	Water	−158
Punch Bowl Spring	Water	−158
Hurricane Vent Spring	Gas	−500
Daisy Geyser	Gas	−278
Punch Bowl Spring	Gas	−240
Iron Creek	Gas	−240
Kaleidoscope Geyser	Gas	−247

The ^{18}O concentration in spring waters, compared with meteoric waters, is shifted toward depletion in this isotope. The authors have explained this with isotope exchange between water and carbon dioxide, which follows the reaction $C^{16}O_2 + 2H_2^{18}O \leftrightarrow C^{18}O_2 + 2H_2^{16}O$. The fractionation factor of this reaction at a temperature of 25°C is equal to 1.04. This means that $^{18}O/^{16}O$ in the carbon dioxide exceeds easily and quickly. Therefore, if the ratio of $^{18}O/^{16}O$ for carbon dioxide were lower than in water, then the former should tend toward enrichment in heavy oxygen at the expense of the water.

Thus, the lower the ratio of the amount of water to carbon dioxide involved in the reaction the more the ^{18}O content of the water decreases. The water from the well in Perdgine (Arezzo), where the maximum deviation of ^{18}O content from the initial value has been found, provides an example of such a case. The water from the Perdgine well has the maximum gas–water ratio and is used in obtaining carbon dioxide. The process of oxygen exchange between carbon dioxide and water may be supposedly considered as being predominant over the opposite process of oxygen exchange between water and rocks reduced at low temperatures.

Friedman (1953) has investigated thermal waters and gases in the Yellowstone National Park (USA; see Table 9.3).

It follows from Table 9.3 that the deuterium content in gases is markedly lower than in waters, although there are comparatively low concentrations of this isotope in certain gases. The results of measurements obtained from Hurricane Vent Springs are distinguished among the other gas samples. Hydrogen-bearing gases (0.5% of molecular hydrogen and 20% of methane) were also detected in this spring, whereas there were no molecular hydrogen gases in the water from any other studied springs. According to Friedman (1953), the same mechanism of interaction between water and gaseous phases was proposed for the Larderello thermal waters are evident. Assuming that gases from all the springs were in isotopic equilibrium with water (taking no account of methane from the Hurricane Vent Spring), a minimal temperature of about 400°C, corresponding to the isotope exchange reaction $H_2O + HD \leftrightarrow HDO + H_2$, has been calculated using hydrogen.

White et al. (1963a) investigated the thermal Steamboat Springs (USA). The isotopic composition of the springs exhibits slight seasonal variations. The content of heavy isotopes rises markedly in June, reaches a maximum in August, and drops to a normal level in October. The maximum corresponds to the warmest period with the most intensive evaporation from the surface waters including a nearby lake. The nonequilibrium character of the evaporation process from the open reservoir, which

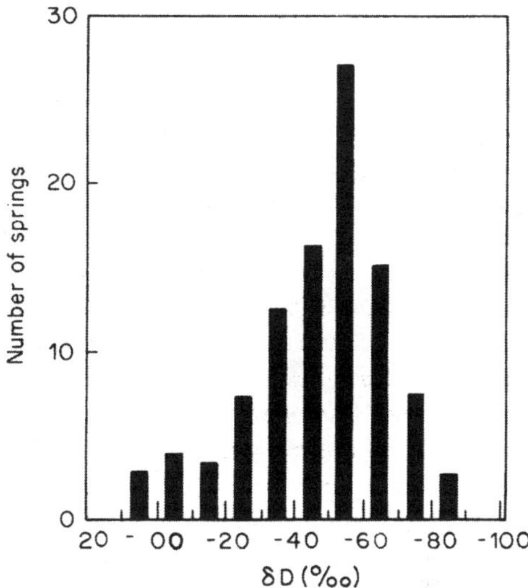

Fig. 9.2 Histogram of deuterium content in groundwater of Japan. (After Kobayakava and Horibe 1960)

is very sensitive to the vapor–liquid system isotopic composition, manifested itself in the course of the spring water isotope analysis. The mean relative deuterium content in water was −90‰, and ^{18}O was −11‰. At the same time in the other springs (e.g., the Galena Spring) δD and $\delta^{18}O$ were about −113 and −15.3‰, respectively, that is, the heavy isotope content follows the relationship $\delta D = 8\delta^{18}O + 10$, which is a feature of atmospheric precipitation. In hot springs, the authors detected the effect of oxygen exchange between silicates and water, being about 2.0 to 3.5‰. The majority of water appears to be absent, if there is any, it amounts to less than 5% and is out of the range of the technique's accuracy.

Deuterium content investigations have been carried out in the thermal springs of Japan (Kobayakava and Horibe 1960). At first the aim of this measurement was to discover the springs whose water, rich in deuterium, could be used in heavy water production. As Kobayakawa has pointed out a tendency of trying highly productive springs with the highest temperatures (these conditions are the most favorable for technological schemes of deuterium extraction from water) was not completely successful from the geochemical viewpoint. About 230 measurements of deuterium concentrations in various samples of water were carried out. The deuterium content varied in them from −103 to +3‰ relative to the SMOW standard. In 70% of the samples, deuterium concentrations ranged between −80 and −40‰. The histogram of deuterium content in these samples is shown in Fig. 9.2.

On the basis of the results obtained for thermal waters of individual groups of springs, some trends are discernible, for example, the deuterium concentration in all the springs of the Ibusika group is proportional to the concentration of chlorine ions and inversely proportional to the temperature. Besides, deuterium content in the water of these springs is markedly higher than in the Tokyo tap water standard (0.0148 at.%). According to the location of these springs it is reasonable to assume that their waters are diluted by sea water. However, if there has been just dilution of the surface waters with sea water, the deuterium concentration in the waters of these springs should fit the line, reflecting the different proportions of sea and inland waters. However, in fact deuterium concentrations in waters of these springs are higher than the corresponding chlorine ion concentrations. Thus, one may assume either that only the chlorine ions are removed from the sea water in underground conditions or that only that portion of water which originally had a higher deuterium content appears on the surface. Since the process of chlorine ion extraction is unlikely, the authors consider that water discharging at the surface had a higher deuterium concentration before mixing with sea water. In support of this statement, it has been pointed out that condensational water from the Hombosudzigoku spring has rather a high deuterium concentration (−40‰).

The groundwaters in the region of the Tamagawa group spring, used for potable supply, have a deuterium content of −73‰ and the thermal springs in this group, the Abuki and Hisogawa, have a deuterium content of water of −50 and −53‰, respectively. The condensational water of fumaroles has approximately the same composition. The water, formed as a result of the mixing of groundwater and fumarole water, exhibits an intermediate deuterium content, equal to −68‰.

Finally, Kobayakawa has pointed out that on the basis of his results it is impossible to draw definite conclusions. For the complete study of the hot spring water's genesis one should investigate the chemical composition, deuterium content, and heavy oxygen content in groundwaters and spring waters.

Friedman et al. (1963) studied various natural waters in Iceland among which were hydrotherms and waters of deep boreholes. On the basis of interpretation of deuterium concentrations in hot and ground waters and in precipitation, it was found that water from the borehole and geyser near Reykjavik is formed from the glacier located a considerable distance northward of them. The same conclusions have been drawn by Craig who measured deuterium and ^{18}O concentrations in hot springs and surface waters around a power station. It has been found that the catchment areas of the thermal and cold springs are different. The lowest deuterium content was detected in two boreholes in the Langaland, being −115 and −117‰. Rather high deuterium concentration, which were detected in some boreholes (up to −8‰), may result from the dilution of hot water with oceanic water.

The results of investigations by Arnason and Sigurgeirsson (1967) indicate that during the movement of thermal waters from the catchment area to the area of emergence, deuterium content remains unchanged. It was also found that deuterium content in hot springs differs markedly from that in precipitation in the region of these springs. The researchers concluded that thermal waters in south-western Iceland are recharged in a region located at a distance of more than 50 km from the place of their emergence at the surface.

The possible applications of the isotope analysis of hydrotherm waters were indicated by Craig (1966) while studying thermal brines in the process of borehole drilling around the Salton Sea in California. The temperature of these highly mineralized (332 g/l) brines, obtained by extrapolation of the bottom of a 1400 m borehole, were 340°C. When the brines were first discovered in the first borehole in this region White et al. (1963b) assumed they were ore-bearing magmatic fluids (i.e., juvenile waters).

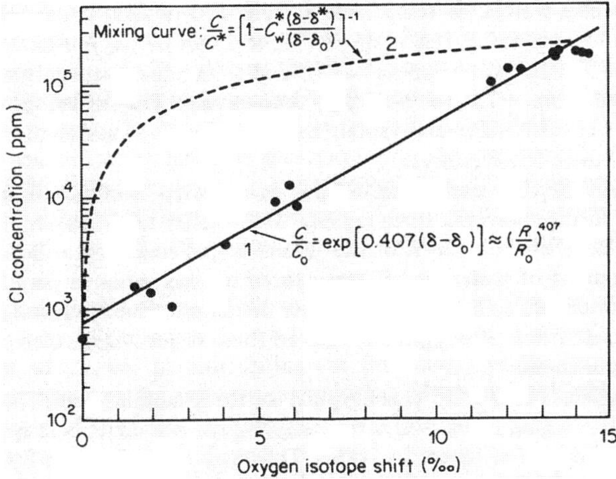

Fig. 9.3 Relationship between chloride concentrations and oxygen isotope shift (δ–δ_0, $\delta_0 = -11$ ‰) in the Salton Sea geothermal brines: (*1*) salt leaching by surface water curve; (*2*) mixing of surface water and hypothetical deep water curve. (After Craig 1966)

Craig has determined deuterium and ^{18}O content in these brines and other surfaces and groundwaters in the region. The isotope ratios obtained (see Fig. 7.1), illustrate the process of oxygen shift observed in nature. In this case, at a constant deuterium content equal to that in local meteoric waters, as usually occurs in regions of hydrotherm formation, the ^{18}O concentration increases in parallel not only with temperature but also with salinity. This fact may be explained in two different ways. Firstly, it may result from simultaneous enrichment of water in ^{18}O and chlorine due to the interaction of surface waters with the water-bearing rocks and, secondly, as a result of deep water (i.e., hypothetical ore-bearing fluid) mixing with the downward moving surface waters. Craig has pointed out curves corresponding to both of these cases. These curves are shown in Fig. 9.3. However, the experimental data shown in the figure indicate conclusively that in fact only the first case holds. Therefore, formation of brines in this case is related to the dilution of salts of water-bearing rocks by local surface waters. Craig (1966) in a note to this work has pointed out that White, after having acquaintance with his results, has agreed that his "ore-bearing fluid" is of meteoric origin.

The results do not contradict Smirnov's conclusion (1971) that the mineral component of bribes, studied by White and coauthors and Craig, is genetically and indissolubly connected with rocks of a salt-bearing basis, at high temperatures and pressures. Craig has also pointed out that surface brines of the Salton Sea Lake differ markedly in isotopic composition of water (solvent) from that of local meteoric waters and thermal underground brines. They are genetically of the same type as waters of Mead Lake, though subjected to a higher degree of evaporation concentration (see Fig. 9.3).

Among the most interesting geothermal regions on the Earth New Zealand is one. Some works (Banwell 1963; Giggenbach 1971) are devoted to the study of isotopic composition of hydrotherms in this region. Banwell (1963) has reported data on deuterium and ^{18}O content of thermal waters in the Wairakei region, which is situated in the north-western part of the North Island. The relative deuterium concentrations in its different springs range from −55 to 0‰ and those of ^{18}O vary from −5 to +4.5‰. The largest surface reservoir in this region is Taupo Lake, δD for which is −30‰ and $\delta^{18}O$ is −5.2‰. Banwell, citing earlier works by Craig and coauthors, inferred that all the hot springs in this region are recharged by local surface waters.

Giggenbach (1971) investigated the isotopic composition of the Broadlands, New Zealand, geothermal waters which are of meteoric origin. It has been found that the borehole waters, at temperatures from 170 to 235°C, have a deuterium content ranging between −41.8 and −33‰ and that of ^{18}O from −4.4 to −3.6‰. The water vapor has δD and $\delta^{18}O$ values varying from −45.2 to −38.7‰ and from −7.0 to −6.0‰, respectively. The difference in isotopic composition between water and vapor corresponded to equilibrium between the two phases at these temperatures.

The authors of the present book have investigated isotopic composition of hydrotherms in the following three regions: in the valley of the Paratunka River, in the Uzon caldera, and on the Kunashir Island at the Kamchatka. Deuterium and ^{18}O content of waters in all these regions varies in parallel (if these regions are considered as a whole): the lowest deuterium concentrations correspond to lower concentration s of ^{18}O . Each of these regions has its own peculiarities in isotope abundances.

The largest values of deuterium and ^{18}O are typical for the Kunashir Island. The deuterium concentrations, ranging from −80 to −32.9‰, do not differ here from those in the waters of a number of sedimentary basins. Thus, may result from the following two reasons. The first and the main reason follow from the fact that precipitation, falling on the island, is a condensed vapor, evaporated from the sea surface. The second reason is a latitudinal effect.

The deuterium content for hydrotherms of the Paratunka River valley varies between −123.4 and −105.7‰ and for hydrotherms of the Uzon caldera ranges from −109 to −68.4‰. The observed difference in isotopic composition of waters from these two regions may be explained by the recharge of the first system from surface waters of higher references.

Relatively high ^{18}O content for hydrotherms of the Kunashir Island, $\delta^{18}O$ in which varies from −11.9 to −4.2‰, may be explained by its marine origin and not by the oxygen shift, the value of which is negligible. The $\delta^{18}O$ content in hydrotherms of the Paratunka valley varies from −16.9 to −12.6‰ and in those of the Uzon caldera ranges from −13.4 to −5.22‰, which results both from the initial isotopic composition of meteoric waters recharging hydrotherms and, partially, from oxygen isotopic exchange between water and rocks. On the basis of the isotope shift, one may come to the conclusion that during the formation of hydrotherms a minimal depth of water circulation is observed on the Kunashir Island, a greater depth for the Paratunka River valley, a maximum depth for the Uzon caldera.

The thermal waters in the Kurile–Kamchatka volcanic area were investigated by Baskov and Vetshtein (Baskov et al. 1973; Vetshein et al. 1971; Meniaylov et al. 1981), found that in this region ^{18}O ranges from −12.9 to +3.4‰. The maximum value corresponds to a condensate of fumarole gases. The relative deuterium values detected on the Kurile Islands gave −51 and −88‰ and those in the Kamchatka region range between −55 and −113‰.

Many researchers, while studying individual questions concerning the genesis and conditions of groundwater formations, at the same time tried to estimate general isotope criteria for the "juvenility" of groundwaters.

Urey (1957) reported that the Earth's gravitational field is insufficiently strong to retain hydrogen and that our planet gradually loses hydrogen, which dissipates into interplanetary space. However, due to the difference in atomic weights of hydrogen isotopes, protium dissipates more easily than deuterium.

In Rankama's opinion, this process must result in the accumulation of deuterium in atmosphere and surface waters over geological time. Quoting Harteck and Suess, he has reported that juvenile water will be depleted in deuterium compared with vadose water, although different mechanisms of isotopic fractionation and dilution of the magmatic waters with vadose water can make the actual picture rather complicated.

From logical considerations, it follows that the hydrogen of water involved in the magmatic activity of the Earth should in principle reflect the isotopic composition of hydrogen which was present during the final stages of the formation of the Earth's crust. It seems, the isotopic composition of water should also "remember" individual stages of the Earth's sedimentary shell formation. Thus, it is of interest to study the deuterium content of parental seated brines and in crystallized water of minerals in the Earth's sedimentary shell, which record data throughout hundreds of millions of years.

There is no conclusive opinion on deuterium content in 'juvenile water. Friedman (1953), Godfrey (1962), and Ferrara et al. (1965) stated that the value of deuterium content in "juvenile" waters should be lower than in surface waters. Kobayakawa and Horibe (1960) reported that "juvenile" waters should be enriched in deuterium. Godfrey also gave the absolute value of deuterium content (obviously an averaged one) obtained on the basis of interpretation of isotopic composition of water in magmatogenic minerals, equal to about −110‰.

A wide range of possible deuterium variation in juvenile waters from −130 to −25‰ has been reported by Ferrera et al. (1965). They have quoted Craig that though the isotopic composition of juvenile waters is unknown, in some ways it can be determined on the basis of reasonable assumptions. This value should be based on deuterium variations in minerals.

It is noteworthy that the discussed problem of juvenile waters has a purely philosophical character because of the problem of the origin of the Earth itself has not yet solved.

Craig (1963) has carried out comparative analysis of the isotopic composition of water and vapor samples, from the main geothermal regions of the world. Figure 9.4a shows isotopic data, obtained in the most famous geothermal regions, which correspond to the volcanic vapor only, typically neutral or slightly alkaline hot springs with pH = 5–9. Deuterium content in these waters and vapor correspond to that in the local meteoric waters. The ^{18}O content indicates a typical enrichment which results from oxygen shift, as the author stated. The range of ^{18}O variations relative to meteoric waters is from 0‰ in the New Zealand springs to 14‰ in the Salton Sea region. The last value is the greatest among all detected data for oxygen shift in hydrotherms. In this case, shown in the Fig. 9.4a for vapor and chloridized waters, the character

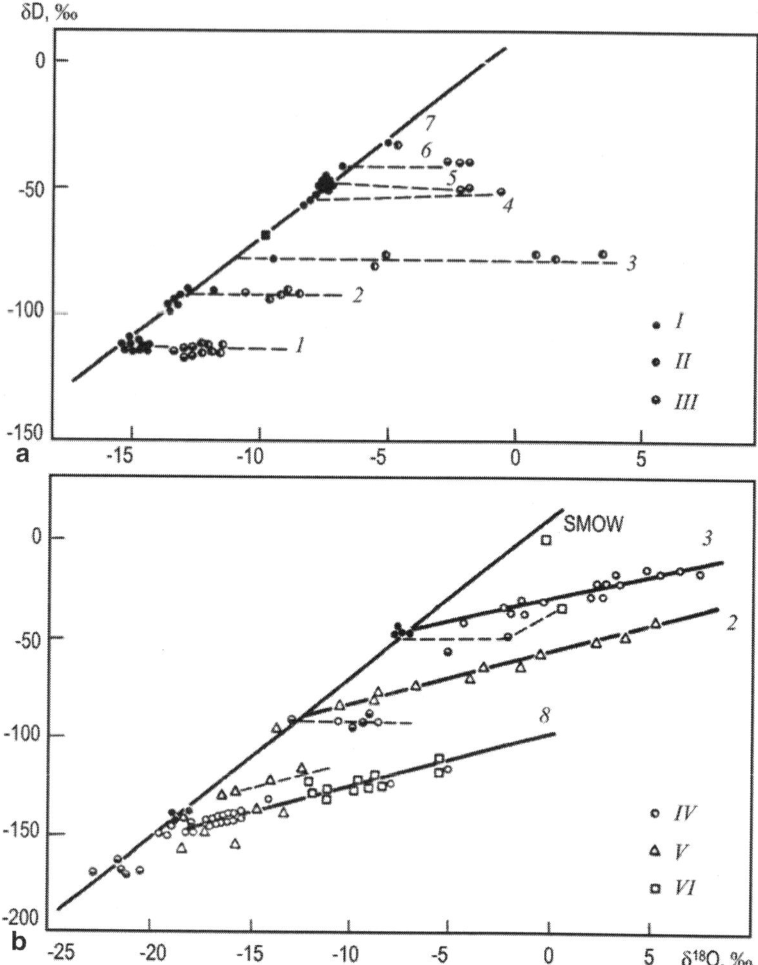

Fig. 9.4 Observed isotope variations in meteoric, near-neutral chloride type geothermal waters and in steam (**a**), and in acid, alkaline type hot springs, in near-neutral waters and in steam (**b**) of some geothermal areas: (*1*) Steamboat Springs (USA); (*2*) Lassen Park (USA); (*3*) Salton Sea (USA); (*4*) Hekla(Iceland); (*5*) The Geysers (USA); (*6*) Larderello (Italy); (*7*) Wairakei (New Zealand); (*8*) Yellowstone Park (USA); (*I*) local meteoric waters; (*II*) geothermal waters; (*III*) geothemal steam; (*IV*) acid type geothermal waters; (*V*) Lassen Park; (*VI*) Yellowstone Park. (After Craig 1963)

of the relationship between isotopic ratios for oxygen and hydrogen is similar. This picture is only observed for vapor at high pressures and temperatures. Craig shows that the observed oxygen shift in vapor and water may result from oxygen exchange between meteoric water and rocks during the motion of water. As to juvenile water, it is not detected in the above-mentioned regions according to Craig and its amount lies out of the range of the accuracy of isotopic investigations.

Table 9.4 Systems of thermal water in Japan. (From Sakai and Matsubaya 1977)

System type	Geology	Chemistry	Example
Arima	Preneogene plutonic and volcanic complex and metamorphic rocks	Na–(Ca)–Cl–HCO_3 High saline, $0 < Cl < 44$ g/kg $t = 20 - 97.5$°C	Arima Ishibotoke
Green Tuff	Green Tuff formations Miocene age	Na–Ca–Cl–SO_4–HCO_3 $0.5 < Cl < 2$ g/kg; $t = 31 - 62$°C	Tottory Owani
Coastal	Quaternary volcanic rocks at ocean cost	Na–Ca–Cl $3 < CL < 20$ g/kg	Ibusuki Shimogama
Volcanic	Quaternary volcanic rocks breed	H–SO_4 H–Cl–SO_4 NaCl HCO_3 Mixed	Hakone Noboribetsu Tamagawa Satsuma Iwojima

The relationship between the isotopes of neutral, alkaline, and acid waters in the previous regions and complementary in the Yellowstone Park is shown in Fig. 9.4b. In acid waters, the following fact is discernible. The straight lines plotted on the basis of experimental data for individual regions do not intersect at one point corresponding to the isotope content of juvenile waters. These lines pass almost parallel to each other, reflecting the relation between deuterium and ^{18}O, is evidence in favor of the absence of isotope equilibrium during the evaporation of meteoric water sampled in regions of the springs and in individual experiments, involving the addition of some chloridized water and vapor.

In the studied geothermal regions, Craig distinguished two different types of vapor by isotopic composition. In all regions, vapor indicates an oxygen shift without any marked deuterium shift relative to meteoric waters. Enrichment of vapor in ^{18}O ranges from 0 to 3‰ in the Wairakei region (New Zealand), where the shift is not detected due to the short residence time of the water. Another type of vapor is found in Yellowstone Park and Steamboat Springs, where deep circulation of water is fixed and vapor has been found to be in isotopic equilibrium with thermal waters. The studied samples from this region show that their isotope content corresponds to vapor-liquid isotope equilibrium in the system at corresponding temperature.

Sakai and Matsubaya (1974; 1977) have used isotope and hydrochemical techniques while studying hydrothermal system in Japan. They distinguished four types of isotopically and chemically different thermal water systems, which are described in Table 9.4.

In waters of the Arima type the presence of a prevailing amount of solvent of meteoric origin has been discovered with the help of isotope data (Fig. 9.5). A considerable oxygen shift is also a feature of these waters. In some brines, the $\delta^{18}O$ values are as much as +8‰, however, one cannot conclude, as the authors reported, that these brines are of magmatic origin, since thermal waters exhibit higher $\delta^{18}O$ values compared with when they were at equilibrium with granite magma. The oxygen isotopic composition of these waters obviously reflects isotopic exchange with carbonate rocks at about 100°C. Sakai and Matsubaya have pointed out that these waters are similar both chemically and in isotopic composition to the formation

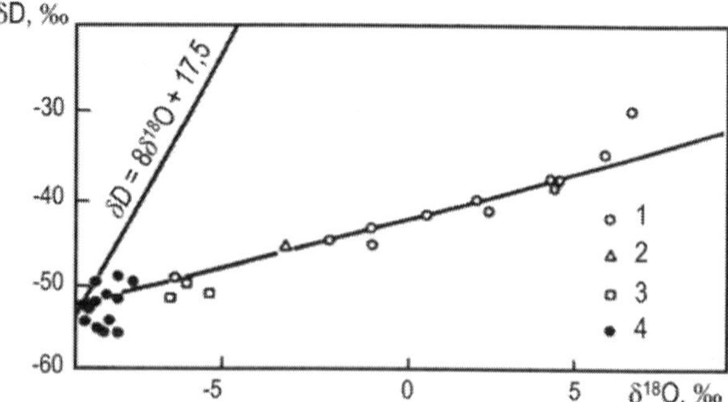

Fig. 9.5 δD and δ^{18}O plot of the thermal and mineral waters at Arima (*1*), Takarazuka (*2*), Ishibotoke (*3*), and for meteoric waters (*4*) of Japan. (After Sakai and Matsubaya 1977)

waters of sedimentary basins in North America, which were studied earlier by Clayton et al. (1966) and Hitchon and Friedman (1969). (The works of these authors were cited earlier).

The water of Green Tuff formation has been found to be mainly of meteogenic origin. Among other places, similar waters were found in mines while passing the thickness formation of the green tuff in the Seikan submarine tunnel were found to be a mixture of marine and local meteoric waters. From Fig. 9.5 one can see that concentrations of deuterium and chlorine ions increase in parallel, which may be explained by the mixing process of the two types of water. Thermal waters of the green tuff type are characterized, in a number of regions, both by the oxygen shift and some deuterium enrichment. On the basis of data obtained for the Seikan tunnel, the authors concluded that, in a number of hydrotherms of the green tuff formation, formation waters are present. These waters are a mixture of marine and local precipitation waters. The increasing deuterium content in waters of this type is in accord with an ordinary process of mixing and the δ^{18}O increase results from the oxygen isotope shift (Fig. 9.5). It follows from this example that the process of isotopic exchange have no influence on heavy hydrogen content in thermal waters of this region but result in considerable enrichment of the thermal waters in ^{18}O.

Coastal thermal waters, as a rule, represent a mixture in certain proportions of marine and meteoric waters. Their chemical composition in a number of gases is markedly metamorphized as a result of the interaction between heated waters and volcanic rocks. In particular, a small amount of magnesium (compared with marine waters), according to Sakai and Matsubaya, may be explained by the following reactions:

The volcanic rocks (glass) $++2Mg^{2+} + H_2O \rightarrow$ Mg–chlorite $+Ca^{2+} + K^+$ $+ H^+ + H_4SiO_4$. The sulfates may be released due to the formation of gypsum:

$$Ca^{+2} + SO^{2-} \rightarrow CaSO_4 \downarrow .$$

As a result of this reaction the waters become of the Cl–Na–Ca type.

The volcanic type of hydrotherm is likely to be formed from meteogenic waters and volcanic vapor mixed in different proportions. This conclusion is illustrated with the data shown in Table 9.5.

The data in Table 9.5 include two models of the formation of thermal waters in the Higashi region (see the footnotes to Table 9.5). Sakai and Mitsubaya prefer model 2,

Table 9.5 Mixing models for acid thermal waters at Higashi. (Satsuma–Iwojima volcanic island in the south of Japanese Archipelago). (From Sakai and Mitsubaya 1977)

Locality	δD (‰)	δ^{18}O (‰)	Cl (mmole/l)	SO_4 (mmole/l)
Meteoric water	−37.0	−6.0	2	1
Sea water	0.0	0.0	535	28.1
Volcanic gas	−25.0	+6.6	370	533
Higashi-1	−32.7	−5.3	52.0	72.0
Higashi-2	−35.0	−2.0	43.7	–
Model 1[a]	−32.7	−5.3	63.9	8.3
Model 2[b]	−35.4	−4.3	52.0	7.8

[a]Model 1 has 88.4% meteoric water +11.6% sea water
[b]Model 2 has 58% meteoric water +42% volcanic gases

but in any case the portion of meteoric waters in thermal volcanic springs is rather large.

The origin of volcanic gases on this island is of specific interest due to their unusual isotopic and chemical composition. The δD values of the fumarole gases are considerably higher than in local groundwaters and precipitation. The oxygen isotopic data is evidence in favor of the isotope equilibrium of the water vapor andesites ($\delta^{18}O = +6.6 \pm 7.0$‰) at magmatic temperatures. This fact, together with temperature of fumaroles (up to 835°C), supports the conclusion that a marked part of these gases is magmatogenic (this word the authors put in quotation marks). However, the mean values of these "magmatic waters" (δD = −25‰) are considerably higher than the δD values observed in waters of the volcanic gases of Surtsey Island (Iceland), in oceanic basalts not far from the Hawaiian Isles, and in hydrogen-bearing minerals originating in the upper mantle, which range from −40 to −60‰ and are typical of the "primary magmatic fluids" (i.e., of juvenile waters). On the other hand, Sakai and Matsubaya have pointed out, the condensates of the volcanic vapor from White Island (New Zealand) during high activity are very similar, in number of parameters, to condensates of the region under study. The observed δD and δ^{18}O values in the New Zealand volcanoes range from −10 to −20‰ and from 0 to +4‰, respectively. In the condensates of these volcanoes, identically to the Japanese fumaroles, the chlorine content reaches the value of 5 g/kg. According to one of the models of Stewert and Hulston (1975), the high δD values in the volcanic vapors of the White Island are related not to discharge of "juvenile" waters but result from the interaction of andesite magmas with marine waters near the surface. This assumption does not contradict the data of Ohmoto and Rye (1974) who studied the isotopic composition of waters of fluid intrusions in pyrite and halkopyrite from a number of deposits in the Kuroko region (Japan). On the basis of variations of δD and δ^{18}O (from −26 to −18‰ and from −1.6 to −0.3‰, respectively), they reported that ore-forming fluids of Kuroko were formed from the waters of marine origin, which is in accord with geological evidence. The admixture of magmatic or meteoric water in these fluids does not exceed 25%.

Therefore, it has been found from the results obtained in the course of groundwater isotope studies in different geothermal regions of the world that meteoric waters are prevalent in the composition of thermal waters such as are found at Larderello (Italy), Wairakei (New Zealand), Hekla (Iceland), Carupano (Venezuela), Yellowstone and Lassen Parks, Steamboat Springs, and Salton Sea (USA), The Kurile–Kamchatka region (Russia), and other regions. This conclusion can be drawn from the following reasoning. If thermal waters contained a considerable amount of juvenile waters, no difference in their isotopic composition would be observed and their composition would be independent of their geographic location. In this case, the difference in the isotopic composition of the studied thermal waters and, moreover, the quantitative correspondence to local surface waters, shows the close genetic relationship between thermal and meteoric waters. On the assumption of the above-mentioned authors, the abundance of juvenile water does not exceed 5%.

As a whole, the isotopic composition of thermal waters, in practically and other case, corresponds to that of surface waters, taking into account the ^{18}O change resulting from the influence of exchange reactions with rocks, and may insignificantly change under the influence of shallow, young intrusive bodies due to an input of the juvenile component.

Finally, it must be noted that the supposed range of deuterium variations in juvenile waters should be narrowed to that in basalts. Moreover, apparently, the isotopic composition of thermal waters is a less reliable criterion of the juvenile component than, for example, distinguishing waters of marine and meteoric origin using isotope composition analysis.

9.2 Isotopic Geothermometers

Base temperature measurement, using ‘isotope’ geothermometers using the temperature dependence of the fractionation factors of carbon, oxygen, and hydrogen stable isotopes between the dissolved gases and water, has been carried out in the Larderello and some other geothermal system by Panichi et al. (1977; 1979; 1983). Practically in all hydrothermal fluids, together with water vapor, gaseous components such as CO_3, CH_4, H_2, H_2S, N_2, etc., are present. In accordance with the Fischer–Tropsch reaction

$$4H_2 + CO_2 \leftrightarrow CH_4 + H_2O,$$

the following isotopic exchange reactions should proceed in hydrothermal systems:

$$^{12}CO_2 + {}^{13}CH_4 \leftrightarrow {}^{12}CH_4, \quad \Delta^{13}(CO_2 - CH_4);$$

$$1/2C^{16}O_2 + H_2{}^{18}O \leftrightarrow 1/2C^{18}O + H_2{}^{16}O, \quad \Delta^{18}(CO_2 - H_2O);$$

$$1/4CD_4 + 1/2H_2O \leftrightarrow 1/4CH_4 + 1/2D_2O, \quad \Delta^{D}(H_2O - CH_4);$$

$$1/2D_2 + 1/4CH_4 \leftrightarrow 1/2H_2 + 1/4CD_4, \quad \Delta^{D}(CH_4 - H_2);$$

$$1/2D_2 + H_2O \leftrightarrow 1/2H_2 + D_2O, \quad \Delta^{D}(H_2O - H_2).$$

Table 9.6 Equations for the base temperature calculation

System	Equation
$\Delta^D(CH_4–H_2)$	$10^3 \ln \alpha. = -90.888 + 181.269(10^6/T^2) - 8.949(10^{12}/T^4)$
$\Delta^D(H_2O–H_2)$	$10^3 \ln \alpha. = -201.6 + 391.5(10^3/T) - 12.9(10^6/T^2)$
$\Delta^{13}(CO_2–CH_4)$	$10^3 \ln \alpha. = -9.01 + 15.301(10^3/T) - 2.361(10^6/T^2)$
$\Delta^{18}(CO_2–H_2O_{vapor})$	$10^3 \ln \alpha. = -10.55 + 9.289(10^3/T) - 2.659(10^6/T^2)$
$\Delta^{18}(CO_2–H_2O_{fluid})$	$10^3 \ln \alpha. = -3.37 + 4.573(10^3/T) - 2.708(10^6/T^2)$

Here Δ denotes the difference in isotopic composition of the components A and B, $\Delta = \delta A - \delta B \approx 10^3 \ln\alpha$ (see Chap. 2).

The time taken to attain isotopic equilibrium ranges from one year to several years, expect for oxygen isotope exchange which proceeds very quickly in the carbon dioxide–water system. Therefore, during the elevation of overheated water up the well or in the course of natural discharge, the isotopic ratios of the components involved in the exchange reactions fix based temperatures in hydrothermal systems even if the water becomes cool at the output of the system.

Panichi et al. (1977; 1979), which calculating base temperatures with the help of data on the equations given in Table 9.6, taken from the works by the authors.

The geothermometers $\Delta^{13}(CO_2–CH_4)$, $\Delta^D(CO_2–CH_4)$ and $\Delta^D(CO_2–H_2O_{vapor})$ have shown the base temperatures of a corresponding hydrothermal system to be equal to 341 ± 37, 314 ± 30, and 254 ± 25°C, respectively. These values surpass the temperature measured directly at the well head and equal to (216 ± 25°C). The geothermometer $\Delta^{18}(H_2O–CO_2)$, by the virtue of its quick attainment of isotopic equilibrium, is indicative of the water temperature at the well head. The geothermometer $\Delta^D(H_2O–CH_4)$ shows barely comparable temperatures resulting from the slow process of isotopic equilibrium in the water–methane system and also from the process of accessory reactions accompanying isotopic exchange.

As mentioned above, Friedman (1953) used the hydrogen-water geothermometer for the estimation of base temperatures for the hydrothermal system drained by the Hurricane Vent Spring (Yellowstone National Park, USA). The base temperature of the system of about 400°C has been evaluated by using the hydrogen isotopic composition of water and molecular hydrogen.

In the Wairakei geothermal region (New Zealand), Hulston (1977) has determined the base temperature to be equal to 260°C, which coincided with the temperature measured directly in the well by using the water-hydrogen isotope thermometer. Similar investigations have been carried out by Arnason (1977a) for geotherms of Iceland.

While studying hydrothermal systems, the geothermometer based on oxygen isotope exchange between water and dissolved sulfate ions is in common use (for more details see Chap. 2). This geothermometer is based on theoretical and experimental studies reported by Lloyd (1968), Mizutani and Rafter (1969), Mizutani (1972), McKenzie and Truesdell (1977), in turn based upon fundamental work by Urey and coworkers (1951). The geothermometer has been used for studying geothermal regions in Wairakei, New Zealand (Rafter and Mizutani 1967), Otake, (Mizutani

1972), Larderello, Italy (Cortecci 1974), Tuscany (Longinelli 1968), Shimogana, Japan (Sakai and Matsubaya 1974; 1977), and other geothermal regions of Japan (Mizutani and Hamasuna 1972).

McKenzie and Truesdell (1977) have used the sulfate–water geothermometers in order to determine base temperatures in the geothermal systems of Yellowstone Park (Wyoming), Long Valley (California), and Ralf River (Idaho). They have used the following dependency of the oxygen isotope fractionation factor on temperature (100–350°C) in the sulfate–water system:

$$10^3 \ln\alpha = 2.88(10^6/T^2) - 4.1,$$

obtained earlier by Mizutani and Rafter (1969) for temperatures ranging from 100 to 200°C. The time required for isotopic equilibration of the system depends on the temperature and pH of the system. It reached at pH = 7 and temperature 300°C equilibrium in 2 years and at 200°C in 18 years.

On the basis of the oxygen isotope analysis of sulfates and water, McKenzie and Truesdell (1977) have determined the base temperatures in a number of springs located in the above-mentioned regions of the USA. In particular, one of the springs in Yellowstone Park with a water temperature of 8°C is hydraulically connected with a hydrothermal system exhibiting a base temperature ranging from 115 to 135°C. As a whole, for most hydrothermal springs with temperatures between 70 and 90°C at the output, the base temperatures of the system, calculated by the sulfate geothermometer, exceeded 200°C.

Chapter 10
Hydrogen and Oxygen Isotopic Composition of Minerals of Magmatic and Metamorphic Rocks and Fluid Inclusions

10.1 Role of Water in Hydrothermal Alteration of the Rocks and Minerals

The results of stable hydrogen and oxygen isotope studies in minerals of magmatogenic rocks and fluid inclusions are in common use in geology mainly when the problem of hydrothermal ore deposition is concerned. Valuable evidence concerning the genesis of ore-forming fluids and mineralization sources of hydrothermal solutions, temperature conditions of the formation of mineral constituents, magmatogenic, metamorphic, and hydrothermally altered rocks and ore minerals may be obtained from isotope investigation (Taylor 1974, 1978; Hall et al. 1974; Rye et al. 1974; Robinson 1974; Sheppard and Taylor 1974; Ohmoto and Rye 1974; White 1974; Javoy 1978; Taylor and Silver 1978; Matsuo et al. 1978; Borshchevsky 1980; Shukoliukov 1980; Vlasova et al. 1978).

Of indubitable scientific and practical interest in this field is the study of water genesis of ore-forming hydrothermal fluids. As White (1974) has pointed out, the most authentic data on the fluid genesis may be obtained by studying the isotopic composition of fluids and accompanying minerals of the ore deposits against a background study of natural waters and, in the first place, of hydrothermal system waters. The data of water isotopic composition provide valuable information on the conditions of ore deposition and the role played by groundwaters of different origin. This question has been studied in a number of works by various authors (see, for example, a special issue of the journal *Economic Geology*, Vol. 69, No. 6, 1974).

White (1974) has distinguished six major types of natural waters involved to some degree in hydrothermal processes: (1) meteoric waters; (2) sea waters of the modern ocean; (3) connate waters of ancient oceans (the waters of sedimentary basins); (4) metamorphic waters associated with rocks in the course of their metamorphism; (5) magmatic waters coming out of magmatic rocks independently of their origin; (6) juvenile waters evolved during degassing of the mantle which were not previously in the hydrosphere.

The possible variations of isotopic composition of the majority of these water types have already been considered. We shall dwell here only on the results of

DOI 10.1007/978-94-007-2856-1_10,

isotopic composition studies of metamorphic, magmatic, and juvenile waters which were obtained by some researchers in their investigations of hydrogen and oxygen isotopic composition of magmatic, metamorphic, and hydrothermally altered rocks of various geologic formations.

A vast data review on hydrogen and oxygen isotope content in magmatic and metamorphic rocks was reported by Taylor in his review paper (Taylor 1974) and is shown in Figs. 10.1 and 10.2, which were composed according to the results of numerous authors.

The calculation of hypothetical isotopic composition of metamorphic and primary magmatic (according to the classification of Sheppard et al. (1969) or juvenile waters is based, as was pointed out earlier, on the usage of data on hydrogen and oxygen isotope content in minerals of metamorphic and mafic rocks, taking into account the experimental evidence of isotopic fractionation factors in the mineral-water system at various temperatures. For qualitative consideration, one can take into account that $\Delta\delta D$ and $\Delta\delta^{18}O$ values for the mineral-water systems are similar to the values of these parameters in the following systems: feldspar-water, biotite (chlorite)-water, muscovite (phlogopite)-water, hornblende-water, and serpentine-water. The estimations of Δ-values at different temperatures are shown in Table 10.1 in accordance with data obtained by Taylor (1974), Suzuoki and Epstein (1976); and Sakai and Tsutsumi (1978).

We shall point out once more that the data presented in Table 10.1 are tentative. These data were obtained by extrapolation in the range of temperatures above 800°C and below 400°C. For the chlorite-water system, only one experimental point has been reported by Suzuoki and Epstein (1976), obtained at 400°C ($\Delta\delta D = -20‰$). Therefore, the values presented in the table for biotite (chlorite)-water system were obtained analogously with the help of the fractionation factor dependence on temperature for other hydroxide-bearing minerals.

Pugin and Khitarov (1978) assumed that during the passage of the mantle ultramafic substance up to the Earth's surface, with the accompanying decrease in temperature and pressure besides the solid phase transformations, the main process is that of serpentinization, resulting in radical changes of its phase composition and the loosening of its substance. There occurs together, not on such a large scale, the metasomatism change of ultramafic rocks which is manifested in the form of amphibolization, mica formation, and talc formation. All these processes are carried out with the participation of water, for example by the following reactions:

$$\begin{aligned} 2\,\mathrm{Fo\,(forsterite)} + 3\,H_2O &= \mathrm{Srp\,(serpentinite)} + \mathrm{B\,(brusite)};\\ 5\,\mathrm{En\,(enstatite)} + H_2O &= \mathrm{T\,(talc)} + \mathrm{Fo\,(forsterite)};\\ 6\,\mathrm{Fo\,(forsterite)} + 9\,H_2O &= 5\,\mathrm{Srp\,(serpentinite)}. \end{aligned}$$

As Pugin and Khitarov (1978) have pointed out, at present there are many reasons to suppose that water involved in the serpentinization of ultramafic substance can be both of juvenile and nonjuvenile character. According to Dmitriev (1973), who studied isotope abundances of sulfur and boron in serpentines sampled from the ocean bottom and the copper content in them, the juvenile waters play the main role in processes of rock serpentinization at great depths. Pugin and Khitarov do not share

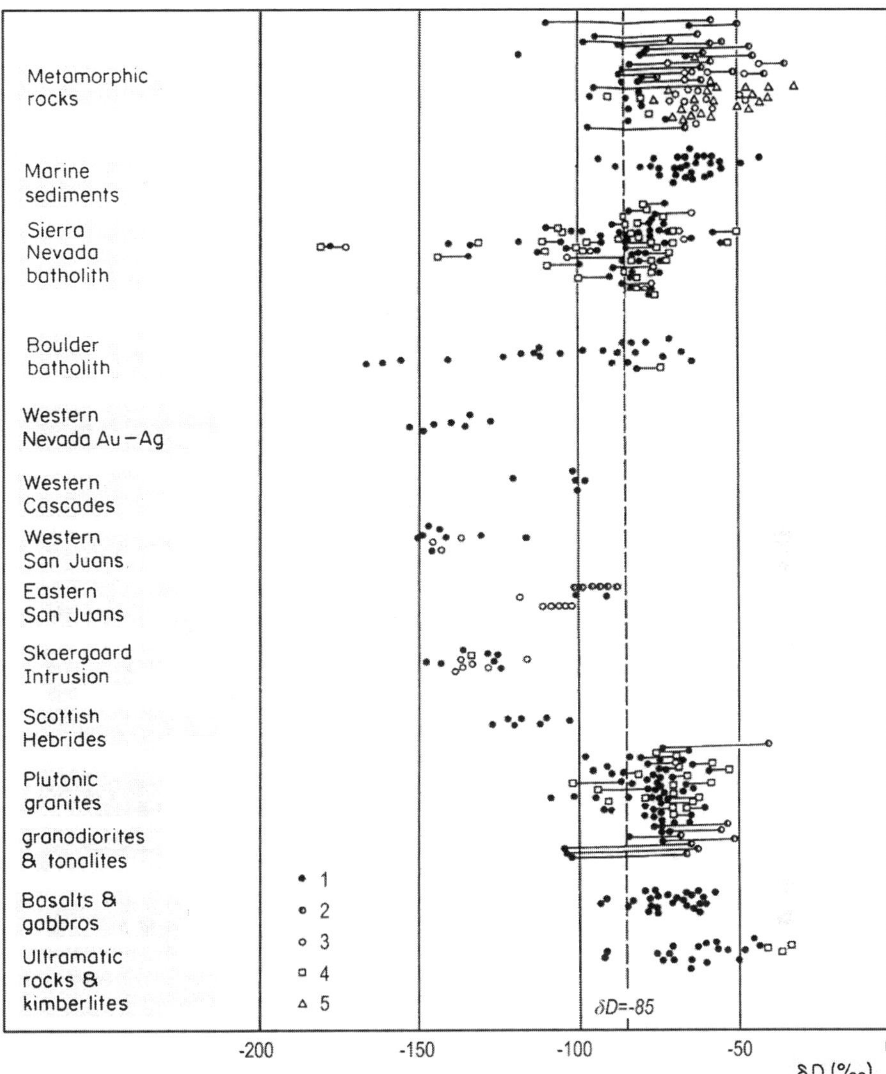

Fig. 10.1 A compilation of δD analyses of minerals from magmatic, metamorphic, and hydrothermally altered rocks from variety of locations and obtained by different authors: (*1*) biotite and whole rock; (*2*) muscovite; (*3*) chlorite or talc; (*4*) hornblende; (*5*) antigorite. (After Taylor 1974)

this view, stating that the amount of water in the mantle's substance does not exceed 0.025–0.1%. The amount of water in serpentinized rocks is dozens of times as great, amounting to several percent. It is difficult to prove that such large-scale processes take place in natural conditions, taking into account juvenile (mantle) waters only, on the basis of these arguments.

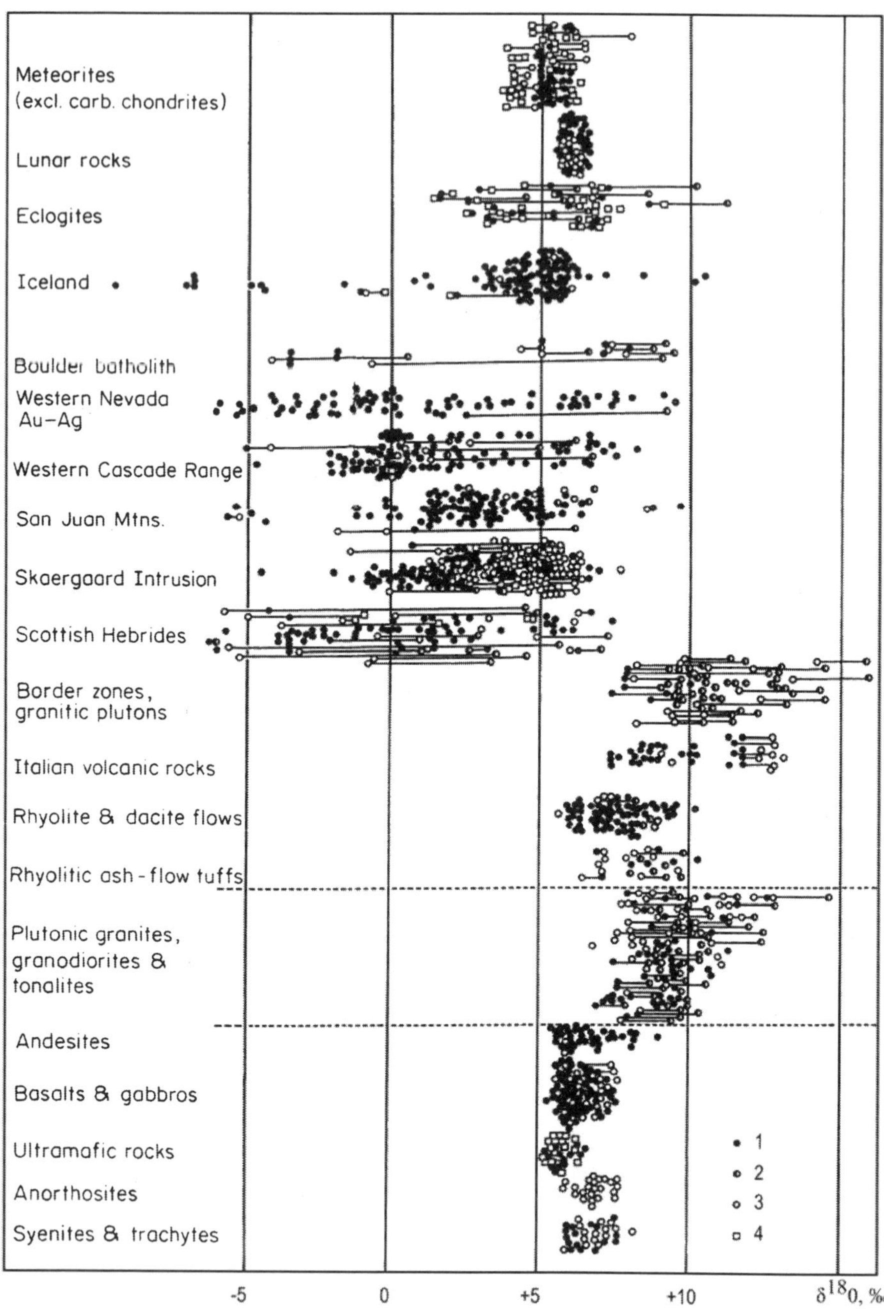

Fig. 10.2 A compilation of ^{18}O analyses of minerals and whole-rock samples of magmatic rocks from variety of locations and obtained by different authors: (*1*) whole rock; (*2*) quartz; (*3*) feldspar; (*4*) pyroxene. (After Taylor 1974)

Table 10.1 Approximate Δ-values for various temperatures. (© IAEA, reproduced with permission of IAEA)

Δ(‰)	Temperature (t°C)					
	1,100	900	700	500	300	200
$\Delta\delta^{18}O_{feldspar-water}$	0	0	+0.5	+0.1	+4.5	+8.5
$\Delta\delta^{18}O_{serpentine-water}$	0 (?)	0 (?)	0–2 (?)	0–2 (?)	0	+1
$\Delta\delta D_{biotite-water}$	−12	−18	−25	−38	−45	−50
$\Delta\delta D_{muscovite-water}$	0	0	−5	−20	−25	−30
$\Delta\delta D_{hornblende-water}$	−5	−10	−19	−32	−40	−45
$\Delta\delta D_{serpentine-water}$	0 (?)	0 (?)	−20	−18	−10	−3

Analysis of the hydrogen and oxygen isotope composition of more than 150 samples of serpentinized rocks taken from the pre-Cambrian formations of various regions of the former Union of Soviet Socialist Republics (U.S.S.R.) have been carried out by Polyakov. The results of these studies show that δD values of the minerals' hydroxide groups vary over a wide range from −50 to −150‰ with an average value of δD = −95‰. Since the processes of serpentinization in nature take place at temperatures of 300–350°C (Pugin and Khitarov 1978), then, taking into account that $\Delta\delta D_{serpentinite-water}(300°C) = -10‰$, it is possible to obtain the δD of water involved in the serpentinization of ultramafic rocks which varies from −40 to −140‰. A wide range of δD variations shows that main role in serpentinization has been played by meteoric waters but not juvenile fluids.

This assumption is confirmed also by the data for oxygen isotope measurements for water extracted at 800°C by thermal destruction of the hydroxide group, which range between −10 and +8‰ with an average value of −2‰. There is reason to believe that serpentinization of rocks occurs at conditions of isotopic equilibrium and that in this case the oxygen isotope composition of the water also reflects the isotopic composition of serpentine. Making such an assumption, it is possible to conclude that either serpentine at isotope equilibrium with forsterite at a temperature of about 500°C or lower has been depleted in oxygen-18 by 7–9‰ compared with forsterite or in the course of serpentinization large amounts of water depleted in ^{18}O have been involved. The latter conclusion seems to be more reasonable. Thus investigations of minerals of the serpentine group cannot give reliable evidence on variations of the isotopic composition of original magmatic or juvenile water.

Taylor discussed the problem of primary magmatic water from a theoretical viewpoint. He pointed out that magma exists over a restricted temperature range from 700 to 1000°C. At these temperatures, the isotopic composition of water released from magma minerals in isotopic equilibrium with it should be practically identical to mantle minerals in oxygen isotope composition and heavier in hydrogen by a value ranging from 0 to 20‰ (on average 10‰). It is known that the majority of plutonic and volcanic rocks have $\delta^{18}O$ ranging from +5.5 to +10‰ (Fig. 10.2) and δD in 95% of cases ranging from −50 to −85‰. At these values of δD and $\delta^{18}O$, taking into account fractionation at $t > 700°C$, the primary magmatic water should have $\delta^{18}O = +5.5$ to $+10‰$ and δD = −40 to −75‰. Taylor with his co-authors, in their earlier work (Sheppard et al. 1969), for the primary magmatic water accepted values of $\delta^{18}O = +7$ to $+9.5$‰ and $\delta D = -50$ to -80‰.

Similar data estimating the hypothetical isotope composition of primary magmatic (juvenile) water have been reported by the other researchers (Epstein and Taylor 1967; Rye 1966; Sheppard and Taylor 1974). White (1974) has pointed out that the δD value for the hypothetical juvenile waters genetically related to the mantle should most likely be close to −50‰. This value of δD is similar, for example, to eaters of the fumarole condensates from the Surtsay volcano, Iceland, for which $\delta D = -53$‰ (Arnason and Sigurgeirsson 1967). For the glassy border of juvenile pillow lavas, emerged at high pressures on the ocean floor, it has been found that $\delta D = -60$‰ (Moore 1970).

According to Sheppard and Epstein (1970), who studied the hydrogen isotopic composition of phlogopites of ancient ultramafic rocks genetically related with the mantle substance, the average value of $\delta D = -58 \pm 18$‰. Taking into account the fractionation factor of deuterium between water and phlogopite at $t = 700°C$ ($\Delta\delta D_{phlogopite-water} = -10$‰ at this temperature according to the author's accounts), the value for juvenile water in contact with rock has been estimated at -40 ± 20‰ ($\delta^{18}O = +7 \pm 2$‰). The hypothetical juvenile water involved in the transformation of ultramafic and mafic rocks of the Kitakami granite massif, northeastern Japan (Kuroda et al. 1974) lie in the range of $\delta D = -29$ to -37‰ as do majority of water in Western Turkmenia, in which the δD and $\delta^{18}O$ values range between −28 and −65‰ and +3.7 and +5.0‰, respectively (Seletsky et al. 1973). The δD values in primary amphiboles of high-temperature peridotites in a number of regions of Morocco and France range from −48 to −88‰, which indicates, according to Javoy (1978), the existence of a deep (juvenile) hydrogen source.

The above-mentioned estimates of the isotopic composition of hydrothermal juvenile (or primary magmatic) water are based on a number of assumptions which do not exclude other considerations. For example, from the viewpoint of plate tectonics one can assume that in the zones of ocean plate movement (zones of oceanic trenches), large amounts of pelagic and other oxygen-bearing minerals which were formed under surface conditions are involved in the mantle. In the course of dehydration of these minerals at high temperatures, water with an isotopic composition ranging from −60 to −80‰ may be released. This possibility is discussed in Chap. 2. It is obvious that a 'juvenile' appearance may be obtained by meteoric waters as a result of isotopic exchange during the metamorphism of rocks. In fact, metamorphic waters are characterized by a wide range of $\delta^{18}O = +4$ to $+25$‰ and a comparatively narrow range of $\delta D = -20$ to -65‰ (Taylor 1974; Magaritz and Taylor 1976a, b).

The hypothetical isotope composition of juvenile waters might possibly be explained from the viewpoint of the condensational theory of the formation of planetary bodies which is developed by the authors of the present book (see Chap. 20). According to this theory, at the stage of 'cold' condensation of siliceous compounds ($T \leq 537°C$), their partial hydration could occur with the formation of a great number of hydroxide-bearing minerals. If condensation occurs at equilibrium conditions between the solid siliceous phase and water vapor, then the siliceous minerals formed in accordance with fractionation factors become enriched in oxygen-18 by 5–10‰ and depleted in deuterium by 25–50‰. Large amounts of captured inertial gases, and in particular 'primary' helium enriched in ^3He isotope, are evidence in favor of a low temperature at the beginning of the condensation of planetary bodies. In the course of the warming of the protoplanetary cloud during its gravitational contraction, the released juvenile water should have an isotopic composition $\delta D = -25$ to -50‰ and $\delta^{18}O = +5$ to $+10$‰. The hypothesis of condensational origin results in initial inhomogeneity of juvenile waters' isotopic composition since, in this case, the processes of isotope exchange between water and condensing minerals occur in the closed system at varying temperatures.

10.2 Meteoric Water in the Processes of Hydrothermal Formation of Minerals

Let us consider now the results of some experimental investigations concerning hydrogen and oxygen isotope studies and liquid-gaseous inclusions which provide valuable information about the evolution of the Earth's core. The majority of investigations on this question, initiated by Friedman (1953), have made it possible to conclude that the main role in the transformation of the upper lithosphere belongs to the hydrosphere's water and a large part of metamorphic and magmatic waters which we are studying now has more than likely passed through a cycle of meteoric circulation.

Friedman and Smith (1958), studying the deuterium content of igneous rocks, found that δD ranges from -80 to -150‰ in obsidians from five geographical regions. The relationship between these concentrations and those of local meteoric water has not been found, whereas in the accompanying perlites this relationship is evident. Therefore, it has been acknowledged that the latter are secondary formations. In the course of biotite and hornblende investigations in rhyolite lavas (New Mexico, USA), the same authors (Friedman et al. 1963) found that the relative deuterium content in 11 samples varied from -33 to -14‰. It has been found that the inverse relationship holds between the amount of water in the sample and the deuterium content. A similar relationship has been obtained by Friedman et al. (1963) under laboratory conditions in the course of heating biotite up to 700°C. It has been stated on this basis that the deuterium content in the investigated samples is a record of the history of rhyolite glass cooling during which iron reacted with water and oxidized and a light isotope of hydrogen had been preferentially lost. Therefore, the less water that remained in the glass, the more it was enriched in deuterium.

Analyzing numerous experimental data on the hydrogen and oxygen isotope composition of granite plutons, Taylor (1978) pointed out the following: The δD values of biotites and hornblende in granite batholites range everywhere between -50 and -80‰. These data are similar to typical δD values detected in metamorphic and marine sedimentary rocks and also in the majority of rocks which are products of weathering in the moderate climate. On this basis, Taylor has stated that primary water involved in the process of granitization of igneous rocks were not juvenile waters. Typical $\delta^{18}O$ values for the majority of granitic rocks range from +7 to +10‰. He concluded that the formation of batholiths was accompanied by the formation of a gigantic meteoric hydrothermal convection system.

Studying different igneous rocks and minerals, Kokubu et al. (1961) found that the relative deuterium content in their water ranges from −25 to −160‰. No correlation has been found between the content of water and deuterium in these specimens. The same authors analyzed several samples of liquid inclusions in basalts from the Japanese Islands, where δD values were found to range from −33 to −60‰. The authors, using results of chemical analyses carried out earlier by other researchers, concluded that these liquid inclusions represent 'juvenile' water. The ^{18}O content was measured simultaneously in the samples. Later on, Craig (1963) showed that the experimental D and ^{18}O values of that water fit the straight line of local meteoric water.

According to Godfrey (1962), the relative deuterium content in amphiboles (35 specimens), biotites (36 specimens), and chlorites (8 specimens) from different regions in the USA range from −203 to −76‰.

It should be pointed out that long before mass spectrometric measurements upon water and minerals, Vernadsky et al. (1941) reported data on deuterium and oxygen-18 content in two water samples from metamorphic rocks of igneous origin using a method based on water density and refraction measurements. It was found that only heavy oxygen is enriched in the water of the rocks, whereas it becomes depleted in deuterium. These data obtained in the 40s of the previous century are in good agreement with later conclusions derived by other researchers and are related to the way in which ^{18}O enrichment may reflect, to some degree, the Earth's magmatic activity.

The authors of this book have measured the deuterium content of two perlite specimens sampled from the Kecheldagsky ore deposits location in the central region of the North Caucasus. The water was extracted under two temperature regimes: up to 600°C (the first stage of burning with extraction of 'perlite water') and in the range 600–1,000°C (second stage of burning with extraction of 'obsidian water'). Relative deuterium content in these specimens was −82 and −88‰, which is in accordance with the data of other researchers obtained for this type of rock.

Interesting investigations were carried out by Hall and Friedman (1963) on the determination of chemical composition and deuterium content in primary liquid inclusions in ore, lead-zinc vein, and fluorite deposits on the Upper Mississippi and Cave-in-Rock (USA). It was found that in liquid inclusions of different minerals (white, yellow, and blue fluorite, quartz, calcite, barite, etc.) of the fluorite Cave-in-Rock deposit, relative deuterium concentrations range from −64.6 to −32.9‰. The ancient sedimentary waters in this region had δD values from −31.7 to −26.6‰ and in one case this value was −147.5‰. The following principle has been obtained: liquid inclusions of more ancient minerals have a higher content of both salts and deuterium compared with the younger ones. A similar picture is also evident for liquid inclusions of minerals (calcite, galenite, sphalerite, etc.) in the lead-zinc deposition in the Upper Mississippi. The δD values there range between −101 for ancient minerals and −25.4‰ for younger ones.

Hydrogen and oxygen isotope techniques, applied to hydrothermal altered rocks and their liquid inclusions, have made it possible to elucidate the role of groundwater of different origin in the formation of ore deposits. In the review papers by Taylor (1974) and White (1974), analysis of isotope data and the role of groundwaters of different type in the metamorphism of rocks and accompanying ore deposition process in hydrothermal deposits of various types are shown. Besides the elucidation of this role of groundwater in ore formation, isotope techniques are useful in the determination of ore formation temperatures as isotope thermometers involving coexisting minerals (see Chap. 9). They are also useful for mean ratio estimations of ore-forming fluids to rocks.

We shall not discuss this problem in detail; only note that the existence of isotope investigations of hydrothermal alteration of ores and the formation of ore deposits of various geological ages makes it possible to conclude that the main role in these processes belongs to groundwaters of meteoric genesis. Thus, Taylor (1974) has pointed out that, from data on hydrogen and oxygen isotopic composition studies, the most widespread types of water in hydrothermal systems of the subsurface layer of the continental crust (3–6 km in thickness) are meteoric and 'connate' formation waters or a mixture of the two. Further, the author has pointed out that though the aim of this work was to prove the importance of magmatic and metamorphic waters, the latter, as Taylor pointed out, are the most important components of hydrothermal solutions, being waters separated from magmatogenic rocks during their evolution. In this case, the terms 'magmatogenic' and 'metamorphic' correspond not only to water circulation in the crust-hydrosphere system. As mentioned earlier, experimental evidence suggests that the water content in the mantle rocks is minimal and insufficient for regional transformations of rocks. This has been proved experimentally, for example, by the laboratory studies of the serpentinization processes of ultramafic rocks (Pugin and Khitarov 1978).

In relation to this problem, White (1974) has also reported that meteoric waters play the dominant role in the majority of hydrothermal systems. He has also added that the present accuracy of measurements and slight variations of isotope composition of the thermal and meteoric waters in each of the studied regions suggest the presence of 5–10% of water of nonmeteoric origin and other isotopic composition in the studied thermal systems. The assumption may be confirmed by a number of experimental results which have appeared in the last few years. For example, Vlasova et al. (1978) examined the origin of hydrothermal solutions associated with trap volcanic activity of the Siberian Platform, on the basis of hydrogen, oxygen, and carbon isotope techniques. They found that δD variations, ranging from −75 to −54‰, in waters in equilibrium with basalts at the temperature of 200–300°C indicate the common meteoric-marine origin of the solutions responsible both for the hydrothermal ore deposition and the chloritization of basalts.

At present, attempts are being made to widen the circle of studied isotope ratios applied for the determination of the genesis of hydrotherms and hydrothermal ore-forming fluids. Considerable success has been achieved in the investigations of isotopic composition of noble gases ($^{3}He/^{4}He$, $^{36}Ar/^{40}Ar$), in hydrothermal liquid inclusions (Naydenov et al. 1978; Kamensky et al. 1976; Fedorov 1999), in strontium isotopes (Faure and Powell 1972; Hedge 1974; Lepin et al. 1975), in strontium and lithium isotopes (Sandimirova et al. 1978; Plyusnin et al. 1978a, b), in silicon (Ustinov et al. 1978), and also in isotopes of other elements such as sulfur, carbon, and lead.(Doe and Stacey 1974; Ohmoto and Rye 1974).

In conclusion, we should like to note that isotope studies of water inclusions in ancient rocks and minerals are of unique interest for understanding the history of water itself and for the elucidation of both the conditions of rock metamorphism and formation of hydrothermal deposits of minerals. Further investigations involving not only stable isotopes of water as the solvent but also dissolved substances will provide new facts which will be of value in solving important scientific and practical problems.

Chapter 11
Other Stable Isotopes in the Hydrosphere

Besides hydrogen and oxygen-18 other stable isotopes, which are present in the hydrosphere, are of use in studying natural waters. They are the isotopes of helium $^{3}He/^{4}He$, boron $^{10}B/^{11}B$, carbon $^{13}C/^{12}C$, nitrogen $^{15}N/^{14}N$, sulfur $^{34}S/^{32}S$, chlorine $^{37}Cl/^{35}Cl$, and argon $^{36}Ar/^{38}Ar$. We discuss here briefly only some aspects of geochemistry and distribution in natural compounds of carbon and sulfur in order to understand dynamics and genesis of natural waters. These isotopes found wide application for solving practical problems in hydrology, hydrogeology, climatology, and oceanography. Information about the other isotopes one can find in the works of Fritz and Fontes (1980), Hoefs (1973), and others.

Carbon and sulfur form a number of mobile, simple compounds in nature (CO, CO_2, CO_3^{2-}, SO_2, SO_4^{2-}, SO_3, H_2S, etc.) which take part in the intensive circulation of substances in the geosphere are in close relationship with water of the hydrosphere. The study of the principles of the distribution of carbon and sulfur isotopes provide important data concerning the origin and sequence of the evolution of the Earth's hydrosphere.

11.1 Stable Isotopes of Carbon

Natural carbon occurs in the form of the two stable isotopes ^{12}C (98.88%) and ^{13}C (1.12%), participating in numerous transition cycles together with radiocarbon. In the course of these transitions carbon isotope fractionation occurs. The carbon isotope composition in different objects depends on their genesis (Craig 1957; Galimov 1968). It is characterized by $\delta^{13}C$, value, defined as

$$\delta^{13}C(‰) = \left[\frac{\left(^{13}C/^{12}C\right)_{sample} - \left(^{13}C/^{12}C\right)_{std}}{\left(\delta^{13}C/^{12}C\right)_{std}}\right] \times 1000\ (‰),$$

where $^{13}C/^{12}C$ is the ratio of the carbon isotopes in the specimen and standard.

V. I. Ferronsky, V. A. Polyakov, *Isotopes of the Earth's Hydrosphere*,
DOI 10.1007/978-94-007-2856-1_11, © Springer Science+Business Media B.V. 2012

The international standard of stable carbon in calcium carbonate is for the belemnite mollusc shell from the Peedee formation (PDB-1 standard). The true ratio of $^{13}C/^{12}C$ is equal to 1.12372 at. % (Craig 1957).

The carbon dioxide exchange between the atmosphere and the oceans involves a number of reactions, resulting in carbon isotope fractionation:

$$^{13}CO_2(r) + {}^{12}CO_2(p) \overset{K_1}{\longleftrightarrow} {}^{12}CO_2(r) + {}^{13}CO_2(p);$$

$$^{13}CO_2(r) + H^{12}CO_3{}^-(p) \overset{K_2}{\longleftrightarrow} {}^{12}CO_2(r) + H^{13}CO_3{}^-(p);$$

$$^{13}CO_2(r) + {}^{12}CO_3{}^{2-} \overset{K_3}{\longleftrightarrow} {}^{12}CO_2(r) + {}^{13}CO_3{}^{2-}(p);$$

$$H^{13}CO_3{}^-(p) + {}^{12}CO_3{}^{2-}(p) \overset{K_4}{\longleftrightarrow} H^{12}CO_3{}^-(p) + {}^{13}CO_3{}^{2-}(p);$$

$$^{13}CO_3{}^{2-}(p) + {}^{12}CO_3{}^{2-}(r) \overset{K_5}{\longleftrightarrow} {}^{12}CO_3{}^{2-}(p) + {}^{13}CO_3{}^{2-}(r).$$

As it was shown in Chap. 2, the processes of the stable isotope fractionation in a system $CO_2(r)$–$CO_2(p)$ can be described by the isotopic exchange equilibrium constant K or by the fractionation factor α.

If the symmetry numbers of the isotopic differences in molecules during the isotope exchange reaction are not changed then $\alpha \equiv K$ (see Chap. 2) and this is true for the above equations. At 25°C $K_1 = 1.009$, $K_2 = 1.0077$, $K_4 = 1.0038$ (Craig 1957). In a number of cases, fractionation of isotopes is convenient to describe by a value of $\varepsilon = \delta_A - \delta_B$, i.e., by the difference in isotopic content between the components A and B, which are in thermo-dynamical equilibrium. The error of the value $\varepsilon \approx (\alpha - 1) \times 1000$ is not worth of 0.1%, for example, the hydrocarbonate, being in isotopic equilibrium at 25°C, is enriched in ^{13}C by about 7.7‰ compared with the gaseous CO_2, i.e., $\varepsilon = 17.7‰$. The ε values for the carbon isotope fractionation between gaseous CO_2 and components of the natural water carbonate system, by the data of different authors, are presented below (Fontes and Garnier 1979; Wallick 1976):

$$\varepsilon_1 = \delta^{13}C_{CO_{2p}} - \delta^{13}C_{CO_{2r}} = -0.5‰;$$

$$\varepsilon_1 = \delta^{13}C_{CO_{2p}} - \delta^{13}C_{CO_{2r}} = -1.18‰ + 4.64 \cdot 10^{-3} t°C;$$

$$\varepsilon_2 = \delta^{13}C_{CO_{2p}} - \delta^{13}C_{CO_{2r}} = 10.2‰ - 0.064 t°C;$$

$$\varepsilon_2 = \delta^{13}C_{CO_{2p}} - \delta^{13}C_{CO_{2r}} = 10.78‰ - 0.11 t°C;$$

$$\varepsilon_2 = \delta^{13}C_{CO_{2p}} - \delta^{13}C_{CO_{2r}} = 9.9‰ - 0.1 t°C;$$

$$\varepsilon_3 = \delta^{13}C_{CO_{2p}} - \delta^{13}C_{CO_{2r}} = 10.3‰ - 0.108 t°C;$$

Table 11.1 Stable carbon isotope variations relative to the PDB-1 standard in various natural objects. (Craig 1953; Galimov 1968)

Object	$\delta^{13}C$ (‰)					
	-80	-60	-40	-20	0	+60
Carbon dioxyde						
Oceanic						
Hot springs						
Atmosphere						
Fresh water						
Wells						
Organic matter						
Bitum						
Coal						
Crude oil						
Carbonates						
Meteorites						
Marine non-organic						
Marine organic						
Carbonatites						
Fresh water						
Hydrothermal						
Gaseous caps (Eocene)						
Sicilian sulphurous						
Limestone						
Saline domes						
Aragonites						
Methane						
Paleozoic						
Mesozoic						
Tertiary (Pliocene, Miocene)						
Quaternary						

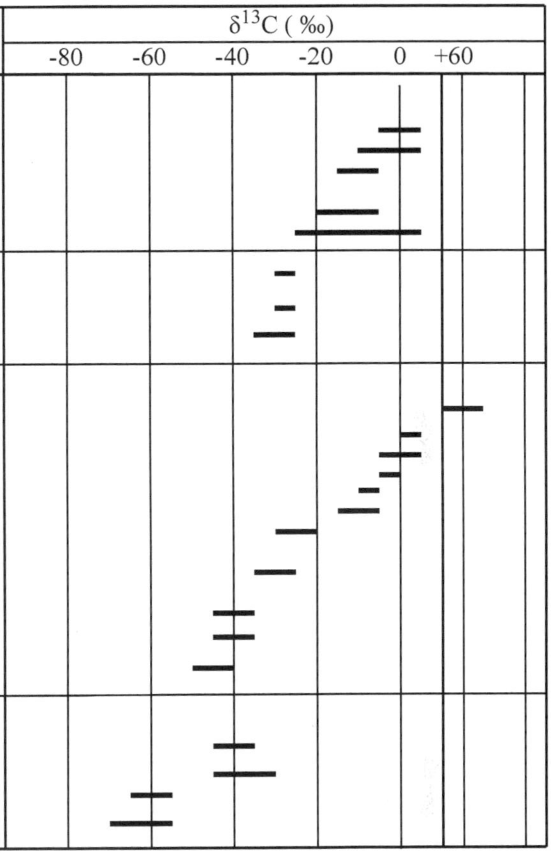

$$\varepsilon_4 = \delta^{13}C_{CO_{2p}} - \delta^{13}C_{CO_{2r}} = 11.2‰ - 0.072t°C;$$

$$\varepsilon_4 = \delta^{13}C_{CO_{2p}} - \delta^{13}C_{CO_{2r}} = 12.38‰ - 0.10t°C;$$

Some differences in ε values, reported by the authors are defined because of the experimental errors and specific conditions of the experiments.

Let us consider distribution in the main Earth's reservoirs of the carbon isotopes and their fractionation in nature (Table 11.1).

It is known that atmospheric CO_2 together with carbon dioxide dissolved in the hydrosphere's reservoir represent one whole physical-chemical volumetric system. The partial pressure of CO_2 in the atmosphere corresponds to the equilibrium state of such a system.

From the other side, atmospheric CO_2 is a course of carbon for the biochemical synthesis. Atmospheric CO_2 represents a common chain of the organic and inorganic cycle of carbon on the Earth. The most important processes of carbon isotope fractionation correspond with those two branches of carbon cycle in the nature. The biological fractionation leads to considerable enrichment of biological substances by light carbon isotopes compared with atmospheric CO_2 and fractionation in the system atmosphere-hydrosphere leads to enrichment of the carbonate-ion by heavy carbon isotopes.

Relative content of ^{13}C in the atmospheric CO_2 as a whole is equal to −7‰. At the same time, marine air masses are less subjected in variation of $\delta^{13}C$ values compared to continental ones. The diurnal, yearly, and other cycles of variation of the isotopic composition are observed. These cycles correlate with the process of photosynthetic activity of the organisms. Mean value of the continental $\delta^{13}C$ value is about −10‰ (Galimov 1968). Because of isotope fractionation, carbon isotopic composition of living organisms depends on the ecologic system and the type of photosynthesis and varies between −7 and −32‰. For the continental plants of the Calvin photosynthetic cycle those value is about −25‰ (Parck and Epstein 1960).

The process of carbon isotope fractionation in the system atmosphere-ocean results in the enrichment of the seawater bicarbonate up to an average value of $\delta^{13}C = -2$‰. Here the value lies within the narrow ranges from −1.3 to −2.9‰ which can be accounted for by the high rate of exchange of atmospheric carbon dioxide with the ocean. Variations of the heavy carbon in the sea sedimentary rocks (limestones) range from +6 to −9‰, but 70–80% of all the carbonates have values ranging from +2 to −2‰ with an average value close to zero. Note that theoretical ideas concerning the above-mentioned processes of carbon isotope fractionation in nature do not contradict the experimental data (Epstein 1969; Galimov 1968).

The fractionation of carbon isotopes is also observed in the course of the decomposition of organic material. Attempts were undertaken to study the variations of the carbon isotope composition in the vertical section of bottom sediments with the simultaneous determination of their age by the radiocarbon method. It has been found that the organic material in sediments is enriched in light isotopes of carbon compared with living organisms (Eckelmann et al. 1962). The accumulation of ^{12}C increases from young to old ooze layers. It is assumed that enrichment in the light carbon isotope occurs as a result of the presence of a great amount of the heavy isotope in carboxyl groups of amino acids, which are less stable and more subjected to decomposition, or due to the preservation of the lipid fraction, which contain an excessive amount of the light carbon isotopes (Epstein 1969).

Besides the above-mentioned carbon isotope fractionation process there are also some other mechanisms of its fractionation in nature, including some high-temperature processes. These mechanisms only have regional importance. Among these processes we shall note only the Fisher–Tropsh reaction (Lancet and Anders 1970), which is as follows:

$$nCO + (2n + 1)\,H_2 \rightarrow C_nH_{2n+2} + nH_2O,$$

$$2nCO + (n + 1)\,H_2 \rightarrow C_nH_{2n+2} + nCO_2.$$

Both oxidized and reduced products are formed by this reaction. The major portion of the oxygen contained in carbon monoxide CO is transformed into H_2O and small proportion (several tenths of a percent) into CO_2. Among the reduced products methane is predominant. The content of hydrocarbons volatile at 400°C amounts to ~30% to the end of the reaction. About 1.5% of the polymerized organic compounds remain on the catalyst at this temperature.

The Fisher–Tropsh process is usually considered as the most probable model of the abiogenic synthesis of organic compounds in nature. It is assumed that this process causes the formation of organic compounds, for example, in the carbonaceous chondrites (Lancet and Anders 1970). It is thought by some researchers that the hydrocarbons and bitumes of the igneous rocks are formed by the Fisher–Tropsh reaction (Galimov 1973).

The question of estimation of the average value of the relative content of heavy carbon in the terrestrial crust as a whole, and the comparison of this value with $\delta^{13}C$ value of the other Earth's shells and meteorites, are of importance.

According to the data of Craig (1953) the $\delta^{13}C$ value in the Earth's crust ranges from −7 to −15.6‰ with an average of −12 ‰. Galimov (1968) has determined a somewhat different range of variations of this value between −3 and −8 ‰ with an average value of −5‰. Epstein (1969), using data on the carbon content in carbonates of the sedimentary rocks, which are the representative reservoir of carbon in the upper sphere of the Earth (73%, $\delta^{13}C = 0$), and data on the carbon content in shales, representing a second considerable reservoir of carbon (26%, $\delta^{13}C = -27‰$), has found that in the Earth's crust the average value of $\delta^{13}C = -7‰$.

Thus, despite of some difference in estimation of the average $\delta^{13}C$ value, obtained by different authors for the Earth's crust, all agree that the Earth's crust is enriched in the heavy carbon isotope compared with meteorite and solar material.

11.2 Stable Isotopes of Sulfur

The main geochemical cycle of sulfur on the Earth is related to the existence of the oceans. Despite the continuous process of sulfur contribution to the ocean together with river runoff its concentration and isotope composition remains constant in the ocean due to sulfur reduction. During the cycle the relatively light continental sulfate ($\delta^{34}S = +5‰$) becomes enriched up to the oceanic average value of $\delta^{34}S = +20‰$ and the excess of this isotope is bonded in diagenetic sulfides.

Sulfur from the oceans is emitted mainly in the form of dimethilsulfide and accounted by the value of 1.5×10^{10} kg of the sulfur per year. In the atmosphere the dimethilsulfide is oxidized to SO_2 and SO_3. A mean value of the $\delta^{34}S$ of these forms is $+15.6 \pm 3.1‰$. Biogenic emission of the sulfur through decomposition of the terrigenic biomass is accounted by the value of about 4×10^{10} kg per year. The volcanic emission gives about 9×10^{10} kg per year ($\delta^{34}S = +5‰$). All the sources provide to the atmosphere about 3×10^{10}kg per year of sulfur with mean value of $\delta^{34}S = +16‰$.

The isotopic composition of sulfur in oceanic sulfate is a sensitive indicator of its dynamic equilibrium in the cycle and constancy of the ocean salt composition on the whole. This equilibrium is determined by the rate of biogenic sulfate reduction and therefore by the total amount of biomass on the Earth. On the other hand, the content of sulfur in the oceans is closely related to its total salinity.

While studying sulfates in the Phanerozoic evaporites, it has been shown (Ault and Kulp 1959; Thod and Monster 1964; Vinogradov 1980) that the sulfur isotopic composition has no time-ordered changes. There occur only slight variations of the $\delta^{34}S$ value from the average value corresponding to the sulfur isotope composition of the present oceans (+20‰). The only exceptions are the Permian evaporites, which always exhibit depletion down to the value of $\delta^{34}S = +10‰$. Vinogradov (1980) has explained this phenomenon in terms of the paleogeographical peculiarities in the accumulation of Permian evaporites. These peculiarities consisted, in his opinion, in an increase in the role of the inland sulfate component in the recharge of salt basins in the transition from the sulfate to the haloid accumulation of salts. On the whole, the isotope composition of sulfur of sulfates in ancient evaporites during the whole Phanerozoic has remained constant, and close to the modern composition of the oceans. This circumstance is evidence in favor of the constancy of the ocean's salinity, amount of the biomass, and concentration of oxygen on the Earth during the whole Phanerozoic.

While studying the Precambrian metamorphic rocks from East Siberia, the Pamirs, and South Africa, which exhibit salt-forming features by a number of minerals (skapolite, apatite, lasurite, carbonates), it has been found (Vinogradov 1980) that these rocks contain relatively high concentrations of sulfide, sulfate, and native sulfur. The isotope composition of sulfur in sulfate of the metasomatic minerals is characterized by the high enrichment in the heavy isotope (from +13 to +45‰) and the identical content of $\delta^{34}S$ relative to sulfates and sulfides of the sulfur-bearing carbonate rocks. On this basis, the author has come to a reasonable conclusion regarding the sedimentary origin of the initial sulfates. These sulfates were accumulated at the steady dynamic cycle of sulfur and served as the initial source of sulfides. In the metasomatic minerals of the Archean carbonate rocks in Aldan (East Siberia) and rocks of the Swaziland system(South Africa) up to 3.5×10^{9}years old, widely developed regional scale sulfates of sedimentary origin were found, with, $\delta^{34}S = +6‰$, which are not typical for sedimentary sulfates.

On the basis of the experimental data analysis, it has been found that the sources of sulfur, participating in the metasomatism and metamorphism processes, were the metamorphic sedimentary thicknesses themselves containing sulfur in the form of sedimentary sulfide. From the existence in the section of sedimentary rocks of layers of dolomites enriched in sulfate-sulfur, it has been concluded that the process of the sedimentation of sulfates has taken place in the normal facial conditions in the course of the salinization of the sea basin. However, the sulfur cycle in the basin has not attained dynamic equilibrium and the sulfate-sulfur in a basin has not yet been enriched in the heavy isotope, but in individual sites of a basin such enrichment of the sulfate-sulfur of metasomatic minerals has occurred and the $\delta^{34}S = +20‰$ has been found there. This has led the author (Vinogradov 1980) to conclude that the establishment of the dynamic equilibrium of the sulfur, at a level which approximates to the modern one, coincides with the age of the studied rocks, i.e., occurred about $(3–3.5) \times 10^9$ years ago. On this basis Vinogradov has concluded that the emergence of the main mass of sulfur from the interior to the upper shell of the Earth, and also the formation of the oxygen composition of the atmosphere and the salinity of the ocean, which approximate to modern levels, ended not later than 3.5×10^9 years.

Let us consider now the comparative analysis of the sulfur isotope composition of the upper shell of the Earth. The balance estimations show (Grinenko and Grinenko 1974) that the major sulfur reservoir in the Earth's crust is the platform sedimentary thickness of the continents, characterized by an average value of $\delta^{34}S$ of about +14‰. In the geosynclinal areas containing up to 30% sulfur the value of $\delta^{34}S$ is close to zero. The ocean water contains 15% sulfur in the form of dissolved sulfates, enriched in the heavy isotope up to 20‰. The sulfur of the ocean sediments, amounting to about 10% of that in the sedimentary thickness, exhibits $\delta^{34}S = +7.7‰$. The average $\delta^{34}S$ value in the abyssal clays, limestone, and siliceous sediments, is equal to amounting to about +17‰. The ultramafic oceanic and inland rocks are characterized by an average value of $\delta^{34}S = +1.2‰$. For basic rocks the value is +2.7‰ and for acid rocks it is equal to +5.1‰. Therefore, the outer sphere of the Earth together with the ocean is enriched in $\delta^{34}S$ by 5.5‰ and the terrestrial crust on the whole by 3‰ relative to meteorite material. It is noteworthy that as one moves from the basaltic sphere toward the granite sphere and further to the sedimentary continental layer and the oceans, the amount of sulfur increases with a simultaneous enrichment in the heavy isotope.

Some researches (Grinenko and Grinenko 1974) have assumed that such global processes as degassing of the mantle, crystalline differentiation, and metamorphism of rocks have the same effect. This effect manifests itself as increases in the amount of sulfur in sedimentary rocks from the Archean to the Proterozoic, Phanerozoic, and Cenozoic with simultaneous increases of the sulfate-sulfur enriched in ^{34}S.

It should be pointed out that such an approach to the observed global principle of increases in the amount of sulfur with enrichment in ^{34}S from the ultramafic rocks to the acid and to the oceans contradicts the evidence given above in favor of the constancy of the salt and isotopic composition of the oceans during the last $(3 – 3.5) \times 10^9$ years. On the other hand, the above-mentioned principle including the relationship between the gross content of the element with its isotopic composition is typical for the other elements stated: oxygen, carbon, and hydrogen. For all these elements we have observed enrichment in heavy isotopes while moving from ultramafic rocks to the acid rocks and to the ocean. The only exception is the oxygen isotopic composition of the ocean. In order to explain this general principle, while considering the isotopic composition of different elements, different researchers employ assumptions which are often contradictory when compared.

In addition to the above-mentioned principles of the distribution of the isotopes of the volatile elements H, O, C, S, already considered, one should bear in mind one more cosmochemical principle. This principle states that the Earth's crust is, in general, enriched in the heavy isotopes of these elements relative to meteorite substances. The available experimental data on boron isotope distribution, despite being limited, show that for isotopes of this element the above-mentioned principles also hold.

It should be pointed out that the data on the isotope distribution of noble gases (He, Ne, Ar, Kr, Xe) in the upper sphere of the Earth and in meteorite substances indicate the applicability of the above principle to this group of elements also.

Part II
Cosmogenic Radioisotopes

Chapter 12
Origin and Production of Cosmogenic Radioisotopes

12.1 Composition of Cosmic Radiation in the Earth's Atmosphere

The Earth's atmosphere is penetrated by a continuous flux of charged particles, constituting of protons and nuclei of various elements of cosmic origin. Consequently, a great variety of radioisotopes, referred to as cosmogenic, are produced due to the interaction of these particles with the atomic nuclei of elements which constitute the atmosphere. Transported by air masses, radioisotopes are abundant over the whole gaseous sphere of the Earth. Being mixed with atmospheric moisture, a proportion falls over the Earth's surface, to enter the hydrological cycle as components of surface waters, soil-ground moisture, and groundwaters. Another proportion becomes a component of ocean and inland basin waters through exchange at the surface of water reservoir. Finally, the Earth's biosphere plays an active role in exchange processes, which are of great importance for some cosmogenic isotopes.

Cosmic dust is another source of cosmogenic isotopes, as are meteorites which are continually falling on the Earth's surface. Having been in cosmic space these meteorites have been subjected to bombardment by cosmic radiation. Nuclear reactions accompanying the process produce many radioisotopes.

Cosmic radiation plays the main role in the origin of cosmogenic radioisotopes. Understanding the nature of cosmic radiation has been important in development of the Earth sciences. In particular, the solution for a number of hydrological and hydrogeological problems, related to natural water dynamics, their genesis, and age, has become possible due to investigations of abundances of cosmogenic radioisotopes in the hydrosphere, i.e., isotopes, produced by cosmic radiation.

Cosmic rays of solar and galactic origin are distinguishable. The nature of nuclear particle flux of solar origin is related to fusion reactions in the solar interior. The origin of galactic and metagalactic cosmic radiation bombarding the Earth's atmosphere is still uncertain.

In the practical studies of cosmic radiation there have been cases when particles of galactic origin with energies of 10^{19}–10^{20} eV[1] have been detected. Some idea of

[1] eV, electron-volt is a unit of energy used in nuclear physics and equal to 1.6×10^{-19} Joules.

V. I. Ferronsky, V. A. Polyakov, *Isotopes of the Earth's Hydrosphere*,

DOI 10.1007/978-94-007-2856-1_12,

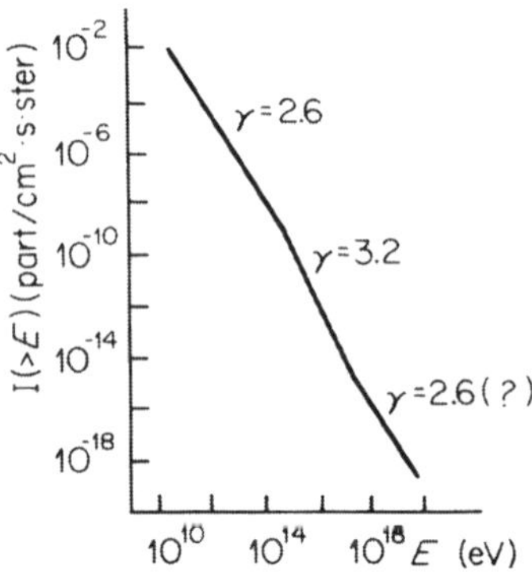

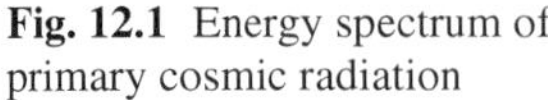
Fig. 12.1 Energy spectrum of primary cosmic radiation

the spectrum of cosmic radiation with energy greater than $E = 10^{10}$ eV is given by Fig. 12.1.

Cosmic radiation is made up of about 90% protons, about 9% helium nuclei (α-particles), and about 1% other nuclei. Table 12.1 shows specific abundances of nuclei in the solar system, cosmic space, and in cosmic radiation (Webber 1967).

Except for hydrogen and helium, the composition of primary cosmic radiation is poorly understood. The estimations of specific abundances of carbon isotopes show that in primary cosmic radiation $^{13}C/^{12}C \approx 1$. Measuring this ratio with the help of photo-emulsion techniques, a value close to unity was obtained.

As for deuterium, no sufficiently reliable measurements of its content in primary cosmic radiation have been carried out so far. Those measurements which were carried out in the upper atmosphere (for atmospheric depths 2–4 g/cm^2) gave values of $^2H/^1H = 0.05$–0.12. The large discrepancy in these data results from the fact that the measurements correspond to different energy intervals and latitudes. However, on the whole, the data are in agreement with satellite data ($^2H/^1H \leq 0.06$ or $\varepsilon = 25$–80 MeV/nucleon), where the effect of the Earth's magnetism and atmosphere are excluded (MeV = 10^6 eV).

The ratio of deuterium abundance to proton abundance in the Universe is of the order of 1.4×10^{-4} (Webber 1967), which may be accounted for by its disintegration. According to spectroscopic measurements, the ratio $^2H/^1H$ is also small in the solar atmosphere, amounting to 4×10^{-5} (Kinman 1956). Only in the atmosphere of magnetic stars does the ratio increase up to 10^{-2}. The ratio of $^2H/^1H = 10^{-5}\rho$, where ρ (particle/cm^3), is the average density of substance through which cosmic radiation has passed (Singer 1958).

While studying the helium isotope composition in cosmic radiation (Appa Rao 1962), the ratio $^3He/(^3He + ^4He) = 0.20$–$0.30$ has been obtained in the energy range of $\varepsilon = 160$–360 MeV/nucleon. Attempts to measure this ratio for higher energies have failed.

Besides nuclei, α-particles, and nuclides of various elements and their isotopes, primary cosmic radiation also contains gamma-ray protons, neutrons, electrons, and

Table 12.1 Nuclei abundances in the solar system, cosmic space, and in primary cosmic rays relative to carbon. (©IAEA, reproduced with permission of IAEA)

Element	Abundances			Element	Abundances		
	Solar system	Cosmic space	Primary cosmic rays		Solar system	Cosmic space	Primary cosmic rays
He	?	400	38	Al	0.004	0.005	0.03
Li	≪0.001	≪0.001	0.27	Si	0.063	0.13	0.11
Be	≪0.001	≪0.001	0.19	P	–	0.002	0.01
B	≪0.001	≪0.001	0.43	$16 \leq Z \leq 19$	0.050	0.02	0.02–0.05
C	1.0	1.0	1.0	Ca	–	≪0.001	0.026
N	0.16	0.27	0.46	Ti	–	0.006	0.017
O	1.7	2.3	0.61	Ci	–	0.06	0.030
F	≪0.001	≪0.001	0.09	Fe	–	0.06	0.080
Ne	?	0.80	0.18	Ni	–	0.008	0.015
Na	0.004	0.006	0.08	$Z>20$	–	≪0.001	0.01
Mg	0.005	0.12	0.15	$Z>30$	–	~ 0.001	<0.004

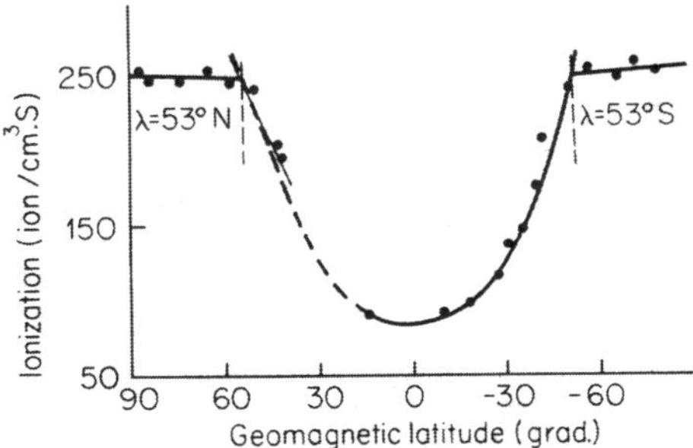

Fig. 12.2 Latitude effect of cosmic radiation for the depth of atmosphere at 50 g/cm^2

positrons. The source of gamma-ray protons and neutrons in primary cosmic radiation is supposed to be related to supernova.

Primary cosmic radiation interacts with the Earth's magnetic field, which results in a relationship between the intensity of radiation and geomagnetic latitude. All available data concerning geomagnetic effects of cosmic radiation are in good agreement with the idea that the isotropic flow of charged particles coming from the Universe is deviated by the terrestrial magnetic field. Analyzing the effect of the Earth's magnetic field on the motion of charged particles, favored and unfavored directions of charged particle motion to a given point of the Earth's surface have been found.

The interaction of the Earth's magnetic field with primary cosmic radiation results in latitude variations of its intensity I (Fig. 12.2). This effect, referred as the latitude effect, is quantitatively expressed by the ratio $[I(90°)–I(0°)]/I(90°)$.

It has been found that the intensity of cosmic radiation coming to the Earth varies with time. Four different types of these time-dependent variations are known (Fireman 1967):

1. Variations related to the 11-year cycle of the solar activity. With increasing intensity of the solar particle flux the galactic radiation intensity decreases.

2. Heliocentric variations of galactic radiation, the radial gradient of which in the range of 1.0–1.5 a.u. of length (1 a.u. $\approx 1.5 \times 10^8$ km), is about +9.6% per 1 a.u.
3. The secular variations of radiation were detected by measuring ^{14}C content variations in the atmosphere through different centuries and recorded in the annual rings of the trees emitting radiations at various ages.
4. The sporadic flows of nuclear particles of low energies emitted during solar flares.

The intensity of the flux of protons of solar origin, with energies greater than 10 MeV near to the Earth, is equal to 100 protons/cm^2 s during a typical solar cycle (relative to the Earth's surface) (Lal et al. 1967). This value was obtained using experimental data on the rate of ^{26}Al production during a time interval of about 10^5 year. The intensity of protons in the solar flux is higher than the galactic intensity by about one order of magnitude, i.e., the latter is characterized by a value of about 10 protons/cm^2 s, but while the solar protons have energies of about several dozens of MeV, the protons of galactic origin have energies two orders greater on average. The intensity of α-particle flux, both of solar and galactic origin, is lower than of protons by one order of magnitude. The velocity of nuclear particles of solar origin near the Earth is about 300 km/s.

12.2 Composition and Steady-state Abundances of Cosmogenic Radioisotopes in the Outer Shells of the Earth

In the course of the interaction of high-energy cosmic radiation with the atmosphere the major part of its energy is absorbed and scattered by the Earth's atmosphere. This leads to the production of secondary low-energy radiation composed of mesons, gamma-ray photons, positrons, and other particles of various energies. This secondary radiation is mainly composed of less energetic protons and neutrons which play the main role in nuclear reactions, resulting in the production of cosmogenic radioisotopes in the Earth's atmosphere. The energy threshold of these reactions ranges between 10 and 40 MeV.

The distribution of secondary neutrons and protons in the atmosphere varies both in latitude and altitude. Figure 12.3a shows the experimental data of energy distribution in the neutron flux, characterized by energies lower than 20 MeV, obtained in 1966 during an experiment in Sicily during a quiet Sun period (Boella et al. 1968). A similar thermal neutron distribution was obtained by Korff using a thermal neutron counter in the Princeton region (USA). The data of these measurements are shown in Fig. 12.3b. One can see from the figure that a total flux of neutrons at first increases with altitude and then decreases with atmospheric density due to their escaping from high atmospheric layers. With the transition in latitude from the equator to the poles the density of neutron flux increases. Figure 12.4 shows the experimental data of latitude dependency of neutron flux at an altitude of 1,000 m, obtained by Simpson (Libby 1967).

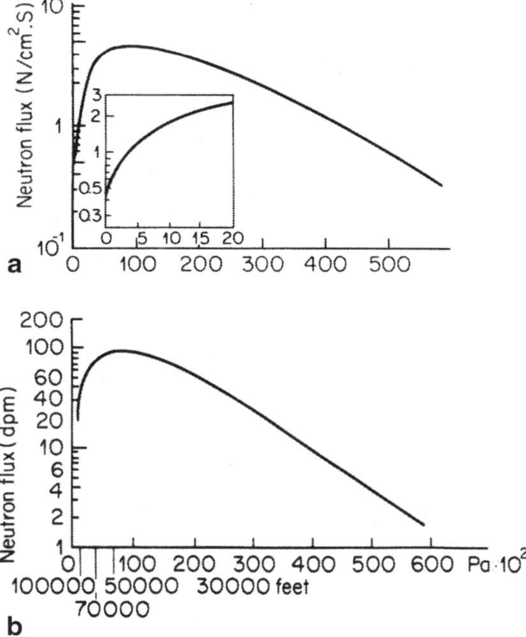

Fig. 12.3 Relation of integrated neutron flux from atmospheric depth at 0–600 mbar in Sicily for neutron energy lower than 20 MeV (**a**) and for thermal neutrons in Princeton region (**b**)

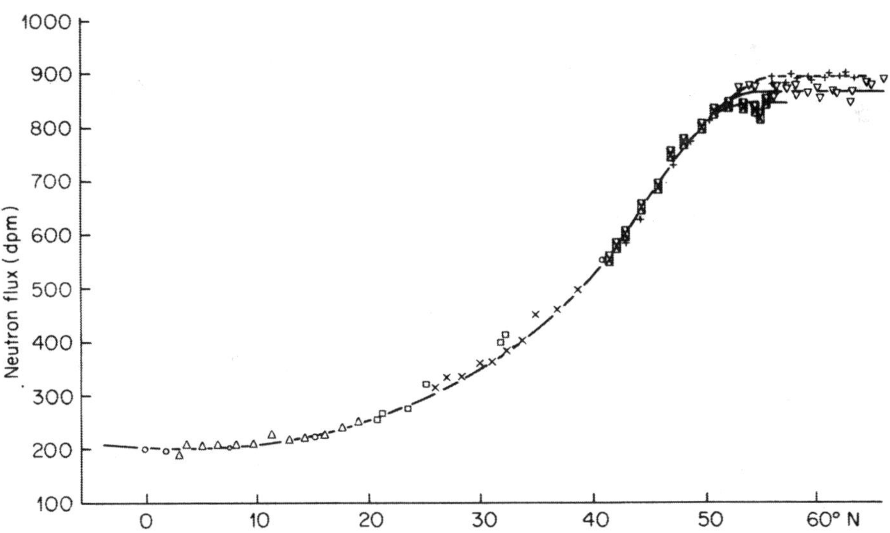

Fig. 12.4 Latitude effect in distribution of neutron flux at an altitude of 1,000 m for the North American continent. (After Libby 1967)

The effect of the production of cosmogenic isotopes in the Earth's crust is negligible even in the upper layers. Nuclear transformations occur here mainly due to the penetration of primary nucleons of light energies, thermal neutrons, and $\overline{\mu}$-mesons. The dominating component of this flux varies with depth. At a depth of several meters most nuclear transformations are provided by quick $\overline{\mu}$-mesons, and at depths where pressure is greater than 400 kg/cm^2, by the interaction with neutrino flux of both primary and secondary origin (Lal and Peters 1967).

According to Lal and Peters, at depths characterized by pressures greater than 0.7 kg/cm^2, the majority of nuclear transformations occur due to interaction with $\overline{\mu}$-mesons and 10% of them are the effect of negative meson capture.

The main proportion of cosmogenic radionuclides is formed in the atmosphere. The main components of atmospheric air are: nitrogen (78.09%), oxygen (20.95%), argon (0.93%), carbon dioxide (0.03%), and neon (0.0018%). In the course of interaction between cosmic radiation and atmospheric elements nuclear transformation occur with nuclei of neon, oxygen and argon, which play a major role in the production of cosmogenic radioisotopes. The most typical reactions are: (n, p), $(n, 2n)$, (n, α), (n, γ), (p, n), $(p, 2n)$, (p, pn), $(p, 2p)$.

Isotopes of cosmogenic origin, which are of interest for the investigations of water circulation patterns in the hydrosphere, are: ^{3}H, ^{3}He (stable), ^{7}Be, ^{10}Be, ^{14}C, ^{22}Na, ^{24}Na, ^{26}Al, ^{32}Si, ^{32}P, ^{33}P, ^{36}Cl, ^{37}Ar, ^{39}Ar, and ^{81}Kr. The irradiation of cosmic dust falling down in amounts of about 10^9 kg/year on the Earth's surface, by the cosmic radiation results in the production of such radioisotopes of great importance as ^{10}B, ^{14}C, ^{22}Na, ^{26}Al, ^{36}Cl, ^{39}Ar, ^{53}Mn, ^{59}Ni.

The rate of production of cosmogenic isotopes varies with altitude and latitude and depends on the intensity of secondary neutron and proton flux in the atmosphere. At the same time the rate of their formation remains constant over time. As a first approximation the ratio of various isotopes produced may be considered to be independent of altitude and time. According to estimations by Lal and Peters (1962) and Young et al. (1970) the rate of isotope production per gram of air, for example, at 46°N, increases exponentially at an altitude of about 21 km is three orders of magnitude higher than at the surface of the Earth. Further on it decreases with altitude due to escaping neutrons from the upper atmospheric layers. The isotope production rate increases by about one order of magnitude with transition from the equator to the poles.

Estimations of the production rate of cosmogenic radioisotopes in the atmosphere and in cosmic dust, carried out by a number of authors (Bhaudari et al. 1969; Craig and Lal 1961; Lal and Paters 1962; Lal and Venkatavaradan 1967; Lal et al. 1967; Schell 1970), used a value of average density of proton flux with energy $E > 10$ MeV at the upper boundary of the Earth's atmosphere, equal to 100 protons/cm^2 s. The latitude and altitude effects of nuclear reactions were taken into account in the course of the calculations.

In order to carry out estimations of steady-state amounts of individual isotopes on the Earth, it was assumed that the flux of cosmic radiation has remained constant over a long period of time (being not less than the half-life of a given radioisotope).

With a constant cosmic flux and a certain atmospheric composition the production rate of isotopes will also be independent of time.

The steady-state amount of an individual isotope on the Earth follows from the equation

$$nS = q\lambda,$$

where n is the production rate of a radioisotope relative to the Earth's surface (atom/cm^2 s); S is the area of the Earth's surface, cm^2; q is the steady-state amount of isotopes or atoms; λ is the decay constant of a given nucleus.

The results of the estimations carried out by the above-mentioned authors and clarified by experimental research, related to the production rates and steady-state abundances of various cosmogenic radioisotopes and also their main physical properties, are presented in Table 12.2.

The total steady-state amount of cosmogenic isotopes on the Earth depends on the balance between their production rate in the atmosphere, accumulation over their life, and reduction due to radioactive decay.

The main proportion of isotopes (except of the noble gases) is oxidized immediately after production. Among these oxides only carbon dioxide and tritium occur in the atmosphere in a free form. The other isotopes are absorbed by aerosols after a short time after of their formation. Radioisotopes, contained in aerosols, are removed from the atmosphere fairly quickly by condensation of moisture in low tropospheric layers, whereas those isotopes which remain constant gaseous components of the atmosphere (CO_2, Ar, Kr) are removed from it far more slowly by molecular exchange at the boundary between atmosphere and oceans.

The ability of cosmogenic isotopes to the study of hydrological and hydrogeological processes is restricted by the condition of correspondence between their lifetimes and the duration of a considered process, and also by the principles of their displacement in geospheres. The major proportion of cosmogenic isotopes ($\sim$70%) is formed in the upper atmospheric layers, about 30% of them in the troposphere. There sequential redistribution is caused by large-scale processes involving the motion of air masses in the troposphere and the precipitation of atmospheric moisture on the Earth's surface, together with cosmic dust or in the form of aerosols.

The absolute amount of an individual isotope, or its ratio relative to another radioactive or stable isotope within a natural reservoir, is used for dating or reconstructing events which have taken place in the past and also for elucidating the nature of prevailing physical, chemical, and biological processes. Resolution of the investigation of these processes depends largely on the information available concerning the source and rate of production of a radionuclide.

As pointed out earlier, the production rate of a cosmogenic isotope in the atmosphere depends on latitude and altitude. Comparison of the expected production rate with those amounts actually observed in the air for isotopes with adequate lifetimes provides a basis for studying the principles of large-scale circulations and tropospheric fallout. The abundance of an isotope in the hydrosphere depends on its lifetime, its biochemical role, and, finally, on the nature of oceanic circulations.

Table 12.2 Physical parameters and steady-state amounts of cosmogenic radioisotopes on the Earth

Isotope	Production reaction	Half-life	Radiation type and energy product (MeV)	Decay product	Production rate (atom/cm^2 s)	Steady-state amount on the Earth (g)
^{3}H	^{14}N(n, ^{14}C)^{3}H ^{16}O(P, ^{14}H) ^{3}H	12.32 year	β−; 0.018	^{3}He	0.25	3,500
^{3}He	N, O(^{3}He)	stable		–	0.2	3.2 × 10^{9}
^{7}Be	^{14}N(n, 3p5n)^{7}Be ^{14}N(p, 4p4n)^{7}Be ^{16}O(n, 5p5n)^{7}Be	53 day	γ; 0.48	^{7}Li	8.1 × 10^{-3}	3.2
^{10}Be	^{14}N(p, 4pn)^{10}Be ^{16}O(p, 5p2n)^{10}Be	1.6 × 10^{6} year	β−; 0.55	^{10}B	4.5 × 10^{-2}	4.8 × 10^{8}
^{14}C	^{14}N(n, p)^{14}C ^{16}O(p, 3p)^{14}C	5,730 year	β−; 0.156	^{14}N	2.5	7.5 × 10^{7}
^{22}Na	^{40}Ar(split)^{22}N	2.6 year	β+; 0.54 γ; 1.28	^{22}Ne	8.6 × 10^{-5}	1.9
^{24}Na	^{40}Ar(split)^{24}N	15 hour	γ; 2.75 β−; 1.4	^{24}Mg	1.2 × 10^{-4}	–
^{26}Al	^{26}Mg(p, n)^{26}Al ^{26}Si(p, 2p)^{26}Al	7.4 × 10^{5} year	β+; 2.77	^{26}Mg	1.4 × 10^{-4}	1.2 × 10^{6}
^{28}Mg	^{40}Ar (split)^{28}Mg	21.3 hour	β−; 0.42	^{28}Al	5.2 × 10^{-5}	–
^{32}Si	^{40}Ar (split)^{32}Si	~450 year	β−; 0.1	^{32}P	1.6 × 10^{-4}	1,400
^{32}P	^{40}Ar (split)^{32}P	14.3 day	β−; 1.7	^{32}S	8.1 × 10^{-4}	0.4
^{33}P	^{40}Ar (split)^{33}P	25 day	β−; 0.25	^{33}S	6.8 × 10^{-4}	0.6
^{35}S	^{40}Ar (split)^{35}S	87.4 day	β−; 1.67	^{35}Cl	1.4 × 10^{-3}	4.5
^{36}Cl	^{40}Ar (split)^{36}Cl	3.0 × 10^{5} year	β−; 0.714	^{36}Ar	1.1 × 10^{-3}	1.5 × 10^{6}
^{37}Ar	^{37}Cl (p, 2p)^{37}Ar ^{40}Ca (n, α)^{37}Ar ^{36}Ar (n, γ)^{37}Ar	35 day	K-capture	^{37}Cl	8.3 × 10^{-4}	–
^{39}Ar	^{40}Ar (n, 2n)^{39}Ar ^{39}K (n, p)^{39}Ar ^{38}Ar (n, γ)^{39}Ar	270 year	β−; 0.565	^{39}K	5.6 × 10^{-3}	–
^{53}Mn	^{53}Fe (p, 2p)^{53}Mn ^{56}Fe (p, α) ^{53}Mn	3.7 × 10^{6} year	K-capture	^{53}Cr	< 10^{-7}	–
^{59}Ni	^{59}Co (p, n)^{59}Ni ^{60}Ni (p, pn)^{59}Ni	7.5 × 10^{4} year	K-capture	^{59}Co	< 10^{-7}	–
^{81}Kr	^{82}Kr (p, n)^{81}Kr	8.1 × 10^{5r} year	K-capture	^{81}Br	1.5 × 10^{-7}–10^{-5}	–

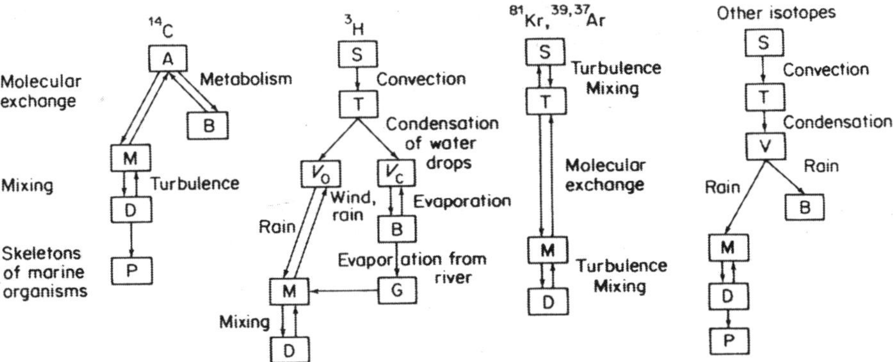

Fig. 12.5 Schematic description and migration of cosmogenic isotopes in the geosphere: (*A*) atmosphere; (*B*) biosphere; (*S*) stratosphere (90–340 g air per 1 cm^2, 1–8 year exchange time); (*T*) troposphere (720–940 g air per 1 cm^2, 30–90 day exchange time); (*V*) water vapor (3–7 g water per 1 cm^2, 4–14 day exchange time); (*M*) mixing layer (75–100 m, 20 year exchange time); (*D*) deep layer (3,500 m, 500–3,000 year exchange time); (*P*) marine sediments; (*G*) soil water. (After Lal and Peters 1962)

Experimental study of abundance of isotopes, characterized by different lifetimes and chemical properties, provide an opportunity for understanding the principles of mass-transfer of substance and, in some cases, helps in understanding the geochemistry of an element's behavior. Isotopes which have got into the biosphere, into oceanic floor sediments, and into some other objects, thus becoming isolated from the transitional dynamics in the cycle (which is characterized by a continuous process of mixing of substance), can be used to determine the time elapsed since the moment they escaped from the cycle.

12.3 Distribution of Cosmogenic Radioisotopes in the Exchange Reservoirs

While studying the nature of geophysical and geochemical processes, which effect the distribution of various isotopes, it is convenient to divide the atmosphere and upper layers of the Earth into a number of zones or reservoirs, characterized by homogeneous pressure, temperature, and character of mass-transfer. A schematic description of isotope migration in natural reservoirs is presented in Fig. 12.5 (Lal and Peters 1962).

The rate of isotope production caused by cosmic radiation in ocean waters is negligible. The detectable concentrations of all the isotopes found in these waters are the result of exchange between the atmosphere and the oceans. The upper layer of the ocean, receiving cosmogenic isotopes from the atmosphere, is characterized by vigorous movements and quick mixing. It is due to these processes that the geographical inhomogeneity of the cosmic radiation is markedly smoothened. Therefore, as

Table 12.3 Steady-state inventories and decay rates of cosmogenic radioisotopes in the exchange reservoirs relative to $1\,cm^2$ of the Earth's surface. (From Lal 1963. ©IAEA, reproduced with permission of IAEA)

Radioisotope	Exchange reservoir				
	Stratosphere	Troposphere	Mixed oceanic layer	Deep oceanic layer	Ocean sediments
^{3}H	6.8×10^{-2}	4×10^{-3}	0.50[a]	0.43	0
^{7}Be	0.60	0.11	0.28	3×10^{-3}	0
^{10}Be	3.7×10^{-7}	2.3×10^{-8}	8×10^{-6}	1.4×10^{-4}	0.999
^{14}C	3×10^{-3}	6×10^{-2}[b]	2.3×10^{-2}	0.91	10^{-2}
^{22}Na	0.25	1.7×10^{-2}	0.62	0.11	0
^{26}Al	1.3×10^{-6}	7.7×10^{-8}	5×10^{-5}	10^{-4}	0.999
^{32}Si	1.9×10^{-3}	1.1×10^{-4}	5×10^{-3}	0.96	4×10^{-2}
^{32}P	0.60	0.24	0.16	1.5×10^{-4}	0
^{33}P	0.64	0.16	0.19	10^{-3}	0
^{35}S	0.57	8×10^{-2}	0.34	6×10^{-3}	0
^{36}Cl	10^{-6}	6×10^{-8}	2×10^{-2}	0.98	0
^{37}Ar	0.63	0.37	0	0	0
^{39}Ar	0.16	0.83	2×10^{-4}	3×10^{-3}	0

[a] The value includes amount presents in biosphere and humus.
[b] Includes amount present in the continental waters.

a first approximation, the source function of the ocean is equal to zero at depth and constant at the surface. This approximation is accurate for ^{14}C dissolved in water and mixed together with water masses. However, there are some other isotopes, such as ^{32}Si and $^{32,33}P$, which are not displaced by water masses. The major portion of them is assimilated by fauna in the upper layers of the ocean. These isotopes enter deep parts of the ocean partly in response to water mixing and partly in the form of organic remains which interact with the environment before sedimentation at the ocean floor. Therefore, the source function of ^{32}Si and $^{32,33}P$ isotopes depends on latitude, longitude, and depth of the ocean (Lal and Peters 1962).

The use of cosmogenic isotopes in the solution of practical problems is more easily realized if the theoretical data related to their abundances and decay rates in the major reservoirs of the Earth, where their migration occurs, are available. For these calculations which were carried out by Lal (1963), the oversimplified box-model (see Fig. 12.5) was assumed. The model consists of six reservoir exchange system: (1) stratosphere ($170\,g/cm^2$), (2) troposphere ($860\,g/cm^2$), (3) continental reservoir (biosphere + surface waters), (4) the upper mixed ocean layer (75 m), (5) deep oceanic layer (3,500 m), and (6) ocean sediments. The calculations were carried out by using the following simplified constants: the average time of air exchange in the stratosphere is equal to 2 years and the average time of water exchange in the deep oceanic layer is equal to 1,000 years (Table 12.3).

The distribution of an isotope in individual geospheres depends both on its half-life and chemical behavior. The usefulness of an isotope in hydrological and hydrogeological process studies depends on its half-life, distribution in exchange reservoirs,

Table 12.4 Specific activity of cosmogenic radioisotopes in the oceans and atmosphere. (From Lal 1963. ©IAEA, reproduced with permission of IAEA)

Radio-isotope	Average specific activity in oceans		Radio-isotope	Average specific activity in atmosphere, disintegration $min^{-1} kg^{-1}$ air	
	Disintegrations min^{-1} t^1 water	Disintegrations min^{-1} t^1 water		Stratosphere	Troposphere
^3H	36	3.3×10^{-4}	^3H	6	7×10^{-2}
^{10}Be	10^{-3}	1.6×10^{-3}	^7Be	17	0.63
^{14}C	260	10	^{22}Na	5×10^{-3}	6.7×10^{-5}
^{26}Al	2×10^{-5}	2×10^{-3}	^{32}P	0.17	1.4×10^{-2}
^{32}Si	2.4×10^{-2}	8×10^{-3}	^{33}P	0.15	7.6×10^{-3}
^{36}Cl	0.55	3×10^{-5}	^{35}S	0.28	7.8×10^{-3}
^{39}Ar	2.9×10^{-3}	5×10^{-3}	^{37}Ar	0.19	2.1×10^{-2}

and the technical ability required to measure the expected activity. The specific activities of a number of cosmogenic radioisotopes in the two principal exchange reservoirs, those of the ocean and atmosphere, were evaluated by Lal (1963) and are presented in Table 12.4.

It follows from Tables 12.3 and 12.4 that the major portion of such isotopes as ^{10}B and ^{26}Al is accumulated in the ocean sediments whereas ^3H, and ^7Be and many other isotopes are absent. The greatest portions of ^7Be, ^{32}P, ^{33}P, ^{35}S, ^{37}Ar and ^{39}Ar are concentrated in the stratosphere and troposphere. The surface ocean layer and inland waters contain the majority of cosmogenic isotopes in significant amounts. Their concentration in ground waters depends on the conditions of interrelation between surface waters and groundwaters. Only those isotopes whose lifetimes are longer than that of infiltrated surface water recharge groundwaters. The varying proportions of concentrations of a given cosmogenic isotope in the water-bearing layer and its steady-state amount in precipitation, surface waters, or in the overlying water-bearing layer, provide a basis for studying processes of water circulation and the hydrochronology of groundwaters.

An additional source of isotopes in the atmosphere is cosmic dust, the terrestrial accretion rate of which is about 10^8–10^9 kg/year. The use of some cosmogenic isotopes, being components of the cosmic dust, provides information on the accretion rate of the Earth in the past. Such isotopes are ^{53}Mn and ^{59}Ni, which are not produced in the Earth's atmosphere, and ^{26}Al, the production rate of which in the atmosphere is less than the contribution due to cosmic dust.

Changes in the isotopic composition of cosmic dust and in the outer shells of meteorites result mainly from bombardment by low-energy cosmic particles of solar origin. The outer shell of a meteorite is usually melted and ablated during movement through the atmosphere. Therefore, while studying the total flux of solar cosmic rays, and particularly low-energy protons, the most convenient object is cosmic dust accumulated in the ocean floor sediments and polar pack ices.

It should be pointed out that during the last two decades the concentration of the steady-state amounts of cosmogenic radioisotopes in nature has been broken due to the additional production of these isotopes during the course of nuclear and thermonuclear tests in the atmosphere. An intense flux of neutrons is produced at the moment of explosion, which interact with the atmospheric constituents and result in the production of identical radioisotopes to those produced by the interaction between the atmosphere and cosmic rays (Styro 1968).

The International Atomic Energy Agency (IAEA Bulletin 1973) has reported that from 1945 to 1973, 936 nuclear tests were carried out, of which 422 took place in the atmosphere. The majority of those tests were before 1963. During the last decade, 43 tests took place in the atmosphere. The most powerful output of bomb radioisotopes in the atmosphere took place during 1958–1959, i.e., related to the most frequent and most powerful thermonuclear tests. The concentrations of some radioisotopes have increased compared with the prethermonuclear steady-state values by one or two orders of magnitude. Thus, the major portion of ^{3}H, ^{14}C, and ^{22}Na present in the atmosphere at present is of bomb origin. Steady-state level of concentration of a number of short-lived isotopes, such as ^{35}S and to a lesser extent ^{7}Be and ^{32}P have been distributed but at present their concentration has returned to normal. The distribution of ^{14}C, and to a great extent ^{36}Cl and ^{81}Kr was also distributed.

The effect of these bombs will be manifested as a distinctive 'mark' for a long time in those reservoirs where the processes of water mixing and dilution are slow, such as in the groundwater reservoir. These marks may serve as a good indicator of groundwater motion and also of the individual water-bearing layers between each other and with surface waters.

Chapter 13
Tritium in Natural Waters

Among the environmental radioisotopes, tritium is the most attractive to those researchers who are studying the principles of water circulation in nature. It is a constituent of water molecules and, therefore, is a perfect water tracer. The interest in application of tritium for hydrological and meteorological purposes was increased greatly during the period of thermonuclear tests in 1953–1962, and also subsequently when a large amount of this artificially produced isotope had been injected into the atmosphere. The bomb-tritium, injected into the atmosphere by installments after each nuclear test, is a kind of fixed time mark of water involved in water cycling.

13.1 Properties of Tritium and Sources of Its Occurrence

Tritium is produced in the atmosphere by the interaction between secondary nuclear particles of cosmogenic origin, mainly neutrons and protons, and nitrogen and oxygen nuclei. Neutrons produced by cosmic radiation originally have energies of about several dozens of MeV. Then, due to inelastic scattering on nitrogen and oxygen nuclei, they are slowed down. At energies greater than 1 MeV the prevailing nuclear reaction is

$$^{14}\mathrm{N} + n + {}^{12}\mathrm{C} + {}^{3}\mathrm{H}.$$

The cross-section of this reaction is about 0.01 barn. Only 3–5% of all the neutrons generated by cosmic rays in the Earth's atmosphere take part in the production of tritium (Table 13.1).

Besides the above reaction, tritium may produced to other reactions, the main ones of which are presented in Table 13.2.

It follows from Table 13.2 that the first two reactions of interactions of ^{14}N with medium energy neutrons, fission of ^{14}N and ^{16}O nuclei by protons at energy higher than 100 MeV have the highest cross-section. The production rate of tritium by protons with energies ranging from 10 to 100 MeV has been estimated to be 0.01 atom/cm^2 s because of an absence of reliable experimental data. The contribution of tritium production by the other components of cosmic radiation is 0.1–0.2

V. I. Ferronsky, V. A. Polyakov, *Isotopes of the Earth's Hydrosphere*,
DOI 10.1007/978-94-007-2856-1_13, © Springer Science+Business Media B.V. 2012

Table 13.1 Main reactions with neutrons in the atmosphere

Reaction	Absolute rate, (neutron/cm^2 s)	Relative rate
Radiocarbon production	4.0	0.56
Tritium production	0.13	0.02
Other reactions	2.2	0.31
Loss	0.8	0.11
Total	7.13	1.0

Table 13.2 Reactions of cosmic ray production of tritium in the atmosphere

Reaction	Energy of particles (MeV)	Cross-section of reaction (mbarn)	Production rate (atom/cm^2 s)
$^{14}N(n, ^3H)\ ^{12}C$	>4.4	11 ± 2	0.1–0.2
$^{16}O(p, ^3H)\ ^{14}O$	>100	25	0.08
$^{14}N(p, ^3H)\ ^{12}N$	–	–	–
$^{16}O(p, ^3H)\ ^{14}O$	10–100	–	0.01
$^{14}N(p, ^3H)\ ^{12}N$	–	–	–
$^{14}N(p, ^3H)\ ^{12}N$	<10	–	0.05
N, O(γ, ^{3}H)	–	–	10^{-5}

atom/cm^2 s. This value is less than that actually observed, which is 0.3 atom/cm^2 s. An additional tritium input to the atmosphere may take place during intense solar flares. It is most probably formed in the course of the reaction ^{4}He $(p, 2p)^3$H in the chromosphere of the Sun.

The steady-state amount of tritium on the Earth, formed by cosmic radiation, varies from 3 to 10 kg. The major part of tritium (~93%) stays in the hydrosphere and only about 7% is in the atmosphere (see Table 12.4). Due to insignificant amounts in natural objects, tritium is commonly expressed in tritium units (TU). A TU corresponds to 1 atom of tritium per 10^{18} atoms of protium, which is equivalent to 7.2 disintegrations per minute per liter of water, or 0.119 Bq/kg. Tritium is a soft β-emitter, characterized by a maximum energy of β-particles equal to 18 keV and half-life of 12.43 years. The final product of tritium decay is the stable isotope of ^{3}He.

Shortly after production tritium is oxidized and forms molecules of water HTO. Since the masses of tritium and protium differ, fractionation occurs during phase transitions of water from gases to solid states and vice verse (Fig. 13.1).

It is for this reason that inhomogeneity is observed in tritium distribution between hydrogen-bearing systems in the tritium–protium exchange reactions. Experiments involving different types of clays (kaolinite, montmorillonite, and silty clays) have shown that in the course of their interaction with water labeled by tritium a marked exchange reaction between tritium and protium is observed. The protium constitutes the clay minerals and hydroxides (Stewart 1965). This effect may be considered as significant in groundwater dating and the large time-scale involved in the investigation of water motion in rocks based on tritium labeling.

Before the first thermonuclear tests in the atmosphere (1952), the majority of the tritium in nature resulted from cosmic ray production. At that time only a few

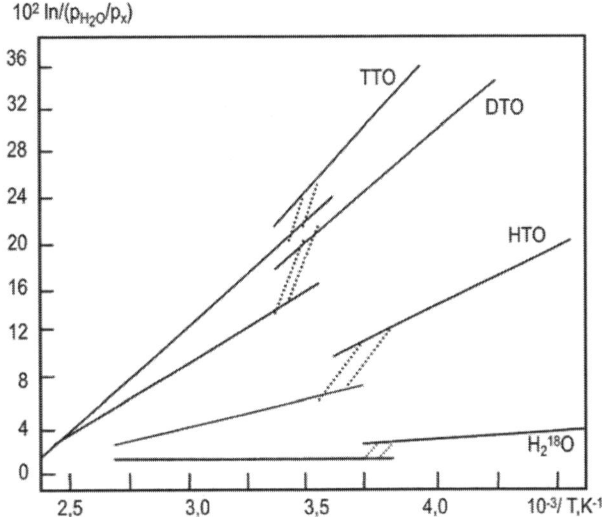

Fig. 13.1 Experimental relationship from vapor pressure of heavy and light water molecules and temperature. (After Van Hook 1968)

measurements of natural tritium on the Earth had been carried out. According to the data of Libby (Kaufmann and Libby 1954), who was the first to study its applicability in hydrology and carried out measurements in Chicago, the average content of environmental tritium in precipitation was about 8 TU. Brown (1961) measured tritium concentrations in the Ottawa Valley (Canada) and found that the mean level of tritium is 15 TU. According to the calculations of Lal and Peters (1962), this value corresponds to 6 TU. Later on, when some principles of distribution and fallout of corresponding amounts of tritium on the Earth were established, it become clear that it's content varies within a large range both in space and time. This range may be estimated as being equal to 0.1–10 TU for both hemispheres at a constant rate of tritium production of about 30 atoms/cm^2 min relative to the terrestrial surface (Suess 1969).

Using data obtained by different authors, Burger (1979) has reported the data of tritium distribution in individual geospheres given in Table 13.3. Estimations were made of tritium fallout on the Earth with the cosmic dust and micrometeorites

Table 13.3 Distribution of tritium in the geospheres. (From Burger 1979. © IAEA, reproduced with permission of IAEA)

Geosphere	HTO		HT		CH_3T	
	Ci	TE	Ci	TE	Ci	TE
Ocean, top 100 m	9×10^8	10–20	(1–10) $\times 10^3$	?	–	–
Troposphere, (3.8 $\times 10^{18}$ kg air)	(2–8) $\times 10^6$	–	(8–18) $\times 10^6$	(3–7) $\times 10^6$	(6–20) $\times 10^6$	$\sim 5 \times 10^4$
Stratosphere, (1.3 $\times 10^{18}$ kg air)	(0.6–5) $\times 10^8$	(2–8) $\times 10^7$	(1–4) $\times 10^4$	–	$<10^5$	$\sim 5 \times 10^4$

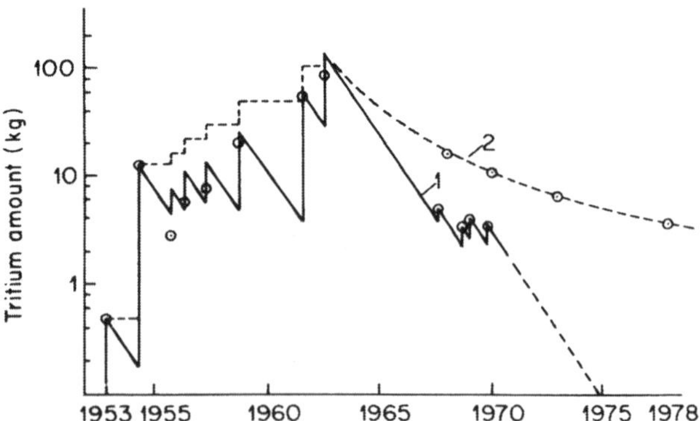

Fig. 13.2 Growth of thermonuclear amounts in the atmosphere taking into account its fallout (*1*) and the stratosphere taking into account its natural decay (*2*). (After Ferronsky et al. 1975. © IAEA, reproduced with permission of IAEA)

(Fireman 1962, 1967). It was shown that in stone meteorites, the tritium activity equals to 200–400 disintegrn/kg min and for iron meteorites this value was 40–90 disintegrn/kg min. Thus, the tritium component contained in meteorites falling down on the terrestrial surface is insignificant and is less than 10^{-5} atoms/cm^2 s.

The results of measurements of tritium concentrations in lunar rocks, carried by spacecrafts Apollo-11 and Apollo-12, gave 270–300 disintegrn/kg min (Bochaler et al. 1971) and appeared to be similar to those in meteorites. The production of tritium in lunar rocks and meteorites occurs due to spallation reactions between cosmic high-energy protons and nuclei or rock-forming elements, such as Fe, Si, Al, etc.

Thermonuclear tests in the atmosphere, carried out since 1952, represent another source of atmospheric tritium. The output of tritium released during a thermonuclear explosion averaged from 0.7 to 5 kg/mt of thermonuclear fission and 0.07 kg/mt of nuclear fission (Miskel 1973). A diagram of tritium injection to the atmosphere from thermonuclear explosions since their beginning is shown in Fig. 13.2 plotted on the basis of data obtained by Eriksson (Schell and Sauzay 1970). Curve 1 accounts for the residence time of tritium in the stratosphere, the main reservoir of accumulation. According to the data obtained for precipitation, this residence time is about 1 year. The residence time in the reservoir is the time required for one half of the tritium present at the beginning to remain in the reservoir. Curve 2 shows the annual variations of the total amount of tritium in the stratosphere.

Up to 1957, the energy of thermonuclear explosions of megaton energy was carried out by the USA in August 1958 at an altitude of 4–7.5 km. In 1957–1958, at a lower altitude, there were 8 mt tests carried out by the UK. Some portion of tritium produced during these explosions has moved in the stratosphere. The most powerful megaton explosions were made in 1961–1962 by the USA and the former USSR in various places of the globe and at high altitudes. As a result, a large amount of tritium (up to 400 kg) has been stored in the stratosphere (Ostlund and Fine 1979), and its concentration in individual places on the Earth (e.g., White Horse, Canada) in the spring–summer months reached 10,000 TE (Thatcher and Payne 1965). It was found later on, after the interdiction of nuclear tests in the three spheres that estimation of the residence time of tritium in the stratosphere, which was accepted to be about 1 year, appeared to be imprecise. Its concentration in precipitation has decreased up to the present time but far more slowly than was assumed theoretically. Thus, the tritium concentrations in precipitation should have reached the natural level by 1970. However, in Western Europe in the summer months of 1968–1970 the levels were still fixed as they had been at the end of 1960, amounting to about 30 TU.

During the last few years up to 1970, some increases in tritium content of the atmosphere were observed due to thermonuclear explosions carried out by France and China. However, the values of these tritium injections are insignificant compared with the previous ones (see in Fig. 13.4 the maxima corresponding to 1967–1970).

Atomic industries (power and research reactors, plants of nuclear fuel reprocessing, etc.) are also sources of environmental tritium. The output of tritium during uranium fission in different types of reactors depends on the choice of fuel, energy spectrum of neutron flux, and a number of technological factors. Depending on the type of reactor tritium is produced in the course of the activation of boron, lithium, and deuterium atoms by neutrons. In the controlled thermonuclear reactor, which is now under construction, tritium will be the main radionuclide. In this case, the major portion of tritium will be ejected from nuclear plants into the environment in the gaseous state (HT, DT, T_2), and partly in the liquid phase in the form of HTO. The gaseous tritium ejected into the atmosphere oxidized quickly and forms water molecules.

According to data obtained by Sehgal and Remport (1971), in the course of uranium and plutonium fission 0.8 atom of tritium is formed per 10^4 acts of ^{235}U fission, 0.9 atom of tritium for that of ^{238}U, and 1.8 atom of tritium for that of ^{239}Pu. Fluss and Dudey (1971) have studied the dependency of tritium production on the energy of neutrons for ^{235}U. According to them, when the energy of neutrons changes from 175 to 630 keV the yield of tritium increases from 2 to 3.4 atom per 10^4 acts of uranium nuclei fission.

In the slow-neutron reactors during ^{235}U fission, the yield of tritium amounts to $8.7 \times 10^{-3}\%$ (Taylor and Peters 1972) and in the fast-neutron reactors it corresponds to $2.2 \times 10^{-2}\%$ (Dudey et al. 1972). This efficiency of yield corresponds to the VVER and RBMK reactor which provides 1.1×10^{-2} Ci/day Mw (t) and for the fast-neutron reactors -2.8×10^{-2} Ci/day Mw (t).

During reactions proceeding in the control rods of the reactors, tritium is ejected in accord with the reactions $^{10}B(\alpha, 2\alpha)^3H$; $^{11}B(n, ^3H)^9B$; $^{10}B(n, \alpha)^7Li$; $^7Li(n, n\alpha)^3H$. The cross-section of these reactions increases with the energy of neutrons. Therefore, the yield of tritium in fast-neutron reactors is considerably higher than in reactors of the other type. Lokante (1971) has reported that the tritium yield corresponding to fission reactions amounts to 11,000 Ci and for boron reactions is about 1,380 Ci in the 3,500 Mw boiling water reactor. In the 300 Mw breeder reactor, the tritium yield is 1,670 Ci for the fission reaction and 3,980 Ci for the boron reaction. The tritium output from the heat-generating elements to the heat carrier depends on the material of the shell. Stainless steel passes up to 60–80% of the produced tritium and zirconium only 0.1% (Lokante 1971).

Table 13.4 Tritium production in various types of reactors. (Bonka 1979. © IAEA, reproduced with permission of IAEA)

Nuclear reaction	Tritium production in Ci/Mw(e) per year					
	BWR[a]	PWR[b]	HWR[c]	AGR[d]	HTR[e]	FBR[f]
Efficiency	0.33	0.33	0.32	0.41	0.41	0.41
Fuel element						
Fission	18	18	20	15	12	30
^{6}Li in fuel (0.05 ppm)	0.3	0.3	0.8	1	0.2	0.1
^{10}B in fuel (0.05 ppm)	4×10^{-5}	4×10^{-5}	3×10^{-5}	5×10^{-5}	1×10^{-5}	2×10^{-5}
^{6}Li in graphite	–	–	–	–	0.5	–
^{9}Be in graphite	–	–	–	–	1×10^{-4}	–
^{10}B in graphite	–	–	–	–	3.5×10^{-3}	–
Coolant						
^{1}H in water	8×10^{-3}	8×10^{-3}	–	–	–	–
^{2}H in water	4×10^{-3}	4×10^{-3}	150	–	–	–
^{10}B in water	–	0.8	–	–	–	–
^{3}He in helium	–	–	–	–	1	–
^{6}Li in sodium	–	–	–	–	–	2
^{9}Be in sodium	–	–	–	–	–	0,01
^{10}B in sodium	–	–	–	–	–	0,01

[a]BWR is boiling-water reactor
[b]PWR is pressurized water reactor
[c]HWR is heavy-water reactor
[d]AGR is advanced gas-cooled reactor
[e]HTR is high-temperature reactor
[f]FBR is sodium-cooled fast breeder reactor

According to Golubev et al. (1979) and Broder et al. (1979), the tritium exhausts from typical VVER-440, VVER-1000, and RBKM-1000 reactors are equal to 0.6, 1.6, and 2.28 Ci/day, respectively. At the Novovoronezhskaya nuclear power station, for example, about 55% of the total amount of tritium is ejected into the atmosphere, 27% into surface waters, and 13% into groundwaters. According to the data for yearly observations, the tritium concentrations at 1 km downstream of the river are higher by one order of magnitude than in water upstream of the river.

As pointed out above, tritium is produced at nuclear power stations (reactors) both due to the process of uranium fission and due to interactions of neutrons of various energies with the constructional materials and coolants. The following substances are some of those used as coolants: light and heavy water, noble gases, melted metallic sodium. The main nuclear reactions, in the course of which tritium is formed, are the reaction of fission of enriched uranium (X) leading to the formation of the fission products $X(n,f)^3H$; $X(n,f)^6He \rightarrow {}^6Li(n,\alpha)\,{}^3H$; ${}^{10}B(n,2\alpha)\,{}^3H$; ${}^2H(n,\gamma)\,{}^3H$; ${}^9Be(n,2\alpha)\,{}^3H$ and so on.

In Table 13.4 data on the tritium yield in different types of reactors due to the above-mentioned reactions are given. In Table 13.5, data on tritium input into the atmosphere

Table 13.5 Tritium emission rates from nuclear power reactors under normal operation and reprocessing plants without tritium retention. (Bonka 1979. © IAEA, reproduced with permission of IAEA)

Nuclear facility		Emission rate (Ci/year)	
		Atmosphere	Surface water
Reactor (1,000 Mw(e))	BWR	30	150
	PWR	20	900
	HTR	10	900
	FBR	100	200
Reprocessing plant (40,000 Mw(e) full load)	BWR and PWR	7×10^{-5}	1,000
	PWR	7×10^{-5}	1,000
	HTR	6×10^{-5}	1,000
	FBR	6×10^{-5}	1,000

and surface waters for various nuclear reactors and nuclear fuel processing plants are presented.

In the report made by the National Council on Radiation Protection and measurements of the USA (Eisenbund et al. 1979) the following data on the tritium yield due to diverse sources are given. The global amount of tritium produced by cosmic radiation equals to 70 MCi (1 kg of tritium is equivalent to about 10 MCi), that is, its production rate is about 4 MCi/year. In 1963, an amount of tritium, estimated at 3,100 MCi, was injected into the atmosphere and hydrosphere as a direct result of nuclear and thermonuclear tests. Thus, the natural level of tritium (70 MCi) may be attained as a result of a decay process up to 2030. The production of tritium in nuclear reactors of the PWR type due to fission reactions ranges from 12 to 20 Ci/day per Mw of the thermal power. The activation of the light elements gives an additional yield of tritium which averages from 600 to 800 Ci/year per Mw of electrical power. These values are equal to 63 Ci/year for the reactors of the BWR and PWR types with light-water cooling. The average residence time of the tritiated water vapor in the troposphere ranges from 21 to 40 days.

The main residence time of HTO molecules in the mixing ocean layer (50–100 m thickness and equal to 75 m on average) is approximately 22 years.

The average time of half-removal of tritium from the human body depends upon individual biological features and is equal to hundreds of days. For two arbitrarily chosen and absolutely healthy men of middle age, the time of half-removal was found to be 340 and 630 days.

Some portion of tritium is injected into the environment from research centers, medical institutions, and industrial plants dealing with works involving the application of artificial tritium. According to the data reported by König (1979), tritium activities of $(1.2–2.1) \times 10^3$ Ci/year have been released into the atmosphere from the Nuclear Research Centre in Karlsruhe (Germany) since 1969. Krejči and Zeller (1979) have reported that a large amount of tritium is ejected into atmosphere from the luminous compound industry, producing tritium, gas-filled light sources and tritium luminous compounds. The urine of the operators working on one of the plants

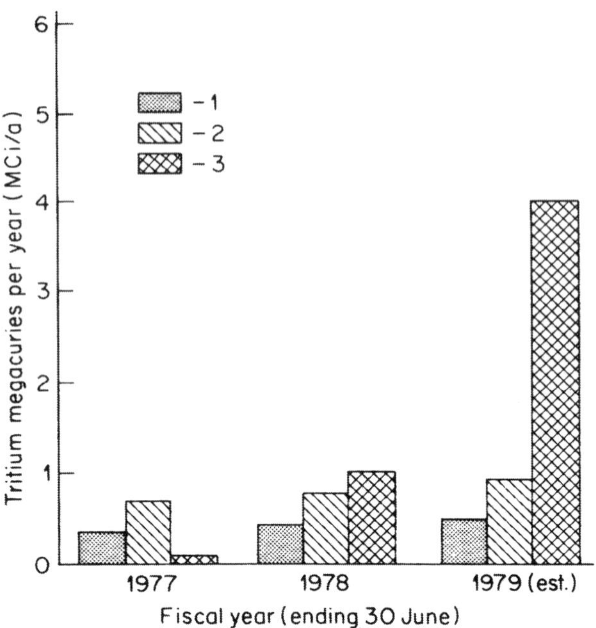

Fig. 13.3 Quantities of tritium produced annually in the USA for digital watch lighting (*1*), by power reactors (*2*), and for other commercial products (*3*). (After Combs and Doda 1979. © IAEA, reproduced with permission of IAEA)

producing luminophor contains about 25 Ci/l of tritium (1 TU = 3.2 pCi/l). In the waste water within the plant area, the concentration of tritium is about 0.3 Ci/l and in water at the exit from the cleaning installations the concentration is 0.004 Ci/l. In precipitation at a distance of 50 m from the ventilation system of the tritium department, the concentration of tritium is 0.1 Ci/l, at a distance of 200 m it amounts to 0.02 Ci/l, and at 2,000 m it is equal to 0.001 Ci/l. A large amount of tritium is now used for the production of liquid crystal displays for digital electronic readouts. The annual production of tritium by different industries in the USA is shown in Fig. 13.3 reported by Combs and Doda (1979). According to their estimations, the amount of tritium used for the production of backlighted digital watches will reach (and it has reached) 4 MCi in 1979.

A considerable amount of tritium in the environment originates from the nuclear fuel reprocessing industry. Daly et al. (1968) have shown that the nuclear fuel reprocessing plant situated in New York State ejects about 200 Ci of tritium per day, 25% of which is released into atmosphere, 65% is contributed in liquid form to the river, and 10% goes into the soil.

Taking into account the modern trend of development of nuclear power stations in most countries of the world, it is easy to estimate that by the beginning of new twenty-first century the production of technogenic tritium, which will be continuously ejected into the environment, will overcome the amount of cosmogenic tritium produced in the atmosphere.

In fact, the tritium production rate from all the nuclear plants (power reactors and nuclear fuel reprocessing plants) of the world at 2000 is overcome by four times the rate of its natural production by cosmic radiation. However, the release of tritium into the environment is negligible since its major portion is collected and buried as radioactive waste. Besides, the tritium produced by nuclear plants cannot reach the stratosphere, where it would be subjected to global redistribution. Therefore, its ejection into the precipitation, surface, and groundwaters is of a local character, related to the neighborhood of the organization of studies of surface and groundwaters involving tritium measurements.

According to Katrich (1990), some amount of tritium was injected into the environment as the result of Chernobyl accident. Concentration of tritium over the European part of Russia in May 1986 increased by 2–5 times compared with May 1985, however, in June–July, the level of contamination was dropped to the normal because the tritium was reached only troposphere.

At least some portion of tritium will be released into the atmosphere due to nuclear explosions used for peaceful purposes (such as the performance of underground oil, gas, and water capacities, excavation of rocks in the course of construction works, etc.). However, the major portion of tritium precipitating in the hydrological cycle in the near future will be, as previously, the bomb-tritium released in the period from 1952 to 1962.

13.2 Global Circulation of Tritium Water

At present the total amount of tritium on the Earth exceeds its prebomb level only by 1.5–2 times. This situation is explained by continuous decay of tritium and the isotopic exchange with the ocean's waters.

Figure 13.4 demonstrates general scheme of the global circulation of tritium water in nature. From the atmosphere, which is the only source of the natural and thermonuclear tritium, the tritium water molecules together with the air flows enter the troposphere where they form isotopic composition of the tropospheric moisture. The other effects which determine tritium content of the tropospheric moisture is the evaporation from the ocean surface and the molecular exchange between the surface ocean layer and the atmospheric moisture. The continental atmospheric precipitation of high tritium content forms river run-off, lake and groundwaters, and also glaciers. Some part of precipitation is evaporated. Concentration of tritium in the river water is close to that in precipitation and in the lakes it depends on residence time of water: the longer residence time, the less content of tritium because of its decay. The same relates to the groundwater basins. In glaciers, especially in polar latitudes (Greenland, Antarctic), water looks like conserved and in the deep layers tritium is completely decayed. The nondecayed part of the HTO molecules from the rivers, lakes and groundwater comes to the oceans together with the surface and underground run-off.

The oceans' waters are divided in two layers: (1) the upper, well mixed with depth of several hundred meters and (2) lower, divided from the upper by the thermocline,

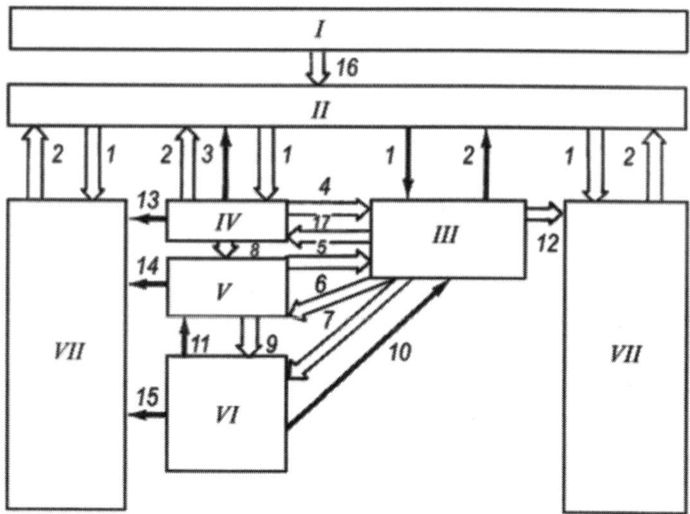

Fig. 13.4 Scheme of tritium cycling in hydrosphere's reservoirs: (*I*) stratosphere; (*II*) troposphere moisture; (*III*) continental surface water; (*IV*) soil moisture; (*V*) shallow groundwater; (*VI*) deep groundwater; (*VII*) ocean water; (*1*) atmospheric precipitation; (*2*) evaporation and molecular exchange; (*3*) transpiration; (*4*) soil run-off discharge; (*5*) underground run-off discharge; (*6*), (*7*) surface run-off discharge; (*8*) infiltration of soil water; (*9*) infiltration of groundwater; (*10*) discharge of deep groundwater to surface water; (*11*) discharge of deep groundwater to groundwater; (*12*) surface run-off to the oceans; (*13*) soil water run-off to the oceans; (*14*), (*15*) submarine discharge to groundwater (*16*) output of tritium from stratosphere; (*17*) discharge of surface water to the soil layer

with water exchange time of several hundreds and even thousand years. The thermocline may absent in the Polar regions and in this case the most favorable conditions for vertical water mixing appear. In the upper layer, maximum concentrations of tritium in waters are observed, which at immersion to the depth and mixture long time lose tritium at its decay.

Thus, the stratosphere is the source of tritium in the hydrologic cycle and the deep ocean waters and glaciers and the deep groundwaters are the reservoir of tritium run-off where it decays. It follows from here that the definite regularities should be expected in distribution of tritium in all chains of the hydrosphere.

13.2.1 Tritium in Atmospheric Hydrogen and Methane

Except the water, which is the main object for the study of tritium distribution on the Earth, there are two other hydrogen-bearing compounds: (1) The molecular hydrogen (H_2) and (2) Methane (CH_4). The study of their behavior is important for the understanding of geophysical and physical–chemical processes in the atmosphere. Of a special interest are H_2O, H_2, and CH_4 cycles, which have close relationship in the atmosphere. Their passage from one form to another may be used as the tracer for determining the residence time of hydrogen stay in its compounds, for estimation of the exchange rate between the hemispheres, for study of the air exchange between the troposphere and stratosphere, and for understanding the nature of the compounds origin.

At present the concentration of H_2 in the nonindustrial regions amounts to 0.575 ppmv in the northern hemisphere and 0.550 ppmv in the southern hemisphere (Schmidt 1974). On a global scale, 50% of the H_2 is of anthropogenic origin. The most significant natural source of H_2 in the biochemical process occurring in the ocean, where waters are enriched with hydrogen by a factor of three. The process of water molecule dissociation and hydrogen photosynthesis occurring in the atmosphere (Romanov and Kikichev 1979) also contribute significantly to H_2 concentration.

The majority of tritium in atmospheric hydrogen is of cosmogenic origin. The principal reaction leading to the production of HT molecules, according to the estimations made by Harteck (1954) for tritium generated by cosmic rays, is the recurrent photodissociation of TO_2 and the subsequent exchange reaction of the form

$$\mathrm{T} + \mathrm{H_2} \rightarrow \mathrm{HT} + \mathrm{H}.$$

Only 0.1% of the tritium produced by cosmic rays exists in the form of the HT molecules and 99.9% is in the form of the HTO molecules. The bulk of the HT molecules are formed at altitudes ranging from 10 to 40 km. The total mass of the HT molecules of cosmogenic origin in the atmosphere is about 5 g (Rowland 1969).

Besides the release of tritium from the device itself during thermonuclear explosions, tritium is produced according to the following main reaction:

$$\mathrm{D} + \mathrm{D} \rightarrow \mathrm{T} + \mathrm{H} + 4\mathrm{MeV}.$$

It is considered that approximately the same relative content of the bomb-tritium is contained both in the molecules of H_2O and H_2, which is the result of isotopic exchange between H_2O and H_2 in the expanding and cooling thermonuclear sphere. In this case, the main reactions are:

$$\begin{aligned}
\mathrm{T} + \mathrm{O_2} &\rightarrow \mathrm{TO_2}, \\
\mathrm{TO_2} + \mathrm{O} &\rightarrow \mathrm{TO} + \mathrm{O_2} + 61\mathrm{kcal}, \\
\mathrm{TO} + \mathrm{H_2} &\rightarrow \mathrm{HTO} + \mathrm{H} + 17\mathrm{kcal}, \\
\mathrm{HTO} + \mathrm{H_2} &\leftrightarrow \mathrm{H_2O} + \mathrm{HT}.
\end{aligned}$$

In the course of underground thermonuclear tests, the increase of tritium content in atmospheric hydrogen has not been accompanied by an increase of tritium content in the atmospheric moisture. This is likely to be due to the lack of conditions necessary for oxidizing reactions in the medium where explosion takes place.

The main sources of technogenic tritium, as pointed out earlier, are nuclear power plants, which release a considerable amount of tritium in the form of HT molecules.

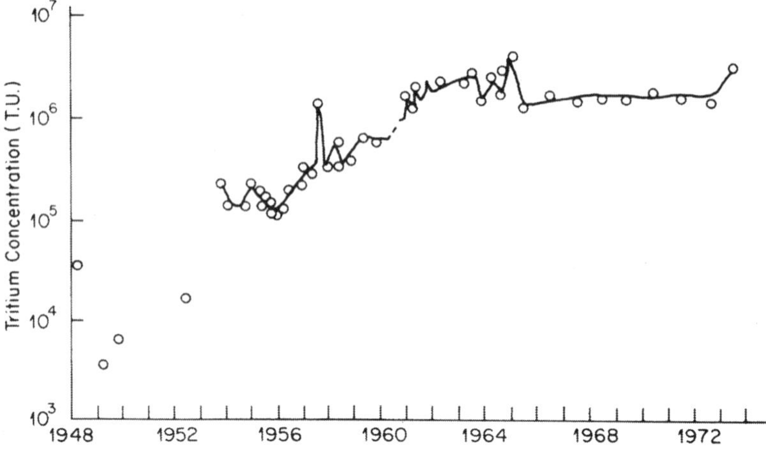

Fig. 13.5 Tritium concentration increase in atmospheric hydrogen from 1948 to 1973 over continents. (After Ehhalt 1966; Martin and Harteck 1974; Östlund and Mason 1974)

Tritium was first measured in atmospheric hydrogen in 1948 near Hamburg, where its concentration was found to be equal to 4×10^3 TU (Faltings and Harteck 1950). Later on, due to thermonuclear tests, the concentration of tritium in atmospheric hydrogen sharply increased (Fig. 13.5). It can be seen from the figure that the concentration of tritium increased from 4×10^3 to 4×10^6 TU from 1948 to 1973 (Ehhalt 1966; Martin and Harteck 1974; Östlund and Mason 1974). In the tropospheric HT, its concentration reached maximum values with a delay of about 2–2.5 years. Ehhalt (1966) has assumed this to be the effect of removal of HT molecules from the stratosphere, which is the main reservoir of tritium accumulation during thermonuclear explosions, into the troposphere after a certain long time, corresponding to that which was observed.

After thermonuclear tests were stopped in the three media, tritium content in the H_2 remained approximately unchanged from 1963 to 1973 at the level of $(2–4) \times 10^6$ TE. The constancy of the HT concentration in the atmosphere can only be explained by ejection of tritium into the atmosphere from some sources in order to maintain the corresponding partial pressures of the HT molecules and compensate the natural losses caused by radioactive decay and other processes of removal.

According to existing estimations (Martin and Hackett 1974), the total anthropogenic release of tritium into the atmosphere should amount to 1.2×10^6 Ci/year in order to maintain the average concentrations of tritium in atmospheric hydrogen at the level of about 80 atom/mg of air. These sources of anthropogenic tritium release are likely to be atomic industrial plants and underground tests.

In contrast to the distribution of the HTO, the spatial distribution of tritium in atmospheric H_2 is characterized by a high homogeneity in the whole atmosphere. According to the data of Östlund and Mason (1974), the concentration of tritium was about 50 atom/mg of air in 1971–1972. Only at high latitudes (60°N and higher)

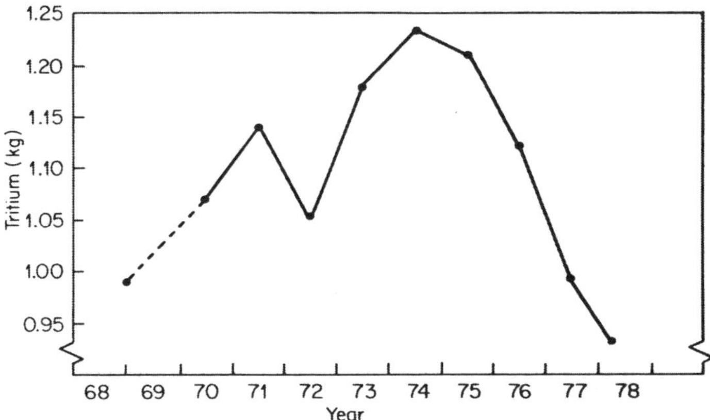

Fig. 13.6 Inventory of the global atmospheric HT for 1968–1978. (After Mason and Östlund 1979. © IAEA, reproduced with permission of IAEA)

does the concentration of tritium increase up to 80 atom/mg of air. According to the limiting data, tritium concentration in the stratosphere decreases with altitude following the barometric law. Thus, in North Alaska at 70–75°N the tritium content in the lower stratosphere varies with altitude from 80 to 30 atom/mg of air (Östlund and Mason 1974). The exceptions to these principles are the regions of anomalous release of H_2 of industrial origin with zero concentration of tritium, and also the regions where nuclear industries are located, characterized by raised tritium content. The global content of HT molecules in the atmosphere during the period 1968–1978 is shown in Fig. 13.6 (Mason and Östlund 1979).

According to the data of Ehhalt (1974), the CH_4 early production is accounted by $(5.4–10.6) \times 10^{14}$ g, from which 80% have biogenic origin. Tritium concentration there is more than 10^4 TU (Ehhalt 1974). Creation of CH_3T molecules in the atmosphere occurs as a result of nuclear and exchange reactions between HT and CH_4. However, as investigations show, these reactions have small efficiency (Begemann and Friedman 1968). It is assumed that the main sources of tritium in CH_4 are research laboratories and institutions of atomic industry, technology of which relates to tritium (Burger 1979).

Molecules CH_3T of biochemical reactions have the same T/H ratio as the environmental water. It is obvious that atmospheric HT takes part in biogenic CH_4. As a result of this process, tritium content in the CH_4 is correspondingly increases. The exchange time of CH_4 in the atmosphere is to about 4–7 years.

13.2.2 Tritium in Atmospheric Water Vapor

As pointed out above, the upper layer of the atmosphere, its stratosphere (15–17 km), is the reservoir where the bulk of natural tritium is accumulated. Despite the small amount of stratospheric moisture is the main source of the tropospheric tritiated water falling to the Earth's surface as precipitation. It will be shown that the stratosphere is also a reservoir accumulating thermonuclear tritium.

Martell (1963), using a supposed production rate of natural tritium of 0.3 atom/cm^2 s, obtained a value of tritium concentration of about 10^6 TU. The first measurements of stratospheric tritium in the air above Minneapolis in 1955–1958 at an altitude of 14–28 km carried out by Hagemann et al. (1959), had shown that the tritium content was equal to 1.1×10^6–1.52×10^7 atom/g of air. On the basis of these measurements and on the ratio of T/C^{14}, it was found that the amount of tritium in the stratosphere is equal to 6×10^{23} atoms (6 kg). Later, Scholz et al. (1970) obtained a tritium concentration of 2.2×10^6–8×10^7 TU using their own experimental data.

The most complete studies of tritium distribution in the troposphere were carried out by Ehhalt (1971). Measurements were conducted from November 1965 to January 1967 at continental (Scottsbluff, Nebraska) and oceanic (near California) stations up to an altitude of 9.2 km. The results of Ehhalt's measurements have shown that concentrations in water vapor increase with altitude. The lowest concentrations were found at an altitude of 2,300 m above the sea level in the spring–summer seasons (1,200 TU) and maximum values at an altitude of 9 km (26,000 TU). It has been found that the altitude of the seasonal variations at amplitude of 7.5–9 km is greater by a factor of 10 than the variation of tritium concentrations at the Earth's surface.

Detailed data concerning the distribution of HTO molecules with height were reported by Mason and Östlund (1979). Water and hydrogen samples were taken with the help of a molecular trap, placed on a special aircraft, up to a height of 13 km. The distribution of HTO in the troposphere and stratosphere during the flights over Boulder, Colorado, is presented in Fig. 13.7. Figure 13.8 shows the distribution of the HTO in the stratosphere below 19.2 km (T-atoms/mg air), discovered during flights in August 1976 and July 1977. Differences in the HTO distribution both with altitude and latitude were observed between 1976 and 1977, caused by an atmospheric thermonuclear test conducted by the People's Republic of China on 17 November 1976, which resulted in the release of a large amount of tritium into the stratosphere. According to the estimations of Mason and Östlund, the inventories of atmospheric tritium were about 1 kg of the HT molecules and 5.3 kg of the HTO at the end of 1977. The major portion of the HTO (about 5.1 kg) has been stored in the stratosphere and 0.2 kg has been in transit to the ocean surface through the troposphere.

Bradley and Stout (1970) carried out individual measurements in order to obtain tritium distribution profiles in atmospheric moisture in Illinois State (USA) up to the amplitude of 5 km. They obtained three different types of distribution of tritium with altitude. The first type is characterized by an increase of tritium concentration with altitude, the second type by a constancy of tritium concentration, and the third type by a decrease of tritium concentration with altitude up to 2.5 km and then by a subsequent increase in concentration. These tritium distributions were explained by Bradley and Stout as the result of different conditions of formation and mixing of atmospheric moisture in the lower troposphere and also in terms of different sources of tritium.

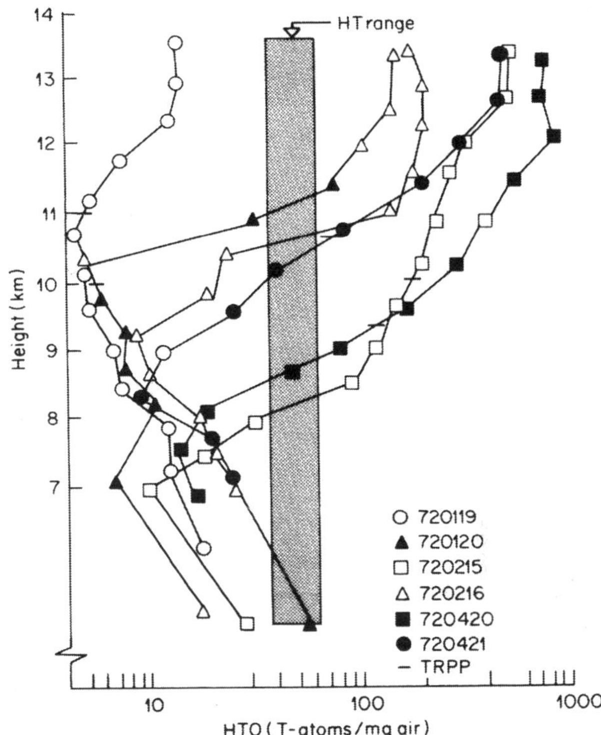

Fig. 13.7 Vertical HTO/HT mixing ratio profiles over Boulder, Colorado, 1972. (After Mason and Östlund 1979. © IAEA, reproduced with permission of IAEA)

In mountainous region, the vertical distribution of tritium in the atmosphere can be estimated using the data of precipitation measurements at different altitudes. These studies have been carried out by Romanov (1978) in the Caucasus near the Aragats Mountain (Table 13.6). It was found that the average annual concentrations of tritium in precipitation, sampled at an altitude of 850–3,500 m in 1971 and 1972 increased by factor three. Assuming the equilibrium conditions of condensation of atmospheric precipitation, it may be assumed that such relationships reflect the actual vertical distribution of tritium in water vapor.

The obtained data by the same author on tritium concentration in the annual layers of the Pamir glacier at an altitude of 4,500 m (Table 13.6) found to be lower. This was explained by difference in the origin of the atmospheric moisture (Indian Ocean), which forms the sampled precipitation.

13.2.3 Tritium in Precipitation

The applicability of environmental tritium as a tracer of air mass circulation in the atmosphere and the formation of precipitation and discharge on the continental surface and in groundwaters, is based upon the experimental data of tritium content in

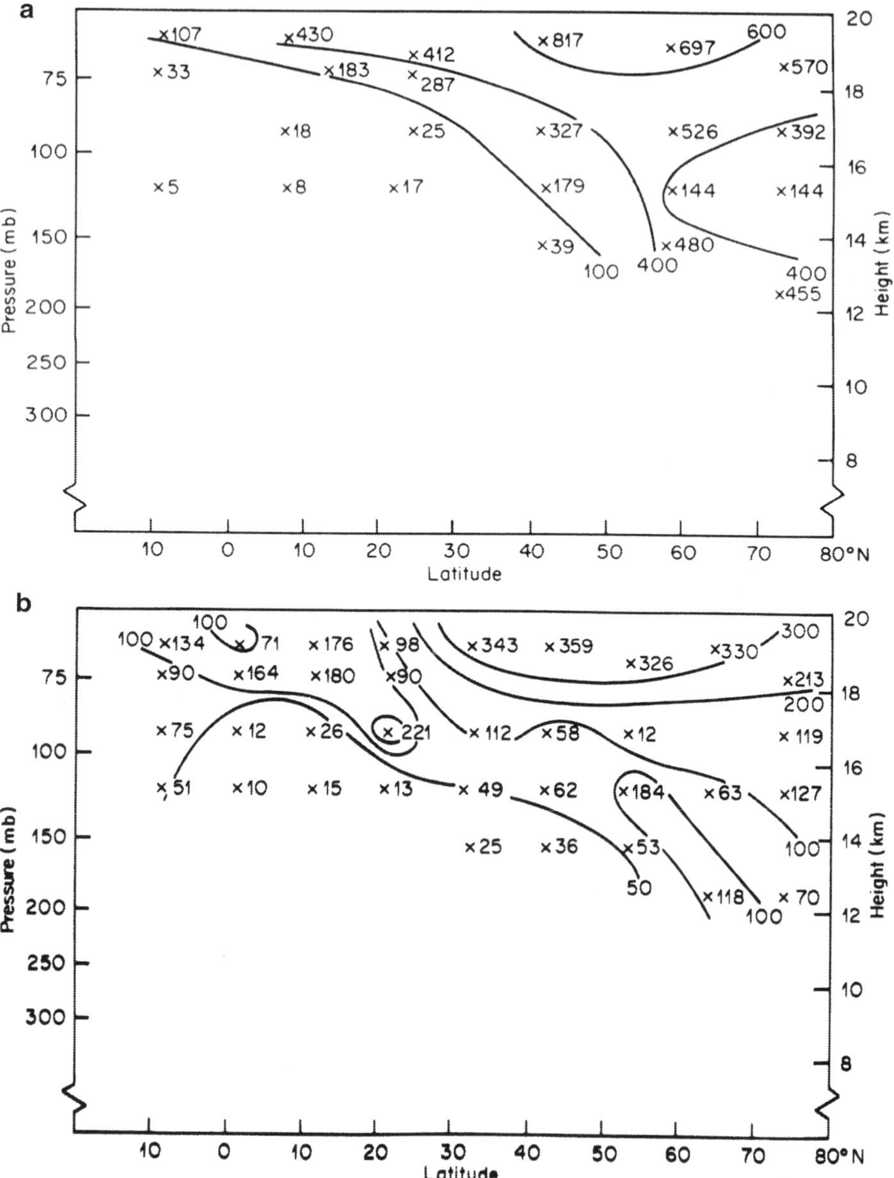

Fig. 13.8 Stratospheric HTO mixing ratios, T-atoms/mg air for May–June, 1976 (**a**) and for August 1976 (**b**). (After Mason and Östlund 1979. © IAEA, reproduced with permission of IAEA)

Table 13.6 Relationships of tritium concentrations from altitude in Aragats Mountain (Caucasus)

Sampling place	Altitude (m)	Mean annual concentration in precipitation (TE)	
		1971	1972
Oktemberian	850	37	82
Garnovit	1,100	114	162
Aragatz	3,238	143	201
Pamir (Abramov glaciar)	4,500	65	72

precipitation on a global scale. This work was initiated in 1961 by the IAEA and the WMO jointly. In order to detect the tritium, deuterium, and oxygen-18 (^{18}O) content in precipitation, more than 100 meteorological stations linked to the WMO, located in different countries, were involved (see Fig. 4.5). The ocean samples were collected on islands and weather ships. Thus, the network of stations included the most characteristic points of the globe both on the continents and oceans in both the northern and southern hemisphere.

In 1965, in connection with the International Hydrological Decade program (1965–1974), the network of stations included additional stations for water sampling from rivers. However, the majority of rivers fell out of this network of stations and therefore, a representative river network was not established. Up to present, the network includes more than 100 stations plus many national points of observation at which the tritium content is measured and measurements of tritium content in precipitation continue.

Sampling and analysis has been performed according to the techniques developed by the IAEA. Samples of atmospheric waters, taking every month, corresponds the monthly average tritium content in precipitation. Tritium content in water samples was measured in low-level counting laboratories in the IAEA and in Canada, Denmark, India, Israel, New Zealand, Germany, Sweden, USA, USSR, and other countries.

All these data are being collected by the IAEA and after processing together with the results of analysis of the stable isotopes (deuterium and ^{18}O) content, measured in the same samples, and also together with meteorological data, are published in special issues of Environmental Isotope Data (IAEA 1969–1994).

The tritium concentrations in precipitation may have substantial difference in individual fallouts, depending on their origin and trajectory of motion. However, in principle, distribution of the fallouts depends on mechanism of circulation of the atmosphere. The seasonal and annual variations are observed. Seasonal variations are related to the strengthening of the air masses exchange between spring and summer stratosphere and the troposphere. The effect leads to occurrence of the so-called spring–summer maximum in the annual tritium distribution. Weakening of this process in the winter and autumn leads to occurrence of the autumn–winter minimum.

The long-term variation of tritium concentration of the natural origin can be related to phases of the solar activity. This is because it occurs at antiphase with intensity of the galactic cosmic rays, which are accepted as a source of tritium.

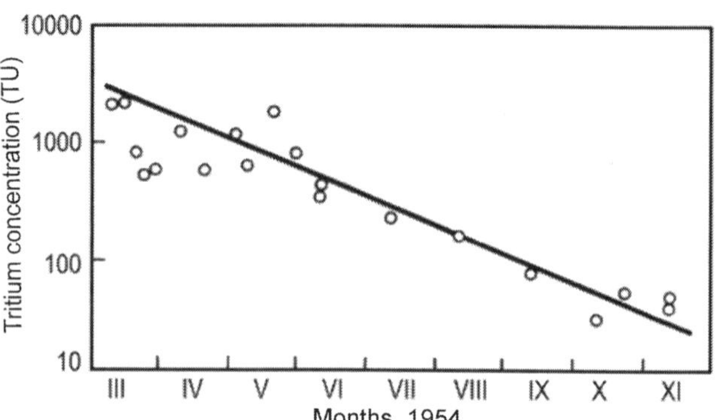

Fig. 13.9 Changes of tritium concentration atmospheric precipitation over Ottawa (Canada) after thermonuclear test in 1 March 1954. (After Brown 1961)

Some researchers tried to study correlation between the long-term tritium variations, which were observed in Greenland glaciers formed before 1952, with solar activity. However, single-valued result was not found, for example, Begemann (1959) in the Greenland glaciers discovered negative correlation between the tritium concentration and solar maximum activity. Ravoire et al. (1970) in Antarctic snow (1950–1957), found this correlation is positive and Aegerter et al. (1967) discovered both types of correlation.

During atmospheric thermonuclear tests tritium in the form of HTO occurs both in the stratosphere and in the troposphere. Proportion of its amount depends on the height and power of the explosion. Tropospheric component of HTO must have the residence time compared with that of the tropospheric moisture, i.e., equal to several weeks. Experimental data proved this conclusion (Buttlar and Libby 1955). Figure 13.9 demonstrates changes of tritium concentration over Ottawa, Canada after thermonuclear test in 1 March 1954 (Brown 1961). Period of removal of half of the tritium from the troposphere here was 45 days. The close figure to this value was obtained by Buttlar and Libby by measurements of tritium in precipitation in Chicago.

The stratospheric tritium part is removed substantially lower. A velocity of this process corresponds with the velocity of exchange between the tropospheric and stratospheric air and have seasonal cyclic character. The annual tritium concentration changes in the Ottawa River water are shown in Fig. 13.10 (Brown 1970). Figure 13.11 demonstrates the annual means of tritium concentration changes over Moscow and in Moscow River (Russia) during the period of 1953–1969, obtained by the authors.

It follows from Figs. 13.10 and 13.11 that the entering of tritium into the atmosphere occurs during the 1954, 1956, 1958 tests. In the period of the moratorium from 1959 up to its interruption in September 1961, the tritium concentration within Moscow region has dropped from 760 to 200 TU. The period of removal of half tritium value for this time interval was equal to about 1 year. In September 1961, the thermonuclear tests were renewed and continued up to December 1962. In that

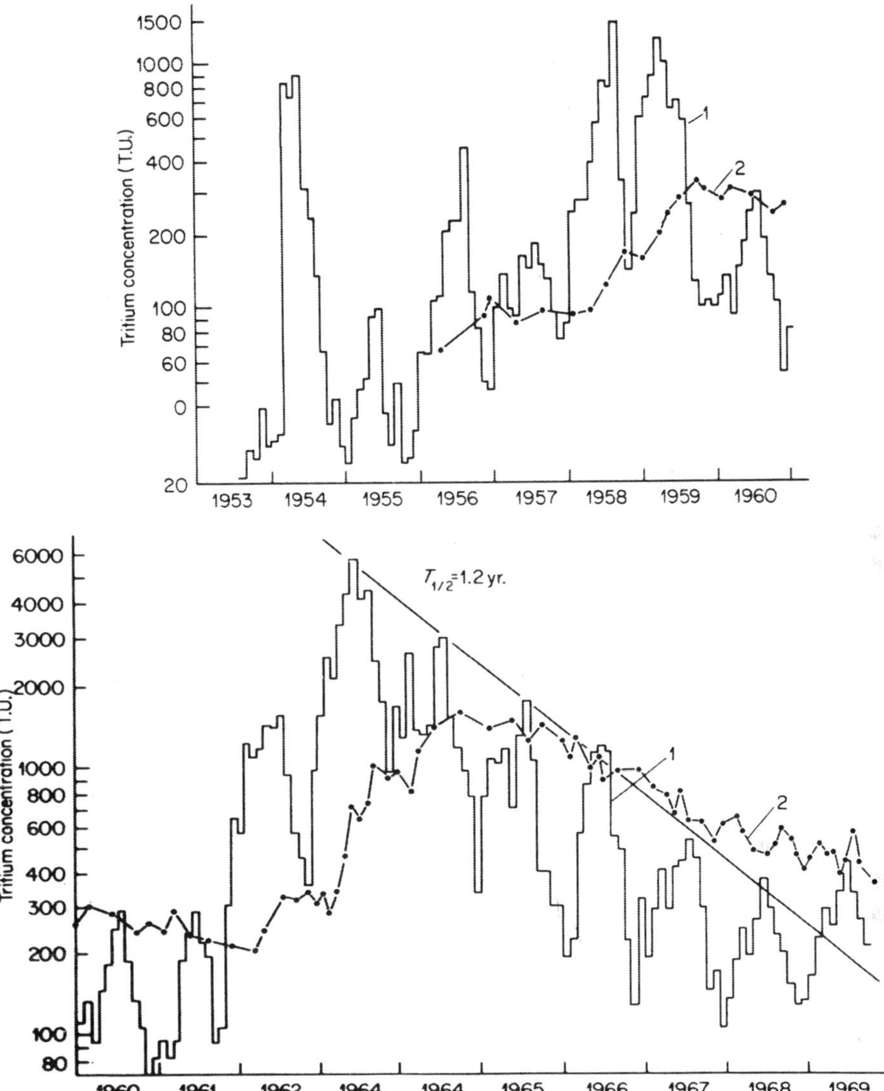

Fig. 13.10 Tritium concentration change during 1953–1969 in precipitation (*1*) and in Ottawa River water (*2*). (After Brown 1970. © IAEA, reproduced with permission of IAEA).

period, the main part of bomb-tritium has accumulated in the stratosphere, which is observed up to now.

The maximum yearly means of tritium concentrations over Moscow region were reached 3,900 TU (see Fig. 13.11). After the thermonuclear test ban treaty in the three spheres had come into force, the stratospheric tritium reserve started to decrease with the period of 1.2 years up to 1967–1968, after that the decrease was slowing

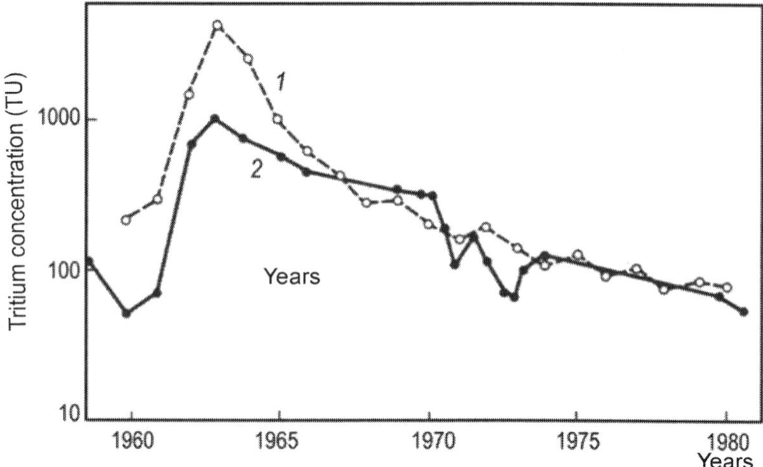

Fig. 13.11 Mean annual distribution of tritium concentration in atmospheric precipitation over Moscow region (*1*) and in Moscow River water (*2*) 1958–1981

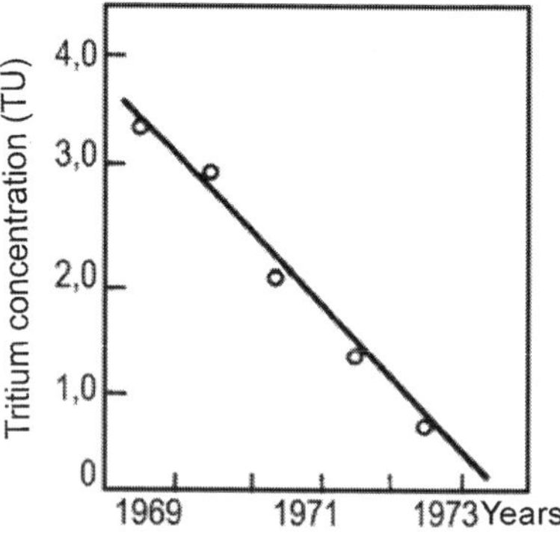

Fig. 13.12 Changes of mean annual tritium concentration by the data of 18 stations over the USSR area

down. During 1969–1974 that period becomes equal about 3 years (see the graph on Fig. 13.12). This value was obtained by the authors on the basis of data over vast territories and, therefore, may be considered adequately authentic. One more explanation of the above phenomenon can be redistribution of tritium in stratosphere between the northern and southern hemispheres. During recent years this transfer was decreased due to approaching the quasiequilibrium state.

It was shown in the work of Weiss et al. (1979) that, starting from 1970, over the Central and Western Europe, technogenic tritium plays a notable role in the formation of its occurrence in precipitation. From here it follows that the sampling stations should be placed in appropriate distance from the industrial plants and institutions.

Variations of tritium concentrations in precipitation during spring and summer are determined by specific conditions of the mass air exchange between the stratosphere and the troposphere, resulting from their easier connection (see Fig. 13.9).

Table 13.7 Seasonal distribution of tritium concentrations in precipitation for stations of the North hemisphere

Month	Yearly means of concentration in precipitation (TU)					
	1964	1965	1967	1968	1969	Mean value
0–20°N						
January	1.09	1.13	1.26	0.94	0.99	1.07
February	1.17	1.29	1.18	1.06	1.04	1.12
March	1.17	1.28	1.20	1.24	1.23	1.19
April	1.30	1.07	1.27	1.09	1.03	1.13
May	1.31	1.23	1.07	1.16	1.04	1.13
June	1.34	1.40	1.30	1.10	1.09	1.22
July	1.44	1.24	1.13	1.27	1.34	1.35
August	0.96	0.84	0.92	1.09	1.24	0.99
September	0.71	0.66	0.83	0.88	0.92	0.78
October	0.51	0.62	0.66	0.72	0.77	0.64
November	0.51	0.62	0.68	0.75	0.66	0.63
December	0.49	0.62	0.51	0.70	0.64	0.58
20–90°N						
January	0.89	0.63	0.73	0.46	0.63	0.70
February	1.09	0.98	0.90	0.80	0.76	0.89
March	1.18	1.20	1.09	1.04	1.04	1.08
April	1.46	1.46	1.22	1.19	1.11	1.29
May	1.67	1.54	1.60	1.42	1.49	1.57
June	1.70	1.73	1.58	1.59	1.62	1.64
July	1.41	1.51	1.36	1.44	1.47	1.45
August	0.99	1.10	1.17	1.28	1.29	1.18
September	0.56	0.60	0.88	0.86	0.80	0.74
October	0.43	0.43	0.54	0.60	0.63	0.53
November	0.31	0.35	0.48	0.54	0.50	0.44
December	0.32	0.36	0.46	0.77	0.64	0.50

The peak of the tritium concentrations in precipitation for the northern hemisphere is observed, as a rule, in June, and in the southern hemisphere in September. However, deviations from this rule are happened very often due to the meteorological peculiarities of differed years. This peak is occurred in a pure sight very seldom. In this connection, it is more correct to obtain average data for a number of regions with more or less identical physical-geographical characteristics at the same time interval. Tables 13.7 and 13.8 are an example of such an average. The tables were prepared on the basis of the data published by the IAEA (1969–1994) using more than 250 stations of the Earth's globe. For the former USSR area the author's own data were used.

Figure 13.13a shows the normalized seasonal variations of tritium in continental precipitation of the northern hemisphere. In Fig. 13.13b, the analogous data are shown for the latitudinal belt 20–90°S and in Fig. 13.14 the same for area of the former USSR. It is seen from the figures that in the equatorial belt between 20°S–0°–20°N the picture of tritium variation does not observed. There are a number of maximum peaks. However, values with minimum concentrations have less dispersion. In the northern hemisphere, those occur in November and in the southern are in May.

Table 13.8 Seasonal distribution of tritium concentration in precipitation for stations of southern hemisphere

Month	Yearly means of concentration in precipitation (TU)					
	1965	1966	1967	1968	1969	Mean value
0–20°S						
January	1.26	1.01	0.94	1.22	1.15	1.15
March	0.96	1.10	0.89	0.88	0.91	0.95
April	0.70	1.20	0.86	0.82	0.86	0.93
May	0.75	0.89	0.71	0.74	0.76	0.77
June	0.94	0.76	0.73	0.88	0.93	0.85
July	0.93	0.98	1.22	1.04	0.93	1.02
August	1.03	1.06	1.03	0.98	1.11	1.04
September	1.16	0.69	0.94	1.21	1.12	1.02
October	1.03	0.71	1.30	0.99	0.99	1.02
November	0.94	1.28	1.06	0.91	0.87	1.01
December	1.00	1.14	1.17	0.98	1.10	1.08
0–20°N						
January	1.06	0.94	1.02	1.00	1.05	1.01
February	1.01	0.87	0.79	1.29	0.89	0.97
March	0.67	0.90	0.78	0.67	0.67	0.74
April	0.77	0.64	0.86	0.70	0.70	0.73
May	0.63	0.73	0.79	0.71	0.73	0.86
June	0.72	0.59	0.70	0.71	0.75	0.69
July	0.82	0.88	0.98	0.86	0.84	0.88
August	1.43	1.18	1.27	1.38	1.43	1.34
September	1.56	1.66	1.35	1.54	1.60	1.54
October	1.29	1.43	1.16	1.22	1.27	1.27
November	0.97	1.06	1,02	1.00	1.,06	1.02
December	1.06	1.10	1.28	0.93	0.90	1.06

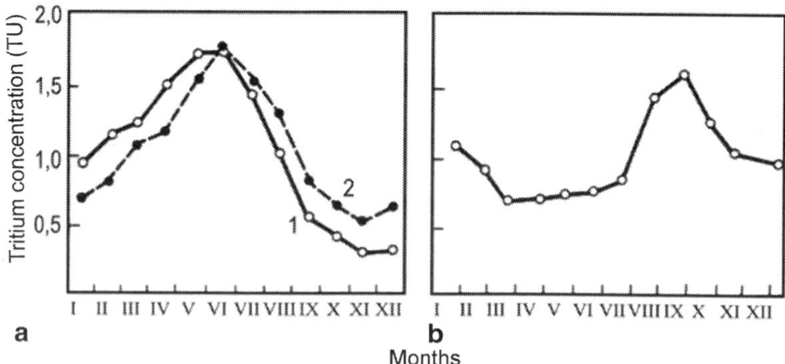

Fig. 13.13 Seasonal change of ratio between mean monthly and yearly means of tritium concentration in precipitation: (**a**) over the continents of the northern hemisphere in 1964 (*1*) and in 1969 (*2*); (**b**) over the southern hemisphere in 1969

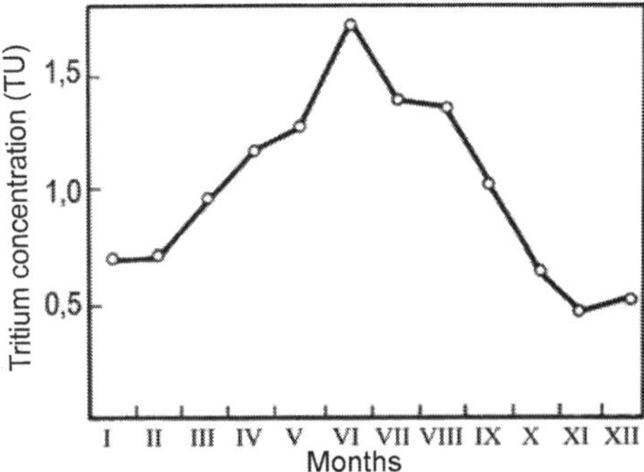

Fig. 13.14 Seasonal change of ratio between mean monthly and mean yearly tritium concentrations for 19 stations of the former USSR area in 1969–1974

Sometimes a substantial deviation in mean monthly tritium values from these for seasonal variations is observed. In this connection, it is interesting to draw attention on the situation observed in a number of stations located in East European and South-Asian regions in December 1969. The tritium concentration in precipitation here was accounted by 0.64 compared to its seasonal mean value (Table 13.9). It seems, this was caused by the air intrude with high tritium content from the upper troposphere into narrow meridian direction on Moscow–Tbilisi–Teheran.

13.2.4 Global Distribution of Tritium

A general picture of global tritium distribution over time (Fig. 13.15) shows that although the degree of decrease of tritium content depends upon the site of a given place, the general features of its distribution remains unchanged. The observed constancy in the character of tritium content variations in precipitation with time indicates the global principle of increases of tritium concentration with latitude, both for oceanic and continental regions. This may be observed while plotting general pictures of global tritium distribution in time. Figure 13.15 gives such pictures for 1964, 1969, and 1975 prepared by the authors. The figure gives general idea

Table 13.9 Tritium concentrations in precipitation for some European and Asian stations in December 1969

Station	Tritium concentration (TE)		Norm for December 1969	Ratio of the norm in December 1969 to the yearly means
	December 1969	Yearly means		
Arhangelsk	133	194	0.69	1.07
Odessa	134	207	0.68	1.01
Vien	119	210	0.57	0.88
Ankara	54	217	0.25	0.39
Teheran	265	133	1.99	3.11
Moscow	500	233	2.24	3.50
Tbilisi	720	218	3.30	5.16

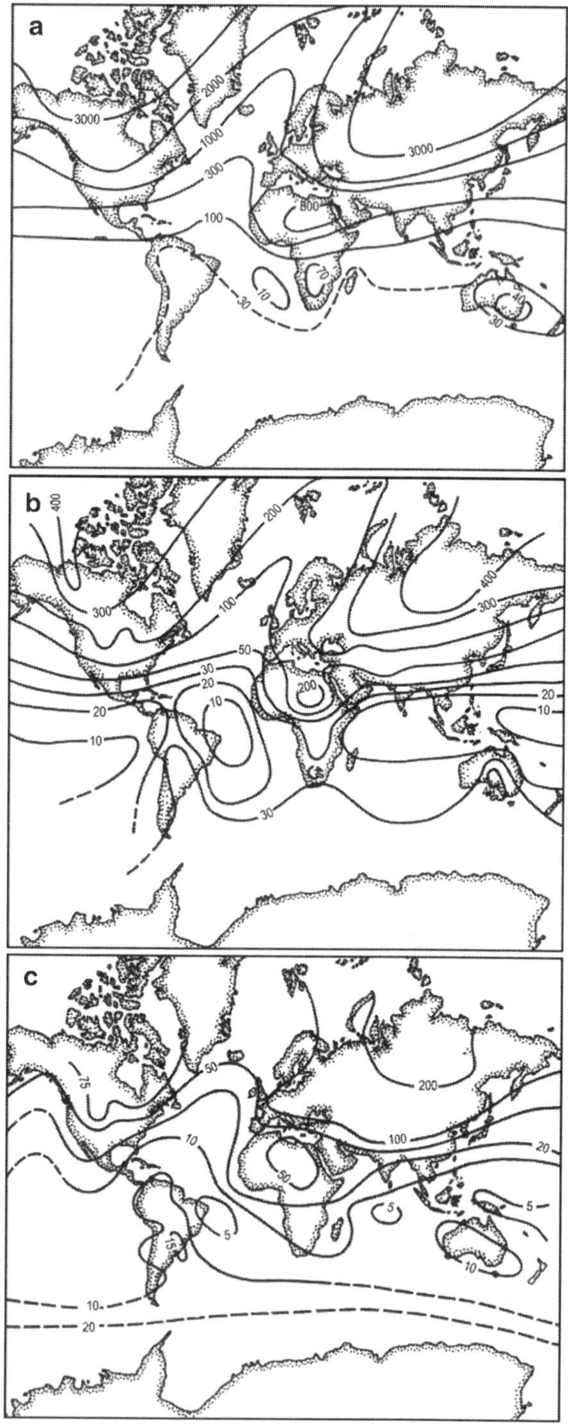

Fig. 13.15 Isolines of yearly weighted tritium means in precipitation of the Earth's globe in 1964 (**a**), 1969 (**b**), and 1975 (**c**). (After Romanov 1978)

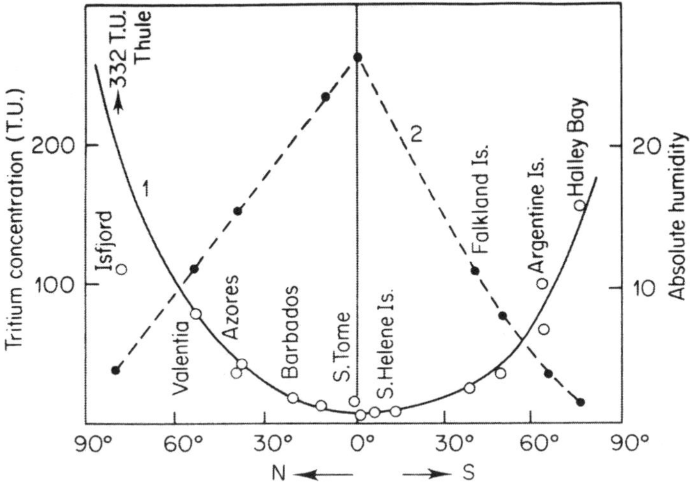

Fig. 13.16 Tritium concentration profile in precipitation (*1*) and atmospheric moisture (*2*) for Atlantic Ocean island stations; average over 1969. (Romanov 1978)

about global tritium distribution in precipitation and changes in time of the distribution due to tritium transfer into the oceans and interhemispheric redistribution in the stratosphere. The similarity of the isolines of the tritium concentrations during 1964–1975 is the evidence of constancy and regular character of its distribution in precipitation.

The latitude effect of the meridional distribution of tritium is its second main regularity, which means decrease of the tritium concentration with latitude both for the oceans and the continents. Figure 13.16 demonstrates this effect for precipitation over the Atlantic Ocean. The mean values of the absolute air humidity are although shown on the graph. Analogous picture for the Pacific Ocean island stations in Fig. 13.17 is presented.

Continental area has a considerable effect on global tritium distribution in precipitation. The total continental area of the northern hemisphere is greater than that of the southern hemisphere. Figure 13.17 also represents the effect of tritium redistribution between the northern and southern hemispheres, which resulted in the relative equality of the tritium content in both of the hemispheres in 1969.

The next characteristic peculiarity, observed in the general picture of tritium fallout, is the influence of continents upon its distribution. At the same latitude, raised tritium concentration is observed above the continents where active enrichment of atmospheric moisture with tritium in the air masses exchange between the troposphere and stratosphere takes place. The continental effect and effect of the relief heights is demonstrated in Fig. 13.18.

Here from 35 to 40°N a number of huge mountain ridges up to 7,000 m in height stretch latitudinally (the Himalayas, Hindu Kush, Tien Shan, and Pamirs). These mountainous ridges stop the motion of atmospheric moisture enriched with tritium over the European–Asiatic continent to the Indian Ocean. At the same time, the same

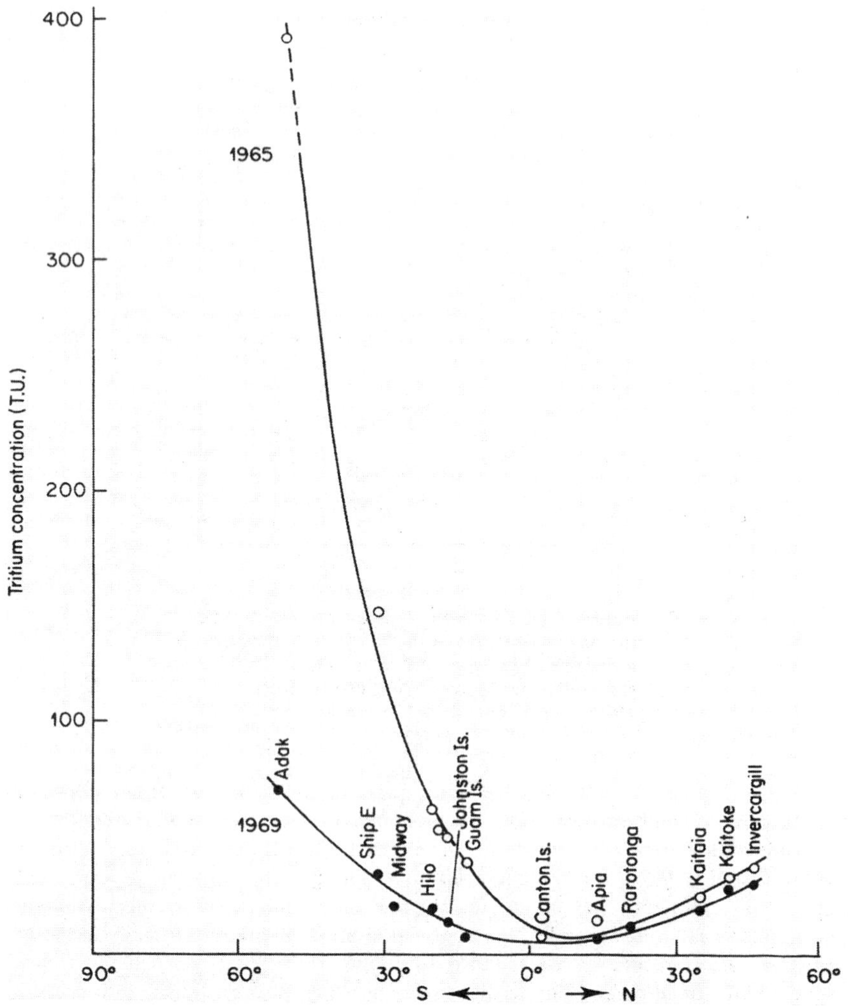

Fig. 13.17 Tritium concentration profile in precipitation for Pacific Ocean island stations; average over 1965 (*1*) and over 1969 (*2*). (After Romanov 1978)

mountainous system prevents the transfer of atmospheric moisture depleted in tritium from the Indian Ocean and Persian Gulf to Siberia and Middle Asia. These factors cause a large gradient of tritium concentration in precipitation over this mountainous region. In the Asian plain, in the region located above 45°N the tritium content in precipitation is governed by the latitude effect.

Figure 13.19 shows the curve of the average tritium content in precipitation for the latitudinal belt 45–60°N, plotted on the basis of data obtained during the second half of 1969 (Ferronsky et al. 1975) and demonstrated also the continental effect.

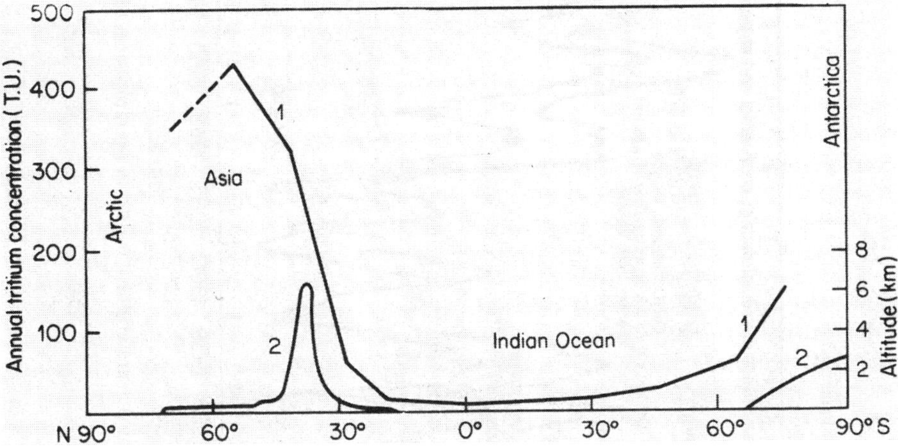

Fig. 13.18 Latitudinal variation of tritium concentration in precipitation at 80–90°E (*1*) relative to relief heights (*2*); average over 1969. (After Romanov 1978)

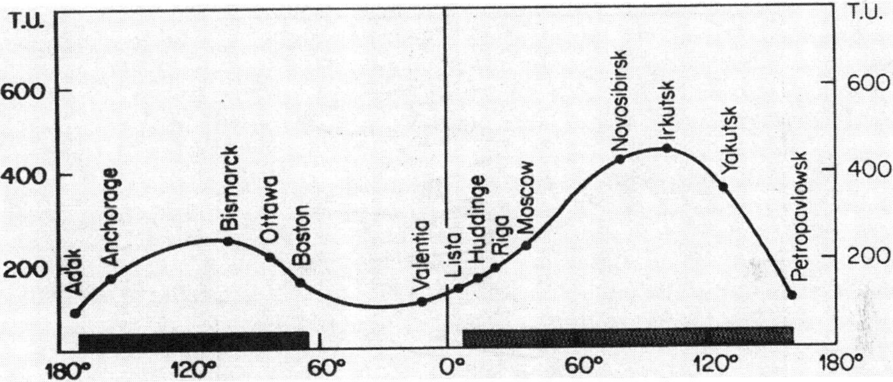

Fig. 13.19 Variation of tritium concentration in precipitation for stations at latitudes 45–60°N; average over July–December, 1969. (After Ferronsky et al. 1975)

One can see the enrichment of atmospheric moisture with tritium in the course of its motion across the European–Asiatic and North American continents, and decreases of the tritium content during the formation of atmospheric moisture above the Atlantic and Pacific Oceans.

13.3 Regional Distribution of Tritium in Precipitation

It follows from the global distribution of tritium in precipitation, presented in Fig. 13.15 that, beside some general features, there are some regional effects. These effects were analyzed by Thatcher and Payne (1965) and Romanov (1978), from whom the following information has been gathered.

13.3.1 The North American Continent

There are more than 30 stations located in this continent, aimed at systematic measurements of the tritium content in precipitation. At the station located near Ottawa (Canada), systematic measurements were made from 1954 (see Fig. 13.10). Tritium variations with time found at this station are typical for the whole continent. Maximum tritium content occurs in the central part of Canada (Edmonton, Fort Smith), where, in April 1963, the concentration of tritium reached 10,000 TU.

Precipitation falling out above the North American continent is formed mainly over the Atlantic Ocean and the Gulf of Mexico (see Fig. 13.15). The mountain chain of the Western Cordilleras and the warm Golf Stream current prevents the movement of Pacific air masses. The great gradient of the tritium concentrations at the west coast is evidence of this mountain effect. However, the contribution of Pacific Ocean moisture coming from Alaska results in a deep bend of concentration isolines toward the continent (Fig. 13.15). Both the continental and latitudinal effects are evident in this region.

13.3.2 The European–Asiatic Continent

There are about 60 observation stations in the Scandinavian countries and Western and Central Europe, 18 stations in European and 19 in Asian part in the former USSR, 10 stations in Near East countries, and about 10 stations in South-East Asia and the Far East.

The Atlantic Ocean is the main source of precipitation over the major part of the European–Asiatic continent. Here, the preferred direction of air mass movement is from west to east. Thus, moisture depleted in tritium while passing over the Atlantic coast of Europe gradually becomes enriched in the course of inland movement. Valentia (Ireland) and Vienna (Austria) (Fig. 13.20) are typical points, corresponding to the Atlantic coast and continental West Europe, with a time-dependent distribution of tritium in precipitation.

The distribution of tritium in the southern part of Europe is affected markedly by moisture coming from the Mediterranean Sea. In addition to continental effect, industrial injections of tritium are significantly increased tritium concentrations over the continental Europe. The studies of Weiss et al. (1979) demonstrate this conclusion (Fig. 13.21).

The long-term picture of tritium distribution in precipitation in principle looks like the Central European picture (see Fig. 13.11). A total yearly means of tritium concentrations decrease over the USSR territory to the end of 1980s of the last

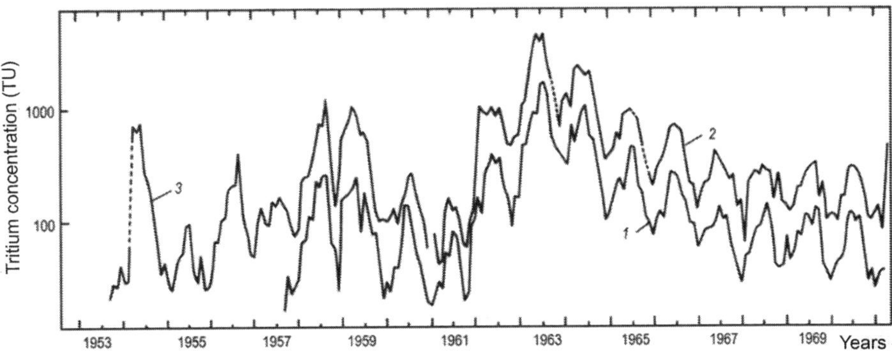

Fig. 13.20 Variation of tritium concentration in precipitation at Valentia (*1*), Vienna (*2*) and Ottawa (*3*) in 1953–1971. Correlation between Vienna and Ottawa is 17%

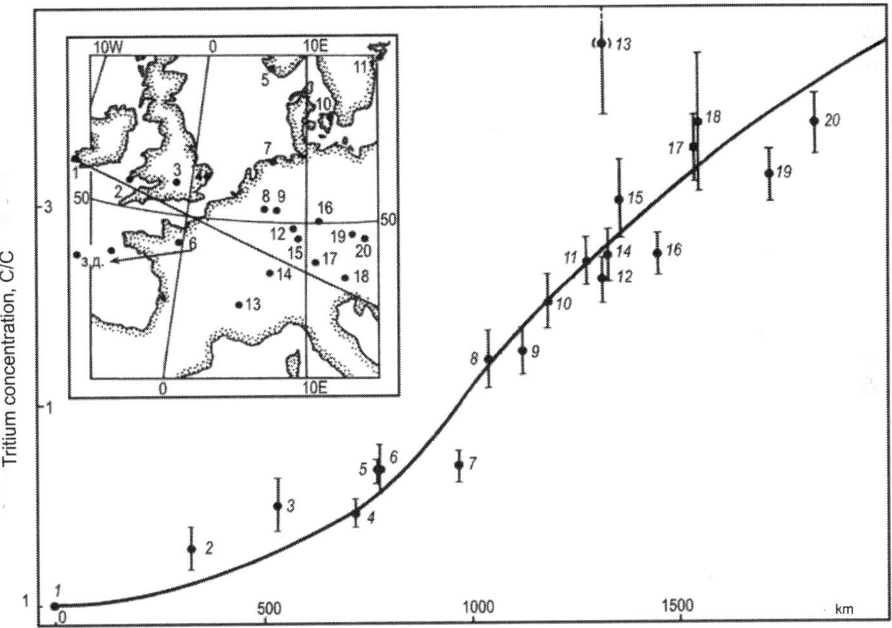

Fig. 13.21 Continental effect of tritium content in precipitation over Western and Central Europe relative to those at the reference station Valentia, taken from observations up to 1970. (After Weiss et al. 1979. © IAEA, reproduced with permission of IAEA)

century, without taking into account the 1975 and 1978 maximums, is described by the equation:

$$C = 2.26 \exp(-0.096t),$$

where C is the ratio of tritium concentration for a particular year to the concentration in 1979.

Table 13.10 Normalized monthly means of tritium concentrations in precipitation over the former USSR for 1972–1979

Month	Normalized monthly means of concentrations				
	1972	1975	1976	1978	1979
January	0.91	0.62	0.74	0.54	0.79
February	0.72	0.82	0.77	0.89	0.82
March	0.95	1.08	0.88	1.03	1.01
April	1.18	1.55	0.84	1.07	1.11
May	1.20	1.68	1.22	1.37	1.30
June	1.63	1.31	1.46	1.47	1.41
July	1.49	1.23	1.55	1.39	1.35
August	1.12	1.05	1.41	1.28	1.12
September	0.86	0.92	1.06	0.98	0.94
October	0.71	0.63	0.86	0.65	0.82
November	0.62	0.53	0.64	0.75	0.65
December	0.60	0.58	0.57	0.57	0.68

The correlation coefficient of this equation is equal to 0.96. Seasonal effects of tritium distribution in time are very clearly expressed (see Fig. 13.14). The monthly means of tritium concentrations for 1972–1979 are given in Table 13.10.

The peculiarity of the European–Asiatic region of the former USSR is the result of high mountainous chains in the south, preventing the arrival of atmospheric moisture from the Indian Ocean. The preferred direction of moisture movement over this territory is toward the east. This effect is manifested particularly in the autumn–winter period. In the summer, when the cyclonic activity becomes stronger, air masses from the South China Sea break through East Siberian and reach Central Asia up to Irkutsk. Pacific air masses depleted in tritium also reach Irkutsk. Thus, it was found that at the Far East stations of Skovorodino, Khabarovsk, and Holmsk the range of tritium variation from month to month may be rather great, depending on the source of moisture coming either from the Atlantic Ocean or the Pacific Ocean (Figs. 13.22 and 13.23).

The Baltic, Black, and partly Mediterranean Seas make substantial influence on the European yearly means of tritium concentrations. This conclusion is proved by the isoconcentration lines of the near Baltic, Black Sea, and Caucasus regions where tritium values are lower compared to the neighboring areas.

Being averaged during the 6 years of observation, the data of Fig. 13.22 are representative and indicate features of the annual mean water transference in the atmosphere over the USSR area. The highest mean annual tritium concentrations in precipitation were detected near Irkutsk. In 1979 (Fig. 13.23), the shape of isolines extends along meridian plane direction and the maximum gradient of concentrations is observed in the domain of 130–145°E. It is obvious, that such a configuration determined the border of influence of the air masses which transfer moisture from the Pacific. The bend of the isolines to the continental depth here along meridian of to 90°E can be considered as a consequence of the Atlantic air masses intervention in this region with farther transfer to the south–east direction. To the contrary of the Atlantic air masses, those from the Pacific move in wide front without a direction privilege. Analyses of tritium distribution in precipitation on the European–Asiatic

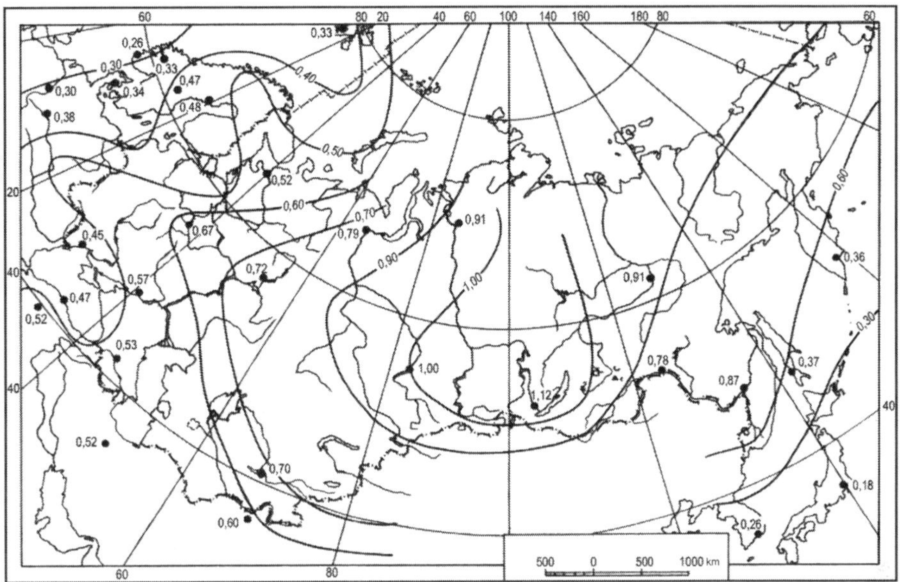

Fig. 13.22 Distribution of tritium concentration in precipitation for the former USSR, normalized relative to 200 TU, and averaged over 1969–1974

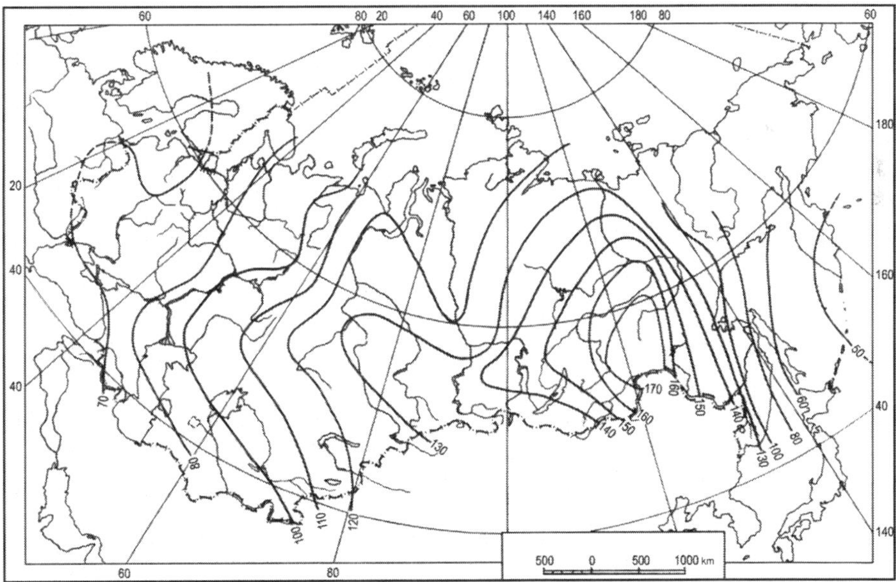

Fig. 13.23 Distribution of yearly (1979) means of tritium concentration in precipitation for the USSR area

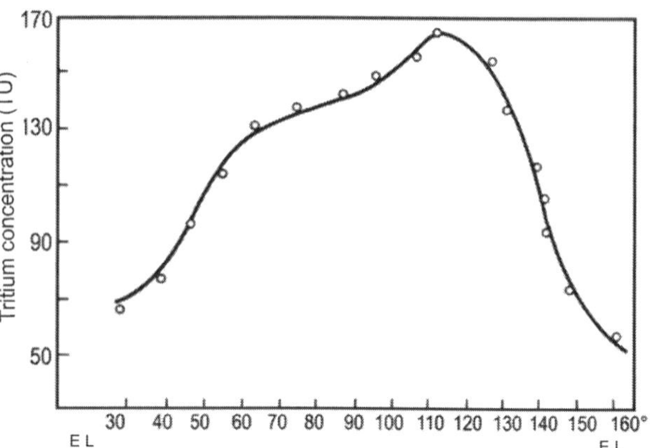

Fig. 13.24 Annual longitudinal means of tritium concentrations in precipitation along 55°N over the former USSR territory in 1979

continent have shown that tritium fallout is very sensitive instrument to dynamics of the atmospheric moisture.

The continental effect over the former USSR territory is developed very clear (Fig. 13.24).

South-East Asia and the Far East may be considered as independent regions from the viewpoint of general principles of tritium distribution in precipitation. The former is a region of monsoon activity with downpours of rain occurring from June to September. During this period the tritium content in precipitation arriving from the Indian Ocean declines sharply. The region is presented by the stations Bahrain, Karachi, New Delhi, Mumbai, and Singapore.

In the Far East region the Pacific moisture plays an important role in the formation of precipitation. When the Pacific moisture is with the continental moisture, the tritium concentration in precipitation decreases. As a result, the tritium concentration in the spring–summer season never reaches maximum value. The region is presented by the stations Hong Kong, Pohang, Tokyo, Habarovsk, and Holmsk. Figure 13.25 shows the yearly (1979) means of tritium concentrations of this region with its distribution effects.

13.3.3 The African Continent

The conditions of the formation of tritium are different in the northern and southern parts of this continent, relative to the equator, because of differences in the formation of atmospheric circulation. This effect was particularly evident during the period of thermonuclear tests (see Fig. 13.15). The annual mean concentration of tritium in precipitation was 125 TU in 1963 in Entebbe (0° 05′ N). At the same time in Dar-es-Salaam (6° 52′ S) it was only 17 TU, which reflects the restricted connection between the hemispheres relative to air mass circulation. Up to 1968 the tritium content in precipitation measured in the northern and southern parts of the continent

Fig. 13.25 Annual means of tritium concentrations in precipitation in 1979 for stations: Pohang (*1*), Tokyo (*2*), Hong Kong (*3*)

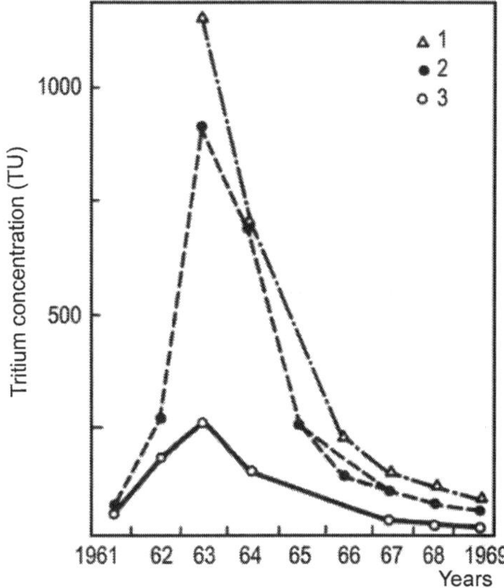

was approximately the same. However, at the same time, the gradient of tritium concentrations between the equator and the poles was great (Fig. 13.26), which is explained by the continental effect and the hemispheres gravitational potential restriction.

The results of yearly measurements of tritium content in precipitation at four stations located in North Africa are presented in Fig. 13.27. As in the hemisphere as a whole, the maximum concentrations were observed in 1963, and the decrease

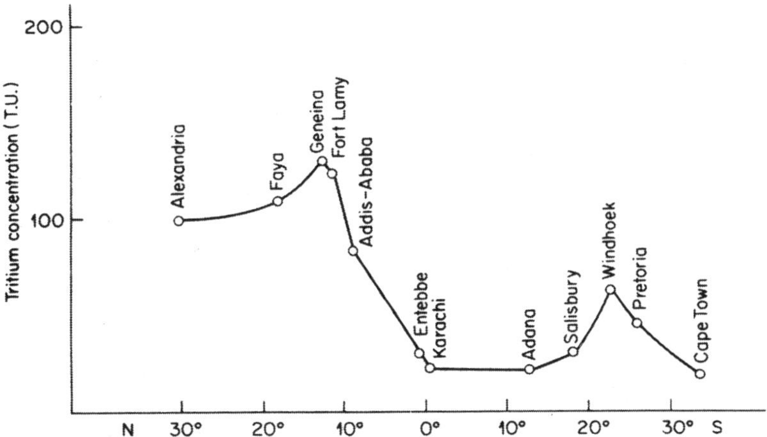

Fig. 13.26 Meridional distribution of tritium concentration in precipitation over the African continent, averaged over 1968. (After Romanov 1978)

Fig. 13.27 Variation of annual tritium concentration in precipitation over North Africa for Alexandria (30° 20′°N) (*1*), Khartoum (15° 60′°N) (*2*), Fort Lamy (12° 13′°N) (*3*), Entebbe (0° 05′°N) (*4*). (After Romanov 1978)

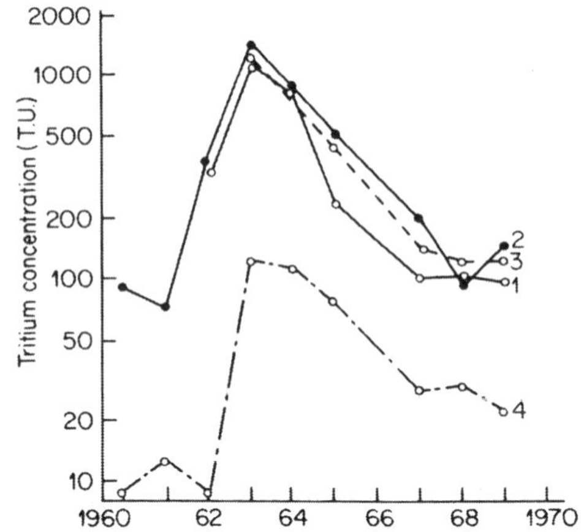

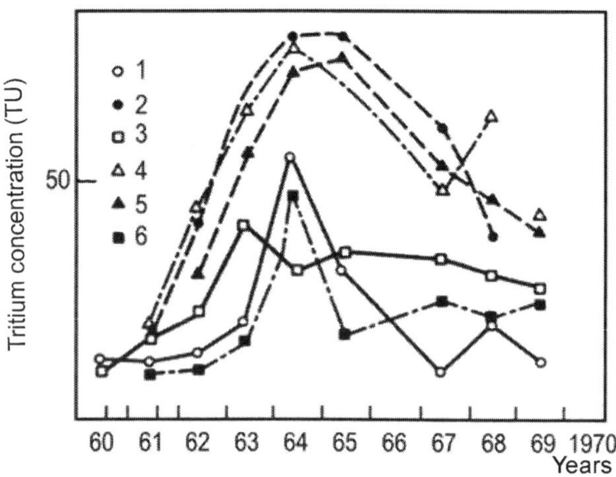

Fig. 13.28 Variation of annual tritium concentration in precipitation over Equatorial Africa for Dar-es-Saalam (*1*), Kinshasa (*2*), Harare (*3*), Windhoek (*4*), Pretoria (*5*), Cape Town (*6*). (After Romanov 1978)

of concentrations with time slowed up to 1967. The half-removal period of tritium estimated for Alexandria region, locater in the north of the continent, was 1.1 years. For Khartoum and Fort Lamy (12–15°N) this period is equal to 1.5 years and for Entebbe (0° 05′°N) is 2.2 years.

The results of yearly measurements of tritium concentration in precipitation at equatorial African stations are shown in Fig. 13.28. The same effects as for southern stations are observed here.

Maximum tritium concentrations were observed in South Africa in 1964 and decrease in tritium content occurred slowly over the period of half-removal equal to 5–6 years. The seasonal variations of tritium concentrations correspond maximum to September and minimum to March. The latitudinal and continental effects are clearly evident.

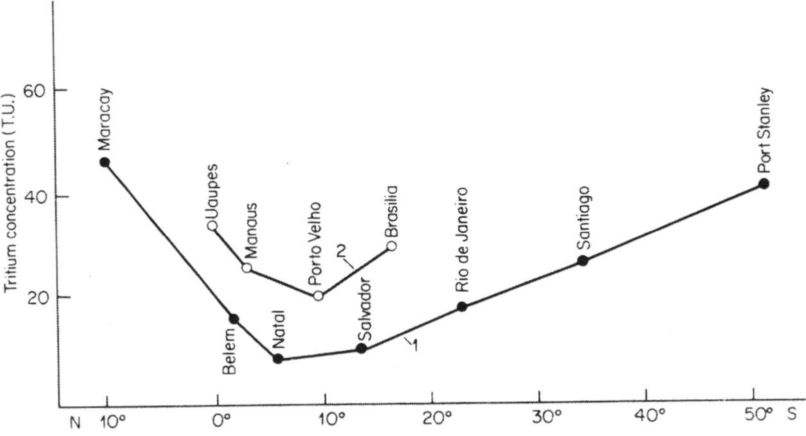

Fig. 13.29 Variation of tritium concentration in precipitation over the South American continent for costal (*1*) and inland (*2*) stations, averaged over 1967. (After Romanov 1978)

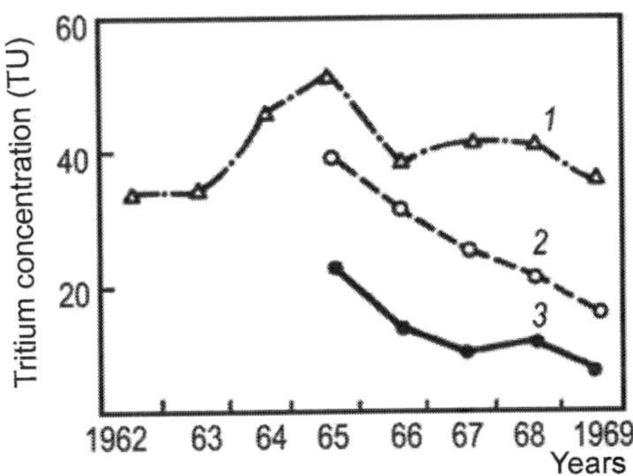

Fig. 13.30 Annual means of tritium concentration in precipitation over the South America for stations: Stanley (*1*), Manaus (*2*), Salvador (*3*)

13.3.4 The South American Continent

There are about 20 stations of IAEA/WMO. Figure 13.29 shows the meridional distribution of tritium concentration in precipitation during 1967 for the continental and costal stations.

The continent as a whole is characterized by relatively low values of tritium with a maximum observed near the Falkland Islands in 1965, according to yearly observations. The effect of the mixing of Atlantic and Pacific moisture is the bending of tritium isolines toward the continent to the north-west. The Amazon River basin in the northern part of the continent and the southern part of the Caribbean Sea basin represent an equatorial region with characteristic features. Minimal values for coastal and inland stations were observed in the region of 10°S with maximum values in 1965 (Fig. 13.30).

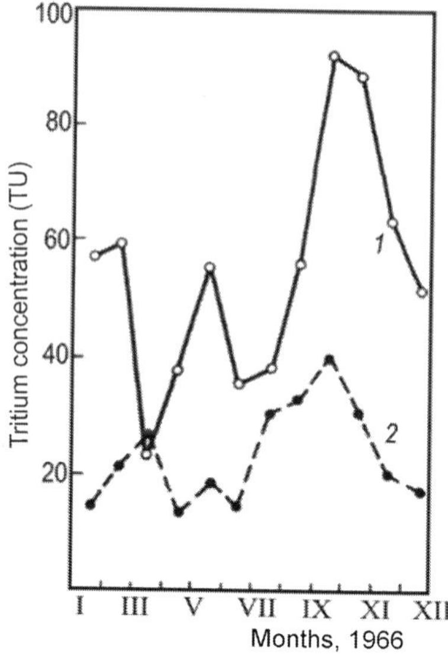

Fig. 13.31 Variation of annual tritium concentrations in Melbourne (*1*) and Kaitaia (*2*) for 1962–1968. (After Romanov 1978)

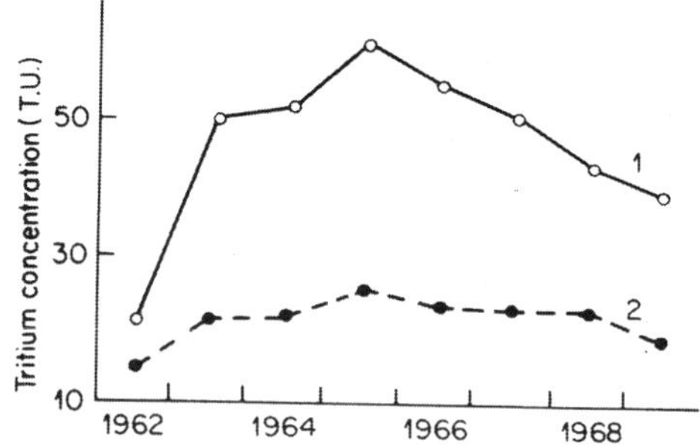

Fig. 13.32 Seasonal variation of tritium concentration in precipitation over Australia (*1*) and New Zealand (*2*) for 1966. (After Romanov 1978)

13.3.5 Australia and New Zealand

The continental effect of tritium distribution in precipitation can be clearly demonstrated in the small Australian continent and the New Zealand Islands. Figures 13.31 and 13.32 show data for annual and seasonal variation of tritium concentration in the stations Melbourne and Kaitaia. The maximum yearly concentration occurred in 1965, i.e., 2 years later than a similar maximum in the northern hemisphere. The seasonal variations of tritium concentrations are typical with maximum in September and minimum in March.

From the spatial picture of tritium distribution averaged during a number of years of systematic observations, presented in Fig. 13.33, it follows that the region of the maximum tritium concentrations lies in the south-eastern part of the continent (the Melbourne region). This indicates the preferential continuous transition of atmospheric moisture from the north-west to the south-east. This is also evidenced by tritium content in precipitation over the southern part of New Zealand. The tritium distribution picture permits the application of balancing equations for accounting

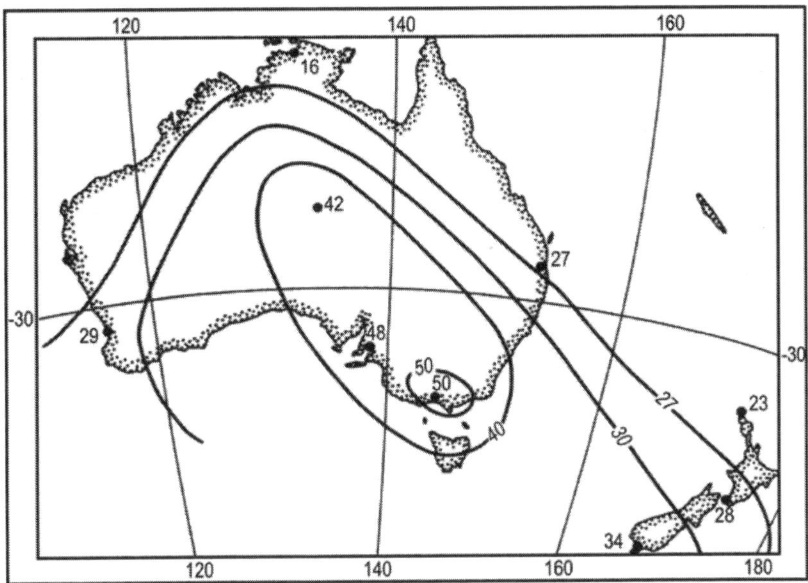

Fig. 13.33 Spatial distribution of tritium concentrations over Australia and New Zealand, averaged over 1963–1969. (After Romanov 1978)

tritium and moisture transfer in order to estimate atmospheric moisture release over the continent.

13.3.6 Antarctic

Jouzel et al. (1979) presented the most complete results on the tritium formation of snow precipitation over the Antarctic. In accordance with their data, which does not differ from other researchers, before the thermonuclear tests tritium concentration were not exceeded 34 TU. The thermonuclear tritium traces after 1952 USA tests were found in a snow layer dated by 1954. The following later on thermonuclear tests (1954–1958) were made tritium peaks delayed also with 2 years. However, the most powerful explosions in the northern hemisphere in 1961–1962 occurred in the southern hemisphere after 2–4 years. The thermonuclear tests done by France in southern hemisphere also accompanied with increase of tritium concentrations.

The only regional IAEA/WMO station on systematic sampling for isotope analysis was Halley Bay. The maximum of tritium concentration within long time observation was fixed in 1969 (Fig. 13.34). The seasonal peak is observed in August–September.

The above authors explained by winter exchange between the stratosphere and atmosphere in the polar region and by direct fallout of precipitation from the overcooled low stratosphere during arctic winter that the high level of tritium concentration is up to 1,800 TU per year.

The continental effect in Antarctic is markedly observed.

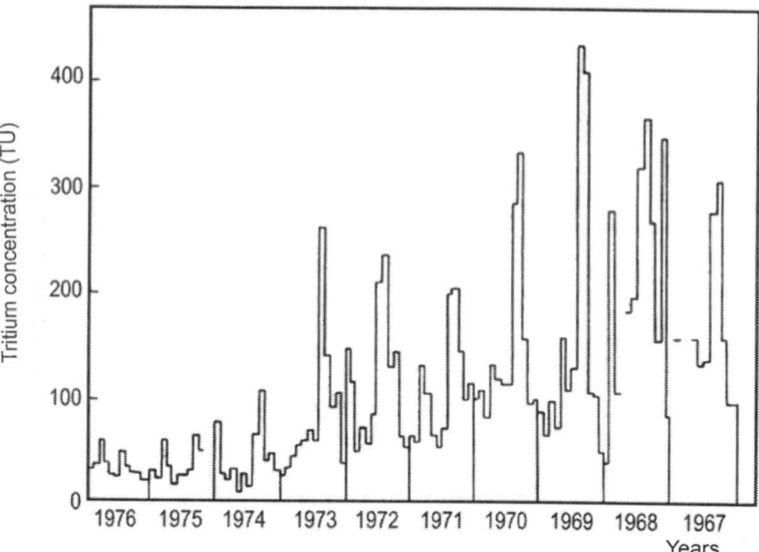

Fig. 13.34 Variation of tritium concentration in precipitation for Halley Bay (Antarctic) during 1967–1976

13.3.7 Tritium in Precipitation over the Oceans

A general picture of the tritium distribution in precipitation over the oceans can be seen in Fig. 13.15. There are three regions characterized by minimal concentrations: (1) The Pacific Ocean (10°N–25°S; 160°E–100°W), (2) The Atlantic Ocean (5°N–20°S; 40°W–0°E), and (3) The Indian Ocean (15°N–25°S; 60°E–100°W).

Such a distribution of tritium in precipitation can be accounted for by intense water exchange between the atmosphere and the oceans, which results in the intense dilution of atmospheric tritium with moisture which is depleted in tritium. Tritium content in atmospheric precipitation gradually increases to the north and to the south of these regions, reflecting the main features of the moisture exchange between the atmosphere and the ocean. Relationship between the atmospheric moisture and the tritium concentration is shown in Fig. 13.16.

13.4 Formation of Tritium Concentrations in the Atmosphere

General principles observed from the picture of tritium concentration, related to latitudinal distribution and seasonal cycles, indicate the existence of a strict mechanism which governs the process of tritium distribution and concentration to precipitation.

In the period of intensive thermonuclear tests the release of tritium in the stratosphere occurred periodically from individual points where the explosions took place. At the same time, the average concentration of tritium fallout does not depend upon the place of injection but is related to latitude and time of the year. This fact has resulted in the conclusion that the upper layer of the atmosphere (its stratosphere) is the reservoir where the accumulation and latitudinal redistribution of tritium occurs on a global scale and from which seasonal tritium releases to the lower atmospheric layer (troposphere) take place. Precipitation is formed in the troposphere. Such a reservoir exists for each hemisphere independently and the relationship between them is rather restricted. The mechanism of formation of tritium concentrations is presented as follows. The oceanic water, evaporated from its surface and having low tritium concentration, moves by the rising up air flows with velocity of about by 3–5 m/s. In the case of unstable thermal atmospheric stratification the velocity may overreach 10 m/s. The created water vapor during the day and night is roused several kilometers in high. On the other hand, the process of mixing of the stratosphere and troposphere leads to interference to the last of enriched by tritium water vapor.

The vertical distribution of tritium shows that its interference into the upper layers of the troposphere begins in December and continues up to June–July. In winter time, the tritium transfer is made by the eddy flows and in spring and summer by high tropic cyclones. In Antarctic region, in winter time, a direct condensation of vapor from the overcooled lower stratosphere is possible (Jouzel et al. 1979). The process of the tritium moisture enrichment in the clouds is continued due to the molecular exchange.

The idea of the accumulation of radioactive products of thermonuclear tests in the stratosphere, with their subsequent redistribution and injection into the troposphere, where precipitation is formed, was first suggested by Libby in 1956 (Libby 1963). Later on, attempts were made to develop this idea and to construct the box model of interacting exchangeable reservoirs: the stratosphere, the troposphere, and the ocean. Investigations were mainly aimed at determining the residence time of a radioactive tracer in each of the exchangeable reservoirs. The final goal of these studies was to fix a relationship between the residence time of a tracer in the stratosphere and the motion of air masses in the stratosphere and troposphere.

In estimating the residence time of tritium in the stratosphere, it was assumed that the release of tritium into the troposphere is exponential. Thus, the residence time of tritium in the reservoir was taken, by the analogy with its half-life, to be the time required for one half of the amount to be discharged from the stratosphere.

On the basis of experimental data concerning tritium concentrations in precipitation measured after 1953 at the two stations of Vienna and Valentia (see Fig. 13.20), which represent typical continental and coastal regions in terms of fallout, it was found that the residence time of tritium in the stratosphere is equal to 1 year (Schell and Sauzay 1970). According to the data of Brown (1970) for the Ottawa River Valley it is equal to 1.2 years. As a result of the observed residence time of tritium in the stratosphere, its total storage (see Fig. 13.2), and the half-life removal, it can be estimated that by 1970 the tritium concentration in precipitation should already have been close to the natural concentration. At the same time, according to experimental data obtained for Western Europe (see Fig. 13.20) and North America (see Fig. 13.10) from the beginning of 1967, the tritium content in precipitation in the northern hemisphere remains constant at a level of 150–200 TU and, by Romanov's estimations (Romanov 1978), decreases with a half-life removal time of about 3 years. This tritium content is an order of magnitude greater than natural tritium concentrations in precipitation. Different causes of this phenomenon were considered, particularly those related to the injection of tritium from thermonuclear tests conducted in the atmosphere during the preceding years. However, in accordance with the measures carried out at a number of stations (e.g., in Tokyo), the increase in tritium concentration in fallout is insignificant because these explosions were not very powerful and provided only small concentrations of tritium to the stratosphere (see Fig. 13.2), where the maximum concentration occurred between 1967 and 1970. Another source of tritium contribution to the atmosphere may be the various installations of the atomic industry. This idea has been well demonstrated by Weiss et al. (1979). However, as pointed out earlier, the possible input of tritium into the stratosphere from this source should be negligible and this tritium should have only a local effect upon sampling stations located near to nuclear plants.

Therefore, the residence time of tritium in the stratosphere (1–3 years) does not remain constant and the distribution and mixing of tritium in the atmosphere is a more complicated process than that described by the model of exchangeable reservoirs.

The questions of interpretation of global tritium distribution in precipitation were studied by a number of researchers (Eriksson 1965a; Libby 1963; Smith 1966; Taylor 1968). In one of the works (Schell et al. 1970), an attempt was made to develop a general model of tracer release from the stratosphere into precipitation in which the injection, exchange, evaporation, and condensation of the water vapor would be related to the meteorological parameters of the troposphere.

The model is based on the following physical reasoning. The exchange of air masses between the stratosphere and troposphere occurs mainly due to powerful air streams being thrown down periodically into the troposphere as a kind of trough or trench. This process coincides with the observed spring maximum of radioactive falls in precipitation at mid-latitudes of the northern hemisphere. Therefore, in the general mechanism of air masses exchange between the two reservoirs, the process of stratospheric air transfer to the troposphere due to diffusion is a secondary factor. The periodic rushes of the dry stratospheric air into the troposphere which contains excessive concentrations of radioactive substances and have large potential velocities determine the mechanism of stratospheric–tropospheric exchange of air masses.

Another process responsible for the ejection of highly radioactive stratospheric air pertains to powerful convective storm fluxes rushing into the stratosphere. These air fluxes, containing a large amount of water vapor, are mixing with radioactive substances and return with a high radioactive content as subsequent precipitation.

It has been found that the greatest concentrations of various radioactive nuclei, e.g., ^{90}Sr, ^{14}C, and T, in precipitation always occur in the spring–summer period. However, the maximum T-values appear with a delay of 1 month compared with that of Strontium-90 (^{90}Sr) and 1–2 months earlier than the maximum ^{14}C-values at the same latitude. The reason for this lies in the differences in their masses and physicochemical properties relative to those of the other atmospheric constituents. In the stratosphere, the behavior of tritium a constituent of water molecules HTO, is the same as that of ^{90}Sr and ^{14}C since gaseous T is usually transferred through the air by eddy diffusion.

It has been found that the $T/^{14}C$ ratio in the lower stratosphere exceeds that in the humid air layers and is equal to 0.4. While approaching the humid layers of the troposphere, stratospheric HTO molecules may exchanged with water at the ocean surface, reevaporated, and transferred in groundwater by infiltration.

The water circulation process in nature is as follows. Water with some tritium evaporates from the ocean surface. The water vapor rises and reaches the temperature of condensation by cooling. Dry air from the lower stratospheric layers with a high tritium content exchanges tritium with the rising ocean water vapor, which contains low concentrations of tritium. The time of exchange appears to be an important factor, the longer the vapor stays in the atmosphere, the greater is the probability that it will become enriched in tritium. At some thermodynamic and meteorological conditions precipitation is formed which is a mixture of ocean and stratospheric water vapor.

The water balancing equation is based on the relationship between radioactive fallout and meteorological parameters for a given atmospheric volume. Near the Earth this equation is:

$$E - P - \Delta F = 0, \tag{13.1}$$

where P is precipitation; E is evaporation–transpiration; ΔF is the derivation of the water flux in the considered air volume.

In order to establish the relationship between the equation and the real conditions, it has been assumed that water transfer is caused by winds characterized by velocities which vary in space. The balancing equation at a concentration of the indicator C, and for a vertical column of air relative to the Earth's surface, is as follows (Schell et al. 1970):

$$\iint (EC_E - PC_P - Q_t C_t)\mathrm{dxdy} = \iiint \left[\frac{\partial(QC_v \cos\theta)}{\partial x} + \frac{\partial(QC_v \sin\theta)}{\partial y}\right]\mathrm{dxdydz}, \tag{13.2}$$

where Q_t is an upward loss of moisture; C_E, C_P, C_t and C_v are the tritium concentration in evaporation, precipitation, and atmospheric moisture at the border of air volume accordingly; Q is the flux of moisture; θ is the angle of wind direction.

The equation may be simplified for computation:

$$\langle \bar{E}\bar{C}_E\rangle - \langle \bar{P}C_P\rangle - \langle \bar{Q}_t\bar{C}_t\rangle = \frac{1}{L}\int_z (Q_2 v_2 C_2 \cos Q_2 - Q_1 v_1 C_1 \cos Q_1)dz,$$

where L is the distance between the observation stations, and indexes 1 and 2 for the parameters Q, v and C relate to those stations.

The above model has been applied for estimate the relationship between the content of T, ^{90}Sr, and ^{14}C in precipitation and meteorological parameters characteristic for a number of sites in Western Europe. Those calculations produce reasonable results.

In order to demonstrate the effects of the global distribution of tritium, Romanov (1978) made a number of assumptions in order to obtain a simpler expression describing the transfer of the atmospheric moisture. The balancing equation then becomes:

$$\frac{dW}{dt} = E - P, \tag{13.3}$$

where W is the moisture content in the atmosphere; E and P are the amounts of moisture in evaporation and precipitation per unit of time.

The balancing equation of HTO in the atmospheric moisture above the ocean is

$$d(WC_A) = qdt - C_A P dt + C_0 E dt - M(C_A - C_0)\, dt, \tag{13.4}$$

where C_A is the concentration of tritium in the atmospheric moisture; q is the rate of tritium release from the stratosphere; t is the residence time of an air mass over the ocean; C_0 is the tritium concentration in the surface oceanic layer; M is the rate of the eddy transfer of the atmospheric moisture to the surface oceanic layer. The fractionation of tritium during the phase transition is not accounted for.

The left hand side of the equation expresses the change in the tritium content of atmospheric moisture over the time period *dt*. In the right hand side, the first term defines the discharge of tritium from the stratosphere, the second term corresponds to the removal of tritium in precipitation, the third term to the release of tritium from the ocean surface layer by evaporation, the fourth term to the injection of tritium into the ocean due to molecular exchange. The last term in Eq. (13.4), according to experimental data obtained by Romanov, is proportional to the gradient of the tritium concentration in the system: atmospheric moisture–ocean surface layer together with the water vapor.

In writing Eq. (13.4) it has been assumed that the tritium content in precipitation is equal to its concentration in the whole of the upper atmospheric moisture layer and that the tritium concentrations in the ocean surface layer are little affected compared with concentrations in the atmospheric moisture and the rate of tritium injection from the stratosphere is constant during the whole time of air mass transfer.

Using (13.3), Eq. (13.4) can be rewritten in the form of

$$\frac{dC_A}{dt} + C_A \frac{E+M}{W} = q + C_0 \frac{E+M}{W}. \tag{13.5}$$

The solution of Eq. (13.5) is

$$C_A = C_0 + \frac{q}{E+M}\left\{1 - \exp\left[-\frac{E+M}{W}t\right]\right\}. \tag{13.6}$$

The integration constant can be defined from the initial conditions: $C_A = C_0$ at $t = 0$.

It follows from Eq. (13.6) that the tritium concentration in the atmospheric moisture is greater when the moisture content W is lower and when the interrelationship between the ocean and the atmosphere, expressed by parameters E and M, is weaker. This principle can be observed in nature by analyzing the experimental data on global

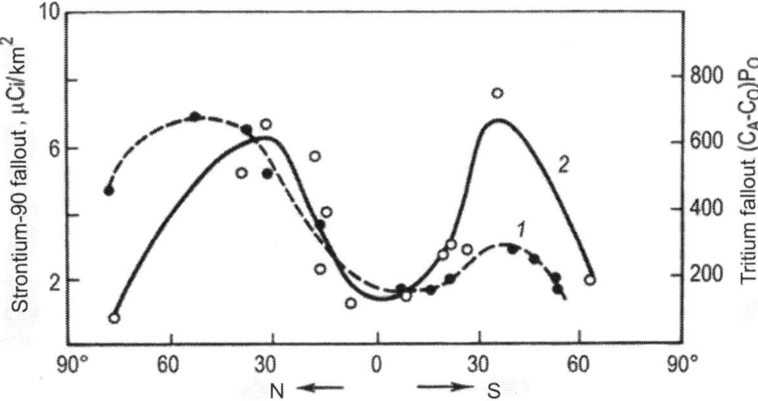

Fig. 13.35 Latitudinal fallout of tritium (*1*) and ^{90}Sr (*2*) from the stratosphere over Atlantic Ocean in 1969

tritium distribution in precipitation. The lower values of moisture content, evaporation, and molecular exchange in high latitudes result in the observed latitudinal effect of tritium concentration exchange in precipitation.

Taking into account that the rate of water evaporation is proportional to the humidity gradient (p_0–p), the rate of molecular exchange is proportional to atmospheric humidity p, and that both processes are identically related to wind velocity, Eq. (13.6) may be rewritten in the form:

$$C_A = C_0 + \frac{q}{kp_0}\left[1 - \exp\left(-\frac{tkp_0}{W}\right)\right], \tag{13.7}$$

where p_0 is the pressure of the saturated vapor at the ocean surface temperature t; k is the parameter determined from the relationship $M = kp$.

If value t is large enough, then

$$C_A = C_0 + \frac{q}{kp_0}. \tag{13.8}$$

From Eq. (13.8) the velocity q of injection of tritium from the stratosphere can be obtained. The plot of dependence of the velocity from the latitude is presented in Fig. 13.35. The values of T and ^{90}Sr content in the atmosphere are placed on the plot. It is seen that in both cases the maximum is located in the belt of 40–50°N, which evidences on the common nature of injection to the troposphere the bomb-tritium and ^{90}Sr.

Let us determine the relationship between tritium concentration and the residence time of the air mass above the continent. The balancing equation in this case is:

$$d(WC_A) + qdt - C_APdt + C_EEdt, \tag{13.9}$$

where C_E is the tritium concentration in the evaporating continental water.

Taking into account the insignificant difference between tritium concentrations in the surface continental water and in precipitation one can assume that $C_E \approx C_A$. In this case, using the balancing equation for the atmospheric moisture (13.3), a solution of (13.9) is obtained in the form:

$$C_A = \frac{q}{W}t + const. \tag{13.10}$$

The integration constant can be determined by the initial conditions $t=0$ at which the tritium content in the atmospheric moisture is equal to its content in moisture transported from the ocean. Then

$$C_A = C_0 + \frac{q}{W}t. \tag{13.11}$$

The last equation accounts for the observed continental effect. In fact, the longer the air mass moves above the continent, the greater the concentration C_A becomes and the better the function $C_A(t)$ may be approximated by a linear dependence, which is confirmed by experimental data obtained in many regions of the world.

Further development of the models establishing the relationship between tritium concentrations in precipitation and meteorological parameters requires a deeper understanding of the natural principles governing air mass circulation in the atmosphere. Experimental data already obtained on tritium falls and their analysis will, in turn, lead to better understanding of the natural principles and also to a more reliable and well-grounded hypothesis.

13.5 Tritium in Ocean Waters

The oceans are the main reservoir of the hydrosphere and the main source of atmospheric moisture on the Earth. From continental run-off, direct falls of precipitation and exchange with the atmosphere, the oceans receive the majority (about 90%) of natural and bomb-tritium. Therefore, the oceans are the main reservoir of the tritium accumulation on the Earth. The distribution of tritium in the surface and deep ocean layers is of interest while studying the principles of water circulation of the oceans together with atmospheric moisture and, in particular, of the ocean itself.

Before the thermonuclear tests the tritium concentrations in the ocean water, measured at different sites, had characteristic values from ~0.5 TU (Kaufmann and Libby 1954) to 1 TU (Begemann and Libby 1957; Brown and Grummit 1956). After the thermonuclear tests in March 1954 tritium concentrations increased at an average to 1.9 TE (Giletti et al. 1958). It was difficult to measure such a concentration by the existing that time techniques with an appropriate accuracy.

After 1980, the systematic measurements of tritium concentrations in the oceans were started. The obtained results allowed Östlund and Fine (1979) to calculate the approximate amounts of tritium in different ocean regions (Table 13.11).

Tritium content in the ocean waters is defined by the effect of interaction between the ocean surface and atmospheric moisture, which is developed in precipitation, by

Table 13.11 Tritium inventory in the oceans. (From Östlund and Fine 1979)

Ocean	Tritium (kg)
North Atlantic	66
South Atlantic	7
Arctic Basin	6
North Pacific	59
South Pacific	14
Antarctic	6
Indian Ocean	6
Total	164

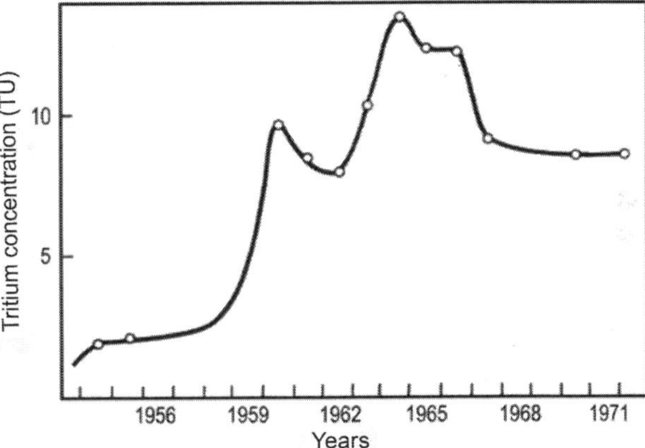

Fig. 13.36 Latitudinal variation of tritium concentration along 30°N in Atlantic Ocean in 1954–1972

the evaporation and molecular exchange, by life time of the surface layer existence, depending on the mean time of vertical mixing of water, and by interaction of water masses having different origin and tritium concentrations.

The general character of tritium content variations in the surface waters of the ocean reflect the picture of tritium input into precipitation including the rise of tritium content during the period of thermonuclear tests and seasonal cycles in each year. At the same time, it follows from the data that the input to the surface layer is delayed and the amplitude of tritium concentration is decreased because of dilution in larger water volume. Figure 13.36 shows variation of tritium concentration in surface water layer along 30°N in Atlantic Ocean plotted by the data of a number of researchers (Dockins et al. 1967; Giletti et al. 1958; Münnich and Roether 1967; Östlund and Fine 1979; Rooth and Östlund 1972).

The corresponding changes in tritium content because of the series of thermonuclear tests in 1958 and 1961–1962 is obvious. However, the peaks are shifted on 2–3 years compared with concentrations in precipitation. The maximum of the amplitudes with higher tritium concentrations were observed in the middle and high latitudes (Östlund and Fine 1979) (Fig. 13.37). The effect is strongly smoothed in the equatorial latitudes.

Tritium content variations in the surface layer differ seasonally. In the summer time, surface concentrations are higher because of its high content in the atmospheric

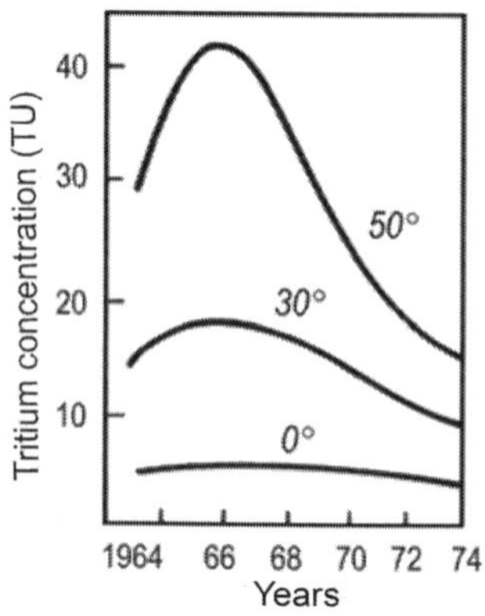

Fig. 13.37 Maximum changes in tritium concentrations in surface layer of Pacific Ocean. (After Östlund and Fine 1979. © IAEA, reproduced with permission of IAEA)

moisture, higher velocity of the molecular exchange and lower dilution of surface layer by the deep low-active oceanic water due to the thermocline effect. Dockins et al. (1967) studied tritium concentrations in Pacific Ocean between 14°S and 52°N during 1959–1966. The highest concentrations correspond to the summer maximum in the northern part of the ocean which marked smoothed in the low latitudes. In the southern hemisphere, the tritium concentrations in the surface layer decline quickly, which is in agreement with the general picture of tritium falls in precipitation.

Münnich and Roether (1967) studied tritium variations in the surface and depth profiles of the Atlantic Ocean. As for the Pacific Ocean, the latitudinal distribution of tritium in the surface water has maximum concentration in the mid-latitudes of the northern hemisphere. The minimum concentrations of tritium were observed in the equatorial region (not more than 1–3 TU), being obviously close to the natural level observed before the bomb tests. The depth of the tritium mixing layer also increases with latitude from 100 m near the equator to several hundreds of meters in high latitudes. The absolute values of tritium concentrations in the Pacific Ocean are higher than those in the Atlantic Ocean due to the more intensive vertical mixing.

Rooth and Östlund (1972) studied the distribution of tritium in the North Atlantic surface waters during 1963–1968. They confirmed the latitudinal increase of tritium content there. Longitudinal variations in concentration are small. They proposed a model applicable for calculation of the vertical turbulent transfer through the ocean thermocline, including advection.

Bainbridge (1963), Michel and Suess (1975) and other researchers, using the ratio between tritium content in the surface ocean layer and in atmospheric moisture carried out assessment of residence time in that layer. For this purpose, they applied a box model where the ocean was divided on two reservoirs: first was the surface well-mixed layer, and second was deep layer with lower tritium concentration compared with the surface one. Thus, it was assumed that tritium input is provided only through the dividing border between the atmosphere and the ocean and the tritium loss results by the radioactive decay in the well-mixing layer. Horizontal advection is refused here because the horizontal tritium gradient in the open ocean is, as a rule, very small.

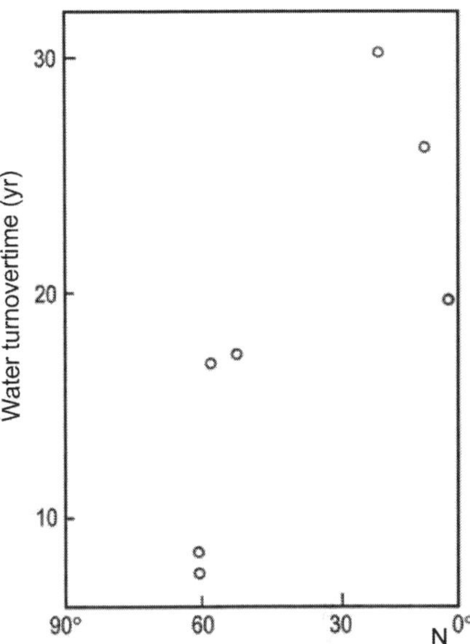

Fig. 13.38 Relationship between water exchange time and latitude for surface layer of the Pacific Ocean

Applying the box model, Michel and Suess (1975) has found that water exchange time for 12 stations in the Pacific Ocean for the period of 1967–1973 is equal to 7.5–26 years. At the same time, they obtained maximum velocities in the subtropical latitudes and minimum in the equatorial zone (Fig. 13.38). Values close to the above, but calculated by a modified methodology, were obtained by Romanov (1982) for the Black and Caspian Seas (10–15 years) and for the margin seas of the Arctic Ocean.

The vertical tritium distribution in ocean water has its specific features related to its circulation on the globe. A typical picture of the vertical tritium distribution is shown in Fig. 13.39 (Östlund 1973). It is specified by presence of the density jump at the thermocline, which confined the surface layer with relatively high level of tritium content and the deep layer with almost zero tritium content. The two layers are divided by an intermediate thin layer where the vertical turbulent mixing of tritium takes place. Such conditions are characteristic for the equatorial and low latitudes.

Applying the tritium gradient of concentration, it is possible to calculate the coefficient of vertical turbulent diffusion K_z. Roether et al. (1970) used this method and for the northern part of Pacific Ocean obtained $K_z = 0.15\ \text{cm}^2/\text{s}$. Rooth and Östlund (1972) for the Sargasso Sea obtained the value of K_z close to the calculated value. In order to obtain this parameter, the authors compile an equation which takes into account the processes of the turbulent diffusion and the vertical advection. In this case, the change of tritium concentration $C(x, y, z, t)$, as a conservative radioactive indicator, in time t in any point can be written in the form:

$$\frac{dC}{dt} = \frac{K_z \partial^2 C}{\partial z^2} + K_H \left(\frac{\partial^2 C}{\partial x^2} + \frac{\partial^2 C}{\partial y^2} \right) - \bar{\mathrm{V}} \nabla \mathrm{C} - \lambda \mathrm{C}.$$

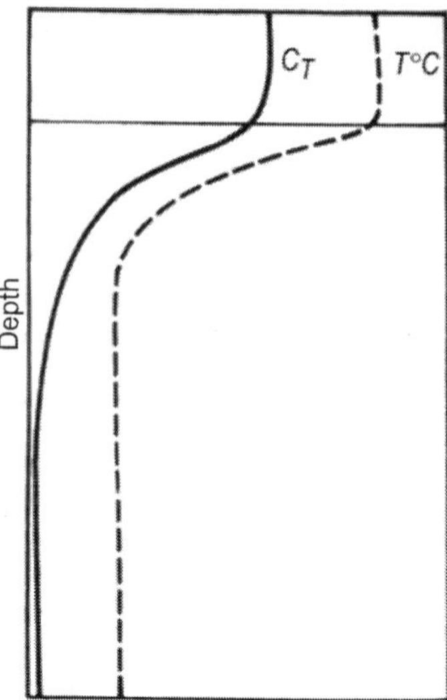

Fig. 13.39 Typical vertical tritium distribution in oceanic waters

The coefficient of turbulent diffusion K_H in the horizon plane (at x and$y$) is taken the same. The vector of velocity of advective transfer $\bar{V}$ appears to be the vectors sum on all the three space coordinates.

Romanov (1982), in order to determine the coefficient of vertical diffusion for the Black Sea, applied a simplified equation in the form:

$$\frac{dC}{dt} = \frac{K_z \partial^2 C}{\partial z^2} - \lambda C.$$

Its solution for nonstationary conditions is

$$C = \frac{1}{2} C_n \left[\exp\left(-\sqrt{\frac{\lambda}{K_z}} z\right) \left(1 - erf \frac{z - 2\sqrt{K_z \lambda t}}{2\sqrt{K_z t}}\right) \right.$$
$$\left. + \exp\left(\sqrt{\frac{\lambda}{K_z}} z\right) \left(1 - erf \frac{z - 2\sqrt{K_z \lambda t}}{2\sqrt{K_z t}}\right) \right],$$

where C_n is the tritium concentration in the surface layer of marine water.

As the tritium concentration C_n in the surface layer does not expressed in analytical form due to the occasional character of thermonuclear tests, then a numerical solution with presentation of C_n in a histogram form is used. After that the value of C for each step is found. This way was used for the interpretation of Black Sea data, where the value of $K_z = 0.1\ \text{cm}^2/\text{s}$ was obtained.

In high latitudes, where the process of air masses vertical mixing is more powerful, vertical tritium distribution appears to be more complicated. The same picture of tritium distribution for the regional concentrations and divergence of ocean water is observed.

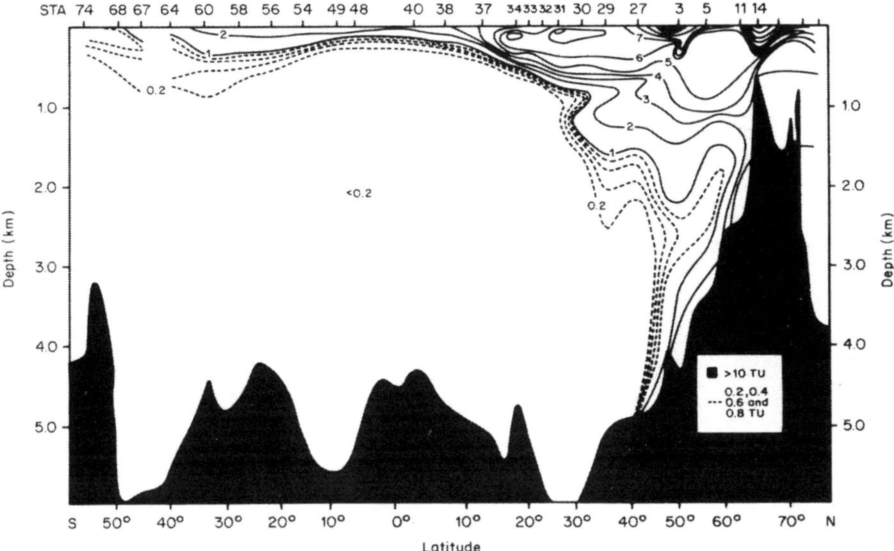

Fig. 13.40 Meridional distribution of tritium concentrations in Western Atlantic according to GEOSECS studies in 1972. (After Östlund and Fine 1979. © IAEA, reproduced with permission of IAEA)

In 1972, tritium content studies at the deep water stations in the western Atlantic Ocean from 3 to 74°N were carried out as a part of the GEOSECS program (Östlund 1973; Östlund and Fine 1979; Östlund et al. 1974). The results obtained (Fig. 13.40) were mainly in agreement with those obtained by Münnich and Roether (1967) and supplemented their profiles to the north and south. In the north latitudes, the penetration of the bomb-tritium—with concentration more than 0.2 TU—was observed up to the depth of 3,500 m, whereas in the equatorial region a sharp decrease in content was observed at the depth of only 200 m.

Analogous profiles, obtained by the same authors for the Pacific Ocean (Fig. 13.41), shows that here bomb-tritium has not fell so deep as it was in Atlantics at the north, and most of it was in the upper layers which circulate in opposite sides in both hemispheres.

The northern and southern currents divide these systems. The deepest tritium penetration (up to 1,000 m) in the northern part of Pacific Ocean is observed. The most complete mixing, especially up to 500 m, in the region of 20–40°N is reached. The maximum tritium concentrations in the near-equatorial current (8–20°N) at the depth of 200 m and at the surface are observed. This is because being formed at higher latitudes and moving along constant density, oceanic water are dropped. Analogous results and conclusions were obtained in their earlier works by Michel and Suess (1975).

Asymmetric distribution of tritium concentrations in both oceans relative to equator is explained by prevailing fallout in the northern hemisphere.

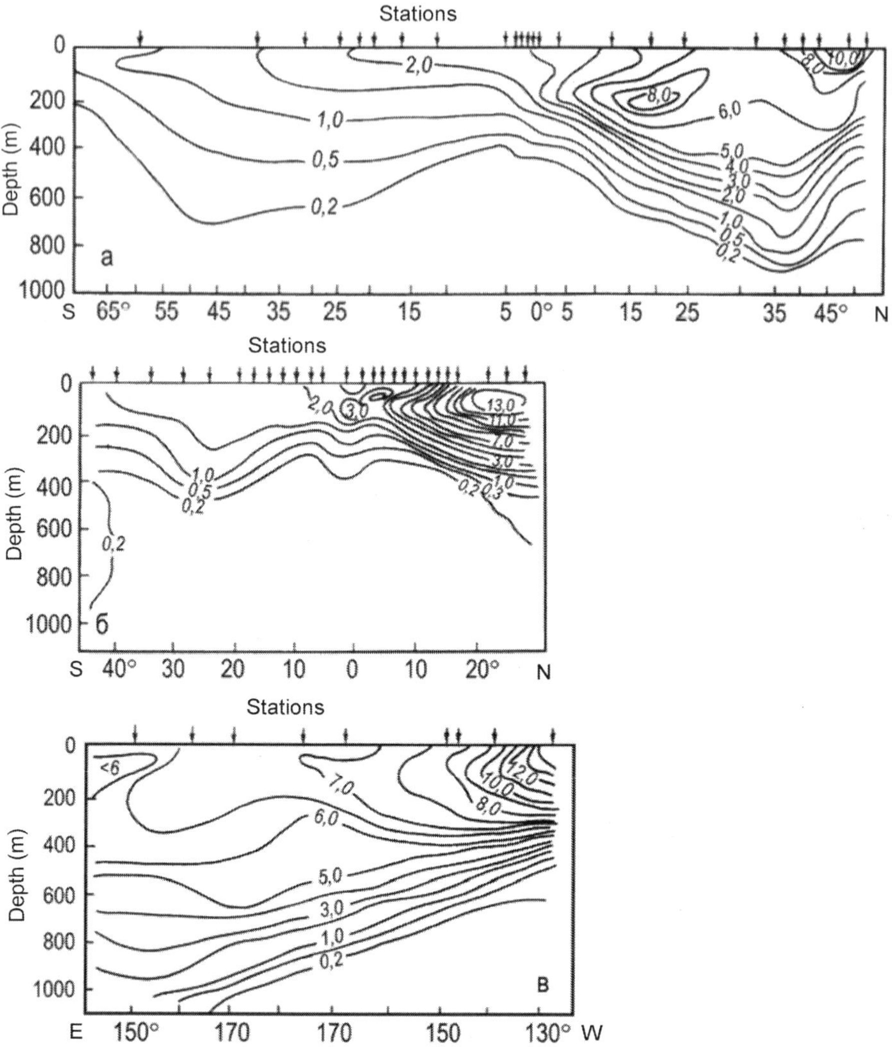

Fig. 13.41 Distribution of tritium concentrations (TU) in waters of western (**a**), eastern (**b**) and northern (**c**) parts of Pacific Ocean for 1973–1974. (After Östlund and Fine 1979. © IAEA, reproduced with permission of IAEA)

Note that the marked influence on distribution of tritium in the surface oceanic layer in a near-shore and continental regions the river run-off effects. There is also a good correlation between the tritium concentrations and the salinity and temperature of the sea water.

13.6 Tritium in Continental Surface Waters

The continental surface waters, together with the precipitation, are an important transport chain of the global water circulation on the Earth. At evaporation they markedly affect on isotopic composition of atmospheric moisture. The surface continental waters are the main source for the groundwaters recharge and in significantly degree determine conditions of formation of isotopic composition and salinity of the marginal seas. A study of isotopic composition of the surface waters helps in obtaining valuable information about the parameters of water dynamics of the river and lakes basins. Thus, the study of regularities in tritium distribution of surface waters is an important scientific problem.

13.6.1 Tritium Content in River Water

The factors which determine tritium content in the river waters are tritium concentrations in precipitation over area of the river catchment basin, residence time of the infiltrating precipitation water in the soil through which it discharged into the groundwater and the river.

Let us express the analytical relationship between the tritium concentration in river water and the water exchange velocity in a river basin. Assume that river water in a basin represents the surface run-off and groundwaters which are well mixed. The age spectral function of river water can be written as $a(t) = \varepsilon \exp(-\varepsilon t)$, where $a(t)$ is the relative portion of precipitation in river water with age t; ε is the exchange velocity (the ratio of total annual recharge to the water volume of the catchment basin). Here value $1/\varepsilon$ has dimension of time (in years). Therefore,

$$\int_0^\infty a(t)dt = 1.$$

The tritium concentration in the river water C_r can be expressed by the equation

$$C_r = \int_0^\infty C_p(t)\varepsilon e^{-\varepsilon t} e^{-\lambda t} dt,$$

or

$$C_r = \varepsilon \int_0^\infty C_p(t) e^{-(\varepsilon+\lambda)t} dt, \tag{13.12}$$

where λ is the tritium decay constant; $C_p(t)$ is the tritium concentration in precipitation at the time moment t.

At $C_p(t) = \text{const} = C_p$ (e.g., during the prethermonuclear period) one obtains

$$C_r = \frac{\varepsilon C_p}{\varepsilon + \lambda}. \tag{13.13}$$

In order to obtain the solution of Eq. (13.12) when $C_p(t) \neq \text{const.}$ (after the beginning of the tests) the function $C_p(t)$ should be rewritten in the form of a histogram. Then

$$C_r = \frac{\varepsilon}{\varepsilon + \lambda} \left\{ \sum_{i=1}^{n} [C_{p_{i=1}} - C_{p_i}] \exp[(\varepsilon + \lambda)(t_i - t_{i-1})] \right\} + \frac{\varepsilon}{\varepsilon + \lambda} C_{p_n}, \tag{13.14}$$

where C_{p_n} is the concentration of tritium in precipitation for the time interval $(t_i - t_{i-1})$; n is the number of the time intervals in a period t.

Eriksson (1963) proposed another model for the interpretation of tritium data. It is based on the assumption that groundwaters, which are recharged by precipitation, move in parallel to the watershed. Then the precipitation waters, which fall at different distances from the watershed, do not become mixed. According to this model, $a(t) = \varepsilon$ for t ranging from 0 to $1/\varepsilon$, equal to the residence time of infiltration water during its course from the watershed to the bed of the river. Then, the concentration of tritium in the river is as follows:

$$C_r = \varepsilon \int_0^{1/\varepsilon} C_p(t) e^{-\lambda t} dt. \tag{13.15}$$

At $C_p(t) = \text{const.} = C_p$ one has

$$C_r = \frac{C_p \varepsilon}{\lambda} (1 - e^{-\lambda/\varepsilon}). \tag{13.16}$$

Note that the ratio of the C_p-values, obtained from the expressions (13.13) and (13.16), is equal to 2.3.

Putting the expression for $C_p(t)$ written in the form of the histogram in (13.16), one obtains

$$C_r = \frac{\varepsilon}{\lambda} \left\{ C_{p0}(e^{-\lambda t} - e^{-\lambda/\varepsilon}) + \sum_{i=1}^{n} C_{p_i} e^{-\lambda(t - t_i)} (1 - \exp[-\lambda(t_i - t_{i-1})]) \right.$$
$$\left. + C_{p_n}(1 - e^{-\lambda t}) \right\}. \tag{13.17}$$

Here the value of t should be less than that of the maximum residence time $1/\varepsilon$.

However, a more reliable method for the estimation of water exchange rates in river basins is a balance method, proposed by Eriksson (1965b). In somewhat simplified form, it is as follows (Romanov 1978):

The spectral function of the water's age can be represented by a number of fixed values $a_0, a_1, \ldots, a_i$, where index i corresponds to the portion of infiltration water characterized by the age t_i. Thus, the portion of precipitation falling in the year of observation is denoted by a_0, that in the previous year by a_1 and so on.

The balancing equation of water is

$$\sum_{i=0}^{\infty} a_i = 1. \tag{13.18}$$

An expression for tritium concentration in river run-off can be written as

$$C_r = \sum_{i=0}^{\infty} a_i C_{a_i} e^{-\lambda t}. \tag{13.19}$$

Before the thermonuclear tests, n years ago, $C_a = \text{const.} = C$, then

$$C_r = \sum_{i=0}^{\infty} a_i C_{a_i} e^{-\lambda t} + C \sum_{i=0}^{\infty} a_i e^{-\lambda t}. \tag{13.20}$$

While studying river basin, the second term on the right-hand side of Eq. (13.20) can be neglected, so that

$$C_r = \sum_{i=0}^{\infty} a_i C_{a_i} e^{-\lambda t}. \tag{13.21}$$

One should write n equations, such as (13.21), for the solution of the problem. In practice, the upper limit of summation can be bounded, for example, in the work of Soyfer et al. (1970) the value of $n = 1/\varepsilon$ has been determined beforehand from the conditions $i = 3$, i.e., for the coefficients a_0, a_1, a_2. The other coefficients were put equal to each other. Then

$$\begin{aligned}
C_{r_0} &= a_0 C_{a_0} + a_1 C_{a_1} e^{-\lambda} + a_2 C_{a_2} e^{-2\lambda} + [1 - (a_1 + a_2 + a_3)] \\
&\quad \exp\left\{-\frac{1}{2\varepsilon} + 1{,}5\right\} \frac{1}{1/\varepsilon - 2} \sum_{3}^{1/\varepsilon} C_{a_i} e^{-i\lambda}, \\
C_{r_2} &= a_0 C_{a_2} + a_1 C_{a_3} e^{-\lambda} + a_2 C_{a_4} e^{-2\lambda} + [1 - (a_1 + a_2 + a_3)] \\
&\quad \exp\left\{-\frac{1}{2\varepsilon} + 1{,}5\right\} \frac{1}{1/\varepsilon - 2} \sum_{3}^{1/\varepsilon} C_{a_{ii+2}} e^{-i\lambda}, \\
C_{r_i} &= a_0 C_{a_i} + a_1 C_{a_2} e^{-\lambda} + a_2 C_{a_3} e^{-2\lambda} + [1 - (a_1 + a_2 + a_3)] \\
&\quad \exp\left\{-\frac{1}{2\varepsilon} + 1{,}5\right\} \frac{1}{1/\varepsilon - 2} \sum_{3}^{1/\varepsilon} C_{a_{ii+1}} e^{-i\lambda}.
\end{aligned} \tag{13.22}$$

Treating the system of equations (13.22), one can obtain the values of a_0, a_1, a_2. In the similar way, any number of equations can be written and solved.

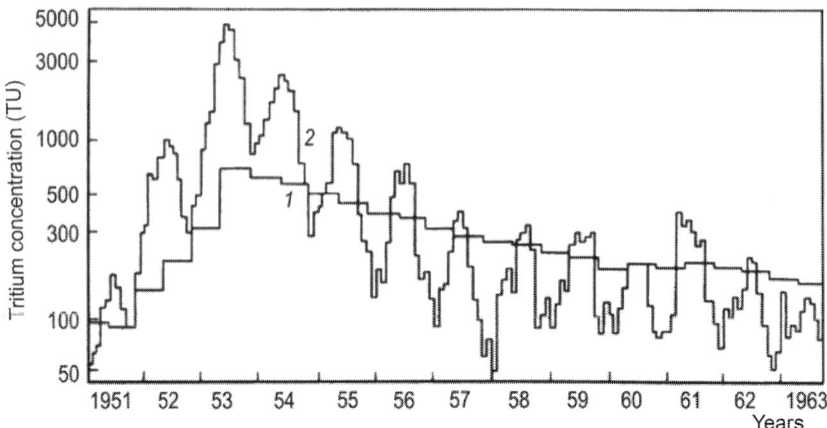

Fig. 13.42 Variation of tritium concentrations in Rheins River (*1*) and in precipitation (*2*). (After Weiss and Roether 1975. © IAEA, reproduced with permission of IAEA)

Applying the balance method, Eriksson (1965b) has reestimated the partition function of the age of run-off in the Ottawa River. Using his solution and the box model, Romanov (1978) has reported that the value of ε for the Moscow River in 1964–1965 is 0.195. This value coincides with the results obtained on the basis of the laminar flow model, following from Eq. (13.17). In this case, the values of the coefficients are a_0 = 0.25; a_1 = 0.18; a_2 = 0.14; $a_i = 0.07$ at $i = 3$–8.

The tritium content in river water, as well as in the atmosphere, is varied in time and space. Long-term distribution of tritium concentrations is governed mainly by that of tritium content in precipitation. Figure 13.11 shows annual means of tritium content in precipitation over the central part of the former USSR territory and in the Moscow River water from 1958 to 1981 (Romanov 1982). The maximum tritium values in river water were observed in 1963 and are accounted by 0.25 part compared with precipitation. Decrease of HTO values in the river water was more smoothed than in precipitation. After 1971, the tritium level in river and precipitation were equalized.

Weiss and Roether (1975) were carried out analogous observations, starting in 1957, for Rheins River (Fig. 13.42). A study in the Ottawa River, as well as in precipitation, started in 1953 (see Fig. 13.10). Prethermonuclear concentrations were measured only in two tributaries of the Mississippi River near St Louis (Illinois, USA): in January 1952 it was 5.6 ± 0.6 TU, and in August 1952 it was 1.15 ± 0.08 TE (Stewart 1965).

Figure 13.43 shows variations of tritium content in the Colorado, News, Arkansas, and Potomac rivers during 1963–1964 according to USGS data (Stewart 1965). One can see in the figure that the tritium peak of 1963 is presented in precipitation (1) and in waters of all the rivers (2). The tritium peak in rivers appears with an average delay ranging from several days of even hours for the mountain regions (see Fig. 13.43a) to a year or even more for the plain regions (Fig. 13.43b), depending on the geological and

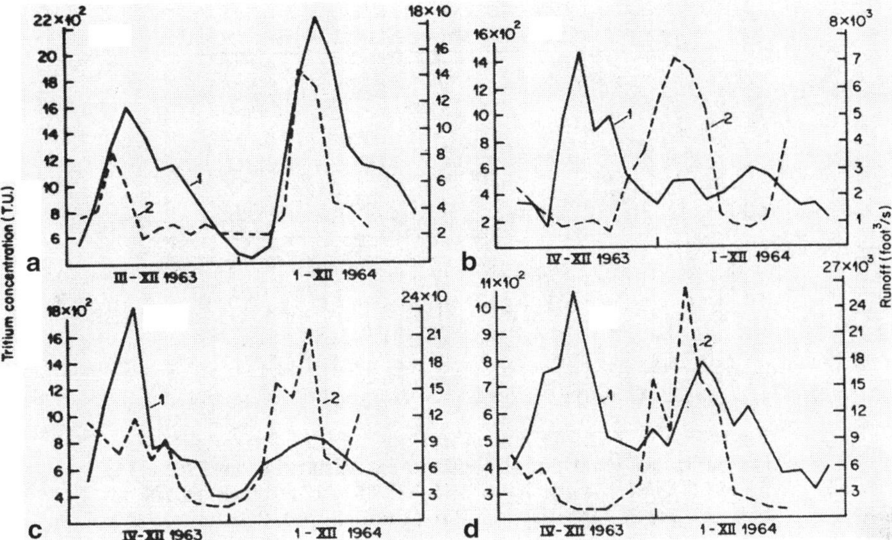

Fig. 13.43 Variation of tritium concentrations in precipitation (*1*) and in river water (*2*) of Colorado (*a*), News (*b*), Arkansas (*c*) and Potomac (*d*). (After Stewart 1965. © IAEA, reproduced with permission of IAEA)

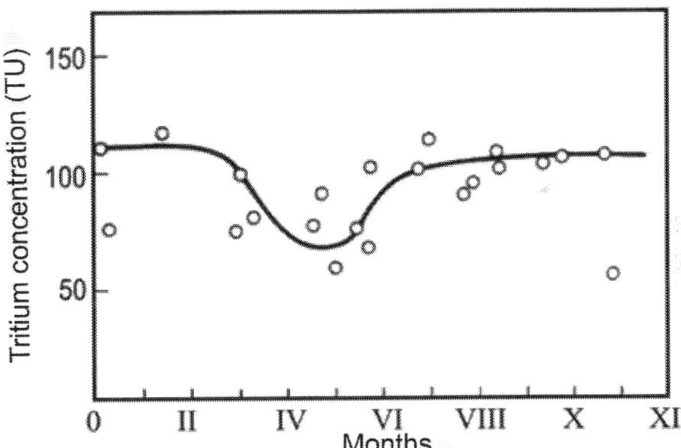

Fig. 13.44 Seasonal means of tritium distribution in Pechora River 1976–1979

geographical conditions of their recharge. The travel time taken by a tracer to move from the catchment area to the river bed is an important parameter, characterizing drainage properties and the capacity of a basin. This parameter, and also the general character of tritium variations in the river water and precipitation over time, permits estimations of the residence time of tritium in the drainage can be gained.

The seasonal variations in tritium concentration are markedly lower than in precipitation. The inner annual maximum values of tritium content for the majority of rivers in the former USSR territory in the 1980s of the last century were shifted to the period of winter time. The seasonal distribution of tritium content for Pechora River during 1976–1979 is shown in Fig. 13.44. Analogous picture was received for Moscow River in 1973–1974.

The minimum tritium values are observed in the flood period, when the river recharge is mainly provided by melted winter snow, where tritium concentrations are by 1.5–2 times lower of the mean annual (Romanov 1982).

After flood pass, the rivers are recharged by the groundwaters of the catchment, where tritium accumulates by the previous fallouts in precipitation. Such a result was obtained in the Upper Angara River in autumn 1973, where after precipitation with about 250 TU of tritium content the river water activity contained about 350 TU (Romanov 1982).

The seasonal, annual, and long-term variations of tritium concentration were sucessfully used by many researchers for the calculation of hydrographs in different river basins, for the determination of residence time of water and water exchange time. For example, Brown (1970), using the box model, found for the Ottawa River (see Fig. 13.10) that about 75% of the river volume of water represents the surface run-off with mean residence time about 1 year and the volume rest of about 25% is the underground run-off which has residence time of about 6 years.

Weiss and Roether (1975) separated three components in the surface run-off on the basis of tritium concentrations in the Rheins River water and precipitation during 1961–1973 and by appropriate modeling. It was found that 25% of the run-off has the value of residence time less than 1 year, 35% have 5 years and 40% do not involve tritium at all. It seems that was old groundwater.

In addition to the time variation, there are space changes in tritium content of river water. Table 13.12 demonstrates tritium data obtainer in river waters and precipitation over a number of river basins in 1979, where also some hydrological characteristics of basins are shown (Romanov 1982).

The values of run-off coefficients of the tritiated water for the river basins were obtained by those data. The coefficients are equal to the ratio of tritium value discharged by the river to the sea and amount of that isotope fell out with precipitation over the catchment. The run-off coefficient is always equal to the run-off coefficient of the water, but some times it is rather different from it. An attempt was undertaken to compare the ratio of these values and also the ratio of the annual mean value of tritium concentration and the coefficient of the underground run-off. Figure 13.45 shows dependence of these values (relative to tritium concentration) on the underground run-off coefficient, which were taken from the work of Domanitsky et al. (1971).

Relationship between these values may be expressed by the regression equation in the form of $C_r/C_p = 1.4K_n + 1.65$, where C_r and C_p are the tritium concentration in the river water and precipitation over the catchment area of a corresponding river; K_n is the coefficient of the underground water run-off of the river basin. The correlation coefficient of the regression equation is $r = -0.762$.

The geographical distribution of the relative tritium concentration in river water, for the European part of the former USSR territory, is characterized by a lower or close to unit value. At the same time, for the Siberian and the Far East regions that value is within 1.05–1.69 values and is increased toward east direction. This tendency is well expressed for the northern rivers (Fig. 13.46). The relative tritium concentrations in waters of the north rivers located between the North Dvina and

Table 13.12 Tritium concentrations in river waters and precipitation and some hydrological characteristics of the basins for the former USSR territory in 1979

River	Catchment area $\times 10^3$(km^2)	Run-off (km^3)	Precipitation (km^3)	Coefficient of surface run-off (K_n)	Concentration of tritium (TU)		Coefficient of surface tritium run-off, K_T	K_T/K_n	Coefficient of groundwater run-off
					River	Precipn			
Pechora	322	130	229	0.57	98	105	0.53	0.93	0.19
Northern Dvina	357	110	255	0.43	83	90	0.40	0.92	–
Onega	57	16	41	0.39	86	79	0.42	1.09	0.32
Neva	281	80	212	0.38	141	72	0.74	1.96	–
Svir	66	19	52	0.38	143*	67[a]	0.81	2.13	–
West Dvina	88	21	68	0.31	67	68	0.30	0.99	0.35
Dniester	72	8.7	42	0.21	81	72	0.24	1.12	0.43
Yuzhny Bug	64	3.4	34	0.10	68	73	0.09	0.93	0.28
Dnieper	504	54	343	0.16	64	74	0.14	0.86	–
Don	422	30	240	0.12	65	80	0.10	0.81	0.56
Volga	1,360	243	900	0.27	90	90	0.27	1.00	–
Oka	188	29	212	0.24	51	78	0.16	0.65	0.62
Moscow	5	1	3.6	0.28	70[a]	66[a]	0.30	1.06	0.24
Kama	522	124	363	0.35	105	100	0.37	1.05	0.30
Ural	229	11.6	102	0.11	92	110	0.09	0.84	0.56
Ob[b]	2,450	390	1,650	0.24	126	125	0.24	1.01	–
Yenisey	2,620	623	1,360	0.46	142	120	0.54	1.18	0.30
Angara	1,045	138	455	0.30	55	140	0.12	0.39	–
Upper Angara	21.6	7.8	11.9	0.65	145*	137[a]	0.69	1.06	0.46
Lower Tunguska	483	116	251	0.46	182	130	0.64	1.40	0.29
Lena	2,490	508	1,145	0.44	199	155	0.56	1.28	0.21
Indigirka	360	57	117	0.49	220	130	0.83	1.69	0.09
Kolyma	665	129	270	0.45	116	70	0.75	1.66	0.18
Amur	1,855	343	1,113	0.31	207	155	0.41	1.34	0.15
Sir-Darya	219	14.5	100	0.14	111	105	0.15	1.06	–

[a]Data related to 1980
[b]Hydrological data do not valid for the mouth profiles

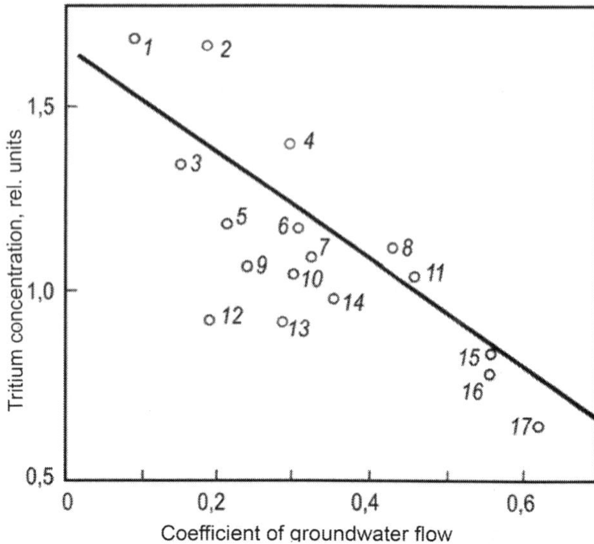

Fig. 13.45 Dependence on tritium concentration in river water from coefficient of the underground water run-off for the rivers: Indigirka (*1*), Kolima (*2*), Amur (*3*), Lower Tunguzka (*4*), Lena (*5*), Enisey (*6*), Onega (*7*), Dnepr (*8*), Moscow (*9*), Kama (*10*), Upper Angara (*11*), Pechora (*12*), Southern Bug (*13*), Western Dvina (*14*), Ural (*15*), Don (*16*), Oka (*17*)

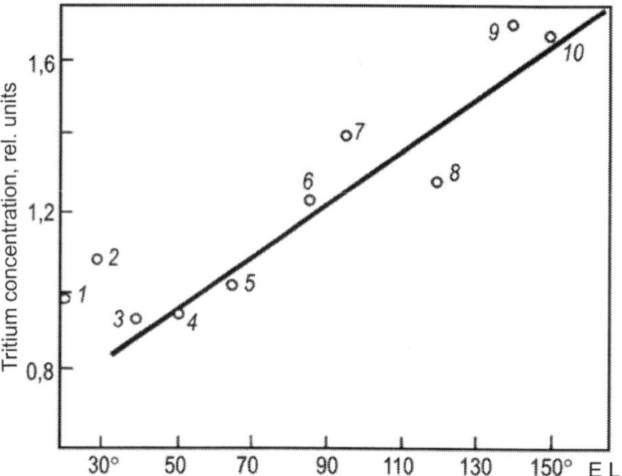

Fig. 13.46 Variation of relative tritium concentration in river waters depending on their longitude: Western Dvina (*1*), Onega (*2*), North Dvina (*3*), Pechora (*4*), Ob (*5*), Evisey (*6*), Lower Tunguzka (*7*), Lena (*8*), Indigirka (*9*), Kolima (*10*)

Kolima are described by the regression equation like $C_r/C_p = 7.15 \times 10^{-3}\lambda + 0.56$ at the correlation coefficient $r = -0.95$ (λ is the degree of the eastern longitude).

The space distribution of the relative tritium concentration in river waters can be explained by the groundwater discharge where its value is lower of unit. In the case when the relative concentration is higher of unit, then the modern precipitation water displaces the groundwater with lower tritium content.

There are experimental data which permit to assess tritium concentrations in groundwaters which take part in the river recharge by use of the equation of isotope balance like $C_r = C_p(1 - K_n) + C_nK_n$, where C_n is the tritium concentration in groundwaters. Some results of the corresponding calculations are in Table 13.13.

Table 13.13 Tritium concentrations in groundwaters taking part in river recharge

River	Tritium concentration in Groundwater (TE)	C_r/C_p
Pechora	68	0.65
Onega	101	1.28
Western Dvina	65	0.96
Dniester	93	1.29
Yuzhny Bug	55	0.75
Don	53	0.66
Oka	34	0.44
Moscow	83	1.26
Kama	117	1.17
Ural	78	0.71
Yenisey	193	1.61
Upper Angara	154	1.12
Lower Tunguska	309	2.38
Lena	364	2.35
Kolyma	325	4.64
Amur	502	3.24

The calculated tritium concentrations in groundwaters are very high and correspond to its content in precipitation fallout 10–15 years ago. In this connection, it is interesting to note that the appropriate groundwater tritium measurements in Yakutia, carried out earlier, just showed high tritium concentrations like that (Afanasenko et al. 1973). At the same time, tritium content in groundwaters of the European part of the former USSR, by multiple studies, very seldom overtopped its value in precipitation (Romanov 1982).

The attempts in finding relationship between tritium content in river waters and some other hydrological parameters have failed.

From the general picture of space distribution in tritium concentration in river waters (see Table 13.12) the data for Svir, Neva, and Angara are of special interest. This is because water of the above rivers represents mainly run-off from the large lakes like Onega, Ladoga, and Baykal where formation of isotopic composition is defined by some other conditions.

Let us estimate the total amount of tritium fallout over the former USSR territory and compare it with the discharge by the rivers. It is seen from Table 13.14 that the run-off coefficient for tritiated water for all the investigated territory is higher of that for the river run-off (0.45 against 0.35). This is the evidence that the tritium has up to now washed out from the basins. In addition to this, about 65% of total tritium is washed out by the rivers to the North Arctic Ocean, about 24% to the North Pacific Ocean, and only about 11% to all other seas.

13.6.2 Tritium in Lakes and Reservoirs

While studying water exchange in lakes and water reservoirs with the help of the tritium tracer, a number of peculiarities should be taken into account which are affecting

Table 13.14 Tritium fallout and run-off to the sea basins of the former USSR territory (data of 1979–1980)

Basin	Catchment area $\times 10^3$(km^2)	Run-off (km^3/year)	Precipitation (km^3/year)	Coefficient of water run-off	Fallout over basin (Bq $\times$ 10^{15})	Run-off (Бк $\times$ 10^{15})	Coefficient of tritium run-off	Annual tritium run-off (% from total)
Barents	1,260	420	890	0,48	9,6	4,0	0,44	6,0
Baltic	661	171	510	0,34	4,6	2,1	0,46	3
Black, Asov	1,347	159	880	0,18	7,9	1,3	0,17	1,8
Caspian	2,927	300	1,430	0,21	17,2	3,3	0,19	4,5
Kara	6,630	1,331	3,700	0,36	58	21,6	0,38	29,4
Laptev	3,670	777	1,650	0,47	29,6	18,5	0,62	25,2
East Siberian	1,326	233	460	0,51	4,4	2,8	0,63	4
Okhotsk	2,561	626	1,600	0,39	24	17,5	0,73	24
Central Asia (non recharged region)	2,420	125	730	0,17	8,7	1,6	0,18	2,1
Total	22,802	4,142	11,850	0,36	164	72,9	0,45	100

tritium distribution patterns in them. Among them are the regime of the recharge, the ratio of the catchment area, and that of the reservoir itself, the existence or absence of run-off, the temperature regime, and many others. All these peculiarities determine whether a reservoir has steady state thermocline or a seasonal one, characterized by a complete mixing of water during cold times. Saline and some tropical reservoirs are of the first type. The majority of other lakes correspond to the second type. For the latter the balancing equations of water and tritium under a study regime can be written in the form:

$$\frac{dV}{dt} = \sum_{i=1}^{n} R_i + P - E \pm U - A, \tag{13.23}$$

where dV is the lake volume change; R_i is the river run-off; P is the precipitation; E is the evaporation; U is the groundwater discharge; A is the surface run-off from the lake.

At a steady state regime ($dV/dt = 0$), then equation of tritium balance is written as

$$\frac{d(CV)}{dt} = \sum_{i=1}^{n} C_i R_i + PC_p - EC_p - M(C - C) - \lambda CV - AC \pm UC_U. \tag{13.24}$$

Here C, C_i, and C_p are the tritium concentrations in the lake water, river run-off and precipitation; M is the rate of the turbulent exchange between the atmospheric moisture and the surface water of the lake (Östlund and Berry 1970); C_U is tritium concentration in the groundwater discharge; λ is the tritium decay constant.

Omitting the value $\pm U = 0$ due to small groundwater tritium discharge in Eqs. (13.23) and (13.24), one obtains

$$\frac{dC}{dt} = \bar{C}_i \bar{R}_i + PC_p - \bar{R}C + PC - M(C_p - C) - \lambda CV, \tag{13.25}$$

where

$$\bar{R} = \sum_{i=1}^{n} R_i;$$

$$\overline{C_i} = \sum_{i=1}^{n} C_i A_i.$$

Introducing the relative balance components $\bar{r} = \bar{R}/V$; $p = P/V$; $m = M/V$, we obtain the first-order differential equation:

$$\frac{dC}{dt} + C(\bar{r} - p - m - \lambda) = \bar{C}_i \bar{r} + C_p P. \tag{13.26}$$

Table 13.15 Tritium concentrations and some of hydrological characteristics for the lakes of the former USSR territory (1979–1980)

Lake, reservoir	Water volume (km^3)	Annual recharge (km^3/year)	Tritium concentration		Water exchange time (year)	C_i/C_p
			Lake C_i (TU)	Precipitation C_p (TU)		
Ladoga	908	79.8	122	66	11.4	1.85
Onega	295	13.6	138	67	21.6	2.06
Sevan	58.5	1.26	80	64	44	1.26
Issyk-Kul	1,730	5.25	35.5	115	330	0.31
Baykal	23,000	70.2	40	128	330	0.31
Valday	0.55	0.044	110	62	12.5	1.78
Aral	1,023	60.3	217	100	17	2.17
Ribinsk Reservoir	25.4	36	86	76	0.71	1.13

In the integral form the last equation becomes:

$$C = \exp[-(\bar{r} - p - m - \lambda)\,t] \int_0^t (\bar{C}_i \bar{r} + C_p p) \exp[-(\bar{r} - p - m - \lambda)\,t]\,dt + \text{const.} \tag{13.27}$$

The solution of Eq. (13.27) can be obtained by the summation of the integrand if the variables $\bar{C}_i \bar{r}$ and $C_p p$ are written in the form of a histogram.

The estimation of the water cycle time in the Baikal Lake, the largest in the world, carried out on the basis of the analysis of tritium patterns in the lake water, influents, and precipitation, has shown that this problem can be treated with the help of the box model. Using the balance method and corresponding experimental data on the tritium content in the water sources, it has been shown that complete water exchange in this lake lasts 330 years, whereas that in the river basins is 2–3 years (Romanov et al. 1979). These results are in good agreement with hydrological estimations (Afanasyev 1960).

Table 13.15 demonstrates the main hydrological characteristics and the data on tritium concentrations in lake waters of the former USSR territory. In order to exclude the influence of geographic differences on tritium content in precipitation over the lake and river catchment areas, the value of relative tritium concentration is used. This value is equal to the ratio of tritium concentration in water and that over the catchment are of the lake or river. This value is used for comparison with the value of time exchange of the lake or river water (see Table 13.15).

The results of tritium study of lakes on the former USSR territory are shown in Fig. 13.47. Here maximum relative tritium concentrations in reservoirs with the water exchange time about 17 years (the Aral Sea and Onega Lake) were found. It is interesting to note that the maximum fallout took place 17 years before the lake water was sampled.

The box model can be applied for the treatment of experimental data for open water reservoirs. The conditions for such treatment are as follows: (1) The equality of the

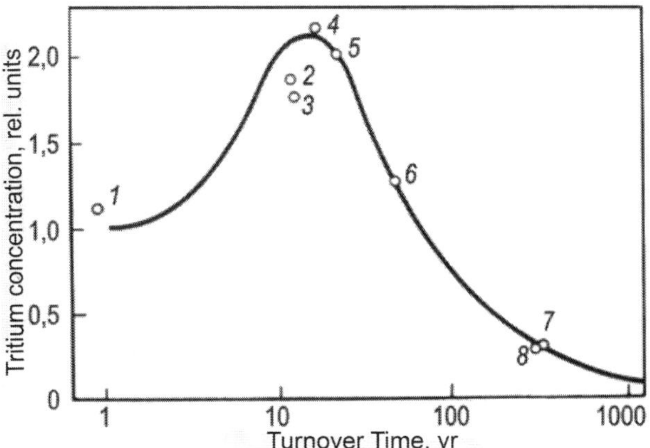

Fig. 13.47 Dependence of relative tritium concentration on the water exchange time for: Ribinsk Reservoir (*1*); lakes Ladoga (*2*); Valday (*3*); Aral (*4*); Onega (*5*); Sevan (*6*); Issyk-Kul (*7*); Baykal (*8*)

mean value tritium concentration in the system itself and in the discharged water and (2) The constancy of volume of the reservoir in time. In fact, the reservoirs with the constant water level and with well-mixed water are satisfied to these conditions.

The box model was used in calculation of water exchange time for river basins, lakes, and reservoirs by many authors (Brown and Grummit 1956; Begemann and Libby 1957; Romanov and Soyfer 1968; Romanov 1978).

13.7 Tritium in Groundwaters

The main source of tritium in groundwaters is precipitation water. However, as pointed out by Andrews and Kay (1982), some amount of tritium is created by nuclear reaction (n, α) on nuclei of ^{6}Li just in the aquifers, especially those represented by acid rocks. The source of neutrons, in this case, is the spontaneous natural uranium and thorium decay. By Andrews and Kay calculations, the above reactions in the pure groundwaters produce tritium concentrations up to 2.5 TU, which can be reached in a granite massif.

The shortest transition time of precipitation and surface water into groundwaters is observed in the case of a direct hydraulic connection between them, which occurs in the regions of tectonic fractures, fissures, and karstic rocks and gravel–pebble sediments. The tritium half-life, equal to 12.43 years, is very often lesser than the groundwater exchange time. Therefore, the tritium radioactive decay is a significant factor which is decreasing its concentration. The intensive exploitation of groundwaters for economic purposes is accompanied with development of the depression funnels and washing of the rocks, intensifying the degree of relationship between the surface and groundwaters and affecting on their tritium content.

In the recent years, when the natural tritium content was disturbed by injection into the atmosphere of large amounts of bomb-tritium, the possibilities of the absolute dating of groundwater with the help of tritium were lost. At the same time, the wide-scale investigations of the tritium falls in precipitation on a global scale allow us to use the obtained data for studying the motion of groundwaters from a somewhat different viewpoint.

Tritium releases into the atmosphere during the decade of nuclear weapon tests occurred in the form of individual pulses, which correspond to a single powerful or a series of moderate explosions. The tritium falls during the test period, and the subsequent period of time, mirrored its injection into the atmosphere in the form of individual pulses differing in magnitude over the yearly cycle. The knowledge of atmospheric water patterns in the underground part of the hydrosphere together with tritium dating information provides the basis for the solution of different time-dependent problems while studying groundwater dynamics.

The study of problems dealing with the transition of tritium 'marks' in groundwaters over time is based on systematic measurements of tritium concentrations in the precipitation of a studied region. The occurrence of tritium in groundwaters depends on their recharge conditions.

The most typical case of recharge of the tritium from the surface aquifer is percolation of surface water through the unsaturated zone. Here, as a rule, the spring–summer component of the annual precipitation, containing the maximum tritium concentration, does not reach the aquifer. This portion of annual precipitation is lost mainly through evaporation–transpiration and partly by surface reservoir recharge. Some portion of the groundwater storage is similarly lost. During the autumn–winter period and in the early spring, when tritium content in precipitation is minimal, the evaporation and transpiration of precipitation water is also reduced. During this period groundwaters are replenished. Thus, in this case tritium concentrations in groundwaters are lower than the annual average concentration in precipitation.

The seasonal variations in tritium concentration were observed (Andersen and Sevel 1974). The delay time of the peak annual concentrations was ranged within about 1–3 months, and the peak itself compared with the tritium peak in precipitation was considerably smoothed. The tritium concentrations in the water of the unsaturated zone were changed in depth. Some times, the maximum concentration, which corresponds to the infiltrating precipitation of the peak fallout, on a definite depth was observed. This depth was depended on the filtration properties of the soil of that unsaturated zone (Andersen and Sevel 1974; Atakan et al. 1974; Morkovkina 1978). Such a tritium distribution was used for the determination of the infiltrated water velocity in that unsaturated zone.

According to the experimental data of Münnich et al. (1967) and Atakan et al. (1974), who carried out their studies on the alluvial plain of the Rhine River, bomb-tritium dating of shallow groundwaters can be used with a sufficient accuracy if the unsaturated soils and the aquifer itself are homogeneous in composition and properties. Tritium change in depth profiles over time, in an unsaturated zone composed of loess-loam soil and in an unconfined aquifer consisting of fine to medium-size and coarse sand were studied by the authors. According to the tritium dating techniques, the average recharge rate of groundwaters in the fine to coarse sand material

Table 13.16 Tritium content in natural waters of Yakutia. (From Afanasenko et al. 1973)

Object	Tritium concentration (TU)
Water from the spring of Upper Nekharan	860
Water from the spring of Yust Nekharan	523
Water from the river mouth Nekharan	382
Snow	252
Precipitation	337

is 200 mm/year which is in good agreement with that measured on the basis of routine hydrologic techniques.

By the tritium distribution in time, the mean residence time of the water is determined in the aquifer. Taking into account that the aquifer water is drained by rivers then the portion of groundwater discharge to rivers can be calculated (Brown 1970) or if this value by some other method is obtained, then by balance relations the mean tritium concentrations discharged to rivers are obtained. The attempt was undertaken to obtain these values for the large northern river basins of the former USSR (Romanov et al. 1982). The results of these calculations, presented in the previous paragraph, show that tritium concentration in the groundwaters is increased from west to east of the basin location and also, for rivers, located to east behind the Ural, their concentration is higher than in modern precipitation. This is because of the continental maximum over the catchment area, and possible due to permafrost affect of the Siberian surface rocks which save tritium concentrations accumulated in the period of maximum fallout in 1962–1966. This idea is proved by the observational data of tritium in groundwaters, rivers and precipitation obtained in Yakutia (Table 13.16).

The aquifers of fractured and karstic rocks have intensive recharge from precipitation. In this case, there are good conditions for infiltration of water and restrictions for its evaporation. Here irregular distribution of tritium concentrations due to the existence of a number of hydrologic subsystems or different ways of water transit is observed.

Devis (1970) studied groundwaters in noncarbonate fractured rocks on volcanic island CheYu in South Korea; substantial variations in tritium content, discovered there. According to the author, the mean residence time of the water in the range of 1–8 years to be found, this is the normal "age" for groundwaters. A similar study was done in Aragatz Mountain in Armenia by Vlasova et al. (1978). The transit time for precipitation between the points of recharge and discharge in the summer time was found to be 2–3 months and in the cold period it was 7–8 months.

The tritium concentrations in such waters are, as a rule, high and close to precipitation, for example, high tritium concentrations in karstic waters with substantial seasonal variation were found in Southern Turkey (Dinçer and Payne 1971).

A number of researchers found good relationship between karstic and surface waters. The velocity of water movement in karstic rocks was found to be from several to hundreds of meters per hour (Fontes 1976).

The anthropogenic activity influence on the groundwater exchange rate is the problem of a special interest, for example, concentration of tritium in the water of the Middle Carboniferous rocks in the Moscow artesian basins, which groundwaters are intensively exploit, in 1978, was equal in average about 50 TU. At the same time, in marginal parts of the basin, which are closer to the recharge area, the values very seldom exceed 10 TU (Zlobina et al. 1980). This shows that anthropogenic activity may lead to substantial disturbance of the natural circulation of groundwaters.

Detailed studies of groundwater motion in saturated and unsaturated zones have been carried out by a number of researchers (Allison and Hughes 1974; Andersen and Sevil 1974; Atakan et al. 1974; Morkovkina 1979; Münnich et al. 1967; Verhagen et al. 1979).

13.8 Dating by Tritium

In view of the interpretation of isotope data and the solution of different problems elucidating groundwater dynamics, different models have become widely used. The mathematical ground for the construction of these models is the balancing equation of water masses and isotope tracers, together with the water dynamics in the system under investigation. Let us now consider some of the hydrological models.

One of the common problems in groundwater studies is the estimation of the residence (exchange) time of water in hydrological system, or, as a sometimes said, the age of the water. It has been suggested that several models applicable for the interpretation of experimental results involving tritium tracers for the determination of the age of groundwaters (Maloszewski and Zuber 1996; Nir 1964; Zuber 1994).

13.8.1 Piston Flow Model

This model is based on the assumption that portions of water coming into the system follows each other along the flow and do not intermix. The model underestimates the residence time in hydrological systems since water mixing does occur in nature. However, it is useful for estimations of the minimal residence time in a system.

According to this model the concentration of radioactive isotope C at a sampling point located at a distance x_0 from the recharge zone is defined as

$$C = C_0 \exp\left(-x_0/vT\right) = C_0 \exp\left(-t/T\right), \tag{13.28}$$

where C_0 is the concentration of radioactive isotope in recharge water; v is the rate of groundwater motion; T is the lifetime of an isotope ($T = T_{1/2}/\ln 2$).

Introducing a dimensionless parameter $k = t/T$, one obtains:

$$C/C_0 = \exp(-k). \tag{13.29}$$

Thus, the residence time of water from the recharge region of a basin to the sampling point is determined by the ratio C/C_0. Equation (13.28) can be readily transformed into one which corresponds to the common exponential law of radioactive decay:

$$C = C_0 \exp(-\lambda t), \tag{13.30}$$

where t is the age of the water, λ is the decay constant of an isotope $\lambda = 1/T = \ln 2/T_{1/2}$.

Therefore, the age of water at a sampling point for the piston flow model is

$$t = \frac{T_{1/2}}{\ln 2} \ln \frac{C_0}{C_t} = \frac{1}{\lambda} \ln \frac{C_0}{C}. \tag{13.31}$$

In the framework of this model, it has been admitted that C_0 is a constant and that intermixing of waters of different ages does not occur in the system.

13.8.2 *Dispersive Model*

According to this model, the hydrodynamic dispersion and intermixing of waters, entering the system at different time, results in a Gauss' distribution of transition times along the flow. As in the previous case, assuming the spatial distribution of isotope tracer being dependent on one coordinate x, one has

$$C(x_0, t) = \frac{C_0}{(4\pi Dt)^{1/2}} \exp\left[-\frac{(x_0 - vt)^2}{4Dt}\right] \mathrm{d}x, \tag{13.32}$$

or

$$C(x_0, x) = \frac{C_0}{(4\pi D_m x)^{1/2}} \exp\left[-\frac{(x_0 - x)^2}{4D_m x}\right] \mathrm{d}x, \tag{13.33}$$

where $D_m = D/v$ is the coefficient of hydrodynamic dispersion being characteristic for a given hydrogeological system.

Expression (13.33) represents the concentration of tracer $C(x_0, t)$ at a distance x_0 from the source of recharge. When the tracer passes the average distance $x = vt$, the amount of tracer remains unchanged in time and is equal to $C_0\mathrm{d}x$. The average lifetime of tracer is T and the concentration at a point x can be estimated from the equation

$$C(x_0, x) = \frac{C_0}{(4\pi D_m x)^{1/2}} \exp\left[-\frac{(x_0 - x)^2}{4D_m x} - \frac{x}{vT}\right] \mathrm{d}x. \tag{13.34}$$

If the concentration of tracer at the input of the hydrogeological system varies within time, which it does for natural isotopes, the variation of the concentration at the output will depend mostly on the input parameters of the tracer.

Let us consider a hydrological system of volume $V(t)$, having an inflow of water $a(t)$ and an outflow $q(t)$. In this case, the water age t is

$$t = V(t)/q(t)\,.$$

For the steady hydrodynamic state, the input concentration of tracer in the discrete form can be given as follows (Martinec et al. 1974):

$$C(t) = \sum_{t=0}^{\infty} p(t)C_a(\theta - t)e^{-\lambda t}, \tag{13.35}$$

where θ is the year of sampling; t is the age of the water; λ is the constant of tritium (0.056 year^{-1}).

The distribution function of the water's age $p(t)$, in a hydrogeological system, is given in fractions of annual replenishment at the output of a system at the moment of sampling.

In the case of the dispersive model the distribution function of the water's age for a semiinfinite aquifer is as follows (Burkhardt and Fhöhlich 1970; Maloszewsky and Zuber 1996; Martinec et al. 1974):

$$p(t) = \frac{2}{\sqrt{\pi Dt}} \exp\left[-\frac{(t-t_0)^2}{Dt}\right] - \frac{2\exp(4\pi t_0/D)}{D} erfc\left(\frac{t+t_0}{\sqrt{Dt}}\right). \tag{13.36}$$

The result of the above expression depends on two parameters, having dimensions of time: $t_0 = x_0/v$ (t_0 is not equal to the average residence time of water in the system) and $D = 4D_m/v^2$, (x_0 is the coordinate of a sampling point), where D_m is the hydrodynamical dispersion. Using a radioactive isotope tracer, one should introduce in Eq. (13.36) a term to account for its decay to the moment of sampling.

Dispersive models have been used by a number of researchers for estimating the residence time of water in hydrological systems (Burkhardt and Fhöhlich 1970; Martinec et al. 1974; Maloszewsky and Zuber 1996; Zuber 1994; Zuber et al. 2001).

13.8.3 Complete Mixing Model

According to this model water input at different times mixes quickly. It is not possible to account for water flow lines, as has been done in previous models, only to speak of water residence time in a system obeying exponential distribution. The model, as a rule, gives higher values of the residence time in a system if the distribution function of time is expressed by a continuous function.

In the interpretation of the observed results of tritium concentration changes in the Ottawa River basin, Brown (1961) used a box model which assumes complete mixing of the meteoric and groundwater in the bed flow. In this case, at the output of a hydrological system the relative fraction of water of a certain age is expressed as an exponential dependency of the form:

$$p\,(t) = (1/\tau)\exp\left(-t/\tau\right), \tag{13.37}$$

where τ is the average value of residence time of water in a system; $1/\tau = \varepsilon$ is the water exchange rate, i.e., the ratio of the total annual inflow of water to the volume of the basin.

Due to the function of water distribution (13.37), this model in literature is often called 'the exponential model'.

The distribution of tritium concentration in water tritium distribution at the output of the system, in the discrete form and in accordance with Eq. (13.35), is

$$C(t) = \sum_{1953}^{\theta} \alpha C_a(\theta - t)p(t)e^{-\lambda t}. \tag{13.38}$$

where α is the statistical distribution of the input function, determined as a function of the seasonal precipitation, participating in the recharge of a system compared with the annual amount of precipitation.

During the prethermonuclear era when $C_a = \text{const.}$ we have expression (13.13) and when $C_a \neq \text{const.}$ we have expression (13.14).

In practice it is not always convenient to estimate $C(t)$ by a method of successive approximations. Expression (13.38) can be simplified by letting $\mathrm{e}^{-1/\tau} = k$. In this instance, at $1/\tau \ll 1$, expanding the function $k = \mathrm{e}^{-1/\tau}$ in a power series of $1/\tau$, one obtains

$$k = 1 - \frac{1}{\tau}, \tag{13.39}$$

or

$$\tau = 1 - \frac{1}{k}. \tag{13.40}$$

Substituting the last expression into the age function $p(t)$, coming into Eq. (13.38), one obtains

$$P(t) = (1 - k)k^t, (k < 1). \tag{13.41}$$

Then, taking $\alpha = 1$, Eq. (13.38) becomes (Dinçer and Payne 1971)

$$C(t) = \sum_{1953}^{t=n} C_t(\theta - t)(1 - k)k^t e^{-\lambda t}. \tag{13.42}$$

Here, as a rule, the summation is carried out over all the years since the beginning of thermonuclear tests, i.e., 1953, and $C_t(\theta - t)$ is accepted as an average tritium concentration in precipitation for a corresponding year. For practical purpose the values C_t estimated at different values of k (0.1, 0.2, 0.3, etc), are indicated in the plot together with the experimentally measured values of tritium in the aquifer. Then, comparing the theoretical and experimental curves, the most probable value of the averaged residence time of water in a basin is determined.

The complete mixing model has been used for estimating the residence time of groundwater in the karstic region of the Anatolian coast in Turkey (Dinçer and Payne 1971), for studying aquifers sited near Vladimir (Polyakov and Seletsky 1978) for studying the groundwater discharge characteristics of the Aragatz Mountain region (Vlasova et al. 1978) and for studying the water dynamics in Moscow artesian basin (Zlobina et al. 1980).

It should also be pointed out that the age distribution of water, described by Eq. (13.37), does not always imply the existence of an underground or surface reservoir with good mixing. This model is also advantageous for studying aquifers drained by the aquifer thickness.

13.8.4 Symmetrical Binominal Age Distribution Model

According to this model the probability of the appearance of water characterized by an age t at the output of the hydrogeological system has the form (Martinec et al. 1974):

$$p(t) = \frac{1}{2N}\binom{N}{t}, \quad t = 0, 1, 2, 3\ \ldots, \tag{13.43}$$

The parameter N in this case is dependent upon the mean residence time of water in the hydrogeological system and is given in the form: $N = 2\tau - 1$ or $\tau = (N + 1)/2$.

The numerical coefficient N/t represents the binominal coefficient defined by the binominal theorem

$$(a+b)^N = \binom{N}{0}a^N + \binom{N}{1}a^{N-1}b + \binom{N}{2}a^{N-2}b^2 + \ldots + \binom{N}{N-1}ab^{N-1} + \binom{N}{N}b^N.$$

The following properties of this expansion are known: $N/0 = N/N = 1$, and the sum of all the binomial coefficients is equal to 2^N. Therefore, it is evident from the discrete binomial distribution that in the case of the continuous variation of the parameter t a normal distribution results with dispersion and average residence time depending upon the parameter N.

Figure 13.48 indicates some models of age distribution corresponding to the groundwater basin of the Dischma River (Switzerland).

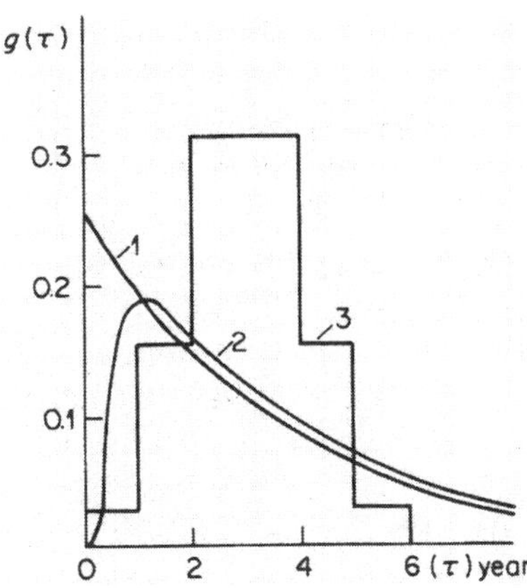

Fig. 13.48 Models of age distribution for groundwater basin of the Dischma River: (*1*) experimental data, $\tau = 4$ years; (*2*) dispersive, $\tau = 4.8$ years; (*3*) binominal, $\tau = 3$ years. (After Martinec et al. 1974. © IAEA, reproduced with permission of IAEA)

13.8.5 Model of Mixing Waters of Different Ages

This model accounts for the case when discrete differences in the distribution of residence times are observed, for example, young waters in an aquifer may mix with very old waters. If the volume of the whole system is V and the volume of an individual constituent is V_i, then their relative contribution is

$$p_i = \frac{V_i}{V}, \tag{13.44}$$

where $V = \sum V_i$ and $\sum p_i = 1$.

If the true age of an individual component is t_i, then the age of a mixture is

$$t_{\Sigma} = \sum p_i t_i = \bar{t}, \tag{13.45}$$

where $\bar{t}$ is the mean weighted age.

If the initial concentration of each constituent is C_{0I}, then at the moment of mixing its concentration is

$$C_i(t_i) = C_{0i} e^{-\lambda t}. \tag{13.46}$$

After mixing in a system its concentration becomes

$$C = \sum p_i C_i = \tau = \sum p_i C_{0i} e^{-\lambda t}.$$

While estimating the age of a water mixture, e.g., using the piston flow model, the formal application of expression (13.31) gives the age:

$$t' = \frac{1}{\lambda} \ln \frac{C_0}{C} = \frac{1}{\lambda} \ln \Big[\sum p_i \exp(-\lambda t_i) \Big]^{-1}. \tag{13.47}$$

Comparing expressions (13.45) and (13.47) we find that actually $t \neq \bar{t}$. Therefore, it follows that the true age of a mixture of waters of different ages is equal to the average age of its components. Further, the isotope age of the water mixture does not normally equal to the true age and its theoretical value has nothing in common with that for a real system in the framework of the piston flow model. In principle, the problem of the mixing of waters of different ages can only be solved while taking into consideration the behavior of several radioisotopes and counting out the time from the moment the waters become mixed. In this way, the problem is reduced to the solution of the system of equations

$$C_k = \sum^{m} p_i C_{0ik} \exp(-\lambda kt), \sum^{m} p_i = 1, \qquad (13.48)$$

where C_{0ik} is the concentration of the kth isotope on the ith constituent of a mixture at $t = 0$; p_i is the contribution of the ith constituent; λ_k is the decay constant of the kth isotope; t is the time elapsed since the moment of mixing; m is the number of constituents in a water mixture.

Using the stable isotopes only, i.e., those for $\lambda_k = \infty$, system (13.48) is reduced to that describing a simple mixing. For more details the reader is referred to Ferronsky et al. (1977).

13.8.6 Complicated Model

Such a model is indicative of most natural systems. In this model, the output function of a system of one type becomes the input function of a system of another type, for example, the groundwater basin with a normal distribution of transition times is connected to a basin characterized by complete mixing, and so on. Studies of the complicated systems, involving a tritium and other tracer, require a detailed knowledge of the geology and hydrology of a basin.

Besides the above-mentioned models other combinations and varieties are used in hydrogeological studies.

A number of authors have carried out measurements of the tritium concentrations in groundwaters up to the depth of several hundreds of meters. These studies were carried out in various hydrogeological conditions: in the Vienna basin sited near the Alps, in the hydrothermal regions of New Zealand and Iceland, in the limestone and dolomitic formations of the Transvaal in South Africa and elsewhere. It has been found that tritium concentrations decrease sharply with depth. Seasonal variations of tritium in the depth of an aquifer are not observed, indicating the continuous replenishment of the aquifer during the year with a constant rate of recharge. Considerable variations of tritium content exist for various boreholes located within a basin at short distances from each other, indicating the various conditions of recharge and rates of inflow. While studying geothermal regions the applicability of the techniques for the estimation of the inflow rate of surface waters to the zone of heating, and their subsequent time of circulation, was demonstrated (Gonfiantini and Panichi 1982; Pinneker et al. 1978; Theodorsson 1967).

Tritium techniques are an effective instrument in investigation of pollution problems, especially the release of radio-nuclides into groundwater from nuclear power plants and atomic wastes. The last problem becomes actual in connection with active construction of the nuclear power plants. In the last decade, these studies are intensified (Polyakov and Golubkova 2007; Sokolovsky et al. 2007b; Tokarev et al. 2009).

In 1969, Tolstikhin and Kamensky (1969) proposed the helium–tritium method of groundwater age determination; by this method water age is calculated as

$$t = \frac{1}{\lambda} \ln \left(\frac{^3\mathrm{He}^*}{^3\mathrm{H}} + 1 \right),$$

where ^{3}He* is the helium-3 concentration in groundwaters which appears as a result of the tritium decay.

It is not necessary in determination of the input function $C_a(\theta - t)$ in Eq. (13.38). This method is used, first of all, for the study of "young" groundwater (Plumer 2005; Schlosser et al. 1988, 2000).

Chapter 14
Radiocarbon in Natural Waters

Carbon has always played an important role in geochemical processes which take place in the upper layers of the Earth and, in the first instance, in the formation of the sedimentary terrestrial layer and the evolution of the biosphere. Radioactive carbon ^{14}C is often used as a tracer of various natural processes such as the circulation of natural waters, their redistribution between the natural reservoirs, water dynamics of the hydrosphere and its elements, and is applicable for estimating the age of such waters. The age of geological formations and groundwater within the time scale up to 60,000 years is of a great interest for modern geology and hydrogeology. The data of radiocarbon distribution in different carbon-bearing natural objects are used for reconstruction of their paleoclimatic changes and for solving astrophysical problems related to variation with respect to time of the cosmic rays. In this chapter the main attention is drawn to the distribution and application of radiocarbon with respect to dynamics of natural waters.

14.1 Origin and Distribution of Radiocarbon in the Nature

As it was shown in Chap. 1, the radiocarbon is formed in the atmosphere in the course of the interaction of secondary neutrons generated by cosmic rays, mainly with nitrogen-14, according to the reaction as shown in Eq. 14.1:

$$^{14}N + n = {}^{14}C + p + 0.626\ \text{MeB}. \tag{14.1}$$

The cross-section of the reaction is 1.81 ± 0.5 barn. Table 14.1 shows the main reactions responsible for radiocarbon production in the atmosphere, but their contribution to the total ^{14}C balance compared to the reaction (Eq. 14.1) is insignificant.

The radiocarbon produced is usually oxidized to $^{14}CO_2$ after several hours in the atmosphere, which is characterized by approximately the same carbon isotopic composition and takes part in general global circulation of the carbon dioxide. The total equilibrium amount of radiocarbon on the Earth (in the atmosphere, hydrosphere, and biosphere) can be theoretically estimated. According to Libby (1955), it is equal

V. I. Ferronsky, V. A. Polyakov, *Isotopes of the Earth's Hydrosphere*,
DOI 10.1007/978-94-007-2856-1_14,

Table 14.1 Reactions of radiocarbon production in the atmosphere by the secondary neutrons' action. (© IAEA, reproduced with permission of IAEA)

Reaction	Reaction energy (MeV)+, exothermic; −, endothermic	Abundance relative ^{14}N	Relative rate production in atmosphere
$^{13}C(n,\gamma)^{14}C$	+8.17	0.23×10^{-5}	1.1×10^{-9}
$^{14}N(n,p)^{14}C$	+0.626	1.0	1.0
$^{15}N(n,d)^{14}C$	−7.98	0.37×10^{-2}	3.7×10^{-5}
$^{16}O(n,{}^{3}He)^{14}C$	−14.6	0.269	2.7×10^{-3}
$^{17}O(n,\gamma)^{14}C$	+1.02	0.99×10^{-4}	2.3×10^{-5}
^{20}Ne, ^{21}Ne (split)^{14}C	–	0.12×10^{-4}	1.2×10^{-7}

to 81 t and the estimations of Lal and other researchers (see Chap. 11) give 60–75 t which is equivalent to an activity of about 3×10^{8} Ci.

Despite considerable variations in the secondary neutron flux from the equator to the poles, which differs by a factor of 3.5, the ^{14}C isotope is sufficiently homogeneously distributed on the Earth. This effect was well studied by the artificial bomb-radiocarbon which gives evidence of the high rate of mixing of the atmosphere. Experimental data show that variations with latitude and altitude do not exceed 3–5%.

Some amount of ^{14}C can be received on the Earth together with the meteoritic matter, where it is produced by interaction with cosmic rays. The lunar soil and rock studies showed that the carbon content in the sample No 14163 is 109 ± 12 g/t (109 ± 12 ppm together with correction on the Earth's contamination (Fireman and Stoenner 1982)). The radiocarbon activity in two fractions, extracted by heating in oxygen at $T=1{,}000°C$, was 31.2 ± 2.0 disintegration min^{-1}/kg for the fraction of >53 μ and 11.2 ± 2.0 disintegration min^{-1}/kg for the fraction of <53 μ.

The above mentioned authors assume that the radiocarbon derived in that way was produced by action of the solar wind. The carbon activity is approximately the same for both fractions (19.2 ± 2 and 21.0 ± 15 disintegration min^{-1}/kg). This is explained by interaction of cosmic rays with the lunar rocks and direct partial production of the radiocarbon.

The studies of dynamics of the carbon exchange between the natural reservoirs, based on ^{14}C tracer and theoretical investigations of the radiocarbon redistribution between the natural reservoirs (boxes), their number is arbitrarily accepted from one (Gray and Damon 1970; Ralf 1972) to six (Craig 1957; Plesset and Latter 1960; Bacastrow and Keeling 1973; Ekdahl and Keeling 1973). However, the majority of studies were done using two or three box-models (Damon 1970; Lingenfelter and Ramaty 1970; Yang and Fairhall 1972; Keeling 1972).

Figure 14.1 shows schematic two-box model used by Sternberg and Damon (1979). In accordance with this model, the radiocarbon, being produced in the upper parts of the atmosphere with the rate $Q(t)$ atom cm^{-2} sec^{-1}, is immediately moved into the reservoir A composed of the atmosphere, biosphere, and mixing layer of the ocean. Between this reservoir and the reservoir B of the deep ocean waters, the radiocarbon exchange results with the rates K_{as} and K_{sa}. Due to the radioactive decay, the loss of radiocarbon from the reservoirs A and B occurs. The exchange rate K_{as}, as

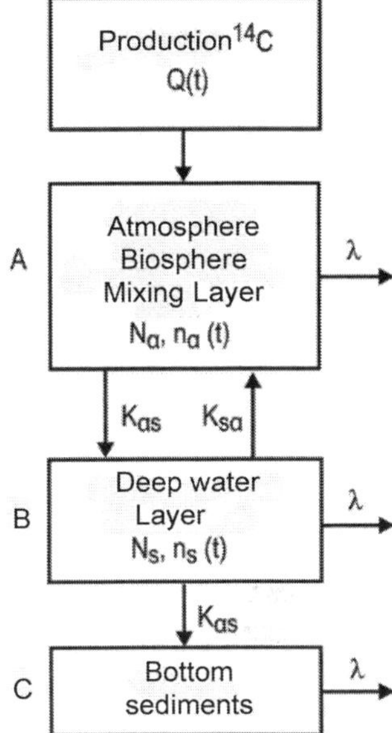

Fig. 14.1 Two-box radiocarbon exchange model. (After Sternberg and Damon 1979)

well as N_a and N_s, which are the total amount of carbon per cm^2 of the Earth surface as well as in the reservoirs *A* and *B*, are accepted constant in time. The amount of ^{14}C atoms/cm^2 in the reservoirs *A* and *B* ($n_a(t)$ and $n_s(t)$ accordingly), is the function of $Q(t)$.

The two-box model can be described by two differential equations

$$\frac{dn_a(t)}{dt} = Q(t) - K_{as}n_s(t) + K_{sa}n_s(t) - \lambda n_a(t), \tag{14.2}$$

$$\frac{dn_s(t)}{dt} = K_{as}n_a(t) - K_{sa}n_s(t) - K_{as}n_s(t) - \lambda n_s(t). \tag{14.3}$$

Note that the reverse value of the exchange rate is equal to the mean residence time *t* of radiocarbon in the reservoir. Taking into account that the specific activity of radiocarbon A_i in the ith reservoir is $\lambda n_i/N_i$, the Eq. 14.2 can be written as:

$$\frac{d(\lambda n_a(t)/N_a)}{dt} = \frac{\lambda Q(t)}{N_a} - K_{as}\frac{\lambda n_a(t)}{N_a} + \frac{K_{sa}}{N_a}\lambda n_s(t) - \frac{\lambda^2 n_a(t)}{N_a}, \tag{14.4}$$

$$\begin{aligned}\frac{dA_a(t)}{dt} &= \frac{\lambda Q(t)}{N_a} - K_{as}A_a(t) + \frac{K_{sa}N_s}{N_a[\lambda n_s(t)/N_s]} - \lambda A_a(t) \\ &= \frac{\lambda Q(t)}{N_a} - K_{as}A_a(t) + \frac{K_{sa}A_s(t)N_s}{N_a} - \lambda A_s(t)\end{aligned} \tag{14.5}$$

Analogously for the reservoir *B*, one has

$$\frac{dA_s(t)}{dt} = \frac{K_{as}A_a(t)N_a}{N_s} - K_{sa}A_s(t) - K_{as}A_s(t) - \lambda A_s(t). \tag{14.6}$$

In equilibrium conditions, the process of the carbon exchange between the reservoirs A and B is satisfied to equality

$$K_{as}N_a = K_{sa}N_s \tag{14.7}$$

or

$$K_{as}/K_{sa} = N_s/N_a = \nu, \tag{14.8}$$

where ν is the ratio of the carbon content in the reservoirs A and B.

It follows from Eq. 14.8 that the carbon exchange rate is in reverse proportion to the total carbon reserve, and vice versa, the residence time t is proportional to the carbon amount in the exchange reservoirs. The effect of carbon isotope fractionation at its redistribution between reservoirs should be taken into account while studying the parameters K_{as} and K_{sa}. As shown by Craig (1957), the real value of the fractionation factor $\alpha_{s/a}$ is equal to about 1.012.

On the basis of experimental data, by Eqs. 14.5–14.8, one may calculate values of ^{14}C distribution and assess the effect of dynamics of carbon exchange between the reservoirs. For these calculations, the values A_q, A_s, Q should be taken for the steady state condition. The Q parameter, as a rule, is determined by means of the decay rate I of radiocarbon, which relates to 1 cm^2 of the Earth's surface. In general case, I relates to $Q(t)$ by equation $I = -\lambda \int_0^\infty Q(t)e^{-\lambda t}dt$. It is obvious, if one considers that $Q(t)$ does not depend on time (which is not entirely correct), then $I = Q$. Equating the left hand side of Eq. 14.5 to zero replacing $K_{sa}N_s/N_a$ by K_{as}, for the stationary state condition one obtains

$$K_{as} = \frac{\lambda(Q/N_a - A_a)}{(A_a - A_s)}, \quad t_a = \frac{(A_a - A_s)}{\lambda(Q/N_a - A_a)}. \tag{14.9}$$

where t_a is the residence time of radiocarbon in the reservoir A.

On the basis of experimental data, Sternberg and Damon (1979) accept the following mean values: $\bar{A}_a = 14.5 ppm/g(C)$, $\bar{A}_s = 12.6 ppm/g$, $Q = I = 108$ ppm/cm^{-2} and $N_a = 0.361$ g (C)/cm^2. In this case, the carbon residence time in the reservoir A is about 50 years. In a similar way, on the basis of more number of reservoirs (boxes), the residence time of carbon in the atmosphere exchanging with the fast mixing ocean layer can be estimated. For this case, Sternberg and Damon accept the following parameters: $\bar{A}_a(1.890) = 13.8 ppm/g(C)$, $\bar{A}_{\Pi.c.} = 0.965\bar{A}_s$ and $N_a = 0.125$ g(C)/cm^2. In this case, one has $t \approx 5$ years. The similar estimates of the carbon residence time in the biosphere, in the fast mixing layer of the ocean and its deep layers give mean values of 60, 10 and 1,500 years respectively. These are the approximate results because of low accuracy in the parameters calculated. The values of carbon in a number of exchange reservoirs used for theoretical calculations are given in Table 14.2.

The possibility of ^{14}C production in nitrogen-bearing objects (in the wood, for example) *in situ* by reaction with neutrons, generating by the cosmic rays or occurring spontaneously in rocks at nuclear reactions, were considered.

Table 14.2 The amount of carbon in some exchange reservoirs

Reservoir	Carbon amount (g/cm^2)	
	Libby (1955)	Rubey (1964)
Ocean carbonates	7.25	6.95
Ocean dissolved organic substance	0.59	–
Biosphere	0.33	–
Living organisms and non-decomposed organic substance	–	0.775
Atmosphere	0.12	0.125
Total	8.29	7.85

Radnell et al. (1979) considered a possibility of the radiocarbon accumulation with respect to time in woods by the thermal neutrons of cosmic origin irradiation. It was shown that in nuclear reaction with nitrogen-14 (fast neutrons) in bristlecone pine, the activity of about $(1.7 \pm 0{,}6) \times 10^{-3}$ ppm/g (C) during 8,000 years can be reached, i.e., this is only 0.03% of natural ^{14}C radioactivity in wood 8,000 years of age.

Zavelsky (1968) has shown that possibility of ^{14}C generation in situ is limiting the upper value of the measured radiocarbon age. According to calculation, such a limit for a number of objects can be age of 80,000–100,000 years.

The theoretical specific activity of ^{14}C in the modern carbon-bearing samples should be about 17 ppm/g (C) (Libby 1955). Numerous comparable studies in different regions have shown that the specific activity of ^{14}C in biosphere varies from 14 to 16 ppm/g for the inland specimens and from 13 to 17 ppm/g for the ocean specimens. The average activity is 15.3 per gram of carbon. For the ocean carbonates, this value averages around 16.0 ppm/g, i.e., approximately 5% higher than for the biogenic ^{14}C. Later on, these figures were refined.

The major proportion of carbon, which participates in the cycle in the form of dissolved carbon dioxide, carbonates, and bicarbonates ($H_2CO_3 - HCO_3 - CO_3^{2-}$), is in the oceans. If one accepts CO_2 contents in the atmosphere as $N_a = 0.62 \times 10^{12}$ t(C), then the ocean contains 65 N_a (C), and the biosphere 2.4 N_a (C), from which 90% is contained in the oceanic plankton (Oeschger and Siegenthaller 1979). The organic substances are subjected to decomposition after death. Carbon is cycling through the biosphere about every 300 years. This process can be observed, for example, in the biogenic ocean carbonates, the major portion of which is dissolved while precipitating on the ocean floor.

Besides this, the carbon present in the common exchangeable reservoir, which is contained in the sedimentary strata, representing its main terrestrial storage, takes part in the dissolution and admixing of the cosmogenic radiocarbon. Sedimentary carbonaceous rocks are, on the one hand, being continuously formed and, on the other hand, are being constantly disintegrated. In this form of solutions and suspended particles, the disintegrated rocks are carried out into the ocean. The amount of carbon contained in sedimentary rocks is estimated to be 2×10^{22} g. If the process of the formation of sedimentary carbonaceous rocks has taken place over the last 3 billion years with a variable rate of formation in various epochs, then, according to different estimations, less than 3% of the ^{14}C participates in this process at any one time.

An important question in radiocarbon dating concerns the efficiency of ^{14}C mixing in the main reservoirs over time. The homogeneity in ^{14}C distribution in a reservoir can only be attained if the mixing time is short compared with the lifetime of ^{14}C. The mixing time for the biosphere is not more than 300 years and in the atmosphere 10 years (Libby 1955). The mixing time of the Atlantic Ocean does not exceed 2,000 years, that of the Mediterranean Sea is about 100 years, and for the Black Sea it is about 2,500 years. Another fact in agreement with the assumption of complete mixing of the ocean is the magnitude of the heat flux from the oceanic floor, which is equal to that of the terrestrial crust at 30 cal/cm^2/year (Libby 1955). If this value is correct, the absence of a heat inversion near the bottom suggests good mixing of the ocean in the framework of a radiocarbon scale of time.

The constant rate of radiocarbon production and the constancy of the amount of stable carbon in the exchangeable reservoir is of importance in the problem of radiocarbon dating. The constancy depends upon (Stuiver 1965): (a) the variation of intensity of the cosmic radiation due to solar activity; (b) the variation of magnetic dipole and field of the Earth; (c) the climate change of the Earth. Libby (1955, 1967) has pointed out that considerable corrections in view of these factors should not be made since the ^{14}C lifetime is relatively small ($T_{1/2} = 5{,}730$ years). However, in the cold periods of glacial times, the stable carbon content in the ocean may drop and the specific activity of ^{14}C may increase by 5–10%. The amount of the living organic material has no effect on the specific activity of carbon (since its ratio in nature has always been small).

During the last 100 years, the content of CO_2 has markedly increased in atmosphere of the Earth as a result of the industrial burning of fossil fuels (coal, oil, and gas). This effect is known as 'the industrial effect' or 'Suess effect' consisting of a certain decrease (by about 3% for the northern hemisphere) of the ^{14}C (Houtermans et al. 1967; Oeschger and Siegenthaller 1979). However, as a whole, the natural equilibrium in the ^{14}C content has settled during the last two decades due to thermonuclear explosions conducted in the atmosphere. Due to these, the ^{14}C content in the northern hemispheric atmosphere has more than doubled and became higher in the biosphere, and in the surface oceanic layer increased by about 20% (Nydal et al. 1979). In the wooden rings of 1963–1965, the ^{14}C increased up to 180–190% compared with pre-bomb level (Cain 1979).

14.2 Natural Variations of Radiocarbon in the Atmosphere and Biosphere

The specific activity of radiocarbon in the atmosphere and, as a consequence, in the biosphere is mainly governed by variation of the cosmic rays' intensity at the Earth's surface. The natural ^{14}C variations can be divided between the short-periodic—governed by the Sun modulation of the galactic cosmic rays—and long-periodic—connected with the geomagnetic Earth's field and climate variation (Dergachev and Kocharov 1977; Sternberg and Damon 1979). Long-periodic variation of ^{14}C level

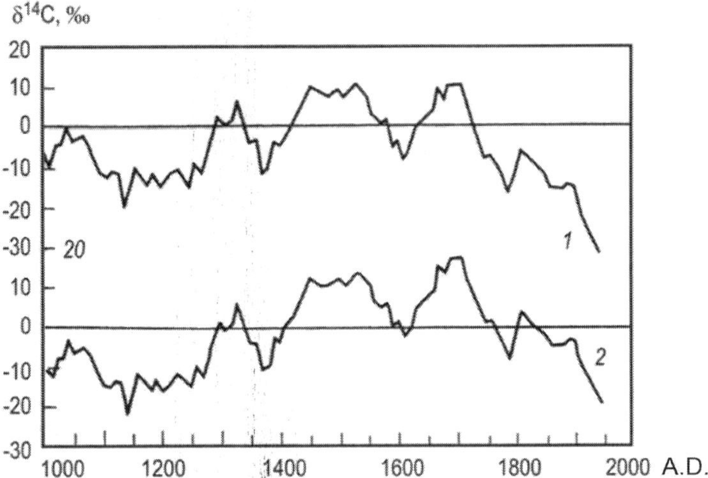

Fig. 14.2 ^{14}C variation in the atmosphere in the last 1,000 years: (*1*) non-corrected and (*2*) corrected on the basis of the long-periodic geomagnetic field change. (After Stuiver and Quay 1981)

can also be effected by the corpuscular radiation from the super-nova flashes and, by low probability, by occurrence of neutron flows during annihilation of the meteoric matter. The last problem has a more exotic than practical meaning (Sternberg and Damon 1979). As Dergachev and Kocharov pointed out, the degree of correctness of the solar activity (Wolf's numbers W) is high starting after 1749. The attempts were undertaken to extend the time scale up to 1610 and even to 648 before present (Dergachev and Kocharov 1977), but the correct measurements of 11 years solar cycle by W numbers were achieved to be only since 1749. The last time of the 80 years (century) cycle is tried to be checked, but the difficulty has arisen in connection with the short period of observations (230 years). In accordance with the calculations, the amplitude of specific ^{14}C activity for the 11 years' cycle of the solar activity is counted by 5%, and for the century cycle about 1%. Registration of the 11 years' variation in the yearly wooden rings' radiocarbon content is a difficult instrumental task. Stenhause (1979) used two high-stability proportional counters of quartz and metallic for its solution. This allowed obtaining relative error in the experiments equal to 6.4‰. In yearly oak rings over 1840–1890, he found ^{14}C peaks relating to 1851, 1869, and 1880 (at the level of about 1%). ^{14}C fluctuations on the level of 0.3–0.4% have not been reached, but it was sure that the 2–3% variations have not been found. The total ^{14}C activity from 1840 to 1890 has decreased on an average by 0.03% per year. After 1890, an increase of ^{14}C concentration in the atmosphere fixing in yearly wooden rings was observed. The radiocarbon variations in the past and changes in the solar activity during the last 300 years are considered in the study of Dergachev (1975). ^{14}C variation in the atmospheric carbon dioxide during the last 1,000 years are discussed by Stuiver and Quay (1981). Their results are presented in Fig. 14.2.

It was observed from the experimental data that natural reservoirs of the carbon do not stay in equilibrium relative to the atmospheric ^{14}C level. Before 1890, variation of the radiocarbon content resulted mainly due to the solar activity changes. Starting since 1890, the ^{14}C activity decrease was connected mainly with dilution of the atmospheric carbon dioxide by the "dead" CO_2 coming from burning the fossil fuel, but as it was shown by Stuiver and Quay, ^{14}C content in atmospheric CO_2 during twentieth century is not distinguished by something specific (see Fig. 14.2, (*1*)). These changes resulted due to long-periodic variation of the geomagnetic field. If one makes correction of ^{14}C long-periodic changes by means of sinusoidal curve of several thousand years period, then the ^{14}C level in atmospheric CO_2 (and biosphere) in twentieth century at least during the last thousand years will be minimal (see Fig. 14.2, (*2*)).

These data prove the anthropogenic influence on the increase of total carbon content in the atmospheric reservoir which has led to a decrease in the specific activity of radiocarbon in the atmosphere, biosphere, a mixing ocean layer. By the data of Stuiver (1980) with reference to Douglas, ^{14}C level in 1954 differed from the 'standard modern level' (0.95 activity of the NBS standard) by $24 \pm 2‰$.

The study of ^{14}C variation with respect to time by the yearly rings of the eighteenth and nineteenth century woods is convenient because at that time the anthropogenic impact on the atmosphere was weaker, the level of solar activity (W-number) change was well known, and the climate variations by the synoptic data can be restored. At present, a lot of data on ^{14}C concentration measurements in the yearly rings of woods from that time period are collected. By measurements of Stuiver (1965) over the time period from 1700 to 1870, a decrease in ^{14}C content by 2% from 1700 to 1790 is noticed. The maximum difference in ^{14}C activity within the studied period reaches 2.8%. It is noteworthy that due to small diameter of rings, Stuiver measured the ^{14}C activity in several rings simultaneously. In addition, the studies were performed on the ring slices of four trees from different locations of growth. This may lead to some errors depending on local conditions.

Analogous measurements on ^{14}C changes in yearly wooden rings of pines growing in Lithuania were done by Dergachev and Kocharov (1977). Figure 14.3 shows their results of a series of measurements on 5-year rings over 1707–1859 period of time.

It is seen that the ^{14}C change curve has a sinusoidal form with the amplitude of 1.5%. The period of ^{14}C variations for the time interval from 1780 to 1840 is equal to about 60 years. The curve itself shows an inverse proportionality in relationship between the solar W-numbers and radiocarbon content. The time shift, equal to about 10 years, is specified by the mean residence time of the carbon in the exchangeable reservoir (see Fig. 14.1 A) and depends on the period. The following expression of the changes in rate of ^{14}C generation depending on the W-number for the considered solar activity cycle, is proposed by Dergachev and Kocharov:

$$\Delta Q(W)/Q_0 = 0.4 - 0.01W,$$

where Q_o is the equilibrium rate of the ^{14}C generation.

Suess (1970) discovered in the yearly bristlecone pine rings a century ^{14}C variation with the period of 181 years within the time period of 7,000 years (from the middle of twentieth century up to 5000 years before present). The ^{14}C variation amplitude

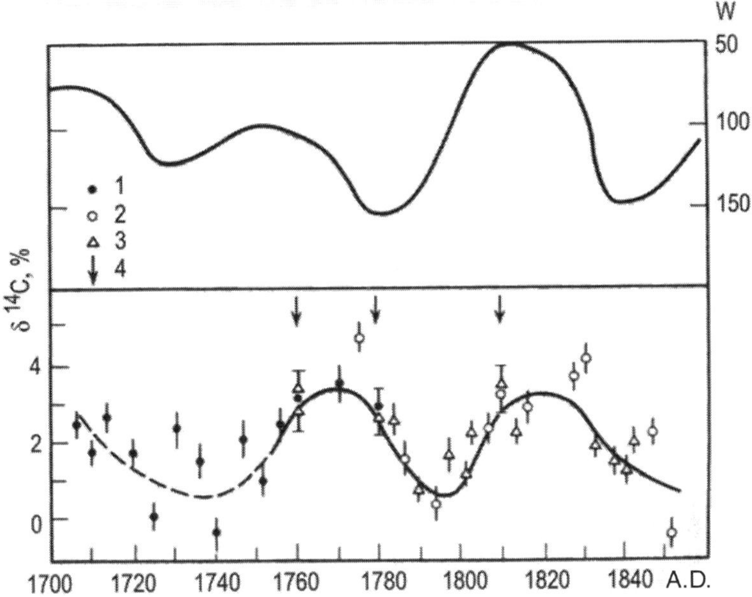

Fig. 14.3 Relationship between δ^{14}C variations in pine rings and solar W-number over 1707–1850: (*1* and *2*) measured by proportional counter; (*3*) measured by scintillation counter; (*4*) calibration points. (After Dergachev and Kocharov 1977)

is about 2–3% and the mean value of the ^{14}C activity is changed with respect to time by a sinusoidal law. In connection with solar activity changes and the corresponding climatic variation, it is noreworthy that Dansgaard et al. (1969) in ice core of 404 m length from the Greenland Camp Century covering the time interval from 1970 to 1200 years by the Fourier treatment derived the δ^{18}O values equal to 78 and 181 years (cited by Dergachev and Kocharov 1977). Obviously, the climatic changes are resulted by the solar activity and the 180-year period corresponds to the Suess data.

It seems the record in observation of the periodic climate changes based on the solar activity and registered by isotope data belongs to Libby and Pandolfi (1979). They studied distribution of deuterium and oxygen-18 in the yearly rings of Japanese cedar (*Cryptomeria Japonica*) within about last 1,800 years and derived there eight cycles with the periods of 58, 68, 90, 96, 154, 174, 204, and 272 years. The shorter cycles like 11 and 21 years were not discovered due to averaging of the measured wooden rings within 5 years, and also the 174th cycle of the authors corresponds to 183-year Suess period.

For the last relationship of δD and δ^{18}O in the Japanese cedar with radiocarbon in the bristlecone pine—measured by Suess for the same time intervals—Libby and Pandolfi discovered inverse correlation. To the lower values of δD and δ^{18}O there correspond the higher ^{14}C values. These relations have the form

$$\delta D = 0.677\delta^{14}\mathrm{C} - 75.5(\mathrm{SMOW}), \quad r = 0 - 0.62,$$

$$\delta^{18}\mathrm{O} = 0.0613\delta^{14}\mathrm{C} - 22.5(\mathrm{PDB}), \quad r = 0 - 0.77.$$

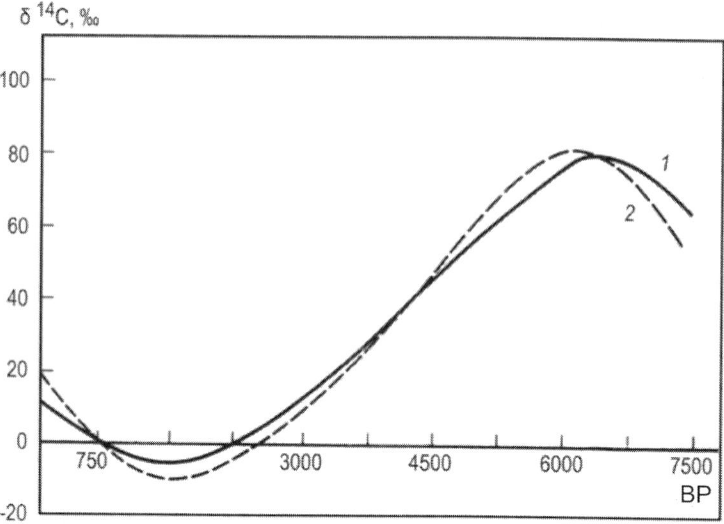

Fig. 14.4 Experimental and calculated data of ^{14}C variation in yearly wooden rings: (*1*) experimental data approximated by the polynomial of the fourth degree; (*2*) the curve obtained by calculation. (After Sternberg and Damon 1979)

The δD value was measured in relation to the SMOW standard and $\delta^{18}O$ in relation to the PDB standard. The above authors explain—by the climate change due to the solar activity—the relationship between the stable hydrogen and oxygen-18 isotopes and the radiocarbon in wooden rings with the ^{14}C amplitude of about 1%. The relationship between the rate of radiocarbon generation in the atmosphere and the solar activity variation seems to have physical basis, but the problem of the influence of the solar activity on the galactic cosmic rays' modulation and on the Earth's climate change has not been yet studied in detail. The search of cyclicity in the long- and short-periodic processes always attracted attention of the researchers, but as Gribbin (1980) pointed out, the cyclic attraction conceals in itself the risk to discover it there, where there is distribution of random values. To prove his words, Gibbin demonstrates the known graph, drawn by computer, generating random figures from 0 to 9, where deviation of the sliding average value from the mathematical expression appears to be equal to 4.5. On the graph of the random values distribution, a quasi-sinusoidal short-periodic fluctuations and long-periodic 'temporal' trend is traced.

The long-periodic variation of ^{14}C content in the atmosphere and correspondingly in the biosphere was reliably fixed by the radiocarbon concentration measurement in the long-living woods. Figure 14.4, taken from the Sternberg and Damon (1979) study and drawn on the basis of results obtained by many laboratories in the world, traces sinusoidal changes in the wood ring's ^{14}C concentration during about 7,500 years before present. The averaged data show that the minimal ^{14}C concentration (−5.5‰ from the level of 1890) comes approximately on 1,400 years and the maximum (+85‰) on about 6500 before present.

Sternberg and Damon assumed that the observed ^{14}C variation is related to the geomagnetic dipole moment changes. The relationship between the rate Q(t) of ^{14}C generation and the strength of the geomagnetic dipole moment has a reverse proportional dependence as

$$\frac{Q(t)}{Q_0} = \left(\frac{M(t)}{M_0}\right)^{-a}, \tag{14.10}$$

where Q_o is the equilibrium rate of generation of the magnetic moment M_0; α is the coefficient changing within 0.4–0.6.

It was also found by the data of paleomagnetic study that the magnetic dipole moment varies with respect to time within (4–12) × 10^{25} Gs.cm^3. It changes on quasi-sinusoidal law and during the last 10,000 years can be described by equation

$$M(t) = M_0 + M_1 \sin \omega(T - t + \theta),$$

where M_0 is the mean value of the dipole moment; M_1 is the amplitude variation; T is the period of oscillation; θ is the phase; $\omega = 2\pi/T$; t is the time.

Denoting $T - t + \theta$ by t', Eq. 14.10 can be rewritten in the form

$$Q(t) = Q_0[1 + (M_1/M_0) \sin \omega t']. \tag{14.11}$$

Applying the two-box model for the carbon exchanging reservoir (see Fig. 14.1), Sternberg and Damon reached good accordance between the experimental and calculating data on the long-periodic variation of the radiocarbon change in the atmosphere for the following conditions: T = 8,500 years; M_{max} = 12.5 × 10^{25} Gs cm^3; $T_{M max}$ = 2,500 years; α = 0.45; τ_a = 75 years (see the residence time of carbon in a reservoir in Fig. 14.1); $K_{as} = 5 \times 10^{-4}$. The calculations were done by Eqs. 14.5, 14.6 and 14.11. The value of ^{14}C (‰) was calculated by the equation $\delta A_a(t) = [(A_a(t) - A_0)/A_0] \times 1000$, where A_0 is the activity of the modern carbon standard.

Figure 14.5 presents the compared experimental and calculated long-periodic changes of radiocarbon in the Earth's upper atmospheric layers (Sternberg and Damon 1979). The authors note that a delay of several hundred years in the phase between the geomagnetic field change and the radiocarbon variation in the exchange reservoirs is observed. This is due to the mean residence time of carbon in the reservoir A, the ratio of exchanging carbon funds in the A and B reservoirs, and the period T of the geomagnetic field variation.

In addition to the astronomical and climatic factors' effect on distribution of the natural ^{14}C concentration in the atmosphere and biosphere, the dilution of the atmospheric carbon dioxide by the endogenic CO_2 may have influence on local radiocarbon content changes which is typical to the volcanic and breaking Earth's crust regions. It was certain that in the modern volcanic regions, the ^{14}C content in the biosphere is lower than the present day level. This is because of throwing out of the endogenic carbon dioxide in the period of volcanic activity and CO_2 entering from the Earth's crust breaks (Sulerzhitsky and Forova 1966; Olsson 1979; Karasev

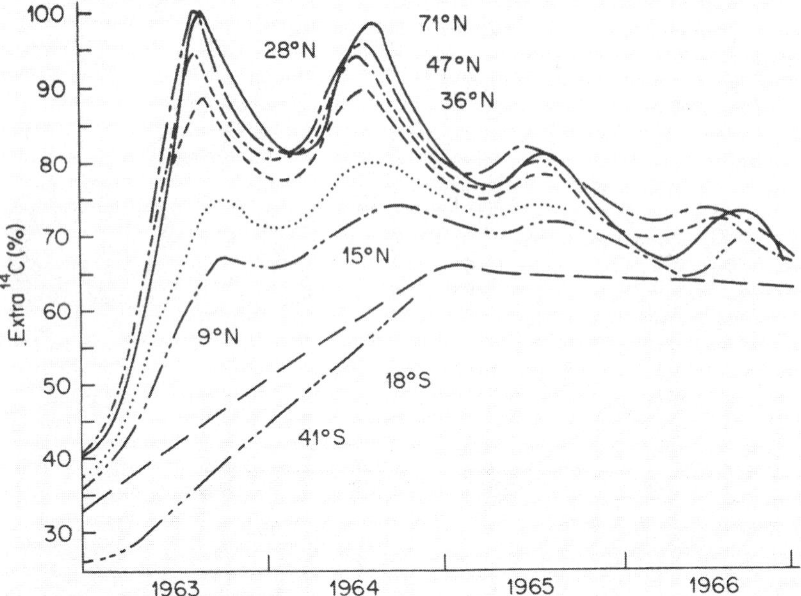

Fig. 14.5 Variation with respect to time of ^{14}C concentrations in tropospheric CO_2. (After Fairhall and Young 1970)

et al. 1981a). On this basis, Karasev et al. proposed a method to break the discovery in active tectonic zones by radiocarbon measuring in plant samples. As it was shown by the authors, in a break zone $\delta^{14}C = -50\%$ and at a distance of about 10 m $\delta^{14}C$ increases up to +27%.

14.3 Natural Radiocarbon in the Oceans

A number of studies were devoted to the analysis of radiocarbon in the oceanic water's carbonate system and the water-bearing organic matter (Fonselius and Östlund 1959; Broecker et al. 1960; Broecker and Olson 1961; Bien et al. 1963; Fairhall and Young 1970; Fairhall 1971).

Starting in 1940–1950s of the last century, the studies on ^{14}C distribution in the ocean water were later on strained due to the bomb radiocarbon entering into the oceans. The artificial mark appeared by culminate tests in 1961–1962 allowed detailed further study of dynamics of the CO_2 between the atmosphere and oceans exchange, but became an irresistible obstacle for further study of the ^{14}C space distribution in the oceans and seas.

Radiocarbon enters into the oceans as a result of the carbon exchange between the dissolved marine carbonates and the atmospheric CO_2. The radiocarbon entered into the surface ocean layer is rather quickly and uniformly distributed in exchangeable layer (h $\approx$ 100 m, r $\approx$ 10–15 years). From there, by the vortex diffusion and pelagic biogenic carbonate sedimentation, the radiocarbon enters the deep water layers, where its residence time (t), by different estimates, is equal to 1,500–2,000 years. As Craig (1957) notes, CO_2 exchange between the atmosphere and oceans is accompanied by the isotopic effects. The mean isotopic content of stable carbon of the dissolver—hydrocarbonates of sea water—has a value of $\delta^{13}C = -7‰$ (here and further relative to the PDB standard). Thus, as a result of exchange processes, the carbon of the hydrocarbonates becomes heavier by 7‰. Enrichment of the oceanic ^{13}C can be described by means of the fractionation coefficients $\alpha_{13C} = R_0/R_a$, where R_o and R_a are the ratio of $^{13}C/^{12}C$ in the ocean and atmosphere.

The value R can be presented as $(1 + \delta^{13}C/1{,}000)$. Then the expression for α is rewritten as

$$\alpha_{13C} = \frac{1 + \delta^{13}C_0/1000}{1 + \delta^{13}C_a/1000}.$$

Analogously, for ^{14}C one has

$$\alpha_{14C} = \frac{1 + \delta^{14}C_0/1000}{1 + \delta^{14}C_a/1000} = \frac{(^{14}C/^{13}C)_0}{(^{14}C/^{13}C)_a}.$$

It is known (for example, Galimov 1968), that the fractionation constant increases by the square dependency on each additional neutron to the isotopic nucleus. Then, $\alpha_{14C} = \alpha^2_{13C}$, and applying the approximate equality $(1 + x)^2 = 1 + 2x$ at $x \ll 1$, one obtains

$$\alpha_{14C} = \frac{1 + \delta^{14}C_0/1000}{1 + \delta^{14}C_a/1000} = \frac{1 + 2\delta^{13}C_0/1000}{1 + 2\delta^{13}C_a/1000}. \tag{14.12}$$

Using the last expression, it is possible to show that if ^{13}C of the ocean bicarbonate is enriched by 7‰, then at such enrichment, the ^{14}C should be equal to about 14‰. In other words, $\varepsilon_{14C} = 2\varepsilon_{13C}$, where $\varepsilon_{13C} = \delta^{13}C_0 - \delta^{13}C_a$. This task is discussed in more details later on.

Thus, the theoretical consideration of the problem of CO_2 exchange between the atmosphere and the ocean shows that the surface oceanic water should be enriched in ^{14}C isotope by about 14‰ compared with the atmospheric carbon. Analogously, it is possible to show that the biosphere should have ^{14}C in deficit by about 3% ($\delta^{13}C_a = -25‰$) compared with the atmosphere. As it was shown earlier, the normalization of ^{14}C content is provided for unification of the radiocarbon measurements (−25‰ is the normalizing value for the studied samples). The formula for normalization from Eq. 14.12 was obtained. It was pointed out by Libby (1955) that the oceanic carbonates have ^{14}C activity 5% higher than the biogenic carbon. His statement coincided with theoretical consideration of the natural exchange processes, but later the studies have not proved the conclusion. In particular, it was shown that the hydrocarbons of the mixing layer in the Atlantic, Indian, and Pacific oceans to the north from 40°S have a practically constant ^{14}C content close to that in the biosphere (without the isotope correction) (Fonselius and Östlund 1959; Broecker et al. 1960; Broecker and Olson 1961; Bien et al. 1963; Bien and Suess 1967). This fact is explained by increase of ^{14}C content due to isotope fractionation in the exchanging processes eliminated at mixing with the deep layers.

Stuiver (1980), on the basis of studies of Broecker, Östlund, Craig, and other researchers compiled a short report on ^{14}C distribution in ocean waters during the pre-bomb period. It is evident from this summary that in nine surface water samples taken in the North Atlantic in 1955, the $\Delta^{14}C = -49 \pm 2‰$. The data from 13 stations of the South Atlantic (to the north from 40°S.), sampled during 1956–1957, have little lower values, namely, $\Delta^{14}C = -57 \pm 2‰$. Up to that time of measurements, the decrease of the pre-industrial ^{14}C level in the surface oceanic waters was about 12‰. From this, we can accept that $\Delta^{14}C$ for the surface Atlantic waters before the intensive thermonuclear tests is be equal to −40‰ from radiocarbon in the biosphere. It means that this value can be accepted as characteristic for the mixing layer over the thermocline. In the depth from 100 to 600 m, the content of radiocarbon decreases by exponent from −40 to −100, and at more depth it remains constant. The value −110‰ was characteristic for the Atlantic deep layers in the pre-thermonuclear epoch. The studies of Stuiver (1980) show that after thermonuclear tests, the 'natural' distribution of radiocarbon in the Atlantic ocean in 1970s was preserved only below 3,500 m.

Giant amounts of experimental data were collected by dating of mollusk shells living in the normal salinity marine water. Investigations of samples collected in the period between the mid-nineteenth century to 1950 near the shore areas of different oceans provides the basis for conclusion that the radiocarbon content (without isotopic correction) in the mixing ocean layer during pre-thermonuclear epoch was also lower of biogenic level determined as 0.95 activity of the NBS oxalic acid. By the data of Gillespie and Polach (1979) who studied the radiocarbon distribution in mollusk shells from near shore of the oceans in 1840–1950, the conclusion follows that ^{14}C content in the mixing layer varies in the natural conditions from +8 to −11% relative to the modern standard, and without taking into account the effects of the fresh river water recharge and the Suess effect, the variation range is narrowing. In average, the carbon is depleted by ^{14}C (without correction) by about 15‰. By the age and being corrected by 25‰, the value of $\overline{\nabla}^{14}C = -64.5‰$ (0.95 NBS). Then the "apparent age" of the shells related to the reservoir effect in the mixing layer appears to be 535 years. For the Australian shore area, the mean value of $\overline{\nabla}^{14}C = -55 + 4‰$ (non-corrected value is ~5‰), which corresponds to the "apparent age" as 450 ± 35 years.

For the shells from the North and South American shore, taken in 1878–1940, the non-corrected value of $\delta^{14}C$ varies from +1 to −8%, and the averaged value is about −3% (−30‰). The true ^{14}C value for the mixing layer in the pre-thermonuclear epoch is of great practical interest for radiocarbon dating using mollusk's shells. The ^{14}C differences in mollusk shells are hardly defined by the metabolic processes. They more likely reflect radiocarbon variations in off-shore ocean layers, which affect the river runoff and discharge of the continental groundwaters. Sternberg and

Table 14.3 Distribution of natural and thermonuclear ^{14}C at the end of 1962. (Fairhall and Young 1970)

Reservoir	Total amount of carbon (g)	^{14}C before 1950 (atom/g × 10^{10} C)	Natural ^{14}C, (atom × 10^{27})	Thermonuclear ^{14}C, (atom × 10^{27})
Atmospheric CO_2	6.8×10^{17}	6.07	41	54
Continental biosphere	3.1×10^{17}	5.85	18	<1
Humus	1.1×10^{18}	<5.8	<64	<1
Oceanic living biosphere	3×10^{15}	6	0.2	<0.1
Oceanic non-decomposed rests	2×10^{16}	6	1	<0.1
Dissolved organic matter in the oceans	8×10^{17}	5.3	42	<1
Non-organic matter in the oceans				
Upper 100 m	1×10^{18}	5.8	58	~3
Below 100 m	3.8×10^{18}	5.1	1,940	~1
Total	42×10^{18}		2,160	~60

Damon (1979), after analyses of the large amount of collected experimental data, for theoretical calculation accepted that the mean radiocarbon activity in the mixing layer for the pre-thermonuclear epoch is equal to 0.965 from that in the biosphere in 1890 equal to 13.8 disintegrations $min^{-1} \times g^{-1}$ C.

The ^{14}C activity in different exchangeable reservoirs before 1950 can be estimated using the data from the study of Fairhall and Young (1970) shown in Table 14.3.

On the basis of the abovementioned Fairhall and Young data, it is possible to conclude that the ^{14}C activity of the mixing ocean layer in the pre-thermonuclear epoch should be 1% lower compared to the biosphere and by about 3% than in the atmospheric CO_2. A number of researchers (for example, Oeschger and Siegenthaller 1979), in their theoretical calculations, accept ^{14}C in the surface ocean layer before thermonuclear tests to be 95% from the atmospheric level in the mid-nineteenth century. The carbon distribution in the deep layers, as a result of the thermonuclear tests, inconsiderably changed due to the very long mean carbon residence time in this layer.

By the data of Fairfall (1971), who summarized his own results and the results obtained during 1958–1970 expeditions by Broecker, Bien, and Rafter before 1969 in the Atlantic, Indian, and Pacific oceans, the non-disturbed picture of radiocarbon distribution below 500 m was observed, and also its relative uniform distribution was fixed in the deep ocean layer. Radiocarbon comes to the deep layers from the mixing layer by the vortex diffusion (Oeschger and Siegenthaller 1979) and partly due to sedimentation of mollusk shells, which dissolved in deeper layers. By the Fairfall's calculations, the ^{14}C absolute concentration in the ocean water before 1950 was about 1.4×10^9 atoms per liter for the entire depth, for which the ^{14}C concentration depth gradient is equal to zero or correspondingly about 84% of specific radiocarbon activity in the atmosphere (see Table 14.3). Using the Fairfall's data, we may estimate the carbon mean residence time in the ocean deep layers. For this case, the following material balance can be used: $M\mathrm{d}A_2/\mathrm{d}t = q_1A_1 - q_2A_2 - MA_2\lambda$, where M is the carbon mass in the exchangeable layer; q_1 and q_2 are the carbon mass income; A_1 and A_2 are the specific carbon activities in the exchangeable and deep layers; λ is the ^{14}C decay constant. In the stationary state, the left hand side of the equation is equal to zero, and $q_1 = q_2$. Taking into account that $M/q = \tau$, then one has $\tau = (A_1 - A_2)/A_2\ \lambda$.

Accept $A_1 = 0.965$, $A_2 = 0.835$ and $\lambda = 1.21 \times 10^{-4}$, then one obtains $\tau = 1{,}300$ years. If $A_1 = 0.99$, which corresponds to many experimental data for the Atlantic ocean (Fairfall 1971), then $\tau \approx 1{,}500$ years, which corresponds to Libby's (1955) estimation.

14.4 Technogenic Radiocarbon in the Atmosphere and Oceans

According to data obtained by different authors (Fairhall and Young 1970), 6×10^{28} atoms of ^{14}C have been released into the Earth's atmosphere during thermonuclear tests. Since before the tests the ^{14}C content in the atmosphere was estimated to be 4.1×10^{28} atoms, the total amount of the carbon atoms has increased by 2.5 times. Compared with the total equilibrium amount of the isotope on the Earth (2.2×10^{30} atoms), the bomb component amounts to 2.5% (see Table 14.3).

On an average, at a thermonuclear, test 3×10^{26} atoms of ^{14}C per Mt are produced, which is equivalent to 7 kg of the radiocarbon. By the radiocarbon thrown into the atmosphere, over 200 Mt of total explosion power was done up to 1962, which injected about 1.4 t of artificial radiocarbon into the atmosphere.

Numerous measurements (up to a thousand specimens) were carried out in order to determine ^{14}C content variations in the troposphere and stratosphere both with latitude and altitude (Fairhall and Young 1970; Münnich and Vogel 1963; Fairhall et al. 1969; Hagemann et al. 1959). The most representative results of these studies according to the data of different authors for the troposphere CO_2 generalized from 1963 to 1966 are shown in Fig. 14.5.

There are distinct seasonal variations of ^{14}C concentrations in the northern hemisphere and a considerable latitudinal gradient of concentrations in 1963 indicated in the Fig. 14.5 but the gradient quickly decreased to zero by 1967. Both of these effects are a consequence of seasonal variations in the release of bomb ^{14}C from the stratosphere into the troposphere and the longitudinal mixing which occurs in the troposphere. The most intensive release of ^{14}C into the troposphere, as in the case of tritium, is detected in spring and early summer. At the same time, the most effective mixing of this isotope occurs in the longitudinal direction.

In contrast to the thermonuclear dust, which is removed as a rule quickly from the troposphere with precipitation, CO_2 has a rather long residence time in the troposphere. Therefore, the levels of bomb ^{14}C in the troposphere air were highest in the middle latitudes of the northern hemisphere and reached maximum concentrations late in the summer.

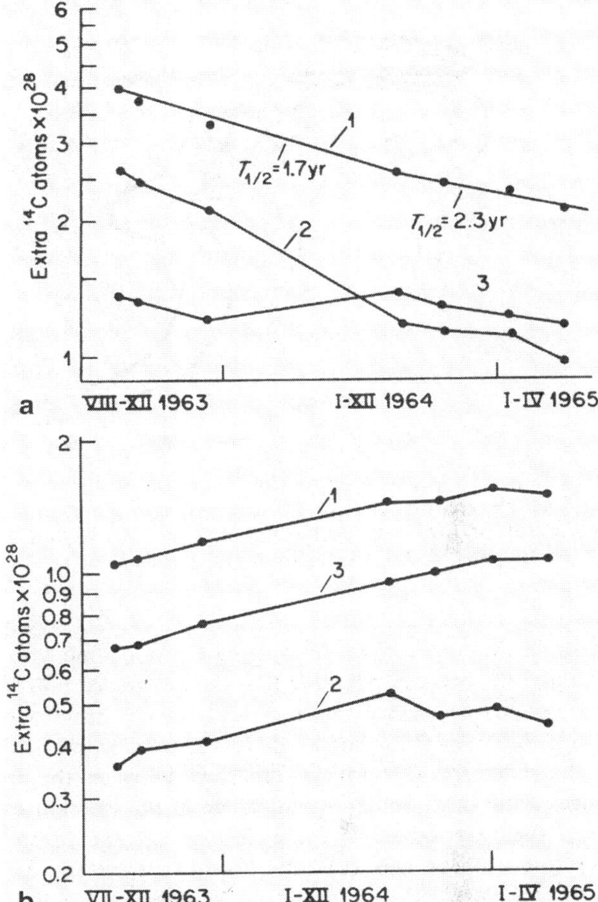

Fig. 14.6 Variation of total amount of bomb ^{14}C in northern (**a**) and southern (**b**) hemispheres for the atmosphere as a whole (*1*), the stratosphere (*2*), and the troposphere (*3*). (Fairhall and Young 1970)

The longitudinal mixing of ^{14}C extends southwards. Annually, from September to February, the level of ^{14}C in the northern latitudes decreased, and in the southern latitudes increased, up to the middle of 1966. Such a process took place until 1966 when the whole troposphere became homogeneous with respect to the ^{14}C.

It has been pointed out (Fairhall and Young 1970) that the relative amounts of tropospheric air participating in the circulation at various latitudes should be taken into account when comparing the latitudinal variations of the ^{14}C concentrations (Fig. 14.5). The convergence of the meridians and the descent of the tropopause at high latitudes result in a sharp decrease in the volume of tropospheric air with increasing latitudes, compared with the equatorial region. Thus for the same ^{14}C ejection from the stratosphere, its activity in high latitudes will be considerably higher than in the lower ones.

Figure 14.6 shows the variations with time of the absolute amounts of the bomb-radiocarbon in the troposphere, stratosphere, and atmosphere as a whole for the northern and southern hemispheres.

It is observed from the Fig. 14.6 that with the decrease in the ^{14}C concentrations in the northern hemisphere, the opposite process has taken place in the southern hemisphere during 1963–1965. The theoretical estimation of the total amount of the ^{14}C in the atmosphere based on experimental data shows that the time of half-removal of this isotope is equal to 3.3 years. In the subsequent years, the rate of the ^{14}C removal from the atmosphere decreased. Its amount in the troposphere up to 1970 compared with 1963 decreased by two times and by 1977 only one third of it was left (Berger 1979). Before 1984, the ^{14}C amount in the troposphere numbered about 125% with respect to pre-thermonuclear level. The residence time of CO_2 in the troposphere calculated by decrease of ^{14}C activity during period of 1963–1976 is about 10 years (7/0.693), and of 1970–1982 this time was about 15 years, which is close to the earlier estimates made by Arnold and Anderson (14–30 years) (Miyake 1969).

It is assumed that the rate of removal of bomb ^{14}C from the stratosphere is proportional to the difference (gradient) between its concentration in the stratosphere and the surface (mixing) ocean layer. While decreasing the gradient, the velocity of the radiocarbon removal from the troposphere is dropping. On this basis, the conclusion follows that during the 'peak' injections of radiocarbon (and tritium as well) into the atmosphere, the ^{14}C concentration changes between the atmosphere and hydrosphere take place in $^{14}CO_2$ but not in general CO_2. The sharp drop of ^{14}C half-removal from the atmosphere in the first years after nuclear tests stopped is shown in Fig. 14.6a. It is most probably, that the ^{14}C half-removal, measured by its activity before 1966, characterized in reality not only CO_2 exchange process between the atmosphere, ocean, and biosphere, but also ^{14}C activity decrease in the northern hemisphere as a result of its transfer with air masses into the southern hemisphere.

The altitudinal distribution of ^{14}C specific activity in the atmosphere is not uniform. In 1962, for example, ^{14}C activity in CO_2 of the lower troposphere was about 140% with respect to the natural level. In April 1962, before the USSR thermonuclear tests, at the height of 12.2 km at 36°N, the ^{14}C activity was measured to be three times higher than the natural one (Fergussen 1963). These measurements have shown that the considerable amounts of ^{14}C were accumulated in the stratosphere as a result of the nuclear tests. This fact is proved by the summer maximum of its coming into the troposphere (see Figs. 14.7 and 14.8).

Figure 14.8 shows the summarized Troncheim Laboratory data on radiocarbon concentration changes in the troposphere and the surface waters of the Pacific and Atlantic oceans (Nydal et al. 1979).

By the data of Fairhall and Young (1970), the radiocarbon concentrations in the troposphere and atmosphere are practically equal; however, the observations in 1968–1974 summer maximums gives evidence against this assumption (Berger 1979). In order to find the correct answer, the CO_2 stratospheric sampling at 18 and 21 km high was carried out during 1975–1977 in the framework of the NASA program. The results of this experiment are shown in Table 14.4.

As shown in Table 14.4, the spring peak of ^{14}C concentration has appeared in the lower stratosphere which looks like the summer maximum in the troposphere. In this connection, one may assume that during the summer the ^{14}C enriched air breaks the troposphere and comes there from the lower stratosphere. By Berger's opinion, the mean residence time of CO_2 in the stratosphere is about 15–20 years.

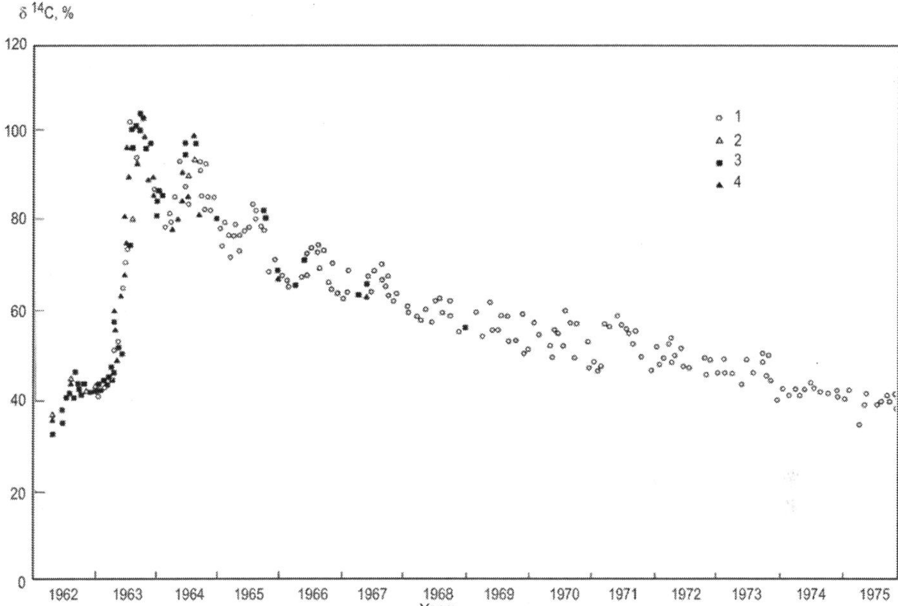

Fig. 14.7 Radiocarbon concentration changes in troposphere of high latitudes by data from the stations: (*1*) Nordcap, 71°N, 24°E; (*2*) Spitsbergen, 78°N, 19°E; (*3*) Troncheim, 63°N, 10°E; (*4*) Lindesnes, 58°N, 7°E. (After Nydal et al. 1979)

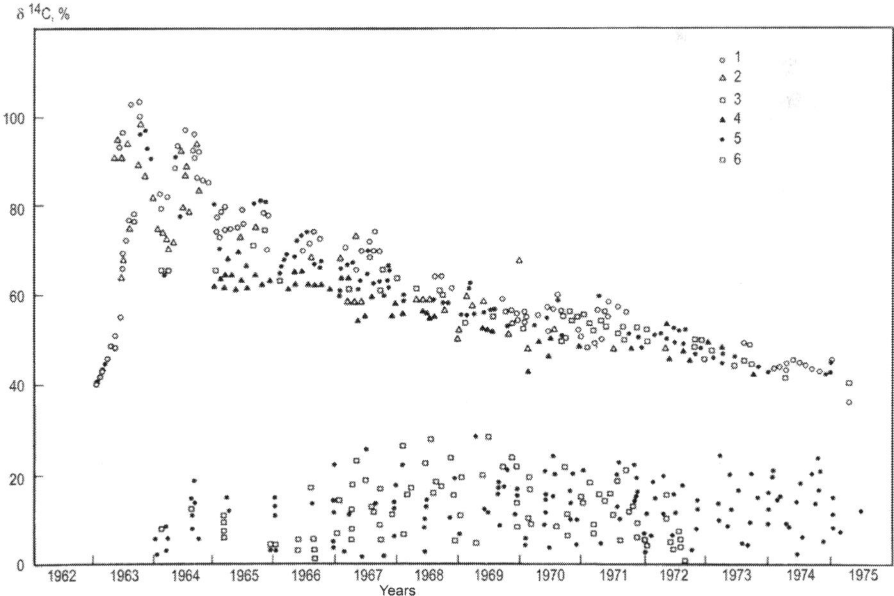

Fig. 14.8 Summarized radiocarbon content data on troposphere and ocean surface waters: (*1*) Nordcap; (*2*) Gran Canaria; (*3*) N'Djamena; (*4*) Madagascar; (*5*) surface waters of the Pacific Ocean; (*6*) surface waters of the Atlantic waters. (After Nydal et al. 1979)

Table 14.4 ^{14}C content in stratosphere. (From Berger 1979)

Sampling date	Sampling height	Sampling place	$\delta^{14}C$ (‰)
13.06.75	19.8	Northern Sierra	+127.4
25.09.75	19.8	California	+90.7
03.12.75	18–21	San Francisco—Phoenix	+79.1
02.06.76	18–21	San Francisco—Salt Lake City	+85.6
30.03.76	19.8	San Francisco—Oregon	+107.0
23.09.76	19.8	San Francisco—Denver	+99.2
22.03.76	19.8	Pacific Ocean—Los Angeles	+199.6

In 1963, the USSR, USA, and GB have signed the treaty on cessation of the nuclear weapon tests in the atmosphere, oceans, and on the Earth surface. Since that time, the exchange reservoirs were insignificantly replenished by the artificial ^{14}C during the French and Chinese explosions. The peak ejections of radiocarbon into the atmosphere, which created in the atmosphere and stratosphere some kind of marks, allowed a more detailed study of the mass exchange between the atmosphere, oceans, and biosphere, and also between different layers of the atmosphere and oceans. The wide program on the study of distribution of ^{14}C was caused by the bomb-tests, carried out in the Trondheim Laboratory starting 1962. In particular, the scientists who took part in the program, since 1963 published their results on the study of CO_2 exchange between the atmosphere and oceans. The study on ^{14}C redistribution between the northern and southern hemispheres on a profile Spitsbergen–Madagascar was also initiated. Before 1967, the ^{14}C concentrations in the lower atmosphere of the Earth were practically equal. This was allowed in calculations of carbon exchange cycles without taking account the time in different parts of the atmosphere. The obtained results are shown in Figs. 14.8 and 14.9.

The decrease of bomb ^{14}C in the atmosphere occurs due to its absorption by the terrestrial biosphere and CO_2 exchange between the atmosphere and the oceans. The latter reservoir is most important since more than 80% of the atmospheric CO_2 is exchanged with the ocean and 20% is replaced through the inland biosphere. In view of the role of the oceans in the absorption of bomb-^{14}C, considerable investigations have been carried out aimed at the measurement of its concentration in surface and deep oceanic waters (Bien and Suess 1967; Fairhall 1971; Nydal 1967; Fairhall et al. 1969). In contrast to the atmosphere, characterized by quick ^{14}C mixing (less than 10 years), the process of mixing in the ocean is more complicated and requires more time. The radiocarbon technique is advantageous in studying such oceanographic processes as mixing of the oceanic waters and their global circulation.

Figure 14.10 illustrates in isolines the picture of ^{14}C variation in the surface waters of the Pacific Ocean (from 140°W to 170°E), plotted by Bien and Suess (1967) on the basis of the data obtained by different authors. The main features of the character of the ^{14}C variations in the atmosphere of the northern and southern hemispheres are in principle present in the surface waters of the Pacific Ocean, but the effect of the motion of oceanic waters is also reflected, for example, regions of upwelling of deep waters to the surface are indicated, characterized by low ^{14}C content, near the equator and at 40°N.

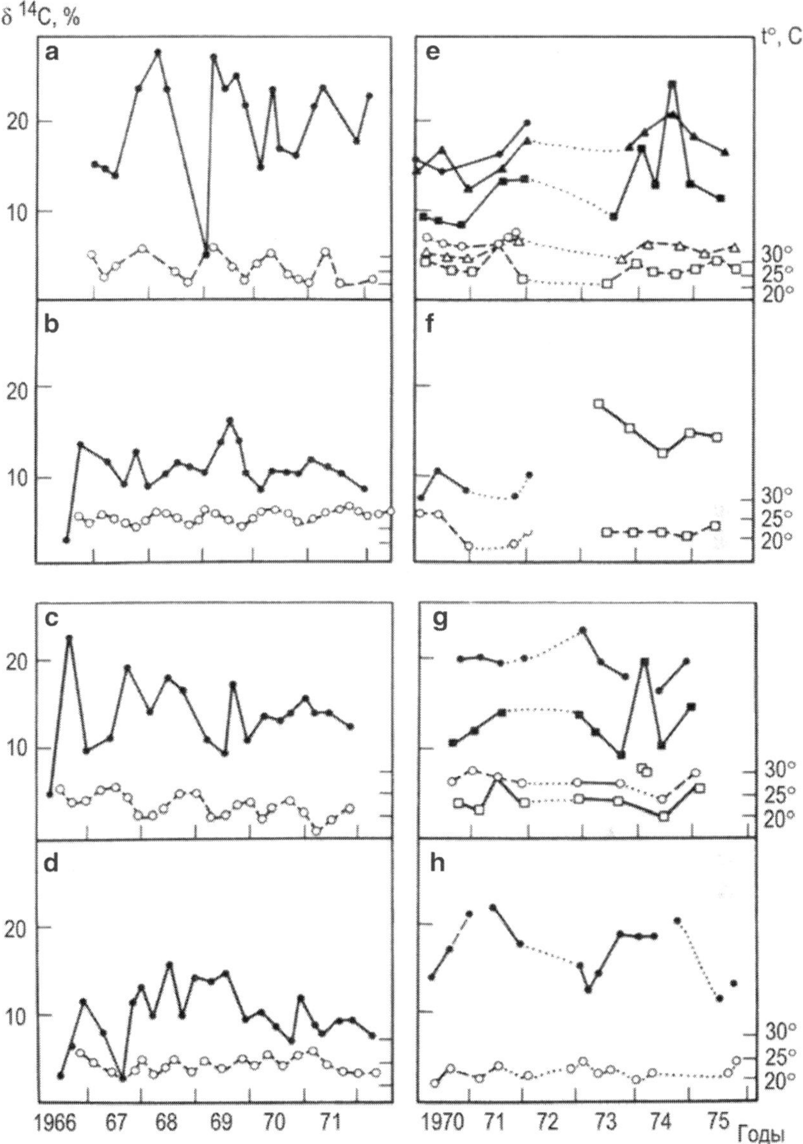

Fig. 14.9 ^{14}C concentration changes in surface of water oceans: (*a–d*) Atlantic ocean: (*a*) 31–33°S, 50–52°W; (*b*) 4–7°S, 31–33°W; (*c*) 20–30°N, 15–18°W; (*d*) 17–18°N, 20–22°W; (*e–h*) Pacific Ocean: (*e*) 25–36°S, 120–180°W; (*f*) 14–20°S, 110–120°W; (*g*) 28–34°N, 150–150°E, (*h*) 19–26°N, 119–126°W. (After Nydal et al. 1979)

Figure 14.11 shows the variation of the relative ^{14}C concentrations in a depth profile of the Pacific Ocean at high latitudes in the southern hemisphere. The experimental data, corresponding to the low latitudinal depth profile in the ocean, are shown in Fig. 14.12 (Fairhall 1971).

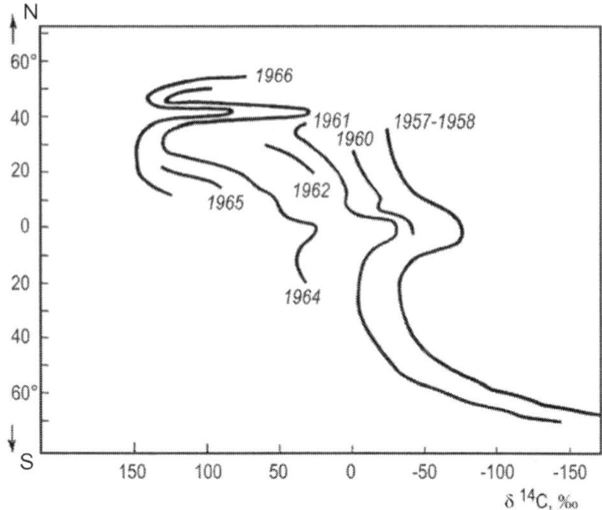

Fig. 14.10 Variation of ^{14}C content (relative to standard) in surface water of the Pacific Ocean between 140°W and 170°E in 1957–1966. (After Bien and Suess 1967. © IAEA, reproduced with permission of IAEA)

The spatial ^{14}C distribution in depth of the Atlantic Ocean was studied in framework of the GEOSECS Program (Stuiver 1980). Figure 14.13a shows location of the stations and also the stations of the research ship 'Meteor' (Germany). The ship's route was adjusted to GEOSECS Program (Roether et al 1980). The $\delta^{14}C$ values' distribution at the GEOSECS stations in 1972–1973 is presented in Fig. 14.13b, 14.13c. It is seen in Fig. 14.13c that the two main fields of the surface waters' immersion into deep layers are observed. The first field is located between 36° and 40°S and the second between 20° and 30°S. The upward flow of the deep waters is traced around the equator. The two more localized upward flows are traced between 40°–50°S and 40°–50°N. It is observed from the Fig. 14.13b, 14.13c that significant amount of the bomb-radiocarbon was entered into the surface ocean layer and reached the deep layers in 1972. Analogous picture on the longitudinal profile, obtained during 23rd route of the 'Meteor' in 1971, is observed (Fig. 14.14; Roether et al 1980).

As it was pointed out above, the ^{14}C mark, occurred in the atmosphere after the thermonuclear tests during a short period of time, is used to study the atmosphere and oceans dynamical characteristics. The carbon concentration increase in the Earth's atmosphere has a head form of δ-function at the well-known change of the carbon concentration in the atmosphere and oceans. This gives the possibility for estimation of the exchange parameters of the reservoir, the main of which are the atmosphere, oceans, and biosphere. Each reservoir can be presented by a number of 'boxes' (Dergachev 1977). For example, the oceans can be considered as two boxes reservoir that is the upper one over the thermocline (75 ± 25 m) in which intensive water mixing occurs, and lower, weakly mixing 'box' in which the carbon concentrates due to the diffusion process. The continuous carbon exchange between the deep ocean layer and the ocean sediments takes place. The natural border in the atmosphere is the tropopause which is at the height of 11–12 km. The tropopause divides the troposphere into the layer, well mixing in vertical direction, and the layer where the

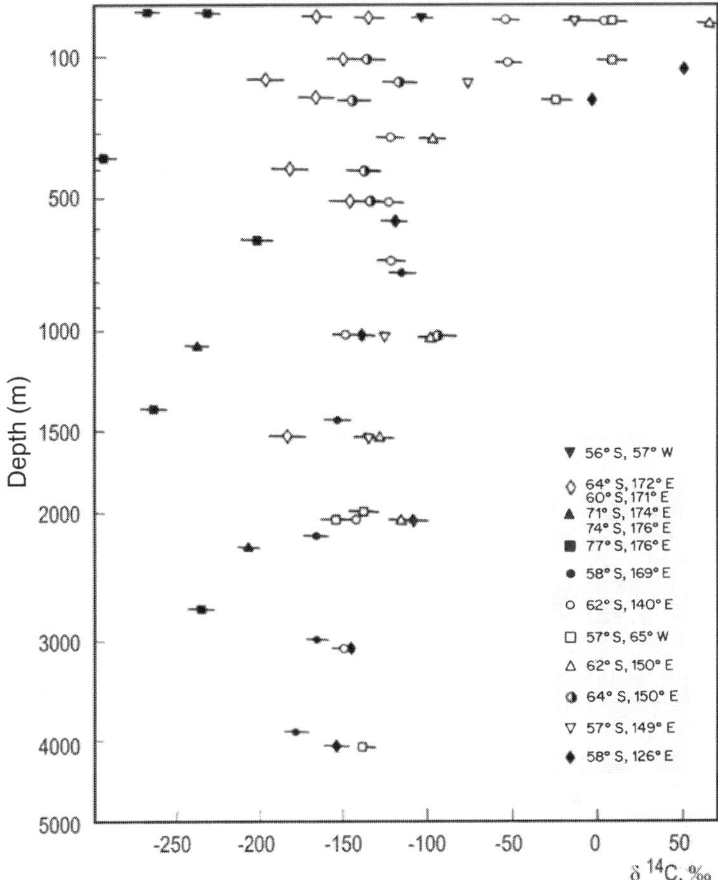

Fig. 14.11 ^{14}C distribution in a depth profile of the Pacific Ocean for high latitudes of the southern hemisphere in 1958–1969. (After Fairhall 1971)

meso-longitudinal exchange is prevailing. The use of the box models, the possibilities of which were discussed earlier, on the basis of experimental study of the ^{14}C distribution in reservoir in space and time, allows the rate of exchange processes and the carbon mean residence time in reservoirs to be estimated.

Dergachev (1977), after analysis of many studies on this problem, came to conclusion that the values of the exchange parameters, calculated on the basis of the artificial ^{14}C distribution, are too scattered. The scattering has two causes: first, due to incorrectness in determination of the reservoir borders of the exchange reservoir, and secondly, because of confusing the meaning of 'half-exchange time' and 'half-life time' of the ^{14}C in either part of reservoir. The following values of the above parameters have been obtained by Dergachev: the exchange time between the troposphere and stratosphere is equal to 1.5–2 years; the ^{14}C mean residence time in

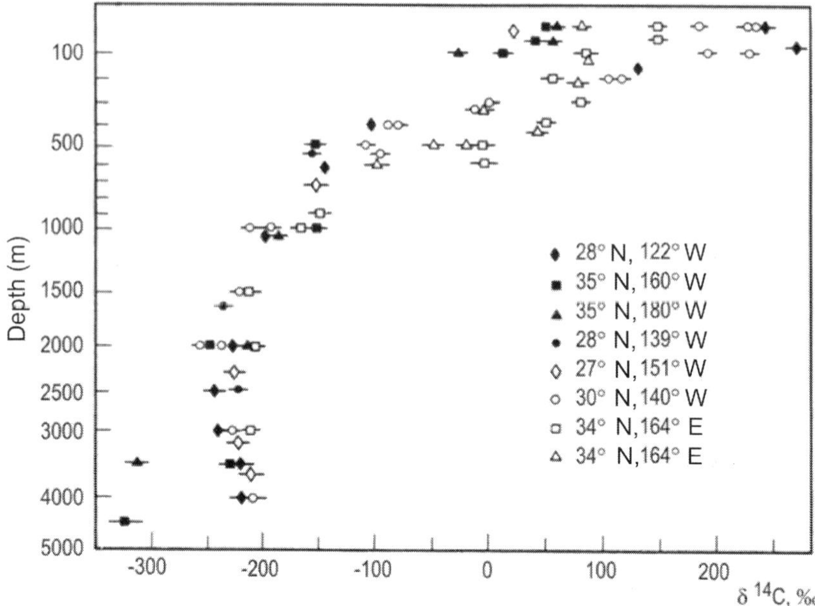

Fig. 14.12 ^{14}C distribution in a depth profile of the Pacific Ocean for low latitudes of the northern hemisphere in 1958–1969. (After Fairhall 1971)

the stratosphere is 3.5–4 years, and in the troposphere 1.5–2 years; the mixing time of the atmosphere between the hemispheres through the equator is about 1.5 years. The mean values for the both hemispheres are:

$$\tau_{ma} = 9.1;\ \tau_{ma} = 7.1;\ \tau_{ma} = 8.7;\ \tau_{md} = 4.3;\ \tau_{md} = 2.8;\ \tau_{md} = 3.8 \text{ years};$$

where *m* is the mixing layer; *a* is the atmosphere; *d* is the deep layer.

As Dergachev points out, the small values τ_{md} contradict to the results of calculations by box models, by which $\tau_{md} \approx 10$ years. This is possible because of dependence of exchange rate between the atmosphere and different ocean layers on the ^{14}C concentration gradient in exchangeable reservoirs. In this case, the decrease of the rate of the radiocarbon removal from the troposphere must lead to an increase in the mean residence time in the atmosphere and mixing ocean layer. In favor of such assumption, the experimental data on ^{14}C distribution in the atmosphere and oceans obtained in the recent years are evident. The authors of this book, applying the Trondheim Laboratory data (Nydal et al. 1979), on the basis of the one-box model, estimated the residence time of radiocarbon in the mixing and deep ocean layers. Table 14.5 shows the function (A) of the radiocarbon change in the atmosphere and the ^{14}C concentration increase in the mixing ocean layer H_m used in the calculations.

The data for the atmosphere before 1963 were taken on the basis of ^{14}C measurements in alcohols of the Georgian and Portuguese wines (Burchuladze et al. 1977; Lopes et al. 1977). For the mixing layer, the ^{14}C concentration before 1963 was

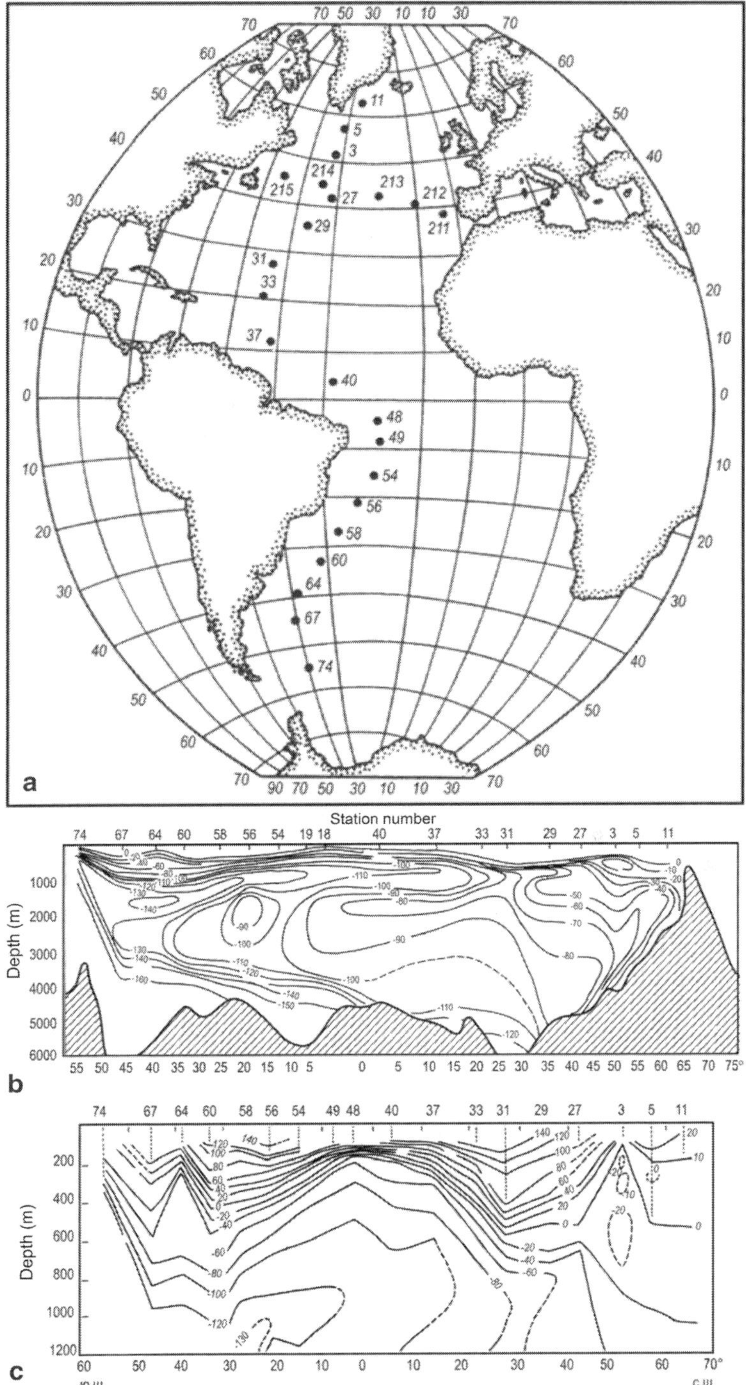

Fig. 14.13 Location of measurement and sampling stations for GEOSECS program (1972–1973, No No5-74) and 23rd 'Meteor' route (1971, NoNo 211-215) (*a*) (Stuiver 1980; Roether et al. 980); and $\Delta^{14}C$ vertical distribution in western part of Atlantic Ocean in 1972–1973 (*b*) (Stuiver 1980); the same for the depth up to 1,200 m (*c*))

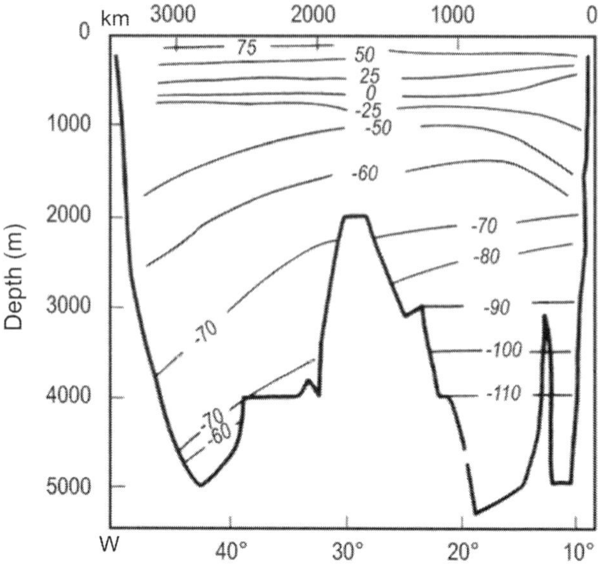

Fig. 14.14 $\Delta^{14}C$ vertical distribution obtained by ship 'Meteor' stations, 23rd route, 1971. (After Roether et al 1980)

Table 14.5 ^{14}C (%) content from the standard of modern carbon

Year	Reservoir	
	A	Hm
1957	104	98
1958	112	98
1959	123	99
1960	122	99
1961	116	100
1962	137	100
1963	193	102
1964	188	105
1965	170	108
1966	165	110
1967	160	112
1968	158	114
1969	155	118
1970	152	116
1971	150	116
1972	148	115
1973	145	115
1974	142	114
1975	140	113

obtained by extrapolation of the data published in the studies of (Bien et al. 1963; Broecker et al. 1960). The calculations were carried out by the formula analogous to that used for the tritium data interpretation (see Chap. 12):

$$A_t = \overline{A} + \sum_{1957}^{1975} \delta A_\theta \frac{1}{\tau} e^{-t/\tau-},$$

where A_t is the radiocarbon concentration in the mixing layer at t years after 1957; $\delta A_\ominus$ is the difference (%) between the biogenic radiocarbon level before 1950 (100%) in the calendar year $\ominus$; $\bar{A}$ is the ^{14}C concentration in the mixing layer (98%).

The best coincidence of the calculated and experimental data is achieved at value $\tau \approx 15$ years. The radiocarbon half-life was not taken into account because of the short time interval compared with the ^{14}C half-life. Certainly, the one-box model used for estimation of the dynamical parameters in the atmosphere-ocean system (Gray and Damon 1970; Ralf 1972) gives an approximate result, but the carbon constant of the reaction transfer rate from the deep to mixing layer by substitution of the value $\bar{A}$ was taken into account. The carbon exchange time of the mixing layer by Eq. 14.7 can be estimated. Taking into account that $N_{\Pi.c.} = 1.3N_a$, one obtains $\tau = 13$ years, if τ for the atmosphere is equal to 10 years. As in this case the carbon exchange in the system mixing-deep layers is not taken into account, the more realistic τ value for the mixing layer seems to be 10 years. This does not contradict to the three-box model of Dergachev (1977). The ^{14}C distribution in the surface layer of the central ocean parts is of interest. The observations on the Atlantic and Pacific Ocean stations show (Nydal et al. 1979) that ^{14}C concentrations in local areas do not stay constant with respect to time but vary within 10‰ (and even up to 20‰). A positive correlation between radiocarbon content in the surface layer and the surface ocean temperature is observed.

In the study of Rafter and O'Brien (1972), the ^{14}C distribution in the Pacific surface waters after 1968 is discussed. Figure 14.15 shows their results, where two belts of increased concentrations ($\delta^{14}C \approx 20\%$) near 27°N and 27°S and minimal values around the equator ($\delta^{14}C \approx +5\%$) are observed.

While considering the ^{14}C distribution in the surface Atlantic waters, Dergachev (1977) derived two maximum values: $\delta^{14}C \approx +7\%$ (70°N) and $\delta^{14}C \approx +19\%$ (27°N). It is interesting to note that the location of the above maximum values is unchanged up to the depth of 500 m (Fig. 14.16). The nature of this phenomenon is not explained. The only thing clear is that the maximum salinity of the oceans coincides with those from Fig. 14.15. The water zones are characterized by higher evaporation of the marine water and have a relationship with global circulation of the air and water masses. It is obvious that before the thermonuclear tests, the ^{14}C concentration in the surface ocean waters was not constant and changed under the same effects as after the tests.

The artificial radiocarbon from the mixing layer enters to the deep ocean layers. According to Ferhall (1971) calculations, the process of equilibrium state between the atmosphere and oceans will continue for 45 years (starting from 1970). Experimental study of this process will help in understanding the oceanic currents' nature and the mixing time of ocean waters.

14.5 Forecast of Carbon Dioxide Increase in the Atmosphere

The carbon dioxide increase in the atmosphere, in addition to the ^{14}C specific activity dilution in the exchangeable reservoirs, plays the own independent and important role. Increase of the carbon dioxide content may lead to a considerable change in

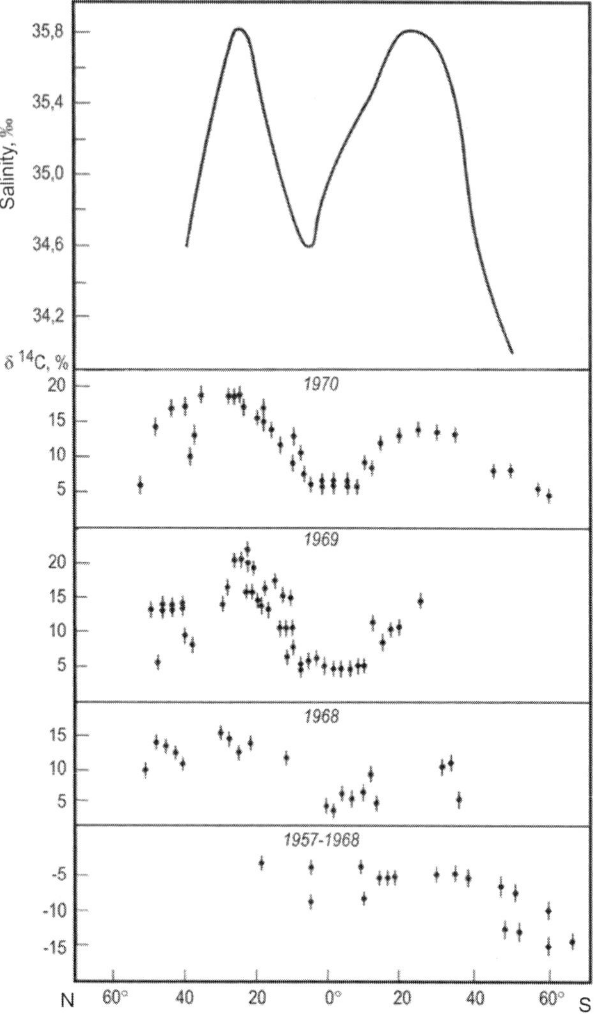

Fig. 14.15 ^{14}C concentration change in the Pacific Ocean surface layer over 1957–1970 (Rafter and O'Brien 1972) and salinity of the ocean surface layer. (After Miyake 1969. © IAEA, reproduced with permission of IAEA)

the mean yearly temperature on the Earth, which is a consequence of the 'green house' effect. The carbon dioxide is practically transparent for the solar radiation, but it absorbs the thermal radiation of the Earth in a number of lines of the infrared spectrum. If the water and carbon dioxide are absent in our planet, then the infrared radiation leaves it for the space and the lower layers of the atmosphere to lose the warmth. The carbon dioxide concentration growth leads to the stratosphere cooling and the near-earth air growth (Table 14.6).

Oeschger and Siegenthaller (1979) used CO_2 to change with respect to time parameters for the main reservoirs, obtained on the basis of the artificial radiocarbon distribution, and to predict the possible increase of CO_2 content in the atmosphere. They proposed models which based on the following assumptions.

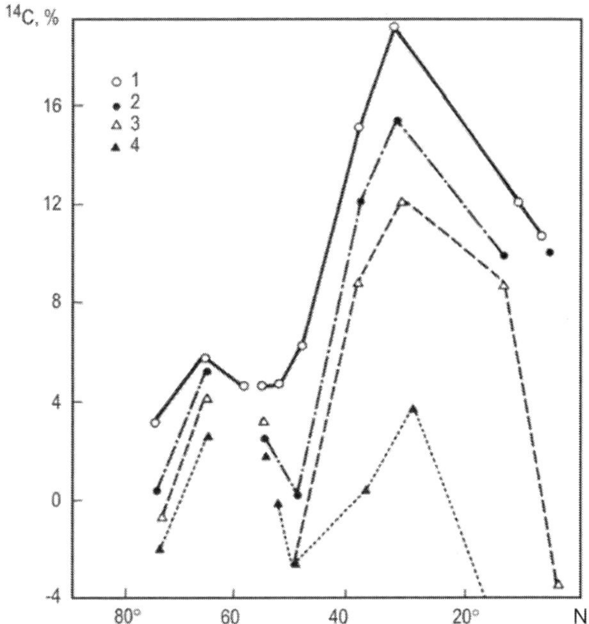

Fig. 14.16 Radiocarbon concentration change in upper layers in the North Atlantic Ocean: (*1*) surface layer; (*2*) at depth 100 m; (*3*) at 200 m; (*4*) at 500 m)

Table 14.6 Assuming CO_2 growth in the atmosphere. (From Kellogg 1980)

CO_2 content change in in atmosphere (%)	Assuming year of the change (year)	Growth of near-earth temperature (°C)
+25	2000	0.5–1.0
+100	2050	1.5–3.0

1. The study of ^{14}C content in deep Pacific and Atlantic oceanic waters over 1957–1959 has shown that distribution of radiocarbon in the deep waters practically was not disturbed by the thermonuclear tests.
2. As a result of the thermonuclear tests, the ^{14}C activity in the surface ocean waters is increased by 95% with respect to the present carbon content (its 'pre-industrial' value for the atmosphere is equal to 100%) in 1957 and by about 112% in 1970.
3. About 19% with respect to 'pre-industrial' amount of CO_2 has been injected into the atmosphere before 1970, but its concentration decreased only by 10%.
4. Because of dilution by carbon dioxide originated at combustion of the fossil fuel, the ^{14}C specific activity in the atmosphere before 1950 has decreased by about 2%.

Oeschger and Siegenthaller (1979) assumed for their model that the carbon transfer from the mixing layer to the deep waters results by the vortex (turbulent) diffusion (Fig. 14.17). Any increase in the partial pressure in CO_2 leads to its redistribution between the atmosphere and oceans. Such redistribution can be described by a parameter called the 'buffer-factor'. For example, if the CO_2 partial pressure in the

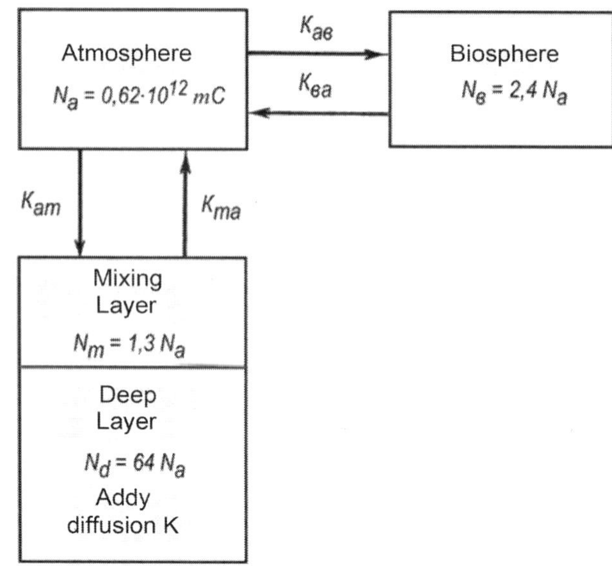

Fig. 14.17 Reservoirs of the carbon exchangeable system which takes into account diffusion transfer to the deep ocean layers. (After Oeschger and Siegenthaller 1979. © IAEA, reproduced with permission of IAEA)

atmosphere increases by $\alpha\%$, then the total CO_2 concentration in the ocean waters, to be in equilibrium with the atmosphere, increases in waters only by $\alpha/\xi\,\%$. The biosphere in the model is represented by the well mixed reservoir, the amount of carbon in which is 2.4 times more than in the atmosphere and its residence time is about 60 years. The partial pressure increase of CO_2 in the atmosphere leads to the photosynthetic activity of the plants described by the 'growth-factor' ε. If CO_2 pressure in the atmosphere increases by $\alpha\%$ then the carbon flow from the atmosphere to the biosphere increases by the value of $\varepsilon\alpha\%$. According to (Oeschger and Siegenthaller 1979), the ε value varies for different conditions from 0 to 0.4. For prediction of the CO_2 increase in the atmosphere, the following independent parameters were used (see Fig. 14.17):

N_a	CO_2 in atmosphere	Pre-industrial level	0.62×10^{18} g
N_b	CO_2 in biosphere		2.4 N_a
N_m	CO_2 in mixing layer		1.3 N_a
N_d	CO_2 in deep water layer		64.2 N_a
K_{am}	Exchange coefficient in system atmosphere—mixing layer		1/7.7 years
K_{ab}	Exchange coefficient in system atmosphere-biosphere		1/25 years
K	Coefficient of vortex diffusion		3,987 $m^2\ sec^{-1}$
ξ	Buffer-factor of CO_2 absorption by the ocean		10
ε	Growth-factor connecting with CO_2 absorption by biosphere		0.2

The authors give a prediction on the CO_2 content increase in the atmosphere at different input functions. For example, if single CO_2 concentration in the atmosphere increases by $\alpha\%$, then a new equilibrium is reached characterized by $\alpha_\infty\%$. The ratio α_o/α_∞ can be obtained by Equation,

$$\alpha_0 N_a = \alpha_\infty N_a + (\alpha_\infty/\xi)(N_m + N_d) + \alpha_\infty \varepsilon N_b,$$

or

$$\alpha_\infty/\alpha_0 = N_a/[N_a + (N_m + N_d)]/\xi + \varepsilon N_b.$$

At $\xi = 10$, $\varepsilon = 0.2$, $\alpha_o/\alpha_\infty = 0.125$, i.e., one eighth part of the total CO_2 input into the atmosphere. Here possible decrease of α_∞ value as a result of oceanic carbonates dissolution and weathering of the rocks has not been considered, but because of too slow CO_2 transfer into deep ocean layers these processes will continue for a long time. If the CO_2 inflow into the atmosphere stopped in 1970, then $\alpha \approx 10\%$, in 2000 α value to be 7%, and after several centuries it would reach equilibrium value, that is, $\alpha_\infty \approx 2.3\%$. The carbon content in the fossil fuel (oil, coal, and gas) recalculated by CO_2 is 11.64 times more than CO_2 in the atmosphere at pre-industrial epoch. According to the present-day knowledge, the CO_2 production rate can be described by the equation as

$$P(t) = \frac{d}{dt}\left[\frac{11.65 N_a}{1 + 61\ \exp(-t/22)}\right],$$

where t is the time in years ($t = 0$ in 1970).

The production rate P(t) after 1970 comprises 4.5 N_a per year (the mean velocity over 1960–1970 is about 5% per year). The calculations show that at this rate of the fuel combustion more of 2,000% with respect to pre-industrial level of CO_2 will be accumulated in the atmosphere to 2,050. This amount of carbon dioxide may lead to an increase of near-ground temperatures from 3 to 4°C (Fig. 14.18), but this is unlikely. In twenty-first century, the fossil fuel use decreased because the other sources of energy are now involved. In addition, according to the recent scientific data on heat and mass exchange between the atmosphere and oceans, the green house effect is by orders of 3–4 overestimated (Sorokhtin 2002).

By the Kellogg's (1980) data, in this case, the near-ground temperature will increase only by 1–1.5°C.

14.6 Principles of Radiocarbon Dating

If any system exchange with the atmospheric carbon dioxide is finished, then the accumulated ^{14}C amount is decreased with respect to time by the radioactive decay law which is

$$A_t = A_0 e^{-\lambda t}, \tag{14.13}$$

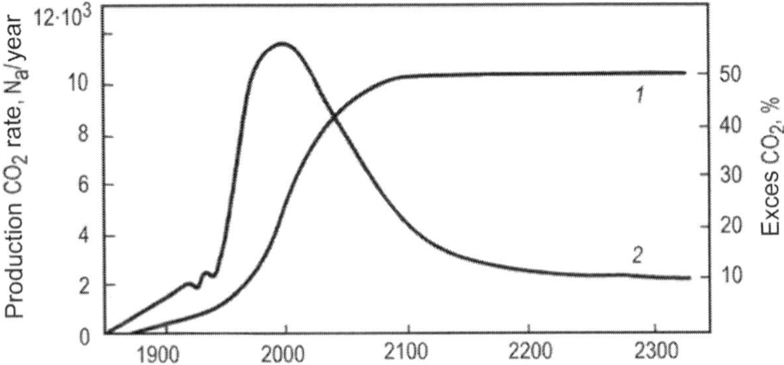

Fig. 14.18 Dependence of CO_2 variation in the atmosphere on present fossil fuel combustion (*1*) and on 30% reduction of its present combustion (*2*). (After Oeschger and Siegenthaller 1979. © IAEA, reproduced with permission of IAEA)

where A_t and A_0 are the radiocarbon activity (or concentration) at time t and $t = 0$; λ is the radiocarbon decay constant; $\lambda = \ln 2/T_{1/2} = 0.693/T_{1/2}$; T is the period of the ^{14}C half-life.

If the exchange processes stop, for example due to death, then the Eq. 14.13 can be used for calculation of the time passed after the process has broken off (the object age):

$$t = (1/\lambda)\ln(A_0/A_t) = 8033\ln(A_0/A_t). \tag{14.14}$$

The half-life period equal to 5,568 years was used in the Eq. 14.14. In order to pass the 5,730-year period, all the age values calculated by Eq. 14.14 should be multiplied by the coefficient 1.03. The method of obtaining radiocarbon dates on organic specimens has been developed by Libby (1967).

Radiocarbon dating is based on a number of assumptions; the most important of which are (Olsson et al. 1970): (1) the intensity of cosmic radiation and, as a consequence, the ^{14}C concentration in the atmosphere, remained constant at least during the radiocarbon dating scale (0–80,000 years); (2) the time of exchange of atmospheric carbon with terrestrial carbon is considerably smaller than the half-life of carbon-14 and does not change with time; (3) secondary processes do not affect the isotopic composition of the studied specimens after sampling. This is equivalent to the assumption that the radiocarbon content in a specimen decreases only because of radiocarbon decay.

In natural cases, the above-mentioned assumptions do not hold strictly, although in most cases the age determined on the basis of radioactive dating are in good agreement with those obtained using other techniques (Libby 1967).

In the case of ^{14}C variations in atmospheric CO_2 and in biosphere (see Fig. 14.4) over long time periods starting from about 2,500 years, radiocarbon and dendrochronological timescale of the organic specimens becomes different (Fig. 14.19; Mook 1977).

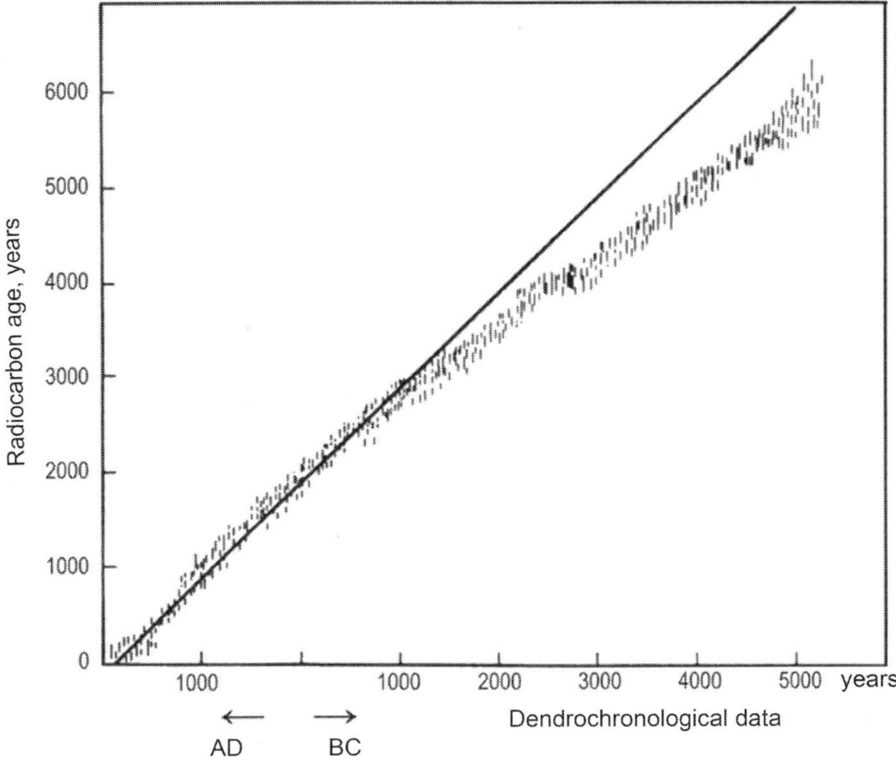

Fig. 14.19 Relationship between dendrochronological and radiocarbon ages of *Sequola gigantea* and *Pinus aristata*. (After Mook 1977. © IAEA, reproduced with permission of IAEA)

In order to reconcile the radiocarbon results with those obtained by the dendrochronological ones within 0–8,000 years, a number of correlations were proposed. According to Wendland and Donley (1971), the age correlation can be done by the third degree polynomial:

$$T_{corr} = 112 + (0.710T_{14C}) + \left(1.610 \times 10^{-4}T_{14C}^{2}\right) - \left(1.50 \times 10^{-8}T_{14C}^{3}\right).$$

For this purpose, there are also other correlation equations (Ralf and Klein 1979).

Radiocarbon dating on the basis of organic fragments has a wide application in the archeology, geology, geography, oceanography, and other Earth's sciences (Suess 1979).

Isotopic composition of the stable carbon in a carbonate system is used for correction of the radiocarbon age of groundwater. The processes of the carbon isotope fractionation were considered in Chap. 10.

As it is observed from the Eq. 14.14, the most stable carbon fractionation is between gaseous CO_2 and hydrocarbonate-ion (HCO_3^--ion is enriched in ^{13}C by about 10% at 0°C compared with the carbon dioxide). It is also obvious that the complete fractionation of the carbon isotopes in the gas-solution and gas-solid phase

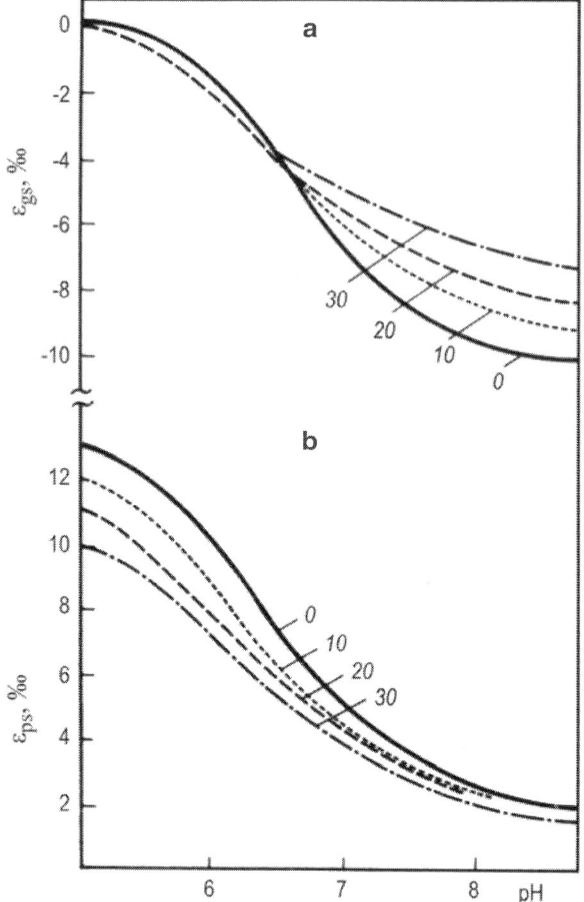

Fig. 14.20 Equilibrium fractionation of carbon isotopes versus pH and temperature for the systems: (*a*) gaseous CO_2—dissolved carbon components; (*b*) carbon sediments—dissolved carbonates. (After Wigley et al. 1978)

system should depend on the medium pH. A graphical relation of such dependence is shown in Fig. 14.20 (Wigley et al. 1978).

Fractionation of the carbon isotopes in the thermodynamically equilibrium processes, stipulated by the kinetic factors (for example, in biochemical reactions), leads to the nonsteady state in carbon isotopes of the carbon-bearing matter of different natural objects.

Figure 14.21 indicates the isotope variation limits in carbon contained in various objects (Stiel et al. 1979). It is observed from this figure that the carbonates precipitating from the oceanic water (mean values of $\delta^{13}C = 0$) are the most ^{13} Cenriched and relatively constant by isotopic composition. The PDB-1 isotope standard is composed of these carbonates. Compared with the standard, the atmospheric CO_2 is enriched in the light isotope of carbon by −7‰. Plants are even more enriched in the light isotope. The average value for ground plants is $\delta^{13}C = -25$‰; close to this value is the soil CO_2. A large value of the shift ($\delta^{13}C = -30$‰) is detected in oil and maximum enrichment in ^{12}C is observed in natural methane formed in sedimentary rocks ($\delta^{13}C$ up to −90‰).

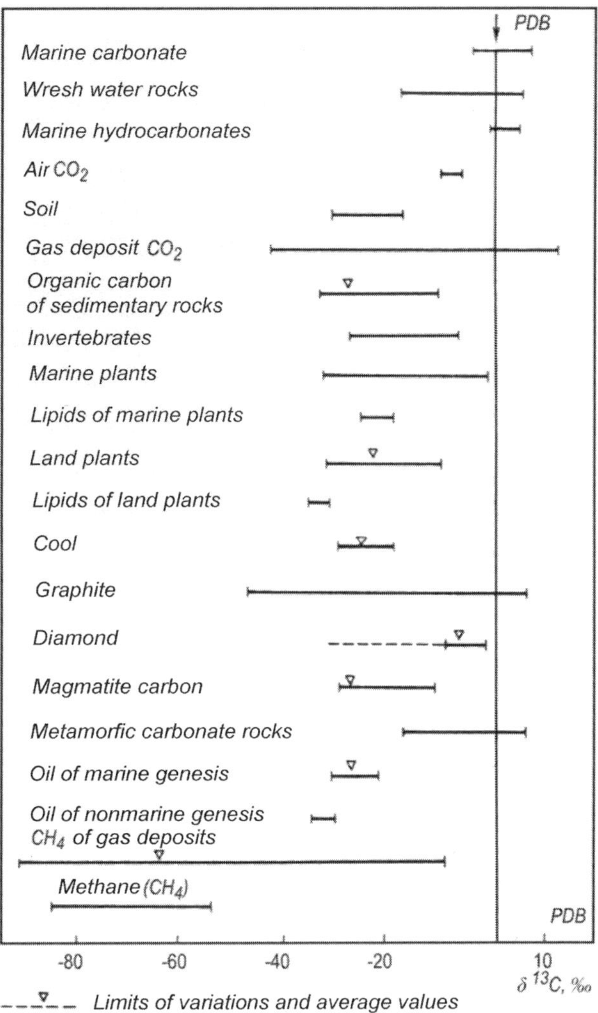

Fig. 14.21 Stable carbon isotope variations relative to the PDB-1 standard in various natural objects. (After Stiel et al. 1979)

Since the isotope fractionation factor increases by its square with the addition of each neutron, the corresponding isotope shift in ^{12}C–^{14}C isotope fractionation will be twice as great. Therefore vegetation should be depleted in ^{14}C compared with the atmosphere and have $\delta^{14}C = -50‰$.

14.7 Radiocarbon Dating of Groundwater

At present, for groundwater age determination, most of the researchers apply the piston model. It was earlier pointed out that the radioactive isotope concentration in a sampling point *A*, located at a distance x_0 from the recharge, is calculated by the expression:

$$A = A_0 \exp[-(x_0/v)\lambda] = A_0 \exp(-\lambda t), \tag{14.15}$$

where A_0 is the isotope concentration (activity) in the recharge area; v is the velocity of groundwater motion; λ is the ^{14}C decay constant; $\lambda = \ln 2/T_{1/2}$; t is the water age, determining as the transit time, during which the isotope reaches the distance from the recharge area to the sampling point.

The time t from Eq. 14.15 can be obtained by the formula:

$$t = (T_{1/2}/\ln 2)\ln(A_0/A) = (1/\lambda)\ln(A_0/A). \tag{14.16}$$

As it was noted earlier, the radiocarbon method of groundwater age determination was applied for the first time by Münnich (1957, 1968). The main assumptions of this method are as follows:

1. Radiocarbon concentration in the recharge zone is known and does not change at least over the range of the radiocarbon time-scale.
2. Radiocarbon does not enter the carbonate system of groundwater from outside the recharge zone.
3. If ^{14}C content in an aquifer decreases both due to radioactivity decay and other processes, then the effect of these processes can be taken into account.
4. The leakage of water from one aquifer into another (the mixing of water) is negligible.
5. The rate of movement of dissolved carbonates is equal to the rate of groundwater movement.

The reality of the above conditions is discussed in the later part.

It is worth to stress that the term 'age' of groundwater has a great indefiniteness compared with, to say, 'age' of a solid sample. In general case, the age of a groundwater can be defined by a certain period of time after some hydrogeological event has happened (Dubinchuk 1979). Such an event in radiocarbon dating is discontinuation of the water exchange with the soil air or the atmosphere. In framework of this definition, in the radiocarbon scale the natural waters appears to be modern if they are found in continuous exchange with the soil air or atmosphere, where radiocarbon concentration $A_0 = 100\%$.

In general case, the water 'age' is more easily described in the form of distribution function of residence time in the aquifer, applying corresponding models for a hydrogeological system, which satisfy to its real structure (Ferronsky et al. 1977; Dubinchuk 1979). At the present time, for the isotope data interpretation and for obtaining the groundwater's age characteristics, in addition to the piston model, there are the following models for the mass transfer in hydrological systems: (1) model of complete mixing (exponential or box-model); (2) dispersion model: (3) mixing model of different age waters.

According to the complete mixing (exponential) model (Nir 1964), water input at different times mixes quickly. The relative portion of water of t age in the system is expressed as an exponential function in the form

$$p(t) = (1/\tau)\exp(-t/\tau), \tag{14.17}$$

where τ is the average residence time of water in a system; $1/\tau = v/q$ is the water exchange rate, that is, the ratio of the total annual inflow of water to the output of the system. Concentration of the isotope in the system ($A_{system} - A_{output}$), in general case can be written as the following sum in the discrete form:

$$A_{system} = \bar{A}/(1+\lambda\tau) + \sum_{\theta=1952}^{1952+t} [A(\Theta - t) - \bar{A}]/p(t)e^{-\lambda t}, \tag{14.18}$$

where $\ominus$ is the current year of; $(\ominus - t)$ the time of input into the water system with concentration of the isotope (^{14}C and others.) $A(\ominus - t)$, which was happened by t years from the sampling moment; $\bar{A}$ is a steady state concentration before the thermonuclear tests; other definitions are previous.

Applying the Eq. 14.18, by observed ^{14}C concentrations in the system (A_{system}), the mean residence time of water can be obtained if the function of the isotope distribution $A(\ominus - t)$ at the system entrance is known. The exponential model is used for interpretation of the tritium data, but it can also be used for estimation of dynamic characteristics of groundwater, applying the radiocarbon, because the function $A(\ominus - t)$ (radiocarbon input at the hydrogeological system entrance) is known (Ferronsky et al. 1977). The ^{14}C concentration change in groundwater in the time calculated by the exponential model is shown in Fig. 14.22 (Geyh 1974).

The Figure shows that the effect of ^{14}C concentration change induced in groundwater by the thermonuclear tests so much as the mean residence time in the system is less. If the function $A(\ominus - t)$ is constant the ^{14}C radiocarbon residence time of water $\tau = v/q$ can be determined by the formula:

$$\tau = (A_{input} - A_{output})/A_{output}\lambda = 8{,}033(A_{input}/A_{output} - 1), \tag{14.19}$$

where A_{input} and A_{output} are the radiocarbon concentrations (in % with reference to standard of the modern carbon) at the entrance and exit of the system.

It is not difficult to show that in framework of the exponential model if A_{input} is constant in time, then the water age t is related to the residence time τ by expression $t = (1/\lambda)(1 - \lambda\tau)$. Application of the integral or discrete expressions like Eq. 14.18 allows to estimate the residence time in hydrological system at different forms of the distribution function $P(t)$, but despite the wide application in tritium data treatment, interpretation of the radiocarbon data by such a way is not used.

As a special case, it is necessary to consider the model of different-age water (Ferronsky et al. 1977; Evans et al. 1979). If the hydrogeological system consists of n components of water with the ages T_i, then its age can be determined as a mean-weighted value

$$\bar{T} = \sum_{i=1}^{n} P_i T_i, \tag{14.20}$$

where P_i is the part of water of age T_i.

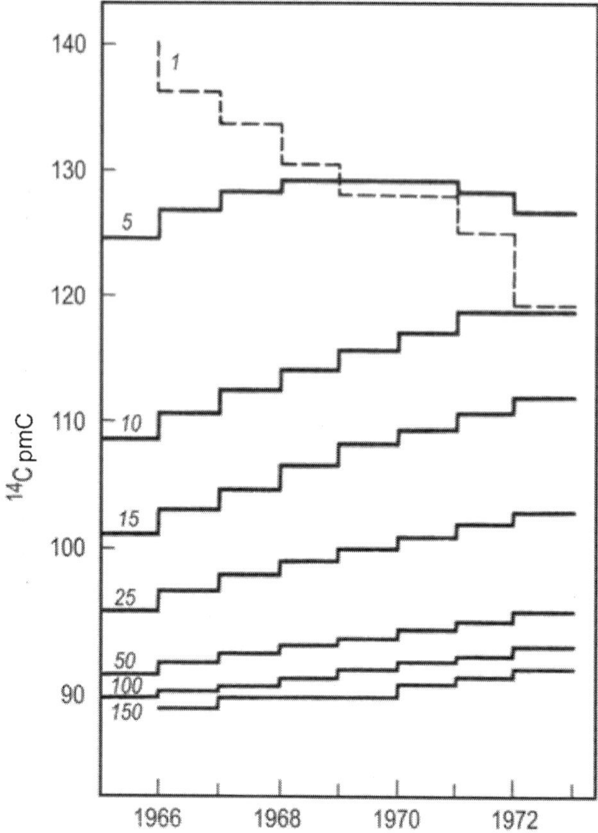

Fig. 14.22 Radiocarbon concentration change in groundwater as a function of residence time calculated by exponential model. Initial ^{14}C concentration was accepted equal to 85% relative to the standard of modern carbon. The figures on curves define residence time in years. (After Geyh 1974)

In such a system, the mean-weighted isotope concentration (for example, ^{14}C) is established:

$$\bar{A} = \sum P_i T_i. \tag{14.21}$$

If the source of the radioactive isotope input is the same, then the values A_i will be different due to partial radioactive isotope decay over the time T_i. Such a condition is satisfied for the radiocarbon concentration of which the recharge area can be accounted constant and equal to A_0. Then the Eq. 14.21 can be rewritten as:

$$\bar{A} = A_0 \sum P_i \exp(-\lambda T_i). \tag{14.22}$$

Determining the mixing age by concentration $\bar{A}$ and using the exponential formulae, we may obtain the value

$$T' = 8003 \ln (A_0/\bar{A}). \tag{14.23}$$

Comparing the Eq. 14.23 and 14.20, we find that $T' \neq \bar{T}$. It is observed from the above that: (1) the true age of mixture of the different waters is equal to the mean-weighted age of the water components; (2) in general case, the isotopic (radiocarbon) mixture of waters is not equal to the true age and its formal determination does not give unique time-information about a system (Ferronsky et al. 1977; Dubinchuk 1979). This problem can be solved correctly only in the case of independent measuring of the date and age for each component. For obtaining such information one may use, for example, the data on distribution of other isotopes (D, ^{18}O, T, and so on). Evans et al. (1979) have calculated the possible change in radiocarbon age for a two-component mixture depending on a portion of the modern component ($A_0 = 100\%$). It follows from their calculations that the groundwater of radiocarbon age equal to 57,000 years may represent a mixture of 99.9% of water with 100,000 years of age and 0.1% of modern water. This result is changes only a little if the present-day component has a 5,000 years of age, but despite this the radiocarbon age of water mixture does not give the true information which can be used to study a relationship between the aquifers, localization of recharge areas, and so on (Borevsky et al. 1981; Sobotovich et al. 1977; Seletsky et al. 1979; Fröhlich et al. 1977; Deak 1974).

14.8 Formation of Chemical and Isotonic Composition of Groundwater's Carbonate System

The consistent and complete analysis of chemical and isotopic composition of the carbonate system of groundwater can only be carried out if all the physicochemical processes involved in the formation of the saline composition of groundwater are accounted for. At present, it is difficult to realize such an analysis since many problems remain unsolved. Even in such simple systems, as the ocean and sea waters where the chemical composition, temperature, and pressure are well known and, as a rule, extra chemical reactions do not occur, the estimation of the carbonates equilibrium state often results in inconsistent conclusions. More complicated physicochemical processes of interaction between solutions and rocks in dispersion systems take place in groundwater. Besides, many of these chemical reactions are conditioned by the participation of micro-organisms. Therefore, the estimations of these system states can be carried out with a less accuracy than those of sea and ocean water systems. Consider only some of the physicochemical reactions resulting in changes of chemical and isotope composition of the carbonaceous system of groundwater.

The hydrocarbonate contained in the soil solutions is formed mainly due to the dissolution of limestone under the influence of carbon dioxide according to the equation (Fontes and Garnier 1979; Mook 1976; Wendt et al. 1967; Münnich 1968)

$$CaCO_3 + CO_2 + H_2O \Leftrightarrow Ca^{2+} + 2HCO_3^-. \qquad (14.24)$$

The following additional reactions are also common: leaching by carbon dioxide of the siliceous rocks

$$CaAl_2(SiO_4)_2 + 2CO_2 + H_2O \Leftrightarrow Ca^{2+} + 2HCO_3^- + Al_2O_3 + 2SiO_2; \qquad (14.25)$$

the decomposition of carbonates in the presence of humites

$$\begin{aligned} &CaCO_3 + 2H^+(hum) \Leftrightarrow Ca(hum)_2 + CO_2 + H_2O; \\ &2CaCO_3 + 2H^+(hum) \Leftrightarrow Ca(hum)_2 + Ca^{2+} + 2HCO_3^-; \end{aligned} \tag{14.26}$$

the reduction of sulfates

$$SO_4^{2-} + 2H_2O + 2C \Leftrightarrow H_2S + 2HCO_3^-; \tag{14.27}$$

the injection of atmospheric carbon

$$CO_2 + H_2O \rightarrow H_2CO_3 \rightarrow H^+ + HCO_3^-; \tag{14.28}$$

the oxidation of organic substances

$$6nO_2 + (C_6H_{10}O_5)_n + nH_2O \rightarrow 6nHCO_3^- + 6nH^+; \tag{14.29}$$

the subsequent interaction of the acid water with calcium carbonates

$$nH^+ + nCaCO_3 \Leftrightarrow nCa^{2+} + nHCO_3^-. \tag{14.30}$$

The list of reactions affecting the formation of chemical and isotope composition of groundwater can be extremely large. The contribution of each reaction to this process has not been completely studied. The contributions change depending upon the common hydrological and hydrochemical conditions of the formation of the groundwater; however, at present it is commonly accepted (Ferronsky et al. 1975; Bondarenko 1983; Giggenbach et al. 1983; Fontes 1976; Fontes and Garnier 1979; Mook 1976; Reardon et al. 1980; Wigley et al. 1978; Geyh 1972; Münnich 1975; Wendt et al. 1967; Tamers 1965) that the major portion of radiocarbon in groundwater is formed by the dissolution of the 'dead' soil carbonates in the unsaturated zone interacting with modern soil carbon dioxide. The modern CO_2 is provided by root respiration and bacterial decomposition of organic substances according to reaction (14.24). In the presence of isotope species of carbon-bearing molecules, this reaction can be written as

$$CaCO_3 + {}^{14}CO_2 + H_2O \Leftrightarrow (H^{14}CO_3^- + HCO_3^-) + Ca^{2+}. \tag{14.31}$$

This equation indicates that the hydrocarbon-bearing ion obtains one half of its carbon from biogenic carbon dioxide and the other half from sedimentary carbonates having no radiocarbon atoms.

The model is based on the assumption that the specific activity of radiocarbon in young groundwaters is equal to 50% of the activity of biogenic carbon dioxide. It follows from the experiments that the radiocarbon concentration of water in the recharge regions normally exceeds those mentioned above and ranges from 50 to 100% of the modern biogenic level. This may result due to two reasons which are the existence of isotope exchange between the biogenic carbon dioxide and the bicarbonate-ion, and the presence of radiocarbon in soil carbonates.

The first assumption is confirmed by the fact that approximately 98% of the biogenic carbon dioxide leaves the soil because this system is open relative to CO_2. In the open system, with isotope exchange occurring under conditions of counter-flow extraction (mass of $CO_2 >>$ mass of HCO_3^-), the ^{14}C content in the dissolved hydrocarbonates, formed by the reaction described by Eq. (14.24), can theoretically reach 102% of the radiocarbon content in the biogenic carbon dioxide, since in this case $\alpha_{HCO_3^- - CO_2}(r) \approx 1.02$, or $\varepsilon \approx 2\%$ (Galimov 1968; Wigley et al. 1978; Pearson and Swarzenki 1974).

It is important that the dead carbonates of the soil layer can contain a considerable amount of ^{14}C. As Geyh (1970) has shown, a higher radiocarbon content in the carbonate system of groundwaters than that theoretically predicted may result from the presence of radiocarbon of 2–15% (and sometimes up to 70%) of the modern level in the soil carbonates . By Geyh's assumption, the radiocarbon is accumulated in the upper part of the soil layer in the summer season when the amount of infiltrated atmospheric water is not high and precipitation of carbonates occurs due to the evaporation processes of salt concentration in the soil layer.

It is obvious that such an explanation does not truly reflect the whole complexity of the formation of pedogenic carbonates. Polynov (1953) has pointed to the process of calcium carbonate formation occurring in the soil itself during the growth of plants. This process produces lime shells of cells, calcium crystals in wood tissue, and scales of calcium carbonate on leaves and stems. Partly during their life and episodically when the plants die, these carbonates enter the upper soil layer, migrate with surface and soil waters, and accumulate in the soils. The formation of thick carbonaceous layers has also been reported during soil-formation processes by other researches (Afanasyeva 1947; Ponomarev and Antipov-Karataev 1947).

While studying the behavior of the stable carbon isotopes in these processes, it is assumed that calcium carbonate, involved in the reaction (14.24), originates from the sedimentary carbonates of sea genesis with $\delta^{13}C_{HCO_3^-} = 0‰$. The soil carbon dioxide in the humid zone of the plants' growth has, according to the Calvin photosynthetic cycle (C_3), $\delta^{13}C_{CO_2} = -25‰$ (Pearson et al. 1978; Galimov 1968; Pearson and Hanshaw 1970). If equilibrium isotopic fractionation in the system CO_2–HCO_3^- does not occur, then, by Eq. (14.24), hydrocarbonate in the solution will have $\delta^{13}C_{HCO_3^-} = -12.5‰$.

In the arid zones, where plants metabolize through a Hatch-Slack photosynthetic cycle (C_4), the soil carbon dioxide has $\delta^{13}C_{CO_2}$ from -12 to $-14‰$ and, therefore, neglecting isotope fractionation in the gas-liquid system, the hydrocarbonates' $\delta^{13}C_{HCO_3^-}$ average value in solution will range between -5 and $-7‰$. The data on the carbon isotope composition of the CO_2 produced by plants metabolizing through a Hatch-Slack cycle are taken from the study of Lerman (1972), who has also reported somewhat different values of $\delta^{13}C_{CO_2} = -27‰$ for the Calvin cycle.

In open systems, in a similar manner to radiocarbon, the isotope composition of stable carbon of dissolved carbonate ions can, in principle, reach the equilibrium value determined by the fractionation coefficient $\alpha_{HCO_3^- - CO_2} = 1.0077$ (25°C) (Vinogradov et al. 1967). In this case, for the Calvin plants $\delta^{13}C_{CO_2}$ equal from -17 to $-19‰$ and for the Hatch-Slack plants $\delta^{13}C_{CO_2}$ is from -4 to $-6‰$.

Typical Hatch-Slack plants are maize, papyrus, camel brushes, and other types of plants typical of arid zones. Further examples are those plants which produce CO_2 characterized by $\delta^{13}C_{CO_2}$ varying from -19 to -34‰, through the Crassulacean acid metabolism (CAM) cycle (Wong and Sackett 1978).

For the zone of semi-arid climate in Arizona, US, the $\delta^{13}C$ value of the soil carbonates is about -3.73 ± 1.31‰, and the ^{14}C content amounts to -23.8 ± 15.3% of the modern standard (Wallick 1976). For the soil carbon dioxide, the $\delta^{13}C$ value ranges from -11.7 to -18.4‰ with an average of 15.12 ± 2.81‰. As the author reported, this is close to the average isotope composition of CO_2, produced by plants through the Hatch-Slack photosynthetic cycle (from -13.9 to -19.8‰).

Experimental study of $\delta^{13}C_{HCO_3^-}$ values and of the specific activity of the hydrocarbonate ion in the north-eastern region of the Germany has shown that the ratio of $\delta^{13}C_{HCO_3^-}/\delta^{13}C_{CO_2} > 0.5$, and is close to 0.6, and the ratio of specific activity of the hydrocarbonate ion to that of the soil carbon dioxide is ~0.8 (Wendt et al. 1967). The specific activity of radiocarbon in the upper aquifers of groundwater varies greatly. Derivation from the average value may reach ±25% and more. In a number of cases, corresponding to Central Europe, the radiocarbon variations are less pronounced and are equal to ±5%. Elevated radiocarbon contents (up to 95% of the modern level) and greater contents of the light carbon isotope $\left(\delta^{13}C_{HCO_3^-} = -15‰\right)$ have been found in water sampled from lysimeter without any plant cover (76% of the modern standard activity and $\delta^{13}C_{HCO_3^-} = -10$‰) (Münnich et al. 1967).

In the Kalahari region (South Africa), the carbon isotope composition of water in the upper aquifer is somewhat different from the European average $\left(\delta^{13}C_{HCO_3^-} = -11‰\right)$, but the radiocarbon content is close to that measured in other regions of the globe (85% of the modern standard) (Vogel 1970). All these facts indicate that a number of problems concerning the formations of the isotopic composition of the groundwater carbonate system should be considered more carefully and new models should be developed.

Wendt et al. (1967) have proposed the idea that equilibrium might occur between the soil solutions and the soil CO_2. The theoretical and experimental data show that the isotope exchanges occurring between the solution and the soil gases do result in changes in $\delta^{13}C$ and the specific activity of the bicarbonates. The experimental and theoretical values agree rather well, butin majority of cases the groundwaters' ^{14}C radioactivity is only 80–85% of the modern standard. This fact also indicates the involved nature of the formation of the isotopic composition of groundwaters, which are obviously related to the kinetics of the process.

It is observed from the consideration of even a simple scheme of radiocarbon contribution to groundwaters that it is impossible to determine the initial radiocarbon activity in groundwaters; one can only approximately estimate the ranges of ^{14}C variations in modern waters. This problem becomes even more complicated due to the fact that in the real conditions occurring in the unsaturated zone, as has been shown earlier, another process different from reaction (Eq. 14.24) may take place, resulting in the enrichment of water with the bicarbonate ion. The effect of these reactions upon the formation of the chemical and isotopic composition of the carbonate system of groundwaters is insufficiently understood. The temperature, type of plant cover, and other factors, depending on the climatic conditions and the geological structure of the unsaturated zone, all affect the procedure of chemical and biological reactions (Bondarenko 1983). The initial amount of ^{14}C also depends upon these factors. The fact that the climate of many regions of the globe has changed sharply during the last 10^4 years leads to uncertainty in choosing the initial ^{14}C concentration of the studied groundwaters. The problem is also complicated by the fact that during recent thermonuclear tests, the ^{14}C content in the atmospheric carbon dioxide has increased by a factor of 2 (see Fig. 14.5) and already, since 1965, thermonuclear radiocarbon has started to recharge into groundwaters (Münnich et al. 1967; Geyh 1974).

It has been found experimentally that the radiocarbon content in fresh groundwater is often equal to about $85 \pm 5\%$ relative to the modern standard. This value is taken by some authors as the initial concentration (Klitzsch et al. 1976; Vogel 1970).

14.9 Corrections in Groundwater Dating by Radiocarbon

The processes resulting in a decrease of the specific activity of dissolved carbonate components during groundwater circulation make the radiocarbon dating of groundwaters difficult. The decrease in the specific activity of carbon in the groundwater carbonate system may be accounted for by the following two main processes: the dissolution of the dead carbonates and the removal of radiocarbon from solution.

The dissolution of the vast amounts of carbonates from the water-bearing rocks is caused by:

1. the presence in the soil moisture which recharges the groundwaters of active carbon dioxide and organic acid;
2. the production of the carbon dioxide in groundwaters due to biological decomposition of the organic substances contained in water and rocks;
3. the change of the alkaline-earth metal cations for the alkali metal cations, for example calcium for sodium;
4. the injection into some aquifers of endogenic carbon dioxide;
5. changes in the thermodynamic conditions (those of temperature and pressure) in the immersed parts of the water-bearing complexes.

A likely cause of the removal of radiocarbon becoming incorporated in the groundwaters' solute is the exchange processes in the solid phase-solution system. The exchange processes do not result in considerable changes of radiocarbon content in the liquid phase at equilibrium conditions, when the dissolved carbonates are in thermodynamic equilibrium with carbonates of the water-bearing rocks. This has been shown by laboratory and natural studies (Münnich et al. 1967; Thilo and Münnich 1970; Vogel 1970). The fact that radiocarbon is present in groundwaters—and that a number of individual studies have produced coincidental ages by radiocarbon dating and from independent hydrodynamical considerations—leads to the conclusion that in many water aquifers the effects of the carbon exchange and decay are small, even during the movement of waters through the carbonate collectors. Thus, Vogel and Ehhalt (1963) have concluded that ^{14}C removal from water should be negligible. This conclusion is confirmed by the general knowledge that the surface of the rocks, which are in direct contact with water, undoubtedly favors isotope exchange on the one hand but on the other hand restricts the removal of radiocarbon from the water. Münnich and co-authors (Münnich and Roether 1963; Thilo and Münnich 1970) have proposed a model of the two reservoirs with the help of which an attempt has been made to show how small the effect of exchange processes in the carbonate collections is upon the change of the carbon isotope composition.

The removal of radiocarbon from the solution due to isotope exchange with carbonate rocks should be limited by the value of the diffusion coefficient of the carbonate ion in $CaCO_3$. According to laboratory data for crashed calcite (Thilo and Münnich 1970), the diffusion coefficient, extrapolated to 10°C, is about 10^{-28} cm^2/sec. According to estimations made by Thilo and Münnich this corresponds to the transition from water into the solid phase of about 50% ^{14}C during 10^4 years at temperature of 10°C. Thilo and Münnich assumed that the observed effect of isotope exchange upon radiocarbon removal is conditioned by the organization of the experiment, particularly by the use of finely-crushed calcite, each particle of which provides a fresh unstable surface. It should be pointed out that under such conditions recrystallization in the finely-crashed particles of calcium carbonate has occurred, resulting in an apparent increase in the diffusion coefficient which is equal to about 10^{-43} cm^2/sec at room temperature (Wendt et al. 1967).

In a discussion on a report by Thilo and Münnich (1970), Hanshaw has pointed out the estimations carried out by Anderson, according to which carbon isotope exchange does not exceed 2–3% during 10^7 years at a temperature below 50°C. In another work, Rye and O'Neil (1968) reported values of ^{13}C corresponding to liquid inclusions of carbonates in calcites, the age of which was estimated to be $n \times 10^7$ years. It has been shown that the value of $\delta^{13}C$ is greater than that of carbon isotope exchange at temperatures > 50°C is small.

Processes of carbon isotope exchange most likely play a decisive role in hydrothermal systems and during the precipitation of the fine-grained calcium carbonate from the solution resulting from the change of the groundwaters' chemical composition, for example, during the alcalinization of gypsum. Thus, while studying thermal waters in the Neocomian deposits of the Rioni Depression (Caucasus), it was found (Polyakov et al. 1979), that the application of ^{14}C dating for local conditions cannot elucidate the ages because isotope exchange processes happen very quickly at high temperatures in the water-rock system. Here, the water circulation time at a number of sites in a hydrothermal system does not exceed 500 years. This water exchange time was obtained by hydrodynamic calculations and also by tritium and uranium isotope studies. But the ^{14}C content in these waters does not exceed 4–5% and the value of $\delta^{13}C \approx -2‰$. The total content of the carbonaceous components at different sites in the aquifer remains practically unchanged ($\sim$150 mg/l).

Isotope studies of groundwater dynamics in the Assel-Kljazmian aquifer (Upper Carbon) near Vladimir (Polyakov and Seletsky 1978) have shown that carbonate precipitation process occurring during the mixing of calcium-hydrocarbonate and calcium-sulphate waters result in considerable radiocarbon removal from solution. This is accounted for by the carbon isotope exchange processes in the fine-grained calcium carbonate precipitating from the liquid phase which is saturated with Ca^{2+} ions. According to the hydrodynamic and tritium data, the residence time of water in the studied aquifer does not exceed several hundred years. But the ^{14}C content decreases sharply from about 60% in the recharge region to 10% in the immersed part of the aquifer. The $\delta^{13}C$ value increases from -10 to -2.5‰, while the total content of carbonates decreases respectively from 300 to 80 mg/l.

Therefore, an account of the possible effect of carbon isotope exchange in the water-rock system upon the ^{14}C decrease in groundwater under specific hydrological conditions should be carried out on the basis of detailed analyses of hydrogeochemical and isotope data. But at present no reliable method has been proposed which takes into account the correlation between ^{14}C content and isotope exchange in the water-rock system.

The decrease in the specific activity of the carbonates in water due to the admixing of 'connate' carbonates to the sedimentary rocks can be estimated both by the increases of the total carbon concentration in the studied groundwaters compared with the waters in the recharge zone and on the basis of the stable carbon isotope studied in the soil carbon dioxide in the carbonate system of groundwaters and water-bearing rocks. In general, if the concentration of the carbonaceous complexes is equal to C_{initial} in the recharge zone, then the correction factor K accounting for changes in the radiocarbon concentration caused by dilution can be estimated by the ratio $C_{\text{sample}}/C_{\text{initial}}$. Since under real conditions the initial radiocarbon concentration can be always estimated, with some error, it was assumed (Tamers 1967, 1969; Tamers and Scharpenseel 1970) that in any carbonate system of groundwaters in agreement with Eq. (14.24) the amount of radoicarbon is equal to only half of the hydrocarbonate ion. According to Tammers the process of radiocarbon concentration change can be described by the equation

$$pA_{\text{initial}} + (1 - p)A_{\text{CaCO}_3} = A_{\text{sample}}e^{-\lambda t}, \tag{14.32}$$

where P is the fraction of the dissolved carbonate arriving from the soil air; A_{initial} is the relative radiocarbon amount in the carbonaceous system, formed due to dissolution of the biogenic carbon dioxide (the latter is taken for 100%); A_{CaCO_3} is the relative radiocarbon amount, released in water due to the dissolution of the calcite (is accepted as 0%); A_{sample} is the relative radiocarbon amount in an altered sample.

From Eq. (14.32), it follows that half of the hydrocarbonate originates from $CaCO_3$. Then

$$p = \frac{[C_{\text{total}}] - \frac{1}{2}[C_{\text{HCO}_3^-}]}{[C_{\text{total}}]}, \tag{14.33}$$

where $|C_{\text{total}}|$ is the total carbon concentration in the water specimen; $|C_{\text{HCO}_3^-}|$ is the hydrocarbonate ion concentration.

From Eq. (14.32), and on account of relationship (14.33), one can estimate the age:

$$t = 8033\left\{\ln\left(\frac{A_{\text{initial}}}{A_{\text{sample}}}\right) + \ln\frac{[\mathrm{C}_{\text{total}}] - \frac{1}{2}\left[\mathrm{C}_{\mathrm{HCO}_3^-}\right]}{[\mathrm{C}_{\text{total}}]}\right\}. \tag{14.34}$$

It follows from Eq. 14.34 that the correction factor by which one should multiply the value of the experimentally determined radiocarbon content in a sample A_{sample} or divide by the value A_{initial} (accepted as 100%), is

$$K = \frac{[C_{\text{total}}]}{[C_{\text{total}}] - 1/2[\mathrm{HCO}_3^-]}. \tag{14.35}$$

It should be pointed out that Tamers' approach is not well grounded from the viewpoint of the carbonate components' geochemistry. It is known (Garrels and Christ 1965; Sokolov 1974) that the carbonate system consists mainly of three components: undissociated molecules of carbonic acid H_2CO_3, hydrocarbonate ions HCO_3^{2-}, and anions of carbonic acid CO_3^{2-}. The forms of weak carbonic acid ate the pH-determining components for a great many types of groundwaters. In turn, the concentration of each of the forms depends on the pH and the ionic strength of a solution. At pH = 6.0 the fraction H_2CO_3 is equal to 69% and that of HCO_3 is 31%. At pH = 7.0, they are equal to 18 and 82%; at pH = 8.0, they are equal to 2 and 98%. Therefore, the chemical correction factor, calculated from Eq. 14.35, depends on the pH and the ionic strength of a solution and for a great many natural waters (at pH $\geq$ 7) is close to 2.

A somewhat different approach for the estimation of corrections, based on the use of stable isotopes, has been suggested by Pearson, Ingerson, and Hanshaw (Ingerson and Pearson 1964; Pearson 1965; Pearson and Hanshaw 1970; Pearson and White 1967). The specimen radioactivity is introduced by the formula

$$A_{\text{sample}} = \frac{\sum_{i=1}^{n} m_i A_{0i}}{\sum_{i=1}^{n} m_i} e^{-\lambda t}, \tag{14.36}$$

where A_{0i} is the initial radioactivity of the ith component; m_i is the mass of the component.

Usually, radioactivity is introduced by one source with mass m_{initial} and radioactivity A_{initial}. Then

$$\frac{A_{\text{sample}}}{A_{\text{initial}}} = \frac{m_{\text{initial}}}{\sum_{i=1}^{n} m_i} e^{-\lambda t}. \tag{14.37}$$

The concentrations of $CaCO_3$ in water after the dissolution of carbonate-bearing minerals, in simple schemes of carbonate generation, are related to the value of $\delta^{13}C$ by the expression

$$\frac{m_{\text{initial}}}{\sum_{i=1}^{n} m_i} = \frac{\left(\delta^{13}C_i - \delta^{13}C_{CaCO_3}\right)}{\left(\delta^{13}C_{\text{initial}} - \delta^{13}C_{CaCO_3}\right)}, \tag{14.38}$$

where $^{13}C_{\text{initial}}$ is the isotopic composition of the soil CO_2; $\delta^{13}C_{CaCO_3}$ is the isotope composition of the calcium carbonate.

From Eq. 14.37, it is observed that the radiocarbon activity of a water sample is inversely proportional to the content of the dissolved carbon in water. The possibility of the introduction of the correction for measured radiocarbon content in a specimen with the use of data on the isotope composition of stable carbon is observed from Eq. 14.38:

$$t = 8033\left[\ln\left(\frac{A_{\text{initial}}}{A_{\text{sample}}}\right) + \ln\left(\frac{\delta^{13}C_{\text{sample}} - \delta^{13}C_{CaCO_3}}{\delta^{13}C_{CO_3} - \delta^{13}C_{CaCO_3}}\right)\right], \tag{14.39}$$

where A_{initial} is the relative ^{14}C content of the soil carbon dioxide; A_{sample} is the measured ^{14}C content in a specimen; $\delta^{13}C_{\text{sample}}$, $\delta^{13}C_{CaCO_3}$ are relative content of ^{13}C isotopes in a sample and carbonate rocks.

Assuming that $A_{\text{initial}} = 100\%$, $\delta^{13}C_{CaCO_3} \approx 0‰$, and $\delta^{13}CO_2 = -25‰$ for zones covered with plants having a Calvin photosynthetic cycle, Eq. 14.39 can be simplified:

$$t = 8033\left[\ln\left(\frac{100}{A_{\text{sample}}}\right) + \ln\left(\frac{\delta^{13}C_{\text{sample}}}{-25}\right)\right]. \tag{14.40}$$

It is observed from Eq. 14.40 that the correction factor by which the measured value of ^{14}C in a sample should be multiplied as in this case $K = (-25/\delta^{13}C_{\text{sample}})$ or in a more general case:

$$K = \frac{\delta^{13}CO_2 - \delta^{13}C_{CaCO_3}}{\delta^{13}C_{\text{Sample}} - \delta^{13}C_{CaCO_3}}. \tag{14.41}$$

As shown earlier, in the case of open systems (unsaturated zone), where the formation of the initial radiocarbon concentration occurs, the $\delta^{13}C = -15‰$ and the ^{14}C content ranges over $85 \pm 5\%$ according to numerous experimental data for dissolved carbonates. Using these data, the correction of the measured ^{14}C content (85%) in a sample, estimated by Eq. 14.41, gives values which are considerably greater than 100%. This fact indicates that more complicated models of the formation of the isotopic composition of the carbonate system should be used, accounting for chemical and exchange reactions occurring in the open and closed hydrogeological systems.

More general models, accounting for the effect of both dissolution of the solid carbonates and exchange in the carbon dioxide dissolved carbonates system, developed for the interpretation of radiocarbon data, are considered by a number of researchers (Fontes and Garnier 1977; Mook 1976; Wigley 1976; Wigley et al. 1978; Wendt et al. 1967; Pearson and White 1967).

Mook has studied three stages of the formation of carbon in the carbonate complex of groundwater. The first stage is the dissolution of soil carbonates while interacting with soil CO_2. The second stage represents isotopic exchange between the hydrocarbonate ion and the gaseous carbon dioxide in the unsaturated zone. The last stage corresponds to the attainment of isotope equilibrium between the dissolved carbon dioxide and hydrocarbonate ion under conditions of chemical equilibrium between the carbonate system components in the groundwaters. The initial ^{14}C concentration (as a percentage of the modern carbon activity), at the suggested formation of the groundwaters' isotope composition, is described by the Eq. 14.42, which uses Mook's notation:

$$A_\Sigma = \frac{1}{\Sigma}\left\{(\Sigma - b)A_{a_0} + 0.5b\left(A_{a_0} + A_1\right) + \left[A_{g_0}\left(1 - 2\frac{\varepsilon_g}{10^3}\right) - 0.5a\left(A_{a_0} + A_1\right)\right]\right.$$
$$\left. \times \left[\frac{\Sigma\delta_\Sigma - a\delta_{a_0} - 0.5b\left(\delta_{a_0} - \delta_1\right)}{\delta_{g_0} - \varepsilon_g\left(1 - \delta_{g_a}/10^3\right) - 0.5a\left(\delta_{a_0} - \delta_1\right)}\right]\right\}, \tag{14.42}$$

where A_Σ, δ_Σ are the general ^{14}C activity (in percent of the modern carbon) and $\delta^{13}C$ (in permil) in the dissolved carbonate dioxide and hydrocarbonate in the aquifer; a, b are the molal concentrations of the total dissolved carbon dioxide and hydrocarbonate; Σ is the molal concentrations of the total dissolved carbon $\left[H_2CO_3 + CO_3^{2-} + HCO_3^-\right]$; A_{a_0}, δ_{a_0} are the ^{14}C (in percent of the modern carbon) and $\delta^{13}C$ (in permil) of the dissolved soil carbon dioxide; A_{g_0}, δ_{g_0} are the ^{14}C activity (in percent of the modern carbon) and $\delta^{13}C$ (in permil) of the soil carbon dioxide; A_1, δ_1 are ^{14}C activity (in percent of the modern carbon) and $\delta^{13}C$ (in permil) of the soil carbonates; ε_g is the difference between $\delta^{13}C$ (in permil) in the gaseous CO_2 and $\delta^{13}C_{HCO_3^-}$ in the dissolved HCO_3^-: $\delta^{13}C_{CO_3^-} - \delta^{13}C_{HCO_3^-}$.

The dependence of the factor ε_g on temperature can be obtained from the below data calculated by Mook (1976):

t(°C)	0	5	10	15	20	25	30
ε_g(‰)	−10.83	−10.20	−9.60	−9.02	−8.46	−7.92	−7.39

The value of A_1 can be set by calculation:

$$A_1 \approx (A_{a_0} + 0.2\varepsilon_k)\frac{\delta_1 - \delta_{10}}{\delta_{a_0} + \varepsilon_k - \delta_{10}}, \tag{14.43}$$

where δ_{10} is $\delta^{13}C$ (‰) of carbonates of sea genesis is equal to about 0‰; ε_k accounts for the effect of isotope fractionation between solid calcium carbonate and dissolved CO_2, equal to $\delta^{13}C_{CaCO_3} - \delta^{13}C_{CO_2(solution)}$.

It should be pointed out that in Eqs. 14.42 and 14.43, $A_{a_0} \approx A_{g_0} \approx 100\%$ and ε_k is practically independent of temperature and is equal to about +10‰. The values of a/Σ and b/Σ can be obtained by the precise measurement of $pH = -\ln a_H$ (a_H is the activity of H^+ in solution):

Table 14.7 Comparison of groundwater ages determined by different methods. (From Mook 1976)

Sampling location	Groundwater age (years)						
	pH	δ_Σ	A_m	Vogel 1970	Pearson 1965	Tamers 1967	Mook 1976
Ezulwini	8.6	−13.6	47.6	4,800	1,100	460	2,370
Lomamba 2	7.9	−14.5	43.6	5,540	2,370	1,390	4,180

Table 14.8 Experimental and calculated values of groundwater parameters. (From Mook 1976)

Sample number	δ_{a0}	pH	δ_Σ	A_m (%)	A_Σ (%)
16	−15.0	7.3	−8.4	53.3	50.4
	−15.1				53.3
17	−15.0	7.4	−7.4	54.7	95.2
	−13.6				54.2
40	−15.0	8.2	−6.2	54.0	126.9
	−12.5				54.1

$$a/\Sigma = a_H^2/\left(a_H^2 + a_H k_1' + k_1' k_2'\right)$$
$$b/\Sigma = a_H k_1'/\left(a_H^2 + a_H k_1' + k_1' k_2'\right), \quad (14.44)$$

where k_1' and k_2' are the first and second dissociation constants of the carbonic acid.

On the basis of Mazor's experimental data (Mazor et al. 1974b), Mook (1976) has compared groundwater age estimated by Pearson's formula (Eq. 14.39), Tammers's (Eq. 14.34), Fogel's ($A_\Sigma = 85\%$), and his own expression (Eq. 14.42). These data are presented in Table 14.7.

Here the water age by Mook is estimated with the use of a typical exponential formula (piston flow model)

$$t = \frac{T_{1/2}}{0.93} \ln \frac{A_\Sigma}{A_m},$$

where A_m is the measured ^{14}C content in a sample as a percentage of the modern carbon; A_Σ is the value obtained from Eq. (14.42), assuming that: $A_{a0} = 100\%$, $\delta_{a0} = -25‰$, $\delta_{10} = \delta_1 = 0‰$.

It follows from the data shown in the table that all four methods give different water ages. Mook considers his technique to be the best, since it accounts for the highest number of factors affecting the formation of the carbon isotope composition of groundwaters.

It should be pointed out that Eq. 14.42 is such that A_Σ depends greatly upon the values of δ_{a0} and δ_Σ which follows from the data presented in Table 14.8, reported by Mook, who cited the experimental data of Gonfiantini et al. (1974).

As a rule the value of δ_{a0} cannot be accurately determined on a regional scale, being dependent on the geochemistry of the landscape and the physicochemical conditions of the soil layer. δ_{a0} also undergoes seasonal and long-term variations (for example, during the climatic changes in Holocene). All these circumstances result

Table 14.9 The groundwater ages from Lincolnshire limestones, United Kingdom. (Evans et al. 1979)

Sampling location	A_{sample} (%)	$\delta^{13}C_{sample}$ (‰)	Percent of modern carbon or the age
Tallington	45.9	−11.7	90.2%
Kates Bridge	60.0	−12.6	109%
Baston Fen, 1	1.9	−2.8	15,200 years
Baston Fen, 2	2.6	−2.8	12,600 years
Six Score Farm	0.4	−2.2	26,000 years
Tongue End Farm	1.1	−2.0	16,800 years
Cuckoo Bridge, 1	2.4	−1.8	9,500 years
Cuckoo Bridge, 2	2.2	−1.8	10,000 years

in considerable uncertainty regarding the determination of parameter A_Σ, estimated from Eq. 14.42.

Wigley (1976) has made an attempt to account for the effect of $CaCO_3$ precipitation from the solution upon the carbon isotope composition of the carbonate system of groundwaters. Wigley's model has been used by Evans et al. (1979) for the interpretation of data on the radiocarbon dating of groundwaters of some aquifers in England. In agreement with this model, the age can be found by the Eq. 14.45 as

$$t = \frac{T_{1/2}}{\ln 2} \ln \frac{50}{A_{\text{sample}}} \left[\frac{k - \delta^{13}C_{\text{sample}}}{k - \delta^{13}C_{\text{initial}}} \right]^{(1+\varepsilon_{13}/1000)}, \qquad (14.45)$$

where the value of $T_{1/2}/\ln 2 = 8{,}267$ (accepted by Evans); A_{sample} is the measured ^{14}C activity in a sample (as a percentage of the modern carbon content); $\delta^{13}C_{initial}$ is the relative ^{13}C content in the carbonate system at the initial stage of dissolution of the soil carbonates; $\varepsilon_{13} = \delta^{13}C_{rock} - \delta^{13}C_{solution}$; $k = \delta^{13}C_{rock} - \varepsilon_{13}$.

An example of groundwater age calculation by Eq. 14.45 is presented in Table 14.9, taken from the study of Evans et al. (1979).

While calculating the data in Table 14.9, it has been accepted that $\varepsilon_{13} = +24‰$; $\delta^{13}C_{rock} = +2.35‰$; $\delta^{13}C_{initial} = -11.5‰$.

It should be pointed out that Wigley's model, as used by Evans, is not quite correct, since it only accounts for the equilibrium carbon isotope distribution between the liquid and soil phases in the one-stage process. In real conditions, the processes of $CaCO_3$ precipitation are permanent (of the fractional crystallization type). In this case, the isotope fractionation factor should be multiplied by the coefficient n (n is the number of theoretical plates), which cannot be determined precisely in the case of real aquifers. In particular, according to the data obtained by Evans (Evans et al. 1979), the isotopic composition of modern and ancient water is identical. For the modern waters in England, $\delta D = -52‰$, $\delta^{18}O = -8.3‰$. For waters 29,000 years old, these values are practically the same ($\delta D = -52‰$, $\delta^{18}O = -8.5‰$), although the formation of groundwaters in the last case should have occurred during the Pleistocene under much cooler climatic conditions than those typical for Europe today.

Fontes and Garnier (1979) have suggested carrying out an interpretation of radiocarbon data based on the chemical and isotope balance of the components of the carbonate system of groundwaters. According to their model, the initial concentration A_{initial}, being used for the age estimation, can be calculated by the equation:

$$A_{\text{i}} = \left(1 - \frac{C_{\text{rc}}}{C_{\Sigma}}\right) A_{\text{s}} + \frac{C_{\text{rc}}}{C_{\Sigma}} A_{\text{rc}} + (A_{\text{s}} + 0.2\varepsilon - A_{\text{rc}}) \times \frac{\delta_{\Sigma} - (C_{\text{rc}}/C_{\Sigma})\,\delta_{\text{rc}} - [1 - (C_{\text{rc}}/C_{\Sigma})]\,\delta_{\text{s}}}{\delta_{\text{s}} + \varepsilon - \delta_{\text{rc}}}, \tag{14.46}$$

where A_s and A_{rc} are ^{14}C concentrations in soil carbon dioxide and in carbonate rocks C_{rc} and C_{Σ} are carbon concentrations from the dissolution of the carbonate rocks and the total concentration of carbonate; δ_s, δ_{rc}, δ_{Σ} are $\delta\,^{13}C$ values (in ‰) of the biogenic carbon dioxide, carbonate rocks, and samples; $\varepsilon = (\delta^{13}C_{CaCO_3} - \delta^{13}C_{soil})$‰.

It is easy to show that this model supplements the results of Tammers on changes in the initial ^{14}C activity due to dissolution of the 'dead' carbonate components of rocks but accounting for the carbon isotopic fractionation in the gaseous phase-solution system. The term K can be written as

$$K = (A_{\text{s}} + 0.2\varepsilon - A_{\text{rc}}) \times \frac{\delta_{\Sigma} - (C_{\text{rc}}/C_{\Sigma})\delta_{\text{rc}} - [1 - (C_{\text{rc}}/C_{\Sigma})]\delta_{\text{s}}}{\delta_{\text{s}} + \varepsilon - \delta_{\text{rc}}}.$$

According to Fontes and Garnier, the concentration of C_{rc} can be calculated, using the experimental data, by equation

$$\text{mC}_{\text{rc}} = \text{mCa}^{2+} + \text{mMg}^{2+} + 1/2(\text{mNa}^{+} + \text{mK}^{+} - \text{mCl}^{+} - \text{mNO}_3^{-}) - \text{mSO}_4^{2-}. \tag{14.47}$$

Wigley et al. (Wigley 1976; Wigley et al. 1978) have proposed the model which accounts for the highest number of factors affecting the carbon isotope composition of the carbonate system. The authors distinguish the following three stages in the formation of the carbonate system: the processes of interaction in the unsaturated zone; congruent dissolution in the unsaturated zone; incongruent reactions occurring in the saturated zone. The carbon isotope composition at each stage of formation of the system can be calculated by the corresponding equations. This system of equations permits the determination of carbon radioactive decay. It is assumed that in the first stage of formation of the system, the carbon isotope composition changes by the equilibrium isotopic exchange process between soil gas and soil solutions under open conditions relative to the CO_2 system. At this stage, the ^{14}C activity and $\delta^{13}C$ can be estimated on the basis of the equilibrium isotope fractionation theory. The degree of carbon isotope fractionation between the solution and carbon dioxide gas at this stage is determined by the relative amount of the dissolved carbon-bearing components. The isotope composition of the dissolved carbon (A_1 for ^{14}C and $\delta^{13}C$ for ^{13}C) at the end of the first stage is determined completely by the isotope composition of the CO_2 and the pH of the soil solution.

At the second stage, congruent (without change of the isotopic composition) dissolution of soil carbonates occurs under closed conditions relative to the CO_2 system. The change of the isotopic composition in this case can be estimated by Eq. 14.39 proposed by Ingerson and Pearson (1964).

At the third stage, the processes occur under conditions of a closed system by an incongruent mechanism, that is, with the involvement of carbonate isotope fractionation effects. At the end of the third stage, the ^{14}C radioactivity (as a percentage of the modern carbon), neglecting radioactive decay (A_{nd}), can be described by the equation

$$A_{nd} = A_1 x_2 x_3, \tag{14.48}$$

where A_1 is the ^{14}C radioactivity (as a percentage of the modern carbon) at the end of the first stage; $x_2 = mC_1/mC_2$ is the fraction of the molal concentrations of the inorganic carbon in the system at the end of the first and second stage;

$$x_3 = (mC/mC_2)^{\beta\Gamma/(1-\Gamma)},$$

where mC is the molal general inorganic carbon concentration at the end of the third stage; β and Γ are constants related to the change of carbon content in a system and the carbon isotope fractionation between precipitation and solution; $\beta = 1 + \varepsilon_{ps}/1{,}000\Gamma$.

The carbon isotope composition after the second stage will be determined as follows:

$$\delta^{13}C_2 = \delta^{13}C_S + x_2(\delta^{13}C_1 - \delta^{13}C_s), \tag{14.49}$$

where $\delta^{13}C_s$ and $\delta^{13}C_1$ are the carbon isotope composition of the solid $CaCO_3$ and soluted carbon at the end of the first stage.

At the end of the third stage, we have

$$\beta\delta^{13}C_{sample} = \delta^{13}C_s - \frac{\varepsilon_{ps}}{\Gamma} + \chi_3\left(\beta\delta^{13}C_2 - \delta^{13}C_s + \frac{\varepsilon_{ps}}{\Gamma}\right), \tag{14.50}$$

where $\varepsilon_{ps} = \delta^{13}C_{precipitation} - \delta^{13}C_{solution}$.

Taking into account Eqs. 14.48–14.50, one obtains

$$A_{nd} = A_1\left(\frac{\delta^{13}C_2 - \delta^{13}C_s}{\delta^{13}C_1 - \delta^{13}C_s}\right)\left(\frac{\beta\delta^{13}C - \delta^{13}C_s + \varepsilon_{ps}/\Gamma}{\beta\delta^{13}C_2 - \delta^{13}C_s + \varepsilon_{ps}/\Gamma}\right). \tag{14.51}$$

Putting A_{nd} into exponential form $t = (1/\lambda)\ln(A_{nd}/A)$ and assuming that $A_1 = 100\%$, one obtains the following expression for age estimations:

$$t = 19035\left\{\lg\left(\frac{100}{A_s}\right) + \lg\left(\frac{\delta^{13}C_2 - \delta^{13}C_s}{\delta^{13}C_1 - \delta^{13}C_s}\right)\left(\frac{\beta\delta^{13}C_s - \delta^{13}C_s + \varepsilon_{ps}/\Gamma}{\beta\delta^{13}C_2 - \delta^{13}C_s + \varepsilon_{ps}/\Gamma}\right)\right\}. \tag{14.52}$$

where A_s and $\delta^{13}C_s$ are ^{14}C activities (in percentage of the modern carbon) and $\delta^{13}C_s$ values (in ‰) in a sample.

Equation 14.52, being too complicated in practice, can be simplified, assuming that $\delta^{13}C_s = 0$ and $\delta^{13}C_1 = \delta^{13}C_{CO_2} + 4.5$ (the isotope fractionation effect of carbon

Table 14.10 Age correction accounting for incongruent dissolution (t_3) and error in the t_3 value when the error of $\delta^{13}C$ determination is equal to 0.2% (δt_3)

$\delta^{13}C$, ‰	−10	−7	−5	−4	−3
t, πeT	−600	−1,900	−4,000	−6,300	−13,000
Δt₃, πeT	60	130	330	690	2,800

in the soil carbon dioxide components system at pH = 6.5). For the plant system fixing carbon dioxide according to Calvin cycle (C_3), we have $\delta^{13}C_{CO_2} = -25‰$, and for the Hatch-Slack cycle (C-4) we have $\delta^{13}C_{CO_2} = -13‰$. The account of the carbon isotopic exchange in the first stage (the open system) results in the difference between the age estimated by Eq. (14.52) and by the Ingerson-Pearson equation (Eq. 14.39), where carbon isotopic fractionation during dissolution of carbon dioxide has been neglected. These differences are 1,600 years and ~3,500 years for ecosystems with C_3 and C_4 plant cycles respectively. Assuming that $\delta^{13}C_s = 0$, $\delta^{13}C_{CO_2} \approx -13‰$, which is characteristic for groundwaters at the second stage in ecosystems with C_3 plant cycle, $\Gamma = 1$, $\beta \to 1$, one can obtain some difference between the age obtained by Eq. 14.39, which takes into account dissolution of the carbonates in the closed system only, and that obtained by Eq. 14.52, accounting isotopic fractionation at noncongruent dissolution. This difference is shown in Table 14.10.

A convenient practical method of introducing corrections on the basis of data on stable and radioactive carbon isotope distribution has been suggested (Karasev 1979; Polyakov and Karasev 1979; Polyakov and Seletsky 1978) as follows.

If the increase in the content of carbonates in water occurs at conditions of underground circulation due to the congruent dissolution of the carbonate rocks of a collector, then by normalizing the total content of carbonate components in water, a simple equation can be written (Ingerson and Pearson 1964; Pearson and Hanshaw 1970):

$$p\delta^{13}C_{sc} + (1 - p)\delta^{13}C_{rc} = \delta^{13}C_{wc}, \tag{14.53}$$

where $\delta^{13}C_{sc}$, $\delta^{13}C_{rc}$, $\delta^{13}C_{wc}$ are the relative ^{13}C contents in the soil carbon dioxide, carbonates of the water-bearing rocks, and in the water carbonate system.

From Eq. 14.53, it is observed that the fraction of the initial soil carbon dioxide, with which practically all the radiocarbon is bound, is equal to

$$p = \frac{\delta^{13}C_{wc} - \delta^{13}C_{rc}}{\delta^{13}C_{sc} - \delta^{13}C_{rc}} \tag{14.54}$$

Assuming that the case of limestones of sea origin $\delta^{13}C_{rc} = 0$, the Eq. 14.54 becomes:

$$p = \frac{\delta^{13}C_{wc}}{\delta^{13}C_{sc}}. \tag{14.55}$$

Similarly, the Eq. 14.53 can be written for radiocarbon, assuming that the effect of radioactive decay is negligible:

$$PA_{sc} + (1 - p)A_{rc} = A_{wc}, \tag{14.56}$$

where A_{sc}, A_{rc}, A_{wc} are the activities of the radiocarbon isotope in the soil carbon dioxide, soil carbonates and carbonate system of the studied waters.

From Eq. 14.56, it is observed that

$$p = \frac{A_{wc} - A_{rc}}{A_{sc} - A_{rc}}. \tag{14.57}$$

At $A_{rc} \to 0$, which is the case for carbonaceous water-bearing

$$p = \frac{A_{wc}}{A_{sc}}. \tag{14.58}$$

Equating the Eqs. 14.55 and 14.58, one obtains

$$\frac{\delta^{13}C_{wc}}{\delta^{13}C_{sc}} = \frac{A_{wc}}{A_{sc}}. \tag{14.59}$$

Hence it follows that

$$A_{wc} = A_{sc}\frac{\delta^{13}C_{wc}}{\delta^{13}C_{sc}}. \tag{14.60}$$

Assuming that $A_{sc} = 100\%$, $\delta^{13}C_{sc} \approx -25‰$, Eq. 14.60 can be simplified to

$$A_{wc} = -4\delta^{13}C_{wc}. \tag{14.61}$$

For the above assumptions, the Eq. 14.61 represents a straight line passing through the origin and the point with coordinates $\delta^{13}C_{sc} = -25‰$ and $A_{sc} = 100\%$.

In a similar manner, assuming that the initial radiocarbon activity $A_{initial} = 85\%$ and the isotope composition of the carbonate system in the recharge region $\delta^{13}C_{initial} = 15\%$, one can derive a second equation relating radiocarbon content in a studied sample A_{wc} to the $\delta^{13}C_{wc}$ isotope composition of carbon:

$$A_{wc} = -5.7\delta^{13}C_{wc}. \tag{14.62}$$

The straight line described by the equation is plotted in Fig. 14.23 (curve II).

If the values of $\delta^{13}C$ and A_c in Eqs. 14.54 and 14.57 are not equal to zero, we obtain a more general form for dependency of A_{wc} from $\delta^{13}C_{wc}$:

$$A_{wc} = a + b\delta^{13}C_{wc}, \tag{14.63}$$

where

$$a = A_{rc} - \frac{(A_{sc} - A_{rc})\delta^{13}C_{rc}}{(\delta^{13}C_{sc} - \delta^{13}C_{rc})},$$

and

$$b = \frac{A_{sc} - A_{rc}}{\delta^{13}C_{sc} - \delta^{13}C_{rc}}.$$

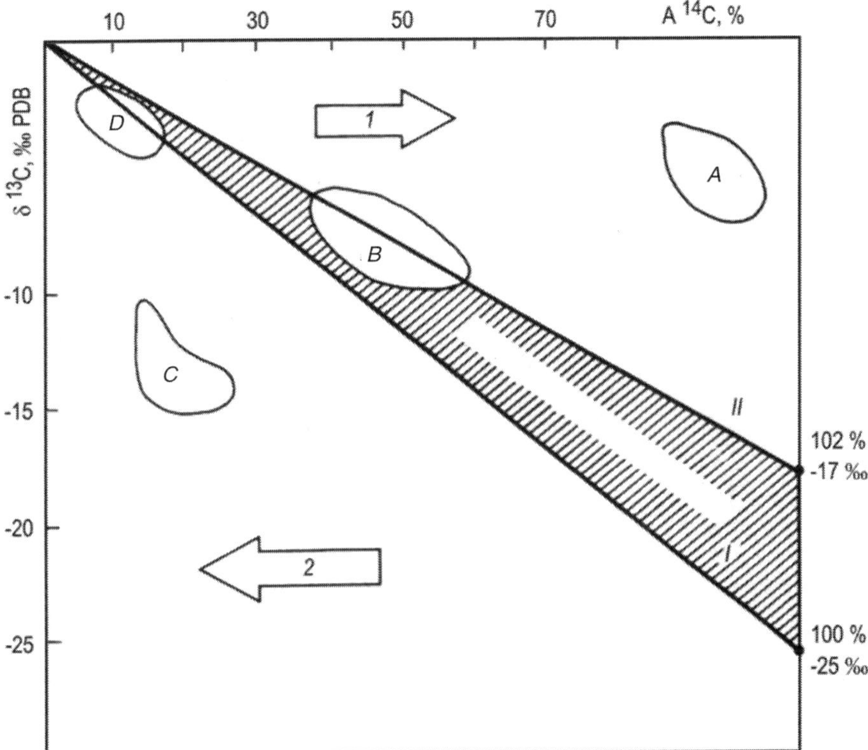

Fig. 14.23 Relationship between parameters ^{14}C (% of the modern carbon content) and δ^{13}C of carbonate system of water: (*I, II*) modern waters ^{14}C of which has decreased due to 'dead' carbon entering ($\delta^{13}C = 0$); (*A*) water from the North Kasakhstan lakes; (*B*) groundwaters from Riony depression; (*C*) thermal waters of Neocomian aquifer of Riony depression; (*D*) groundwaters from Paleozoic aquifer, North Kasakhstan; (*1*) region of post-thermonuclear epoch, or region of 'heavy' carbon entering; (*2*) region of water age determined by Eq. 14.64)

It is difficult to apply Eq. 14.63 to real conditions since the value A_{rc}, A_{sc}, and $\delta^{13}C_{sc}$ cannot be determined experimentally. This follows both from the spatial variability of geological structures of the recharge regions and transition zones, and from the fact that the climatic conditions at the time when water passed from the recharge zone to the aquifer could have been considerably different from those of today.

The Eqs. 14.61–14.63 describe the changes in the radiocarbon activity of the carbonate system of 'young' waters due to the process of normal dissolution of the carbonate water-bearing rocks. At such an approach, the age estimation is determined by formula;

$$T = 8033 \ln(A_{wc}/A_{sc} = 8033) \ln\left(-5.7\delta^{13}C_{wc}/A_{sc}\right), \tag{14.64}$$

where A_{wc} and $\delta^{13}C_{sc}$ are the measured ^{14}C activity in a sample and isotopic composition of carbon in a carbonate system. The term 5.7 accounts conformity between the % and ‰ dimensions.

Curve II in Fig. 14.23 passes through the point with coordinates A = 102% and $\delta^{13}C = -17$‰, being close to the equilibrium state of carbon isotope exchange in the gas-fluid system at fractionation factors ^{13}C and ^{14}C equal to 1.0070 and 1.016, and the isotopic composition of modern soil carbon dioxide characterized by the $\delta^{13}C_{sc} = -25$‰, $A^{14}C_{sc} = 100\%$ (relative to the modern standard). Such an assumption is fully justified for climatic zones where plants with a Calvin photosynthetic cycle exist. It is obvious that the plants lying near the correlation line will correspond to the 'young' waters, in which the decrease in the carbon specific activity is due to dilution with the old dissolved carbonates but not due to radioactive decay. The deviation of the experimental points to the left of the curve may indicate the arrival of ultramodern radiocarbon. The deviation of the points to the right of the curve may be indicative of ^{14}C radioactive decay and also the arrival of the carbon with light isotope content. A similar analysis of radiocarbon data has been used by Deak while studying groundwaters in Hungary (Deak 1979).

Figure 14.23 indicates points corresponding to ground and surface waters sampled in some regions of the former USSR. In those regions where Hatch-Slack plants are typical, the correlation line should pass through the point with the coordinates 102% and from −4 to −6‰, which reflects the isotope equilibrium between coil carbon dioxide ($A_{14C} = 100\%$, $\delta^{13}C$ from −12 to −14‰) and the dissolved components.

The admixing of carbonates during groundwater circulation can be determined for a studied aquifer by the analysis of the dependence of the radiocarbon content upon the inverse value of the total content of dissolved carbonates. In fact, if the initial concentration of carbonates in water of the recharge zone is C_o and the total concentration becomes equal to C_{wc}. Then the initial fraction of the carbonates in groundwaters is:

$$p = \frac{C_0}{C_{wc}}, \tag{14.65}$$

Equating the Eqs. 14.65 and 14.57, one obtains

$$\frac{C_0}{C_{wc}} = \frac{A_{wc} - A_{rc}}{A_{sc} - A_{rc}}, \tag{14.66}$$

hence it follows that

$$A_{wc} = A_{rc} + \frac{C_0(A_0 - A_{rc})}{C_{wc}}. \tag{14.67}$$

As pointed out earlier, if the water receives carbonates from ancient sedimentary limestones, then the term A_{rc} in Eq. 14.67 can be ignored. In the last case, Eq. 14.67 becomes similar and describes the linear dependency of A_{rc} on $1/C_{wc}$ since it is accepted that, for waters of the same age, A_o and C_o are constants. Pearson and Swarzenki (1974) have shown that, under real conditions occurring in some aquifers,

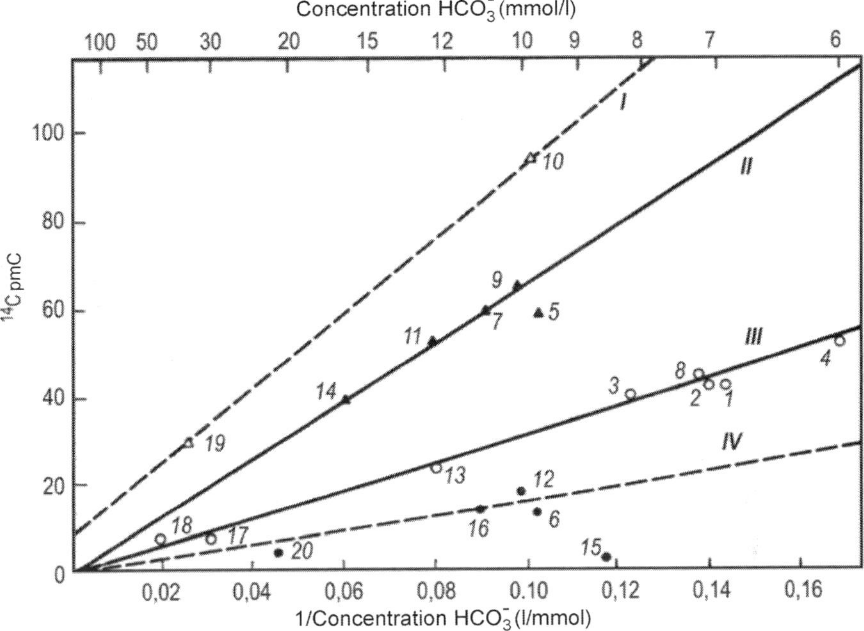

Fig. 14.24 Relationship between the determined ^{14}C content and the reverse value of carbonate component concentration in groundwaters for the Upper Quaternary rocks in the north-eastern province of Kenya. The figures show number of boreholes. (*I*) Modern waters; (*II*) age of water about 2,500 years; (*III*) ~8,000 years; (*IV*) ~14,100 years. (After Pearson and Swarzenki 1974. © IAEA, reproduced with permission of IAEA)

Eq. 14.67 can help to derive waters of the same age even with different radiocarbon specific activities (Fig. 14.24). A similar equation can be derived the dependency of $\delta^{13}C_{wc}$ on $1/C_{wc}$

$$\delta^{13}C_{wc} = \delta^{13}C_{rc} + \frac{C_0\left(\delta^{13}C_0 - \delta^{13}C_{rc}\right)}{\delta^{13}C_{wc}}. \tag{14.68}$$

It should be pointed out that Eqs. 14.67 and 14.68 describes the decrease of ^{14}C activity due to the congruent dilution of the initial carbonate components of groundwater by the dead carbonates of the water-bearing rocks.

Besides such processes as the dissolution of the dead carbonates (resulting in the decrease of the radiocarbon specific activity of the groundwater carbonate system)—and $CaCO_3$ precipitation and isotope exchange in the soil CO_2-hydrocarbonate and hydrocarbonate-rock systems—other processes affecting the reliability of radiocarbon (and tritium) groundwater dating were considered. These processes are:

a. the diffusive transfer of ^{14}C and T isotopes through impermeable thicknesses from one aquifer into another (Matthess et al. 1976; Klitzsch et al. 1976);
b. stratification of the ^{14}C and T concentrations through the aquifer's section and the effect of borehole construction—which facilitates taking the samples of water—upon the obtained radioisotope age (Vogel 1967; Matthess et al. 1976; Dubinchuk 1979; Banis and Yodkasis 1981);
c. differences in the motion rates of water (solvent) and soluted carbonate components containing radiocarbon along the aquifer (Tamers and Scharpenseel 1970; Thilo and Münnich 1970).

Klitzsch et al. (1976) have shown, on the basis of Matthess' estimations (Matthess et al. 1976), that with diffusion coefficient of carbonates through the clay impermeable layer thickness of 1.2×10^{-5} cm^2/sec, the transfer of radiocarbon from an aquifer with a ^{14}C concentration of about 85% (relative to the modern level) can result in decreases in the water age in the underlying aquifer from 40,000 to 18,000, 27,000, and 36,500 years at depth within the impermeable clay layer, lying between modern and ancient waters, of 30, 50 and 70 m, respectively. In the course of estimations, it has been assumed that the porosity of the aquifer containing ancient waters is equal to 0.3 and its thickness is 100 m.

Stratification of radiocarbon through a section of an aquifer in which recharge is distributed down its thickness results in errors in groundwater dating in the framework of the piston model since the samples taken from imperfect boreholes cannot be considered reliable (Dubinchuk 1978).

In Vogel's opinion (Vogel 1967), the water age down the section of an aquifer should vary with $\ln(H/h)$ (where H is the thickness of the aquifer and h is the distance from the base of its bed). Vogel's model is derived from simple geometrical constructions and, therefore, can be far from the conditions occurring in real hydrogeological systems. For example, at $h \rightarrow 0$ the water age, according to the above-mentioned model, tends to infinity independent of the aquifer's thickness, which is in disagreement with hydrodynamical calculations.

The ^{14}C radioisotope becomes incorporated in the dissolved component of the carbonate system of groundwaters. From common knowledge regarding sorption chromatography dynamics (Helfferich 1959; Rachinsky 1964), it is observed that the motion of water and solute at conditions of underground circulation occurs at different rates to those observed in the chromatography section. A similar phenomenon should take place during ^{14}C motion, especially in the course of groundwater migration through the carbonate collectors. The rate of a trace motion (v) is related to the rate of the solvent motion (u) by the equation $u = v\mathrm{R}$, where R is the retardation factor. If $\mathrm{R} > 1$, then the rate of the trace transfer (in our case it is ^{14}C) is in by R times less than the rate of water filtration. In general case $\mathrm{R} = 1 + k_d$ (Dubinchuk et al. 1988), where k_d is ratio between volumetric trace concentration (^{14}C) in solid (C_s) and liqid (C_l) phases; $k_d = [n(1 - n)]k = C_s/C_l$. Here n is the porosity and k is the coefficient of sorption distribution of the trace in solid (m_s) and liquid (m_l) phases; $k = m_s/m_l$. In the volumetric concentration units, one has $m_s = nC_l$, $m_s = (1 - n)C_s$.

Thus, the radiocarbon age of water estimated in the framework of a piston model should differ from the true age by the value of hk_d. The value of this quantity, according to experimental data (Thilo and Münnich 1970), varies for different carbonate rocks from 1 to 0.3. This circumstance in turn results in the fact that the peaks of T and ^{14}C, which occurred, for example, during the thermonuclear tests, will arrive at a sampling point in the recharge region, which both isotopes have reached simultaneously, at different times (radiocarbon will arrive later than tritium). This phenomenon can be responsible for the higher tritium content than of ^{14}C observed in some groundwaters (Tamers and Scharpenseel 1970).

In conclusion of this chapter, we summarize some results of the theoretical and experimental studies of formation the isotope composition in a carbonate system of groundwaters in order to increase reliability of the groundwater age determined by the radiocarbon.

Firstly, the most proposed equations include a number of parameters which practically are not determinable in the field. Moreover, in many cases the stages of the formation and observation of isotopic composition of groundwaters are divided by significant time intervals during which climatic changes affecting the hydrological conditions in unsaturated zone have happened. Secondly, the use of more complicate equations does not mean that they improve the reliability of the determined age. In the end, the experimental values of the parameters, which are substituted into the equations, being determined for a specific water point, do not take into account their space variability. In this connection the age determined by the radiocarbon is called in literature as "radiocarbon age" which has meaning of indeterminancy, characteristic for the method. As a rule, ^{14}C is defined the upper border of the age (i.e., the maximal age).

Chapter 15
The Other Cosmogenic Isotopes in Natural Waters

15.1 Origin of Other Cosmogenic Radioisotopes in the Atmosphere

In addition to tritium and radiocarbon, other cosmogenic isotopes are used in studying natural waters.

While considering the amounts and production rates of primary cosmic rays and secondary nuclear particles with the three main constituents of the atmosphere such as nitrogen, oxygen, and argon were accounted for, but it has been pointed out that the nuclei of neon, krypton, and xenon, and also those of volcanic and meteoric origin, can make a certain contribution to the production of cosmogenic isotopes.

In the course of neon nuclei fission reactions, ^{7}Be, ^{10}Be, ^{3}H, and other isotopes are produced, but the yield of isotopes occurring during these reactions is lower than those occurring during argon fission reactions. In addition, in the atmosphere, the ratio of concentrations of Ne/Ar = 0.02. The content of krypton and xenon in the atmosphere is even lower and, respectively, the contribution of cosmogenic isotopes occurring from reactions involving these elements is also small. Moreover, only reactions of the (n, γ)–type on nuclei of ^{80}Kr, ^{84}Kr, and ^{132}Xe can serve as a source of very small amounts of ^{81}Kr ($T_{1/2} = 2.1 \times 10^5$ years), ^{85}Kr ($T_{1/2} = 10.8$ years), and ^{133}Xe ($T_{1/2} = 5.3$ days). An approximate estimation of the cosmogenic ^{81}Kr yield by the (n, γ) reaction produces a value of 3×10^{-7} atom/cm^2 s.

Bartels (1972) has suggested that the major portion of the cosmogenic isotopes of ^{22}Na, ^{32}Si, ^{32}P, ^{32}P, ^{35}S, and ^{36}Cl may be produced by the reactions of cosmic rays on nuclei of the elements released into the atmosphere in the form of volcanic explosions. His suggestion is based on data of the rate of contribution of s number of elements, together with volcanic material, to the terrestrial atmosphere (Table 15.1).

The conditions in which the production of many cosmogenic isotopes takes place were considered by Lujanas (1975, 1979). He pointed out that the atmospheric concentrations of the elements, indicated in Table 15.1, are much less than those of the main air constituents. Therefore the effect of fission reactions on their nuclei will be negligible, but in a number of cases the cross-sections of reactions on slow neutrons are great. In order to estimate the yield of cosmogenic isotopes in these

V. I. Ferronsky, V. A. Polyakov, *Isotopes of the Earth's Hydrosphere*,
DOI 10.1007/978-94-007-2856-1_15, © Springer Science+Business Media B.V. 2012

Table 15.1 Amount of some elements released into the atmosphere from volcanic explosions

Elements	Released amounts (kg/year)
B	7.3×10^7
F	7.3×10^9
Na	6×10^9
S	1.7×10^{10}
Cl	7.6×10^9
K	2.3×10^9
Ca	7.2×10^9

Table 15.2 Flux density of radioactive isotopes in meteoritic material falling into the atmosphere and the rate of their production by cosmic rays (from Lujanas 1975)

Isotopes	Flux density due to meteorites (atom/cm^2 s)	Rate production by cosmic rays (atom/cm^2 s)
^{3}H	1.5×10^{-8}	0.45
^{14}C	1×10^{-5}	2.5
^{10}Be	2×10^{-4}	4.5×10^{-2}
^{22}Na	8×10^{-10}	8.6×10^{-5}
^{26}Al	1.6×10^{-4}	1.4×10^{-4}
^{36}Cl	1×10^{-5}	1.1×10^{-3}
^{37}Ar	6×10^{-12}	1.7×10^{-3}
^{39}Ar	1×10^{-3}	6.3×10^{-3}
^{53}Mn	1×10^{-8}	–
^{54}Mn	6×10^{-3}	–
^{55}Fe	1×10^{-6}	–
^{59}Ni	8×10^{-6}	–
^{60}Co	2.5×10^{-10}	–

reactions let us write the expression for the production rate P_i of a given isotope in relation to the production rate P_c of ^{14}C isotope in the reaction ^{14}N (n, p) ^{14}C on the slow neutrons

$$P_i = \frac{P_c \sigma_i N}{\sigma_c N} P_i, \tag{15.1}$$

where σ_i and σ_c are the effective cross sections of production of a given isotope and ^{14}C, respectively; N and N_c are number of atoms in the target nuclei and in the atmosphere column with a unit section.

For example, one can find from Eq. (15.1) that in reaction ^{10}B (n, p) ^{10}Be, at $\sigma_i = 4{,}010$ barn, $\sigma_i/\sigma_c \approx 2{,}300$, and $N/N_c = 1.9 \times 10^{-11}$, the value of $P_i = 8.7 \times 10^{-7}$ atom/cm^2 s is less than the rate of yield of the same isotope in reactions on air nuclei, being equal to 9×10^{-2} atom/cm^2 s, by several orders of magnitude. The estimation of the production rates of the other isotopes in reactions on the slow neutrons give even smaller values.

The studies of meteorites falling on the Earth's surface show that they also contain such radioactive isotopes as ^{3}H, ^{10}Be, ^{22}Na, ^{26}Al, ^{36}Cl, ^{37}Ar, ^{39}Ar, ^{40}K, ^{44}Ti, ^{45}Ca, ^{46}Sc, ^{48}V, ^{49}V, ^{51}Cr, ^{53}Mn, ^{54}Mn, ^{57}Co, ^{58}Co, ^{60}Co, and others. Table 15.2 shows data on the flux density of some cosmogenic radioisotopes released into the atmosphere after burning of between 0.5–0.7 of the mass of meteorite material assuming that total amount of meteorite material amounts to 10^9 kg/year. Table 15.2 also shows the cosmic ray production rates of the same isotopes in the atmosphere.

It is evident from Table 15.2 that the flux densities of the isotopes ^{3}H, ^{10}Be, ^{14}C, ^{22}Na, ^{36}Cl, ^{37}Ar, and ^{39}Ar arriving together with meteorites into the atmosphere are lower by several orders of magnitude than the rates of production in the atmosphere due to the irradiation of the atmospheric nuclei of atoms with cosmic rays. At the same time, the ^{26}Al meteorite contribution, for example, is comparable in magnitude to its atmospheric production rate.

15.2 Distribution of Other Cosmogenic Isotopes in the Hydrosphere

Many studies have been carried out on a number of the above-mentioned cosmogenic isotopes to demonstrate their abundance in natural waters and applicability in practice.

The sodium content studies in the upper atmosphere show that at an altitude of 80 km, the atomic sodium present is $\sim 10^{10}$ atom/cm^2. The cross section of the reaction $^{23}Na\,(n, 2n)\,^{22}Na$ is about 10 mbarn. In this case, the ^{22}Na production rate is about 10^{-5}–10^{-6} atom/cm^2 s. Although in the lower atmospheric layers the nitrogen density is greater by several orders of magnitude, the ^{22}Na production rate is of the same order as in the upper atmospheric layers because of a decrease in the intensity of the cosmic ray flux.

Systematic measurements of the ^{22}Na and ^{7}Be isotope concentrations in atmospheric air and fallout were carried out between 1960 and 1973 in the Baltic region near Vilnius (Lujanas et al. 1975) and in Leningrad (Gritchenko et al. 1975). They have shown a number of general principles regarding variations of the studied isotopes with time which are common for all cosmogenic isotopes. Among others, seasonal variations of ^{22}Na and ^{7}Be concentrations have been detected with a spring–winter minimum. The maximum concentrations of the isotopes related to the thermonuclear tests were detected in 1963. The results of these observations for ^{22}Na are shown in Fig. 15.1.

As Lujanas et al. (1975) have pointed out, there were no observations in Vilnius during 1967 (Fig. 15.1a) and the data obtained in 1970 correspond to winter measurements, the period of minimal concentrations of ^{22}Na. From the above data it is evident that during 1962–1963, considerable amount of ^{22}Na had been stored in the atmosphere due to thermonuclear tests and its content in fallout during 1962–1966 was many times greater than the ^{22}Na falls of cosmogenic origin. From the above-mentioned data, the half-removal period of ^{22}Na from the stratosphere can be estimated, which in this case is equal to 9.6 months (Fig. 15.1a) and about 12 months (Fig. 15.1b). In the experiments described above, the ^{7}Be maximum—equal to 79 atoms/g of air—was obtained in May 1965, and the minimum—equal to about 2.7 atoms/g air—in January 1971.

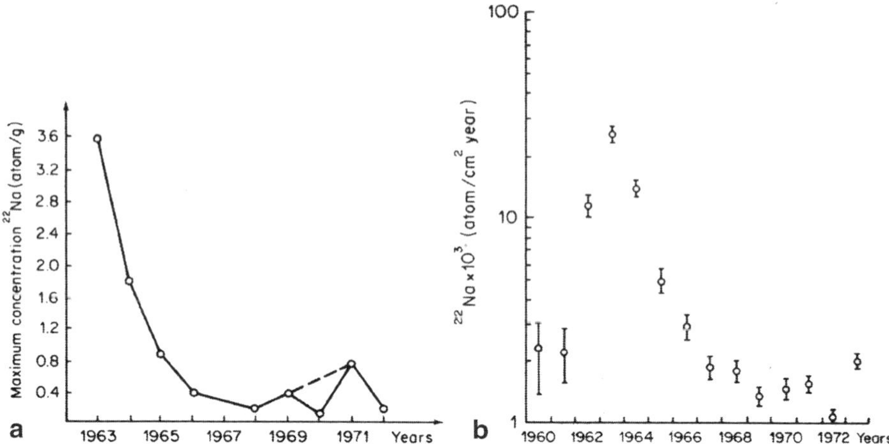

Fig. 15.1 Annual maximum fallouts of ^{22}Na **a** near Vilnius and **b** averaged annual near Leningrad in 1962–1973. (Lujanas et al. 1975; Gritchenko et al. 1975)

One may estimate the average expected concentration of a cosmogenic isotope if one assumes that tropospheric air is fixed rather quickly. If at the moment $t=0$, the concentration of the ith isotope $q_i(0)=0$, then

$$q_i(t) = \frac{\sigma_i}{\lambda_i} S_t(1 - e^{-\lambda_i t}), \tag{15.2}$$

where S_t is the average tropospheric air production rate of isotopes due to cosmic rays per gram of air; σ_i is the mean cross section of the reaction, characterizing the isotope yield; λ_i is the decay constant.

A similar equation can be written for the stratosphere. The theoretical estimations of the equilibrium tropospheric and stratospheric concentrations are equal to 98 and 22,000 atoms/g of air for ^{7}Be and 0.112 and 8.3 atoms/g of air (the residence time τ of the isotope in the stratosphere being equal to a year).

The ratio of the concentrations of cosmogenic isotopes can be estimated by the equation

$$\frac{q_i(t)}{q_j(t)} = \frac{\sigma_i \lambda_i [(1 - e^{-\lambda_i t}) + K(S_s/S_t)(1 - e^{-\lambda_i \tau})e^{-\lambda_i t}]}{\sigma_j \lambda_j [(1 - e^{-\lambda_j t}) + K(S_s/S_t)(1 - e^{-\lambda_j \tau})e^{-\lambda_j t}]}, \tag{15.3}$$

where S_t is the mean production rate of isotope in the stratosphere; τ is the average residence time of the isotope in the stratosphere; K is the dilution factor of stratospheric air with tropospheric air.

A comparison of experimental and theoretical data shows, for example, that in 1972, the ratio of concentrations was ^{22}Na/^{7}Be $= 10^{-2}$, under conditions when the stratospheric air has been present in the troposphere without being washed out for about 10–15 days and with a dilution coefficient equal to 0.2.

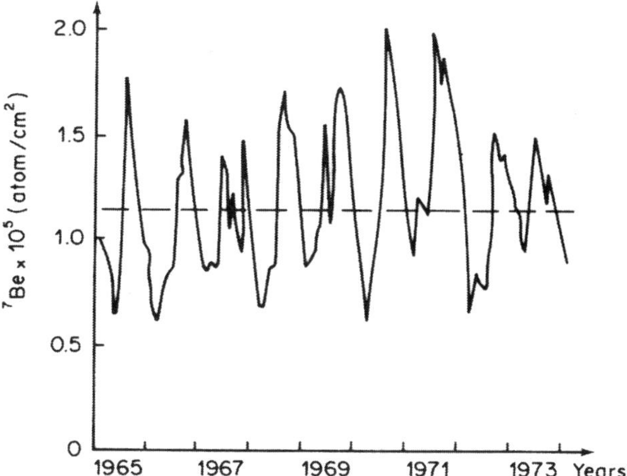

Fig. 15.2 Annual fallouts of ^{7}Be in 1965–1977 near Leningrad (After Gritchenko et al. 1975)

The behavior of ^{22}Na cosmogenic isotope does not differ from the behavior of the other cosmogenic isotopes or those produced in the course of thermonuclear tests and released into the stratosphere. It is evident from the above observations, carried out over a period of seven years, that about 1.6×10^3 atom/cm^2 year of the cosmogenic ^{22}Na isotope falls near Leningrad. Assuming this value to be representative of the whole northern hemisphere, it corresponds to about 1.4×10^{22} atoms or 0.5 g of ^{22}Na per year. The same estimates, in the case of the thermonuclear ^{22}Na fallout, give near Leningrad $\sim 6 \times 10^4$ atoms/cm^2 year, which corresponds to $\sim 1.5 \times 10^{23}$ atoms or 5 g of ^{22}Na falling over the whole northern hemisphere annually.

After the test ban treaty, observations of ^{7}Be (Fig. 15.2) have shown that its concentrations, as designated by a dotted line in the figure, are characterized by considerable greater constancy compared to the other isotopes of thermonuclear origin.

As shown in Fig. 15.2, the annual equilibrium amount of ^{7}Be in fallout during 1965–1973 was equal to 1.2×10^5 atom/cm^2 year. Observations of the ^{7}Be isotope fallout during 1960–1973 have shown that the average ratios of the ^{7}Be atoms to those of ^{22}Na in atmospheric precipitation is equal to 200–500 in the case of cosmogenic isotopes. In the case of thermonuclear isotopes, this ratio decreases sharply due to preferential fallout of ^{22}Na.

15.3 Use of Radioisotopes as Tracers in the Hydrological Cycle

The ^{22}Na isotopes of cosmogenic and thermonuclear origin are in common use in the dating of surface and slightly mineralized groundwaters (Fleishmann et al. 1975). In this case, the balance equation of the isotope content—on the assumption that it holds in groundwater (the amounts of sorption, exchange of ion, and biological fixation are assumed to be negligible)—is

$$\frac{d}{dt}(VA) = RA'_0 - RA - VA\lambda, \tag{15.4}$$

where V is the volume of the reservoir; $A'_0 = A_0Q/R$ is the isotope concentration in the recharge water corrected to evaporation from the reservoir; A and A_0 are the average values of the isotope concentration in the reservoir and inflowing water; Q is the annual water discharge (recharge); R is the runoff of water; λ is the decay constant of the isotope.

K is designated by Q/V, the rate of water exchange in the reservoir and K_1 by R/V, the rate of water exchange (conditioned by the runoff) which determines the rate of exchange of the dissolved substance in the reservoir. Then the inverse values of $T = 1/K = V/Q$ and $T_1 = 1/K_1 = V/R$ correspond to the age of water and that of the solute. At $Q \approx R$, i.e. when evaporation is absent, then $T = T_1$. If $d(VA)/dt = 0$. Now it is evident from expression (15.4) that

$$T_1 = \frac{1}{\lambda}\left(\frac{A'_0}{A_0} - 1\right). \tag{15.5}$$

The ^{22}Na concentration in the fallout A_0 is changed. Due to the thermonuclear tests the atmosphere has become highly contaminated with artificial ^{22}Na. Later, its concentration started to decrease. At $d(VA)/dt \neq 0$, from Eq. (15.4) it is evident that

$$\begin{aligned} A(t+\Delta t) &= A(t)\ \exp\left[-(K_1+\lambda)\Delta t\right] \\ &\quad + \frac{K_1}{K_1+\lambda}A'_0(t+\Delta t)\left\{1-\exp\left[-(K_1+\lambda)\Delta t\right]\right\}. \end{aligned} \tag{15.6}$$

If the dependency $A_0(t)$ is known, the value of $A(t)$ as a function of K_1 can be estimated and the age T_1 determined. If atmospheric precipitation is the source of the inflowing water, then

$$A'_0(t) = \frac{N(t)}{M}, \tag{15.7}$$

where $N(t)$ is the variation with time of the intensity of the isotope in atmospheric fallout per square unit of reservoir; M is the mean annual modulus of the water runoff, tabulated in a number of hydrological works.

The values of $N(t)$ for Leningrad, obtained by Fleishman et al. (1975) during 1960–1973, are shown in Table 15.3.

The standard deviations of the data, indicated in Table 15.3, are equal to about 50% for 1960–1961 and about 10% for 1962–1973.

In view of the obtained values $N(t)$ by formulae (15.6) and (15.7), the ^{22}Na concentrations were determined in natural waters dependent upon the values of T_1. These data are plotted in Fig. 15.3, with the value of A_0 in 1963 being taken to be 100%.

Table 15.3 Values of $N(t)$ (dpm/m^2 year) for Leningrad in 1960–1973 (from Fleishman et al. 1975)

Year	$N(t)$
1960	12
1961	11
1962	59
1963	133
1964	72
1965	25
1966	–
1967	9.5
1968	8.9
1969	7.1
1970	7.6
1971	8.0
1972	5.0
1973	10.4

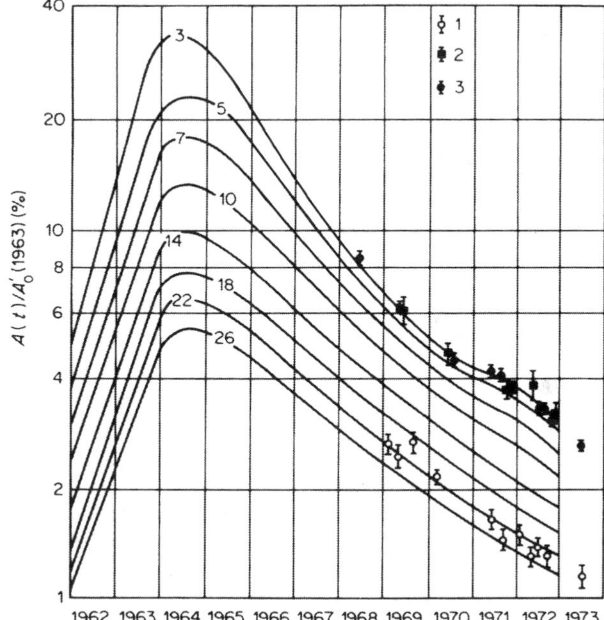

Fig. 15.3 Variation in ^{22}Na concentrations in natural waters: Theoretical isolines for some fixed points and experimental data for Neva River–Ladoga Lake (*1*); Volkhov River–Ilmen Lake (*2*); Alexeyevskoe Lake (*3*) (After Fleishman et al. 1975)

Table 15.4 indicates the values of the water ages (the time of complete water turnover) for a number of lakes in Eastern Europe, obtained on the basis of data on ^{22}Na isotope concentrations. The upper limit of ^{22}Na dating is about 10 years at concentrations of the stable sodium isotope less than 5 mg/l. The upper limit of dating decreases with increasing mineralization of water.

Attempts were undertaken in using the ^{32}Si cosmogenic isotope for groundwater dating (Fröhlich et al. 1977; Jordan et al. 1983). This isotope is formed in the atmosphere due to the cosmic rays' interaction with the nuclei of ^{40}Ar (see Chap. 12)

Table 15.4 Age (water exchange time) for some large lakes of Eastern Europe determined by ^{22}Na concentrations (from Fleishman et al. 1975)

Lake (River)	Sampling time	$A(t)$ (dpm)	T_1 (year)	V/R (year) (hydrologic data)
Ladoga	23.01.69	11.7 ± 1.0	22 ± 3	12
(Neva River)	07.09.72	5.8 ± 0.4	23 ± 3	
Onega	21.10.69	12.2 ± 1.0	16 ± 2	16
(Svir River)	06.10.71	8.3 ± 0.6	18 ± 2	
Sayma	16.08.69	18.8 ± 1.6	10 ± 2	4
(Burnaya River)	02.08.72	9.7 ± 0.7	11 ± 2	
Ilmen	02.06.69	35.7 ± 4.0	4 ± 2	0.2
(Volkhov River)	18.12.72	18.0 ± 2.5	4 ± 2	
Chudskoe	–	–	–	–
(Narva River)	12.06.70	15.4 ± 2.6	4 ± 2	2

and migrates with natural water in the form of anions of silicon acid. The period of ^{32}Si half-life (450 years) allows to cover the age gap between the tritium and radiocarbon scale, i.e., to determine the groundwater age in framework of the piston flow model from 50–2000 years (Fröhlich et al. 1977). The ^{32}Si activity in natural waters is very small, which is evident from the experimental data of different researchers (see Table 15.5).

As it is evident from Table 15.5 and Figs. 15.4 and 15.5, the ^{32}Si concentration in atmospheric precipitation is not higher than 0.4 disintegrations/min $\times$ 10^3 l and in average of 0.25 disintegrations/min $\times$ 10^3 l. The ^{32}Si in groundwaters is even lower. Because of this, for the activity measurements, the isotope is concentrated from about 20 m^3 of water sample.

The methods of precipitation by silicon acid on the hydroxide iron or ion-exchange are used for the isotope concentration (Fröhlich et al. 1977). The low-level semiconductor detectors are used for the ^{32}Si activity measurements, as a rule, by ^{32}P daughter product, which is a more ridged β-emitter ($E_{\beta Si} \approx 0.1$ Mev, $E_{\beta P} = 1.7$ Mev). The ^{32}P period of half-life is equal to 14.2 days and in about one month, the radioactive phosphorus is accumulated enough for measurement. Phosphorus is separated radiochemically from the sample and the obtained ^{32}Si concentrate can be measured many times.

Table 15.6 demonstrates comparative ages of the same sample of water obtained by the radiocarbon and radiosilicon methods. The results were noncorrected on dilution factor, exchange effects, and so on (IAEA 1968).

Rama (1968), whose data are shown in Table 15.6, has assumed that the corrections needed for the age of water calculation by ^{32}Si are less compared to the radiocarbon method, but that assumption needs to be checked because silicon-32, as well as radiocarbon, is one of the dissolved components. Therefore, the interaction processes in the system of water-rock should considerably affect the silicon acid concentration in the dilution and on the ^{32}Si specific activity as well. The ^{32}Si natural level in 1960s was notably changed by the thermonuclear tests. In this connection, its initial activity on entering the hydrological system is known at present time with great

Table 15.5 ^{32}Si activity in natural waters (from Fröhlich et al. 1977)

Sampling location	Sampling time	Activity (disintegrations/ min in 1,000 l)
Atmospheric precipitation		
Glostrub (Denmark)	June–September 1965	1.1
Bombay (India)	July–September 1966	0.31 ± 0.04
Freiberg (Germany)	May–September 1972	0.24 ± 0.04
	October–November 1972	0.13 ± 0.03
	May–July 1973	0.34 ± 0.01
River waters		
Ganga at Haridvar (India)	March 1964	0.14 ± 0.01
Godavari at Nasik (India)	May 1964	0.11 ± 0.01
Freiberg mold at Weissenborn (Germany)	August 1972	0.19 ± 0.03
Groundwaters		
Tubewell at Banaras, sample from 100 m (Uttar Pradesh, India)	October 1964	0.15 ± 0.015
Tubewell at Dabla, 220 m (Rajasthan, India)	November 1964	0.03 ± 0.01
Tubewell in mottled sandstone, 150 m (Thuringiya, Germany)	June 1972	0.30 ± 0.08
Tubewell in Saksonean chalky deposits (region of Elbean sandstones, Germany)	November 1972	0.05 ± 0.01
Tubewell in crystal deposits (Erzgeburge, Germany)	October 1972	0.08 ± 0.02

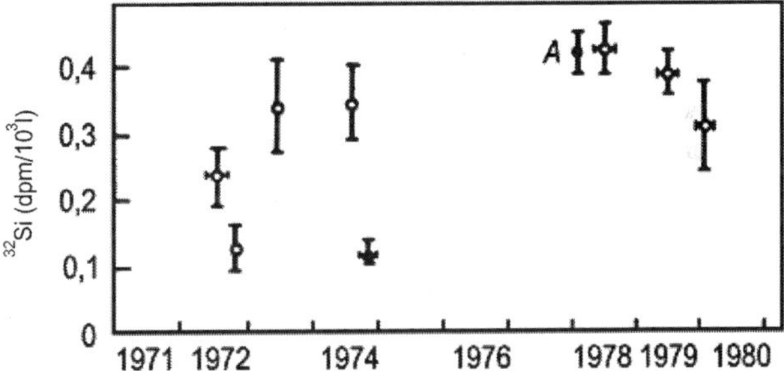

Fig. 15.4 ^{32}Si activity in atmospheric precipitation near Freiberg (Germany). Point *A* shows activity in the Antarctic liquid water (After Jordan et al. 1983)

inaccuracy and also the period of ^{32}Si half-life is not yet identified with high accuracy. The literature data varied from 105–710 years. This also reduces accuracy of the calculated age values obtained by the radioactive decay formula.

Only the experimental studies can answer the above questions. The present day half-life period of ^{32}Si is accepted as 450 years.

Among the cosmogenic isotopes in common use for the dating of natural waters, the ^{39}Ar isotope is of interest to researches. Although its specific activity is very low (0.1 dpm/l argon), its application has the following advantages. Firstly, the half-life period of ^{39}Ar ($T_{1/2} = 269$ years) allows dating of water samples characterized by

Table 15.6 Comparisons of age of a number of groundwater types in India determined by ^{32}Si and ^{14}C

Sampling location	Depth (m)	Noncorrected age (years)	
		^{32}Si	^{14}C
Palana tubewell, Rajasthan	168	2,000	5,000
Palana open well, Rajasthan	66	2,000 ± 1,000	5,000
Vijapur tubewell, Gujarat	110	1,500	2,280
Bhalod tubewell, Gujarat	65	600	2,600
Ropar open well, Punjab	30	900	300
Amritsar tubewell, Punjab	90	1,000	800
Neyvelli tubewell, Tamil Nadu	130	500	540

ages ranging from 50–1000 years, which fills the gap in dating left by restrictions of using tritium and radiocarbon. Secondly, ^{39}Ar—being a noble gas—does not interact with water-bearing rocks.

A group of scientists from Physics Institute in Bern University (Oeschger et al. 1974; Loosli and Oeschger 1979) has studied the problem of ^{39}Ar application to hydrology for a number of years. The physical reasons behind ^{39}Ar usage in groundwater dating remain the same as in the case of tritium and ^{14}C. ^{39}Ar is produced in the atmosphere by the reaction ^{40}Ar(n, 2n) ^{39}Ar. The half-life of ^{39}Ar is greater than the mixing time of the atmosphere.

According to the data obtained by Loosli and Oeschger (1979), the ^{39}Ar activity in the Zurzach thermal springs, Switzerland, is 3.8 times greater than that in the atmosphere. The authors assumed that this phenomenon can be associated with ^{39}Ar production in ground conditions by the reaction ^{39}K(n, p) ^{39}Ar, initiated by neutrons formed by (α, n) processes. If this is the case, the ^{39}Ar concentration in groundwaters depends considerably on the content of uranium, thorium, and potassium in the water-bearing rocks. In order to estimate the upper limit of the ^{39}Ar production rate in the ground, Loosli and Oeschger (1979) made the following assumption. In rocks containing 2% of potassium, about 3,000 neutrons/kg of rock are produced annually. The cross section of the reaction ^{39}K(n, p) ^{39}Ar is equal to 30 mbarn and that of the neutron capture reaction is 2,000 mbarn. All the produced atoms of ^{39}Ar diffuse into pore water. With these assumptions, it has been found that the ^{39}Ar activity in the ground, equal to 0.4 dpm/l Ar, should be greater than its activity in the atmosphere by four orders of magnitude. The maximum ^{39}Ar contribution from thermonuclear tests and ejections of the atomic industry does not exceed 12% of its cosmogenic content. According to Oeschger et al. (1974), the ^{39}Ar activity in the atmosphere during the last thousand years has changed within ± 3% (Fig. 15.5).

It should be pointed out that the method of ^{39}Ar measurement in natural waters is rather laborious. In order to obtain 3 l of argon, required for its activity measurement, at least 20 tons of water has to be processed. This procedure consists of three stages: (1) extraction of gases from the water; (2) separation of argon from the gaseous mixture, during which special attention should be paid to the separation of the ^{85}Kr admixture ($T_{1/2}$ = 10.8 years), the activity of which is rather high; (3) measurement of the ^{39}Ar activity with the help of the proportional counter.

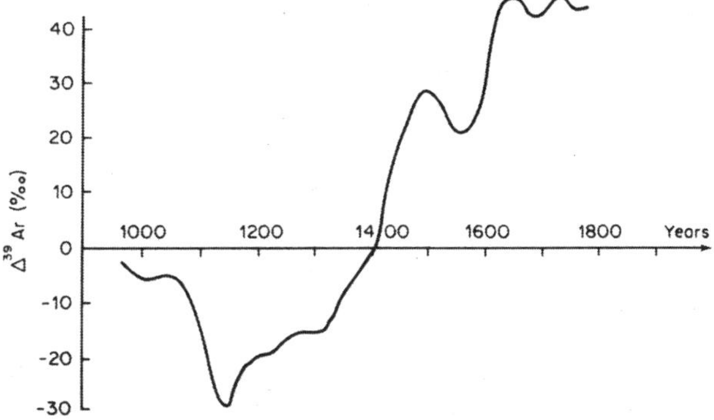

Fig. 15.5 Calculated deviations of the $\Delta\ ^{39}$Ar function in the atmosphere for the last 1000 years (After Oeschger et al. 1974. © IAEA, reproduced with permission of IAEA)

According to the data of (Oeschger et al. 1974), the argon age of water from the Zurzach thermal spring has been found to be 70 years, which is much less than the radiocarbon age (12.4% of the modern radiocarbon activity). Loosli and Oeschger (1979) have reported the following reasons for the discrepancies in groundwater dating by ^{39}Ar and ^{14}C. Firstly, the error in the age obtained by ^{39}Ar can be caused by admixture of gases remaining after their separation and also by a considerable contribution of the $^{39}K(n, p)\ ^{39}Ar$ reaction, proceeding in ground conditions. On the other hand, there could be errors in the radiocarbon age determinations due to the admixture of dead radiocarbon. These discrepancies in the argon and radiocarbon ages of water have remained even after more careful experiments. One reason for obtaining ages which are too young in ^{39}Ar dating might be the possible ^{39}Ar isotope exchange between the aquifer and the atmosphere but the solution of the diffusion equation at a diffusion coefficient equal to 1.2×10^{-4} cm^2/s indicates that at the obtained flow rates in the aquifer the diffusion process cannot account for young argon age of water. An attempt is made to explain the observed discrepancy in the ^{39}Ar and ^{14}C water ages in terms of the dilution of water of different ages in the aquifer. The theoretical estimations, based on the mixing model of modern and ancient waters, the exponential model, and the piston flow model, have not given an answer to this question.

Among the cosmogenic isotopes that have found their application in natural water studies, ^{85}Kr has a number of advantages. The unique, well-mixed natural reservoir, where ^{85}Kr is formed, is the atmosphere. The ^{85}Kr isotope concentration in the atmosphere is known very accurately. As a natural gas, krypton does not react chemically with the water-bearing rocks. The natural level of ^{85}Kr activity in the atmosphere is conditioned by the reaction $^{84}Kr(n, \gamma)\ ^{85}Kr$ in which the secondary neutrons of the cosmic rays take part. The production of some additional amounts of ^{85}Kr nuclei occurs in the course of spontaneous fission of uranium and thorium nuclei.

The determination of the nuclear industry resulted in a considerable release of technogenic krypton into atmosphere. The increase in the ^{85}Kr concentrations in the atmosphere during the period of 1950–1978 is shown in Fig. 15.6. The average ^{85}Kr concentration in the southern hemisphere is about 85–90% of that in the northern

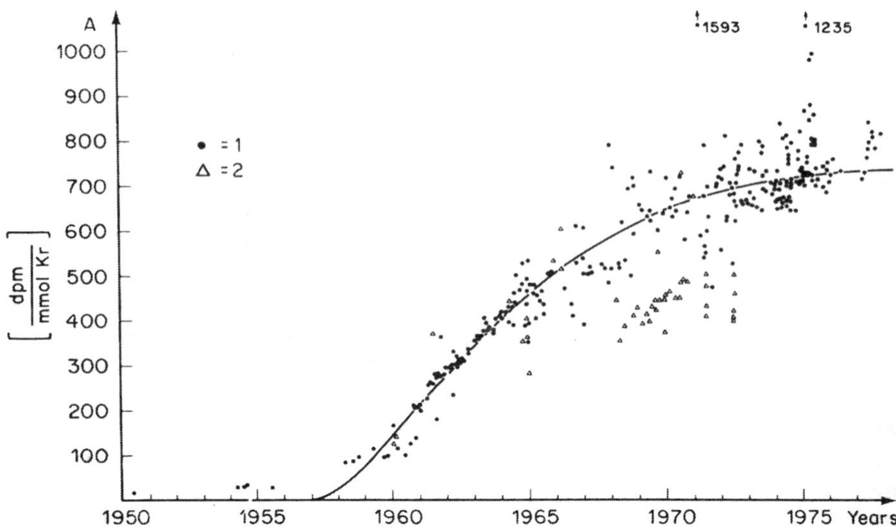

Fig. 15.6 ^{85}Kr activity in the atmosphere for the (*1*) northern hemisphere and for (*2*) the southern hemisphere (After Rozanski and Florkowski 1979. © IAEA, reproduced with permission of IAEA)

hemisphere, where its activity is ∼780 dpm/mmol Kr. The contribution of various sources in the total ^{85}Kr production is shown in Table 15.7.

In recent years, the annual production of ^{85}Kr into the atmosphere has decreased from 85 ± 5 dpm/mmol Kr in the 1960s to the 40 ± 5 dpm/mmol Kr in 1976 (Heller et al. 1977). Only 2.8% of the total amount of krypton is contained in the hydrosphere, the major portion being present in the atmosphere. The mass transfer of krypton between the atmosphere and hydrosphere occurs together with atmospheric precipitation and also due to molecular exchange in the atmosphere-surface water system.

According to estimations made by Rozanski and Florkowski (1979), the annual average transfer of ^{85}Kr from the atmosphere into the hydrosphere is 0.1% of its general content in the atmosphere. The krypton concentration in the atmosphere is equal to 1.11×10^{-4} vol %, and its solubility in water at 15°C is about 6.9×10^{-6} vol %. Thus, the content of krypton in water does not exceed 0.07 cm^3/m^3, which corresponds to a krypton activity of about 2.6 dpm.

The method of the ^{85}Kr measurement (Rozanski and Florkowski 1979) is as follows. The water sample is pumped into a volume of about 360 l, filled with nitrogen before sampling in order to avoid atmospheric contamination. After that, by means of the gas extraction system, approximately 7 l of gases are evolved from the water at a temperature of about 90°C. Then the oxygen and nitrogen gases are removed from the water-gas mixture. The krypton content in the mixture is determined by means of mass spectrometry and its activity is measured by a proportional counter. The measurement time for a single groundwater sample is about 100–200 h. The ^{85}Kr,

Table 15.7 ^{85}Kr contribution into the atmosphere from various sources at the end of 1976

Source	^{85}Kr activity in the atmosphere (mCi)
Natural production	1.4×10^{-5}
Nuclear explosion	2.0
Nuclear reactors	28.3
Plutonium production	30.3

^{14}C, T data obtained for 14 water samples taken from surface and groundwaters have shown good results while making a comparison using the complete mixing model.

Considerable success has been achieved in the application of such cosmogenic isotopes as ^{10}Be ($T_{1/2} = 1.6 \times 10^6$ years), ^{26}Al ($T_{1/2} = 7.16 \times 10^5$ years), ^{36}Cl ($T_{1/2} = 3.01 \times 10^5$ years), and ^{129}I ($T_{1/2} = 1.57 \times 10^7$ years). These isotopes have already proved to be of use while studying the conditions and rates of accumulation of bottom sediments in reservoirs and the ocean and also in surface and groundwater dating (Alder et al. 1967; Bentley et al. 1986; Fröhlich et al. 1977; Kocherov 1975; Lal et al. 1970; Lehmann 1993; Möller and Wagner 1965).

Part III
Radiogenic Isotopes

Chapter 16
Production and Distribution of Radiogenic Isotopes

At present, more than twenty long-lived radioisotopes of heavy elements are known to exist in the Earth's crust. These are evidence of the gigantic processes which resulted in the formation of chemical elements in our Galaxy. The main characteristics of these elements are shown in Table 16.1.

Because of the considerable difficulties involved in deriving and measuring very small amounts of radioisotopes, only ^{40}K, ^{87}Rb, and isotopes of the uranium-thorium series are of use in practice. But in future, with the perfection of analytical methods, an increasing amount of natural radioisotopes will be used to solve the practical problems of isotope geology, hydrology, and hydrogeology. Among them the elements of the uranium-thorium series (with atomic numbers from 81 to 92 in the Periodic System) are of practical interest. Undergoing numerous sequential nuclear transformations, these elements give rise to three radioactive series (Figs. 16.1–16.3). The existence in nature of each of these series is determined by the existence of the primary substances, the half-lives of which are comparable to the Earth's age. In the uranium-radium series, the parent radioactive isotope is uranium with an atomic weight of 238 and a half-life of 4.51×10^{10} years. Uranium-235, characterized by a half-life of 7.13×10^{8} years is the originator of own series, and thorium-232 with a half-life of 1.39×10^{10} years is the originator of the thorium series. The final decay products in each series are the stable isotopes of lead, ^{208}Pb, ^{207}Pb, and ^{206}Pb, respectively.

If there were no separation of elements and isotopes in natural interacting systems, the radioactive series would be at a state of radioactive equilibrium, and the content of each element of a series in the subsystems would be determined strictly by the content of the originator of a corresponding series, in accordance with the equation of radioactive equilibrium; but in natural systems a state of radioactive equilibrium is rare. Chemical processes and radioactive decay itself result in the displacement of equilibrium. A system taken out of the radioactive equilibrium state has a tendency to attain this state again. The observed effect can be employed in studying the temporal characteristics of the processes resulting in this shift. The specific conditions of the formation of isotopic content of the heavy radioelements can be of use to solve a number of genetical problems. These problems are related to the origin and

V.I. Ferronsky, *Isotopes of the Earth's Hydrosphere*,
DOI 10.1007/978-94-007-2856-1_16,

Table 16.1 Long-lived radioactive isotopes. (From Voytkevich 1961)

Parent elements	Daughter elements	Half-life (years)	Decay type	Relative abundance (%)
^{40}K	^{40}Ar	1.3×10^9	β (88%)	0.0118
	^{40}Ca	1.3×10^9	K-capture (12%)	0.0118
^{50}V	^{50}Ti	6.0×10^{15}	K-capture	0.24
	^{50}Cr	6.0×10^{15}	β	0.24
^{87}Rb	^{87}Sr	4.7×10^{10}	β	27.85
^{115}In	^{115}Sn	5.9×10^{14}	β	95.67
^{123}Te	^{123}Sb	1.2×10^{13}	K-capture	0.87
^{138}La	^{138}Ba	1.1×10^{11}	K-capture (70%)	0.089
	^{138}Ce	1.1×10^{11}	β (30%)	0.089
^{142}Ce	^{138}Ba	5.0×10^{15}	α	11.7
^{144}Nd	^{140}Ce	2.4×10^{15}	α	23.8
^{147}Sm	^{143}Nd	1.0×10^{11}	α	15.1
^{148}Sm	^{144}Nd	1.2×10^{13}	α	11.35
^{149}Sm	^{145}Nd	4.0×10^{14}	α	14.0
^{152}Gd	^{148}Sm	1.1×10^{14}	α	0.205
^{156}Dy	^{152}Gd	2.0×10^{14}	α	0.057
^{174}Hf	^{170}Yb	4.3×10^{15}	α	0.163
^{176}Lu	^{176}Hf	2.2×10^{10}	β	2.588
^{187}Re	^{187}Os	4.0×10^{10}	β	62.93
^{190}Pt	^{186}Os	7.0×10^{11}	α	0.0127
^{207}Pb	^{200}Hg	1.4×10^{17}	α	1.4
^{232}Th	^{208}Pb	1.39×10^{10}	6α + 4β	100
^{235}U	^{207}Pb	7.1×10^8	7α + 4β	0.715
^{238}U	^{206}Pb	4.51×10^{10}	8α + 6β	99.28

characteristics of the shift processes or other geological processes. Let us consider the main chemical and geochemical properties of the radioelements constituting the uranium-thorium series.

16.1 Geochemistry of Radiogenic Elements

16.1.1 Uranium

Uranium is the heaviest element in the Earth's crust, and consists of three α-emitting isotopes: ^{238}U, ^{235}U, and ^{234}U. The isotopes ^{238}U and ^{235}U are the parent elements of the corresponding radioactive series. Among all the members of the series only the last two isotopes have remained in the terrestrial crust from the beginning of its formation. Uranium-234 is the decay product of uranium-238. In equilibrium the ratio of the uranium-234 content to that of uranium-238 is 5.5×10^5. Table 16.2 shows the main properties of the uranium isotopes (Hyde et al. 1971).

Uranium is a member of the actinide group and therefore, belongs to Group III of the Periodic System, but due to a strict shielding of the nucleus with electrons,

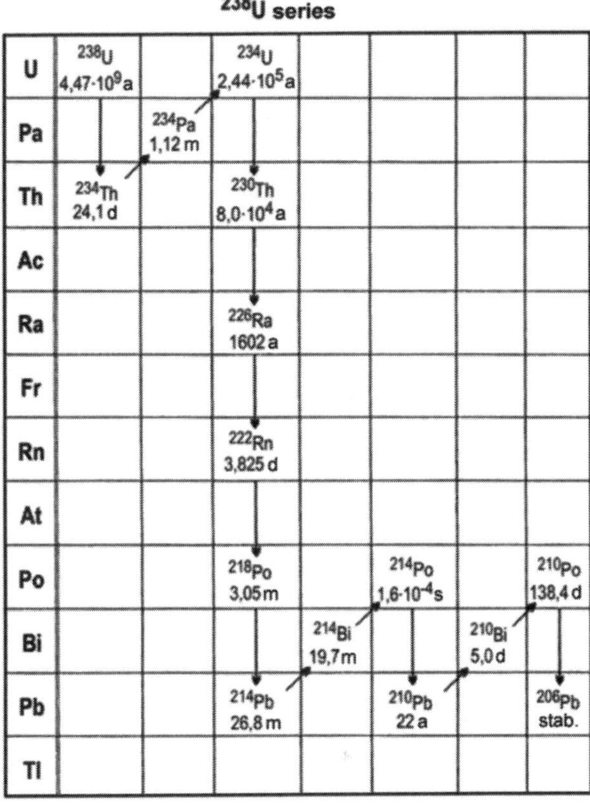

Fig. 16.1 Radioactive uranium-238 series

the valence electrons are not only the electrons of the outer shell but also those of P and Q shells, providing a maximum equilibrium valency equal to six. In natural conditions uranium occurs in the form of the four- and six-valent compounds. The three-valent compounds of uranium can be obtained under laboratory conditions only. The U^{6+} ion is energetically unstable. The compounds of the six-valent uranyl cation $UO_2{}^{2+}$. The four-valent uranium occurs in the form of the U^{4+} cation forming salts such as UCl_4 and $U(SO_4)_2$ (Vinogradov 1963). The complex compounds with carbonic anion-uranyl-carbonates, which are easily dissolved in water, are typical of the cation $UO_2{}^{2+}$. The possibility of uranium transference is determined by the chemical properties of its behavior in natural waters of a certain composition. The ion concentrations being negligible (10^{-10} g/l at pH = 4 and 10^{-22} at pH = 7), the hydroxide of the four-valent uranium has a very poor solubility product ($\sim 10^{-52)}$) even in weak acid solutions and all the more so in neutral and alkaline solutions (Vinogradov 1963). The uranium concentration, in the case of the six-valent uranium, can reach considerable values even in weak alkaline solutions, but in the presence of an insignificant amount of weak reductants the uranyl ion is reduced readily and precipitates in the form of hydroxide $U(OH)_4$. Natural waters become enriched in

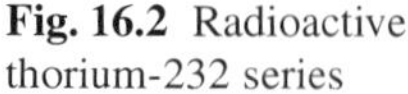

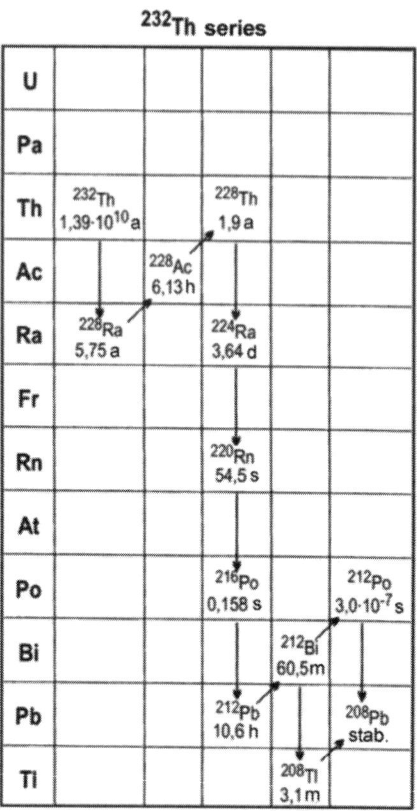

Fig. 16.2 Radioactive thorium-232 series

elements, while interacting with water-bearing rocks. The main factors determining the content of an individual radioelement are the content and chemical forms of its occurrence in rocks of the water-bearing complex. According to Vernadsky, there are three forms of uranium occurrence in rocks: (1) in the form of individual minerals (90% of which are secondary uranyl minerals), together with the uranyl group $U_2{}^{2+}$ (there are few minerals containing U^{4+}); (2) in the form of isomorphic admixtures to the crystalline lattices of the other minerals (accessory rock minerals); (3) in the dispersed state (in the form of the uranyl group $UO_2{}^{2+}$) or in the dissolved form, occurring in liquid inclusions, in the minerals and in capillary waters. In the last form, uranium is most mobile and can be leached easily both by acid and alcaline-carbonate solutions. The total fraction of uranium bonded to the uranium-bearing minerals is insignificant.

In summary, the degree of uranium mobility is determined both by the concentration of the solutions, reduction-oxidation conditions, and the form of its occurrence in nature. There are various forms of uranium transfer in natural waters, depending on the composition of pH (Kazhdan et al. 1971) of the solution: (1) in form of the uranyl-carbonate complex $UO_2(CO_3)_2{}^{2-}$ with sodium, calcium, and magnesium at

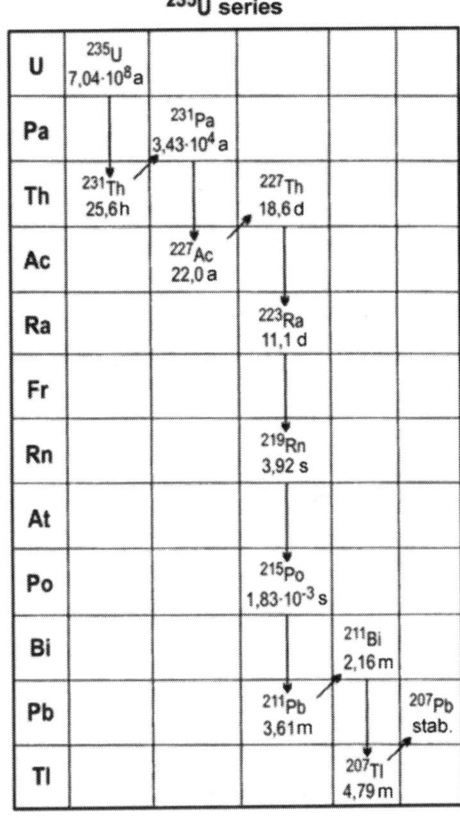

Fig. 16.3 Radioactive uranium-235 series

elevated levels of the $(HCO)^-$ and $(CO_2)^{2-}$ ions at pH 4.5–6.5; (2) in the form of the $UO_2{}^{2+}$ and $UO_2(OH)^+$ at pH between 4.5 and 7.5; (3) in the form of sulfate compounds in strong acid solution at pH < 4 (its migration occurs over small distances only due to the instability of the $UO_2(SO_4)_m$ complex); (4) in the form of organic compounds in weak alkaline and acid waters.

The precipitation of uranium, occurring with changes of the chemical composition of waters at increasing reduction capacity of rocks, is accompanied by the reduction of U^{6+} to U^{4+} at the absorption and coagulation of colloids. The main processes leading to the precipitation of uranium compounds are (Tokarev and Shcherbakov 1956):

1. Hydrolysis and coagulation of uranium hydroxides; when the pH is close to 4.2 for $UO_2(OH)_2$ and 4.1 for $Al(OH)_3$, uranium is assimilated predominantly by clays.
2. Absorption processes; uranium sorption increases in the following order: sandstones-limestones-clays-limestones-phosphorites-clayey rocks and coals.

Table 16.2 Isotopic composition and main properties of natural uranium

Isotope	Half-life (years)	Relative content (%)	Energy of α-particles (MeV)	Output of α-particles of energy (%)
^{238}U	4.5×10^9	99.2741	4.195	77
			4.147	23
^{234}U	2.48×10^5	0.0055	4.768	72
			4.717	28
^{235}U	7.13×10^8	0.7204	4.20	4.2
			4.30	85.6
			4.58	10.2

3. The destruction of the complex ions, causing dissolution of the uranium components, which results in the formation of the simple and poorly soluble compounds.
4. The reduction of the four-valent uranium to the six-valent one and the related hydrolysis of the four-valent uranium compounds.
5. The formation of poorly soluble salts such as vanadates, arsenates, carbonates, and silicates. Uranium which can be readily extracted from rocks is usually called 'mobile' or 'leaching'.

The sources of readily leached uranium can be (Vinogradov 1963):

1. Primary uranium minerals, related to oxides (uranite and nasturan) and complex oxides, because they contain, alongside four-valent uranium, six-valent uranium, especially oxides (uranite and nasturan), are substituted in the oxidation zone with secondary minerals (phosphates, vanadates, silicates, etc.), in which uranium is six-valent. All the secondary minerals of uranium, being formed in the hyper genesis zone, are also readily dissolved with weak acids.
2. The non-uranium minerals, where uranium is presented in the form of an isotopic admixture, if they are in the isomorphic condition.
3. Absorbed uranium.
4. Uranium occurring in the form of an isomorphic admixture to such minerals as zircon, apatite, orthite, etc., cannot be practically subjected to oxidation process.

16.1.2 Thorium

Six isotopes of thorium occur in nature: ^{234}Th and ^{230}Th (Io) in the uranium-238 family, ^{231}Th and ^{227}Th in the actino-uranium series, ^{232}Th and ^{228}Th (RdTh) in the thorium series. ^{232}Th, Io and RdTh are of importance only in practice. All these isotopes are alpha-ray emitters. The main properties of these isotopes are given in Table 16.3 (Hyde et al. 1971).

Thorium, like uranium, is a lithophilic element, being concentrated predominantly in the upper parts of the lithosphere. In rocks thorium is located in a dispersed form and in the thorium-bearing accessory minerals: monazite, ortite, zircon, thorite,

Table 16.3 The main parameters of long-lived thorium isotopes

Isotope	Half-life (years)	Energy (MeV)	Output of *respective* energy (%)
^{232}Th	1.39×10^{10}	4.00	76
		3.94	24
^{230}Th (Io)	8.0×10^{4}	4.68	75
		4.61	25
^{228}Th (RdTh)	1.9	5.42	72
		5.34	28

apatite, pirochlorine, sphene, thorianite, etc. Almost all the minerals containing thorium are stable in natural conditions. Chemical processes have no effect on the redistribution of thorium and do no not lead to its accumulation in the form of secondary minerals.

In aqueous solutions, only the four-valent thorium occurs. Th^{4+} ions form many water-insoluble compounds. Thorium hydroxide, fluorite, iodate, and phosphate are poorly soluble compounds. They are only dissolved in those media where thorium forms complex compounds. Thorium carbonate is poorly soluble in water, but is readily dissolved in an excess of an alkali carbonate forming a stable complex compound. In aqueous solutions as $pH < 5$, thorium is most likely to be present in the ionic, dispersed, and molecular forms. At $pH > 5$ the colloidal form of its existence is predominant. The element has a strong tendency toward hydrolysis and absorption and is liable to form complex compounds (Bernat and Goldberg 1969). The main form of migration is transfer of thorium in the form of the clastic products of destruction of chemically stable minerals, where it is bonded to a crystalline lattice (Kaplan et al. 1960; Nuclear Geology 1955). The transfer of thorium by groundwaters occurs in the colloid-suspended form and in the form of complexes of anionic character, most probably of organic origin. In summary, the problem of thorium transference is far from being solved.

16.1.3 Protactinium

In nature there are two isotopes of protactinium: ^{234}Pa, which has half-life of 1.175 min and a member of the uranium-238 family, and ^{231}Pa, being a member of the actino-uranium family. ^{231}Pa is of great importance in geochemistry, being alpha-emitted with half-life of 34,000 years and having a complicated alpha-spectrum (4.66 MeV: 1–3%; 4.72 MeV: 11%; 4.84 MeV: 3%; 4.94 MeV: 25%; 5.00 MeV: 47%; 5.04 MeV: 11%). Its principle oxidation state is the +5 state. Most of the protactinium compounds in aqueous solutions are readily hydrolyzed to form poorly soluble hydrolysis products. In the absence of complex-forming substances protactinium is readily absorbed by suspension being present in a solution. In acid solutions it is captured by many insoluble compounds and precipitates together with MnO_2, $CaCO_3$, SiO_2 etc. (Kuznetsov et al. 1966). In practice it can be extracted from alkaline solutions together with any precipitate. Protactinium exhibits geochemical properties similar to those of thorium; their transference forms are also likely to be the same. In rocks, it is abundant in accord with the occurrence of its parent element uranium-235, if active chemical processes are absent.

16.1.4 Actinium

There are two isotopes of actinium in nature: ^{227}Ac, a number of the ^{235}U (action-uranium) series, which is a beta-ray emitted with a boundary energy of spectrum of about 40 keV (Hyde et al. 1969), and a half-life of about 22 years, and ^{228}Ac with a half-life of about 6 h, which is of no practical importance. The main oxidation state of actinium is equal to +3. Actinium hydroxide precipitates in aqueous solution. Actinium, by its chemical properties, is analogous to lanthanum. As with all the other rare-earth elements actinium forms a poorly dissolved fluoride, phosphate, carbonate, silicate fluoride, oxalate, and hydroxide. The geochemical properties of this element have not yet been studied.

16.1.5 Radium

There are four natural isotopes of radium: ^{226}Ra, which is an alpha-ray emitter (E = 4.78 MeV: 94.3%; E = 4 > 59 MeV: 5.7%) with a half-life of 1,610 years in the uranium-238 series; ^{223}Ra is an alpha-ray emitter of complicated energy spectrum and with a half-life of 11.2 days in the actino-uranium series; ^{228}Ra is a beta-ray emitter with a boundary energy of 0.053 MeV and a half-life of 6.7 years; and ^{224}Ra is an alpha-ray emitter with a half-life of 3.64 days, a member of the thorium series. Among these natural isotopes of radium only ^{226}Ra and ^{228}Ra isotopes, have relatively long half-lives and practical value as well.

By its chemical properties, radium is a typical alkaline-earth metal, which exhibits only one oxidation state, +2 and has no independent minerals. Radium nitrate, chloride, sulfite, and bromite are dissoluble in water and radium carbonate, sulfate, iodate, and oxalate are insoluble. Radium is a decay product which gives it specific particularities with regard to its behavior in natural conditions. Usually, radium does not form part of the mineral crystalline lattice since its chemical and crystallochemical properties differ from those of the elements constituting the uranium minerals. It therefore escapes readily into the capillaries and crystalline micro-imperfections from where it can be washed out and transported to natural waters. The radium content in natural waters is usually so less that it can never reach the solubility limits. In weak mineralized waters radium can be present independently of chemical composition (Cherdyntsev 1969), but this state is unstable and if a microcomponent precipitates out of a solution (calcium from carbonate waters, barium from the sulfate ones) then radium co-precipitates together with it. An unstable state of radium in all natural waters (besides chloride waters) leads to its sorption onto the surface of a solid phase.

Table 16.4 The main properties of radon isotopes

Isotope	Half-life (years)	Energy (MeV)	Output of *respective* energy (%)
^{222}Rn	3,825 d	5.48	100
^{220}Rn (Tn)	54,5 c	6.28	100
^{219}Rn (An)	3,92 c	6.43	8
		6.56	11
		6.82	81

16.1.6 Radon

The radon isotopes ^{222}Rn in the uranium-238, ^{220}Rn (Tn) in the thorium-232 series and ^{219}Rn (An) in the actino-uranium series are radioactive gaseous substances. All these gases are alpha-ray emitters. Their characteristics are shown in Table 16.4 (Cherdyntsev 1969). Radon isotopes are typical of the inert-gas subgroup. They form chemical compounds only in exceptional cases. In nature they occur in the form of individual atoms. Originating in rocks from radium isotopes, they either remain inside the mineral structure or are accumulated in pores, capillaries, and fissures. The releases of radon into natural waters of a region are determined by the processes of emanation and diffusion.

Release of emanation into the air, water, or any other medium from its sources, containing radium, is quantitatively determined by the coefficient of the emanating power (K_{em}), which is defined as the ratio of the fraction of emanation into the air or any other medium to the total amount of emanation, in equilibrium with radium or its isotopes, contained in a sample.

It has been found that emanation depends on the structure and density of a mineral crystalline lattice, the branching of capillaries, and the migration of parent substances. The radioactive recoil and diffusion determining emanation are also dependent on the radium distribution in the crystalline lattice. Questions concerning emanation are discussed in detail in a number of monographs (Serdyukova and Kapitonov 1969; Starik 1961; Cherdyntsev 1969; Gudsenko and Dubinchuk 1987). The emanation power varies from several percent to 100% for different minerals. Table 16.5 shows the Rn, Tn and An emanation powers and their ratios for some minerals (Cherdyntsev 1969).

Considerable differences in the K_{Tn}/K_{Rn} ratio from unity can be explained by the inhomogeneity of the distribution of uranium and thorium atoms in the crystalline lattice. A K_{An}/K_{Rn} ratio close to unity indicates that the main cause of the emanation of atoms into fissures and capillaries is the result of the radioactive recoil phenomenon, in contrast to all the other processes where a large difference between the lifetime of radon and action should occur. The emanation power of rocks containing primary minerals is usually determined by the emanation power of minerals. With secondary minerals being present in rocks, the emanation power is usually affected by the radium leached from minerals and absorbed by rocks. In fact, according

Table 16.5 Emanation coefficients of Rn, Tn and An

Minerals	K_{Rn}	K_{An}	K_{Tn}	K_{An}/K_{Rn}	K_{Tn}/K_{Rn}
Primary minerals					
Uranite	0.21	0.20	–	0.95	–
Zircolite	12.0	10.4	38	0.80	2.9
Malacon	4.2	4.4	44	1.05	10.0
Tantalite	0.61	0.60	1.20	0.98	2.0
Tantalite	2.03	2.1	18	1.03	9.0
Kolumbite	1.00	1.0	5.4	1.00	5.4
Sphen	2.8	3.0	2.8	1.07	1.00
Hornblende	4.2	4.4	17	1.05	4.0
Molibdenite	2.08	2.3	3.6	1.11	1.7
Molibdenite	6.6	8.0	–	1.21	–
Secondary minerals					
Khrizokolla	68	57	–	0.84	–
Khrizokolla	64	57	–	0.89	–
Copper bituminous ore	8.6	10	16	1.16	1.9

to the data of Starik and Melikova (1957), rocks containing secondary minerals exhibit an emanation power equal to 30–65%, whereas the individual minerals have a similar emanation power (otenile 6%, torbernite 14.6%, carnotite 32%). According to Serdyukova and Kapitonov (1969), the emanation power of unaltered igneous and dense sedimentary rocks is insignificant. Loose sediments possess the greatest emanation power because of the presence of the mechanical or salt halo of scattered radioactive elements.

According to Baranov (1955), the average emanation power of soils for radon is 41% and that for thoron is 45% with a variation ranging from 18 to 100% for the different soils. Radon solubility in water is very high. The solubility coefficient at 15°C ranges from 0.25 to 0.30, i.e., if the volumes of water and air are equal then the radon concentration in water will be four times than that in the air at equilibrium between the liquid and the gas phases. With increasing temperature, the solubility factor decreases by the law

$$\alpha = 0.1057 + 0.405\mathrm{e}^{-0502t}.$$

The solubility factor decreases with increasing mineralization of water. Table 16.6 indicates the variation of the radon solubility factor with the concentration of some salts (Serdyukova and Kapitonov 1969).

16.2 Separation of Radiogenic Elements and Isotopes

Consider now the main processes resulting in the disequilibrium of radioactive series. According to the idea of Starik (1961), two processes occur at the interaction of solutions with rocks: dissolution, resulting in the extraction of radioisotopes with

Table 16.6 Dependence of the solubility factor of radon upon salt concentration

Salt water solution	Density (g/cm^3)	Solubility factor	Salt water solution	Density (g/cm^3)	Solubility factor
NaCl	1.008	0.239	NaCl	1.203	0.042
NaCl	1.021	0.202	$CuSO_4$	1.042	0.194
NaCl	1.039	0.163	$CuSO_4$	1.087	0.147
NaCl	1.096	0.096	$CuSO_4$	1.110	0.131
NaCl	1.121	0.077	$CuSO_4$	1.093	0.075

the destruction of the crystalline lattice; and leaching, resulting in the extraction of radioisotopes without the destruction of the crystalline lattice.

Isotopes extracted by dissolution into natural waters preserve the isotopic composition typical for a given water-bearing layer and the separation of isotopes does not occur in the process, but the separation of elements does occur, since they are released in accordance with the power of formation of soluble compounds. When the radioelements transfer into the solution, the following factors usually affect the solubility: the chemical composition of water, concentration of the pH-ion, the amount of the free oxygen, temperature, etc. Uranium and thorium, the parent elements of the radioactive series, are accumulated during magmatic differentiation in the acid rocks, where their chemical properties are highly similar. The ratio of thorium to uranium varies over a small range. The four-valent ions of uranium and thorium have approximately equal ionic radii, but in the hypergenesis zone, a sharp difference between the properties of uranium and thorium is observed. Uranium forms readily soluble compounds in water at the last oxidation stages, whereas thorium compounds are practically insoluble and practically no migration of thorium occurs. Therefore, all the secondary formations in the hypergenesis zone are usually enriched in uranium and have little thorium and transference of this element only occurs in a suspended form. Radium is practically stable in chloride waters, and there is almost no evidence concerning the ability of actinium transference by natural waters.

Extraction of a radioelement by leaching does not result in the destruction of the mineral crystalline lattice. The content and the form of its occurrence in the crystalline lattice are therefore the main factors governing the abundance of an individual element in natural waters. The forms of the parent elements and the decay products present in the mineral crystalline lattice are essentially different. If ^{238}U, ^{235}U, and ^{232}Th, the originators of radioactive series, were involved in all the processes (from nuclear fusion to formation of the Solar System and the Earth), and, together with stable elements, had taken part in processes which had resulted in the differentiation of the Earth's substance (formed their individual minerals or are present in them as isomorphic admixtures), then the content of the decay products would have been largely dependent on the content of parent elements of a corresponding series and they would have been present in the crystalline lattice in another state. On obtaining recoil energy during radioactive decay, a nucleus leaves the crystalline lattice and reaches the microcapillaries and microimperfections of the natural crystals where it can be readily washed out by solutions. They are abundant in rocks in a dispersed form, which is more favorable for extraction by aqueous solutions. On leaching, the solution obtains some portion of the parent isotopes, which make their way into the microimperfections and capillaries in the course of the destruction of the crystalline lattice and subsequent diffusion. According to Starik, leaching is the main form of transference of the radioelements in natural waters and therefore the decay products should be predominant there. Separation of both the elements and isotopes can occur during the process of leaching. The rate of leaching is largely dependent on the condition of the crystalline lattice (the degree of which it is destroyed). The extraction rate of an element from capillaries and imperfections of the crystalline lattice depends on the chemical properties of an element and on the composition of the leaching solution as well as on the dissolution process. It should be pointed out that the processes of leaching and dissolution do not occur in nature independently. The isotopes of radon, which is a noble gas, are most mobile. In particular cases the radon emanation power reaches 100%, i.e., almost all the emanated radon is released into natural waters. Numerous experimental data on leaching, obtained both in the laboratory and in the course of studying the content of radioisotopes in natural waters, have shown (Starik and Melikova 1957; Cherdyntsev 1955, 1969) that the separation of radioelements during their transference in natural waters occurs in the following order: Rn > Ra > U > Th > Pa.

16.2.1 Separation of Uranium Isotopes

Among the three natural isotopes of uranium (^{238}U, ^{235}U, ^{234}U) the ratio of abundance of ^{238}U and ^{235}U primordial isotopes is essentially constant. Uranium-234 is a decay product of uranium-238 and, therefore, it has another chemical state in the mineral's crystalline lattice compared with the primordial ^{238}U. In view of this, it should be expected that during leaching uranium-234 will preferentially escape into natural waters. This phenomenon was experimentally found by Cherdyntsev and Chalov in 1953 (Cherdyntsev 1955). Numerous determinations of uranium content in natural waters, and minerals deposited from waters, have shown that the natural waters exhibit an excess of uranium-234, i.e., the ratio is $^{234}U/^{238}U > 1$. During continuous leaching the mineral can become markedly depleted in uranium-234 and then the ratio becomes $^{234}U/^{238}U < 1$. Natural waters, interacting with these minerals, can have uranium isotopic ratios less than one, though these waters are very rare. Uranium leached out of uranium minerals usually displays the equilibrium isotopic composition. Syromyatnikov (1961) has accounted for this effect in terms of isotopic exchange. In the uranium minerals, the uranium-234 atoms exchange with uranium 238 atoms and substitute for them after radioactive recoil. With increasing dispersion, this effect should be markedly reduced. In effect, the greatest value of the uranium isotope ratio has been detected in those waters leaching uranium out of minerals in which it occurs in the form of isomorphic admixture. It should be pointed out that shifts in isotope abundances of the radiogenic (daughter) elements are characterized by deviation from their equilibrium values, and their isotope ratios are described in terms of the activity of the daughter and parent isotopes.

Chalov (1959) has shown that uranium isotope separation during transition into the liquid phase can be partly accounted for by uranium-234, which is more favorable oxidized since it is not bounded to the crystalline lattice, but this effect is not widespread. The correlation between the six-valent uranium-238 and uranium-234 isotope ratios in the mineral and the extraction, has been found by Chalov only for natural uranium oxides.

A suggestion proposed by some authors that ^{234}Th, an intermediate decay product between uranium-234 and uranium-238, can play a role in the variation of the uranium isotope composition, is unlikely to prove correct. The weak migration power of thorium does not allow for it to be leached from minerals in significant amounts, although, as a decay product, it should be preferentially released into the liquid phase compared with the parent isotope of uranium-238. In fact, the ^{234}Th/^{238}U ratio in waters of the uranium deposits and ore exposures does not exceed 0.01–0.10. In experiments on comparative ^{234}Th and ^{238}U leaching, Syromyatnikov (1961) has always obtained ^{234}Th/^{238}U$\,<1$, only in a few cases (Cherdyntsev 1955) elevated values of the ^{234}Th/^{238}U ratio have been detected in natural waters, and these were in extractions from the thorium minerals reported by Syromyatnikov (1961); but in the case of the uranium minerals a great number of other thorium isotopes, being carriers of ^{234}Th, are leached.

16.2.2 Separation of Thorium Isotopes

Let us consider now the separation of isotopes of ^{232}Th, ^{230}Th and ^{238}Th, which are of interest from the practical viewpoint. The thorium-uranium fraction of rocks, being on average 3.78 units, varies over limited ranges. At equilibrium conditions, the value of the ^{230}Th/^{232}Th ratio in activity units is equal to 0.79; but usually this ratio is much greater. Experiments concerned with the relative leaching of ionium and thorium with ferrithorite, have shown (Cherdyntsev 1955) that the relative content of ^{230}Th/^{232}Th in the extraction is 30 times as much as in the mineral. Ionium is a decay product, whereas thorium (a primordial element) is incorporated into the mineral's crystalline lattice. Therefore, an elevated ionium content, compared with thorium, should be observed during leaching, which is actually the case in the natural process.

The ^{232}Th and ^{228}Th isotope separation has some peculiarities. Radiothorium is a daughter product of ^{228}Ra, the lifetime of which is 6.7 years. The radiothorium content in natural waters is usually determined by the content of radium, being extremely mobile and an unstable element. In this case, the isotopic shift is conditioned not by the separation of the isotopes, but by the separation of the elements radium and thorium.

16.2.3 Separation of Radium Isotopes

The condition of radium accumulation in natural waters can be written as (Cherdyntsev 1969):

$$n_{Ra} = q_U\ a\left(1 - e^{-\lambda Ra t}\right),$$

where q_U is the content of the uranium parent isotope in water-bearing rocks; a is the coefficient accounting for the fraction of the daughter isotope transition in water.

In first approximation, one can consider that the coefficient a is the same for all radium isotopes. Since all radium isotopes are daughter products and their occurrence in the minerals is similar, then the ratio ^{224}Ra/^{226}Ra becomes:

$$\frac{\text{Th}}{\text{Ra}} = \frac{\text{Th}}{\text{U}}\frac{1 - e^{-\lambda_{Th} t}}{1 - e^{-\lambda_{Ra} t}}.$$

The lifetime of the ^{224}Ra isotope is 3.64 days and, practically, for all the waters the value is $1 - e^{-\lambda(\text{Ra}-224)t} = 1$. For the young waters (with ages less than the radium half-life, being equal to 1,600 years) the value is $1 - e^{-\lambda_{Ra} t} \approx -\lambda_{Ra} t$. Then, the expression given above becomes:

$$\frac{\text{Th}}{\text{Ra}} = \frac{\text{Th}}{\text{U}}\frac{1}{1 - \lambda_{Ra} t},$$

i.e., large values of this ratio should be expected for young waters. Such waters (^{224}Ra/^{226}Ra>1590) were found by Cherdyntsev (1969) as long ago as 1934; but due to the interphase isotope exchange between the water and minerals this ratio is smoothed down quickly to the normal one (Voytkevich 1961). A similar ratio can be written for the radium isotopes ^{228}Ra/^{226}Ra (for the young waters with ages of several years):

$$\frac{^{228}\text{Ra}}{\text{Ra}} = \frac{\text{Th}}{\text{U}}\frac{\lambda_{Ra-228}}{\lambda_{Ra}},$$

i.e., the ratio ^{228}Ra/^{226}Ra for these waters can reach the limiting value of 240, which really occurs in nature (Cherdyntsev 1973). The interphase isotope exchange also smoothes this ratio down to the normal one, equal to the ratio of the activities of the originators of the radioactive series. In ancient waters radium isotope ratios usually corresponds to their parent isotope ratios.

The main factors resulting in the separation of isotopes of radioactive elements in nature were first summarized by Cherdyntsev (1955):

1. The difference in the chemical properties of the elements (the shift of the ^{228}Th/^{232}Th ratio from equilibrium).
2. The bond of the daughter isotopes to the crystalline lattice is less than that of the parent isotopes (the preferential escape of ionium compared with thorium and uranium-234 compared with uranium-238).

Table 16.7 Uranium and thorium content in rocks and stone meteorites

Rock of meteorite	Content (10^{-6} g/g)		Th/U
	U	Th	
Stone meteorites (Chondrites)	0.006–0.03	0.03–0.08	4.0
Basalts	0.59	2.7	4.6
Gabbo	0.96	3.9	4.0
Igneous rocks (averaged values)	1.5	5.4	3.6
Granodiorites	2.0	7.8	3.9
Granites	3.0	13	4.3
Sedimentary rocks	3.0	13.3	4.4
Soils	2.9	9.0	3.2

3. The difference in the half-decay times: the recoil atoms of the short-lived isotopes result in a quicker saturation of natural solutions (the anomalously high ratios of $^{234}Ra/^{226}Ra$, $^{223}Ra/^{226}Ra$, $^{228}Ra/^{226}Ra$).
4. The interphase isotope exchange results in a shift of the radioisotope ratio toward radioactive equilibrium ($^{234}U/^{238}U$ ratio shifts toward the equilibrium one in a mineral deposit).

16.3 Distribution of Radiogenic Elements in Natural Waters

The content of uranium and thorium, the originators of radioactive series in natural waters, is governed by their content and distribution in rocks. Table 16.7 shows their content in the main types of rocks and stone meteorites according to (Cherdyntsev 1969).

Among all radioisotopes of the uranium-thorium series, the most comprehensive evidence of their occurrence in natural waters is available for uranium and radium; somewhat less common are the data concerned with thorium-232, uranium-234, and radon occurrence; an insignificant amount of data is concerned with the isotopes of thorium ^{230}Th and ^{228}Th. There are only a few data on the content of protactinium, mesothorium, polonium, and radioactive plumbum-210. All these are conditioned by considerable methodological difficulties related to the extraction and detection of small amounts of heavy radioelements.

In order to determine the content of an isotope, it should be concentrated out of a large volume of water (of an order of hundreds of liters), and normally in field conditions; only for the determination of the uranium content just 1 l of water is required. A large amount of the available data on uranium-238 content in the hydrosphere was determined in this way.

16.3.1 Uranium Isotopes in Natural Waters

According to Germanov (Vinogradov 1963) the average value of uranium content in atmospheric precipitation is (2–3) $\times 10^{-8}$ g/l. The main carriers of uranium in atmospheric precipitation are dust particles. Therefore, the uranium content in precipitation depends markedly on the regional climatic conditions. Rain waters in arid regions, where a great amount of dust is injected into atmosphere, has a great content of uranium ($n \times 10^{-8} - 2 \times 10^{-6}$ g/l). At present the artificial radioisotopes formed during the thermonuclear tests are present in atmosphere. The uranium concentration reached 73.3×10^{-6} g/g in atmospheric fallout near Vilnius (Stiro et al. 1970). The correlation of high uranium content with that of ^{239}Pu, which is a typical technogenic isotope, in atmospheric precipitation indicates the technogenic origin of uranium. High uranium concentrations are observed in the winter months. The uranium-228 falls in the summer are of local character and are conditioned by dust transport from nearby regions. The uranium content in fallout decreases during this season.

The content of dissolved uranium in rivers (according to Germanov) ranges from 3×10^{-8} to $n \times 10^{-5}$ g/l with the average value of 1×10^{-6} g/l. The climatic conditions have a great effect upon the concentration of uranium. The general mineralization and uranium content in river in arid regions is greater than that of humid regions. The uranium content increases downstream for some rivers (Dniepr, Don, Volga, Syr-Darya, Amu-Darya).

The occurrence of uranium deposits results in an increase of its content in waters of some small rivers and springs up to $n \times 10^{-4}$ g/l. The problem of uranium migration in rivers was studied in details by Baturin and Kochenow (1969) and Baturin (1968). They have carried out a complex study on the uranium content in dissolved and suspended forms. According to Baturin's data the uranium content in the deltaic deposits of 12 rivers in the former USSR is in the range of (0.4–3) $\times 10^{-6}$ g/g. The maximum values are typical for suspensions of the Syr-Darya, Don, and Volga rivers and the minimum values for those of the Severnaya Dvina and Amu-Darya. The average value, determined in relation to the annual solid runoff, is 1.05×10^{-6} g/g, which is less than half of the average uranium content in the Earth's crust. Baturin has found two general principles for the basins of the Black and Caspian seas: (1) the direct dependency of the relative amount of suspended uranium on sediment discharge (the annual average amount of material from a unit area of the watershed) and considerable less evident but direct dependency on the fraction of mechanical denudation; (2) the inverse relationship between average uranium content in water and water discharge (the average annual amount of water from a unit area of the watershed).

A correlation between the content of the dissolved uranium and the total mineralization of the surface waters is also observed. The estimations carried out by Baturin and Kochenov (1969) have shown that the total amount of dissolved and suspended uranium transported by rivers is in the ratio of 1:1, with variations from 24:1 to 1:9 for an average content of dissolved uranium in world river runoff of

(0.50–0.55) × 10^{-6} g/l. The above given data concerning the world river runoff were obtained by the authors on the basis of generalizations of data obtained for 16 rivers in the former USSR.

The most comprehensive data on the uranium content in river discharge were reported by Sackett et al. (1973). According to their data, the average concentrations of uranium for the 10 deepest rivers, giving 40% of the total world runoff into the ocean, are in good agreement with the data obtained by Baturin and Kochenov (1969).

The uranium content in the oceanic water is close to 3 × 10^{-6} g/l (Vinogradov 1967b). Based on the experimental data, Starik and Kolyadin (1957) have shown that in the ocean waters uranium occurs in the ionic-dispersion state in the form of the firm uranyl-carbonate complex $[UO_2(CO_3)_3]^{4-}$. Only at pH > 7.5 and Eh < 0.1 V, which are not typical for oceanic waters, does uranium occur in the form of hydrolysis products which can be absorbed by colloids and larger suspended particles and removed to the floor. The uranium content is somewhat lower in waters of seas fed by rivers flowing in humid regions (for the Baltic Sea (0.8–2.2) × 10^{-6} g/l; Baturin 1968), and uranium content increases in the waters of closed reservoirs (for the Aral Sea (30–50) × 10^{-6} g/l, for the Caspian Sea 10 × 10^{-6} g/l; Kochenov and Baturin 1967). The distribution of uranium over a basin is significantly homogeneous both with depth and area and only in the near bottom layers of the Black Sea does the uranium content markedly decrease down to $n \times 10^{-7}$ g/l (Baturin et al. 1966). They have accounted for the decrease of the uranium content by the processes of absorption of uranium from water during contact with sediments in highly reducible media. One of the main factors, in their opinion, is the occurrence of organic substances which can capture uranium during sedimentation from the water thickness contaminated with hydrogen sulfide. The uranium content in suspension in the Indian Ocean (Kuznetsov et al. 1967) ranges from 0.1 × 10^{-6} to 2 × 10^{-6} g/g, but the amount of uranium in suspension contained in 1 l of water is (0.1–1.2) × 10^{-6} g/l, i.e., more than 99.9% of uranium in the oceanic water is in the dissolved form.

The uranium enters the sea and oceans together with suspended and dissolved portions of river runoff. At present the mechanism of precipitation of dissolved uranium on the oceanic floor is not completely understood. To a great extent this is a result of the inadequacy of experimental material. A correlation exists between the content of organic substances and uranium in precipitation. In Baturin's opinion, the major portion of the dissolved uranium precipitates after the suspended material over shelves and continental slopes. The sedimentation of uranium occurs by its extraction with the organic components of sediments. Since the fractions of dissolved and suspended uranium are on average equal in river runoff, and the total content of uranium in the suspended and dissolved material corresponds to its content in the rocks of the Earth's crust, then in modern sediments there occurs a regeneration of the uranium content which finally approximates to its content in rocks of the Earth's crust.

The rate of precipitation of dissolved uranium on the floor can be estimated by its residence time in a basin, which is usually determined as the quotient of the element's total amount in a basin divided by its average annual input as a component of river runoff (see Table 16.8).

Table 16.8 Residence time of dissolved uranium in the water of some seas and the oceans as a whole. (From Baturin and Kochenov 1969; Nikolaev et al. 1966)

Basin	Average depth (m)	Dissolved uranium content (kg)	Annually recharged dissolved uranium (kg)	Residence time of dissolved uranium in water (years)
Oceans	3,800	4×10^{12}	$(17–20) \times 10^6$	$(200–300) \times 10^3$
Black Sea	1,200	$(1-1.65) \times 10^9$	$2 \times 63 \times 10^5$	4,000
Caspian Sea	188	4×10^8	3×10^5	1,300
Baltic Sea	54	3×10^7	2×10^5	150
Aral Sea	16	4×10^7	$(3–4) \times 10^5$	100–130
Azov Sea	6,8	7×10^5	4.85×10^4	14.5

It follows from Table 16.9 that the uranium residence time in a basin is closely related to the basin's depth. A number of additional sources of uranium have not been accounted for in the estimations (groundwater discharge, atmospheric precipitation, aeolian deposits due to the coastal erosion).

A question of ultimate importance in nuclear geochronology is that of the constancy of the uranium content in the ocean in the Tertiary, since practically all the methods of the nuclear geochronology of the sea sediments, which use the elements of the uranium-thorium series, are based on it. Indirect data can be obtained on the basis of the uranium content in the sea sediments, since the age of the latter is considerably greater than the uranium residence time in the world ocean. A number of data are in agreement with the assumption of its constancy in the past (the approximate constancy of the uranium to carbon ratio in the samples of the modern and Paleozoic shales, the constancy of the uranium distribution in the vertical columns of the sea and ocean sediments). However, conflicting evidence also exists. The ancient molluscs contain more uranium than their modern counterparts. The ^{230}Th content should decrease with age down the section of a column, described by smooth exponential curves, but peaks are found in the ionium curve. In a number of the sea ooze samples an excess of ^{230}Th was found, which does not result from the uranium content in the overlying column of sea water. The problem of the uranium content in the ocean in the past and present is as jet far from being solved.

The uranium content in lakes, according to Germanov (Vinogradon 1963), is largely determined by climatic zonality, and varies from 3×10^{-8} to $n \times 10^{-4}$ g/l with an average value of about $\sim 1 \times 10^{-6}$ g/l. The lowest concentrations of uranium were found in the high mountain regions and in the lakes of the northern latitudes. The highest uranium concentrations are observed in the shallow lakes of the arid steppe zones, in the regions where the concentration of uranium is higher. The balance of uranium has been estimated for two lakes in mid-Asia. The uranium residence time in the Balkhash Lake is 72 years and that in the Issyk-Kul Lake is 5,400 years (Baturin and Kochenov 1969).

The uranium content varies the most in groundwaters. This is accounted for by the difference in the hydrodynamical regimes and differences in the hydrogeochemical situations which condition the transition of the radioactive elements in water. Usually most researchers distinguish three hydrodynamical zones over the range of each artesian basin: those of the intensive, reduced, and very reduced water exchange.

Table 16.9 Uranium, radium, and radon content in some groundwaters. (From Tokarev and Shcherbakov 1956)

Water-bearing rocks	Zone of water exchange	Uranium (g/l)			Radium (g/l)			Radon (10^{-10} Ci/l)		
		Max	Min	Average	Max	Min	Average	Max	Min	Average
Sedimentary	Intensive	2×10^{-7}	8×10^{-6}	5×10^{-6}	1×10^{-12}	6×10^{-12}	2×10^{-12}	1	50	15
	Reduced	2×10^{-8}	6×10^{-6}	2×10^{-7}	1×10^{-11}	1×10^{-8}	3×10^{-10}	1	20	6
Acid magmatic	Intensive (crust of weathering)	2×10^{-7}	3×10^{-5}	7×10^{-6}	1×10^{-12}	7×10^{-12}	2×10^{-12}	10	400	100
	Reduced	2×10^{-7}	8×10^{-6}	4×10^{-6}	2×10^{-12}	9×10^{-12}	4×10^{-12}	8	400	100
Uranium deposit	Intensive (oxidation zone water)	5×10^{-5}	9×10^{-2}	6×10^{-4}	8×10^{-12}	2×10^{-9}	8×10^{-11}	50	50,000	1,000
	Reduced (reduction zone water)	2×10^{-6}	3×10^{-5}	8×10^{-6}	1×10^{-11}	8×10^{-10}	6×10^{-11}	50	3,000	500

The waters with dissolved oxygen, possessing high oxidation-reduction potential, occur in the zone of intensive water exchange. In this zone, the four-valent uranium is oxidized to the six-valent state with transition into solution. In the zones of reduced and very reduced water exchange, characterized by reducing conditions, the waters have no oxygen but are enriched with hydrogen sulfide and organic substances. The waters of these zones contain uranium in small amounts. Table 16.9 (Tokarev and Shcherbakov 1956) indicates the average, minimum, and maximum uranium contents in waters in different water-bearing rocks and in waters of the uranium deposits.

The hydrochemical factors have little effect upon the uranium isotope composition in natural waters, which is determined markedly by the form of the uranium distribution in rocks of the water-bearing complex. At present there are several hundreds of determinations of the uranium isotope ratio $\gamma = {}^{234}U/{}^{238}U$ in natural waters. Despite the fact that the obtained data do not throw light upon all the types of natural waters, some general conclusions can already be drawn. Many works are devoted to the study of the uranium isotopic content in atmospheric precipitation. The precipitation, related to intensive circulation of the marine air masses, falls over the Baltic region in the cold period of the year. The uranium isotope ratio during this period ranges from 1.08 to 1.18 (for sea water the average value is 1.15). In the summer months, the isotope ratio is close to that typical of rocks in the Earth's crust (0.89–1.07) and indicates the continental origin of the uranium isotopes in the atmospheric fall-out. According to Cherdyntsev (1969), the uranium isotope ratio in the rivers is 1.1–1.4 (45 determinations) and does not change with time. In salty rivers in the United States it is greater (Thutber 1965), being equal to 1.3–2.0, and in the mountainous river Williams-Green it even reaches 6.35.

In river deposits in the United States(Thurber 1965), in some cases $\gamma < 1$, which is accounted for by the greater mobility of the uranium-234 atoms during weathering. There are also cases when $\gamma > 1$. Such sediments contain more organic substances. It is possible that, as in the marine sedimentation cycle, sedimentation of the dissolved uranium also occurs in river waters with organic substances. Due to the low sensitivity of the apparatus used, Cherdyntsev (1955) could not find the uranium isotope shift in ocean waters. Thurber was the first to discover this shift and to find that $\gamma = 1.15 \pm 0.5$ (Thurber 1962, 1963). Later on, the value of the uranium isotope shift was defined more precisely by many researchers. Its most probable value for the world oceans is 1.15 ± 0.1 (Cherdyntsev 1969). According to the data of Cherdyntsev, the value of the uranium isotope shift for water of the Red Sea is 1.18 ± 0.1 and 1.2 ± 0.08 for waters of the Azov Sea. In large river mouth regions the uranium isotope ratio can differ considerably from the mean; so, in the Black Sea water near the Bzib River mouth, it is equal to 1.02 ± 0.1 and in the central parts of the sea 1.17 ± 0.1.

Detailed studies of the average uranium concentrations in the open regions of oceans and sediments were carried out by Ku et al. (1977). By measurements of more than 100 samples they found that uranium contents vary from $(3.27 \pm 0.05) \times 10^{-6}$ in Antarctic to $(3.43 \pm 0.04) \times 10^{-6}$ in Arctic. The values of ${}^{234}U/{}^{238}U$ in sediments changed within 0.78–1.07 at the average value of 0.93. In Cherdyntsev's opinion (1969), about 25% of the uranium is leached by sea water from the sediments, leading to a decrease in the uranium isotope ratio in the sea sediments, i.e., the dissolved

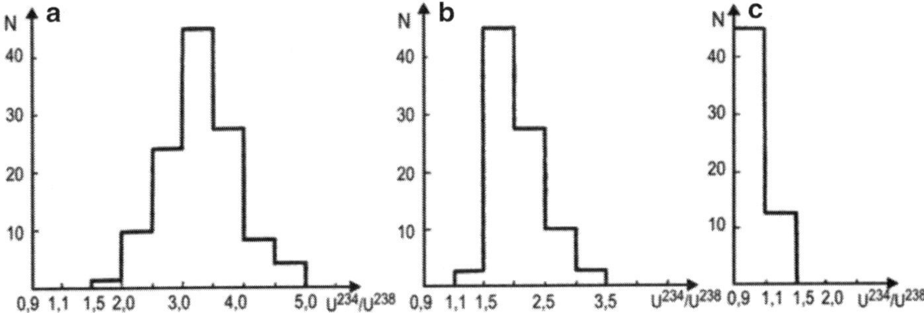

Fig. 16.4 The value of $^{234}U/^{238}U$ ratios in waters from igneous rocks (**a**), sedimentary and metamorphic rocks (**b**), and from uranium deposits (**c**)

uranium does not precipitate to the bottom but is even extracted from the deep sea deposits. With the average uranium content in the ooze being $(2 - 3) \times 10^{-4}$ g/g, the extraction from deposits of 25% of its content releases into water ~1 mg of uranium per 1 cm of sea floor sediment. This corresponds to uranium content in the sea water column several kilometers deep. In summary, the problem of uranium transference in the ocean is far from being solved at present. According to the data of Rona et al. (1965), the value of γ in the sea ooze is 1.16–1.17, i.e., approximately the same as in the sea water with a mean of 1.08 and does not fall below the equilibrium value.

The uranium ratio in lake waters ranges on the whole between the same limits as for rivers, being equal to 1.1–1.4. This is because the sources are the same for both rivers and lakes. The total number of uranium ratio determinations does not exceed several dozen for lake waters. For the Sevan Lake $\gamma = 1.72$. The total amount of determinations of uranium isotopic composition in lake waters does not exceeds several tenths.

The uranium isotope ratio in groundwaters varies greatly depending upon the type of water-bearing rocks. The uranium isotope composition in the Kazakhstan groundwaters has been very intensively studied by Syromyatnikov (1961). The main features of uranium migration were ascertained for waters of the igneous, sedimentary, and metamorphic rocks and for waters of the uranium deposits. Figure 16.4 shows the histograms of the uranium distribution in the above-mentioned types of rocks.

In the igneous rocks 117 wells were studied, where the uranium isotope ratio varied from 2 to 5.5 (80% of the values lie in the range of 2.5–4.0). In waters of the sedimentary and metamorphic rocks from the data for 82 wells, 85% of the values fall in the range of 1.5–2.5. The uranium content in the waters of both types of rocks has varied greatly. The hydrochemical characteristics have also varied over wide ranges but the isotope composition has not varied greatly. In waters of the hydrotherms and sedimentary uranium deposits, uranium is practically at its equilibrium value. At the same time, in the deposits characterized by dispersed uranium distribution, ~90% of the values of the isotope ratio range from 2.5 to 7.0. In waters of the three uranium deposits no dependency has been found between the uranium isotope ratio and the hydrochemical parameters.

Cherdyntsev (1969) reported a summary of the uranium isotope content in some types of waters. In peat soil waters the value of γ varies from 1.12 to 1.78 (12 determinations) and in waters of thermal and carbonic acid springs γ varies from 1.00 to 3.24 (17 determinations). The thermal waters of an active volcanic zone (Kamchatka, Kurile Islands) indicate, on an average of 12 determinations, a uranium isotope ratio of 1.3 (maximum 2.27 and minimum 0.97; Kuptsov and Cherdyntsev 1969).

16.3.2 Thorium Isotopes in Natural Waters

For convenience of interpretation, the thorium isotope content is sometimes given in the form of the ratios ^{232}Th/^{238}U, ^{230}Th/^{232}Th, ^{228}Th/^{232}Th. There are not much data on these isotope ratios because of the technical difficulties involved in thorium extraction and the large amounts of water required for measurements, due to its small content in waters. Thorium is mainly transferred by water in the form of the clastic products of destruction and in a colloid-suspended form. Therefore, the results of measurements are essentially dependent upon the conditions of its extraction from natural waters.

According to the data of Cherdyntsev (1969), the ^{228}Th/^{232}Th ratio in the rivers of the former USSR ranges on the whole from 0.8 to 1.8. Considerable deviations from the average thorium/uranium ratio are observed for rocks of the Earth's crust (0.8 in activity units). An even greater value of this shift has been detected in rivers in the United States (up to 15).The ^{228}Th/^{232}Th ratio in 50% of the studied rivers in the former USSR does not deviate from equilibrium. In all the other cases an excess of ^{228}Th (up to 2) is observed. This ratio is markedly different from the equilibrium value (4.6 and 6.8) for the two studied rivers in the United States. The thorium-uranium ratio varies over sufficiently large ranges (0.01–9.2), although it is small for the majority of rivers (in 50% cases it is less than 0.1). In two cases out of the 27 studied rivers the ratio is greater than at equilibrium (> 1.5 and 9.2).

The results of the thorium content measurements in waters of the Sea of Azov have been reported in the work by Nikolaev et al. (1961). The thorium concentrations range widely between 4×10^{-9} and 219×10^{-9} g/l. These variations are related to the existence of large amount of suspended material in the Sea of Azov. Waters sampled from its central part in calm conditions contain 4×10^{-9} g/l of thorium, which is a feature of the open seas. The maximum thorium content in water sampled from the middle part of the Taganrog Bay, replenished with waters from The Don River, ranges from 218×10^{-9} to 40×10^{-9} g/l. The thorium concentration in the Black Sea is in the range $(2.2–0.2) \times 10^{-9}$ g/l (and remains constant over the studied region; Starik et al. 1959). According to the data reported by Higashi (1959), the thorium content in the Pacific Ocean waters reaches $(2.6–9.1) \times 10^{-9}$ g/l but in filtered water thorium was not detected. Assuming, as Vinogradov (1967) has shown that the average thorium content in ocean water is $n \times 10^{-9}$ g/l and that of uranium is 3×10^{-6} g/l, then in the ocean water the ratio Th/U ≈ 0.003, i.e., lower than in the Earth's crust by 3 orders of magnitude. In summary, the thorium content in water ranges very widely. More accurate measurements of the thorium-232 content in surface ocean waters, carried out by Kaufman (1969), gave a value of 0.07×10^{-9} g/l.

The ^{230}Th content in the open waters of the Black Sea is $(2.5-4.2) \times 10^{-12}$ g/l, which increases to 40×10^{-13} g/l near the shores in the deltaic regions (Starik et al. 1959). In the Pacific Ocean its content is in the range of $(6–40) \times 10^{-13}$ g/l in surface waters and is $\sim 2 \times 10^{-13}$ g/l in deep waters (Higashi 1959). Assuming the average ^{230}Th content is 5×10^{-13} g/l, then the ocean contains only 1% of the amount required for equilibrium with uranium. The mechanism of thorium isotope removal to the bottom has been studied in detail by Kuznetsov (1962) on the basis of an investigation of the distribution of radioelements in the ocean water suspensions-bottom sediments system. The suspensions have been considered to be ocean sediments at the initial stage of their formation. The major portion of thorium is contributed to suspensions as a component of terrigeneous material. In addition, absorption extraction of thorium from sea water is also possible. The main processes of ^{230}Th concentration in suspensions are those of biochemical extraction and sorption to particles constituting the suspension.

The geochemical balance of ^{230}Th and ^{232}Th in the basins of the Black Sea and the Sea of Azov has been considered by Nikolaev et al. (1966). The residence time of thorium in the Black Sea is 56.6 years, i.e., 66 times less than that in the world ocean. The residence time of ^{230}Th in the Black Sea is 365 years. The authors have accounted for the difference in the residence times of ^{230}Th and ^{232}Th in terms of the different forms of their contribution and occurrence both in the river and sea waters, and in the bottom sediments. ^{232}Th is considered to be contained in the terrigeneous component of the river runoff and ^{230}Th in the colloidal form. In the Sea of Azov the ^{232}Th residence time is 160 days and the ^{230}Th 172 days, which is explained by Nikolaev et al. in terms of the shallowness of the basin.

The ratio of ^{228}Th/^{232}Th in waters of the Sea of Azov, according to the data by Nikolaev et al. (1961), is 1.43. In waters of the Pacific Ocean this ratio varies between 10 and 25. It is assumed that the ^{228}Th excess occurs due to ^{228}Ra release during the decay of ^{232}Th, contained in the solid phase (in the ocean sediments and terrigeneous material; Bernat and Goldberg 1969). The cause of deviations is the content of ^{234}Th from equilibrium with ^{238}U (up to 2.8 in activity units) in the upper mixed ocean layers has been considered by Bhat et al. (1979). The deviations from the equilibrium are explained by the mixing of the upper ocean layers and biological activity.

A vast summery of data on the thorium isotope content in groundwater is reported in the work by Syromyatnikov (1961). Usually, the thorium-uranium ratio for groundwaters is considerably lower than that in river and lake waters. Radioelements do not typically occur in the colloidal form in groundwaters. Sixty water samples from igneous rocks were studied along with 40 from sedimentary rocks and 35 from uranium sediments. The thorium-uranium ratio is less than 0.09 for 58% of water from the igneous rocks, 77% from sedimentary rocks, and 97% from the uranium ores. In some cases higher values of the thorium-uranium ratio are also indicated. They are likely to occur in waters in contact with those rocks which have a high thorium-uranium ratio, but it has been shown (Titaeva et al. 1973) that the value of this ratio falls at a distance of several dozens of meters from the point of their emergence to the surface, due to decreases of the thorium concentration.

The value of the $^{230}Th/^{232}Th$ ratio for groundwaters in general is markedly greater than that typical of the rocks constituting the Earth's crust (~0.8 in the activity units) due to greater ^{230}Th leaching compared with thorium-232. In 95–100% of the studied samples this ratio ranges from < 1 to 4.5. In some cases of rocks with a low thorium-uranium ratio the value of the $^{230}Th/^{232}Th$ is > 15 (and sometimes it reaches 100). The waters of peats also exhibit a raised $^{230}Th/^{232}Th$ ratio (Cherdynysev 1969), for which a low thorium-uranium ratio is characteristic. The $^{228}Th/^{232}Th$ ratio for groundwaters is practically always > 1. The majority of waters of igneous rocks exhibit a value of this ratio ranging from 4 to 10 and the waters of sedimentary and metamorphic rocks have a ratio ranging from 1.5 to 4.5 (Syromyatnikov 1961).

The protactinium content in natural waters has not been determined in practice. This is accounted for by the fact that the analytical chemistry of the extraction of protactinium from natural objects has only been developed in recent years for dating sea sediments, therefore, only data on the protactinium content in sea water exists. The content of protactinium in sea water amounts to 3% of its equilibrium amount with uranium (Sackett 1963); in the surface ocean water it is equal to 5%, at a depth of 500 m it is 7%, and of 4000 m it is 5% of the equilibrium (Kuznetsov et al. 1966). According to Kuznetsov's data the correlation coefficient between the content of protactinium and ^{230}Th in the sea sediments is equal to 0.95. Both elements are extracted from sea water by sorption, carried out by poorly soluble particles having highly absorptive surfaces. Protactinium is formed from uranium in sea water and in partly carried there by river runoff.

There are also few data on the actinium content in natural waters. In the Pacific Ocean it amounts to $\sim < 2 \times 10^{-15}$ g/l, and in the Atlantic Ocean it is $\sim 7 \times 10^{-16}$ g/l (when in equilibrium with uranium its content should be 6.4×10^{-12} g/l; Sackett 1963). It is transferred from the ocean water into sediments as readily as ^{230}Th and ^{232}Th. The $^{227}Ac/^{235}U$ ratios in water of three springs in the Caucasus neovolcanic region are markedly less than unity (Kuptsov and Chedyntsev 1969). In the Kamchatka thermal waters the ratios range from 0.003 to 12.9.

There are few data on the radium content in river water (they are mainly related to North American rivers). The average radium content is $(0.03-1) \times 10^{-12}$ g/l, although in some cases it reaches considerably greater values, for example $(1-3) \times 10^{-12}$ g/l in the Mississippi river. The radium content in the Thames (England) is $(0.01-1) \times 10^{-12}$ g/l. In the rivers of Germany the radium content ranges over $(0.07-0.8) \times 10^{-12}$ g/l. Khristianov and Korchuganov (1971) have also reported data on the content of radium in four rivers within the area of the Russian platform (in g/l): the Volga river 0.72×10^{-12}, the Ruza river 0.16×10^{-12}, the Moscow river 0.04×10^{-12}, the Oka river 0.19×10^{-12}. In Vinogradov's (1967) opinion, the average value of radium content in rivers is 0.1×10^{-12} g/l.

The radium content in the Black Sea amounts to $n \times 10^{-13}$ g/l (Grashchenko et al. 1960), ~3.7×10^{-14} g/l in the surface ocean waters (Broecker et al. 1967), and 1.35×10^{-13} g/l (Blanchard and Oakes 1970).

Detailed studies of the uranium content have been carried out (Broecker et al. 1976; Chung 1974b, 1976; Chung et al. 1974; Ku and Lin 1976; Chan 1976; Sarmiento and Feely 1976; Li et al. 1973, 1977). These studies have thrown light upon the main principles of radium distribution in the surface and deep waters of all the oceans. The main factors governing the radium content are water circulation and the release of radium from bottom sediments. Radium is contributed to the surface oceanic waters from coastal sediments and to the near-bottom waters from bottom sediments. Based on the two-reservoir model (the warm surface waters and the cold deep ones), Li et al. (1973) have evaluated the radium flux from the coastal sediments, which amounted to $17 \times 10^{3} 1 \times 10^{-12}$ g/years, and flux from the deep bottom sediments, which was 43×10^{-3} g/years. These processes result in a natural increase of the radium concentrations with depth, which is most noticeable for the Pacific Ocean waters, where the water circulation is retarded compared with the other oceans. In the Arctic and Antarctic waters, where deep waters increase up to 9.3×10^{-14} g/l (Ku and Lin 1976), but concentrations common for surface waters also occur: $(4.2-7.4) \times 10^{-14}$ g/l (Chung 1974). In the surface waters of the Atlantic Ocean the radium content varies over a small range of $(3.0–3.9) \times 10^{-14}$ g/l with an average value of 3.5×10^{-14} g/l. In the near-bottom waters the radium concentration varies over a greater range from 4.4×10^{-14} g/l to 11.2×10^{-14} g/l (Ku and Lin 1976).

There are a number of works in which the behavior of the short-lived radium isotope^{228}Ra in the oceans has been considered (Moore 1969; Kaufman et al. 1973; Knauss et al. 1978). As well as its long-lived isotope, ^{228}Ra is released into surface waters by diffusion from the continental shelf sediments; it is distributed over the whole ocean. During the decay of the ^{228}Ra the thorium isotope ^{228}Th is formed. Biochemical processes in the surface waters of the ocean result in the separation of the ^{228}Ra and ^{228}Th isotopes. The following relationship always holds for surface waters: $^{228}\text{Ra} > {}^{228}\text{Th} > {}^{232}\text{Th}$. The radium content in different type of natural waters and in waters of uranium deposits is indicated in Table 16.9. In natural waters the ^{226}Ra/^{238}U ratio (in activity units) is usually less than unity, although waters do occur in which radium is excess compared with uranium.

Extensive studies of the ^{228}Ra content in different types of natural waters have been carried out by Syromyatnikov (1961). The majority of samples of rock-waters have a ^{228}Ra/^{226}Ra ratio in the range 2–5. Waters of the highly destroyed rocks and terrigeneous sediments are usually enriched with ^{228}Ra isotopes. The value of the ratio $^{228}\text{Ra}/^{226}\text{Ra} < 2$ is usually typical of waters surrounding uranium deposits. Very high values of the ratio ^{228}Ra/^{226}Ra (up to 380) were found in thermal waters of an active volcanic zone (Kuptsov and Cherdyntsev 1968). A value of the ^{228}Ra/^{226}Ra ratio equal to 240, at a thorium-uranium ratio of 3.7, should be expected by the contribution of radium to water by the nuclear recoil mechanism.

Radioactive emanations are released into the atmosphere from the Earth's crust. Their decay products are absorbed by aerosol particles of natural origin and returned

Table 16.10 Concentration of radon in various regions of the Earth's surface. (From Junge 1963)

Region	Concentration of Ra (Ci/cm^3)
Emanation flux, Ci/cm^2 s	$(0.1-25) \times 10^{-17}$ (average 4×10^{-17})
Soils	$(0,5-10,000) \times 10^{-13}$ (average 3×10^{-13})
Near the Earth's surface:	
Continental regions	$(70-330) \times 10^{-18}$
Continental regions with disturbed soil structure	$(400-600) \times 10^{-18}$
South America	$(20-70) \times 10^{-18}$
Antarctica	$(0.2-2.0) \times 10^{-18}$
Oceans	$(0.5-3.0) \times 10^{-18}$
Tn/Rn ratio near the Earth's surface	0.01–0.1

to the Earth together with precipitation. The horizontal redistribution of radon relative to its sources in the terrestrial crust occurs in the ocean and atmosphere (see Table 16.10).

The flux of radon emanation depends on the state of the soil. During rain a decrease in its output of 70% is observed. The temperature and humidity of the soil do not have any effect upon this process. The concentration of radon in the atmospheric layer nearest the Earth depends upon the flux of emanation and turbulent mixing in the overlaying layers. The concentration of radon over inland regions is approximately twice as much as over the ocean, i.e., radon is formed almost completely over the land. The radon concentration in precipitation is about 2×10^{-11} Ci/cm^3 on average (the minimum value is 0.1 and the maximum value is 10×10^{-11} Ci/cm^3).

Radon content in river waters has not been practically determined. In water supply sources the content of radon is about 100 times greater than in water at equilibrium with radium. In surface continental waters the amount of radon is 10^4 times as much (Khristianov and Korchuganov 1971). In summary, the concentration of radon in rivers recharged by groundwaters is one order of magnitude greater than in equilibrium with radium. The saturation of groundwaters with radon is conditioned by its high solubility. The major source of radon in groundwaters is that emerging from rocks during emanation. The other sources of emanation (emanation from the atmosphere together with rain, dissolution from the atmosphere, contribution from bottom sediments, from suspensions contained in water, formation from dissolved radium) are relatively unimportant although in a number of cases they should be accounted for. The Rn/Ra ratio (Khristianov and Korchuganov 1971) for the Volga River (Eltsi village) is 96, for the Ruza River (Ruza town) is 31, for the Moskva River (Zvenigorod town) is 470, and for the Oka River (Shchurovo settlement) is 42.

In the surface ocean layer, 150 m in depth, the radon concentration gradually increases up to the value relating to equilibrium with radium (Broecker and Kaufman 1970). Table 16.11 indicates the concentrations of radium and radon in the surface ocean waters. The principles of deviations in radium concentration are conditioned by two main factors: the rate of the gas exchange between the ocean and the atmosphere, and the rate of vertical mixing. In near-bottom waters the radium concentration decreases with increasing distance from the ocean floor.

Table 16.11 Radium and radon concentrations in the upper (150 m) ocean layer. (From Broecker and Kaufman 1970)

Depth	^{222}Rn (dpm/100 kg)	^{226}Ra (dpm/100 kg)
1	5.3	10.1
1	6.0	–
1	5.3	–
50	7.8	–
60	7.5	–
75	–	9.6
150	–	–

Table 16.12 Variation of excess radon concentration with distance from the sea bottom. (From Chung 1974a)

Distance from the bottom (m)	^{222}Rn (dpm/100 kg)	^{226}Ra (dpm/100 kg)
100	35.3 ± 0.4	10.8 ± 0.8
80	36.6 ± 0.4	13.6 ± 0.9
60	36.0 ± 0.7	19.3 ± 1.1
50	35.6 ± 0.5	20.7 ± 0.9
40	35.0 ± 0.7	23.1 ± 1.0
30	34.3 ± 0.7	20.0 ± 0.9
20	35.6 ± 0.6	30.4 ± 1.2

Bottom sediments with high radium content in the upper layers are important sources of radon, the concentration of which decreases with increasing distance from the sea-floor due to the process of diffusion and mixing (Chung 1974). In Table 16.12, ^{226}Ra and ^{222}Rn concentrations in the near-bottom waters are given on the basis of Chung's data.

The flux of radon from the oceans into the atmosphere is estimated by its relationship with the radium concentration in the surface waters. This flux is rather small, ranging from 11 ± 1 to 260 ± 32 atom/m^2 s (Hoang and Servant 1972). According to these authors, the Rn/Ra activity ratios usually amount to 0.2–0.8 in the surface waters.

Radon is released into ground waters during the emanation of rocks of the water-bearing complex. Diffusion of radon atoms in a solid body is negligible, therefore, it mainly releases when radium is located on the surface of the rock (absorbed on the rock). Emanating collectors are thus formed and the radon content in waters can reach large values. In the case of strongly fractured rocks radon escapes from capillaries located deep inside the rocks. The main factors affecting the emanating power are (Tokarev and Shcherbakov 1956): the degree to which the rock is fractured, the temperature (with increasing temperature the emanation power greatly increases), the humidity of the rocks (with increasing humidity emanation drops), and the air pressure (with the increase in pressure emanation decreases).

Despite large variations of the emanation power for individual rock samples (0.01–100%) the average emanation power in acid magmatic rock ranges from 15 to 30%. The average for sedimentary and metamorphic rocks is 10–25%. The higher values of emanating power in ore deposit formations (32–91%). The radon content in groundwaters is indicated in Table 16.9 (Tokarev and Shcherbakov 1956).

In some rocks with a low content of the radioelements a lower radon content can occur. Thus in the neovolcanic rocks of the Caucasus the average uranium content amounts to 3.5×10^{-7} g/g (Cherdyntsev et al. 1968), i.e., almost an order of magnitude lower than the average value typical of the rocks of the Earth's crust. The average content of radon in thermal, carbonic acid, and fresh waters is extremely low. In the eight water samples studied, it is equal to $(0.08) \times 10^{-10}$ Ci/l (with a maximum value of 0.29×10^{-10} Ci/l and a minimum value of 0.01×10^{-10} Ci/l). An average value of $\sim 1 \times 10^{-10}$ Ci/l for the uranium content is observed for the studied products of the active Kamchatka volcanism and the Kurile Islands (Kuptsov and Cherdyntsev 1969); but the existence of the hot rocks, having high emanating power, and liquid magma hearths lead to the raised radon contents (Kuptsov and Cherdyntsev 1968). The average content of radon for 11 fumarol waters of Kamchatka and Kurile Islands is 4.45×10^{-10} Ci/l (with a minimum value of 0.05 and a maximum of 10.8×10^{-10} Ci/l; Kuptsov and Cherdyntsev 1968). Considerably higher radon concentrations occur in the regions of active volcanism. Thus in the thermal springs of the Japanese volcanic regions radon concentrations vary over the wide range $(0.3 - 1{,}200) \times 10^{-10}$ Ci/l, in New Zealand $(8 - 3{,}200) \times 10^{-10}$ Ci/l, in Iceland $(3 - 2{,}000) \times 10^{-10}$ Ci/l, and in Kamchatka up to $1{,}000 \times 10^{-10}$ Ci/l (Chirkov 1971). The raised radon content can be accounted for by the acid magma hearths being sources of deep emanations. With decreases in temperature the radon content usually drops. The effect of deep emanations is likely to be manifested less in this case.

In the soil air Th/Rn ratios (in equilibrium units) are practically coincidental with the thorium-uranium ratios for water-bearing rocks, since the emanating powers of radon and thoron by rock minerals are approximately equal and the circulation rate of ground air and gaseous currents is not very great. For the eight gas samples from the active volcanic zones of Kamchatka and Kurile Islands the Th/Rn ratio is low, in all cases being below the sensitivity threshold (the lowest value is less than 0.05). In some volcanic gases of Japan the Th/Rn ratio is elevated to about 10 and sometimes even 370, which may account for the formation of the intermediate collectors of ^{228}Ra–^{228}Ra with a high ^{224}Ra/Ra ratio (Kuptsov and Cherdyntsev 1968).

Radioactive lead ^{210}Pb is formed from radon. The ^{210}Pb concentration in the atmospheric layer near the Earth in England, during the winter months, is about 9×10^{-15} Ci/kg of air (Peirson et al. 1966). In rain water its concentration from 1961 to 1964 at different points on the globe has ranged over $(0.2–9.7) \times 10^{-12}$ Ci/l. In the summer months the ^{210}Pb concentration in the near-earth layers of India increase to about 40×10^{-15} Ci/kg of air (Joshi and Mahadevan 1967). In the sea water near Cape Town the ^{210}Pb concentration on average is about 38×10^{-15} Ci/l and varies from 0 to 135×10^{-15} Ci/l. No variation of ^{210}Pb concentration up to a depth of 600 m has been found (Shannou et al. 1970). It was found that ^{210}Pb is delivered to the surface layer of the Pacific Ocean with the mean velocity of $(1-5) \times 10^{-4}$ cm/s (Tsunogai and Nozaki 1967).

More detailed geochemical studies of the behavior of ^{210}Pb in the oceans have been carried out by Somayajulu and Craig (1976); Thomson and Turekian (1976); Nozaki et al. (1976); Nozaki and Tsunogai (1976). The ^{210}Pb isotope arrives at the surface layer of the ocean from the atmosphere, where it is formed during the decay of radon. As Nozaki et al. (1976) have shown, the flux of ^{210}Pb from the atmosphere along the transect from Tokyo to San-Diego decreases gradually from 1.9 to 0.7 dpm/cm^2 years. The additional source of lead is likely to predetermine the raised value of the $^{210}Pb/^{226}Ra$ ratio, but usually this ratio is considerably lower than the equilibrium one, which is accounted for by the rather short residence time of ^{210}Pb in the surface ocean waters.

In thermal waters of the active volcanic zone of Kamchatka and Kurile Islands the $^{210}Pb/^{226}Ra$ ratios are subjected to exceptionally large variations (from 2 to 970+ with a mean of 25). Other data of the ^{210}Pb content are not available. There are practically no data on the ^{210}Po content. In the years between 1961 and 1965, the value of the $^{210}Po/^{210}Pb$ ratio deviated over England in the range 0.05–0.35 (Peirson et al. 1966), and in the sea water near Cape Town in the range $(8-41) \times 10^{-15}$ Ci/l (with mean 20×10^{-15} Ci/l; Shannou et al. 1970).

In ocean waters the $^{210}Po/^{210}Pb$ ratio are usually less than the equilibrium one. The mean value of this ratio in surface ocean waters is equal to 0.5, which corresponds to the average residence time of ^{210}Po in these waters which is 0.6 years (Nozaki and Tsunogai 1976). On the map plotted by Nozaki et al. (1976), the distribution of the $^{210}Po/^{210}Pb$ ratio is indicated for the surface of the Pacific Ocean, this ratio varies between 0.3 and 0.9 with an average of about 0.5.

Chapter 17
Dating of Surface Water, Groundwater, and Sediments

The notion of the age of water is rather ambiguous. The age of water is usually understood to be its residence time in the studied geological object. It is further assumed that either the isotopic composition of the radioactive elements changes only through radioactive decay or that the law which governs their contribution or removal from water with a definite isotopic composition is known. These simple, theoretical suggestions are hard to apply in practice. Therefore, the main criterion of data verification is the comparison of results obtained by different methods, which usually correlate weakly with each other.

17.1 Dating of Closed Reservoirs

Cherdyntsev (1969) made the first estimation of groundwater age using radium and radon. The accumulation of radium in groundwaters can be approximately expressed as:

$$n_{\mathrm{Ra}} = n^0_{\mathrm{Ra}}(1 - e^{-\lambda_{\mathrm{Ra}} t}).$$

When the emanating power is equal to the radium extraction factors, one can put $n^0_{\mathrm{Ra}} = n^0_{\mathrm{Rn}}$.

Then

$$t = \frac{1}{\lambda_{\mathrm{Ra}}} \ln \left(1 - \frac{n^0_{\mathrm{Ra}}}{n^0_{\mathrm{Rn}}}\right).$$

This expression is only true as a first approximation. Radium, as a rule, is absorbed by rocks and radon can be lost to the surface or during the movement of gas currents. But this method is undoubtedly useful in distinguishing young waters. Table 17.1 shows the data about the ages of a number of surface waters, according to Cherdyntsev.

Chalov (1968) has developed a method applicable for the dating of closed basins by disequilibrium uranium. If uranium containing an excess of uranium-234 arrives at some reservoir, then the decay of the excessive amount of the daughter products

V. I. Ferronsky, *Isotopes of the Earth's Hydrosphere*,
DOI 10.1007/978-94-007-2856-1_17,

Table 17.1 Radon and radium content in some natural waters and their ages. (From Cherdyntsev, 1969)

Sampling location and type of sample	Rn 10^{-10} (Ci/l)	Ra 10^{-12} (Ci/l)	n_{Ra}/n_{Rn}	Age
Surface waters				
Caucasus:				
Trachiliparites	386	7.3	1.9×10^{-4}	170 days
Trachiliparites	45	0.5	1.1×10^{-4}	90 days
Trachiliparites	5.1	0.65	1.27×10^{-3}	2.8 years
Paleogenic marl	58.0	2.4	4.1×10^{-4}	0.9 years
Oligocenic lime	18.6	1.3	7.0×10^{-4}	1.5 years
Quaternary sediments	9.8	0.35	3.6×10^{-4}	290 days
Quaternary sediments	54.2	0.5	9.2×10^{-5}	75 days
Kirgizstan:				
Sienites	7.6	0.25	3.3×10^{-4}	280 days
Deep fresh groundwaters				
Caucasus	3.5	0.79	2.3×10^{-3}	5 years
Deep thermal groundwaters				
North Caucasus	11.7	9.5	8.1×10^{-3}	19 years
Dzhermuk, Armenia	0.46	7.0	0.15	380 years
Old Matsesta	5.84	89	0.15	380 years
Old Matsesta	9.9	21	0.21	550 years
Agura	1.27	34	0.27	740 years
Oilfield waters				
Middle Asia	3.0	300	1.0	∞

occurs by the incident component and the value of the isotope shift in a reservoir. If it can be observed at present time, then one can estimate how long a reservoir has been in existence. The disequilibrium uranium arrives at closed reservoirs together with river runoff. In the case of constant contribution of uranium to a reservoir during the whole period of its existence, one can write simple differential equations, reflecting the balance of uranium isotopes, as:

$$\frac{dN_1}{dt} = v(t) - \lambda_1 N_1,$$

$$\frac{dN_1}{dt} = k(t) - \lambda_2 N_2 + \lambda_1 N_1,$$

where N_1 is number of uranium-238 atoms at time t; N_2 is the number of uranium-234 at time t; $v(t)$ and $k(t)$ are the rates of input of the corresponding isotopes into the system.

While solving these differential equations, knowledge of the functions $v(t)$ and $k(t)$ is essential.

During his continuous studies, Chalov has shown that the ratio of uranium isotopes for one and the same source is a constant. As the object of his studies he chose the rivers of the Issyk-Kul basin. These rivers have shown large seasonal variations in water level at reasonable steady average annual water discharge, according to yearly observations. Field studies have been carried out in the spring, summer, and winter, when the hydrodynamic and hydrochemical characteristics have been markedly different. The water discharge on average changed by nine times for average changes in the uranium content by three times, but the deviations of the uranium ratios for a source have been within the accuracy of measurements. The average variations of the product of the water discharge and uranium concentrations (being on average 4.8 times in magnitude) prove the seasonal variations in the uranium content to be related not only to the dilution of some waters involved in persistent circulation during the period of melting of glaciers, but also (mainly) with the expansion of the domain of uranium leaching from rocks. Thus, Chalov has concluded: The uranium isotopic composition of natural waters is only determined by the composition of rocks, subjected to leaching. Therefore, returning to the above-mentioned differential equations we can assume that $v(t)$ and $k(t)$ are alike within a constant factor. In this case these equations can be rewritten as:

$$\frac{dN_1}{dt} = a\varphi(t) - \lambda_1 N_1,$$

$$\frac{dN_1}{dt} = b\varphi(t) - \lambda_2 N_2 + \lambda_1 N_1.$$

In order to solve these equations, one further function, $\varphi(t)$, needs to be known. Chalov has studied four possible cases: (1) increasing input of uranium into the object over time; (2) time-independent input of uranium; (3) a single contribution of uranium to the object; (4) decrease in the uranium input to the object. In Chalov's opinion, the increase in the rate of uranium contribution to the object is unlikely and should be eliminated. The decrease of the rate of uranium contribution to the object can be described by the function $\varphi = ae^{-nt}$, which can describe the two other cases at the following values of the parameter n: at $n \to 0$, which describes the time-independent arrival of uranium, and at $n \gg \lambda_2$, which is the single contribution of uranium to the object. The parameter n cannot be determined experimentally, since we do not know the law according to which input has occurred in the past; but for the interval over which the assumed age has been varying, $t_{\text{min}} = t_{\text{n}} \gg \lambda_2$ and $t_{\text{max}} = t_{\text{n}\to 0}$. Therefore, the age of the studied object can be assumed to be equal to $t = (t_{\text{max}} + t_{\text{min}})/2$ assuming that a single and unchangeable arrival of uranium is unlikely. For the Tertiary period (to 1 million years), the error in the age determinations, i.e., deviations of t from t_{max} to t_{min} increase from 0 to 50% with increase in age. In some special cases one can evaluate the law by which the arrival of uranium occurred and use the individual formula.

Usually uranium is contributed to a studied object from several sources. The mean value is then determined as

$$\gamma_0 = \frac{\sum \omega_i \gamma_i}{\sum \omega_i},$$

where ω_i and γ_i are the mass and value of the uranium isotope ratio for an individual source.

Table 17.2 Uranium isotopic ratio for waters of Chatyr-Kul Lake and its recharged rivers. (From Chalov et al. 1964)

Sample location	Relative mass of water (%)	$^{234}U/^{238}U$	Average (in wt.) $^{234}U/^{238}U$, recharged into lake
Kokaigyr River	60	1.21 ± 0.02	
Turgartsu River	40	1.50 ± 0.02	1.33 ± 0.02
Chatyr-Kul Lake	–	1.07	

The mass is determined by the relative uranium content contributed to the object. The mean annual mass contributed by individual water sources is determined by the average of the seasonal masses P_i

$$\omega_i = \frac{\sum \kappa_i P_i}{\sum P_i}$$

where κ_i is the mass of an individual value of P_i, determined by the lifetime during which it can be considered to be constant.

The main formula for the determination of the age can be written in the following forms:

For a single contribution:

$$\frac{\gamma_i - 1}{\gamma_0 - 1} = e^{-\lambda_2 t},$$

for a time-independent contribution:

$$\frac{\gamma_i - 1}{\gamma_0 - 1} = \frac{1 - e^{-\lambda_2 t}}{\lambda_2 t},$$

where γ_i is the uranium isotopic composition of the studied object; γ_0 is the isotopic composition of uranium arriving at the object.

By this method Chalov et al. (1964) have determined the ages of the lakes Issyk-Kul and Chatyr-Kul, Central Asia. The determined age of the Issyk-Kul Lake is $110{,}000 \pm 40{,}000$ years and that of the Chatyr-Kul Lake is $320{,}000 \pm 50{,}000$ years. These determinations are in good agreement with geological considerations. Table 17.2 shows the values of uranium isotope ratios for waters of the Chatyr-Kul Lake and its river system with which the basin age has been evaluated.

On the whole, the model assumed by Chalov reflects the process of uranium accumulation in a basin, but does not account for some important factors. The main factor is the accumulation of uranium in sediments. The obtained value of the age is not the age of a basin but the residence time of uranium there. Therefore, attempts aimed of age estimation of the Aral Sea and the Balkhash Lake have failed.

The effect of atmospheric precipitation and aeolian material upon the uranium isotope ratio in the Issik-Kul Lake waters has been considered by Alekseev et al. (1973). The contribution of the last two uranium sources to lake waters was not considered in Chalov's earlier works.

17.2 Dating of Groundwater

An interesting model applicable for the determination of the ages of groundwaters has been suggested by Kigoshi (1973). According to his model rain waters are accumulated in a certain reservoir. It is assumed that the rate of contribution of uranium and thorium elements to the liquid phase is permanent during the determined time, and the dissolved elements are removed from the liquid phase at a rate proportional to the concentrations of these elements in the liquid phase. With these assumptions, the change in the elements' concentration in a liquid phase can be written as:

$$\frac{dC_{U_8}}{dt} = E_{U_8} - D_U C_{U_8} - \lambda_{U_8} C_{U_8}, \tag{17.1}$$

$$\frac{dC_{U_4}}{dt} = E_{U_4} + \lambda_{T_4} C_{T_4} - D_U C_{U_4} - \lambda_{U_4} C_{U_4}, \tag{17.2}$$

$$\frac{dC_{T_4}}{dt} = E_{T_4} + Q_{T_4} + \lambda_{U_8} C_{U_8} - D_T C_{T_4} - \lambda_{T_4} C_{T_4}, \tag{17.3}$$

where C is the concentration of nuclei in a liquid phase, in atom/ml; U_4, U_8, T_4, T_8 are uranium and thorium, the figure denoting the last figure in the atomic number; E is the rate of dissolution, in atom/ml; D is the rate of nuclei output from the liquid phase, in atoms/years; Q is the rate of injection of atoms during the α-decay from the solid phase into the liquid phase by radioactive recoil, in atom/ml years.

In Eq. (17.2) expression $E_{U_4} + \lambda_{T_4} C_{T_4}$ characterizes the dissolution of uranium-234 which is in the solid phase during the formation of uranium-234 from thorium-234 in the liquid phase. Assuming that all the thorium decay products, being soluble, contribute to the water, the above given expression can be written as $(\lambda_{U_8}/\lambda_{U_4})E_{U_8}/\lambda_{U_4} + Q_{T_4} + \lambda_{U_8} C_{U_8}$. This assumption is rather robust, since it follows from the model experiments that the absorbed uranium is extremely soluble. Therefore, Eq. (17.2) can be rewritten in the form:

$$\frac{dC_{U_4}}{dt} = \frac{\lambda_{U_8}}{\lambda_{U_4}} E_{U_8} + Q_{T_4} + \lambda_{U_8} C_{U_8} - D_U C_{U_4} - \lambda_{U_4} C_{U_4}.$$

For rather young waters (their age is considerably less than the half-life of uranium-234) the decay process of uranium-234 and uranium-238 can be neglected. Uranium is in a relatively steady state in groundwaters and its removal from the liquid phase can also be neglected. Therefore, Eq. (17.1) and (17.2) become:

$$C_{U_8} = E_{U_8} t, \tag{17.4}$$

and

$$C_{U_4} = \left[\frac{\lambda_{U_8}}{\lambda_{U_4}} E_{U_8} + Q_{T_4} + \lambda_{U_8} C_{U_8} \right] t. \tag{17.5}$$

Thus, it follows that

$$\lambda_{U_4} C_{U_4} - \lambda_{U_8} C_{U_8} = \frac{dC_{U_4}}{dt} = \lambda_{U_4} (Q_{T_4} + \lambda_{U_8} C_{U_8}) t. \tag{17.6}$$

The half-life of thorium-234 is short and therefore its concentration can be considered to be unchangeable, so from Eq. (17.3), the expression for the determination of C_{T_4} is as follows:

$$C_{T_4} = \frac{E_{T_4} + Q_{T_4} + \lambda_{U_8} C_{U_8}}{\lambda_{T_4} + D_T}. \tag{17.7}$$

The D_T value can be determined by measuring the activities of ^{228}Ra, ^{228}Th and ^{232}Th in groundwaters. In fact, the balancing equations for these isotopes are:

$$E_{T_8} + \lambda_{R_8} C_{R_8} = (\lambda_{T_8} + D_T) C_{T_8},$$

$$E_{T_8} = \frac{\lambda_{T_2}}{\lambda_{T_8}} E_{T_2},$$

$$E_{T_2} = D_T C_{T_2}.$$

From the above equations it follows that:

$$\frac{D_T}{\lambda_{T_2}} = \frac{\lambda_{R_8} C_{R_8} - \lambda_{T_8} C_{T_8}}{\lambda_{T_8} C_{T_8} - \lambda_{T_2} C_{T_2}}. \tag{17.8}$$

Assuming that thorium-234 is released into the water phase mainly due to radioactive recoil, $E_{T_4} \ll Q_{T_4}$, Eq. (17.7) may be rewritten as:

$$(D_T \lambda_{T_4}) C_{T_4} = Q_{T_4} + \lambda_{U_8} C_{U_8}. \tag{17.9}$$

Then the expression for the estimation of the age becomes

$$t = \frac{\lambda_{U_4} C_{U_4} - \lambda_{U_8} C_{U_8}}{\lambda_{U_4} C_{T_4} (D_T + \lambda_{T_4})}. \tag{17.10}$$

Here D_T can be evaluated by Eq. (17.8).

Kigoshi used this model for the determination of the age of groundwaters near Tokyo and Kyoto. The obtained data are in good agreement with those obtained by radiocarbon dating. Table 17.3 shows the main results of the analysis of groundwaters near Tokyo on the basis of which their age has been determined.

The model suggested by Kigoshi has some faults; some of its assumptions are disputable. Kigoshi considers that the excess of uranium-234 in natural waters is conditioned by the large contributions to natural waters of thorium-234. Cherdyntsev and Chalov do not share this belief. But the approach to the problem of groundwater dating by using the isotopes of heavy radioelements is of interest and the attempts being undertaken in this direction will lead in future to the construction of perfect model.

Table 17.3 Activity of heavy radioactive elements in groundwaters near Tokyo and water age. (From Kigoshi 1973)

Isotopes	Activity (dpm/80 l water)	Water age by Eq. (17.10) (years)	Radiocarbon age (years)	Radiocarbon age correction (years)
^{234}Th	192 ± 27			
^{232}Th	5.4 ± 7			
^{230}Th	14.7 ± 1.2			
^{228}Th	8.0 ± 0.8	$5{,}000 \pm 600$	$6{,}930 \pm 140$	1,350
^{238}U	4.9 ± 0.7			
^{234}U	8.2 ± 0.2			
$^{234}U/^{238}U$	1.68 ± 0.08			
^{228}Ra	8.0 ± 2.0			
$^{234}U_{excess}$	3.3 ± 0.20			

17.3 Dating of Sediments

The isotopes of heavy radioactive elements are most advantageously used in dating sediments in reservoirs. The characteristic feature of the application of heavy radioactive elements is the possibility of dating over a time range of up to 1 million years, which cannot be studied by classical radiocarbon, potassium-argon, rubidium-strontium, or plumbum methods.

17.3.1 Uranium-Uranium Method

As described above, uranium occurs in natural waters in a disequilibrium state being conditioned by the preferential leaching of uranium-234 as a daughter product. Therefore, it occurs in precipitation in a disequilibrium state. If subsequent uranium migration does not occur, the system attains a radioactive equilibrium state governed by the radioactive law as:

$$\frac{\gamma_t - 1}{\gamma_0 - 1} = e^{-\lambda_2 t},$$

where γ_0 is the initial ratio of the uranium isotopes; γ_t is the uranium isotopes in time t; λ_2 the decay constant of uranium-234.

The main difficulty in dating while using this method lies in the estimation of the initial uranium isotopic ratio and in proving the absence of the migration of uranium isotopes during the time period considered. Thurber (1963) was the first who used this method for the determination of the age of fossil corals. In fact, during their metabolism corals accumulate uranium from the sea water, characterized by a constant isotope ratio, i.e., the initial uranium isotope ratio is known with sufficient precision. By elevating the uranium isotope shift in the studied specimen of coral one can determine its age, but the first determinations of the uranium isotope ratio have shown natural decreases with the depth of sampling (see Table 17.4).

Table 17.4 $^{234}U/^{238}U$ fossil corals of Enivetok Atoll. (From Thurber 1963)

Sampling depth (feets)	$^{234}U/^{238}U$
24–26	1.17 ± 0.01
47–52	1.13 ± 0.03
64–69	1.09 ± 0.01
90–97	1.09 ± 0.01
180	1.07 ± 0.01
700	1.00 ± 0.01

Experimental data (Cherdyntsev 1969) indicate that at least during the last 200,000 years the uranium isotope composition of the ocean waters was remained unchanged. There are no reasons to suppose that it differed greatly in more distant times. After death, the corals form thick homogeneous accumulations in which uranium migration is very improbable, but on the whole the question of uranium migration is not completely understood. According to the data of some authors, corals gradually lose uranium due to the recrystallization of aragonite (Cherdyntsev 1969).

The method of disequilibrium uranium is applicable to the shells of the sea mollusks, but the probability of uranium migration there is greater than in corals since they occur in Tertiary deposits in the form of individual inclusions, where, according to some authors, uranium migrates intensively. The amount of uranium in the shells of living mollusks is small, but after their death it increases over several thousands of years. There is a direct possibility of dating the shells of river and lake mollusks, fossil bone, and thermal water deposits. But it is hard to evaluate the initial uranium isotope ratio in these cases, since river, lake, and groundwaters have lower concentrations than the waters of the World Ocean. Thus Thurber, while determining simultaneously the uranium isotope ratio and the age of carbonates in the Bonneville Lake, has shown that uranium content in the lake waters changed in the past, i.e., its water contained uranium of different isotopic composition (see Table 17.5).

As described earlier, the uranium isotope content is dependent on the composition of the rocks subjected to leaching. Therefore, assuming the water-bearing complex of rocks during the considered interval of time to be unchanged, the uranium isotope composition should also remain unchanged.

Chalov et al. (1966a, 1970) have shown that in closed reservoirs such as the Balkhash Lake and Aral Sea, uranium is carried down to the bottom sediments and securely fixed by them. Knowing the value of the mean uranium isotope ratio contributed by river discharge and the uranium isotope ratio in the lower layers of the bottom sediments, one can estimate their age and therefore the age of the reservoir. The absolute age of the Aral Sea, according to Chalov, is $139{,}000 \pm 12{,}000$ years and that of the Balkhash Lake is $37{,}000 \pm 7{,}000$ years. Table 17.6 indicates the uranium isotope ratio in the bottom sediments of the Balkhash Lake.

The mean value of the $^{234}U/^{238}U$ ratio for waters of the lake is 1.498 ± 0.003. The additional determination of the value obtained for the lower layers of the bottom sediments yields 1.462 ± 0.005. Using these two values the age of the lake has been determined.

The main advantage of the disequilibrium uranium dating method is its wide range, including practically all the Tertiary period. The simplicity of the method

Table 17.5 $^{234}U/^{238}U$ in carbonate material from Bonneville Lake. (From Thurber 1965)

Sampling location	Elevation about lake level (ft)	^{14}C age (10^3 years)	$^{234}U/^{238}U$	$^{234}U/^{238}U_0$	Uranium content (10^{-6} g/g)
Oolites (Tip of cape)	4,200	1.5	2.30 ± 0.001	2.30	5.8
Tufa (Peak of Oakveer)	4,550	12.9	1.08 ± 0.04	2.11	–
Tufa (Big Gully)	4,780	17.7	1.92 ± 0.01	1.95	10
Gastropod (Little Valley)	4,818	46	1.79 ± 0.03	1.90	10.5
Gastropod (Little Valley)	4,858	12.4	1.92 ± 0.04	1.94	1.2
Tufa (Leatington)	5,040	13	1.90 ± 0.02	1.93	–
Gastropod (Little Valley)	4,820	14.8	1.79 ± 0.04	1.82	5.0
Gastropod (Big Gully)	4,800	18.5	1.75 ± 0.05	1.79	6.5
Tufa (Peak of Oakveer)	5,100	15.65	1.71 ± 0.03	1.74	–

Table 17.6 $^{234}U/^{238}U$ in bottom sediments of Balkhash Lake at maximum depth. (From Chalov et al. 1970)

Sampling depth (m)	$^{234}U/^{238}U$
0–12	1.515 ± 0.013
12–22	1.513 ± 0.013
22–32	1.501 ± 0.015
32–42	1.501 ± 0.009
42–52	1.510 ± 0.014
52–62	1.511 ± 0.013
62–72	1.517 ± 0.012
72–82	1.500 ± 0.017
82–92	1.491 ± 0.008
92–102	1.490 ± 0.009
102–112	1.462 ± 0.009

should be also noted. The age is determined by the age of one isotope which markedly simplifies the analysis. The sensitivity of the method depends upon the activity of the studied specimens and is to a large extent restricted by the low accuracy available in the determination of the value of the uranium isotope ratio for the oceanic waters, 1.15 ± 0.01.

17.3.2 Uranium-Ionium Method

Among the methods based upon uranium-thorium isotope dating the most widespread is the uranium-thorium one. The ionium content in natural waters is low, therefore all the secondary formations have practically no ionium at the moment of deposition, but in the course of time the ionium is formed from uranium-234 in deposits by radioactive decay:

$$I\text{o}(t) = I\text{o}(0)e^{-\lambda_{I\text{o}}t} + {}^{238}\text{U}\left\{(1 - e^{-\lambda_{I\text{o}}t}) + \frac{\lambda_{I\text{o}}}{\lambda_{I\text{o}} - \lambda_{^{238}\text{U}}}\left(\frac{^{234}\text{U}}{^{238}\text{U}} - 1\right)\right. \\ \left. \times (1 - e^{(-\lambda_{I\text{o}} - \lambda_{^{238}\text{U}}t)})\right\} \quad (17.11)$$

where I_o (U) is the initial ionium content.

Table 17.7 Uranium and ionium isotope ratios and ionium and uranium-234 ages for fossil corals from the Pacific and Indian oceans. (From Thurber 1965)

Sampling location	$^{234}U/^{238}U$	$Io/^{234}U$	Age (10^3 years)	
			U/U	Io/U
Havaii, Is. Oahu	1.10 ± 0.015	0.81	140 ± 30	140 ± 50
Tuamotu Is., Anaa	1.12 ± 0.015	0.72	110 ± 20	80 ± 50
Tuamotu Is., Niau	1.09 ± 0.015	0.76	120 ± 20	180 ± 60
Cook Is., Mangaia	1.11 ± 0.014	0.73	110 ± 20	110 ± 50
Western Australia	1.11 ± 0.014	0.81	140 ± 30	110 ± 40
Mauritius	1.11 ± 0.014	0.87	160 ± 40	110 ± 40
Seychelles Is.	1.10 ± 0.014	0.80	140 ± 30	140 ± 50
Seychelles Is.	1.12 ± 0.015	0.78	140 ± 30	80 ± 40

The uranium-ionium method is applicable over a time range exceeding that for the uranium-uranium method, but within the same range the accuracy and sensitivity of the former is higher, especially over short time intervals. The upper limit of dating provided by this method is dependent upon the value of the measured activity, in practice less than 400,000–600,000 years. While estimating the age by the uranium-uranium method the determination of the uranium isotope ratio is required, and therefore the age of the studied object can be estimated by the two methods (uranium-uranium and uranium-ionium) simultaneously. A comprehensive summery of the application of the uranium-ionium method has been given by Cherdyntsev (1969) in this Book. Here we shall consider in brief some main objects for which the application of this method is advantageous.

A great number of determinations by the uranium-ionium method have been carried out for sea corals and shells of sea mollusks. In a number of cases the estimations were carried out for one object by two different methods (radiocarbon and uranium-uranium). For the Holocene shells a good coincidence was found between radiocarbon and uranium-ionium methods, but for the ancient samples radiocarbon exhibits the lower age values. Comparisons of ages determined by the uranium-ionium and uranium-uranium methods usually give a good coincidence (see Table 17.7).

Cherdyntsev et al. (1963, 1965) has determined the ages of shells of continental mollusks of the Dnestr River and its terraces. The data obtained are in good agreement with the geological considerations. Extensive studies of mollusk shells in deposits of lakes Lahontan and Bonneville were carried out by Kaufman and Broecker (1965). The age has been determined simultaneously by the radiocarbon, uranium-ionium, and radium-uranium methods (see Table 17.8).

The most suitable material for dating by the uranium-ionium method is travertine which have precipitated from natural waters, and deposits from karstic caves (Cherdyntsev et al. 1965, 1966, 1967). Uranium from carbonaceous waters is transferred into the solid phase with which it becomes strongly bonded. A large number of determinations have been carried out by the authors in the travertines of Mashuk (Caucasus), Tate and Vertesszollos (Hungary), lake tuffs of the Karatan (Kazakhstan), tuffs of the Leninakan region, Arzni, Djermuk (Armenia), travertines of the Ob-I-Garma

Table 17.8 The age of the Bonneville Lake's carbonate materials from the $Io/^{234}U$, $Ra/^{234}U$ and ^{14}C methods. (From Kaufman and Broeker 1965)

Sample	$Io/^{234}U$	^{14}C	$Ra/^{234}U$
Big Gully, Utah			
Gastropod	11.0 ± 1.5	17.6 ± 0.6	11.3 ± 2.2
Tufa	21.4 ± 2.2	17.1 ± 0.3	21.2 ± 4.2
Gastropod	10.2 ± 1.1	17.7 ± 0.6	10.6 ± 1.7
Lamington, Utah			
Gastropod	17.5 ± 1.6	14.1 ± 0.31	19.3 ± 1.8
Gastropod	16.8 ± 2.3	14.9 ± 0.3	17.9 ± 3.0
Gastropod	32.8 ± 4.8	17.4 ± 0.4	38.4 ± 5.8
Little Valley, Utah			
Gastropod	8.7 ± 2.2	11.7 ± 0.3	14.3 ± 3.6
Gastropod	16.7 ± 1.4	15.4 ± 0.3	21.4 ± 3.2
Gastropod	18.1 ± 2.3	15.3 ± 0.4	16.9 ± 2.9

(Tajikistan), and the tuffs of the Pleshcheyevo Lake, and travertines of Elatma. The majority of age determinations are in good agreement with the geological evidence.

The ages determined for travertines and other carbonaceous formations of the karstic caves are in good accord with the archeological considerations. The karstic caves were normally inhabited by primitive man. The dating carried out on the carbonaceous formations of the caves has been duplicated by determinations of the ages of bones by the utanium-ionium method and radiocarbon dating of charcoal from primitive man's hearths.

A somewhat difficult problem during dating by uranium-ionium method is the determination of the initial ionium content. Usually it is evaluated by the Io/Th ratio for modern waters in the studied region, or for modern samples. Sometimes the initial value of ionium is determined from the radiocarbon age of a sample.

Cerrai et al. (1965) have proposed a method of sediment dating where initial ionium content can be estimated. In fact, by writing down Eq. (17.11) for two different types of deposits (two different minerals) the initial ionium content can be estimated, According to Cerrai's assumption, the Io/Th ratio should be the same for the both types of deposits. But it has been shown (Kuptsov et al. 1969; Kuptsov and Cherdyntsev 1969) that in some cases, i.e., for minerals of the fumarol fields, the ratio is different for different types of sediments. Therefore, Cerrai's assumption should be investigated further.

The possibility of determining the absolute age of organic river sediments has been considered by Titaeva (1966). In order to estimate the possibility of secondary migration of radioelements, studies were carried out in the frozen-rock zone of the Lena River terraces. Samples containing no organic substances did not exhibit radioactive disequilibrium. An excess of uranium was observed in the presence of organic substances. In the course of the accumulation of uranium by organic substances it was observed to be in excess relative to the decay product. In order to carry out the dating of sediments, the value of this excessive concentration should be known. The value of thorium concentration in ashes of the upper melted layer of peat had been taken as its initial concentration. The ionium content for the ooze sediments of the flood plain was taken to be that determined in the modern oozes of the lake, located near the point of sampling. Based on these concentrations, the equilibrium concentrations were estimated. This value was subtracted from the values of corresponding contents of uranium and ionium in the studied samples, after which their age was determined. The obtained data are in good agreement with the results of radiocarbon dating.

Among other methods of sediment dating in reservoirs, based on the utilization of heavy element isotopes, the following methods should be mentioned: radium-uranium (Chalov et al. 1966b), ionium-thorium (Hamilton 1968; Starik 1961), ionium-protactinium (Kuznetsov 1962), the method based on the utilization of plunbum-210 (Crozaz 1967; Goldberg 1963), and also the decay product tracks (Berzina 1969; Shukolyukov 1970).

17.4 Isotopes of Radiogenic Elements as Indicators of Hydrologic Processes

The value of the uranium isotope ratio in waters is practically independent of hydrochemical factors. It is mainly determined by the uranium distribution in the rocks of water-bearing complex. Therefore, the uranium isotope ratio serves as a kind of natural indicator of water of a certain water-bearing complex. This fact makes it possible to determine patterns of natural water filtration; to distinguish the rock of the water-bearing complex, to construct models of groundwater circulation, to establish the mixing proportions of waters of different water-bearing complexes, and to examine the interrelations between waters of different water-bearing complexes. The results of regional studies of uranium isotope ratios in groundwaters in an artesian basin located near Tashkent, being the most intensively studied basin in geological and hydrological respects, were reported by Sultankhodzhaev et al. (1970). Based on the discovered principles, three zones were distinguished within a basin which differ from each other by the value of the $^{234}U/^{238}U$ isotope ratios: (1) the zone of infiltration, including the recharge zone and the adjacent area and being characterized by the adequately high isotope ratios ranging from 2.0 to 2.5 (average 2.1); the waters of this zone belong to the active water exchange region according to the hydrological parameters; (2) the transition zone, related to the marginal part of the basin, characterized by uranium isotope ratios ranging from 1.3 to 3.5 (average 1.9) and related water motion; (3) the central part of the basin, being the domain of the majority of slow moving waters and characterized by the lowest isotope ratios ranging from 1.3 to 1.8 (average 1.6).

The above-mentioned zonality in the distribution of the $^{234}U/^{238}U$ ratios corresponds to such hydrochemical indexes as the deuterium concentration, factors of water migration, and groundwater age.

Studies of the uranium isotopic composition of the Florida artesian basin were carried out by Kaufman et al. (1969). In their opinion the value of the uranium isotope ratio can be considered as a regional hydrodynamic parameter characterizing the aquifer. Based on the uranium isotope studies, the regional characteristics of permeability of the rocks have been determined, the sources of water recharge to the different parts of the basin have been found, and the vast domain of the Pleistocene waters incorporated into the Florida aquifers has been mapped.

Changes in the uranium isotope ($^{234}U/^{238}U$) content in water can only occur during the leaching of uranium from the surrounding rock system and in the course of mixing with other waters containing uranium. Osmond et al. (1968) proposed to apply the value of the content and the isotope of uranium for the determination of the mixing proportions of waters which have different uranium isotope concentrations.

In fact, if water 1, mixing with water 2, gives a mixed flow, then the following expressions are true:

$$V_1 + V_2 = V; \quad M_1 + M_2 = M; \quad M_1\gamma_1 + M_2\gamma_2 = M\gamma; \quad M = CV,$$

where V is the volume of water, in l; M is the uranium content, in g; C is the uranium concentration, in g/l; γ is the uranium isotope ratio.

Assuming $V = 1$, one can finally derive the following expression:

$$X = \frac{C}{C_1}\left[\frac{\gamma_1 - \gamma_2}{\gamma_1 - \gamma_2}\right],$$

where X is the fraction of water 1 in the mixed flow.

By measuring the uranium isotope ratios in all the three waters and the uranium content in water 1 and in the mixed flow, one can find the proportions in which water 1 and 2 are mixed. This scheme was applied by Osmond et al. (1968) to the construction of groundwater circulation patterns in the Florida artesian basin.

The uranium isotope studies carried out by the authors in the waters of the Mirgalimsay complex of ore deposits (Kazakhstan) allowed, using adequately simple assumptions, on estimates it has been assumed that the following waters contribute to the mine waters: surface discharge waters of the rivers Bayaldyr, Birisekh, Kantagi, and the water of the Paleozoic shales surrounding the ore body. Analysis of the water mixing proportions was carried out by using the above-mentioned formulas. In total, surface discharge amounted to 80%; in a number of mine water tributaries surface waters were absent.

The uranium isotope composition permits one to solve some problems of the paleogeography. Thus, estimating the age of the Aral Sea (Chalov et al. 1966b) by the uranium disequilibrium method, Chalov has managed to obtain an upper limit of the time when Amu Darya river became a tributary of the Aral Sea (~22,000 years). Estimating the age of the Issik-Kul Lake (Chalov et al. 1964) it was found that the Chu river, one of the main rivers recharging the lake, has not always been contributory. The analysis of the bottom sediments of the lakes Balkhash and Issyk-Kul (Chalov et al. 1973) has shown that they do not have the same γ values. This fact resulted in the conclusion that both lakes originated and existed independently

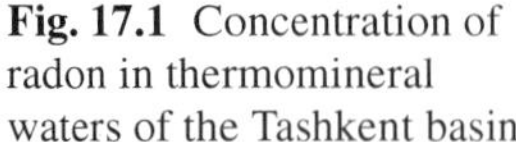

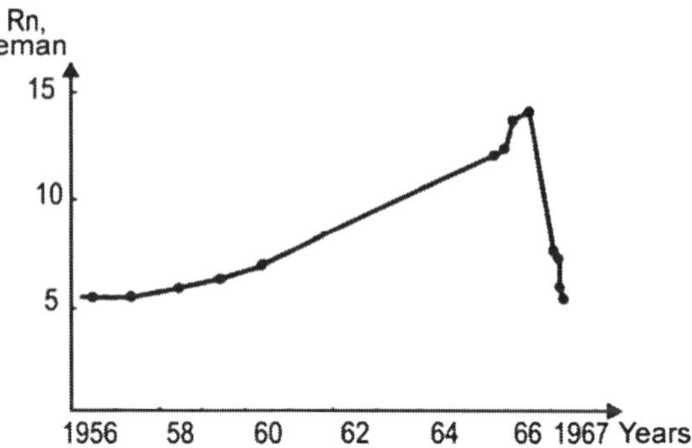

Fig. 17.1 Concentration of radon in thermomineral waters of the Tashkent basin

of each other, despite some geological notions, although the age of the lakes, within the accuracy of the method, is the same.

According to the data obtained by Khristianov and Korchuganov (1971) the excess of the radon concentrations over radium concentrations in river waters has been mainly conditioned by groundwater discharge. The groundwaters contribute the intensive, natural tracer-radon to the river, and therefore studies of the radon content help to solve a number of problems related to deciphering and estimating groundwater discharge into rivers.

The systematic observations of the content of radon in deep groundwaters of tectonically active regions reflect increase in the elastic stresses and changes in the character of the deformations of rocks at great depth. The epicenter of the earthquake represents the focus of the destruction process of rocks when the elastic limit of the medium is exceeded. Besides the main rupture a large number of fissures of various sizes are formed, which result in vibrations of high frequencies. The motion of these elastic waves (of the ultrasonic frequencies) through the thicknesses of hard rocks and liquids results in the escape of radon into the environment (Gorbushina et al. 1967, 1968). The reasons behind radon ejection are: (1) the increase of the emanation factor of rocks with the increasing surface of crushed rocks; (2) the extraction of gases from rocks effecting by the ultrasonic and sound vibrations, occurring during and before the earthquake. The amount of escaping gas depends on the frequency and intensity of the sound waves. One of the reasons behind the ejection of the gas is its transition from a bounded state to a free state. Under the influence of ultrasonic waves the absorption forces holding the gas to the walls of the pores and fissures become weakened and the gas transfers into a free state. Figure 17.1 indicates the curve of radon content in the thermomineral water of the Tashkent basin during 1956–1967. Water samples were taken from the borehole every month. During the 10 years, since the beginning of sampling the radon content has increased by a factor of almost 2.5. The radon content dropped sharply after the earthquake on April 26, 1966. Ten days before the 7-ball shock on March 24, 1967, the radon content increased again by 2.5 times and after that dropped again. Besides the variations in the radon content during the Tashkent earthquake sharp changes were also noticed in the value of the uranium isotope ratio in the waters of the Tashkent artesian basin (Gorbushina at al. 1972; Spiridonov and Tyminsky 1971; see Table 17.9).

Table 17.9 Uranium isotope ratios in groundwaters during and after the Tashkent earthquake. (From Gorbushina et al. 1972)

Sampling well location	$^{234}U/^{238}U$		Ratio II–IV/ 1967–VIII/1968
	II–IV/1967	VIII/1968	
Lunacharsky village (Well № 3)	7.0 ± 0.3	1.4 ± 0.1	5.0
Kibray village (Well № 3)	6.2 ± 0.2	1.6 ± 0.1	3.9
Chinbad sanatorium (Well № 7)	3.4 ± 0.1	1.8 ± 0.1	1.9
Tashkent mineral water (Well № 5)	4.2 ± 0.1	1.5 ± 0.1	2.8
Botanik rest home (Well)	3.0 ± 0.1	1.3 ± 0.1	2.3

According to their opinion, the $^{234}U/^{238}U$ ratios were conditioned by excessive contribution of ^{234}U from rocks to water during earthquake. The elastic vibrations which occur in rocks during the earthquake favor the enrichment of water in ^{234}U. The consequent decrease of the $^{234}U/^{238}U$ ratio after the earthquake is caused by the interphase isotope exchange due to which the unsteady equilibrium between uranium in the liquid phase and the solid phase occurs. Due to this exchange the $^{234}U/^{238}U$ ratio in the waters drops to value (1.3–1.8) characteristic of waters of a sedimentary complex. The possibility of uranium extraction from rocks into solution, initiated by ultrasonic vibrations, was discussed by Andreev et al. (1960).

The elastic vibrations of rocks in the volcanic interiors which precede the earthquake can result in increases of the radon concentrations in the waters of thermal springs of active volcanoes. Chirkov (1971) has attempted to study the relationship between volcanic activity and the activity of radon. Systematic observations of radon content have been carried out on the spontaneous gas of the thermal spring at the foot of the Karymsky volcano, Kamchatka. Figure 17.2 shows the change in the radon content in the spring's gases over time.

All the stages of the volcanic activity are included in the period of observations: the final eruption and the transition to the steady stage, the preliminary stage, and the beginning of the new eruption. A comparison of this plot with the results of observations on the volcano, and data from the Karymskaya seismic station on the incidence of explosive earthquakes, has shown that several days after the appearance of the radon anomaly a related anomaly always appears in the volcanic activity. The study of the radon content allows one to forecast both the resumption of the volcanic activity and changes in the type of eruption during that activity.

One more problem solving by the radiogenic isotopes is the paleotermometry based on fission-fragment tracks. The age of minerals estimated by the fission tracks is usually lower than in reality. The main cause of this is the burning-off within radioactive damage during the geological history. Burning-off is determined by temperature, therefore the content of the fission tracks can provide evidence on the paleothermic conditions in geological objects. The results of the first paleotemperature measurements were reported by Shukolyukov and Komarov (1966). The change in the number of tracks $N(t, E)$ in mineral with time can be expressed as:

$$\frac{dN}{dt} = -\lambda_S {}^{238}\mathrm{U} + Ce^{-E/RT} N(t, E),$$

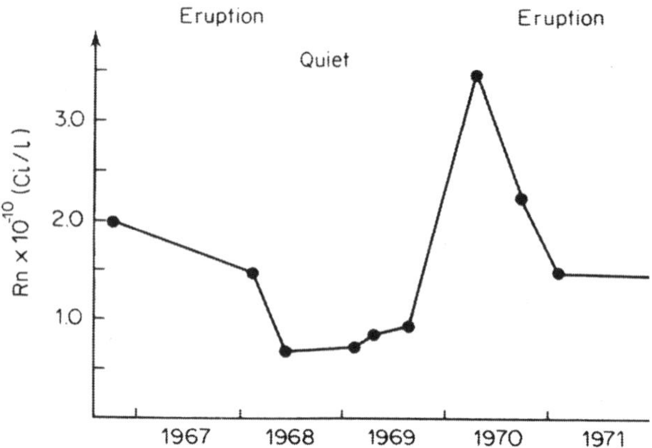

Fig. 17.2 Variation of radon concentration in the thermal spring of the Karymsky volcano

where λ_S is the decay constant of uranium-238; E is the burning-off activation energy; R is the gas constant; C is a constant independent of temperature; ^{238}U is the uranium concentration.

Integration of this equation yields

$$N(t,E) = \frac{\lambda_a{}^{238}U}{\lambda_a - Ce^{-E/RT}}[e^{(\lambda_a - Ce^{-E/RT})t} - 1].$$

where λ_a is the uranium α-decay, which is constant; $N(t, E)$ is the measured density of tracks.

There are three unknowns in the expression: age t, temperature T, and activation energy$E$. In order to determine the activation energy, the dependence of the rate of track activation on temperature should be known. Experiments are carried out at the two different rates of heating. The burning-off activation energy is estimated from the equation

$$E = \frac{R(\ln \alpha_1/\alpha_2 + 2\ln T_2/T_1)T_1T_2}{T_1 - T_2},$$

where T_1 and T_2 are the temperatures at which the track's rate is at a maximum; $\alpha_1$and α_2 are the heating rates, in °C/min.

The value C is determined from the expression

$$\frac{E}{CRd_1t_1^2} = e^{-E/R\alpha_1 t_1}.$$

After the age of the mineral determined by the means of another method (usually by the potassium-argon method), one can determine the heating temperature T in steady thermal conditions. Using two syngenetic minerals, one can determine both the age and the temperature at steady thermal conditions. In this case, one should determine only the value of the activation energy and the frequency factor for each mineral and solve the following two equations (Shukolyukov 1970):

$$t = \frac{2,3}{\lambda_a C_1 e^{-E_1/RT}} \lg \left[\frac{N_1\,(t,E)}{^{238}\mathrm{U}_1} \frac{\lambda_a - C_1 e^{-E_1/RT}}{\lambda_S} + 1 \right],$$

$$t = \frac{2,3}{\lambda_a C_2 e^{-E_2/RT}} \lg \left[\frac{N_2\,(t,E)}{^{238}\mathrm{U}_2} \frac{\lambda_a - C_2 e^{-E_2/RT}}{\lambda_S} + 1 \right].$$

Against a background of constant low temperatures in nature, short heating periods are evident (magmatic activity, high temperature metamorphism, etc.). In this case

$$N_1 = \left\{ \frac{\lambda_S {}^{238}\mathrm{U}}{k - \lambda_a} [e^{-k\Delta t + \lambda_2 t_2} - e^{\lambda_a t_1}] + \frac{\lambda_S}{\lambda a} \mathrm{U}^{238} [e^{\lambda_S t} - e^{\lambda_a t_1}] \right\}$$

$$\times\, e^{-k\Delta t} + \frac{\lambda_S}{\lambda a} \mathrm{U}^{238} (e^{\lambda_S t_1} - 1),$$

where t_1 and t_2 are the beginning and the end of heating; Δt is the duration of heating; $k = Ce^{-E/RT}$.

Experimental studies of muscovite in the Chai River, Mamsky region, have shown that the age determined by the fission tracks (360×10^6 years) is in good agreement with the age determined by the potassium-argon method. Based on the above formulae, one can calculate that muscovite has never been heated over 160°C during a period of 3×10^8 years, or has had temperature less than 220°C being heated during 10^7 years or less than 100°C being heated during the whole period of its existence.

Part IV
Applications

Chapter 18
Applications to the Problems of Dynamics of Natural Waters

The stable and radioactive environmental isotopes are widely applied to solve the practical problems related to studying different aspects of dynamics of natural waters and the hydrosphere as a whole. The solution of such problems by the isotopes as natural tracers is based on natural regularities in their space and time distribution in the hydrosphere, which were discussed in previous chapters. Some examples of practical application of the environmental isotopes in hydrology, hydrogeology, meteorology, oceanography, and climatology are given below.

18.1 Dynamics of Moisture in the Atmosphere

There are three groups of tasks known in the hydrometeorology, which are solved by means of tritium. These tasks are such as the study of parameters of moisture transfer in the atmosphere; the estimation of the water budget in cyclones; the investigation of relationship between the atmospheric moisture and the open water surface. One of the first problems which got solved in this field was determination of the half-removal period of the moisture from the troposphere. This task was solved by Libby with co-workers (Buttlar and Libby 1955; Kaufman and Libby 1954; Libby and Palmer 1960) on the basis of measurement of tritium content in atmospheric precipitation in the atmosphere over Chicago and New York after the thermonuclear tests in autumn of 1952 and in spring of 1954. Analogous studies were carried out by Brown and Grummit (1956) around Ottawa, Canada. The obtained results of both studies were close to each other with the period of half-removal of moisture from the troposphere equal to several weeks.

The direct measurements of the atmospheric moisture transfer on the basis of tritium content study in atmospheric water vapor were carried out by the Water Problems Institute in Moscow (Ferronsky et al. 1980). The sampling method used was based on absorption of water vapor during pumping of air through a drying agent. Synthetic zeolite type NaA was used as the adsorbent. After extracting the moisture from zeolite, a 50 cm^3 sample of the condensate was enriched 10–20-fold

V. I. Ferronsky, V. A. Polyakov, *Isotopes of the Earth's Hydrosphere*,
DOI 10.1007/978-94-007-2856-1_18, © Springer Science+Business Media B.V. 2012

by electrolysis. The tritium concentration in the enriched sample was measured by a liquid scintillation spectrometer. The total measurement error was 10–20%.

The water-vapor samples for analyzing the tritium content were collected in the surface boundary layer (5 m above the ground). Regular sampling of water vapor began in November 1977 and was performed once a month, and after the report in the press on the nuclear explosion in China on March 14, 1978, the frequency of sampling was increased (samples were taken on March 16, 17, 21, 22, 23, and 28, on April 4, 13, 18, 19, and 26, and also during the May and June 1978).

During the two weeks after explosion, the passage of a radioactive cloud was recorded in Japan, Canada, England, and Norway on the basis of higher level of radioactivity in the atmosphere and precipitation (Cambray et al. 1979). It should be noted that a large part of radioactive isotopes, such as ^{137}Cs, ^{90}Sr, ^{95}Nb, ^{140}Ba, etc. (with the exception of ^{14}C, T, and certain isotopes of inert gases) produced in a nuclear explosion in the atmosphere is adsorbed by aerosols, by virtue of which radioactive aerosols are most often studied for determining the content of radioactive isotopes in the atmosphere. The degree of difference in the movement of water vapor and aerosols in the atmosphere is poorly studied, but it is known that the movement of the latter can differ from the movement of air in the lower layers of the troposphere owing to removal of aerosols by clouds and precipitation and their settling on the underlying surface (Styro 1968). Furthermore, it is known from fluid dynamics that the velocities of the fluid-carried and foreign particles are associated by the relationship

$$\nu_{\mathrm{p}} = \nu_{\mathrm{f}} \frac{1}{1+h},$$

where ν_{p} is the velocity of foreign particle; ν_{f} is the velocity of the fluid-carrier; $h = f(m_{\mathrm{p}})$ is the distribution coefficient, which depends on the mass m_{p} of the moving particles. If the airflow transports particles having different masses, the coefficient h will be greater, the greater the mass of the particle, and hence the smaller will be its velocity.

On March 23, the first increase in the tritium concentration in atmospheric vapor to 400 tritium units (TU) was recorded, which was fourfold greater than its concentrations noted in all samples collected starting from November 1977 (Fig. 18.1). In the sample taken on March 28, the tritium content was equal to the normal level for this time of the year (120 TU); on April 4, it again had a peak value (480 TU). A tritium concentration close to this value was observed in samples taken on April 13 and 18; on April 26, the tritium content in water vapor was 336 TU, and thereafter it dropped to the normal level.

The authors estimated possible tritium concentration in atmospheric water vapor of the radioactive cloud from the nuclear explosion on March 14, 1978, for which it was assumed that a minimum value of the tritium yield from the explosion is equal to 0.07 kg/Mt (Miskel 1973). Then for a low-power explosion (up to 10^3 tons), the tritium yield will be 0.7 g, or ~7,000 Ci. A large part of the tritium after explosion forms HTO molecules and at the initial moment is concentrated in a limited volume of the atmosphere, which can be represented in the form of a cylinder with a base of area $S = \pi D^2/4$, where D is the cloud's diameter which, according to Glasston

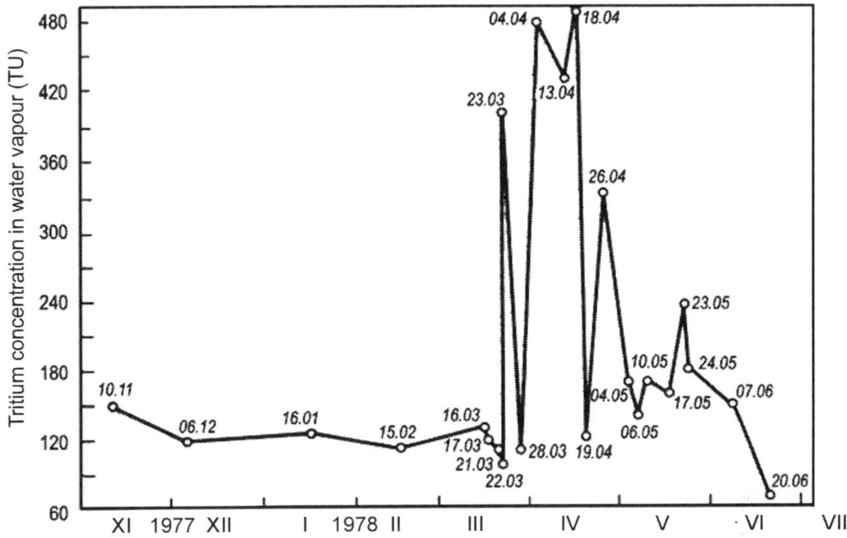

Fig. 18.1 Change in the tritium concentration in atmospheric water vapor from November 1977 to June 1978 (the numerals next to the experimental points indicate the dates on which the samples were taken. © IAEA, reproduced with permission of IAEA)

(1962), is equal to 360 $W^{5/2}$ (W is the energy of the explosion, 10^3 tons). For W = 10 ktons and $D \approx$ 1,000 m, one has $S = 1\ \text{km}^2$.

The content of atmospheric moisture and tritium concentration in the initial cloud was determined. According to long-term observations, the atmospheric water-vapor content (value of precipitate water) in March for north-western China (Drozdov and Grigoryeva 1963) is about 5 mm. Consequently, the amount of atmospheric moisture in the radioactive cloud was equal to 5×10^7 l and the tritium content reached 10^6 TU. This 'package' of moisture is transported by airflows in the atmospheric circulation system, gradually being dispersed in three-dimensional space. Redistribution of the tritium concentration on a global scale occurs within a certain time, the exact value of which has not been established and is the subject of further investigation.

The recorded tritium content in water vapor (~500 TU) is a real fact and, considering its initial concentration in the radioactive cloud (10^9 TU), seems quite probable.

The conditions under which the increased tritium concentrations in water vapor have been noted were determined on the basis of analysis of the synoptic situation in the Northern Hemisphere from March to the end of April 1978. For this, we used the daily constant-pressure charts at upper levels of 1.5 km (850-mb chart), 3 km (700-mb), 5 km (500-mb), 9 km (300-mb), 12 km (200-mb) and 16 km (100 mb).[1] They served as a basis for constructing the trajectories of air particles

[1] The initial meteorological data for calculating the trajectories of air masses were presented by the USSR Hydrometeorological Research Center.

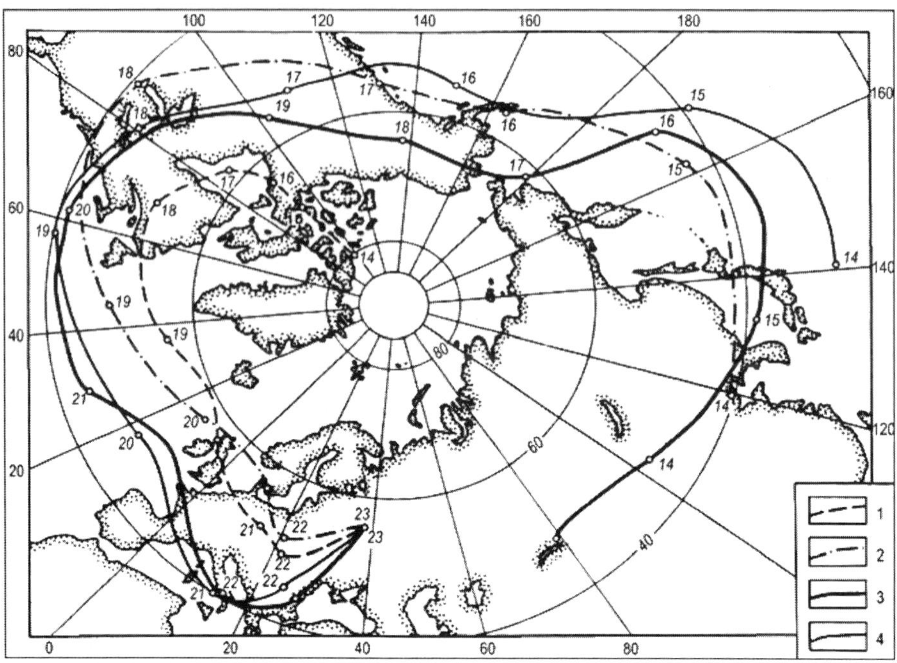

Fig. 18.2 Calculated trajectories and dates of movement of air masses at heights: (*1*) 3 (700); (*2*) 5 (500); (*3*) 9 (300); (*4*) 12 (200) km (the atmospheric pressure. mb, is indicated in the parentheses). Trajectories 2 and 3 are 'tied' to noon and 1 and 4 midnight Greenwich time

in the direction mainly from Moscow, i.e., according to the principle of 'backward movement'. As a result, it was possible to trace the air mass with a certain tritium concentration in water vapor arrived in Moscow.

In certain cases, the 'forward' trajectories of air particles were also analyzed. They helped determine the further trajectories of air masses from Moscow. It should be noted that the products of the nuclear explosion originally remained in the troposphere, since the March 14, 1978 explosion had a low power. Subsequently, the vertical motions of the air masses could have carried these products both to a greater and lesser height (within the troposphere) relative to the height of ascent of the center of the radioactive cloud.

On the basis of analysis of the above results, the following conclusions were made.

The increasing tritium concentration recorded in Moscow on March 23, 1978 was associated with the passage over the city of an air mass which on March 14–15 was over China and passed over Japan, the Pacific ocean, Canada, USA, Atlantic Ocean, Mediterranean Sea, and southeastern Europe (Fig. 18.2). The movement of the air mass with an increased tritium concentration in water vapor was traced at the height of the 300-mb constant-pressure surface (9 km). At lower heights (5 and 3 km), the air masses moved more slowly owing to the greater inhomogeneity of the pressure field and slower rate of travel of the air masses than at a height of 9 km. For instance, at a height of 9 km, the investigated air mass covered the distance from Moscow to Bucharest within 12 h on March 23 whereas at a height of 5 km the air particles covered half this distance.

The analysis of the pressure field shows that on March 23, at heights of 5 and 3 km, the air particles reached Moscow from quite different regions of the Northern Hemisphere than the air particles that traveled at height of 9 km. According to the authors' calculations, the air particles that reached Moscow on March 23 at the 12-km level (200-mb chart) had been over the Pacific Ocean east of Japan on March 14–15, i.e., east of the region of the explosion. However, in the given case it is necessary to take into account the fact that the 200-mb charts are compiled once a day unlike the 300-mb and 500-mb charts, which are compiled twice, at midnight (3 a.m.) and noon (3 p.m.) (Greenwich mean time, in parentheses is Moscow time). Therefore, a greater error in calculating the trajectories of air particles at the 200-mb level is possible.

The constant-pressure charts showed that from March 14 to the end of April, in temperature latitudes of the Pacific, there was a very stable zonal circulation of air masses that prevented meridional dispersion of the radioactive cloud which moved in the strong westerly flow of the air mass that arrived in Moscow on March 23.

Thus, the increase in the tritium concentration in atmospheric water vapor recorded in Moscow on March 23 was due to the arrival of an air mass which on the day of the nuclear explosion was over northwestern China. From there, this mass traveled eastward in the layer between the 5– and 12–km levels along the trajectories shown in Fig. 18.3.

The sample of atmospheric water vapor taken on March 28, 1978 was characterized by a normal tritium concentration (120 TU), since the air mass from which the sample was taken on that day was over the Indochina peninsula on the day of the explosion and moved along a more southerly trajectory than the air mass that formed over China on March 14, 1978 (see Fig. 18.3). The conditions of the atmospheric general circulation (a cyclone over Yakutia and in anticyclone in southern latitudes) did not promote its movement with more northern air masses. About half of its trajectory passed over the Pacific in the latitudinal belt from 20 to 30°N and then over the territory of the USA. The trajectory of this mass over the Atlantic and Europe was little different from the trajectory of the air mass that reached Moscow on March 23 but lagged behind the latter.

Apparently a 'clean' air mass not related to the radioactive cloud arrived in Moscow on March 28 (the tritium concentration in the sample of atmospheric water vapor was negligible). It was assumed that after passage of the 'labeled' air mass in the atmosphere over the Atlantic and Europe, the trail of increased tritium concentration was not preserved and the air masses coming immediately after were not enriched with this isotope.

Of interest were the trajectories of the air particles that reached Moscow on April 4 (see Fig. 18.3) and caused an increase in the tritium concentration in atmospheric water vapor. The air mass that arrived Moscow from China on March 23 passed over northern regions of the USSR and enters the Arctic Ocean region at east of

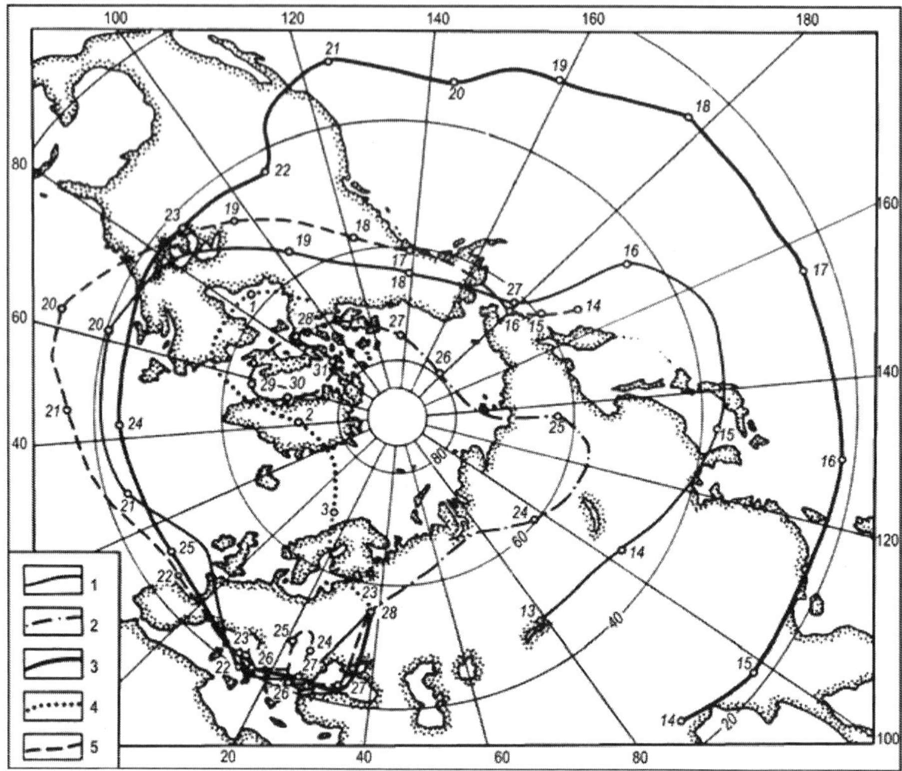

Fig. 18.3 Calculated trajectories and date of movement of air masses that arrived in the Moscow region. At a height of 9 km (300 mb): (*1*) and (*2*) March 23 (trajectory 2 is plotted as 'forward movement'); (*3*) March 28; (*4*) April 4; and also at height of 5 km (500 mb): (*5*) March 28. The trajectories are 'tied' to noon Greenwich time

the Lena River. Then its trajectory turned in the cyclonic vortex near the western coast of Greenland over Baffin Bay. From there, it crossed the Greenland Sea and Scandinavia and by April 4 again arrived to Moscow, where the next peak of tritium concentration was observed.

It is interesting that the concentration recorded on that day exceeded by 80 TU than the concentration measured on March 23. The cause of this increase has still not been deciphered, but preliminarily it can be explained in the following way. Westerly winds prevailed over the region of the explosion from mid-March to the end of April and over Yakutia there was a low-pressure region characterized by ascending air motions, which was clearly traced up to the height of the 100-mb surface (16 km). Thus in the investigated period, the conditions of atmospheric circulation were not favorable for the inflow of tritium from the stratosphere and, probably, the increase in the concentration of this isotope in the investigated air mass cloud have been a manifestation of the 'continental' effect during its passage over Eurasia.

It is also possible that the increased tritium concentration recorded in Moscow on April 4 is a result of the superposition of two processes such as the arrival of an air mass (still enriched with tritium) from the region of Baffin Bay and the arrival of an air mass at the 5-km level from the region of the explosion, where possibly the high tritium concentrations were still preserved by the start of its travel to Moscow.

The peak T concentrations on April 13, 18, and 26 (see Fig. 18.3) are also probably related to the March 14, 1978 explosion, since considerable excesses of the concentrations of short-lived isotopes (such as ^{140}Ba, ^{131}J, ^{95}Zr) in the atmosphere caused by this same explosion were noted at stations of the British AERE network at the end of April (Cambray et al. 1979).

Observations of the tritium label produced as a result of the nuclear explosion in China and plotting of the trajectories of air particles made it possible to outline the path covered by the air mass with increased radioactivity and to determine the time spent on it. This mass arrived Moscow from China nine days after the explosion, covering about 30,000 km during this time. At a height of 9 km, the air particles traveled at a rate of about 3,000 km/day. The air mass needed 12 days to go around the Earth in the latitudinal belt from 40 to 65°N (in individual places the trajectory of the air particles 35°N); in this case, the length of the path covered by it was 33,000 km. The time of circulation of water vapor around the Earth, equal to 12 days, pertains to a height at which very little moisture is usually transported.

The analysis of the movement of the air mass 'labeled' by the nuclear explosion showed that the time of circulation of water vapor around the Earth is not only a theoretical concept but also a natural phenomenon. Long-time existence and movement of the moisture 'package' without their substantial dispersion, at least in the sub-stratosphere, is actually possible under certain conditions of atmospheric circulation. The method used permits tracking the movement of air masses and atmospheric moisture both in cloudy and rainy as well as in clear and dry weather.

The results obtained permit the conclusion that the method of measuring the tritium content in atmospheric vapor can become an effective means of experimental study of global circulation of air masses and atmospheric water vapor, which is extremely important to solve solving the problem of formation of land water resources. In this case, it is not necessary to 'use' nuclear explosions for artificial creation of peak tritium concentrations. For this purpose, tritium can be injected at the required height by ordinary technical means.

Kigoshi and Yoneda (1970) studied daily variation of tritium concentrations in atmospheric moisture in Tokyo region. They found that the increased daily tritium concentrations is an indicator of the continental air masses inflowing from the north and the decreased concentrations attest the entry of oceanic moisture. The daily tritium variation in the atmospheric moisture appears to be useful tool in discovery of its origin. It was shown that daily tritium variation in the water vapor at the Earth's surface connects with tritium entering from the stratosphere due to the vertical convection—with moisture coming from the continent or ocean—and with horizontal air masses advection. The last process was accepted by the authors as a main effect in analysis of their experimental results.

For understanding the nature and evolution of the hurricanes, it is important to study their velocity and role of the local water evaporation from the ocean and the total energy budget of the cyclone. Tritium measurements carried out by Östlund (1967, 1970) and Machta (1969) during studies of the hurricanes Hilda (1964), Betsy (1965), and Fetty (1966) allowed the authors to improve significantly the problem solution. The observed difference in tritium content in the atmospheric moisture and in the surface layer of the ocean water was accepted as a basis of the investigations. Applying the difference in the observed tritium content in water vapor of the eye and periphery of the hurricanes, and taking into account the wind parameters, temperature, and moisture, the molecular exchange and evaporation rate were calculated. It was found that the relative molecular exchange rate inside the Betsy (1965) with radius 100 km is 30–50%. For the Fetty (1966) hurricane, its total water budget of the sea water is 7.5% at the cyclone radius equal to 100 km and 10.5% at the cyclone radius equal to 150 km.

As it was pointed out earlier, tritium concentration in atmospheric precipitation over the continents for the same latitude is higher than that in precipitation over the oceans. This is because of dilution of atmospheric moisture with higher tritium concentration of the surface sea water and vapor which has its lower concentration. Evaporation from a free surface and the molecular exchange are the mechanisms determining dilution in tritium of the atmospheric moisture by the surface sea water. Our laboratory experiments have shown that the molecular exchange rate within 15–30°C temperature interval amounts 10^{-5}–10^{-6} cm/s, which corresponds to daily exchange of water layer of 0.1–1 cm with the atmospheric moisture.

Barry and Merritt (1970) and Brown and Barry (1979) studied the water evaporation rate from lake's surface using tritium. Their method was based on tritium concentration change in the air column of water vapor determining the dead and head shore of wind.

Romanov and Kikichev (1981) have carried out field work on 'Muksun' expedition in 1978 for investigation of the atmospheric moisture dilution by the surface sea water. They did simultaneous sampling of the sea surface water and the atmospheric water vapor in the north-west part of the Black Sea. Water vapor sampling was carried out by the method of absorption on a synthetic zeolite. The tritium measurements were done by the liquid-scintillation method with preliminary electrolytic enrichment. The results of this work are presented in Table 18.1. The tritium concentration measurements were produced with relative accuracy of 10%.

Location of the sampling points is shown in Fig. 18.4. Stations 1–6 were stationary states; sampling on 7–9 stations was carried out during forward motion. The average samples of water vapor and sea water sampling are shown in Fig. 18.4.

It was shown by analysis of the baric topography charts that during the above discussed marine expedition the motion of the air masses in general has taken place from the northwest to the southeast.

The tritium concentration in the surface waters of the Black Sea has had tendency to decrease to the southeast direction from 57 to 28 TU. An increase in tritium content level in the near-shore water area is explained by dilution of the sea water by the Danube River runoff, in which the tritium concentration during investigation reached 91 TU. Analogous results were obtained by the authors in 1975.

Table 18.1 Tritium concentration in the atmospheric moisture and surface water of the Black Sea in 1978

Number of station	Sampling date	Sampling time (h)	Coordinates		Concentration (TU)		Portion of water in water vapor
			N	E	Water vapor	Contaminated water	
1	09.08	12^{50}–17^{30}	44° 25	34° 27	82	28	0.63
2	11.08	15^{25}–18^{25}	46° 00	31° 20	113	42	0.46
3	12.08	00^{50}–05^{15}	46° 36	31° 32	185	33	0.00
4	12.08	20^{55}–23^{55}	46° 00	30° 47	160	50	0.11
5	13.08	02^{25}–05^{30}	46° 00	30° 32	128	57	0.39
6	13.08	14^{40}–17^{55}			164	91	0.00
7	14.08	16^{00}–20^{30}	Vilkov City from 45° 46	Vilkov City to 30° 22	99	–	–
8	14.08	20^{30}–02^{00}	45° 46–44° 42	30° 22–31° 45	84	41	0.68
9	15.08	02^{00}–06^{20}	45° 42–45° 17	31° 45–32° 14	85	33	0.63

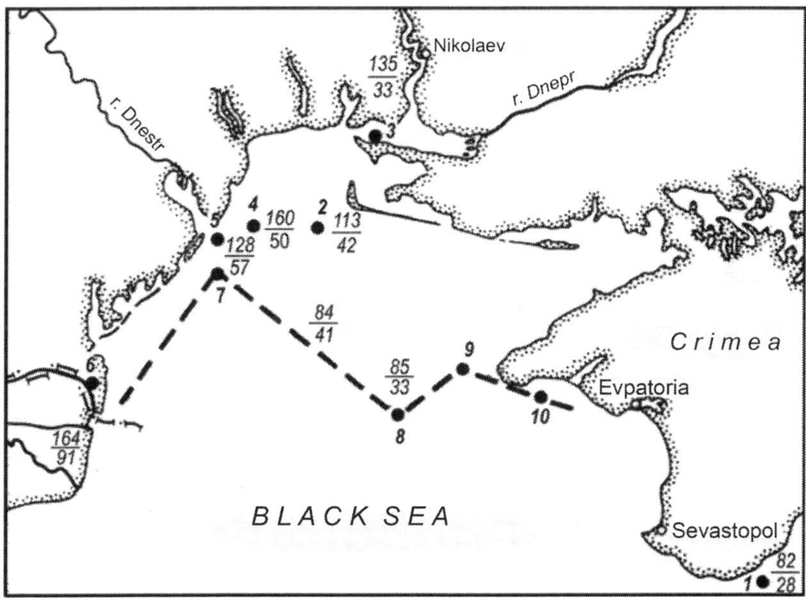

Fig. 18.4 Location of the water and atmospheric vapor sampling points for tritium content measurement (Figures at the station numbers show tritium concentration in the atmospheric water vapor (numerator) and in the surface sea water (divisor))

The following balance equation was used to calculate the tritium content in local sea water and atmospheric moisture: $C_x = C_a(1 - a) + C_w a$, where C_a and C_w are the tritium concentrations in the atmospheric water vapor and sea surface water accordingly; C_x is the tritium concentration in the continental atmospheric vapor; a is the vapor portion of the sea surface water in the atmospheric vapor. In the above equation, the effect of the isotope fractionation is not taken into account.

Three different groups of values of tritium concentrations were observed in the atmospheric moisture. For the stations 3 and 6, located at the mouth of Danube and in Dnieper-Bug Bay, the tritium content level is maximum (164 and 185 TU). The mean value obtained through calculation is 174 TU, which is accepted as a characteristic for the continental atmospheric moisture for the studied region. For the stations 2, 4, 5, and 7, the tritium concentrations have intermediate values from 99 to 160 TU, which corresponds to 11–46% contribution to the sea water. For the station 7, the sea water content in the atmospheric vapor was not calculated because of absence of the tritium content in the sea surface water.

For the stations 1, 8, and 9, located farther in the sea of others, the minimal tritium concentrations (82–85 TU) were obtained, which corresponds to 63–68% of marine water. It is interesting to note that the obtained close results related to sea water portion in the atmospheric water vapor were obtained in precipitation over the Issyk-Kul Lake, whose value reaches 50% (Brezgunov et al. 1980).

This study has shown that the tritium method is a promising tool for the study of atmospheric water exchange for determining the origin of moisture.

18.2 Mixing of River and Sea Waters in Estuaries

One of the near-shore water exchange problem is the marine and river water mixing in the estuaries. Experimental studies of this process by application of the natural tritium and use of other isotopes is a promising direction in the oceanographic investigations. The physical basis for tritium use in the study of interaction between river and sea water appears to be a known global and regional distribution of the isotope and its space and time functions and values of its concentrations in the atmospheric moisture, rivers, and seas.

As it is known (Mamaev 1970), an indicator for study of the processes of the marine water turbulent mixing should satisfy two demands, namely: (1) it should be conservative, i.e., safe at the turbulent transfer and at mixing; (2) it should be passive, i.e., does not affect the character of turbulence. The HTO molecules possess the passivity property, i.e., they do not affect the character of turbulence. The HTO molecule can be accounted as a conservative indicator if the time scale of water mixing is much less than the tritium half-life period, it means less than one year. The mixing river and sea water processes in the estuaries are such kind of phenomenon.

The time scale of vertical mixing of the sea water is close to the half-life period of tritium and it cannot be accounted as a conservative indicator to be used for study of such a process; however, in this case, the other tritium property can be used, namely, the law of the tritium radioactive decay, which is used for investigation of mixing of different ages of water masses by determination of the transfer time. The second property, which is passivity, is completely satisfied. The formation of the salinity and

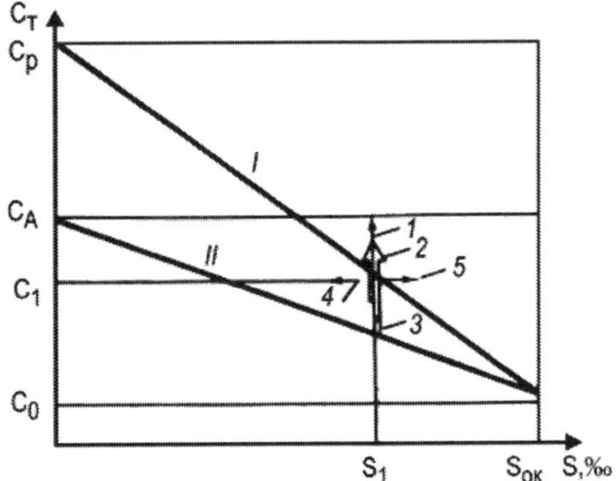

Fig. 18.5 Effects of different natural processes on concentration of surface water, tritium, and salinity: (*1*) molecular exchange; (*2*) 'memory'; (*3*) radioactive decay; (*4*) ice melting; (*5*) ice; (*I*), (*II*) lines of mixing of marine water with river, and precipitation waters

isotopic composition of the surface and near-shore marine waters is schematically shown in Fig. 18.5.

If only the surface and sea waters intermix, then the tritium concentrations of the near-shore waters lie on line *I*. If only the surface and precipitation waters intermix, then the concentration values lie on line *II*. If both sources take part in dilution corresponding to the observable values of tritium and salinity concentrations, then the points lie between lines *I* and *II*. This simple picture of mixing is complicated by a number of factors. Among them are molecular tritium exchange between the atmospheric moisture and the surface layer of water which does not change the salinity but increases the tritium concentration. The fact that tritium concentration in fresh waters has a tendency to decrease in time also shifts the point with co-ordinates C_1 and S_1 upwards. This is because at the moment of observation, tritium can be present being came with fresh waters. The radioactive decay of tritium decreases its concentration depending on its residence time. Finally, ice melting and icy process changes the salinity of water without change in its isotopic composition.

Experimental studies were carried out in the Black and White Sea basins to study the sea and river waters mixing with tritium (Romanov 1982). The considered below results were obtained on the basis of expedition works carried out during 1975–1979 and by measurements of tritium concentrations in precipitation (see also Chap. 12).

Let us estimate the time and rate of exchange of the Black Sea surface and deep layers on the basis of estimation of the total tritium fell out from the atmosphere. The data on tritium concentration in marine water are in Table 18.2.

For determination of the water exchange time of the surface water layer, i.e., the ratio of the surface layer thickness to the value of its exchanging part in unit time, the box model is applicable. The main model admission is equality of concentration of the indicator at the outflow from the reservoir to its mean value in the reservoir itself and constancy of the latter.

Table 18.2 Tritium concentration in Black Sea water

Year of sampling	Number of measurments	Tritium concentration (TU)	Reference
1964	2	107	Romanov (1982)
1970	–	100	Michel and Suess (1978)
1975	19	65	Romanov (1982)
1978	23	41	Romanov (1982)

Equation of water and isotope balance for the surface layer is written as:

$$\left.\begin{aligned} dH &= Pdt - Bdt \\ d(CH) &= C_A Adt - C_B dt - \lambda HCdt \end{aligned}\right\}, \tag{18.1}$$

where H is the thickness of the mixing layer; A is the water inflow per time unit; B is water discharge per time unit; C, C_A, and C_B are the tritium concentrations in the mixing layer, in the inflowing water, and in the discharged part of water accordingly; λ is the tritium decay constant.

With the accepted assumptions for the model $dH = 0$ and $C_B = C$, the equation system (18.1) is rewritten as:

$$\left.\begin{aligned} A &= B \\ HdC &= C_1 Adt - CAdt - \lambda HCdt \end{aligned}\right\}. \tag{18.2}$$

By definition, the water exchange time τ is equal to

$$\tau = H/A \tag{18.3}$$

Then, after a simple transformation of (18.2) and (18.3), one obtains

$$dC = (C_A/\tau)dt - (B/\tau)dt - \pi Cdt$$

or

$$dC/d\tau + C\,(1/\tau + \lambda) = C_A/\tau. \tag{18.4}$$

After solving Eq. (18.4) relative to C, one has

$$C = \int_0^\infty (1/\tau)\,C_A\,(t)\exp[-(\lambda + 1/\tau)\,t]\,dt. \tag{18.5}$$

At the same time, integration is performed with respect to time from the moment of sampling. In order to find time τ, one should know the tritium concentration change in the recharge source $C_A(t)$ within interval t (from 0 to ∞). In the case of constancy $C_A(t) = C_A = \text{const}$ and τ in time, Eq. (18.5) has a form:

$$C = C_A/\tau(\lambda + 1/\tau)\,. \tag{18.6}$$

Isotopic composition of the surface layer of sea water is determined by different types of recharging water. These are atmospheric precipitation, atmospheric water vapor, river water, and also deep marine water. The discharging components include evaporation lost and sinking of the surface water into deep layers. At the same time, the advective transfer of the surface water is neglected because the horizontal gradient of tritium concentration in the open sea is too small. Finally one has

$$\left.\begin{aligned} dH &= Pdt \pm Mdt + \textstyle\sum R_i dt - Edt \pm Kdt \\ d(CH) &= C_A Pdt + MC_A dt - MCdt + \textstyle\sum C_{R_i} R_i dt \\ &\quad - ECdt + KC_K dt - \lambda CHdt \end{aligned}\right\}, \tag{18.7}$$

where P is the precipitation amount; M is the thickness of water layer taking part in the molecular exchange with the atmosphere; $\sum R_i$ is the total river runoff; E is the evaporation; K is the layer thickness of exchange between the surface and deep layer (their values are taken in time units); C_A, C_{Ri},C_K, C are the tritium concentrations in atmospheric moisture, river water, deep marine water, and in surface water layer respectively.

Exchange of P/H, M/H, E/H, $K/$H, and R_i/H by p, m, l, k, r_i respectively. The river runoff $\sum C_{Ri}R_i$ is defined by $C_r r$, where C_r is the mean-weighted value (with respect to the runoff) tritium concentration in river water. Then, assuming $dH = 0$, from the Eq. (18.7) one obtains

$$\left.\begin{aligned} &0 = p + r - e \pm k \pm m \\ &dC/dt + C\,(m + e + k + \lambda) = C_A\,(p + m) + C_r r + C_k k \end{aligned}\right\}. \tag{18.8}$$

The Eq. 18.8 is solved similar to Eq. 18.4. The second equation (Eq. 18.8) can be used directly for calculation of the rate τ of the surface layer water exchange and the rate k of exchange between the surface and deep water layers. In this case, it is assumed that $1/\tau = p + r + m$ at always satisfied condition $|-m| = |+m|$ and assumption that $|-k| = |+k|$, i.e., at absence of the pure water transfer upwards and downwards from the lower border of the surface layer. Then, the tritium concentration change in the surface layer ΔC within the time unit Δt is written as

$$\Delta C/\Delta t = \left(\overline{C}_A - \overline{C}\right) m + \left(\overline{C}_k - \overline{C}\right) k + \overline{C}_r r + \overline{C}_A p - \overline{C}_e - \lambda \overline{C}, \tag{18.9}$$

where $\overline{C}$, $\overline{C}_A$, $\overline{C}_k$, and $\overline{C}_r$ are mean values of tritium concentration within time Δt in the surface sea layer, atmospheric moisture or precipitation, and in the deep marine and river water.

Introduce τ and accept that $p/r = u$; then $r = 1/\tau\,(1 + u)$; $p = u/r\,(1 + u)$. Substituting p and r in terms of u and τ into Eq. 18.9, one obtains

$$\begin{aligned} \Delta C/\Delta t = (\overline{C}_A - \overline{C})m + (\overline{C}_k - \overline{C})k + [u/\tau(1 + u)](\overline{C}_A - \overline{C}_r) \\ + 1/\tau(\overline{C}_r - \overline{C}) - \lambda\overline{C}. \end{aligned} \tag{18.10}$$

Evaluate the role of each of the terms in Eq. 18.10. The molecular exchange appears to be a transfer of the water molecules from the liquid surface to the atmosphere and vice versa, and the difference rate of the water molecules' transfer from one phase to another depending on the sign appears to be the rate of evaporation or condensation. This difference is accounted in equation of the water balance, but because the tritium concentration in the surface water and atmospheric moisture are different, in equation of the isotope balance all the water molecules which take part in the molecular exchange should be taken into account. Experimental studies carried out by Horton et al. (1971) and also the authors of this book have shown that the rate of this process is proportional to the atmospheric moisture. In this connection, the mean velocity value of the molecular exchange m for unit of area is

$$m = 0.0241p, \tag{18.11}$$

where m is defined in m/day and p in Hg mm.

For example, a layer of the exchanged water at 15°C and relative moisture of 70% during one month is accounted by about 65 mm, i.e., by the value related to the fall of precipitation.

The parameter p reflects the role of the atmospheric precipitation and river runoff. In the case of open oceans, $p \gg r$ and for the outskirt seas it is vice versa, i.e., $r \gg p$. In the ocean, the tritium concentration in deep layers is lower than the surface. In this connection, Eq. 18.10 for the ocean can be simplified and written as

$$\Delta C/\Delta t \approx \left(\overline{C}_A - \overline{C}\right) m - \overline{C}k + (1/\tau)\left(\overline{C}_A - \overline{C}\right) - \lambda\overline{C}. \tag{18.12}$$

If $r \gg p$, then

$$\Delta C/\Delta t \approx \left(\overline{C}_A - \overline{C}\right) m - \overline{C}k + (1/\tau)\left(\overline{C}_r - \overline{C}\right) - \lambda\overline{C}. \tag{18.13}$$

The above data at best correspond to the respective calculations done by box model for the case of water exchange time equal to 112 years and diluted by deep water of zero tritium concentration by 3.5 times. The results of this model calculation are shown in Fig. 18.6 (the curve 2 was calculated by Eq. 18.7 for $\tau = 11.2$ years, the curve 3 was obtained from curve 2 by dividing its parameters by 3.5). It is seen that the coincidence of the experimental data with the calculated curve 3 is quite good.

The obtained data allow estimating the mixing layer thickness and the deep water rate to the surface layer. The total amount of water in the form of atmospheric precipitation, river runoff, and molecular exchange put together a layer, thickness of which is 1.7 m (Table 18.3). The deep water contribution amounts $1.7 \times 3.5 = 5.9$ m. If the water exchange time is $r = 11.2$ years, then the thickness of the mixing layer should be $11.2 \times 5.9 = 67$ m, which is in good agreement with the real observed distribution of tritium (Fig. 18.7a). The theoretical curve of tritium concentration in mixing layer, being function of time, gives delay of one year, i.e., experimental values of tritium concentration correspond to mixing of the marine and river waters of one-year age.

Another method can be used for determination of mixing rate of the surface and deep water layers. For this purpose, we apply the isotope and water balance of the surface Black Sea waters, for example, over 1975–1978.

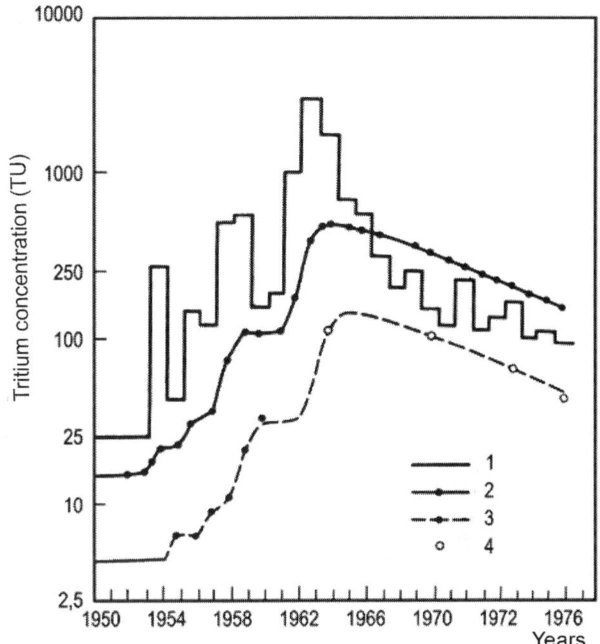

Fig. 18.6 Time changing tritium concentrations for determination of the surface layer water exchange in the Black Sea. Tritium concentration (TU): (*1*) in atmospheric precipitation; (*2*) in the sea surface water at 11.2 years of water exchange; (*3*) at dilution by 3.5 times of the deep water; (*4*) experimental data

In fact, during three years, tritium concentration in the surface layer dropped by 24 TU (by 8 TU per year). This decrease was caused by preferential radioactive decay, evaporation, runoff through the Bosporus and exchange with deep layers compared to the inflow of river runoff, precipitation, and molecular exchange (Table 18.3).

Table 18.3 Mean annual tritium budget of 75 m layer of marine water in the Black Sea during 1975–1978

Budget components	Water volume (m^3/year)	Layer thickness (cm)	Mean concentration (TU)	Amount (TU · cm/year)
Inflow				
Lower Bosporus current	176	42.5	5	200
Inflow from Azov Sea	53	12.8	103	1.330
Atmospheric precipitation	119	26.4	111	2.940
Continental runoff	346	83	103	8.500
Molecular exchange	–	48[b]	111	5.350
Outflow				
Upper Bosporus current	340	83	53	4.400
Outflow to Azov Sea	32	7.8	53	415
Evaporation	332	80	53	4.250
Molecular exchange	–	48[b]	53	2.550
Radioactive decay	–	7.500	53	22.260
Balance			–15.755	(–2.1 TU/year)

[a]Data of Afanasyev (1960)

[b]Estimate calculation of the molecular exchange carried out at mean annual moisture for Tuapse over 1960–1970 equal to 9.3 g/m^3

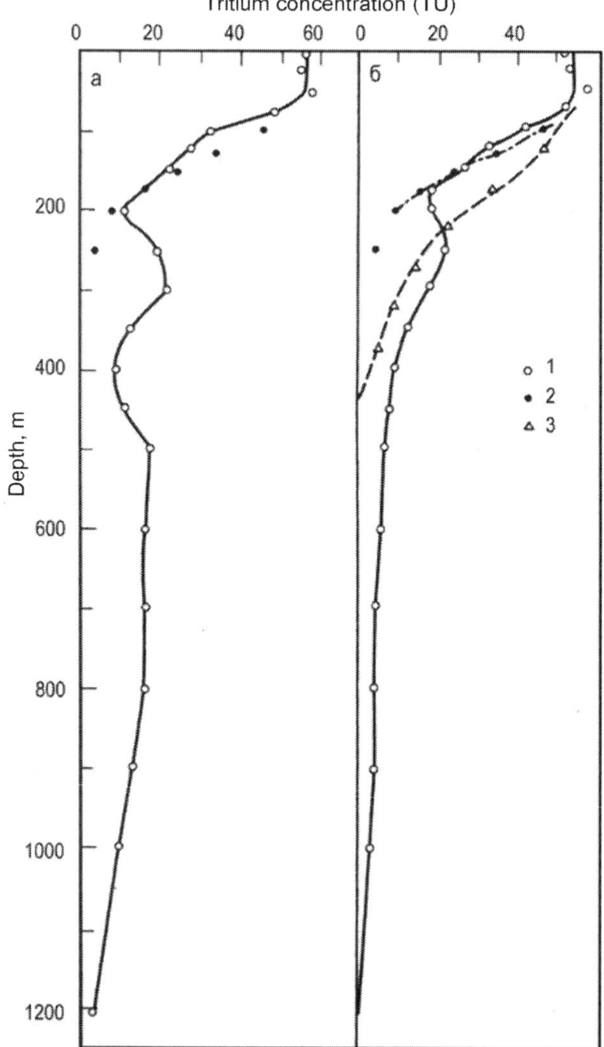

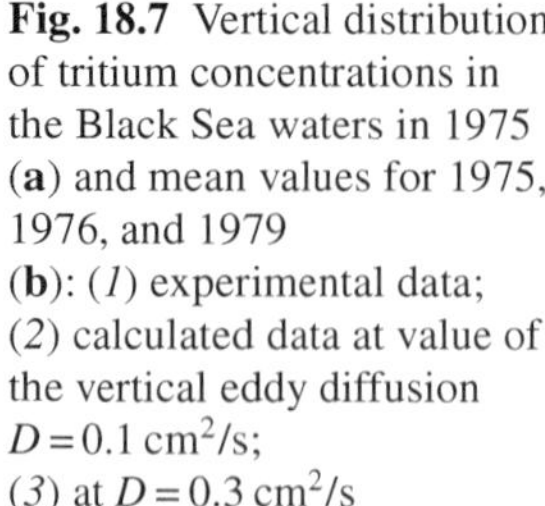

Fig. 18.7 Vertical distribution of tritium concentrations in the Black Sea waters in 1975 (**a**) and mean values for 1975, 1976, and 1979 (**b**): (*1*) experimental data; (*2*) calculated data at value of the vertical eddy diffusion $D = 0.1\ \text{cm}^2/\text{s}$; (*3*) at $D = 0.3\ \text{cm}^2/\text{s}$

Table 18.3 also presents mean values of tritium concentrations in river waters and mean-weighted values of tritium concentration in the runoff waters. For the atmospheric precipitation and atmospheric moisture, the mean values accepted from the Odessa region (northwest of the Black Sea region) and the mean tritium concentrations in the surface layer accepted the data obtained during 1975–1978. Decrease in concentration by 2.1 TU/year is caused by all the balance components except the exchange between deep layers , to which the difference between the calculated and observed data is $8 - 2.1 = 5.9$ TU/year. This corresponds to the layer equal to $5.9 \times 75/53 = 8.3$ m. Thus, the surface water layer of 8.3 m thickness is annually exchanged with the deep layers, which is rather close (within the obtained assumptions)

Table 18.4 Tritium concentrations in waters of the White Sea basin

Water source	Sampling year	Sampling location	Amount of samples	Concentration of tritium (TU)
Atmospheric precipitation	1978	Arkhangelsk	105	12
	1979	Arkhangelsk	121	12
	1979	North Dvina	99	5
	1978	Onega	78	12
	1979	Onega	86	4
	1978	Mezen	84	2
Sea water	1978	White Sea (total area)	55	8
	1979	White Sea (total area)	48	34
	1978	Barents Sea	10	50
	1979	Barents Sea	10	40

to the model results at 5.9 m. This fact gives idea to apply the balance method of the water exchange rate for those reservoirs where the long rows of tritium content observation are absent.

By analogous method of the water exchange, the intensity between the White and Barents seas was applied. In Table 18.4, the data on tritium concentrations in marine and river waters as well as in atmospheric precipitation of the White Sea region are given. The data obtained were on the basis of analysis of samples collected during the expedition studies in 1978–1979.

For calculation of the mean tritium concentration values in the White Sea waters, the data of 20–30% salinity was used. Table 18.5 presents the data of water and isotope balance.

The tritium concentration for the discharged water from the White Sea was determined to be its content in the sea water at Kanin Cape where it is 27.7 TU. The mean value of tritium concentration in the river runoff was accepted to be equal as mean-weighted relative to the North Dvina runoff, which is accounted as half-value of rivers' runoff recharged to the White Sea. The other values of tritium concentration are given in Table 18.5, which are mean-arithmetic values of tritium content over 1978 and 1979.

After a year, the tritium concentration in the White Sea, having volume of 5,400 km^3, decreased by 7.3 TU (see Table 18.5). This value should be equal to difference between the output and input of the balance, such as $7.3 \times 5{,}400 = 3{,}948 + 27.7(X + 218) + 15{,}725 - 19{,}565 - 4{,}407 - 4{,}294 - X \times 10$, from where the recharge from the Barents Sea is $X = 1{,}517$ km^3. According to data of Dobrovolsky and Zalogin (1963), it accounts by 2,200 km^3. The difference between the two values is explained by measurements of tritium content in the sea waters in 1978 and 1979 at different points and also by poor investigation of the tritium vertical distribution. Taking into account that both isotopic and hydrological methods are approximate methods, we account their results as comparable.

The coefficient of the turbulent diffusion for the Black, White, and Barents seas was experimentally determined. It was investigated in more details for the Black Sea basin in 1975–1979 (Table 18.6).

Table 18.5 Water and isotope balance of the White Sea

Balance components	Water volume (km^3/year)	Concentration of tritium (TU)	Tritium amount (TU/km^3)
Inflow			
Rivers	215	91	19,565
Atmospheric precipitation	39	113	4,407
Molecular exchange	38[a]	113	4,294
Inflow from Barents Sea	X	10	X × 10
Outflow			
Evaporation + exchange	76	52[b]	3,948
Outflow to Barents Sea	X + 216	27.7	27.7(X + 216)
Radioactive decay	5,400	52[b]	15,725

[a]Calculated on basis of moisture data for Onega City; X is the inflow from the Barents Sea equal to 1,517 m^3

[b]Mean value for 1978–1979 (see Table 18.4)

It is seen from the data in Table 18.6 that the vertical water circulation covers all the sea water thickness. Up to 500 m, significant tritium concentrations are observed. Lower, up to 1,000 m, such tritium concentrations in the central part of the sea were fixed only in 1975.

In order to compare tritium concentrations in samples taken from different depth and times, the values were reduced to common calculated depths and one (1975) year. This operation is produced by linear interpolation of the data and taking into account the radioactive decay of the tritium by equation

$$C(z, t_0) = \left[\frac{C(z_1, t_0) - C(z_2, t_0)}{z_1 - z_2}(z_1 - z_2) + C(z, t_0)\right] e^{-\lambda(t-t_0)},$$

where $C(z, t_0)$ is the tritium concentration on the depth z and at the time moment t_0; $C(z_1, t_0)$ and $C(z_2, t_0)$ are the tritium concentrations at the depths z_1 and z_2, respectively, at the time moment t_0; λ is the decay constant.

This recalculation of the experimental data makes it possible to have general idea about the scale of the vertical mass mixing within the total basin area.

Figure 18.7 represents graphical dependence of the tritium concentration change with respect to the depth obtained by the above means. It is seen that the mean tritium concentrations in water within the upper 75 m are practically similar. After that, up to the depth of 200–250 m, the concentration little bit increases and then again it decreases with the depth. The corresponding points of tritium distribution with the depth at the coefficients of turbulent diffusion mixing of 0.3 and 0.1 cm^2/s are drawn in Fig. 18.7. The parameters of these points were obtained by equation of the diffusion transfer (see Chap. 12) with the border condition $C_n(t)$—which means the time change of tritium concentration in the surface sea water layer—and graphically presented in Fig. 18.7 (curve 3). This is seen by decreases of the tritium concentration within the depth interval of 75–175 m, which well corresponds to the condition of water mass mixing with the coefficient of turbulent diffusion equal to 0.1 cm^2/s.

Table 18.6 Tritium distribution in the Black Sea section

Coordinates N/E	Sampling depth (m)	Tritium content (TU)	Salinity (‰)	Coordinates N/E	Sampling depth (m)	Tritium content (TU)	Salinity (‰)
15.06.75				01.07.76			
43°34.2	0	60.4	–	44°05.4	200	39.9	–
30°07.8	30	39.4	–	34°25.7	500	0	–
	60	43.6	–		1,000	0	–
	115	31.4	–		1,500	0	–
	210	11.7	–		1,750	0	–
	260	43	–	18.07.76			
	360	2.9	–	42°18.3	0	40.1	–
	400	0.6	–	34°22.0	10	26.1	–
	485	0.7	–		20	34.7	
21.06.75					50	36.0	–
44°05.2	0	52.5	–		100	44.6	–
35°00.0	25	67.3	–		200	22.6	–
	50	74.4	–		300	10.0	–
	75	57.4	–		500	0	–
	100	30.4	–		750	0	–
	150	20	–		1,000	0	–
	200	4.1	–		1,500	0	–
	225	13.8	–		2,150	0	–
	300	15.9	–	10.79			
	500	16.4	–	42°52.5N	0	51	–
	800	15.6	–	40°51.0E			
	989	9.4	–	43°22.0	60	76.5	17.5
	1,300	0	–	40°06.0			
	1,500	5.10	–	43°02.8	128	34.5	20.20
	2,000	0	–	40°25.6			
07.06.76				42°59.0	142	41.4	20.50
43°15.0	0	36.9	–	40°25.5			
30°15.0	50	39.9	–	43°02.7	164	11	20.8
	200	3.9	–	40°22.0			
	750	0	–	42°55.0	180	20.9	20.42
	1,500	0	–	40°25.5			
				42°55.0	200	21.2	20.65
				40°22.0			
				42°52.0	610	9.7	–
				40°47.9			

Analogous calculations of coefficient of the vertical turbulent diffusion were done for the northern part of the White Sea. The coefficient for this basin was found to be equal to 0.07 cm^2/s and for the southern part of the Barents Sea it was equal to 0.24 cm^2/s. The above results have estimative character because they are obtained by the equation where tritium concentration is assumed to be of constant value.

In order to determine the time of the fresh and saline water mixing, let us consider the graphical relationship of tritium in the White Sea waters presented in Fig. 18.8. Because of a very low correlation coefficient between the tritium and salinity concentration, the relationship between two parameters is not a simple linear mixture of the river and marine waters. As it was earlier pointed out (see Fig. 18.5), the deviation from the mixing straight line—drawn through the point with the modern values of

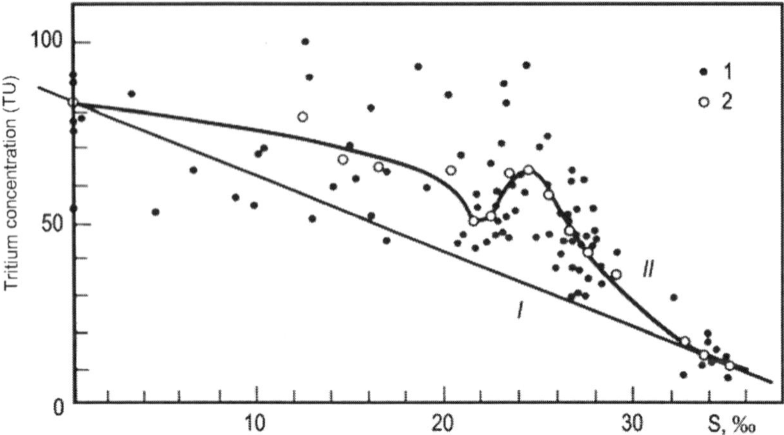

Fig. 18.8 Relationship between tritium and salinity concentrations of the White Sea: (*1*) observed concentrations; (*2*) average; (*I*) mixing line of fresh and sea waters; (*II*) line of ratio of average values of tritium and salinity concentrations

tritium concentration in the fresh and initial saline water of the White Sea—can be initiated by the ice processes, radioactive tritium decay, and effects of memory of the radioactive fallout.

It is seen from Fig. 18.8 that the points are mainly related to the deep layers. The ice process increases salinity of water without changing its isotopic composition, i.e., shifts the points to the right side from the straight line. The 'memory' effect acts in the same direction. It is seen that the observed relationship between the tritium and salinity concentrations in the White Sea waters corresponds to these two process actions. If one averages the values of tritium concentrations up to the intervals of salinity value, then the curve connecting the averaged values of tritium and salinity concentrations occurs. This curve has clear maximum at 24–25‰ of salinity. One may assume that a peak like this corresponds to maximum fallout of tritium in 1963–1964. Thus, the time of mixing of water with such salinity should be 15 years. The value scattering of tritium concentrations shows complication of the process in time during mixing of the river and sea waters. More detailed analysis allows obtaining better results.

Analogous studies on mixing river and sea waters by tritium and oxygen-18 were carried out for the Caspian Sea (Brezgunov et al. 1982; Romanov 1982). Application of both isotopes gives better experimental material for interpretation of the process of river and sea water mixing.

18.3 Water Exchange in the River Basins, Lakes, and Reservoirs

The cosmogenic isotopes have found application to solve the problems related to determination of water exchange rate in the natural reservoirs, relationship between the atmospheric precipitation and the river runoff, and in the hydrograph separation

Table 18.7 Water balance of Lake Baykal

Components of balance	Amounts		Components of balance	Amounts	
	(mm/year)	(km^3/year)		(mm/year)	(km^3/year)
River runoff	1,834	87.77	Angara River runoff	1,933	60.89
Precipitation	294	9.26	Evaporation	294	9.26
Groundwater discharge	99	3.12			
Total	2,227	70.15	Total	2,227	70.15

to estimate the parameters of the groundwater discharge to rivers and reservoirs. As a rule, such studies are combined with use of the hydrogen and oxygen stable isotopes and classical methods. The regularities observed in various natural isotopes in the hydrological cycle—which provide a close relationship between relative proportions in time of atmospheric precipitation in water masses of a reservoir or river basin—create the physical basis for such studies.

The study of the runoff mechanism in the mountain river Dishma (Switzerland) by natural tritium and oxygen-18 was carried out by Martinec et al. (1974). It was found that the main part of water during snow melting in spring is recharged to the groundwater reservoir. The best correlation for interpretation of the isotope data gives the exponential and dispersion models (see Fig. 5.7), according to which the residence times of groundwater of the river basin are 4.5 and 4.8 years respectively.

The other studies for the hydrograph separation of river runoff during snow melting were done by Dinçer et al. (1970), Hermann et al. (1978), Behrens et al. (1979). A successful solution of the problem of hydrograph separation in the study of river runoff from a separate fallout and discharge in the Karstic regions based on tritium and stable isotopes of hydrogen and oxygen were carried out by Schotterer et al. (1979).

Studies in determining the tritium concentrations in waters of Lake Baikal, which with respect to volume of water contained in it is the largest freshwater lake in the world (23,000 km^3, surface area 31,500 km^2, average and maximum depths 730 and 1,620 m), were began in 1963 (Brezgunov et al. 1980; Soyfer et al. 1970). At that time, the water samples were taken and analyzed first time for this purpose at two deep-water stations located in the region of maximum depths. In 1973, the tritium concentration in the surface waters of the lake (along its western shore) was surveyed and the vertical distribution of tritium was obtained in the northern basin of the lake (from Nizheangarsk to the section Kotelnikovsky-Irinda Cape).

The isotopic composition of the lake waters is governed by the interaction of several factors such as the river inflow, amount of precipitation falling on the lake surface, isotope exchange of atmospheric moisture with the water surface, runoff of the Angara River, evaporation, and vertical mixing of water masses of the lake. The role of groundwater flow into the lake at present is still inadequately studied.

Table 18.7, compiled from the data of Afanasyev (1960), demonstrates recharge and discharge components of the lake water balance during 1901–1955. It is seen that the river runoff (more than 82%) represents the main part of the total water balance of the lake.

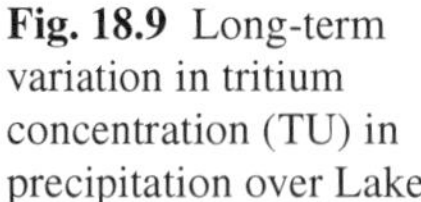

Fig. 18.9 Long-term variation in tritium concentration (TU) in precipitation over Lake

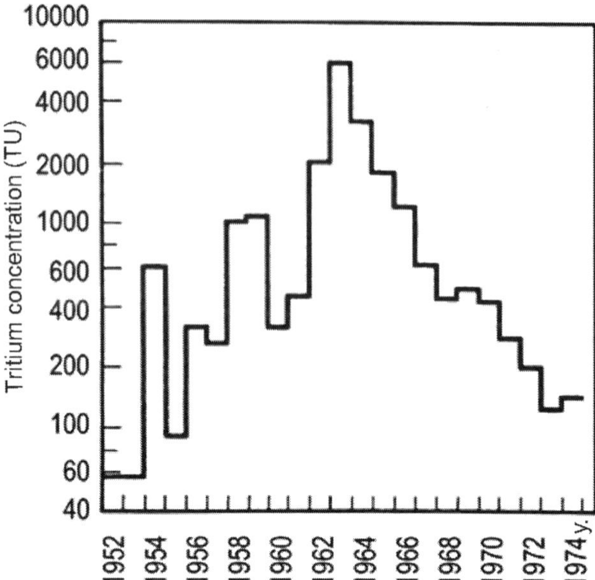

In view of the extensive development of permafrost in the lake region, the depth of the soil drainage by the rivers is small, the water retention time in the soils is negligible, and therefore the tritium content in rivers is little different from that in precipitation. Long-term distribution of tritium concentration in precipitation for one of the lake's section is shown in Fig. 18.9. The missing data (1952–1968) were extrapolated from the data for Ottawa (IAEA 1953–1991).

The Selenga River accounts for half of the river discharge to the lake. Between the rivers of the northern part of the lake, where the experimental studies curried out, the Upper Angara has the greatest runoff (13.6% of the total runoff). The river is recharged mainly by precipitation (71%).

Tritium concentrations are rather high in both the precipitation and river waters of the investigated region. This is explained by the deep inland location of the lake basin. Table 18.8 gives the authors' data on the tritium content in water of the main tributaries of Baikal and in the Upper Angara tributary Dzhelinda.

An attempt was made to correlate the tritium concentrations in river waters with the atmospheric fallout in order to determine the travel time of a rainfall flood. The samples for isotope analysis were taken at the VErkhnyaya Zaimka gauging station (45 km from the mouth of the Upper Angara) and near the Ushikta bank (25 km

Table 18.8 Mean tritium concentration values in lake's river

River	Sampling time	Mean tritium content (TU)
Upper Angara (104 samples)	September 1973	344
Dzhelinda (1 sample)	September 1973	346
Selenga (2 samples)	June 1974	316

Table 18.9 Tritium concentrations and average discharge of the Upper Angara (September 1973)

Site and observation data (September 1973)	Temperature of water (°C)	Debit of water (m^3/s)	Maximum current (m/s)	Tritium content (TU)
Ushikta				
8	12.5	271	1.55	419
13	13.2	234	1.28	311
18	12.3	207	1.18	364
23	8.8	202	1.15	292
Verkhnyaya Zaimka				
9	12.2	302	0.677	359
13	13.4	265	0.607	332
19	11.7	232	0.502	324
20	11.1	238	0.572	416
21	10.1	229	0.507	354
23	8.5	216	0.472	367

Table 18.10 Tritium concentration in precipitation and atmospheric moisture (1973)

Sampling site, date, time	Type of atmospheric water	Tritium content (TU)	Precipitation (mm)
Verkhnyaya Zaimka			
16.09.73, 3^{00}–9^{00}	Rain	272	4.2
8^{00}–15^{00}	Rain	265	–
17.09.73, 0^{00}–5^{00}	Dew	374	–
19.09.73, 0^{00}–5^{00}	Dew	347	–
20.09.73, 0^{00}–8^{00}	Rain	292	9.7
8^{00}–8^{10}	Shower	235	0.3
21.09.73, 0^{00}–5^{00}	Hoarfrost	298	–
16^{30}–17^{30}	Rain	323	4.7
17^{30}–18^{30}	Rain	217	–
18^{30}–20^{30}	Rain	213	–
23.09.73, 7^{10}–14^{00}	Rain and snow	220	5.5
10^{00}–18^{00}	Rain and snow	293	7.4
Uoyan			
20.09.73	Rain	298	–
21.09.73	Rain	311	–
Listvyanka			
27.07.73	Rain	573	–
Solzan			
31.07.73	Rain	464	–

downstream from Verkhnyaya Zaimka). Here the water samples for analysis were taken along the entire course of the river channel and then the average value of the tritium concentration was determined (simultaneously with the discharges). The data obtained are given in Table 18.9.

The tritium concentration in precipitation and atmospheric moisture recorded during the same period are given in Table 18.10.

The gradual decrease in tritium concentration in atmospheric moisture, which originally was equal to its concentration in river waters, reflects the role of

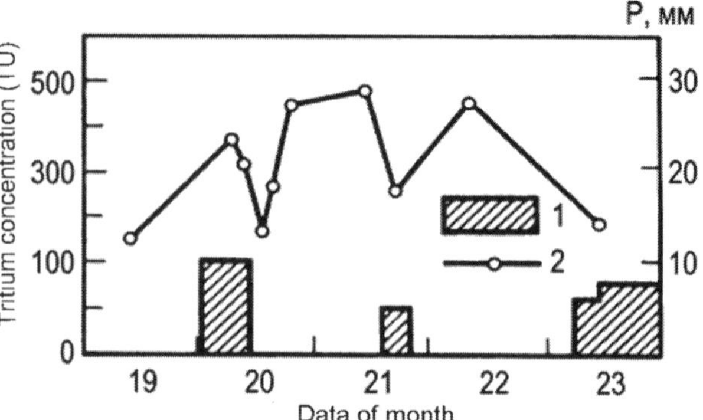

Fig. 18.10 Change of precipitation (*1*) tritium concentration in river water samples (*2*) during the end of September 1973 (the figures on the abscissa axis correspond to the days of September)

re-evaporated water of recent precipitation in which the tritium levels were less than in river waters. The latter fact indicates that at the time of observations, the Upper Angara was draining precipitation that fell in the summer, the amount of which in 1973 was abnormally high. Tritium concentrations of the summer precipitation over nearby points also are given in Table 18.9 and show that their values considerably exceeded the autumn values.

From comparison of the data in Tables 18.8 and 18.9, it is evident that the precipitation that fell on September 19 caused only a negligible increase in discharge (about only 2.5%) at Verkhniaya Zaimka, whereas the change in the tritium content in this case was 28%. The rains on September 21 and 22 in general did not affect the discharge of water in the river, whereas the tritium concentration increased by 13 TU (a small amount of this increase is related to the fact that the sample of water for analysis was taken only 11 h after the start of the rain, whereas on September 20 about 24 h passed after the start of the rain). Thus the time of minimum reaction of river water at Verkhnyaya Zaimka was not less than 24 and not more than 36 h.

An interesting fact is found by comparing the tritium concentrations in river and rain waters. Despite the fact that the tritium concentrations in precipitation during sampling were always lower than in river water, the rainfall effected not a decrease in concentration (as could be expected) but its increase. Presumably, this effect can be explained by the fact that the rainwater displaces from the soil groundwater with a high tritium content stored in the summer of 1973. For a more accurate determination of the travel time of the rainfall flood on September 19–22, samples of the river water were taken more often (several times a day). The results of the measurements are shown in Fig. 18.10. From a comparison of these data with the results of the isotopic analysis of the precipitation that fell simultaneously (see Table 18.9), we can see that the first peak in the ground and river waters after the rain on September 20 was observed 7 h after the rainfall and the second peak reached maximum after 30 h. The rain on September 21 was accompanied by a tritium concentration peak in the river and groundwaters after 20 h.

As already mentioned, the distribution of the tritium concentration was investigated not only in the surface layer of Lake Baikal but also vertically (over the entire water column). The very first investigations have shown that the lake waters

Table 18.11 Vertical change in tritium concentration in waters of Lake Baykal for the Nizhneagarsk–Birakan section

Depth (m)	Tritium concentration (TU)		Depth (m)	Tritium concentration (TU)	
	21.09.73 r.	03.19.73 r.		21.09.73 r.	03.19.73 r.
0	73	54	200	24	35
5	38	–	300	38	33
25	41	–	400	36	33
50	36	39	450	32	–
100	42	42			

are considerably mixed vertically (Soifer et al. 1970). The 1973 investigations confirmed these observations. In this case, five deep-water sections passing from north to south through the following points were investigated: Nizhneagarsk–Birakan, Cape Tiya–Nemnyanka, Slavyanskaya Bay–Khakusy, Baikalskoe–Tompa, Cape Kotelnikovsky–Irinda.

On the first section, the studies were carried out on September 21 and October 3, 1973. During this time, the water temperature on the lake surface dropped from 11 to 6.7°C, which lessened the stability of the surface layer and created conditions for intensification of vertical mixing of water. The latter also affected the vertical distribution of tritium (Table 18.11): its concentration in the surface layer during this period dropped sharply (from 73 to 54 TU), whereas in the subsurface layers it increased slightly.

Figure 18.11 shows the average vertical and surface distribution of the tritium concentration over the entire northern basin of the lake during the period October 3–6, 1973. As one can see, the tritium concentration levels to a depth of 800 m change a little. Thus, if the average concentration on the surface is 47 TU, then for the 0–400 m layer it is equal to 38 TU and for the 400–800 m layer it is 32 TU.

Figure 18.11a also shows the average vertical distribution of the tritium concentration in the southern basin at the end of summer 1970. It is characterized by a greater jump and at a depth of 500 m almost approaches the 1973 level. It also shows that there is an average tritium content in the surface layer in 1973 (late August–early September), which proves to be about the same (88.7) as in 1970.

The distribution of tritium concentration in the surface layers of the lake is nonuniform, though in individual periods it apparently obeys some kind of regularities. A definite trend of the tritium distribution in the surface layer of the lake is traced on the graph plotted from the little data for the end of summer 1973 (Fig. 18.11b). Thus the effect of the runoff of the Selenga and Upper Angara on the tritium concentration is unquestionable. It should be noted that the Selenga runoff is felt more weakly, since the water samples were taken on the western shore of the lake opposite its mouth.

The considerable uniformity of the tritium distribution in the water masses of the lake, its high level in the bottom layers of water, and the passage of the water temperature twice a year below the temperature of maximum density make it possible to use a ‘box’ model having the following main premises for determining the water exchange time of the lake: (1) the tritium concentration in the entire volume of

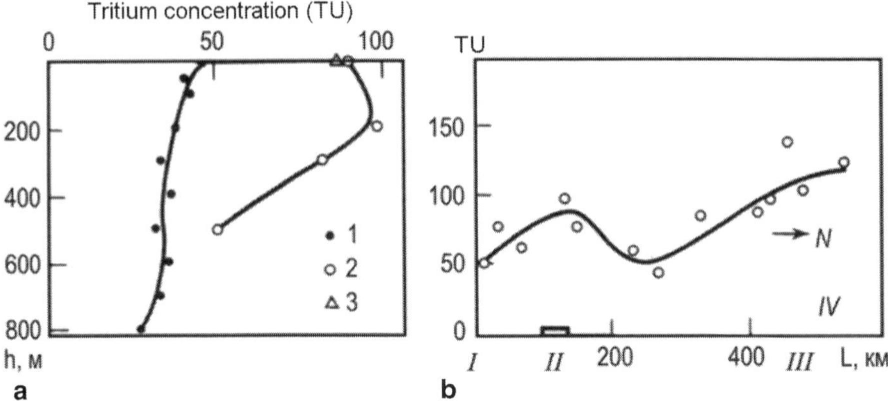

Fig. 18.11 Vertical (**a**) and surface (**b**) distribution of the tritium concentration in Lake Baikal: (*1*) northern basin, October 1973; (*2*) southern basin, August 1970; (*3*) average tritium content in surface layer, late August–early September 1973; (*I*) Listvyanka site; (*II–IV*) the Selenga, Irinda, and Upper Angara rivers, respectively

the lake is about the same since the water in it is mixed during period much less than the half-life of tritium (12.32 years); (2) the runoff from the lake has the same concentration as the average tritium concentration in the lake (in Angara it is equal to 50 TU).

Setting up the hydrologic balance equation on tritium and having tritium concentration data in the recharge sources (precipitation and river runoff) we can calculate the water exchange time of the lake. If the characteristic tritium concentration in precipitation is accepted as basic, then the obtained water-exchange time is equal to 370 years, which slightly exceeds the value obtained by the hydrological data. However, taking into account that the tritium concentration in river water on average is slightly lower than in precipitation (depending on the water-exchange time of the river basins), we obtain that the water-exchange time of the lake is equal to 330 years. For these calculations, we choose a water-exchange time of river basins equal to two to three years, which agrees approximately with the measured tritium concentrations in river waters.

There is a lot of work devoted to studies of water dynamics in lakes and reservoirs based on tritium, radiocarbon, and stable isotopes (Allison et al. 1971; Dubinchuk et al. 1988; Ferronsky and Polyakov 1983; Gonfiantini et al. 1979; Hesslein et al. 1979; Hübner et al. 1979a; Imboden 1979; Pavlov et al. 1981; Quay et al. 1979; Romanov et al. 1981; Weiss 1979).

18.4 Water Dynamics in Unsaturated and Saturated Zone

To solve this problem, tritium is the most applicable tool. The occurrence of tritium in groundwaters is directly dependent upon the conditions and the regime of their recharge due to infiltration of some portion of the runoff through the unsaturated zone and the transition of water from one aquifer into another. The shortest transition time of precipitation and surface waters into groundwaters is observed in the case of a direct hydraulic connection between them, which occurs in the regions of tectonic fractures, fissured and karstic rocks and gravel-pebble sediments.

The absolute tritium dating of groundwaters, characteristic of their complete cycle due to replenishment by precipitation of surface waters and loss due to groundwater discharge, evaporation, and transpiration, was possible before the beginning of thermonuclear tests when its occurrence in nature was determined by equilibrium amount, but at that time work on natural tritium and the hydrological applications have only just began. Therefore, the method of tritium dating based on the application of its decay law has not been used in groundwater studies.

Later on, when the natural tritium content was distributed by the injection into the atmosphere of large amounts of bomb-tritium, the possibilities of the absolute dating of groundwaters with the help of tritium were lost. Nevertheless, tritium dating is still possible for those groundwaters which originated before 1952 and which continues to exchange with natural waters containing pre-thermonuclear amounts of tritium. However, for these studies it is important to know the natural tritium concentrations in precipitation for a studied region before 1952. Such data cannot be precisely obtained. They can be calculated only by extrapolation on the basis of a small number of measurements carried out before the beginning of thermonuclear tests in the atmosphere. At the same time, recent wide-scale investigations of the tritium falls in precipitation on a global scale allow us to use the obtained data for studying the motion of groundwaters from a somewhat different viewpoint.

Tritium releases into the atmosphere during the decade of nuclear weapon tests occurred in the form of individual pulses, which correspond to a single powerful or a series of moderate explosions. The tritium falls during the test period, and the subsequent period of time, mirrored its injection into the atmosphere in the form of individual pulses differ on magnitude over the yearly cycle. The knowledge of atmospheric water patterns in the underground part of the hydrosphere together with tritium dating information provides the basis for the solution of different time-dependent problems while studying groundwater dynamics.

The study of problems dealing with the transition of the tritium 'marks' in groundwaters over time is based on systematic measurements of tritium concentrations in the precipitation of a studied region. The occurrence of tritium in groundwaters depends on their recharge conditions (Dubinchuk et al. 1988; Maloszewski and Zuber 1996; Zuber 1994).

The most typical case of recharge of the tritium from the surface aquifer is percolation of surface water through the unsaturated zone. Here, as a rule, the spring-summer component of the annual precipitation, containing the maximum tritium concentration, does not reach the aquifer. This portion of annual precipitation is lost mainly through evaporation-transpiration and partly by surface reservoir recharge. Some portion of the groundwater storage is similarly lost. During the autumn-winter period and in the early spring, when tritium content in precipitation is minimal, the evaporation and transpiration of precipitation water is also reduced. During this period, groundwaters are replenished. Thus, in this case, the tritium concentrations in groundwaters will be lower than the mean annual concentration in precipitation.

In those regions, where precipitation falls during the cold season of year are the sources of groundwaters recharge (for example in the arid zone), the tritium recharge to the aquifer corresponds to the mean annual values.

Besides the above-mentioned factor, the concentration of the tritium in groundwaters is largely affected by yearly temperature changes and the conditions of the freezing and thawing out of the upper ground layer, the soil composition of the unsaturated zone, the soil cover itself, the type of plant cover, and the regime of vegetation. In view of all these factors determining groundwater recharge, the amount of incoming tritium can vary greatly. Therefore, for the solution of practical problems, an account of the individual conditions of a studied region is indispensable.

According to the experimental data of Münnich et al. (1967) and Atakan et al. (1974), who carried out their studies on the alluvial plain of the Rhine River, bomb-tritium dating of shallow groundwaters can be used with a sufficient accuracy if the unsaturated soils and the aquifer itself are sufficiently homogeneous in composition and properties. Tritium change in depth profiles over time, in an unsaturated zone composed of loess-loam soil and in an unconfined aquifer consisting of fine to medium-size and coarse sand, is shown in Fig. 18.12a, b. According to the tritium dating techniques, the average recharge rate of groundwaters in the fine-to-coarse sand material is 200 mm/year which is in good agreement with that measured on the basis of routine hydrologic techniques. This value was measured using the rate of the tritium peak front movement and the total amount of tritium in the profile.

Detailed studies of groundwater motion in saturated and unsaturated zones, including the works on estimation of level of technogenic contamination, have been carried out by a number of researchers (Allison and Hughen 1974; Andersen and Sevel 1974; Atakan et al. 1974; Morkovkina 1979; Münnich et al. 1967; Polyakov and Golubkova 2007; Sokolovsky et al. 2007a; Verhagen et al. 1979; Zuber et al. 2001).

In the framework of the laminar flow model for a shallow aquifer composed of homogeneous material, the following equations of tritium accumulation can be written (Atakan et al. 1974):

$$M_{\Sigma} = \int_{0}^{z_0} \rho\,(z) C\,(z)\,dz, \tag{18.14}$$

where $M_{\sum}$ is the amount of tritium accumulated in a vertical column of soil with a unit cross-section; $\rho(z)$ is the porosity (a constant under the conditions of the problem; $C(z)$ is the concentration of tritium with depth; z_0 is the penetration depth of the bomb-tritium.

The tritium balance equation between its recharge due to precipitation and accumulation in the aquifer is

$$M_{\Sigma}\,(t) = \int_{1952}^{t} R\left(t'\right) C\left(t'\right) \exp\left[-\lambda\left(t - t'\right)\right] dt, \tag{18.15}$$

where R is the recharge rate into the aquifer; λ is the decay constant of tritium.

Here the exponential factor accounts for the tritium decay between the time of infiltration t' and that of observation t.

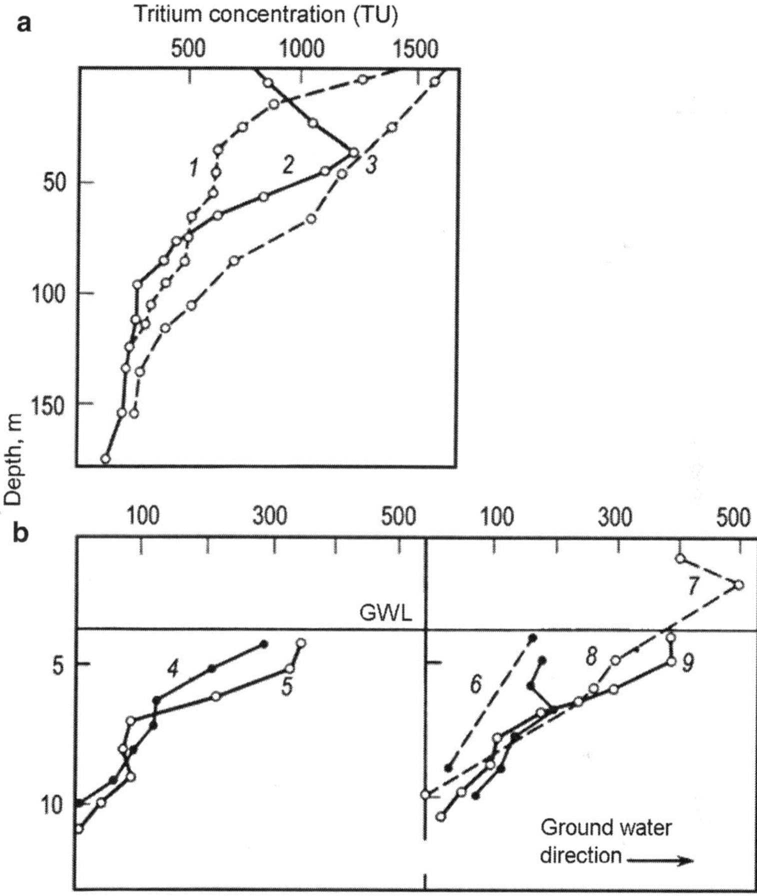

Fig. 18.12 Measured tritium profiles in an unsaturated zone composed of loess-loam (**a**) and in a fine-to-coarse sand aquifer (**b**): (*1*) July 25, 1963; (*2*) Nov. 29, 1963; (*3*) Aug. 28 1964; (*4*) Sept. 1967; (*5*) Dec. 1967; (*6*) July 1965; (*7*) Oct. 1966; (*8*) June 1967; (*9*) Dec. 1967

For the conditions of a year-to-year accumulation of tritium in the aquifer, the Eq. (18.15) becomes:

$$M_{\Sigma}(t) = \overline{R} \int_{1952}^{t} \frac{R(t')}{\overline{R}} C(t') \exp\left[-\lambda(t - t')\right] dt', \tag{18.16}$$

where $\overline{R}$ is the average annual rate of recharge.

Denoting

$$S(t) = \overline{R} \int_{1952}^{t} C(t') \exp[-\lambda(t - t')] dt', \tag{18.17}$$

and assuming $R/\overline{R} = 1$, from the condition of the seasonal recharge of an aquifer due to precipitation, which falls during September to March, Attakan et al. have found that

$$M_{\Sigma} = \overline{R}S. \tag{18.18}$$

In the above equations, the date 1952 is considered to be beginning of bomb-tritium recharge to groundwaters.

Using expression (18.18) and on the basis of the experimental observation on the tritium content in an aquifer profile, Atakan et al. (1974) have estimated the average water recharge rate $\overline{R} = 166\, mm/year$ for an annual average amount of precipitation of 686 mm/year for the Upper Rhine basin during 1967–1971. It has also been found that during this period, the value of M_{Σ} was 500 TU and the depth front of the tritium contribution shifted downwards from 10 to 25 m below the groundwater table.

The distribution of tritium with depth decreases quickly, which is a result of dilution in the aquifer due to hydrodynamical dispersion and molecular diffusion and also due to the natural radioactive decay of tritium. According to the above data (Münnich et al. 1967), at a depth of 5–6 m below the water table, the tritium content in water is reduced by almost two orders of magnitude.

18.5 Recharge and Discharge of Groundwater

Despite some unsolved problems and unanswered questions concerning the reliability of the groundwater radiocarbon dating, the isotope is in common use in hydrogeological studies. As a rule, a combined information, including the data on distribution of the natural isotopes (deuterium, tritium, oxygen-18, carbon-13 etc.) in different elements of the hydrosphere together with hydrodynamical and hydrochemical information related to the object, is used. But it is worth to stress that application of ^{14}C in hydrogeological studies allowed to solve a number of tasks, which in principle have to be solved by the classical methods and even by use of distribution of some other stable or radioactive isotopes. One such problem is determination of groundwater age in the framework of 'radiocarbon time scale' (up to 80,000 years) and also doing paleohydrogeological reconstructions on this basis. All hydrologic and hydrogeologic problem solving by use of ^{14}C can be divided in two groups, namely, qualitative and quantitative. To the first group, the problems related to determination of local areas of water recharge to aquifers, discharge of groundwater and locations of interconnection of the aquifers. As a rule, this complex of problems can be solved by the tritium use, but the radiocarbon data allows to localize not only modern zones of recharge (tritium data are limited by 50 years of time period), but also 'old' occurrence, for example, on the studied area in the end of the Pleistocene and beginning of Holocene, when climate on the planet was different than present day.

The second group of the problems includes estimation of groundwater age, determination of mixing proportions of waters from different aquifers and their complexes, separation of inflow to the mines and excavation of water inflows, determination of regional direction and velocity of filtration of the groundwater flows. The planning and implementation of a real hydrogeologic work should be done in such a way that the optimum amount of information to be obtained on the basis of having the real possibilities. For this purpose a detailed analysis of the geologic and hydrogeologic information and formulation of the main goals and objectives are carried out and also of the natural isotopes and the laboratory possibilities are estimated. Note that the above problems can be solved at the stage of hydrogeologic mapping and exploration of groundwater resources as well.

The study of recharge and discharge of the groundwater is mainly aimed at quantitative estimation of conditions of the aquifer formation. Radiocarbon and tritium was used for study of the Assel-Kljazmian aquifer (Upper Carbon) in the south-eastern part of the Moscow artesian basin (MAB; Polyakov and Seletsky 1978; Seletsky et al. 1979). The study was carried out in the Vladimir region located on the border of Oka-Tsna rivers' billow and Vladimir-to-Shipov crook. The Assel-Kljazmian aquifer as the main subject of the study stratigraphically belongs to the Assel Stratum of the Lower Permian and Upper Carbonic sediments (see Fig. 18.13). The water-bearing rocks are presented by a fissured limestone and marl partially gipsumized. In the aquifer's cover, multicolor clays of Permian-Jurassic age ($P_2 - J_1$) are located and on the foot are the red clays ($P_2 - J_1$), (C_3, šc) up to 15–18 m thickness. Under the clay thickness, the Kasimov aquifer (C_3, ksm) is located. The radiocarbon content in the studied Assel-Kljazmian aquifer varies between 4.8 and 51% of the modern standard. The higher values of ^{14}C were recorded in the surface waters—for the Sudogda River it was 75% and for Oka River it was 94%—and also in the Kasimov aquifer at the recharge zone (see Fig. 18.13).

If, by the obtained radiocarbon data, one calculates the groundwater age, applying the piston model (see Eq. 14.16) even assuming that ^{14}C concentration in the recharge zone is 50% of the modern standard, then for the most testing water points their age is 3,000–15,000 years. Such an estimation of the Assel-Kljazmian aquifer contradicts not only to the data, obtained by the classical hydrogeological methods, but also is not in accordance with the tritium concentrations obtained for a number of wells. It was assumed that the decrease of ^{14}C activity is caused by delusion of the carbonates of the water-bearing rocks not containing ^{14}C and by possible isotope carbon exchange in the system of the dissolved carbonates–carbonate of waterbearing rocks. The possibility of ^{14}C activity decrease due to the dissolved 'dead' carbonates was checked by the Pearson- Swarzenki method (see Eq. 14.67).

In case of the ^{14}C activity decrease due to the dissolved 'dead' carbonates of the water-bearing rocks, a linear dependence of radiocarbon activity in a sample from the reverse value of carbonate components value should be observed. Figure 18.14 shows this dependence of ^{14}C in water of the Assel-Kljazmian aquifer from the reverse value of carbonate ion. Note that at pH of a studied water equal to 75–80 the ion HCO_3- in the carbonate system is dominated.

It is clear that the observed dependence does not fit the classical theory of the carbonate components of groundwater dilution by the 'dead' carbonates of the water-bearing rocks. In fact, at decrease of the hydrocarbon ion in the solution

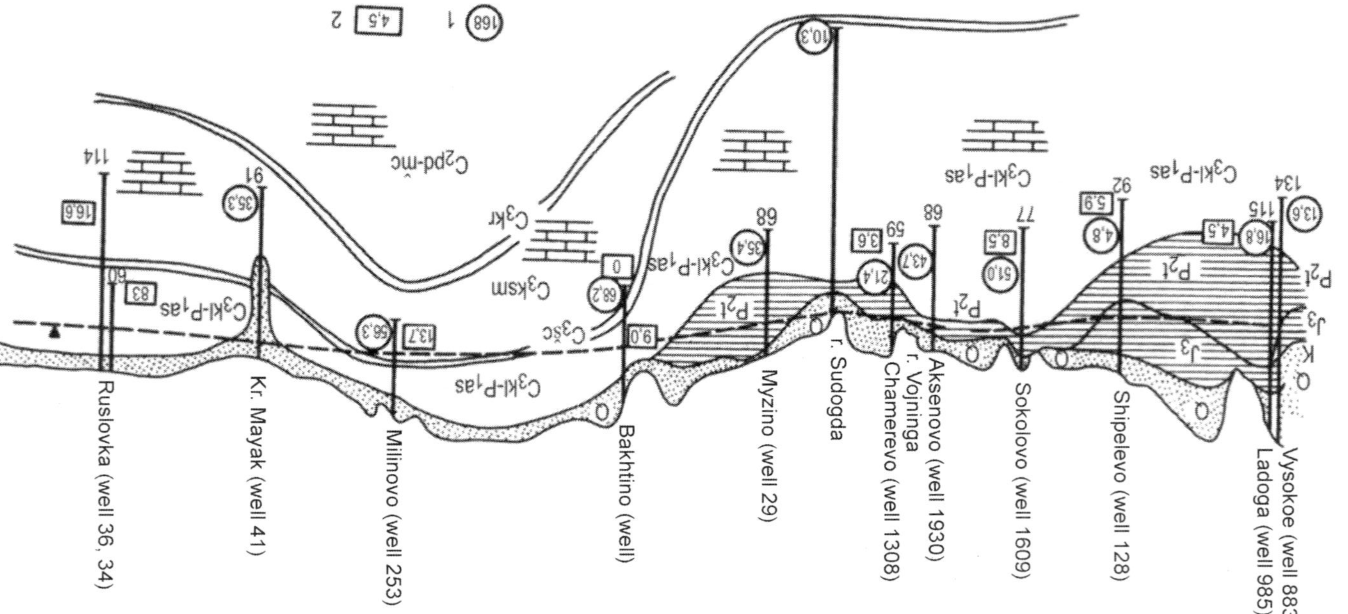

Fig. 18.13 Radiocarbon and tritium concentrations in groundwaters of the Upper Carbon sediments along the section Visoko–Ruslovka (Vladimir region): (*1*) radiocarbon; (*2*) tritium. The depth of wells is in m. (After Polyakov and Seletsky 1978)

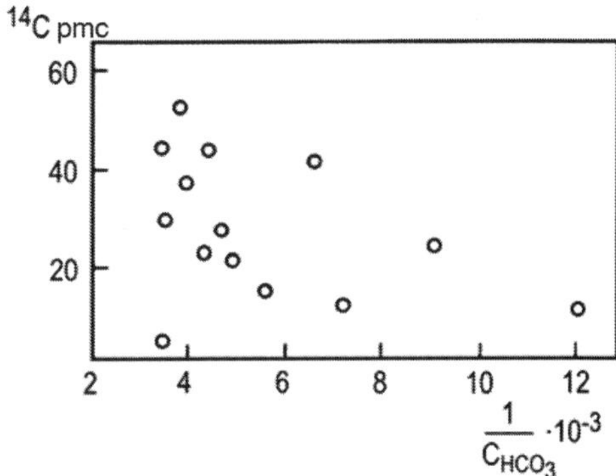

Fig. 18.14 Relationship between radiocarbon concentration and reverse value of bicarbonate-ion in Assel-Kljazmian aquifer of waters. (After Polyakov and Seletsky 1978)

the radiocarbon concentration also decreases. Such dependence has found explanation by the authors (Polyakov and Seletsky 1978) at consideration of groundwater chemistry of the Assel-Kljazmian aquifer. Its water contains the hydro-carbonate calcium component of 0.8 g/l content, as well as the sulfur calcium mineralization up to 4.2 g/l. The latter is adapted to the bottom parts of the aquifer and determined chemical content by leaching out of the clay strata. The hydro-carbonate calcium waters are adapted to the upper parts of the section. The Sudogda River valley appears to be the domain of discharge of the Assel-Kljazmian aquifer. It was proved by the hydrodynamical investigation and using radiocarbon and tritium data. In particular, in the Sudogda River water the ^{14}C and T concentrations are equal to 73% and 41 TU respectively as opposed to 96% and 70 TU in the Oka River.

Because of the vertical component of the underground flow, a mixing of the hydro-carbonate calcium and sulfur calcium waters occurs and leads to abrupt increase of the calcium ions concentration, which shifts equilibrium in the groundwater carbonate system and favors precipitation $CaCO_3$ from the solution. In particular, this phenomenon is fixed by occurrence of the 'dolomite flour'. The isotope exchange occurred on the fine dispersed phase leads to the ^{14}C yield from the solution. All the said relates also to the stable isotope ^{13}C. That is why $\delta^{13}C$ of the carbonate system of the studied mixing zone amounts to $-2.5‰$, i.e., the bicarbonate ion practically occurs in isotopic equilibrium with the solid phase. The correction of the ^{14}C content by the stable carbon isotopic composition, done by Eq. (14.49), gives the value from 80 to 108%, i.e., the water with respect to ^{14}C occurs practically the modern. These data are proved by the study of tritium distribution.

In framework of the exponential model, the mean residence time of the Assel-Kljazmian aquifer water determined by tritium is about 500 years. The qualitative consideration of the radiocarbon and tritium data (see Fig. 18.14), in addition to the recharge area located within Oka-Tsna billow, allows to identify the local 'hydrodynamic windows' near the well 1609. The lower ^{14}C and T concentrations in the well water of the Sudogda River valley are evidences of discharge of the groundwater of the aquifer studied in this area. The higher ^{14}C concentration in the Kasimov aquifer water gives evidence that the 'dry' clay thickness has weak watertight properties.

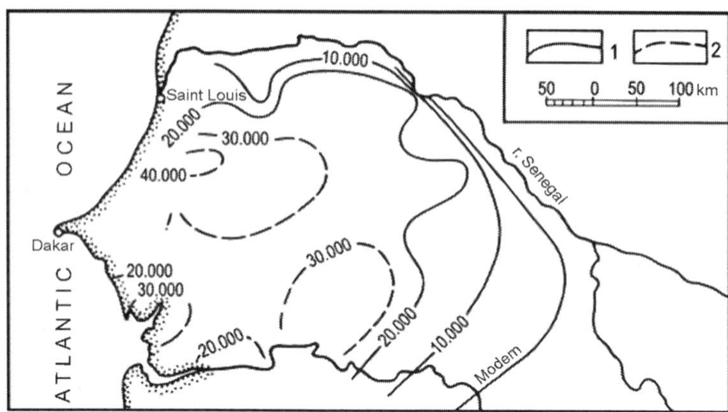

Fig. 18.15 Hydroisochrones of radiocarbon age for groundwater in Maastricht sands, Senegal: (*1*) dynamical and (*2*) stagnant zones. (After Castany et al. 1974. © IAEA, reproduced with permission of IAEA)

The most reliable data about conditions of groundwater formation on the radiocarbon basis is followed to obtain a detailed sampling of a studied aquifer and with subsequent hydroisochrones (lines of the same water ages) plotting. The first such a study was done with the aim of investigation of Alb aquifer in the Paris artesian basin (Evin and Vuillaume 1970). By means of hydroisochrones plotting, the recharge of groundwater conditions from the Alb sediments and their resource replenishment were specified. It was shown that the Seine River in its low current is the regional area of discharge of that aquifer. Also the data of hydraulic connection of the Alb aquifer with the above-lying aquifers were obtained.

Later on, analogous studies by Castany et al. (1974) on the territory of the Senegal, Lebanon, and Morocco were carried out. In particular, in the Senegal, the conditions of the groundwater formation in sandy sediments of the Maastricht stratum, involving clay layers, were studied. The aquifer thickness was practically constant and equal to about 200 m. The depth from the surface was also 200 m, and the area was extended over 150,000 km^2. Zones of the modern recharge on the bare rocks at the west Green Cap Islands were localized. A low hydraulic gradient (from 10^{-4} to 4×10^{-5}) pointed out on a small rate of the groundwater movement. By the pumping tests, the water conductivity of 10^{-4} to 2×10^{-4} m^2/s and average filtration coefficient equal to 10^{-4} m/s were determined. The Maastricht sandy aquifer is covered by the Lower Paleocene-Eocene clay-marl sediments, which create relatively watertight thickness. In this connection, the aquifer contains mainly the pressurized groundwater having the total resources of about 3×10^{12} m^3. The common mineralization of the waters has increased from east to west from 250 to 2,000 mg/l. The water is also changing from the hydrocarbonate-sodium to the sulfur-sodium type.

Figure 18.15 shows the data of radiocarbon distribution in the Senegal Maastrichtian aquifer. The plotted ^{14}C hydroisochrone lines allow the following conclusions to be done. The main modern recharge of the aquifer occurs from the south-east

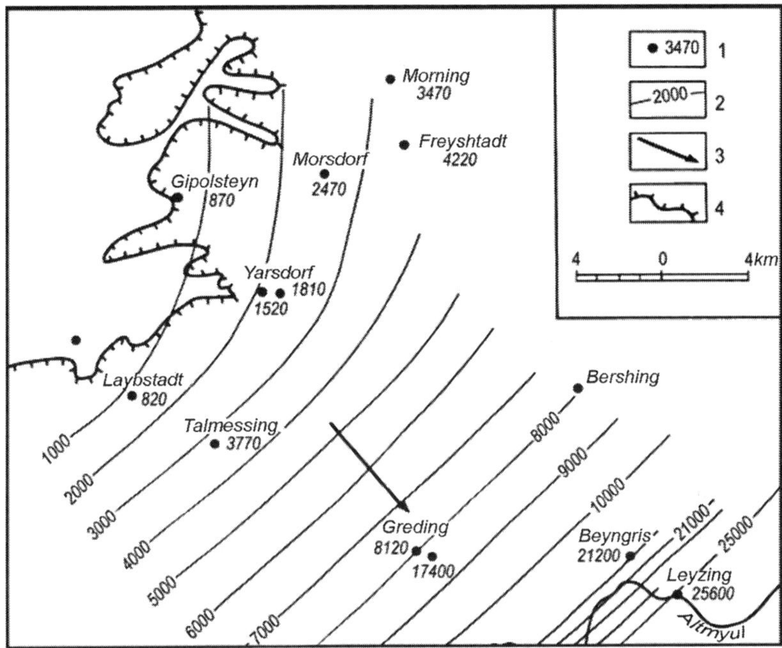

Fig. 18.16 Isochrones of groundwaters for the Alb-multicolor sandstone: (*1*) radiocarbon age of water in the sampling point; (*2*) hydroisochrones; (*3*) groundwater flow direction; (*4*) the borders of the sandstone outlet. (After Geyh 1974. © IAEA, reproduced with permission of IAEA)

and partially along the Senegal River to the east. In the recent past (about 10,000 years ago), the aquifer recharge resulted in the northern part of the region, whereas by the hydrogeological data this area at present is the discharge zone. The authors assume that this is effect of new tectonic shifts which are the cause of substantial changes in the level of the sea-continent system. Comparison of the piezometric map with the isochrone map shows a good coincidence in the estimated rates of the groundwater motion obtained by ^{14}C distribution and by hydrodynamic methods. The water motion rate in the Maastricht sandy aquifer varies from 8 to 30 m/year.

The hydroisochrone plotting method based on the radiocarbon groundwater dating allows in solving a number of common hydrogeologic problems even at absence of the initial ^{14}C concentrations in the recharge area and at the not corrected groundwater age. In this case, in order to obtain quantitative data, some additional conditions should be satisfied (Geyh 1974).

1. The lines of the equal age by ^{14}C (hydroisochrones) must be drawn parallel to the hydroisolines. Otherwise, the velocity and direction of the motion obtained by the isochrones can be incorrect (Vogel 1970).
2. The groundwater recharge should be uniformly distributed within the radiocarbon scale of age. In particular, the climate change of cool and warm epochs in the Pleistocene in radiocarbon age reflects by changes in the recharge conditions in different epochs for the aquifer. Figure 18.16 shows the groundwater isochrones

for the sandy Alb aquifer (Geyh 1974). The last Pleistocene minimum of temperatures in the period of 20,000 to 14,000 years BP is characterized by bunching of the isochrones, which shows lower rate of the groundwater flow in that period. By these data, there isno reason to calculate the groundwater rate but such a study allows to estimate the natural conditions of groundwater formation for the permafrost regions during the last glaciation and to make the paleoclimatic reconstructions.

Geyh (1974) pointed out that the radiocarbon data better reflects natural hydrogeological conditions of an aquifer compared to the pumping method because they give possibilities in estimation of regional filtration characteristics of the aquifer as a whole, but by the pumping method the coefficient of filtration just close to the well is determined.

The studies like the one described above at groundwater exploration were carried out in Hungarian depression (Deak 1974, 1979), in Germany (Fröhlich et al. 1977; Fröhlich 2010; Jordan et al. 1976), in England (Bath et al. 1979; Evans et al. 1979), on African continent (Conrad and Fontes 1972; Geyh and Jäkel 1977; Münnich and Vogel 1962; Sonntag et al. 1978, 1979), in USA (Pearson and Hanshaw 1970; Pearson and White 1967; Winograd and Farlekas 1974), in South America (Tamers and Scharpenseel 1970) and in other regions of the Earth.

18.6 Relationship of Aquifers

Let us consider, as an example, the work of Borevsky et al. (1975), where they investigated relationship between aquifers in the Dnieper-Donets artesian basin with the help of radiocarbon. In this region, the following aquifers were identified.

1. The aquifer in the Eocene sediments represented mainly by sands of 20–40 m thickness. The water-bearing rocks of the Eocene aquifer are located on the relatively watertight marl-chalk Cenomanian-Turonian layer from 3 to 5 m thickness in the west part to 40–43 m in the east part of the region. The marl-chalk layer is absent in the south-western end of Kiev City. Here the Eocene sediments lie down on the Cenomanian rocks, which form a common layer covered by the weakly watertight marls of 25–30 m thickness. In the Dnieper valley, the Kiev marls are washed and the Eocene and alluvial sediments form the common aquifer.
2. The Cenoman-Callovian complex of rocks is represented mainly by sands covered by limy clays with lenses of sandstone and clay. The marl-chalk thickness almost everywhere separates the Cenoman-Callovian rocks of the aquifer from the Eocene sediments. This thickness in general is a serious obstacle for the water overflow between the aquifers.
   The water-bearing rocks of the Cenoman-Callovian complex everywhere lie on the alevrite-clay rocks of the Middle Callovian, which underlined on Bajocian clay sediments and form a stand weakly permeable thickness of 60–100 m. As a result of tectonics, the Bajocian- Callovian thickness were occurred destroyed, which

provides the hydraulic relation between the water-bearing Cenoman-Callovian and Middle Jurassic rocks, which are separated by the thickness.

3. The Cenoman-Callovian aquifer is represented by the sands. Its thickness (as well as the Upper Cenoman-Callovian layer) reaches 20 m and more. The horizon's depth increases from the west to the east. As a result of exploitation, the depression's funnel is formed in the aquifer. The maximal lowering in the Eocene horizon is fixed in the intensive intake structure from the Cenoman-Callovian aquifer. Here the depression funnel is observed on the right and left shore of the Dnieper River. The right hand funnel is limited by river's valley, which is evident from the groundwater recharge from the river. In accordance with the existing version, the Cenoman-Callovian aquifer is mainly recharged at the north-east border of the Dnieper-Donets artesian basin and at the east slope of the Ukrainian Crystal Shield where the Cenomanian rocks under water permeable sediments of the Paleogene are located. The recharge of the underlying Middle Jurassic aquifer results from the same area. The largest depression funnel in the Middle Jurassic horizon is formed and in the northern direction it reaches 60 km. This is evidence of less favorable recharge conditions compared to the Cenoman-Callovian horizon.

The analysis of the exploitation experience of the intake structures and the observational data of piezometric levels and also the investigations in the mathematical modeling, hydrometry, and hydrochemistry made it possible to conclude that the basin is recharged within the area as a whole. The detailed information allowed in comparing the radiocarbon results with the traditional hydrogeological studies. The sampling was provided over a number of profiles which crossed the depression funnel. The samples, as a rule, were taken in the conjugated points from the Cenoman-Callovian and Bajocian aquifers, and also the samples were taken from the alluvial sediments of the Dnieper River valley.

Figure 18.17 shows one of the geological section and the curves of the radiocarbon values which demonstrate a picture of the young water recharge over the basin. General variation of the radiocarbon curves gives evidences about inflow of younger water at the domain of the depression funnels near the Dnieper River which is observed both in the Cenoman-Callovian and Bajocian aquifers. In addition, the younger waters are observed at the south-western wing of the studied area, i.e., at the region of the assumed recharge area on the slope of the Ukrainian Crystal Shield, and also in the eastern part of area where intensive exploitation of water from the Cenoman-Callovian aquifer takes place, whereas on the right hand shore of the Dnieper River the younger waters already flow into the Cenoman-Callovian and Bajocian horizons, on the left shore near Kiev City, where the exploitation of the horizon started later on, radiocarbon concentrations are close to those characteristic for the undisturbed regime. The hydrogeological calculations and the radiocarbon content changes show that younger waters here have not yet reached the intake well structures.

Figure 18.18 presents the data of the radiocarbon content along the section at the water intake structure where water level in the Cenoman-Callovian horizon is higher than in the Dnieper River and the water level in the Middle Jurassic horizon is lower

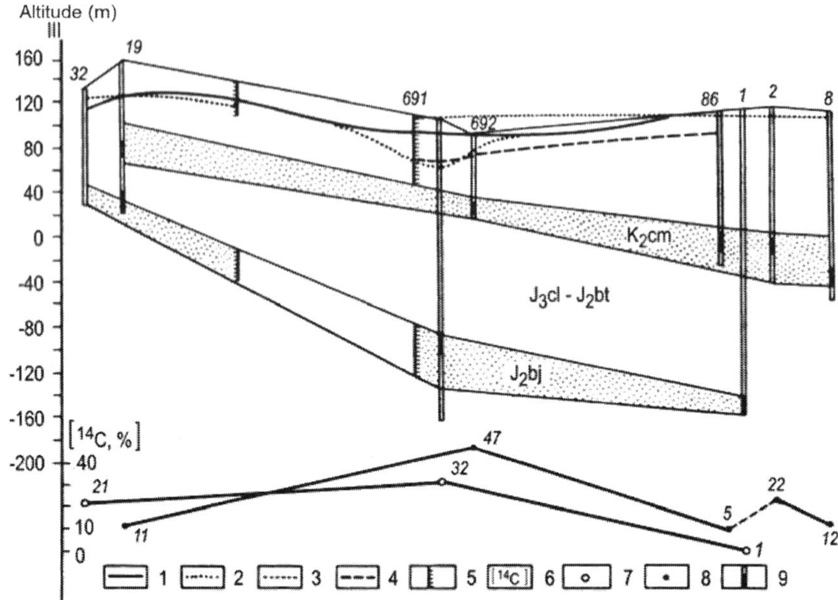

Fig. 18.17 Radiocarbon concentration changes along the section: (*1*) piezometric level in the Cenoman-Callovian horizon in natural conditions; (*2*) after long-term exploitation; (*3*) piezometric level in the Bajocian horizon natural conditions; (*4*) after long-term exploitation; (*5*) the border of water-bearing and weakly permeable sediments; (*6*) radiocarbon concentrations in % of the modern carbon; (*7*) data for the Cenoman-Callovian horizon; (*8*) data for the Middle Jurassic horizon; (*9*) the well filter

than in the Cenoman-Callovian. Thus, one can assume that in this region the water of the Cenoman-Callovian horizon contains the same amount of the radiocarbon which was before exploitation of the Kiev City water intake structure. One may expect that the water inflow occurs from the Cenoman-Callovian horizon to the Middle Jurassic horizon and to the marl-chalk thickness, and actually, the radiocarbon concentration in the Bajocian horizon (49% of the modern carbon) is close to the concentration in the Cenoman-Callovian horizon (7.4%). Because in all the Bajocian horizon wells, which locate on the periphery of the depression funnel, the radiocarbon is found on the limit of sensitivity and accounts about 1% of the modern carbon, the conclusion follows that the water from the Cenoman-Callovian to the Middle Jurassic horizon in this region overflows. Somewhat lower radiocarbon specific activity in the marl-chalk thickness (5.6%) compared to the Cenoman-Callovian is easily explained by the higher concentration of the hydrocarbonate ion in water. One can assumes that an additional amount of the hydrocarbonate ion comes to waters overflowing from the lower horizon due to the excess there of the aggressive carbon dioxide. The waters of the Eocene horizon contain ultra modern amounts of the radiocarbon (109%), which is close to its content in the Dnieper River (107%). This is an evidence of intensive relation of the horizon with the river.

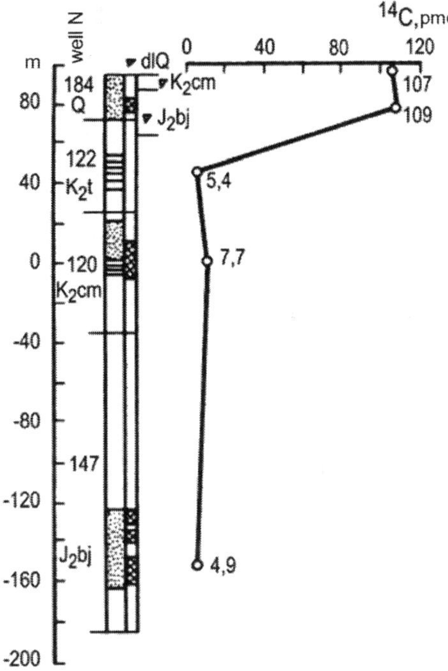

Fig. 18.18 Variation of radiocarbon concentration on depth of the section in Obolon (Kiev City)

The estimation of the aquifers' water age, based on the model of recharge only from the region of the Ukrainian Crystal Shield, shows that its values in the Kiev City's area for the Cenoman-Callovian horizon is equal to about 2,900 years and for the Bajocian about 26,000 years. If one assume that the water of the Bajocian horizon overflows from the Cenoman-Callovian horizon, then its age amounts to about 16,000 years.

The dynamic picture obtained by the radiocarbon studies is in good accordance with the general hydrogeological idea and results of the analogous modeling of the Kiev's water intake structures. For estimation of the groundwater resources in the region by the modeling methods, it is important to obtain direct data about water filtration through the strength thickness of the aleurolite-clayey Callovian-Bajocian sediments. Thus, the radiocarbon studies allowed obtaining unique solution of the problem which was for long time a subject of discussion between the hydgeologists.

The obtained results have a definite interest from the viewpoint of understanding of the hydrochemical migration properties of the hydrocarbonate ion. Because the radiocarbon was found in the Bajocian horizon, where it mainly occurs during filtration through the marl-chalk thickness, this makes it possible to conclude that inconsiderable exchange properties of the horizon take place. If one takes into account that the studied waters can often be a product of mixing of the ancient water of the horizon and young waters of the Dnieper River, then the conclusion about weak role of the exchange processes in this particular case becomes still more convincing.

Borevsky et al. (1981) also used radiocarbon method for study of water leakage through the Neocom-Jurassic clayey thickness within the depression funnel area of the water intake wells of the Bryansk region. The water supply of the Bryansk City is provided by the Upper Devonian aquifer of the fractured-karstic sediments. The exploitation of this complex started at the end of nineteenth century. Intensity of the water pumping had increased. As a result, the wide funnel depression of more than 100 km in diameter, reaching the Orel City in east and the Roslavl City in the west was formed. The funnel depth in the center reaches 65 m. Because of the water pick out was accompanied by decrease of the water level, a danger of the groundwater resources exhausting in the Upper Devonian complex has appeared. In this connection, the problem of the exploration and quantitative estimation of the water resources—taking into account their intensive exploitation roused.

It was assumed earlier that the main source of the groundwater recharge is found in the Jurassic clays which occur on the surface, but regime of the groundwater exploitation gives evidences about existence of an additional source of the recharge. This was presumably connected with the process of water leaching through the Neocom-Jurassic clay thickness. In order to solve the problem, a detailed sampling of the groundwaters for measurement of the natural isotopes including the radiocarbon was doned. The sampling points were located on mutually perpendicular profiles of 250 and 100 km length, which intersect the depression funnel. The Orel–Bryansk–Roslavl section gave the most representative results. The ^{14}C character and values (and also tritium and helium) along the profile gave unique evidence that the water is recharged to the Devonian complex from the above mentioned water-bearing horizon near Orel region. The maximum values of the radiocarbon content (60–80%) were observed at the Jurassic clays absence (about Orel region), where intensity of recharge is maximal. Together with the water-bearing complex submerged under the clayey thickness, the recharge conditions sharply deteriorate. The sharp decrease of ^{14}C (up to 3% of the modern carbon) and tritium (0–4 TU) are the evidences of the above. In Bryansk region, a splash of the ^{14}C concentration (up to 35%) and tritium (up to 20 TU) were fixed. The obtained results are the evidences about the leakage through the Jurassic clays around Bryansk where this clearest fixed in the Desna River valley.

The useful and promising hydrogeological information can be obtained by application of the radiocarbon at investigation of aquifers in the permafrost regions (Karasev et al. 1981b).

The problem of relationship between the under-permafrost aquifers and the surface water sources in the Yakutiyan artesian basin is discussed since long time. Most researchers assume that such a relationship is either absent or limited due to low permeability of the section. Because a lot of observable data about piezometric levels and filtration rock properties are contradicting, the correctness of the above viewpoint is difficult to be checked. The presence of the hydraulic relationship between the under-permafrost and surface waters, found by the well observation, cannot be evidenced about the water movement to the aquifer. The ratio of the ion-salinity components and the dissolved gases, because of going up diffusion of the gases and the existing water exchange with water-bearing rocks, is also not enough for this purpose.

Table 18.12 Radiocarbon content in the central regions of Yakutia

Location of sampling	Interval of sampling (m)	^{14}C concentration (% of modern carbon standard)
North-east of Yakutsk City's suburb	–	< 1
Yakutsk: Worker's colony	–	< 1
Khatassy settlement	–	< 1
Kangalassy settlement	–	< 1
Melting zone Maya	–	3 ± 1
Melting zone Lomtuga		
Well covered by frozen peak	38–60	14 ±
Sub-permafrost waters	200–240	38 ± 3
Melting zone Tungulu		
Lateral melting zone	68–78	4 ± 1
Central part of melting zone	–	18 ± 2
Melting zone Abalakh		
Sub-permafrost waters, acting well	222–225	14 ± 2

By ratio of the surface and groundwater pressure within the central part of Yakutia stands out two regions in which oppositely directed processes of water exchange between the surface and groundwaters are possible. One region is characterized by a low aquifer pressure (with deficit of pressure) of the under-permafrost waters. Its borders in general coincide with the borders of Lena-Viluy rivers' artesian basin. The second region covers the main part of the Aldan River wing of the Yakutia artesian basin, where piezometric levels of the under-permafrost waters are located over the local base of erosion. Over all the considered territory zones of permafrost rocks of 600–700 m thickness—broken by under-river, under-lake, and under-abyssal splashes—are developed. As a result, the first region became less closed for the water pressurized system with water recharge places for the under-permafrost aquifers, the second region have had the local discharge places.

Estimation of the ascending and descending scale of water filtration through the under-river, under lake, and under-abyssal splashes raises difficulties in determination of filtration properties, etc.

The existing materials characterized the region as a very low studied with respect to the hydrogeology. In many cases, the water exchange rates within a region selected for study, cannot be estimated even approximately. In such cases, a qualitative estimation of the relationship between surface and groundwaters as well as estimation of the scale of such a relation and understanding of degree of closing of separate structures is found to be an actual problem.

In this connection, the radiocarbon content in groundwater with the aim of possibility estimation of the ^{14}C content data use over the permafrost territory for estimation of its development and obtaining of information about relationship between the under-permafrost and surface waters (Karasev et al. 1981b).

The studies of the radiocarbon content were carried out on two sites, the data of which are presented in Table 18.12.

The sampling for radiocarbon measurements was carried out from (1) the wells of subbed waters of Lena River in the Yakutsk City region and in the Khatassy and Kangalassy settlements, and (2) from a number of the sub-permafrost melting zones.

The first site was located within the zone of low stratum pressure. In the Khatassy settlement of Yakutsk City, the groundwater levels fill seasonal variation of 1 m amplitude, but their hydrographs have nothing common with the Lena River hydrograph. The piezometric water levels in the well of Kangalassy settlement are fixed lower than the water level in the Lena River.

^{14}C concentrations in the region's groundwaters are in all cases lower than 1% of modern carbon activity which characterizes the pressurized water system as isolated from the surface radiocarbon sources. Despite the hydrodynamical conditions which contribute in infiltration of the Lena River waters to the studied aquifers, this process, according to the radiocarbon data, is not observed. The study of the gases and salinity of the samples from the first site also gives evidences about unfavorable conditions of the subpermafrost water recharge, but the criteria, as was earlier pointed out, cannot give unique answer. The only radiocarbon data can be used as reliable criteria in this case. The results of the radiocarbon study of the under-permafrost melting water gives evidence about this.

The sub-permafrost melting objects selected for study are located in the first (Tungulu, Lomtuga, and Maya) and in the second (Abalakh) sites. The sub-permafrost melting waters are enriched with the radiocarbon (Lomtuga well, interval 200–240 m, Alabakh well) compared to the sub-permafrost melting waters of Lena River zone and a number of sub-thalas (Lomtuga well, interval 38–60 m, Tungulu, interval 68–78 m), and also the Maya zone contains considerable amount of the radiocarbon but much lower than the non-permafrost waters. All these facts gives evidence that at present time the water exchange with the surface in the first under-permafrost aquifers in the Lomguta and Abalakh thalases goes more intensive than in the wells of the under-bed and under-lake melting permafrost sides. The higher radiocarbon concentrations in the Abalakh region, where the piezometric levels are higher than the earth surface, gives evidence about existence of relatively powerful under-permafrost flow, the formation of which is not connected with the radiocarbon entering through the melting zone.

The under-permafrost and melting waters often contain a substantial amount of the sulfur compounds such as the hydrogen sulfide (up to 80 g/l) and hydrosulfide. Therefore, it is possible that the sulfate reduction processes, one of the product of which is carbon dioxide, changes isotopic composition of the carbon. Moreover, it is not completely clear if the scheme of the diffusion and filtration exchange in the melted zones. Isotopic composition of the stable carbon and initial radiocarbon concentration of the possible recharge sources in the studied objects also were not determined. Therefore, it is difficult to make here the age estimations.

The radiocarbon data allowed to find that in beginning of the Holocene, the Baltic Sea appeared to be the area of recharge of the groundwaters of the near-Baltic artesian basin (Sobotovich et al. 1977). Analysis of the radiocarbon use possibilities in studying the relationship between the aquifers in Lithuania was done by Banis and Yodkasis (1981).

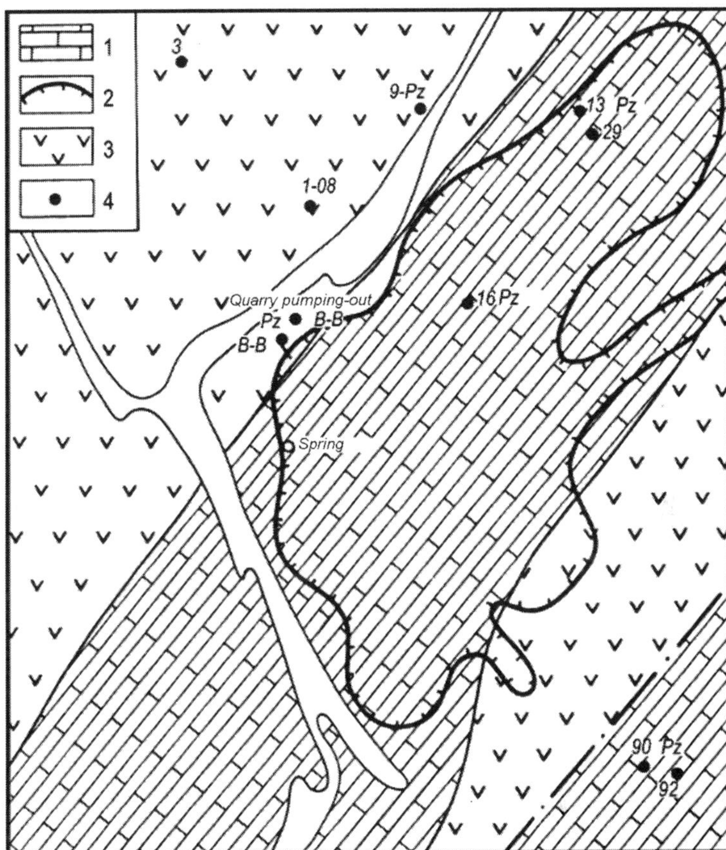

Fig. 18.19 Schematic location of water sampling points around the open-cut mining Ayat-II: (*1*) limestone; (*2*) outline of the open-cut mining; (*3*) porphyrytes; (*4*) sampling points; B–B—water pumping points. (Karasev et al. 1981b)

18.7 Separation of Recharged Water of Different Genesis in Mining

Estimation of the water recharge sources and Tobol River runoff portion to the open-cut bauxite mining in the North Kazakhstan by the radiocarbon method was done (Karasev et al. 1981b). By hydrogeological studies within deposit area, a number of aquifers were identified (Fig. 18.19) such as (a) the aquifer of the Quaternary alluvial deposits; (b) the Cretaceous-Paleogene water-bearing complex; (c) the sporadically distributed bauxite aquifer of the Cretaceous age; (d) the water-bearing complex of the Paleozoic age.

It appeared that the real water inflow exceeds the calculated value by about two times. The assumed portion of the surface water inflow was negligible, and also it was found that mineralization of the water pumping from the Paleozoic limestone

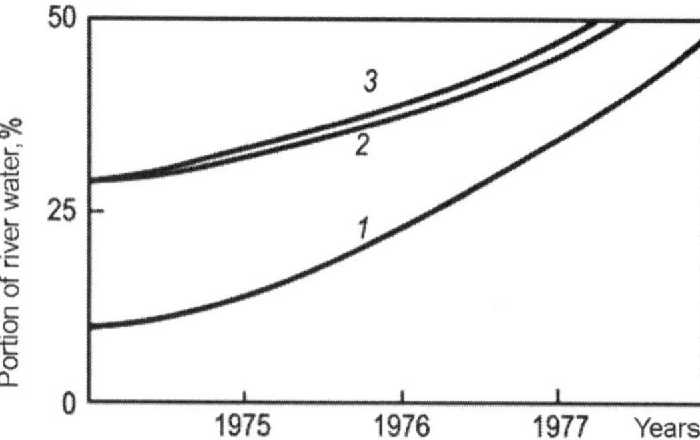

Fig. 18.20 Change in time of surface water portion in the open-cut Ayat-II recharge: (*1*) well 16-I; (*2*) lateral water source; (*3*) mixing water inflow. (After Karasev et al. 1981b)

has decreased. Thus, the problem aroused in overestimation of the open-cut's water inflow components such as the river runoff, groundwater from the alluvial aquifer, Cretaceous-Paleogene and Paleozoic water complexes. Multimineralization of water has not allowed making such estimation by the ordinary hydrochemical methods based on the chemical composition study. The radiocarbon method was applied for this purpose. Figure 18.19 shows the scheme of groundwater sampling from all the water-bearing horizons around the open-cut, from the Tobol River, and water outflows. High values of the radiocarbon content in the open-cut water outflows and in the Paleozoic horizon gives evidence on presence of 'young' waters there. On the basis of radiocarbon data, this can be only water of the alluvial aquifer or from the Tobol River. The samples for radiocarbon (and some other isotopes) measurements were taken in the wells sited on the profile crossed the depression funnel of the open-cut. In the wells 90, 13, 16-I, the data on radiocarbon content in the Paleozoic aquifer and in the depression funnel zone were obtained. For the ^{14}C background value (out of the zone of influence) of this horizon the well 3-E was used. In order to determine interconnection between the alluvial and Cretaceous-Paleogene aquifers, the radiocarbon contents in the wells 1, 29, and 92 were determined. The water from open-cut's outflow, Paleozoic source, and river were sampled. The calculation was done by mixing formula. The data of ^{14}C changes in time shows that the river water portion increases together with the area and depth of the pit, which is shown in Fig. 18.20.

Analogous study was carried out on polymetallic mining (Dubinchuk et al. 1974), located in the south-western slope of the Central Karatau, Kazakhstan. In particular, it was shown that together with the depth of mining portion of groundwaters of the Paleozoic complex has continuously increased reaching about 60%.

The radiocarbon and other isotopes were used for water inflow into the salt mining (Zuber et al. 1979) and into the tunnel Arpa-Sevan (Dubinchuk et al. 1981).

18.8 Determination of Radiocarbon Age of Groundwater

It was earlier pointed out that for determination of the groundwater age, a corresponding model satisfying the real natural conditions of the studied hydrogeological object should be selected. Application of the exponential formula of the radioactive decay (14.15) is justified only for closed basins of 'single filling' or assumption that the water motion in the aquifer on the piston mechanism has occurred. If the radiocarbon comes continuously, which is characteristic, for example, for non-pressurized aquifers with a uniform distribution in area recharge, the exponential expression of type (14.16) is not applicable. The Eq. (14.18) with substitution under the integral sign (or sum), the function of the mean value staying time of the groundwater should be used in this case (Dubinchuk and Polyakov 1983). Most researchers using radioactive dating of the groundwaters apply, as a rule, the exponential formula (14.14) applicable for study of water dynamics in artesian basins (Polyakov 1981) or for determination of age of the 'relict' objects (Dubinchuk et al. 1974; Sobotovich et al. 1977; Vogel 1970). Let us discuss an example of determination of groundwater age studied at the Yaskhan Lens of fresh water in Turkmenistan (Karasev et al. 1975).

The Yaskhan Lens of fresh water is a well investigated hydrogeological object. It is located in the central part of the near-Uzboy Kara-Kum desert. The lens is a localized reservoir of fresh water adapted to the ancient alluvial sediments of the pra-Amu-Dariya River of the Middle Quaternary age (sands with lenses of sandy loam and clay underlying by the Akchagil-Apsheron rocks). The lens of fresh water has in plan elliptic shape extended in the latitudinal direction. The length of the major axis is 65–70 km, width is 30 km, and the area is about 2,000 km^2. The thickness of the fresh water zone is 70 m, which decreases to the periphery. The depth from the surface varies from 10 to 40 m. The lens is 'floating' on the saline water of up to 60 g/l of mineralization. There is transition diffusion zone of about 50 m thick and 25 m in the plan between the fresh and saline waters. There are a number of hypotheses related to the origin of the lens. Two of them are the most probable, such as the infiltration (gradual infiltration of the atmospheric precipitation and condensation of the moisture) and the relict (runoff of the under bed of the pra-Amu-Dariya River water during the last pluvial period). The radiocarbon study appeared to give answer on the problem.

Water sampling for the ^{14}C content measurement was carried out from different depths of the fresh water and diffusion zones in the wells (Fig. 18.21). It was found that the upper lens layer is enriched with radiocarbon compared to its main part, where relatively constant amount of ^{14}C (17–22% of the modern carbon standard) was observed. In the diffusion zone, radiocarbon content was low and amounted 12–14%. It was assumed that the diffusion zone ^{14}C content is formed due to diffusion shift of the fresh water carbonate components and more ancient underlying saline waters. It was shown by study of the chlorine ion distribution in profiles of the diffusion zone that, due to mixing of the fresh and saline waters, ^{14}C concentration in the diffusion zone can decrease by value of 5–10% and thus, the corrected ^{14}C concentration in carbonate system of the diffusion zone is close to that of the main part of fresh water of the lens. The above data prove the hypothesis of simultaneous in a short period of time creation of the main fresh water volume of the lens. The radiocarbon lens age, calculated by using the initial ^{14}C concentration value $A_0 = 100\%$ (river water value) and $A_0 = 53\%$ (water in two wells), is found to be 8,000–13,000 years.

The close to above data for the Yaschan Lens of fresh water (7,000–8,000 years) were obtained by Onufriev et al. (1975), but because of less detailed investigation of the radiocarbon distribution in the fresh and diffuse zone waters, the authors allow

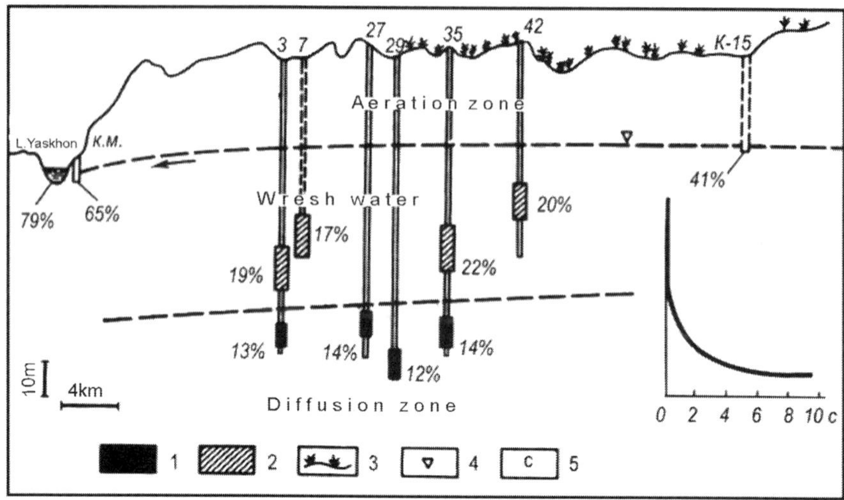

Fig. 18.21 Distribution of radiocarbon concentration in the Yaskhan lens of fresh water: (*1*) filter in diffusion zone; (*2*) filter in fresh water zone; (*3*) sand surface;(*4*) level of groundwaters; (*5*) chlorine-ion concentration along profile (g/l). (After Dubinchuk et al. 1974)

possibility in formation of the lens by intensity of 1.6 mm/year water recharge. The authors' calculations based on the average ^{14}C content in the lens' water is equal to 30%. In the studies done by Dubinchuk et al. (1974), it was shown that this value for the fresh water is equal to $20 \pm 2\%$. In addition, it was found that the lens' body in its section has uniform distribution of ^{14}C content. Such a picture could be not observed in the case of long term layer-by-layer accumulation of fresh water, but the final conclusion of both researcher groups is the Yaskhan Lens of fresh water age equal to, as a minimum, 8,000 years.

Formally the mean residence time in the lens can be calculated by the formula (14.19) in the case of $A_{input} = 50\%$ (which does not contradict to the generally accepted scheme of formation of the isotopic composition of groundwater's carbonate system) and $A_{output} = 20\%$. The time will be equal to about 10,000 years. Taking into account that $\tau = v/q = \alpha\ H/h$, where v is the lens volume in m^3; q is the infiltration discharge, m^3/year; α is the porosity; H is the thickness of the fresh water lens, m; h is the infiltration recharge, m/year; and substituting the values $\alpha =\ 0.3$ and $H =\ 60$ m (Onufriev et al. 1978), one obtains $h = 1.6 \times 10^{-3}$ m/year, or 1.6 mm/year. This value is analogous to that obtained in the above work, but it is worth to stress that such an approach cannot explain the observed ^{14}C distribution in the fresh and salinity lens waters and the most probable origin of the lens' waters is assumption of their relict creation. The higher ^{14}C content in the upper lens layer is pointed out on its modern recharge, which is also fixed by the existing recharge of the Yaskhan Lake. On the other hand, the similar ^{14}C distribution along the profile can be explained by the isotopic exchange of the soil carbon dioxide with the groundwater carbonate system. Such a phenomenon was observed at the study of the ^{14}C distribution in pressurized groundwater of the arid and semi-arid regions, for example, in the Africa (Geyh 1980) and Northern Kazakhstan (Polyakov et al. 1981). The Yaskhan Lens is not a unique object because the relict waters are fixed by the isotope investigations in many arid regions of the Earth globe (IAEA 1980).

The radiocarbon method has opened a new approach in solving the problem of the fresh water formation in the Sahara. Münnich and Vogel (1962) determined the ^{14}C content and hydrogen and oxygen isotopic composition of fresh waters in Eastern Sahara (West Egypt) in locations of some oases. These studies allowed to determine by calculation that the radiocarbon age of fresh waters in the Sahara is equal to 20,000–30,000 years, i.e., their resources recharged in the epoch of the last Pleistocene glaciation when climate on the Sahara territory was cold and humid. These conclusions are proved by the data of hydrogen and oxygen isotopic composition. It is known (Dansgaard 1964) that isotopic composition of the atmospheric precipitation (δD and δ^{18}O) correlates with the mean annual temperature of the region, i.e., in the climatic zones with colder climate, the precipitation of water is depleted in deuterium and oxygen-18 compared to the warmer regions (see Chap. 3). Thus, the hydrogen and oxygen isotopic composition of precipitation and recharged groundwater appears to be indicator of the climatic changes. In the East Sahara, δD and δ^{18}O are on average -85 and -11‰ accordingly. These data are even lower than the δD and δ^{18}O values characteristic for the modern Central European climate. The data on hydrogen and oxygen isotopic composition proved the radiocarbon dating conclusion that formation of the groundwaters in the East Sahara has taken place in the pluvial epoch characteristic for the northern part of African continent during the last glacial period.

According to early work of Ambroggi (1966) the fresh groundwater resources of Sahara were accounted by the value of 15 billion m^3. Later on Gischler (1976) obtained value of 60 billion m^3. If one assumes that this amount of fresh groundwater is uniformly distributed over the all Sahara area (45 million km^2) then the thickness of the hypothetic layer is from 3.3 to 13 m. For comparison, the thickness of the hypothetic global fresh water layer is accounted by value of 55 m (Baumgartner and Reichel 1975), i.e., the average Sahara groundwater resources per unit area are considered one fifth of the average resources per unit area of the continental crust. Such groundwater resources could not be accumulated at the modern Sahara climate (Sonntag et al. 1978) and long period of time is needed for their formation at humid climatic phase. The modern recharge by atmospheric precipitation in the mountain regions of the Sahara periphery (Sahara Atlas, Tibesti, Ennedi, Darfur) occurs. The recharge is accounted by value of 4 billion of cubic meters per year which is from 1/4,000 to 1/16,000 of the total static resources, i.e. mean value of exchange time is within 4,000–16,000 years.

As noted by Sonntag et al. (1978), this value in Ambroggi (1966) calculations was too low and in the central part of Sahara replenishment of fresh groundwaters in conditions of the modern climate has not occurred. It was earlier mentioned that radiocarbon studies of groundwaters in Eastern Sahara were carried out by Münnich and Vogel (1962). Later on the data about the age of fresh water in the Central Sahara and Algeria (Gonfiantini et al. 1974) and Libya (Allemmoz and Olive 1980;

Klitzsch et al. 1976; Salem et al. 1980; Srdoč et al. 1980) have appeared. The results of groundwater radiocarbon dating in Africa (Central and Eastern Algeria, Western Libya, Western Egypt, Sudan, Chad, Nigeria and Senegal) were summarized (Sonntag et al. 1979, 1980).

The largest fresh groundwater reservoir in Africa is the Continental Intercalaire aquifer (Gonfiantini et al. 1974). It covers the main part of the Northern Africa and extends on Tunisian, Algerian, and Libyan territories. The water-bearing sediments of the aquifer represent a sub-horizontal thickness of fractured permeable rocks of the Mesozoic age and occupied 6×10^5 km^2 of the African platform. It extended from the north to south between the Saharan Atlas and Hoggar hills and from the west to east between the Girl-Saura Valley and Libyan Desert. The mountain chain Mzab (the structural bridge between the Saharan Atlas and plateau Tinrert) extended in the meridian direction allows in dividing the aquifer area on the western and eastern complexes. To the west of the mountainous Mzab, the aquifer's thickness is accounted by 200–400 m. The overlying clayey thickness of the Upper Cretaceous (Cenomanian) age is substantially erosive. In these places, the aquifer appears on the surface or covered by water-permeable terrigenous rocks of the Tertiary age and Quaternary sandstones. The aquifer rocks are represented by the sandstones and sandy loams of the Lower Cretaceous age. At the Mzab border the aquifer is in pressurized state. To the west and south-west direction it becomes non-pressurized and less deep. Mineralization of the groundwaters everywhere over the area does not exceed 2 g/l.

To the east from the mountainous Mzab, overlying clayey thickness is uniform and non-dislocated. The aquifer becomes pressurized because of submersion of the Cretaceous rocks from the west to east direction. The thickness of the continental sediments has increased up to more than 1,000 m. At the foot of the aquifer, the Jurassic rocks appear. By piezometric data, the regional flow was directed to the north-east, to the Gulf of Gabes. In the sandstones of the Albian age, the low-mineralized waters (1–2 g/l) overlying the saline waters of the Neocome and Jurassic age are extended. By the hydraulic test data, recharge of the aquifer is provided in its surface occurrence such as in the Saharan Atlas on the north, in plateau Tinrert on the south, and in Dakar region on the east.

In the eastern part of the aquifer, the mean values of deuterium and oxygen-18 practically do not change over the area and have $\delta^{18}O = -8.4 \pm 0.4‰$ and $\delta D = -61 \pm 3‰$. These values are characteristic for the aquifer waters in regions where the leakage from the overlying complexes is absent. The ^{14}C activity is very low and varies from 0 to 2.5% of the modern carbon standard. In the recharge zones ^{14}C values are higher such as on the south of the Saharan Atlas is 54.7% and in Dakar 53.3%. In the plateau Tinrert on south, the recharge region is identified and characterized by ^{14}C content values of 22.8 to 9.9% of the modern carbon standard. Based on the radiocarbon data, the age of the main aquifer's fresh groundwater amount can be estimated by the value of 30,000 years.

Somewhat different picture is observed in the western part of the Continental aquifer. Here the heavy oxygen isotope varies in wider ranges such as $\delta^{18}O$ varies from -9.6 to $-4.0‰$. In the southern part of the aquifer, ^{14}C varies from 4 to 40%

which showsthe presence of the local recharge. As a result, the mixture of ancient groundwater, having age of several periods of the ^{14}C half-life, and waters of the modern infiltration recharge are formed. In the region of the Great Western erg a significant modern local recharge of the aquifer is observed. The ^{14}C ion content at these places reaches up to 60%. The same recharge effect by the piezometric levels was fixed here. The close hydraulic connection of the Continental aquifer in the western part with the non-pressurized Cenomanian aquifer was found by ^{14}C dating. This fact is the consequence that the low thick covered layer of the Upper Cretaceous sediments is considerably washed and does not play the role of a hydraulic screen with respect to the water-bearing Continental aquifer, which in the western part of the mountainous Mzab is a non-pressurized aquifer. The isotope studies also proved the aquifer connection with the Cenoman-Pliocene rocks in the Gulf of Gabes region.

To the east from the mountainous Mzab, one more water-bearing aquifer of fresh waters, adapted to the fractured limestone-marl rocks and underlying the desert, is located. This aquifer lies higher than the Continental Intercalaire aquifer and contains fresh water with 1 g/l mineralization. By the isotope data, the conclusion was made that the modern recharge plays a great role in its recharge in the region of desert sand formation. The heavy hydrogen and oxygen isotope content (their values are -40 and $-3‰$ respectively and the straight line slope in the co-ordinates $\delta D - \delta^{18}O$ is equal to 5.3 ± 0.5) demonstrate significant evaporation processes during filtration of precipitation through the sand dunes.

The researchers also included the second mechanism of the groundwater resources formation in the Upper Cretaceous–Eocene rocks. On the basis of isotope data analysis, it is possible to assume that the horizon is partly recharged by waters of the Continental Intercalaire aquifer and in the sandy ergs region the mixing of the deep and infiltrated waters has occurred. Certainly, the evaporation process from the sand dunes affects the isotopic composition of water, but the chemical content undergoes minimum changes because the Saharan dunes consist of pure quartz sand.

The study of ^{14}C, deuterium, and oxygen-18 content in the Saharan groundwaters made the following conclusions (Gonfiantini et al. 1974; Sonntag et al. 1978, 1979, 1980).

In general, the Saharan fresh groundwaters are of relict origin which has happened in the pluvial epoch between 50,000 and 20,000 years ago. In the last Pleistocene the minimum of temperature (22,000–16,000 years ago) was fixed interruption in recharge of the aquifer recharge, which was connected with the climatic peculiarities of that period. Intensification of the recharge is fixed at about 8,000 years ago, when on the African territory in the early Holocene the phase characteristic for the continent in the late Pleistocene was changed by the short-term humid phase.

It is clear from the above that the radiocarbon method is an important tool to study the paleoclimatic and paleohydrology events.

The groundwaters related to the last Pleistocene glaciation was discovered in the Gdov and Strelnik aquifers (PR_3V) on the southern shore of the Gulf of Finland, St. Petersburg region (Bondarenko 1983; Sobotovich et al. 1977). The Gdov aquifer was opened by a well at a depth of 110–190 m. The water abundance in the aquifer, which contains chloride-sodium water with 1.6 g/l mineralization, related to the sandy stratum thickness which varies from 2.0 to 6.0 m. The pressurized Strelnik aquifer lies under the clayey layers of 4.0–8.5 m thickness, which separate it from the Gdov aquifer. The thickness of the Strelnik aquifer composed from the alternate aleurolites, sandstones, and clays, changing from 11 to 30 m. By chemical composition, the Strelnik aquifer represents chloride-sodium water of 2.6 to 10 g/l mineralization depending on the depth. The modern direction of groundwater motion is oriented to the north-northwest to the Gulf of Finland, which is the region of Strelnik aquifer discharge. The radiocarbon age was calculated by the formula

$$t = 8030 \ln (A_0 \delta^{13} C_{\mathrm{HCO_3}} / A_t \delta^{13} C_{\mathrm{CO_2}}).$$

The measured value of $\delta^{13}C_{\mathrm{HCO_3}} = -20.5‰$ and $\delta^{13}C_{\mathrm{CO_2}}$ was accepted equal to $-25‰$. It is worth to note that the corrections were introduced without possibly taking into account the stable isotope fractionation in the system gas-liquid-solid phase and also without taking into account the isotopic composition of carbon of the carbonates of the water-bearing rocks. Certainly, this reduces reliability of the obtained values but the results do not contradict with the general geological premises. The radiocarbon age of the Strelnik groundwater aquifer varies from 9,000 to 14,000 years. The substantial depletion of isotopic composition of these waters in deuterium and oxygen-18 ($\delta D = -119‰$ and $\delta^{18}O = -18.8‰$) was compared with the modern atmospheric precipitation where $\delta D = -90‰$ and $\delta^{18}O = -12‰$. On the basis of the obtained results, the authors conclude about the glacial origin of the Strelnik aquifer's groundwater. Before 10,000 years, when the common Baltic Lake was covered by ice, the groundwaters were recharged by the glacial melting waters. After the lake melted, the hydrogeological situation changed and the Gulf of Finland became the reservoir of the groundwater discharge. The analogous inversion in hydrogeological situation was found by the radiocarbon data in the Wendian water-bearing complex in North Estonia (Banis and Yodkasis 1981). The calculated radiocarbon age of the groundwater decreases in the direction of the modern region of discharge which is the Gulf of Finland. This is also the evidence of groundwater recharge in the late Pleistocene–Early Holocene by vertical leakage from the Baltic ice lake.

An exotic—with respect to isotopic composition—groundwater in the Cambrian-Vendian rocks was found in the near-shore zone of the Gulf of Finland around Tallinn (Yezhova et al. 1996). The radiocarbon age of this water is found to be 9,000–14,000 years and the values of δD and $\delta^{18}O$ reach $-170‰$ and $-122.2‰$ respectively. By the chemical content, they are the hydrocarbonate calcium waters with mineralization of 0.46–0.84 g/l. The genesis of the waters was defined by the deglaciation of the Baltic basin, which started not earlier than 14,000 years ago (Yezhova et al. 1996).

The natural resources of groundwaters in the arid regions could probably have happened in the Holocene during the short periodic cooling, accompanied by increase in the mean-annual precipitation. The ^{14}C studies of the natural groundwater resources of the main water-bearing horizon of the Neogene age rocks in the north-east Kenyan province were carried out (Pearson and Swarzenki 1974). The main source of fresh water in this part of Kenya is groundwater in the sediments under beds of the Ewaso Ngiro and Larh Dera River Valleys on the area located 200 km south-east of Habaswein City. The lenses of fresh water lie on the saline waters. These waters probably represent the under bed runoffs of the Ewaso Ngiro and Larh Dera Rivers but from hydrogeological conditions their resource replenishment can occur only during periodic strong floods of the rivers. The radiocarbon and tritium sampling campaign in 20 points of the fresh and saline waters from the transition zones was carried out. There was no tritium fixed in any sample and the ^{14}C content changed from 2.8 to 87.7% of modern carbon standard.

The ^{14}C higher values are characteristic for the fresh water, but correlation between ^{14}C concentration and mineralization is absent. Pearson and Swarzenki (1974) performed analysis on the basis of the ^{14}C activity and total carbonate components in the samples (see formula 14.67). If the specific ^{14}C decreases results due to dilution by the 'dead' carbonate components but not by the radioactive decay, then in this case the dependency of the ^{14}C activity (% of modern carbon standard) on the reverse carbonate component concentration by the straight line from the origin is represented. Graphically experimental data of Pearson and Swarzenki are shown in Fig. 14.24. The points are arranged near the straight lines I–IV. The lines II, III, IV go out from the origin and the line I at $^{14}C = 8\%$ (of the modern carbon standard).

As the above authors pointed out, by the sample groups (see Fig. 14.24), the time intervals between the separate intervals of the groundwater replenishment can be determined. It is obvious that the samples with the same age of water, the specific radioactivity of carbon, in which the initial ^{14}C concentration of the 'dead' water-bearing carbonates decreased by dilution, are grouping near each line. In particular, for the water samples 9 and 10, the age difference is accounted to $\Delta t = (1/\lambda)$ ln (A_{10}/A_9), where $1/\lambda$ is accepted by the authors equal to ($T_{1/\lambda} = 5{,}730$ years); A_{10} is the ^{14}C concentration in the sample 10 is 88%; A_9 in the same in sample 9 is 64%. From here one obtains $\Delta t_{\text{II}\ -\ \text{I}} \approx 2{,}600$ years, i.e., between the second and first stage of replenishment of the water resources 2,600 years have passed. By the same procedure, it was found that between the III and II stages 6,300 years and between IV and III stage 5,200 years have passed. The obtained results do not claim to be the absolute truth but their interpretation agrees with the hydrogeological conditions of the territory. One may assume that the main groundwater resource replenishment happens during catastrophic river floors from swamp area of the Lorian region, which appears to be the area of groundwater flow discharge during shallow-water periods of time. The water pressure changes during floods convert the discharge area into area of recharge when the old waters were replaced in periphery of the system by new young water. The obtained results prove this hypothesis. The oldest waters are found on the far places of the Lorian swamps and the young water places aspire just to the swamping site. The samples 9 and 10 were taken from the wells placed near the Ewaso Ngiro River in its upper part where floods seldom happen.

18.9 Determination of Flow Velocity and Direction in Regional Scale

The radiocarbon (as well as tritium) data can be used to study the regional flows of groundwater. As it was earlier pointed out, the regional characteristics of groundwater movement in a studied aquifer can be obtained by plotting of the hydroisochrones (the lines of equal water age). In a number of cases, it is reasonable to operate not by the age but by the reduced values of radiocarbon concentrations. In this case, correlation of the radiocarbon content, using the hydrochemistry of the groundwater carbonate system and the stable carbon isotopic composition is carried out. The hydroisochrones plotting (see Fig. 18.15) by a natural indicator allows determining the groundwater recharge and discharge areas and characterization of direction of the regional flows. It is followed by applying the radiocarbon data in determining real velocities of the groundwater motion and at the same time it is not necessary to calculate the absolute water age. Such assumptions can be done without exact determination of the ^{14}C specific activity in the carbonate system of groundwater in the recharge area (Fritz et al. 1974; Vogel 1970; Pearson and White 1967; Pearson and Hanshaw 1970).

As an example, let us consider the date of the Carrizo Sand, Texas, USA (Pearson and White 1967; Pearson and Hanshaw 1970). The Carrizo basin is located in south-central Texas. This aquifer has been thoroughly studied by conventional hydrogeologic techniques. In particular, good estimates of the hydrodynamic flow rates at different distances from the recharge zone were available. The part of the aquifer studied has a relatively narrow, climatologically constant recharge area, and thus it seems reasonable to assume a constant value of $m\mathrm{C}_{\mathrm{initial}}$ prevails. The authors have measured ^{14}C concentrations and determined the carbonate component content in water of a number of wells. If the radiocarbon content determined in two points located on the line of the flow motion is x, then the velocity of motion v is determined as $v = x/\Delta t$, were Δt is difference of radiocarbon age of water in the sampling points with distance x, which can be obtained from equation $\Delta t = 8{,}033 \ln (A_1/A_2)$. Here A_1 and A_2 are the radiocarbon concentrations (activity) of the groundwater carbonate system in points 1 and 2, but the data obtained by above formula do not coincide with results of the hydrodynamic studies (Fig. 18.22).

Pearson and Hanshaw (1970) points out that the most reliable data on velocity of ground water motion by radiocarbon can be obtained if the ^{14}C activity is corrected by the stable carbon isotopes or on the basis of chemistry of carbonate components (see Eqs. 14.36–14.38). In particular, for determination of the water motion velocity in the Carrizo Sand aquifer the following formula for calculation of Δt was applied:

$$\Delta t = \frac{1}{\lambda} \ln \left[\frac{A_1 \sum mC_1}{A_2 \sum mC_2} \right] = \frac{1}{\lambda} \left[\frac{A_1\, \delta^{13}C_2 - \delta^{13}C_{\mathrm{CaCO_3}}}{A_2\, \delta^{13}C_1 - \delta^{13}C_{\mathrm{CaCO_3}}} \right],$$

where $\sum mC_1$ and $\sum mC_2$ are the carbonate component concentration in water at the sampling points 1 and 2; $\delta^{13}C_{\mathrm{CaCO_3}}$ is the carbon isotopic composition of the water-bearing rocks.

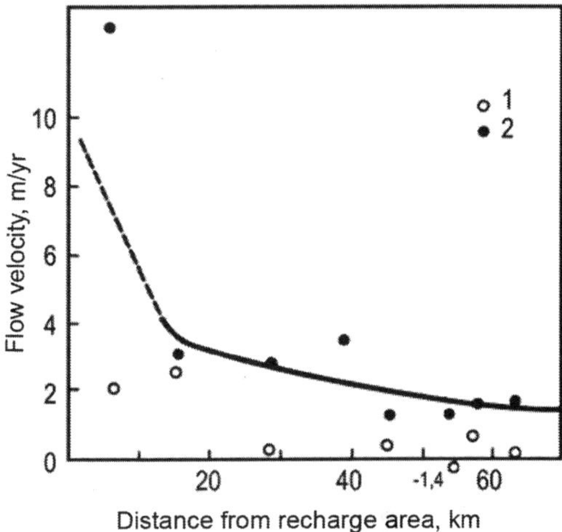

Fig. 18.22 Hydrodynamic velocities of motion of groundwater in the Carrizo Sand, Texas: (*1*) the radiocarbon data without correction; (*2*) the same data with correction. (After Pearson and Hanshaw 1970)

The results obtained by calculating hydrodynamic and radiocarbon data are in good agreement (Fig. 18.22). If the number of sampling points are more than two, then experimental data are easily graphically presented (Vogel 1970) with coordinates as the radiocarbon age and the distance from the recharge area. In this case, the analysis of condition of the uniform groundwater flow motion and the places of its disturbance may be identified, and also the velocity for uniform intervals, for example, by the least square or graphically as inverse value of the angular coefficient of straight line $t = x/v$ can be calculated. These understandings are true for the aquifers not having recharge from the surface or having hydraulic connection with another water-bearing horizon.

A profile through two such similar water-bearing horizons is shown in Fig. 18.23 (Vogel 1970). Most of the area is covered by a thick layer of Kalahari (African arid region) sand. The groundwater is contained in the underlying crystalline and consolidated rock and, according to the water-level controls, is flowing from the east to west. Starting from areas where the pre-Kalahari rocks outcrop, the ^{14}C ages, obtained from several boreholes, increase in the same direction. The studied aquifer is practically non-pressurized. This is practically groundwater flow in the Kalahari sands. The age pattern, therefore, suggests that little or no recharge occurs through the Kalahari Beds and that replenishment takes place only where the covering sand layer is absent or thin. The uniform increase in age with increasing distance from the recharge area shows that the average flow over a distance is remarkably regular. In the far west of the western aquifer ages are encountered which do not conform to the general pattern, but these can be explained by converging flow from the north to south and/or by perched water from local recharge.

The coefficient of filtration or porosity of the water-bearing horizon can be found by the value of flow velocity, because these parameters are connected by the Darcy equation $v\alpha = \mathrm{KI}$ (where I is the gradient of pressure; Vogel 1970). It is natural that in this case by an independent way one of the parameters (α or K) should be obtained.

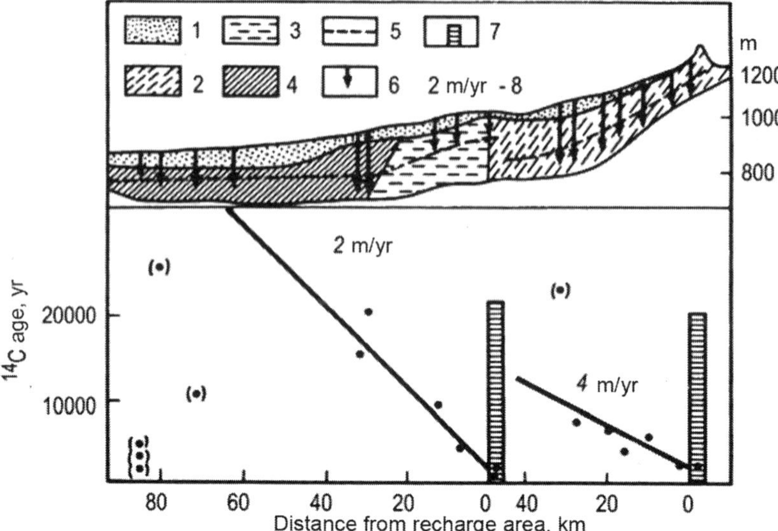

Fig. 18.23 Radiocarbon concentration change as a function of the distance from the recharge area in Kalahari Beds, South Africa: (*1*) Kalahari sands; (*2*) Waterberg quartzite; (*3*) Koling quartzite; (*4*) Dwika tillites; (*5*) groundwater level; (*6*) filter in borehole; (*7*) recharge area; (*8*) groundwater flow velocity calculated by radiocarbon. (After Vogel 1970. © IAEA, reproduced with permission of IAEA)

Many works were devoted to the problem of the velocity of groundwater flow motion determination by the radiocarbon (Castany et al. 1974; Evin and Vuillaume 1970; Fontes 1976; Fritz et al. 1979; Ingerson and Pearson 1964; Jordan et al. 1976; Münnich and Vogel 1962, 1963).

18.10 Paleoclimatic and Paleohydrogeologic Studies

The radiocarbon time scale (0–80,000 years) practically coincides with the last glacial period continuation which allows the radiocarbon method of dating to be successfully used to study the Late Quaternary geological, climatology, and paleohydrogeology (Arslanov 1987; Ferronsky et al. 1993). As it was earlier noted, in the epochs of cooler climate, accompanied in the modern arid regions by pluvial effects, intensive recharge of aquifers occurred, which led to considerable water resources formation. The radiocarbon method in combination with geological, hydrogeological, paleobotanical and other methods for paleoclimatic reconstruction of conditions in the late Pleistocene–early Holocene is used. The radiocarbon dating of samples allows connection of the periods of climatic change and the groundwater age and in combining with distribution of hydrogen and oxygen isotopes gives possibility in reconstruction of the paleohydrogeological conditions of the natural groundwater resources formation.

The hydrogeological regime of an aquifer largely depends on the climate change, which affects the recharge and discharge of groundwater. In general, for the last glacial period in the high latitudes, the low evaporation is characteristic and for the middle latitudes higher precipitation is typical. This is certainly schematic treatment of the meteorological peculiarities in the glacial epochs.

The late Quaternary time for the both hemispheres on the latitude of about 25°, the following paleoclimatic pattern was observed (Rodnon 1976; Rodnon and Williams 1977).

- 40,000–20,000 years before present day: considerable atmospheric precipitation and high lake water levels;
- 17,000–12,000: inner tropical aridity accompanied by dune formation and lake drying;
- 11,000–5,000: plenty of atmospheric precipitation and very high lake water level.

Similar climatic peculiarities related to the mean annual temperatures of the near land air layer and also changes in the air mass circulation caused by creation and degradation of the ice cover in the Northern Hemisphere were observed. The high pressure belt in the glacial maximum shifted to the south. In this connection, the trade zone also shifted in the same direction, which led to its formation in the southern regions; for example, the monsoon precipitation on the modern territory of Sahara. The global climatic changes are in good agreement with the oceans' level. During the late Pleistocenian temperature minimum (which started ~20,000 and completed 14,000 years ago), the oceans' level was 90–100 m lower than the modern (Mörner 1971). In that period, the mean annual temperature in the Central Europe was 12–15°C lower than the modern.

Before the last long interstadial of the Würm-Wisconsin glaciation, known as Stillfried, about 38,000 years ago, the notable cooling which continued for ~2,000 years (Flohn 1980) took place. The global climate warming began ~10,000 years ago after the short-term Alleröde interstadial (11,000 years). This warming continued for several centuries and was accompanied by the ice melting and eustatic rise of the oceans' level. The process continued for ~8,000 years.

During the last 10,000 years (Holocene), the four climatic phases were of main interest from hydrological and hydrogeological viewpoint (Gribbin and Lem 1980). First was so called after-glacial climatic optimum. The culmination phase of that event was between 7,000 and 5,000 years ago. The mean annual temperature in the Central Europe in that period was higher than the modern by 2–3°C and the sea level was higher by ~3 m. The high pressure belt was shifted to the north up to 40–45°N. The trade winds' zone also shifted to the north together with rising of the region of summer monsoon precipitation.

Between 2,900 and 2,300 years ago, the cooling of the 'iron century' occurred. As a result of decrease of the temperature in Europe, the atmospheric precipitation roused, which led to widen the areas of swamps. In Northern Africa, that period of time was somewhat dry but more humid than the present day.

The second after-glacial climatic optimum has taken place about 1,000–1,200 years of present time (or little bit later). In that time, the summer temperatures were about 1°C higher of the modern. In the Central America, Kampuchea, Middle East, and Mediterranean regions that period coincides with the period of high level of humidity.

In the period from 1,430 to 1,850 years, the climate became cooler and so cold that 'the small glacial epoch' has occurred. The mean-annual temperatures on the Central Europe territory were 1–1.5°C lower than the present day (Gagen et al. 2006). In this epoch, the glaciers of Europe, Asia, and North America roused their activity. In the mountain regions of Ethiopia, the snow fall was first registered. The Caspian Sea level was very high, The White Nil level, depending on the amount of equatorial precipitation, was low due to the monsoon shift to the south. Between 1850 and 1900 years, in the Northern hemisphere, the climatic regime characteristic for this region in the small glacial epoch was restored.

The above climatic phases put on the mark on the formation of natural groundwater resources (Ferronsky et al. 1993). As it was noted, replenishment of the fresh groundwaters of Sahara, defined by the radiocarbon data, took place during pluvial epochs, which were characteristic for the period of the Pleistocene glaciation. This conclusion is proved not only by the groundwater age determination by radiocarbon but also by isotopic composition of hydrogen and oxygen of those waters. It is known (Dansgaard 1964) that between deuterium and oxygen-18 in the atmospheric precipitation, there is close relationship described by the equation $\delta\mathrm{D} = \delta^{18}\mathrm{O} + 10$. The free term of the equation characterizes kinetic fractionation of the hydrogen and oxygen isotopes at non-equilibrium evaporation of the ocean water giving rise to atmospheric precipitation. The more the evaporation in non-equilibrium conditions, the more is the value of free term in the above equation (see Chap. 3). For example, for the East Mediterranean it is equal to +22. The value $d = \delta D - \delta^{18}\mathrm{O}$ is called 'excess deuterium' or 'excess parameter'. There is relationship between the local mean annual temperature and the mean weighted isotopic composition (Dansgaard 1964): $\delta\overline{D} = 5.6\bar{t} - 100‰$ and $\delta^{18}\mathrm{O} = 0.7\bar{t} - 13.6‰$. Thus, the stable isotope of hydrogen and oxygen content in the atmospheric precipitation and the recharged groundwater appears to be the function of the local mean annual temperature. This fact can be used as some kind of isotope geothermometer (see Chap. 3). On the basis of distribution of the stable hydrogen and oxygen isotopes in groundwaters, studied at ^{14}C age, it is possible to interpret paleohydrogeological conditions of a territory (Fontes 1981). The pioneers in this field were Münnich and Vogel (1962) and also Degens (1962), who studied the aquifer of Nubian sandstones (Continental Intercalaire) in the Western Egypt. Isotopic composition of hydrogen and oxygen of these groundwater ($\delta\mathrm{D} \approx -85‰$, $\delta^{18}\mathrm{O} \approx -11‰$) is much more light than the modern atmospheric precipitation in the recharged aquifer area. The radiocarbon groundwater age, calculated at assumption that $A_0 = 72.5\%$, lies within 18,000 and 40,000 years.

The groundwaters of the Nubian sandstones were investigated on Sinai Peninsula in the region of the Negev desert (Gat and Issar 1974). The groundwater dating by radiocarbon with correction on the stable carbon in the carbonate system gave the age value from 13,000 to 31,000 years. The deuterium and oxygen-18 content in these waters are somewhat higher (δD from −30 to −65‰, δ^{18}O from −6 to −7‰) than in Egypt's part of the same aquifer. The genesis of these waters connects with the glacier period, which is proved by the radiocarbon age and by the excess parameter. For the modern waters of this region, d = + 22‰ because the atmospheric precipitation relates to evaporation in a very non-equilibrium conditions of the Mediterranean Sea. For the paleowaters, the value d = + 10‰, which characterizes the degree of evaporation from the sea surface at lower than modern mean annual temperatures.

Sonntag et al. (1979) carried out analysis of the paleowater from different regions of Sahara. The results of this work are shown in Fig. 18.24. As it is evident from the figure, the most number of the aquifers in the central part of Algeria have the age from 12,000 to14,000 years, i.e., to the warm post-glacial period with the plenty of convective tropical precipitation. During the last Pleistocene temperature minimum (20,000–14,000 years ago), a cold semi-arid climate was over Sahara territory, which is determined for all the studied Saharan regions having the above age. The period before the last glacial (20,000–40,000 years ago) for Sahara was characteristic of plenty of precipitation in the winter season at the prevailing western air masses transfer (Sonntag et al. 1978). This caused the occurrence in all the studied regions' groundwaters with the age from 20,000 to 48,000 years and more.

For the Central European groundwaters, the analogous analysis done by Vogel (1970) shows that the samples obtained have the age more than 12,000 years. In author's opinion, this is explained by existence of permafrost rock thickness over the European territory, which during the last Pleistocene temperature minimum prevented the groundwater recharge. The same explanation is supported by Geyh (1974) study (see Fig. 18.16).

Considerable decrease in δD and $\delta^{18}O$ values with increase in the age of water, as a rule, is traced in the modern arid and semi-arid regions. For example, in the Cretaceous aquifer of the London's basin (Smith et al. 1976) and in the Lincolnshire's limestone (Downing et al. 1977) of the eastern part of England, the age of water-bearing horizon is changing from the modern to value of about 25,000 years, depending on the distance from the sampling point. At the same time, $\delta^{18}O$ value decreases only by 0.7‰. The water that is 25,000 years old was formed in rather severe climatic conditions which have taken place in the Atlantic region. For example, Fontes (1981) demonstrates data of Dansgaard et al. (1969) from Greenland for the time where the atmospheric precipitation was depleted in ^{18}O by 13‰ compared to the modern values. This can be explained by recharge of the aquifer during some interglacial, when climate was like the modern one or the water-bearing horizon during the Pleistocene glaciation has not recharged at all due to the permafrost effect. In this case, the radiocarbon age occurs from mixture of the ancient and modern waters. Analogous pattern in one of artesian basins in North France was observed by Fontes and Garnier (1979). Here the corrected ^{14}C age of water was accounted by about 15,000 years and during that period the climate in the studied area was colder, but difference in isotopic composition of the modern and ancient water was small. Probably, the statement that the last Pleistocene glaciation, characterized by lower temperatures for all of the globe regions, should correspond with lower deuterium and oxygen-18 values in precipitation is not always correct. Recently, Yapp and Epstein (1977), while studying isotopic composition of deuterium in cellulose of the fossil woods dated by ^{14}C, found that in North America during the glacial maximum mean annual relative content of deuterium in precipitation was 19‰ higher than that

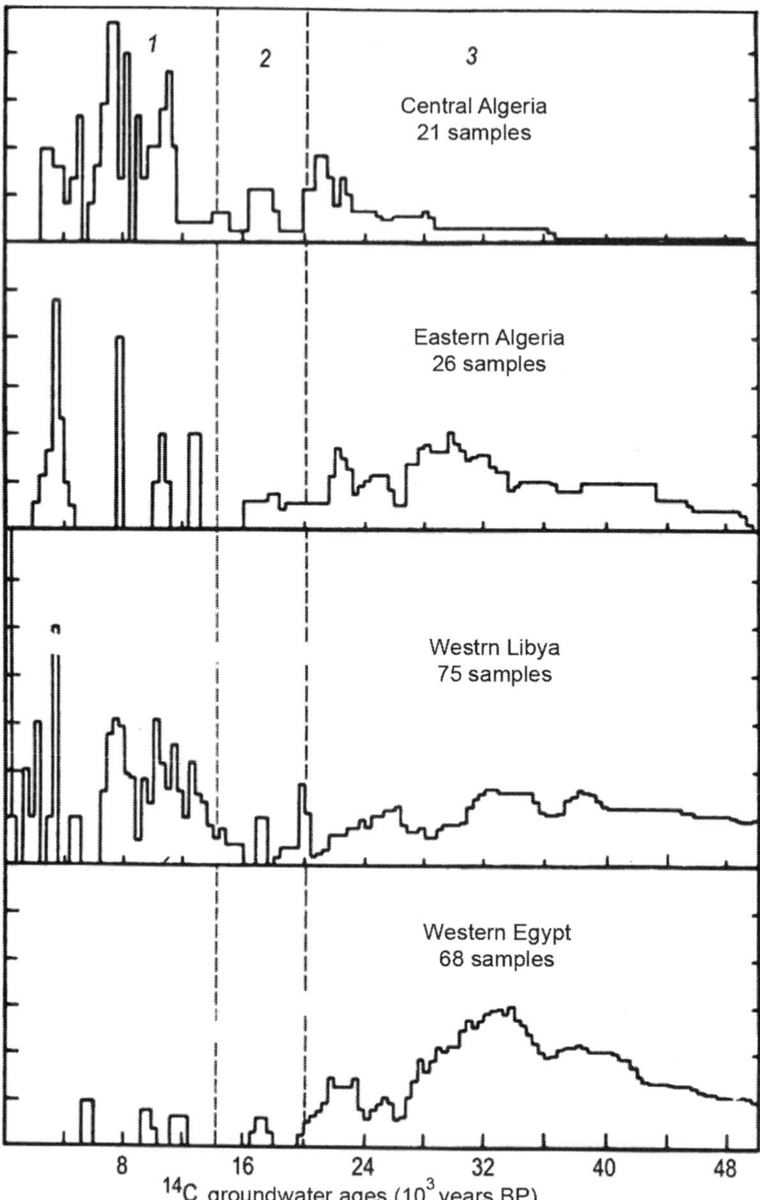

Fig. 18.24 Radiocarbon age of groundwaters of the Sahara: (*1*) transition from humid to arid phase; precipitation of the convective origin; (*2*) semi-arid maximum; (*3*) the period preceded to the last Pleistocene temperature minimum; precipitation of the western transfer. (After Sonntag et al. 1979. © IAEA, reproduced with permission of IAEA)

Table 18.13 Distribution of deuterium and chloride ion along the section of the impervious stratum of clays and marls in the region of Shadrinsk (Southern Urals)

Sampling depth (m)	δD (‰)	Cl^- (g/l)
209.0 – 208.8	−99 ± 3	1.2
177.0 – 176.8	−100 ± 3	1.3
168.0 – 167.8	−107 ± 3	1.2
167.0 – 166.8	−101 ± 3	–
158.0 – 157.8	−103 ± 3	1.3
157.0 – 156.8	−107 ± 3	–
148.0 – 147.8	−115 ± 3	1.6
141.0 – 140.8	−110 ± 3	1.6
138.0 – 137.8	−97 ± 3	–
128.0 – 127.8	99 ± 3	1.6
118.0 – 117.8	−88 ± 3	1.2
114.0 – 113.8	−98 ± 3	–
112.0 – 111.8	−108 ± 3	1.0
99.0 – 98.8	−109 ± 3	0.8

of the modern. The authors explained this by lower temperature gradients between the ocean surface and the land, by higher δD values of the ocean surface waters as a result of concentration of lighter isotopic composition of water in the ice cups, and also by lowering of the kinetic factor at fractionation of isotopes during evaporation of water in conditions of lower temperatures ($d \to 0$). Certainly, the data of Yapp and Epstein obtained only for one region of the land needs to be checked but the arguments are in favor of the observed effect. The authors of this book also found that on the European part of the former USSR, except the near-Baltic region, there is insignificant difference in isotopic composition on the modern and ancient waters having corrected radiocarbon age equal to 30,000 years. Probably this also connects with development of permafrost cover preventing the groundwater recharge. Replenishment of the natural groundwater resources probably has happened after degradation of the permafrost thickness in the conditions close to the modern waters and the groundwaters, ^{14}C age of which is accounted by 20,000–30,000 years, in reality are a mixture of the ancient and modern water. At present, this problem has not found its final resolution.

The Pleistocene stratums of permafrost also existed in the south of the West Siberia and Northern Kazakhstan, where somewhat different picture is observed (Ferronsky et al. 1982). For these regions, a dependence of the hydrogen and oxygen isotopes on age is observed. In the work of Dubinchuk et al. (1974), on the basis of studying the distribution of deuterium and oxygen-18 in pore waters, the possibility of a quantitative estimation of leakage of groundwater through an aquiclude was theoretically justified. The method was tested in the region of Shadrinsk (South Urals) in groundwaters of Pleistocene age (Polyakov et al. 1978). The stratum of marls and clays with a thickness of about 130 m separated Cretaseous and Paleogene water-bearing beds. The content of deuterium and chloride ion was studied in pore waters along the section of the stratum. The results of this investigation are given in Table 18.13 and Fig. 18.25. The stratum occurs in the depth interval 98–220 m from the surface. The deuterium content in waters of the Lower Cretaceous and Upper Paleogene horizons is equal respectively to −99 ± 3‰ and −109 ± 3‰. The chloride ion concentration in waters of the lower horizon varies within 1.4 ± 0.2 g/l

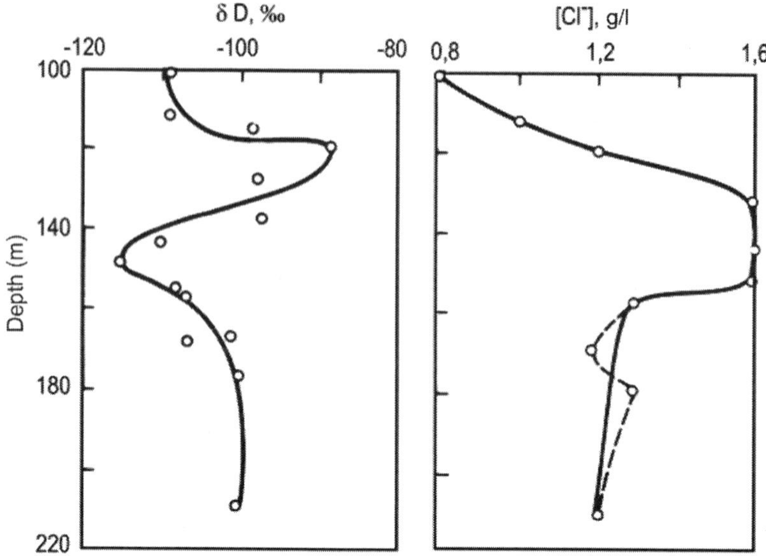

Fig. 18.25 Variations of values of δD and Cl^- concentrations along section of the impervious clay-marl stratum (region of Shadrinsk, Southern Urals). (After Ferronsky et al. 1982)

and in the upper horizon it varies with depth, amounting to 0.80 ± 0.06 g/l. On the curve of the deuterium distribution along the section of the stratum (Fig. 18.25), a minimum depth of 148 m and a maximum depth of 118 m are observed.

The maximum Cl^- concentration is noted at an average depth of 138 m. Taking into account that in the lower horizon there is a gauge pressure of water about 5 atm, we can assume the existence of a flow (leakage) of water from the Cretaceous horizon. In this case, the minimum in the deuterium distribution along the section of the stratum is due to the presence of light water formed at the start of the Holocene in Cretaceous horizon as a result of melting of glaciers and snow in the recharge area confined to the Ural Mountains. The δD maximum can be explained by the low seepage velocities in the glacial epoch during the existence in the recharge area of poorly permeable frozen rocks, which led to diffusion salinization of the Lower Cretaceous sedimentary stratum.

In different parts of the investigated territory are found waters with an age from 9,000 to 13,000 years, determined with respect to ^{14}C, whose isotopic composition ($\delta D = -121 \pm 5‰$, $\delta^{18}O = -12.9 \pm 5‰$, tritium content ~2 TU) considerably differs from the composition of modern waters ($\delta D = -110 \pm 4‰$, $\delta^{18}O = -11.2 \pm 0.3‰$, tritium content ~40 TU). Thus, we can conclude that recharge of the investigated aquifer began 9,000–13,000 years ago. On the basis of data on the deuterium distribution, assuming that recharge of the lower aquifer (K_2) began about 10,000 years ago (Kind 1974), we can easily calculate the rate of leakage, which is 0.7 cm/year ($\sim 2 \times 10^{-5}$ m/day). It is characteristic that the maximum of the Cl^- distribution during this time 'lag behind' the deuterium maximum by about 20 m, i.e., the clay-marl

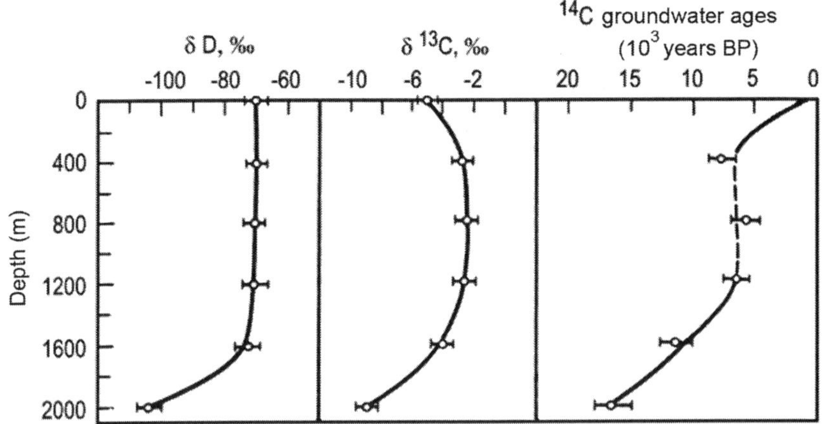

Fig. 18.26 Variations of the values of δD, δ^{13}C and change with respect to depth in the age of water of the Bakharden basin of mineral waters (Southern Turkmenistan)

stratum operates on the principle of a chromatographic column known in laboratory practice and Cl^- cannot be used as an indicator of the movement of water masses in clay strata.

With consideration of the diffusion 'blurring' of the deuterium minimum on the distribution curve, the deuterium content in water that entered the Cretaceous horizon at the beginning of the Holocene can be estimated by the value of $\delta D = -125 \pm 5‰$. Then the conclusion can be made that the mean annual value of the temperature in the Southern Urals was 3–5° lower than that of the present days (+1°C).

Data on the content of hydrogen, oxygen, and carbon isotopes were analyzed during a study of the conditions of formation of mineral waters of the Bakharden deposit confined to the Pre-Kopet Dag deep fault (Southern Turkmenistan; Polyakov and Tkachenko 1980). The results of the analyses are given in Table 18.14 and shown graphically in Fig. 18.26.

The groundwaters of the deposit are confined to a Malm-Neocomian carbonate-terrigenous stratum. The upper aquifers are characterized by values $\delta D = -70‰$ and $\delta^{18}O = -10‰$, which are close to typical modern precipitation in the recharge area. A marked decrease in the deuterium concentration ($\delta D = -104‰$) is noted for waters tapped at a depth of about 2,000 m. These waters probably entered the Malm-Neocomian aquifer system in the epoch of the last Pleistocene glaciation. The radiocarbon age of these waters, calculated with consideration of corrections based on the isotopic and chemical composition of carbon of the carbonate system, is estimated to be about 17,000 ± 3,000 years. It is evident from Table 18.13 and Fig. 18.26 that about 12,000 years ago, a pronounced warming of the climate occurred in the investigated territory.

Lightening of the isotopic composition of carbon of the carbonate system in groundwater with depth can be explained by two factors such as a change in the photosynthetic types of modern vegetation (Hatch-Slack cycle) compared to plants

Table 18.14 Isotopic and chemical composition of groundwaters of the Bakharden

Sampling points	Sampling depth (m)	Temperature of water (°C)	pH	H_2S (mg/l)	HCO_3^- (mg/l)	δD (‰)	$\delta^{18}O$ (‰)	$\delta^{13}C$ (‰)	^{14}C corrected age (years)
Well 11	30	19.5	7.4	–	238	−69	−9.2	−6.0	Modern
Well 16	400	30.5	7.7	–	256	−71	−10.8	−2.8	5,000
Well 17	800	30.3	7.7	–	281	−70	−9.0	−2.8	5,500
Well 18	1,200	38.5	6.7	25	354	−71	−11.4	−2.5	5,500
Well 19	1,600	38.3	7.0	60	342	−73	−8.9	−4.0	12,000
Well 20	2,000	49.0	7.3	80	378	−104	−11.5	−9.0	17,000
Spring Germab	–	19	7.2	–	261	−68	−9.0	−5.0	Modern

Chemical composition

Well 11:$M_{0,5}\dfrac{SO_4 46 HCO_3 46 Cl 8}{Mg35Ca34Na31}$; Well 16:$M_{1,8}\dfrac{SO_4 65 Cl 9 HCO_3 16}{Na59Mg22Ca19}$; Well 17:$M_{2,2}\dfrac{SO_4 65 Cl 8 HCO_3 14}{Na66Mg22Ca12}$; Well 18:$M_{2,4}\dfrac{SO_4 67 Cl 12 8HCO_3 15}{Na57Ca28Mg27}$;

Well 19:$M_{3,7}\dfrac{SO_4 71 Cl 18 HCO_3 10}{Na55Ca28Mg27}$; Well 20:$M_{4,9}\dfrac{SO_4 71 Cl 20 HCO_3 8}{Na54Ca30Mg16}$; Spring Germab: $M_{0,5}$ (SO_4–HCO_3–Ca–Mg)

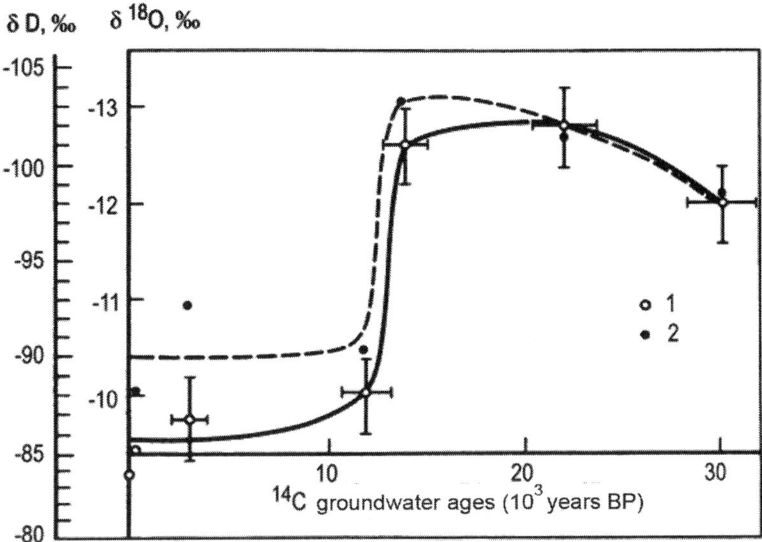

Fig. 18.27 δD–δ^{18}O relationship and changes in the radiocarbon age of groundwater for the region of Southern Kazakhstan (*1*) with correction by isotopic composition of carbonate system, and (*2*) without correction

that grew in the pluvial epoch of the Pleistocene (Calvin cycle); the entry into the carbonate system of light carbon of dissolved organic matter as a result of sulfate reduction processes. Apparently, the second assumption is more substantiated, since a change in the δ^{13}C values occurs symbiotically with an increase in the hydrogen sulfide in the investigated aquifers.

The effect of the pluvial epoch for the territory of Central Asia on the formation of groundwaters is noted in the Sir-Darya, Chu-Sarysu, and Illi artesian basins (Southern Kazakhstan). Figure 18.27 shows change in the isotopic composition of investigated waters confined to the Upper Cretaceous terrigenous-sedimentary stratum with increase in age of the water, calculated on the basis of radiocarbon considering the corrections for the dissolved carbonate components.

As we see in Fig. 18.27, within the age scale from 30,000 to 20,000 years cooling of the climate, which continued until about 14,000 years (from modern time), occurred in the studied area in the Pleistocene. Warming of the climate occurred rather abruptly about 14,000 – 12,000 years ago. Starting 12,000 years back, the climatic conditions in the region became close to the modern. Such a climatic change at the end of the Pleisticene has probably had a global character.

The effect of the last Pleistocene glaciation is displayed in the isotopic and chemical composition of the groundwaters in Moscow artesian basin. During investigation of a hydrogeological section in this basin near Kashin City, some 'lightening' of the water with respect to isotopic composition was observed in the saline and brackish waters of the Upper Carboniferous deposits (depth 180–200 m), which are areas with some decrease in mineralization of the studied groundwaters.

A study of the isotopic composition of groundwaters of the aquifers (Proterozoic) in the Baltic region showed that near Tallin there are waters with an anomalously light isotopic composition ($\delta D = -170‰$, $\delta^{18}O = -22.2‰$). These waters with respect to isotopic composition considerably differ from the modern non-artesian groundwaters ($\delta D = -88‰$, $\delta^{18}O = -9.8‰$) and are close to modern precipitation in the region of Mirniy (Yakutia) ($\delta D = -167‰$, $\delta^{18}O = -19.9‰$). At present time, the mean annual temperature in the Southern Baltic region is +4.5°C, and in Mirniy region it is −8°C. On this basis, it is possible to assume that during recharge of the above aquifer the mean annual temperature in the Baltic region was about 13°C below the modern one. The recharge of this aquifer occurred at the end of the Pleistocene time, i.e., during the period of the Baltic Sea degradation and creation of the Baltic ice lake. Thus, isotopic composition of the Scandinavian ice cover can be calculated using common figures as $\delta D =- 170‰$, $\delta^{18}O = -22‰$), which differ from the value $\delta^{18}O = -30‰$, given by Epstein and Yapp (1976).

The considered material in this section gives general idea of the possibilities of isotopic methods to study paleohydrological conditions in formation of groundwaters. More detailed information on the problem can be found in the IAEA proceedings and respective expert group reports.

Chapter 19
Paleohydrology of the Aral-Caspian Basin

The use of the stable and radioactive environmental isotopes in combination with the conventional classical methods appears to be a most productive way to study the regional natural processes which are developed during considerable intervals of time. The example of such a work based on application of the isotope and classical methods is the paleohydrologic investigation in the Aral-Caspian basin which was carried out for a number of years by the initiative and practical action of the authors.

19.1 Formulation of the Problem

It is accounted that the hydrological, climatic, and ecological changes in the Aral-Caspian basin appears to be an integral effect of the natural climatic changes which happen on the European and Central-Asian territory under action of the atmospheric moisture flow transferred by the powered cyclonic whirlwinds formed in the Atlantic Ocean and moved from the West to East by the Earth's force field. It is established since long time that the integral reaction on the natural climatic changes is in the form of the Caspian Sea level variation which occurs as quasi-periodic oscillations, the amplitude of which reaches tens of meters at periods from tens to hundreds of thousands of years. One such recent increase of the sea level, after its 40-year decrease, had happened from 1977 up to 1995 with amplitude of about 2.5 m.

At the beginning of the twentieth century, Andrusov (1900) substantiated (on the basis of the observational data available at that time) the idea that, among numerous probable causes of the variation in the sea level, climatic changes play the decisive role. Fedorov (1957) provides the following arguments in support of the climatic nature of transgressions of the sea:

- Transgressions and regressions of all the sea coasts in areas of elevation and subsidence occur simultaneously;
- Alterations in the complexes of the mollusk and fauna are always correlated with the regression epochs and the dispersion of new complexes with the transgression epochs;

V. I. Ferronsky, V. A. Polyakov, *Isotopes of the Earth's Hydrosphere*,
DOI 10.1007/978-94-007-2856-1_19,

- Sea transgressions are accompanied by a general relative freshening of water masses and the regressions by an increase in the salinity of the sea water;
- The most essential change in the evolution of the mollusk fauna coincides with the largest Khvalyn transgression;
- The facial character of continental sediments and palynological data testify that the Caspian Sea transgressions have been preceded by climatic cooling and wetting;
- The amplitudes and velocities of tectonic movements are tens and hundreds of times smaller than those of the variations in sea level caused by climatic changes. This indicates that deformations of the Earth's crust play a small role in the transgressions and regressions of the Caspian Sea.

The above reasons prove the leading role of climatic changes in the variations in the sea level and the existence of a functional relationship between the two. The conclusions of Andrusov and Fedorov have been confirmed by the results of numerous studies conducted in the past few decades. Further investigations in this direction aimed at revealing the nature and mechanism of long- and short-period variations in the level of the sea seems to be promising.

Series of instrumental observations for the variations in sea level and data of hydrometeorological observations cover only the last 100–150 years. Reconstruction of these series for the pre-instrumental period is mainly based on studying and interpreting the results of investigations conducted on coastal terraces. However, this information reflects the effect of other factors as well. Therefore, we attempted to use another source of information about the past of the Caspian Sea, namely, the data of deep-sea bottom sediments. Their advantage is that the disturbance of the temporal sequence of events recorded in sediments is not very probable. In addition to the Caspian Sea, it was decided to use the results of the bottom sediment studies in the Aral Sea, Issyk-Kul and Yaskhan Lakes, and the Kara-Bogaz-Gol Gulf. The final results of the collected materials allowed obtaining new, interesting conclusions.

19.2 General Description of Study in the Caspian Sea

To take the cores of bottom sediments, a joint Russian-French expedition to the Caspian Sea was organized in 1984. The French scientists (climatologists from Paris University and several other institutes) are interested in the Caspian Sea problem because the Caspian Sea is an excellent object to study regional and global paleoclimatic changes.

The equipment for coring consisted of a Kullenberg piston corer with an outer diameter of the steel pipe equal to 114 mm. Within the steel pipe, a plastic pipe-envelope with a 90-mm diameter was inserted. It was fitted with the piston system. At the upper edge of the 25-m column of pipes, a 1.5-ton ballast weight was fixed. The process of coring was as follows. The column was hoisted over-board. When the column was about 100 m above the sea bottom, the hoist line was freed and the whole construction fell (due to gravity) to the bottom to sample the core. Then, the column was hoisted on-board the ship. The plastic envelope with the core was extracted from the steel pipe and cut into 1.5-m-long sections. Having their faces closed with cups, these sections were isolated and marked. In this form, the cores were sent for laboratory analysis so that the material and isotope composition of the sediments could be studied. Twenty eight cores were taken during the expedition. They were divided between the partners in order to have the laboratory analysis doubled. The maximum core length in the Middle and Southern sea was about 10 m at the depth of 500 m. Together with sediment cores, the salinity and temperature of water were measured at the coring stations and also the water samples for the chemical and isotopic composition were taken. Seven stations in the south, six in middle and two in north part of the sea were made along the south-north profile.

In 1995 and 1996, two IAEA international Caspian cruises were organized, during which water sampling, temperature, and salinity along the meridian profile were carried out. The deuterium, oxygen-18, and tritium were measured in the water samples. Special samples for measuring helium and uranium-thorium isotope were taken. Also mean-month precipitation and Volga River water were sampled. The carried out measurements made it possible to obtain representative picture of the present-day distribution of isotope and oceanographic parameters of the Caspian water masses on the stage of the seawater level stabilization after almost 20-year period of its rise (1977–1995). The obtained data were also used for the analysis of the current water exchange processes in the basin. The expedition routes and sampling stations are shown in Fig. 19.1.

In 1999, one more expedition was organized to the Kara-Bogaz-Gol Gulf in framework of the European "Copernicus" program for bottom sediments coring. The work was carried out in the north-west part of the gulf. One core of 90 mm in diameter and 5.5 m in length has been taken. The core in the form of 1 m sections in plastic pipe was delivered to the laboratory for analysis.

19.2.1 Laboratory Data of Core Analysis

Two cores of bottom sediments were thoroughly investigated. The core GS04 core, with length of 787 cm, was taken from a depth of 405 m in the southern Caspian Sea at a sampling station with coordinates 38°41′39″ N and 51°36′36″ E; the GS19 core was taken from a station with the coordinates 41°32′38″ N and 51°06′06″ E, its length was 935 cm, and it was raised from a depth of 478 m in the middle of the sea. Two other cores (GS05 and GS20), having a length of about 10 m each and raised in the same site, were analyzed by the French colleagues according to the agreed-upon program. The aim of doubling the analysis was to check the most important results and to interchange the results of additional original analyses. The following laboratory analyses of the sea core sediments for documentation of the Late Quaternary variations were carried out:

- Lithology;
- Oxygen and carbon isotopic composition of bulk carbonates;
- Hydrogen isotopic composition of pore water;
- Radiocarbon dating;

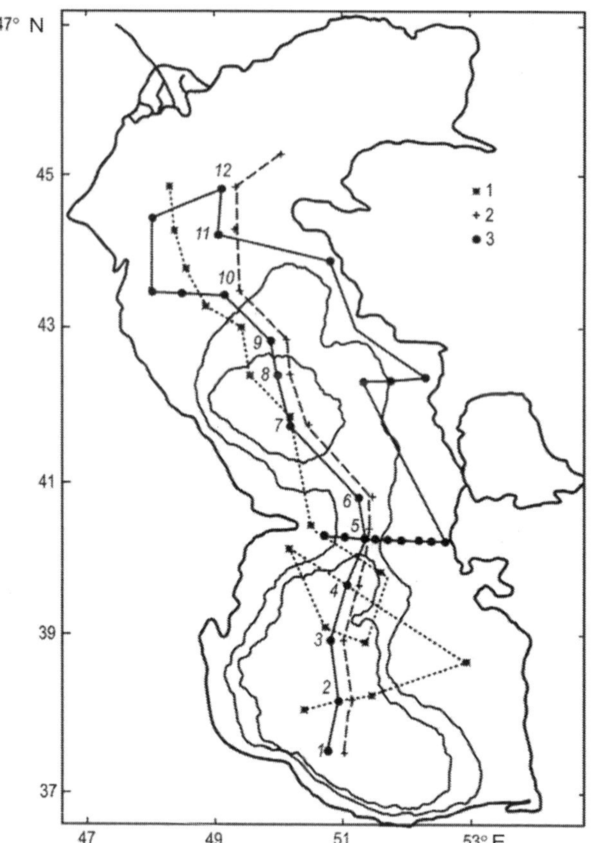

Fig. 19.1 Routes and stations of the core and water sampling during expeditions in (*1*) 1994, (*2*) 1995 and (*3*) 1996 (© IAEA, reproduced with permission of IAEA)

- Granulometry;
- Mineralogy;
- X-ray and difractometry of carbonates and fine crystalites;
- Magnetic susceptibility;
- Pollen;
- Fossil remains of organisms and coarse grained minerals;
- Diatoms;
- Bulk nitrogen;
- $CaCO_3$ and $MgCO_3$ contents;
- Salinity of pore waters;

Laboratory analyses of the Kara-Bogaz-Gol Gulf's core included:

- Lithology;
- $CaCO_3$ and $MgCO_3$ contents;
- Salinity of pore waters;

- Oxygen and carbon isotopic composition of bulk carbonates;
- Hydrogen isotopic composition of pore water;

Laboratory analyses of the modern Caspian Sea system samples included:

- Mean monthly samples of precipitation (Astrakhan, Archangelsk, Pechora);
- Volga River monthly water samples;
- Analyses of stable isotopes of ^{18}O and ^{2}H of water samples;
- Compilation of hydrometeorologic and oceanographic data;
- Modeling of carbonate precipitation and formation of isotopic composition;
- Analysis of the bathymetric topography of the Apsheron Sill.

19.2.2 Structure of Core Cross Section

Figure 19.2 presents the columns of lithologic cross sections and their brief description for the GS04 and GS19 cores with short description in legend from the southern and middle parts of the sea. The description is based on visual examination, the evaluation of particle size composition of sediments, and analysis of clay material and laboratory measurements (Ferronsky et al. 1999; Kuprin et al. 2003).

In the cross section of the GS04 core, the aleurite fraction is predominant, making up more than 50% of the sediment volume; the clay fraction prevails in some places within the depth intervals 0–0.6, 5.2–5.7, and 7.0–8.2 m; the upper part of the cross section contains up to 60% of the carbonate component; within the intervals 0–0.5, 2.2–3.0, and 3.2–4.2 m, the content of sand makes up from 5 to 10%, whereas the sand fraction in the lower part of the cross section (from 4.2 to 7.8 m) does not exceed 3%. Characteristic features of the sediment bedding and changes in the particle-size composition give evidence of the temporal variability of the hydrodynamic conditions of sedimentation.

Unlike the composition of sediments in the core GS19 from the southern Caspian Sea, the core column from the middle sea is characterized by the prevalence of the clay fraction, whose content comes to 60%. The aleurite component only dominates at depths of 6.4–6.6 m. Its content approaches the volume of the clay fraction at the reference marks of 8.0 and 8.8 m. The amount of sand throughout the cross section is $\leq$1% with a content of virtually 0% in the lower part of the cross section and a content of up to 5% at the depths of 1.4, 5.3, 6.5, and 8.8 m.

The above data gives evidence of different conditions of sedimentation and sources of sediment material in the southern and middle basins of the Caspian Sea.

19.2.3 Carbonate Mineral Content of Sediments

To determine the distribution of carbonates along the sections of the two cores, samples of sediments were taken at 50 cm intervals. The samples were processed

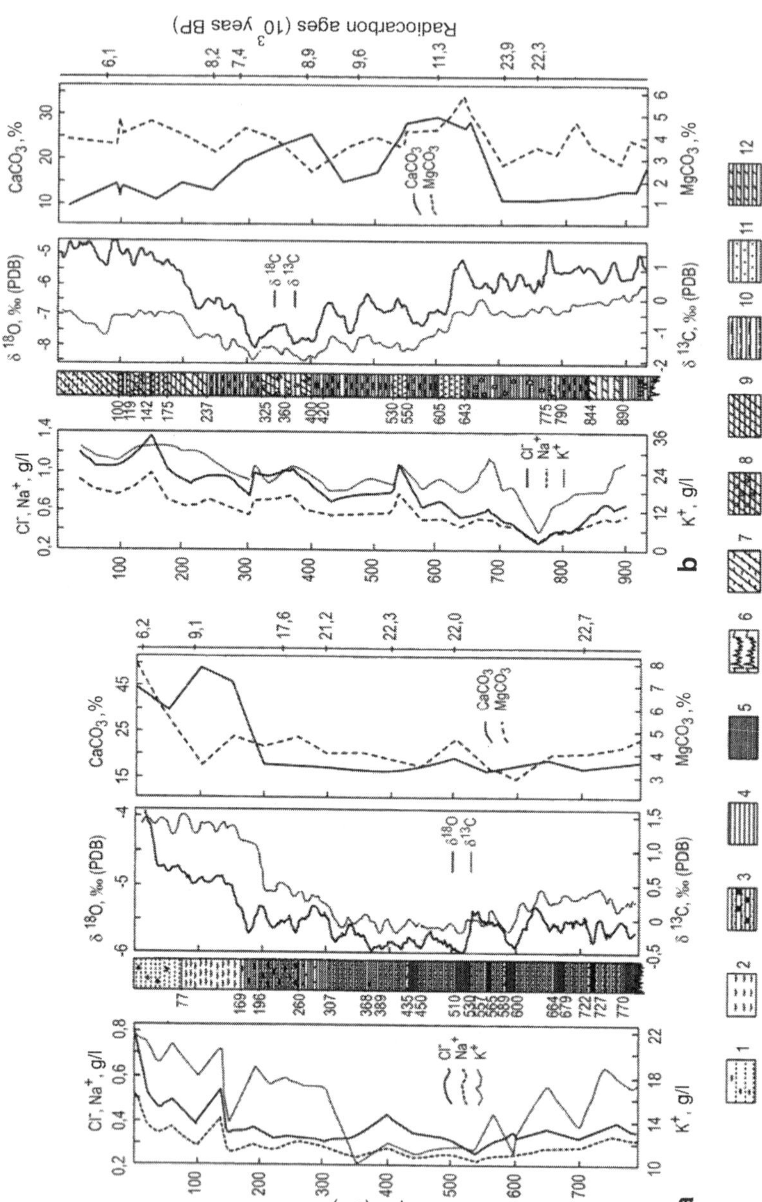

Fig. 19.2 Isotopic composition of oxygen and carbon in bulk carbonate minerals, lithologic description of cross section, content of $CaCO_3$ and $MgCO_3$, values of Cl^-, Na^+, K^+, and results of radiocarbon dating the **a** GS04, and **b** GS19 cores: *1* silts, dark grey clay; *2* grey, highly carbonaceous silt; *3* light grey and greyish brown clays with sand strata; *4* dark grey clays with sand strata; *5* clays without sand; *6* clays with sand; *7* clayey sapropel silt; *8* sapropel clays; *9* troilite clays; *10* light grey marly clays; *11* greenish hydrotroilite clays; *12* greenish clays with carbonaceous sand (© IAEA, reproduced with permission of IAEA)

Table 19.1 Percentage of calcium and magnesium carbonates in sediments from the two cores, % (© IAEA, reproduced with permission of IAEA)

Depth (cm)	$CaCO_3$	$MgCO_3$	Depth (cm)	$CaCO_3$	$MgCO_3$
Core GS04			Core GS19		
0	54	7.98	100	11.8	4.84
50	45.4	5.84	200	14.7	4.04
100	62.9	3.5	250	12.8	3.23
200	21.4	4.3	300	19.2	4.3
250	20.4	4.84	350	22.4	3.77
300	20,1	4.04	400	24.9	2.42
350	17.9	4.04	450	14.4	3.5
400	17.6	3.77	500	16.9	4.04
450	20.4	3.5	540	26.8	3.5
500	24.3	4.84	550	27.8	4.3
550	17.6	3.5	600	29.4	4.3
600	19.5	2.96	640	27.2	5.92
650	22.4	4.03	650	28.8	5.11
700	17.9	4.03	700	11.2	2.7
750	18.8	4.3	750	11.2	3.5
780	19.8	4.84	780	11.5	3.23
			820	12.1	4.57
Core GS19			840	12.5	3.5
20	9.5	3.77	890	13.1	2.7
95	14.05	3.5	915	13.1	3.77
105	13.4	4.04	935	17.3	3.5

with hydrochloric acid to transfer the readily soluble carbonate fraction into solution. The concentration of Ca^{2+} and Mg^{2+} in the solution were determined by titration. The results of analysis are given in Table 19.1. They show that the percent values of carbonate minerals and their distribution along the sections are quite different in the two cores under study.

19.2.4 *Ion-Salt Composition of Water Extracts*

For objective technical reason, we did not manage to sample pore solutions when cutting the cores. Therefore, water extracts from dried samples of sediments were analyzed. In this case, the mineralization of solution is affected by the natural porosity (humidity) of the sediment lithologic composition. To obtain water extracts, distilled water was poured over the dried at $t = 95°C$ samples and at a sample-to-water weight ratio of 1:5. Then, the values of Na^+, K^+ in solution were determined using flaming photometry and that of Cl^- by titration. Cl^-, Na^+, and K^+ in the GS04 and GS19 cores are given in Fig. 19.2.

Note that the humidity of samples from the GS04 core is about 30% lower than that of the GS19 core. Therefore, to correlate the results of measurements, all the values for the GS04 core should be multiplied by 1.3. To reconcile the hydrochemical data in the course of their interpretation, we should assume that the minimum mineralization of water extracts from the cores of the two basins falls within the same time interval. This does not contradict the results of the radiocarbon dating of sediments given above.

19.2.5 Radiocarbon Age of Sediments

The radiocarbon age of sediments was determined by the relevant standard method for the carbon bulk content of carbonate minerals. This method has an error associated with the presence of "dead" carbon in the terrigenous component of carbonates that leads to overestimation of the values determined. The French participants of the experiment made allowance for this error when dating the upper 1.2 m part of the GS05 core from the southern basin of the sea. They assessed the percentage of the course component of terrigenous carbonates by difractometry. These data show that, for the core studied, the above error in age owing to the terrigenous component is 10–20%.

Table 19.2 presents the carbon dates in bulk carbonates determined in two laboratories dealing with the radiocarbon investigations (at the Institute of Geology of Lithuania, Vilnius, and at the IAEA, Vienna), as well as the data for the upper part of the GS05 core corrected by the French researchers (Escudie et al. 1998).

19.2.6 Isotope Record in Carbonates

Figure 19.2 illustrates the results of measurements of the isotopic composition of oxygen and carbon in bulk carbonates separated from the samples of sediments taken along the section of each core. The data have a high resolution in terms of details of the recorded events, for the 1-cm sampling of column was performed uninterruptedly throughout the depth. Mass-spectrometric measurements of the isotopic composition of oxygen and carbon in carbonate minerals was carried out by the standard method with an accuracy of ±0.1‰ for $\delta^{18}O$ and ±0.2‰ for $\delta^{13}C$.

As seen in figure, the sensitivity of the isotopic record in changing conditions of the formation of carbonates is very high. The isotopic characteristics change in a different way along the sections in the southern and middle Caspian Sea. To interpret the above isotope data, it is necessary to consider the conditions of formation of the carbonates and their isotopic composition.

It is known that the index of saturation of the Caspian Sea water with calcite reaches its value of 3. It means that the water is oversaturated by Ca^{2+} and CO_3^{2-} which leads to precipitation of chemogenic carbonatesoccurring in isotopic equilibrium with the seawater. The oversaturation of the water with calcium carbonate is caused by (a) salt concentrating in the surface layer of the water in the process of

Table 19.2 Results of the radiocarbon dating of sediments from different cores (dash means lack of data. © IAEA, reproduced with permission of IAEA)

Depth (cm)	Activity (pmC*)	Age by ^{14}C (yr)	Depth (cm)	Activity (pmC*)	Age by ^{14}C (yr)
GS04					
10	46.07	6230 ± 370^b	600	24,5	11298 ± 560^c
15	32.4	9053 ± 540^c	710	5,5	23905 ± 890^c
80	32.19	17600 ± 700^b	750	6,2	22337 ± 670^c
230	10.5	9100 ± 803^b			
225	11.18	18104 ± 540^c	*GS05*		
295	7.14	21200 ± 1130^b	10	–	4190 ± 100^d
400	6.2	22334 ± 1100^c	45	–	7980 ± 100^d
500	6.5	21957 ± 1050^c	60	–	8720 ± 120^d
700	5.9	22735 ± 680^c	65	–	9290 ± 100^d
			75	–	9670 ± 160^d
GS19			85	–	9920 ± 160^d
70	46.2	6100 ± 140^b	95	–	10090 ± 220^d
245	36.23	8160 ± 330^b	100	–	10330 ± 380^d
280	39.8	7400 ± 590^c	105	–	10600 ± 1360^d
395	33.18	8860 ± 140^b	110	–	11190 ± 1310^d
410	33.6	8761 ± 260^c	115	–	11600 ± 730^d
470	30.27	9600 ± 280^b	120	–	12180 ± 690^d

[a]Percent of the present-day content of ^{14}C. Values of activity in the core samples GS04 and GS19 are not corrected for terrigenous-component content

[b]Measurements were carried out at the Lithuanian Institute of Geology

[c]Measurements were carried out at the IAEA

[d]Measurements were carried out at Paris University. Values of age were corrected for terrigenous-component content (Escudie et al. 1998)

evaporation, and (b) changes in the ionic strength and pH of water in the zones of mixing of river and sea water. Calculations of the ionic equilibrium performed by Solomin at the Institute of Hydrogeology and Engineering Geology (Moscow region) with the use of the MIG-4 program show that a drastic over-saturation of the initially under-saturated Volga River water occurs even at a ratio of the river-to-seawater of 10:1 at a salinity of $\sim 1‰$.

The carbonate mineral content of the Caspian Quaternary bottom sediments varies from 10 to 70%. This is mostly chemogenic and biogenic calcium carbonate. However, many researchers note that, apart from pelagic carbonates forming in the water mass under the condition of isotopic equilibrium, bottom sediments include terrigenous carbonate minerals transported to the sea by river runoff and wind on land. It is evident that this part of the carbonate sediments is not in equilibrium with the seawater in terms of its oxygen isotopic composition.

It follows from the theory of isotope thermometry that the isotopic composition of oxygen in carbonate sediments is the function of two variables: the temperature of water from which carbonates precipitate and its isotopic composition (Bowen 1991). However, the isotope paleoclimatic investigations of the Pleistocene glaciations have shown that the isotope records in the sea carbonates primarily reflect the change in isotopic composition of water (it gets heavier because of the accumulation of glaciers during glacial epochs) and, to a lesser extent, the effect of water temperature (Imbrie and Imbrie 1979).

This is also the case in the Caspian Sea, where the isotopic composition of chemogenic and organogenic pelagic carbonate sediments is correlated not with the isotopic composition of water from which $CaCO_3$ precipitated, but rather with its temperature. The former, in turn, changed significantly in various periods of the life of the Caspian Sea. During transgressions, the rivers carried the water of melting glaciers into the sea, which had a light isotopic composition; during regressions, the Caspian Sea water got heavier.

Similar variations should be observed in the isotopic composition of carbon in the carbonate sediments, i.e., the components of the carbonate system (H_2CO_3, HCO_3, CO_3^{2--}) should be depleted in ^{13}C as compared with the dissolved carbonates of the seawater.

The alteration of the isotopic composition of water in the glacial and interglacial epochs is the most important factor for the paleoclimatic interpretation of data of the organogenic and chemogenic carbonates in continental water bodies, including the Caspian Sea. Alternations of climate cooling and warming led to changes in the isotopic composition of the atmospheric precipitation supplying water to the continental water bodies and catchment basins of river systems. These changes concerning ^2H and ^{18}O are similar to the latitudinal effect and can be described by the equations (Van der Straaten and Mook 1983):

$$\Delta\delta^2H/\Delta t \cong (5.1 \pm 0.9)‰/°C,$$

$$\Delta\delta^{18}O/\Delta t \cong (0.62 \pm 0.10)‰/°C.$$

Thus, during cold climatic epochs, the Caspian Sea was fed by atmospheric precipitation and river-water that had a relatively light isotopic composition. However, by virtue of the fact that the duration of water change (water residence time) in the Caspian Sea, even at the highest and lowest point of the sea level, evidently exceeder hundreds of years, its water mass had to get heavier in isotopic composition owing to kinetic fractionation of oxygen and hydrogen isotopes through evaporation of water and isotope exchange with atmospheric water vapor. This assumption is supported by the research results on the water exchange of the Sevan and Issyk-Kul lakes (Ferronsky and Polyakov 1983).

Data on the isotopic composition of hydrogen in pore solutions sampled in events characteristic of the changing periods of the GS20 core (the middle Caspian Sea) testify to a relatively heavy isotopic composition of the Caspian Sea water in the past (Fig. 19.3). Within the studied time interval, the isotopic composition of hydrogen from -27 to $-6‰$ and never approached the relevant value for the present-day Volga River water ($\delta^2H = -90‰$).

Thus, we conclude that the isotopic composition of chemogenic carbonates formed in the zones of mixing of river and sea water with the dominance of river

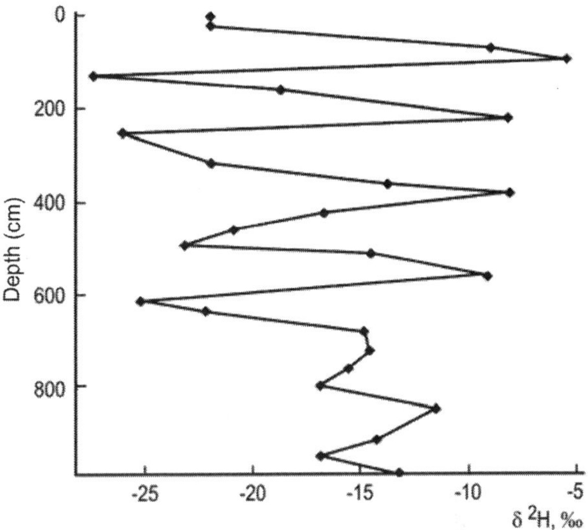

Fig. 19.3 Variations in the deuterium content of water extracts from the GS20 core (© IAEA, reproduced with permission of IAEA)

water. Fine carbonate sediments are carried by the sea currents throughout the middle and southern Caspian Sea and precipitate onto the sea bottom according to the Stokes law. Their maximum storage is to be anticipated in the zones of slowed water movements. The record of the genetic past of carbonate particles, i.e., the isotopic composition of the river water, is well preserved. This is due to the fact that the isotope exchange in the heterogeneous, solid deposit-water system at a temperature below 400 K occurs extremely slowly, and the period of the isotope-exchange reaction through diffusive kinetics is fairly long. With a drastic decrease in river runoff (which is characteristic of the current stagnant conditions in the southeastern part of the Caspian Sea), chemogenic carbonates mostly reflect the isotopic composition and temperature of the sea water.

In addition to the pelagic fraction, carbonate sediments contain the terrigeneous component carried from the land to the sea with river runoff and wind. This component is evidently not in equilibrium with the seawater in terms of the isotopic composition of oxygen and hydrogen. Therefore, it can misrepresent the isotope paleoclimatic records. Unfortunately, nowadays, we have no reliable methods of partitioning the pelagic and terrigeneous carbonate minerals. Preliminary studies have shown that the biogenic component of carbonates in the form of micro-flora, known to be at equilibrium, is negligible in the sampled cores, and its amount is insignificant for routine mass-spectrometric measurements of oxygen- and carbon-isotope composition. Therefore, the isotope paleoclimatic interpretation discussed here is based on studying the bulk samples of carbonate sediments.

To assess the degree of equilibrium of the bulk carbonates used in this study, $\delta^{18}O$ and $\delta^{13}C$ in shells of the mollusks *Cardium edule* and *Didacne trigonoides* and in the carbonate silts in which the values of mollusks occurred were determined. The relevant samples were taken with a bottom-dredger from a depth of 13 m near

Ogurchinskii Island. The length of these silt samples was 45 cm, of which the upper 0 to 5 cm layer was studied. The average isotopic composition of the mollusk's values was equal $^{13}C = +0.98‰$, $\delta^{18}O = -2.25‰$, and that of carbonates in silt sediments was $\delta^{13}C = +1.45‰$ and $\delta^{18}O = -3.27‰$. It is clear that the isotopic composition of silt is enriched in ^{13}C by about 0.5‰ and depleted in ^{18}O by 1‰. Following the refined Urey-Epstein equation, the temperature at which calcium carbonate was accumulated and chemogenic calcite formed was determined, taking into account that the oxygen isotopic composition of water in the site of sampling was equal to −1.32‰. The calculations yielded the following results: the temperature of the shell's growth is ~20°C and that of carbonate sedimentation is ~24°C. At these temperatures, the oxygen of water and sediments occurs in isotopic equilibrium. During sampling (17 August 1994), the water temperature was the same in the surface layer and near the bottom, at 28°C (pH = 8.3, $[HCO_3^-] = 0.254$ g/l, $[Ca^{2+}] = 0.343$ g/l, S = 13.2‰). At the above temperatures, pH, and concentrations of calcium hydrocarbonate-ions, the water is 1.5–2.0 times oversaturated with calcite.

The results obtained led to the conclusion that, in the zone not influenced by surface runoff, chemogenic carbonates are virtually at equilibrium with water at summer temperatures. This is due to the fact that the rate of $CaCO_3$ precipitation increases with increasing water evaporation and, hence, with increasing oversaturation of the water surface layer with calcite. The equilibrium temperature of formation of calcium carbonate of mollusk's shells is 4.6°C lower than that for chemogenic $CaCO_3$. This may be due to the larger range of temperatures at which shells grow. The pattern is inversed for the isotopic composition of carbonates. The cause of this phenomenon is not quite clear, especially considering that, at higher temperatures of formation, chemogenic $CaCO_3$ should have lighter isotopic composition of oxygen and carbon relative to the $CaCO_3$ in valves. This might be a manifestation of the effect of metabolism and the associated isotope exchange of carbon in the valve-mollusk body (lymph) system. It is known that carbon in tissues of live mollusks can be depleted in ^{13}C by 15–20‰ as compared with carbon in the carbonate components of water. Enrichment of carbonate sediments with ^{13}C, as compared with that in valves, was earlier observed by Stuiver and Suess (1996).

Values of $\delta^{18}O$ at t = 26–28°C close to equilibrium are characteristic of the present-day southern Caspian Sea sediments in the 0 to 5 cm layer, where $\delta^{18}O = -3.4‰$ for carbonates and $\delta^{18}O = -1.72‰$ for water. It follows from the above data that, under the present climatic, hydrologic, and lithologic conditions, chemogenic carbonate minerals are virtually in equilibrium with seawater accumulate in bottom silts.

According to the data for the 0–5 cm upper layer of the GS19 core, current carbonate sediments in the middle Caspian Sea have an average value of $\delta^{18}O = -4.7‰$. The water surface layer having $\delta^{18}O = -1.89‰$ and a temperature of 24.6°C, the calcium carbonate here is depleted by about 0.8‰ as compared with the isotopic composition of oxygen in calcite precipitating under the conditions of isotopic equilibrium at t = 24.6°C and $\delta^{18}O_{water} = -1.9‰$. For pelagic chemogenic calcite, the value of $\delta^{18}O$ for conditions of isotope equilibrium must be equal to −3.9‰. The absence of isotope equilibrium between water and sediment can be explained by two factors: the higher temperature of sedimentation (~30°C) and the presence of foreign fraction in bulk carbonates. The first factor is unlikely because the value 30°C far exceeds the average water temperature for the warmest month (~25°C, August) in the middle of the Caspian Sea. It is probable that bottom sediments contain minerals of other origins, which have a lighter isotopic composition of oxygen and carbon with respect to dissolved carbonate components. These may be carbonate sediments genetically associated with suspended matter of the Volga River, for which the value of $\delta^{18}O$ varies about −10‰, and in this case the calcite of river origin (the upper part of the GS19 column) by the simple formula for mixing will have 15–16‰. Carbonate minerals associated with the weathering of rocks (so-called terrigeneous material) are unlikely to be present in these sediments because the values of $\delta^{18}O$ and $\delta^{13}C$ for carbonate rocks of marine origin that compose the mountains of the Caucasus, Iran, and Central Asia are close to zero. The isotopic composition of carbon in bottom sediments also indicates the presence of carbonate material carried out by river-water into the Middle Caspian Sea. This material is traced by the 1.5‰ depletion with ^{13}C as compared with that in current sediments of the Caspian Sea (Fig. 19.2). It should be underlined that a number of researchers use carbonate silts of continental water bodies for isotope paleoclimatic reconstructions. Thus, Fritz (1983) found experimentally that the paleotemperature curves obtained by average isotopic composition of oxygen in carbonate sediments and valves of the mollusks *Valvata sinsera* in the lakes of Canada are cymbate, although they differ in ^{18}O absolute values (the ^{18}O content of mollusk shells is 1–2‰ higher than that in the lake lime). In addition, Fritz mentions using the carbonate sediments of Swiss lakes for paleoclimatic reconstructions. Finally, he concludes that the lake chemogenic lime is suited both for paleoreconstructions using the isotopic composition of oxygen and for dating bottom sediments using radiocarbon methods. Estonian researchers have used lake lime as an indicator of paleoclimatic changes in the Holocene (Martma et al. 1985).

To complete the section, we draw the following conclusions.

- The isotopic composition of oxygen and carbon of calcium carbonate in bottom sediments is governed by the isotopic composition of the water from which the chemogenic calcite precipitates;
- The isotopic composition of the Caspian Sea water has undergone significant alteration owing to the change in the share of river-water with a light isotopic content during transgression and regression phases of the sea;
- In the present-day sediments of the southern Caspian Sea, carbonate minerals are in isotopic equilibrium with the seawater with respect to oxygen. In the middle of the Caspian Sea, there is a certain depletion of carbonate sediments in heavy oxygen isotopes with respect to equilibrium even when the highest probable seawater temperature is observed;
- The phenomenon mentioned above is most likely due to the input into the Caspian Sea of chemogenic carbonate minerals depleted in ^{18}O and formed in the zone of mixing of river and sea water. In this zone, water oversaturation with calcite and dolomite developed and leads to precipitation of fine $CaCO_3$ carried out by the sea currents to the middle Caspian Sea.

Consequently, we can admit that transgressive phases of the sea are fixed by the relatively light isotopic composition of oxygen in chemogenic carbonates and regressive phases are, on the contrary, characterized by the relatively heavy isotopic composition of oxygen and carbon in bottom sediments.

In addition to the river chemogenic carbonates, the terrigeneous carbonate minerals of rocks carried by rivers and wind are likely to be present in the composition of bottom sediments. The river type of chemogenic carbonates strengthens the degree of contrast of the paleoclimatc pattern because, in this case, the isotopic composition of carbonates is genetically bound with isotopic composition of river water. The latter is, in turn, a function of the temperature and humidity, i.e., a function of paleoclimate. On the contrary, the terrigenous type of carbonate minerals smoothes the contrasts of paleoclimatic information. However, this type of suspended matter must be coarser and its migration capability in the sea must be limited by the zone of influence of high velocities of river water. Finally, the version of compensation of the effects of the chemogenic and terrigenous river components on the isotope record on the paleoclimatic pattern in the sea sediments is also probable. As mentioned above, nowadays there are no reliable methods to divide the chemogenic and terrigenous carbonate minerals. Therefore, when interpreting the isotope information in sediments, we can safely single out the phases of the sea transgressions and regressions but cannot obtain quantitative data on the seawater temperature in the past.

19.3 Interpretation of Paleoclimatic Events

As follows from Table 19.2, the columns of bottom sediments of the southern and middle Caspian Sea used in this study cover the time interval up to 24,000 years ago but their mineralogical and pollen composition, the lower layers of the two cores data back to the Lower Khvalyn period $h\nu_1$. Analysis of the complex of data obtained for the isotopic composition of carbonate minerals and of the chemical composition of water extracts and the results of radiocarbon dating and bathymetric investigations of the Apsheron sill enable us to draw conclusions about past events associated with paleoclimatic changes.

19.3.1 Rate of Sedimentation and Change in the Regime of the Northern and Southern Rivers

Figure 19.4 illustrates the data of Table 19.2 on the radiocarbon age of sediments sampled from different depths of the core columns from the southern and middle Caspian Sea. The ratio of the increment of the layer through the column depth Δh to the relevant increment of the time interval of sedimentation Δt is the rate of sedimentation. Figure 19.4 shows the graphs of sedimentation rate $\nu = \Delta h/\Delta t$ for each core. We can see that the graphs of sedimentation rate in the two sea basins are

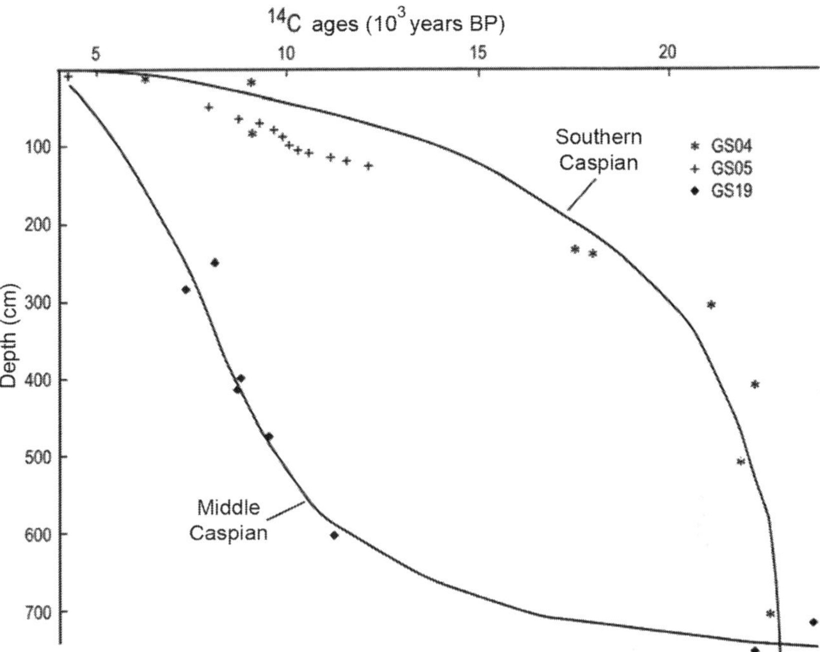

Fig. 19.4 Temporal changes in sedimentation rate based on the data of two cores (© IAEA, reproduced with permission of IAEA)

diametrically opposite: an increase in sedimentation rate in one basin is followed by its decrease in another. Within the time interval from 24,000 to 11,000 years ago, the average rate of sedimentation in the middle Caspian Sea was low, at 0.12 mm/yr, whereas, during the time interval from 11,000 to 6,000 years ago, it was an order of magnitude higher, at 1.2 mm/yr. In contrast, the average rate of sedimentation in the southern Caspian Sea during the period from 23,000 to 17,600 years reached 0.9 mm/yr, and from 17,600 to 6,000 years, it decreased on average to 0.19 mm/yr. This time dependence of sedimentation rate might be explained if we admit that the growth of the river runoff into the middle Caspian Sea started only 11,000–12,000 years ago and that, until that time, water was supplied to the enclosed sea by river runoff from the southern slope of the sea drainage basin.

Supposition about changing volumes of river runoff from the northern and southern parts of the drainage basin is supported by the comparative data on the sea salinity obtained through analysis of water extracts (Fig. 19.2). They show that, within the 5.4–1.5 m section of the GS04 core, the water from the southern Caspian Sea is less mineralized than the water of the middle Caspian Sea. This layer of sediments was forming during the period from 22,500 to 17,000 years ago. The replenishment of the southern Caspian Sea with fresh water could occur only owing to the river runoff from the southern slope. As for the middle Caspian Sea (GS19 core), the

7.5–2.0 m depth interval is characterized by an enhanced salinity of water extracts. This testifies to the excess of evaporation over river runoff in the basin considered. These depths represent the period of 22,500–12,000 years ago. At that time, the river runoff reached the middle Caspian Sea mostly from the east and south through the Apsheron sill and, in lesser volumes, from the north. The peak salinity of seawater is recorded twice. For the middle basin, it is fixed at the 5.8–5.5 m depth of the column of sediments and amounts to 12–15‰, the age of sediments being 12,000–12,500 years. In the column from the southern basin, the salinity starts to grow from the 5.4 m depth and reaches its peak value at the 1.4 m depth, where the sediment age is also 12,000 years but the salinity only 10‰. This is an evidence that, at that time, the southern basin was still fed by river runoff from the southern slope.

The second peak of salinity in the southern basin of the sea was 5,000–6,000 years ago (the 0.5 m level of the relevant core column) and in the middle basin, 6,000–6,500 years ago (1.5 m level of the column). There is a significant correlation of salinity and sediment depth with isotope data.

The curves showing the variations in the isotopic composition of oxygen and carbon in carbonate minerals also prove the idea of the replenishment of southern Caspian Sea by river runoff from the southern slope during the period of sedimentation generally characterized by the largest negative values of $\delta^{18}O$ and $\delta^{13}C$. It is interesting to note that the water of the southern rivers at that time differed slightly in isotopic composition from the present-day Volga River water. This fact indicates that, in the past, the present-day European humid zone was shifted to the Central Asian and Iranian-Caucasian regions of the Caspian Sea drainage basin (Fig. 19.5).

Finally, we consider one more fact to support the concept of the periodical replenishment of the Caspian Sea with river-water flowing from the southern slope of the basin. This is the river-bed entrenchment into the Apsheron sill with an evidently predominant south-to-north direction of water flow. The existence of this entrenchment is confirmed by the relevant documentary proofs.

19.3.2 Variations in the Sea Level

Experimental data obtained in this study made it possible to find reference marks of the sea level position during the periods of its quasi-static stage.

According to the data on the mineralization of porous water obtained through the analysis of water extracts and taking into account the consolidation of sediments, the lowest salinity of the Caspian water equaled 5–7‰. This was recorded at the depth of 5.4 and 7.6 m in the cores from the southern and middle basins, respectively. Here, the age of sediments in both cores is the same and equal to 22,500 years. The above salinity is 2–2.5 times lower than its present-day values. Taking into account the volume of water in the Caspian Sea at present (80,000 km^3) and its average salinity (13‰), the volume of water in the Caspian Sea 22,500 years ago should have been $\sim$200,000 km^3 and the sea level should have been $\sim$70 m above its current mark. It

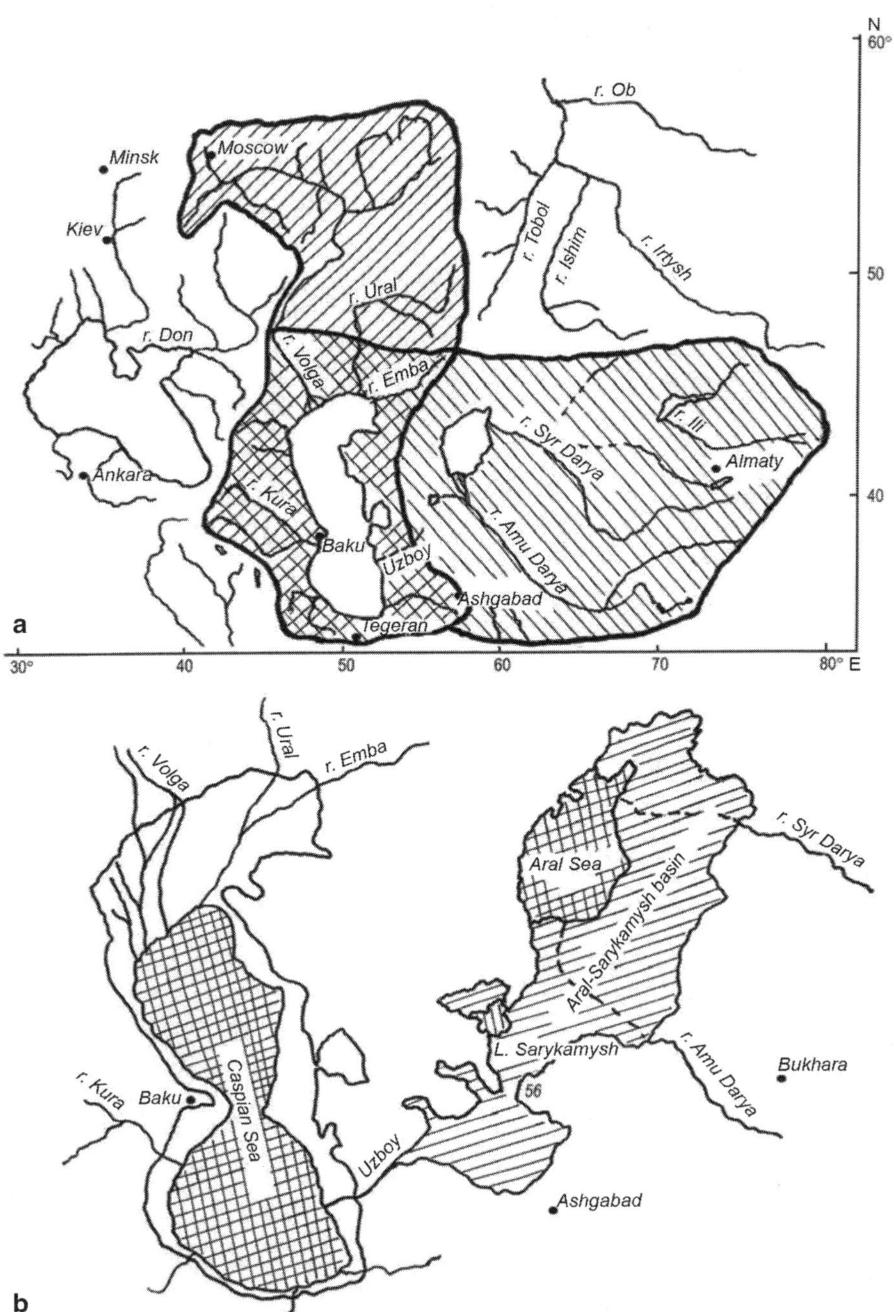

Fig. 19.5 Effect of latitudinal shift of the zone of humid climate of the Aral–Caspian region: **a** drainage basin change from the Volga–Ural to the Amu-Darya–Sarykamysh, **b** regression of the Caspian and transgression of the Aral seas during the transitional period of the climate-zone shift (© IAEA, reproduced with permission of IAEA)

is likely that the ancient Caspian Sea had a hydraulic link with the Black Sea basin through the Kuma-Manych watershed.

Important information can be obtained from the analysis of the bottom topography of the Apsheron sill (Ferronsky et al. 1999; Kuprin et al. 2003). Its hydraulic map and three latitudinal cross sections are shown in Fig. 19.6. One can see that the sill has a river-bed entrenchment with traces of water flow from south to north. Evidence of this is a narrow entrance (several hundreds of meters at the bottom) and a broad mouth outlet (up to 30 km) into the middle of the Caspian Sea. In the uppermost ancient part of the riverbed with a depth of up to 80 m, the western bank is steep and washed up owing to the rotation of the Earth (by the Coriolis' force), whereas the eastern bank is gently sloping. Below 80 m, the eastern bank gets steeper (probably owing to the softer rocks of the sill being washed out and the flow turning in the northeastern direction). As seen from the map, the paleowater way makes a loop in its lower part and comes out again on the western coast of the middle Caspian Sea above 40.5°N. Water flows also from north to south. This is confirmed by a narrow entrenchment from the southern part of the sill to depth of about 200 m (Fig. 19.6d).

Another important piece of information offered by the riverbed entrenchment is river benches distinctly seen in the sections. They indicate the sea-level position of quasi-equilibrium water balance of the sea. We can single out the most ancient main eastern bench situated at a depth of 80 m below the present position of the sea level. There are several more shallow benches indicative of other sea-level positions in the past. On the western slope, below 80 m, we can single out at least three benches formed later. Unfortunately, we have no core material from the bottom of the benches and riverbed entrenchment. This hampers the dating of historical events and the obtainment of other information. The history of this ancient riverbed associated with the history of the Caspian Sea is undoubtedly recorded in sediments and is waiting to be read.

19.4 Study of Water Regime in the Aral Sea

Analysis of bottom sediments in deep depressions in the southern and middle Caspian Sea showed that the present-day runoff regime of the rivers flowing into the Caspian Sea from the Volga–Ural basin have started developing since the onset of Holocene. Until then, the climatic shift of the humid zone toward lower southern latitudes had maintained the conditions under which the river inflow into the Caspian Sea was largely due to the water flowing from the Aral Sea basin through the Uzboi River into the southern Caspian Sea and partially through the Karyn-Zharyk River into the middle Caspian Sea. Therefore, the Aral-Caspian basin should be considered as an indivisible drainage area which consists of two sub-basins: one lying in the arid climatic zone and the other, in the humid climatic zone. When the climate changes, the roles of the sub basins also change.

The Aral Sea drainage basin occupies the Turanian depression and the slopes of the adjoined mountains. In the period when the region lies in the arid climate zone,

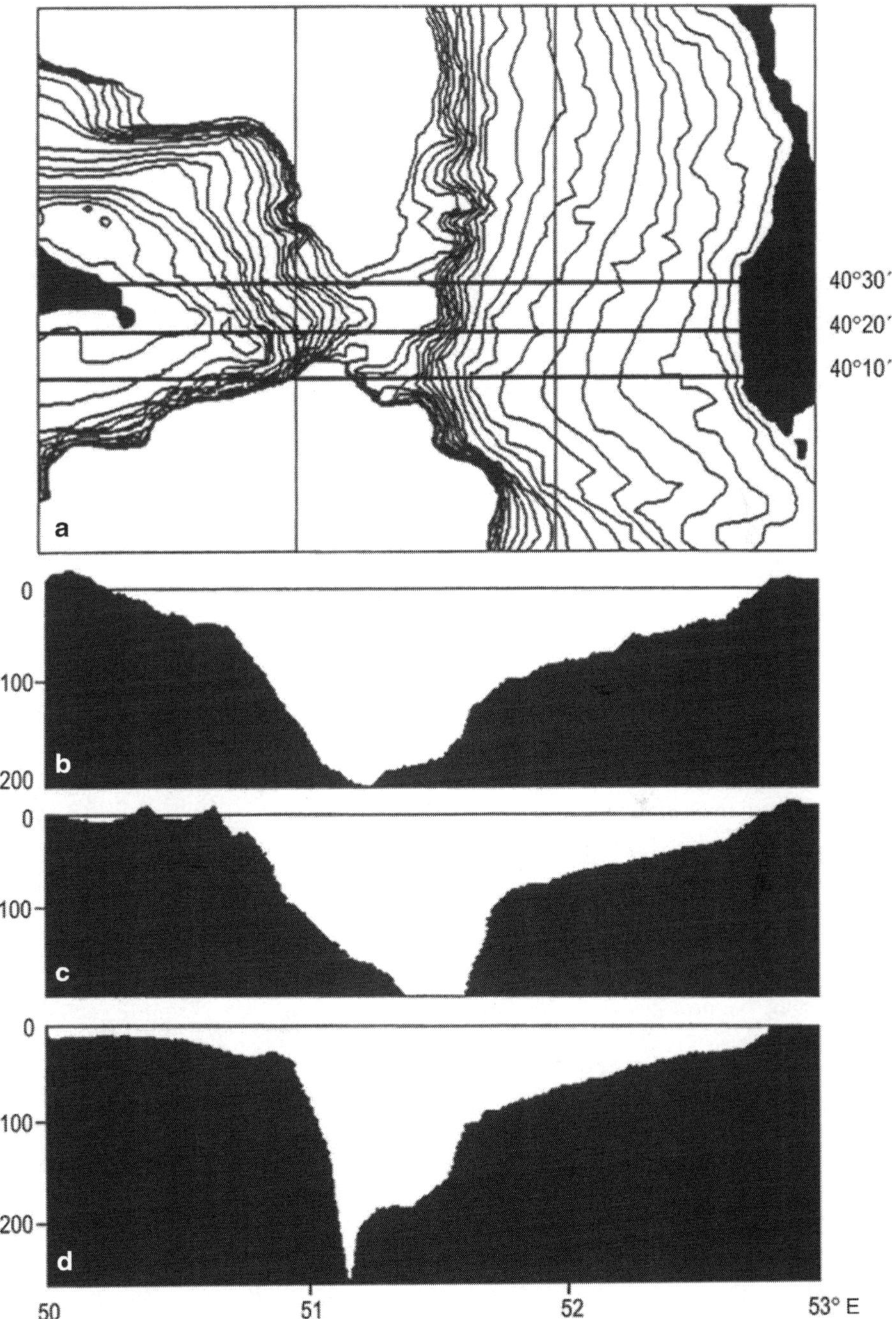

Fig. 19.6 **a** Bathimetric map of the Apsheron sill with a 10 m isobath interval and cross section of the sill along latitudes, **b** 40°10′, **c** 40°20′, and **d** 40°30′ (© IAEA, reproduced with permission of IAEA)

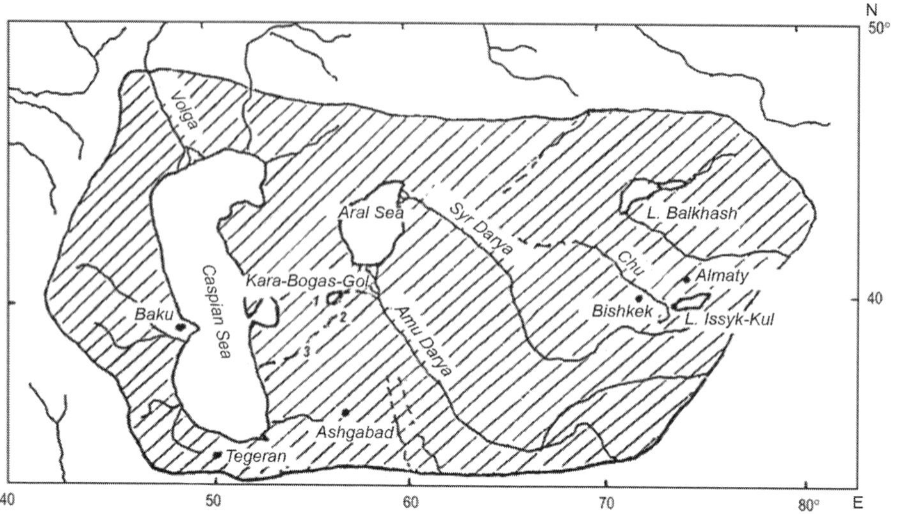

Fig. 19.7 The Aral-Caspian drainage basin and the water bodies under study: *1* Sarykamysh Lake; *2* Kugunek Mount; *3* Uzboi River. Shadow area is the Aral-Caspian drainage basin (© IAEA, reproduced with permission of IAEA)

the rivers, lakes, and the Aral Sea itself are recharged mainly at the expense of melting of mountain glaciers. When the humid zone shifts to these latitudes, the amount of precipitation in mountain regions increases up to 600 mm/yr (the precipitation in mountain regions will be even higher) and the runoff into the Turanian depression increases many times. However, the level in the Aral-Sarykamysh Sea has risen only up to the elevation of 56 m above the absolute sea level, where the threshold at Kugunek Mountain is located; the river passes over this threshold into the Uzboi River channel and farther to the Caspian Sea (Fig. 19.7).

When the climate in Central Asia becomes wetter, the activity of the drainage basin of the Karyn-Zharyk River, which collects the runoff from the Ustyurt and Mangyshlak plateaus, is provided (Dzens-Litovsky 1967). This river's runoff recharges Kara-Bogaz-Gol Gulf (Black Throat Lake) of the Caspian Sea (Fig. 19.8).

The mountainous Issyk-Kul Lake lies at an elevation of about 1600 m in the Tien Shan and collects the runoff of more than 50 mountain rivers and springs, including those originating from glaciers. Despite the relatively small water area, the lake is 700 m in depth and its water volume is almost twice as large as that of the Aral Sea (Table 19.3). In the period of humid climate, the water level in the lake rises only up to the elevation of the Chu River (the western part of the lake). The river discharges the excessive water from the lake into the Aral Sea.

Core sampling with the length of 4–5 m were taken from bottom sediments in the Kara-Bogaz-Gol Gulf, Lake Issyk-Kul and the Aral Sea at depth of 0.5, 250, and 23 m, respectively. The samples were taken in order to study the variations in the hydrological regime in the Aral Sea basin based on variations in the material and

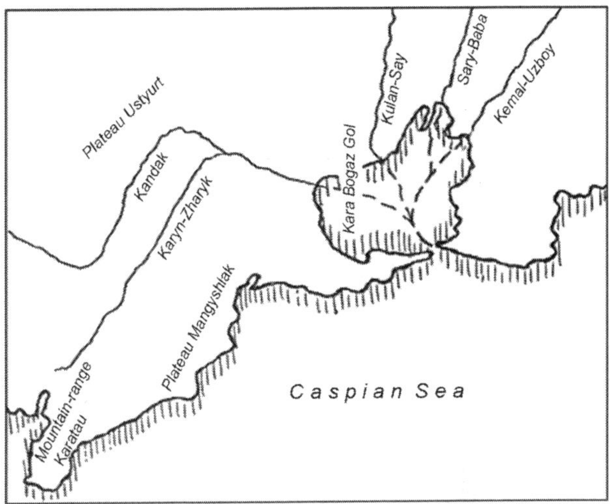

Fig. 19.8 Basin of the Karyn-Zharyk River (© IAEA, reproduced with permission of IAEA)

Table 19.3 Hydrological characteristics of the water bodies under study (© IAEA, reproduced with permission of IAEA)

Description	Aral Sea	Lake Issyk-Kul	Kara-Bogaz Gulf
Area (km^2)	64,500	6330	10,000
Volume (km^3)	1000	1700	200
Watershed area (km^2)	690,000	21,900	50,000 (?)
Depth (m)[a]	20/67	278/702	1/3.5
Mean annual water temperature (°C)[b]	10/3	8/2	6/4
Salinity (‰)	11	5,8	250
River runoff (km^3/yr)	55	3	10
Precipitation (km^3/yr)	6	1,2	1
Evaporation (km^3/yr)	61	4.2	11

[a] The numerator and denominator are the minimum and maximum values, respectively
[b] The numerator and denominator are the values for the surface and bottom waters, respectively

isotopic composition of the bottom deposits. Analysis of the samples allowed the establishment of relationship between climate changes in the region and the water budget of the respective closed water bodies. The analyzed characteristics include: ^{18}O and ^{13}C in the bulk carbonate samples, ^{2}H in the interstitial water, radiocarbon dating, the isotopic composition of oxygen and hydrogen in the present-day lake water, the concentrations of $CaCO_3$, $MgCO_3$ and major salt-forming ions (Na^+, K^+, Cl^-, and SO_4^{2-}) in deposits.

The concentrations of ^{18}O and ^{13}C in carbonates were analyzed by the standard mass-spectrometry relative PDB international standard with the measurement error of ±0.1‰ for $\delta^{18}O$ and ±0.2‰ for $\delta^{13}C$. The concentrations of ^{2}H and ^{18}O in water samples were also determined using the conventional mass-spectrometry relative to the SMOW standard. Radiocarbon dating was carried out using bulk carbonate samples, mollusk shells, and organic matter from peat laminas and dispersed remnants. The concentrations of Na^+, K^+, Cl^-, and SO_4^{2-} were measured in aqueous extracts from bottom sediment samples. To do this, dried samples were covered by distilled water at room temperature in a weight proportion 1:5. Next, the concentrations of Na^+ and K^+ in the solution were determined by flame photometry; Cl^- concentration, by silver nitrate titration; and SO_4^{2-} concentration, by weight method of barium sulfate precipitation. The carbonate minerals were isolated by treating the core sample with hydrochloric acid to transfer the carbonate fraction to solution. The concentrations of Ca^{2+} and Mg^{2+} in solution were determined by trilonometry.

19.5 Interpretation of Bottom Sediments

19.5.1 Kara-Bogaz-Gol Gulf

The experimental data were obtained from a limited number of samples taken from the core in 10–20 cm intervals, and the rate of sedimentation was estimated at 0.1–1 mm/yr. Therefore, the time interval in the events described based on these data amounts to ~1,000 years. Figure 19.9 gives the results of these analyses and lithologic description of the core taken from Kara-Bogaz-Gol Gulf. The deposits are represented by alternating bands of clayey and sandy silts, some of which contain inclusions of salt crystals.

The entire body of collected data allowed recognizing three sedimentation stages with characteristic features (Ferronsky et al. 2003b). Data from the deepest core interval (5.2–4 m) with an age of 9,200–8,500 years suggests that the sedimentation proceeded during the reservoir filling with fresh water originating most likely from the Karyn-Zharyk River and its tributaries. This conclusion is based on a vast body's data. The fact that $\delta^2H = +6‰$ at the depth of 5.2 m suggests that the process took place in a drying-up water body, the water of which is enriched by ^2H due to evaporation. The concentration of ^2H in water decreased to $-6‰$ by the elevation of 4 m, which indicates that river water had found its way into the previously formed drainless salt lake. The time interval considered here coincides with the beginning of an intense input of the paleo-Volga and paleo-Ural rivers into the middle Caspian Sea. At the same time, the decay of glaciers that had formed at the maximum of Valdai glaciation brought about water runoff through the channels of Karyn-Zharyk and Kaidak paleorivers from the Karatau Ridge, Ustyurt Plato, and the southern slope of the Urals (Fig. 19.8). The wide paleochannels of these rivers are distinctly seen now on the Plato Mangyshlak and near Kara-Bogaz-Gol Gulf (Dzens-Litovsky 1967). The concentration of ^{13}C in the carbonate deposits at this stage of the reservoir existence varies from +4.7 to +2‰ (PDB), which shows that the river water inflow into the lake has increased. The concentrations of K^+, Na^+, Cl^-, SO_4^{2-}, $CaCO_3$ and $MgCO_3$ in the elevation interval of 5.2–4 m also confirm the larger amounts of fresh river water emptying into the water body. Thus, the concentrations of ^{18}O and ^{13}C in carbonates confirm the relation between the latter and the continental carbonates

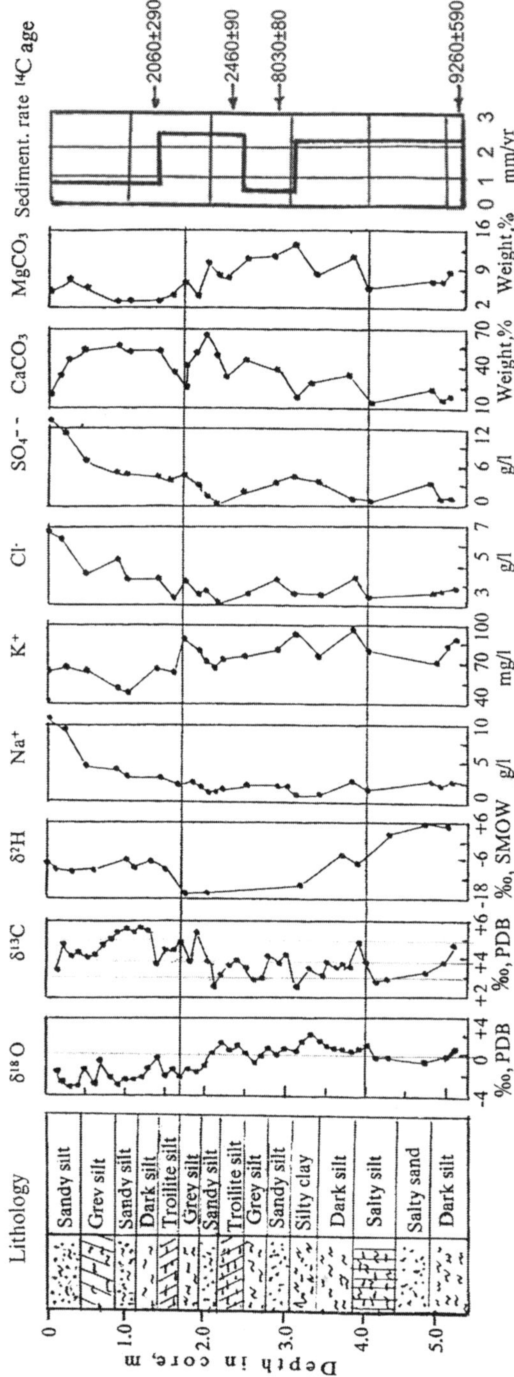

Fig. 19.9 Lithological and geochemical characteristic of core sediments from the Kara-Bogaz-Gol Gulf (© IAEA, reproduced with permission of IAEA)

delivered into the water body by river runoff. The second stage of sedimentation covers the core depth interval from 4–1.7 m and the time interval of 8500–2200 years. Here the values of δ^2H for the pore water decrease from −6 to −17‰. This fact suggests an increase in the river runoff inflow into the water body. The Kara-Bogaz-Gol Gulf as a whole, which is relatively small in volume, was rapidly filled in with water. Thus, about 3000 years ago, water from the former drainless lake started inflowing into the middle Caspian Sea, the level of which was then lower than it is now. The intense river runoff into Kara-Bogaz-Gol Gulf at this stage can be identified by the values of $\delta^{18}O$ in the carbonate fraction of deposits, which indirectly indicate that larger amounts of terrigene carbonate rocks were being carried into the gulf in this period (Fig. 19.9). It is worth noting that the proportion of carbonate fraction in the sediment increases at this stage with the sampling depth decreasing from 4 to 2 m. This appears to account for the radiocarbon age obtained from the analysis of bulk carbonate samples being 2,000–4,000 years higher than the real value (Escudie 1998). The elevated concentration of K^+ in the aqueous extracts also suggests the notable contribution of river runoff in the formation of bottom deposits in Kara-Bogaz-Gol Gulf in the first two intervals. The Na/K ratio in river runoff is known to average 0.4, whereas in the water of marine basins, this ratio decreases to 0.04. The values of $\delta^{13}C$ and the concentrations of Na^+, Cl^-, and SO_4^{2-}decrease, and the values of $\delta^{18}O$ and the concentrations of K^+, $CaCO_3$, and $MgCO_3$ increase to the level, which can be attributed to the steady inflow of river water into the water body. The narrow limits of variations in the characteristics measured in this study allow the conclusion that the excessive water occasionally discharged from the water body into the middle Caspian Sea.

The formation of the third core interval, which covers the depth interval of 1.7–0 m and the time interval from 2200 years to the present time, is associated with the start of water inflow from the middle Caspian Sea into the Kara-Bogaz-Gol Gulf. This process appears to start in the period of the latest Novo-Caspian transgression. It should be mentioned that the site where the core was taken is located near the present-day shore, where the water depth is not greater than 0.5 m. The value of δ^2H in pore water increases from −17 to −6‰ in this interval of sedimentation; this can be attributed to the inflow into the gulf of the Caspian water, which has heavier isotopic composition, and the precipitation of a large amount of gypsum ($CaSO_4 \cdot 2H_2O$) into bottom deposits. The formation of crystalline hydrates of gypsum is known (Gat and Gonfiantini 1981) to be accompanied by a decrease in deuterium in the liquid phase by about 15‰ ($\alpha = 0.985$), which is compared with the crystallization water, and the gay-lussite ($Na_2CO_3 \cdot CaCO_3 \cdot 5H_2O$) precipitation caused approximately the same enrichment of the liquid phase by deuterium ($\alpha = 0.987$). The values of $\delta^{18}O$ in carbonate deposits decrease in this period varies from −3.8‰ to 0. These values are somewhat higher than the average value of ^{18}O in the bulk carbonate deposits of the recent northern Caspian Sea (−4.8 ± 0.2‰). A plausible explanation of this fact is an increase in the coefficient of fractionation of oxygen isotopic composition in the natural carbonate system $CO_{2\,\text{atmosphere}}$–$H_2CO_{3\,\text{plant}}$–$HCO_{3\,\text{plant}}$–$CO_{3\,\text{plant}}^{2-}$–$CO_{3\,\text{solid phase}}^{2-}$ in solutions with a large ionic strength (brines). The concentration of K^+ in the aqueous extracts from the third interval decreases, thus indicating to the marine type of the water body. At the same time, an increase is recorded in the concentrations of chlorides and sulfates, which enter the gulf mainly with the Caspian water. The proportion of the carbonate fraction in sediments decreases at the expense of gypsum precipitation. The rate of sedimentation decreases in this period as compared with the first and second intervals because of a lower dissolved solids discharge.

One may see from Fig. 19.9 that the proportion of magnesium carbonate in bottom sediments in the second sedimentation period increases. In the course of earlier studies of carbonate deposits in Le-Rueret Lake in the Alps, it was found that the concentration of Mg^{2+} in carbonates of mollusk shells increases with decreasing temperature of water inhabited by the mollusks (Lemeile et al. 1983). With this fact taken into account, we can suppose that a relatively rapid drop in water temperature took place in the gulf at this stage of sedimentation at the expense of river water inflow (the stage of a flowing-through lake). Next, as noted above, the part of the gulf under study dried up for a long time, after which a period of water mass warming took place, which appears to be due to the beginning inflow of the Caspian water. This assumption is in agreement with experimental data on the distribution of ^{18}O and ^{13}C in carbonate and other sediment characteristics along the core column. The salinization of the gulf water at the expense of evaporation during the third stage of sedimentation made the carbon isotopic composition in the bulk carbonate fraction of bottom deposits somewhat heavier because of a decrease in the biomass of aquatic microorganisms.

The obtained results show that the Kara-Bogaz-Gol Gulf was a flow-through water body in the early Holocene. The name of the Gulf seems to adequately reflect its state in the past. The relatively small volume, the canal $\sim$11 km in length, which connects it with the sea, the presence of a hard-rock threshold are the features that create the conditions for water exchange with the sea with the direction of flow depending on the sea level. Before 2,200 years, the level in the middle Caspian Sea was lower than the threshold in the gulf and the runoff of the Karyn-Zharyk River was considerable. Under these conditions, the gulf was a flow-through water body. Climate changes affected the gulf hydrological regime, and the gulf came to its current state.

19.5.2 Lake Issyk-Kul

Lake Issyk-Kul is a typical mountain reservoir of tectonic origin. Its level is controlled by the runoff of $\sim$50 rivers and springs recharged by precipitation and mountain glaciers. The evaporation from the lake accounts for about 20% of the precipitation over the lake watershed, the rest of precipitation being due to the water vapor from the Atlantic Ocean (Brezgunov 1991). The lake water level is changing continuously. For example, water level dropped by 130 cm from 1965 to 1985. In the next years, the level somewhat roused. In order to study the paleoclimatic effects of the hydrologic regime, core samples $\sim$4 and 2 m in length were taken from the depth of 250 and 650 m, respectively.

The results of analysis of the isotopic and chemical composition of bottom sediments from the 4 m long core, its texture, and radiocarbon dating based on *ostracode* shells are given in Fig. 19.10, where three characteristic intervals of sedimentation are derived.

The deepest interval, 4.3–2.6 m in the depth and 7800–6000 years in age, has a relatively homogeneous texture, low concentration of organic carbon, monohydrocalcite, $\delta^{18}O$, and Sr/Ca, and slightly elevated concentrations of $CaCO_3$ and $MgCO_3$ at the upper boundary of the interval. These characteristics are in good agreement with relatively stable sedimentation conditions at steady inflow and outflow of cold water from the lake into the Chu River. An increase observed in Mg^{2+}content of calcite in bottom deposits is indicative, as noted above, of a decrease in water temperature, especially near the end of the stage. The parameters of the sediments from the second interval, 2.6–1.7 m in depth and 6,000–4,300 years in age, reflect the sedimentation conditions with a reduced river water inflow, lower water level in the lake, increased salinity, and decreased sedimentation rate. This conclusion is confirmed by an increase in $\delta^{18}O$ in mollusks shells, a decrease in Mg^{2+} concentration in the deposited calcite, and an increase in the concentration of organic matter and Sr/Ca ratio in shells. The ratio of Sr/Ca increases with the temperature of water where the growth of aragonite shells proceeds. The increase in $\delta^{18}O$ appears to be due to increase in ^{18}O in the water because of decrease in the river runoff and an increase in the share of evaporation in the lake water budget. The characteristics of the upper interval of the core from the depth of 1.7 m to the bed surface and from 4,300 years before present day suggest periodic changes in the lake regime with level drops alternating with level rises up to the elevation of the Chu River bed into which the excessive runoff was discharged. In particular, at a depth of 1.5 m and about 4,000 years ago, short-term increase in the water inflow into the lake took place, which can be seen from turbidide flows. A substantial drop in the lake level took place in the lake about 2,900 years ago as demonstrated by the low rates of sedimentation and higher $\delta^{18}O$, Sr/Ca and the organic carbon content of sediments. In this interval of sedimentation, water temperature in the lake steadily increased along with an increase in Sr/Ca and a decrease in Mg^{2+} content of calcite. In such periods, the lake became drainless though the short-term increases on river runoff could cause water discharge into the Chu River. However, these events have no effect on the lithology and the isotopic and chemical composition of the sediments. The lake water temperature likely was controlled not only by seasonal variations in the air temperature, but also by the discharge of numerous thermal springs into the lake depression. The occurrence of monohydrocarbonate ($CaCO_3 \cdot H_2O$), in the third interval of the sediments is perhaps due to the increase in the water temperature at the stage when the lake has been drainless.

Since the height of the threshold, separating the lake from the Chu River channel, is about 12 m and the present-day distance between them is 6 km, the amplitude of the level rise in Lake Issyk-Kul is limited by this value. During the humid climatic events and in the periods, when the accumulated masses of mountain glaciers were melting rapidly, the lake quickly fills up to the river threshold. After that, the excessive runoff into the lake discharges into the river and finally into the Aral Sea. A lake terrace of ~12 m height, which can be distinctly seen all around the lake, supports the above

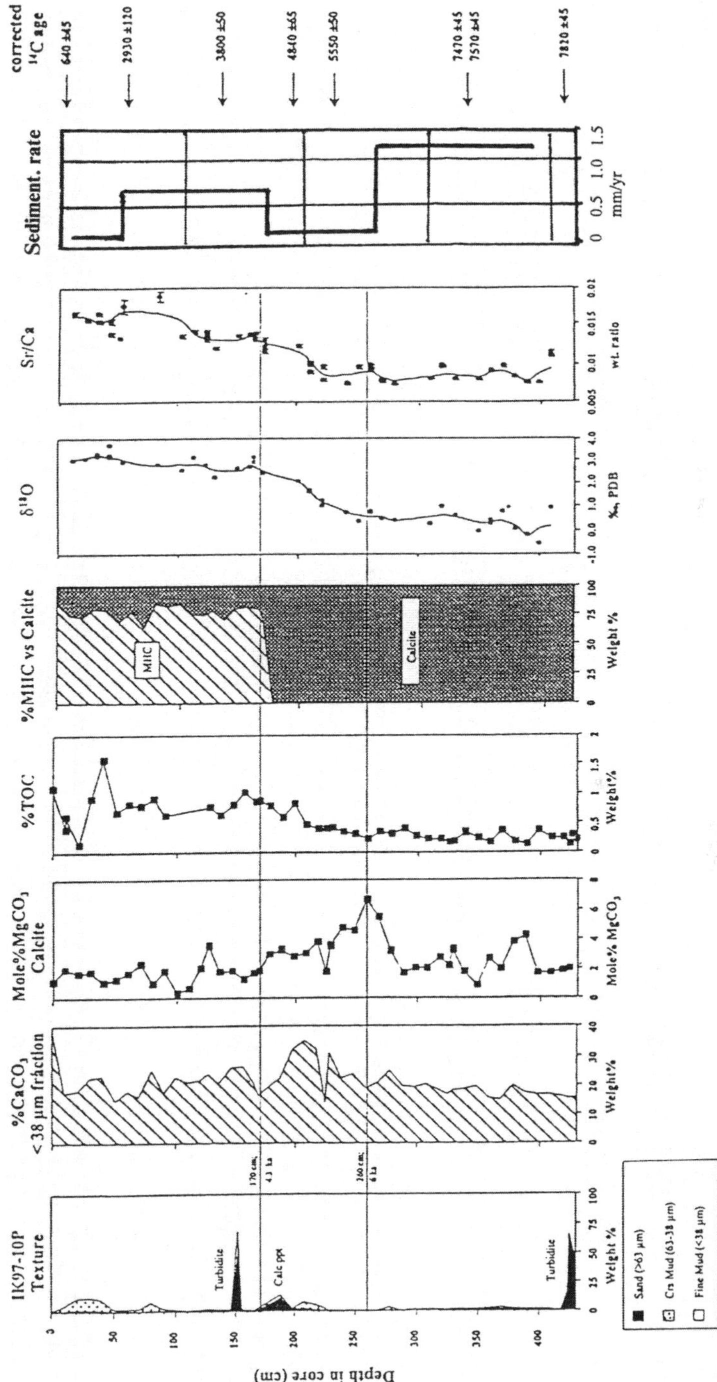

Fig. 19.10 The results of laboratory studies of bottom sediments core from the Lake Issik-Kul (© IAEA, reproduced with permission of IAEA)

conclusion regarding the events that took place more than 6,000 years ago. The cold period between 6,000 and 4,300 years, resulted in a decrease in the rate of glacier melting, river runoff, sedimentation rate, and a drop in the lake level.

19.5.3 Aral Sea

Data on shore terraces and bottom sediments allows us in recognizing four transgressive and three regressive periods in the Holocene history of the Aral Sea (Nikolaev 1995). The level rise in the transgressive periods reached the elevations of 57, 54.5, 53.5 and 53.0 m above the absolute sea level. In the periods of regression, the level dropped down to 44, 43, and 35 m. This means that water flow from the Aral Sea into the southern Caspian Sea took place at least once in the Early Holocene and the major basin of the Aral Sea dried up at least twice in this period. The latter conclusion is confirmed by the fact that two layers of bog peat and one layer of salt were detected in the bottom sediments of the sea at the sea depth of 20–25 m.

In the course of expedition carried out in 1989 by Geographic Department, Moscow State University, a 4-m-long core sample of bottom deposits was taken in the central part of the sea from the depth of 23 m. The study of the core included lithological description of the sediments, analysis of their carbonate content, and radiocarbon dating. Thirty-six dating were made, including the determination of the radiocarbon age by analyzing the *cardium* mollusk shells, the organic matter of peat-bogs, and organic matter scattered in the deposits. The radiocarbon dating of the samples was done in Water Problems Institute, Russian Academy of Sciences (Maev and Karpychev 1999). The objective of this stage of the study was to obtain comparative data on radiocarbon age estimates based on carbon extracted from different natural carbon-bearing compounds. As seen from Table 19.4 and Fig. 19.11, the values of radiocarbon age obtained using mollusk shells and peat-bog organic material are close to each other. The age, determined from the radiocarbon of the dispersed organic matter, exhibits small deviations from the previous values. More significant deviations were recorded when the radiocarbon of bulk carbonates was measured: this seems to be attributable to the effect of terrigenous component of the ancient, dead carbonates, which shifted the results toward higher values. This difference between the age averages ~2000 years and in some cases can be as large as 4000 years. The latter value is mostly typical of sandy-clayey bands, which were formed in the period when the river runoff into the lake was high. The deviations in the radiocarbon age of bottom sediments attain maximum in the core samples taken from the depth of 2.5–3.0 and 0.2–1.0 m. According to lithological data, these intervals are associated with long transgressive phases of the Aral Sea, which fell within periods of 3,500–5,000 and 500–1,300 years ago (Fig. 19.11 and Table 19.4). In these periods, the share of carbonate components in bottom sediments decreased because of higher concentrations of sandy-clayey fractions. These stages also exhibit higher rates of sedimentation equal to 0.8 and 1.1 mm/year, respectively.

In the regressive stages, the difference between the sediment radiocarbon age determined from all the carob-bearing components notably decreased (down to 500–900 years) at the expense of the higher share of pelagic chemogenic and organogenic

Table 19.4 Radiocarbon age obtained by bulk carbonate ($CaCO_3$), carbonate of *cardium* (TOC), and the carbon of bog peat (TOC*) in the Aral Sea bottom sediments (© IAEA, reproduced with permission of IAEA)

Core sample (cm)	Material for dating	Age by ^{14}C (yr)	±σ	Core sample (cm)	Material for dating	Age by ^{14}C (yr)	±σ
0–17	$CaCO_3$	750	120	205–221	TOC*	3450	80
0–17	TOC	460	160	225–238	$CaCO_3$	7670	180
19–30	$CaCO_3$	4370	100	242–260	$CaCO_3$	6390	80
19–40	TOC	1560	100	260–287	$CaCO_3$	9030	150
34–47	$CaCO_3$	3050	100	260–287	$CaCO_3$*	4810	180
34–47	TOC	720	120	288–300	$CaCO_3$	8200	220
72–84	$CaCO_3$	3950	120	288–300	TOC	4990	250
84–98	$CaCO_3$	3690	100	310–324	$CaCO_3$	6890	150
84–98	TOC	1830	300	310–324	TOC	5750	250
98–111	$CaCO_3$	3550	120	310–324	$CaCO_3$*	5610	220
98–111	TOC	1270	250	325–341	$CaCO_3$	8160	100
111–125	TOC	2300	180	325–341	TOC	7960	120
125–130	TOC*	1510	150	342–345	$CaCO_3$	7590	150
141–150	$CaCO_3$	2730	100	342–345	$CaCO_3$*	6090	150
150–164	$CaCO_3$	3600	120	367–377	$CaCO_3$	8030	100
150–164	TOC	1900	250	367–377	TOC	7300	390
164–190	$CaCO_3$	3970	100	382–400	$CaCO_3$	7540	100
205–221	$CaCO_3$	5330	100	382–400	TOC	6650	390

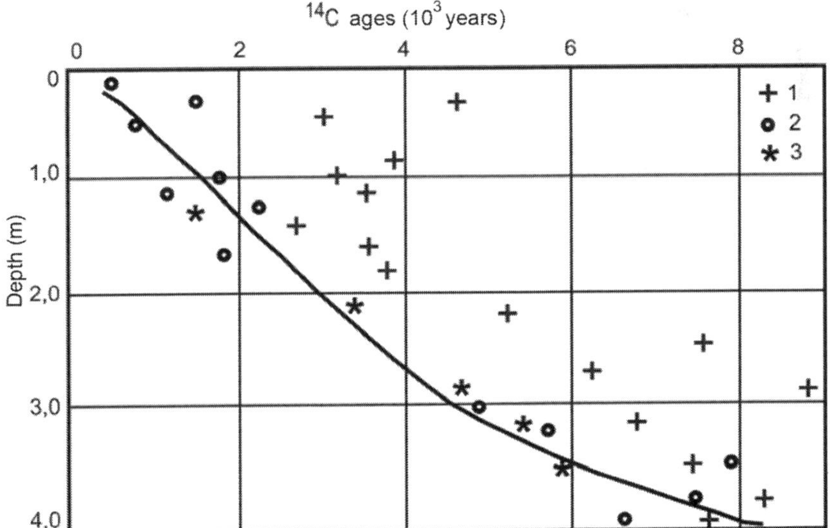

Fig. 19.11 Radiocarbon age of *1* bulk carbonates, *2* total dispersed organic matter, and *3* shells of *cardium* mollusk and bog peat in the sediments from the core sample taken in the Aral Sea (© IAEA, reproduced with permission of IAEA)

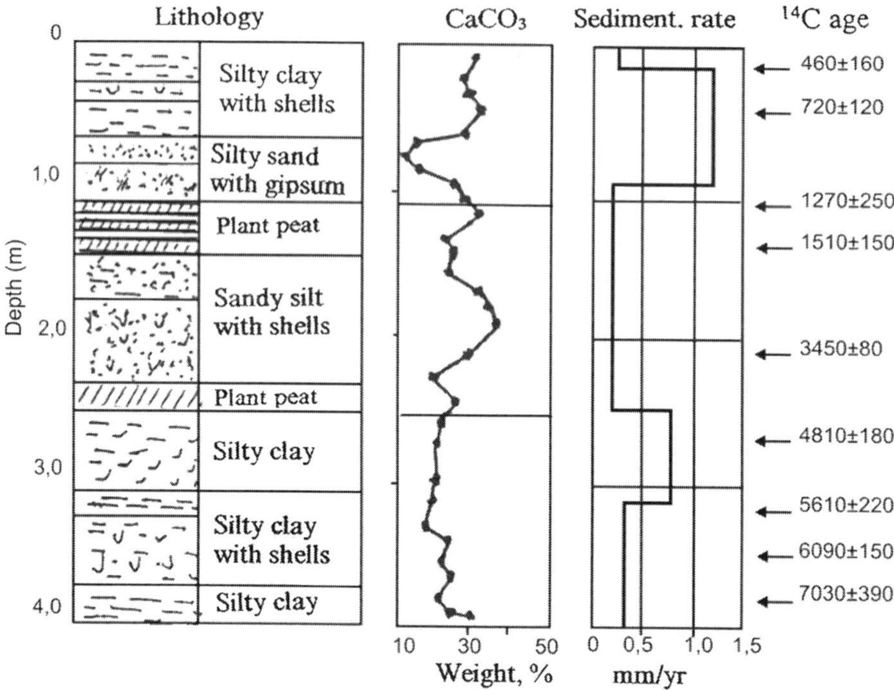

Fig. 19.12 Lithological composition, bulk carbonate content, sedimentation rate, and radiocarbon age of sediments in the core sample from the central Aral Sea at depth of 23 m (© IAEA, reproduced with permission of IAEA)

carbonate minerals in the carbonate fraction of bottom deposits. Long regressive phases can be seen in the core depth intervals of 0.0–0.2 m (0–460 years), 1.0–2.5 m (1,300–3,800 years), and 3.0–3.9 m (5,600–7,300 years). The rate of sedimentation in these periods decreased to 0.2–0.5 mm/yr. There are at least two periods in this time, when the sea in the sampling site was completely dry as can be seen from the bands of bog peat and coastal vegetation remains. It is worth mentioning that the sea regression periods are also associated with an increase in the amount of shells in bottom sediments. This is due to the water mass over-saturation by calcium carbonate at the expense of an increase in evaporation and the sea biological production.

Both these factors facilitate the improvement of the environmental conditions in the mollusk habitat. The amount of carbonate minerals in the depositing minerals found in the core layers where remains of bog vegetation occur, drops abruptly because the pore water of these layers is acid at the expense of humid and fulvous acids. The bog water has pH ≤ 5. Calcium carbonates decompose in acid environment to form difficultly soluble humates, fulvates, and CO_2.

According to Fig. 19.12, three intervals can be derived by the sedimentation regime. The interval from 4 to 2.5 m in depth and from 8000 to 5500 years in age corresponds to the Early Holocene. It formed first at low sedimentation rates

(0.5 mm/yr) and at a drop in the sea level. The subsequent increase in the sea level was replaced by its drying up and overgrowing with swamp vegetation as indicated by a peat layer 0.2 m in thickness at a depth of 0.2 m and ~45,000 years. The subsequent long-term decrease in the sea level is a typical feature of the hydrological regime of the next sedimentation stage, which was considered by a new drying up of the sea about 1,500 years ago. This conclusion is supported by the presence of a series of bog peat streaks in the depth interval of 1.1–1.4 m. The process of sedimentation in the upper interval of the core from 1.1 m to the bed surface with the age from 1,300 years before present time took place at variable hydrological regime with alternating rise and drop in the sea level as can be seen from the interstratifying clayey and sandy silts with inclusion of salt crystals.

The Aral Sea is a shallow water body. Its maximum depth in 1900 was 60 m. By 1900, the sea level has dropped by15 m; the water area has decreased from 65,000 to 40,000 km^2; and the sea water volume, from 1000 to 400 km^3. In this period, the seawater salinity has increased from 10 to 24‰. The general increase in the sedimentation rate from the core bottom to the surface, which is seen in Fig. 19.12, means that the general tendency toward a decrease in the sea level since 7,000 years up to present epoch will be kept.

19.6 Results and Conclusions

The most important results of the studied bottom sediments and water masses in the Caspian and Aral seas, Kara-Bogaz-Gol Gulf and Issyk-Kul Lake related to the nature of the hydro-climatic changes are as follows:

1. The sedimentation rate in the middle and southern Caspian Sea has bright-expressed character. In the time interval from 24,000 to 11,000 years ago, the mean sedimentation rate in the middle basin was 0.12 mm/year. At the same time in the period from 11,000 to 6000 years that value has roused up to 1.2 mm/year. To the contrary, in the southern basin during period from 23,000 to 17,000 years sedimentation rate was 0.9 mm/year and from 17,000 to 6,000 years it dropped to 0.19 mm/year.
2. The oxygen and carbon stable isotopes of the carbonate fraction of Caspian bottom sediments show that in the period of active sedimentation, the light in isotopic composition river water discharged alternately from the Central Asian and European catchment basin.
3. It follows from the chemical content of pore water of sediments that the identical salinity in the two Caspian basins was fixed only about 22,000 years ago when intensive water discharge from the Central Asian catchment basin has taken place. During that period of time the water level in the sea was 70 m higher than that of the present day and the Caspian Sea has had hydraulic connection with the Black Sea basin.

4. The analysis of the coarse grained and detrital material of 1 to 0.63 mm size allows to divide 8 regression-transgression cycles that occurred during last 20,000 years in the Caspian Sea.
5. The dynamic sea study by isotopic composition of hydrogen and oxygen carried out in 1995–1996 by the IAEA expedition shows that the fixed sea level rise up to 2.5 m during 1978–1995 has not taught the structural changes of water body lower of 300 m. In 1996 it was fixed on the basis of tritium studies that the changes occurred only on the southern station of sampling. It appears that the tritium indicator is most sensitive to such kind of environmental effects (Brezgunov and Ferronsky 2005).
6. The study of the Kara-Bogaz-Gol Gulf bottom sediments shows that during period from 9,200 to 8,500 years ago, this water reservoir has been over-drying and the sedimentation has gone by the solid runoff of the Karyn-Zharyk River and its tributaries. This conclusion is based on the fact that δ^2H value of the pore water is equal to +6‰. This value shows that the reservoir this time was drying. Concentrations of Na^+ and Cl^- equal to 2.5–3 g/l show on absence of connection between the gulf and middle Caspian Sea. About 8500 years ago, the δ^2H value dropped −6‰, which is an evidence of filling in the saline reservoir by the river water. Within the period of 8500 and 2200 years the δ^2H value in the pore water decreased to −17‰, which indicates increasing of the river runoff. The lake has fast filled in by water and about 3000 years ago a periodic discharge of water to the middle Caspian Sea has started. After the New-Caspian transgression 2200 years ago, the Karyn-Zharyk River and its tributaries started drying and since that time the gulf is recharged by water from the Caspian Sea. All other studied parameters prove this conclusion.
7. By the data of the shore terraces and bottom sediments study, in the Holocene history of Aral Sea it was fixed four transgression and three regression phases. The sea level was roused up to absolute marks of 57, 54.7, 53.5, and 53 m and dropped to 44, 43, and 35 m, accordingly. The sea level on the mark 57 m indicates the event when water discharged from the Aral Sea to the south Caspian basin through the Uzboi River. The two peat layers of 0.2 and 0.4 m thickness and 4500 and 1300 years old, respectively, in the bottom sediment core have been found. Since 1300 years and up to the present days, the process of sediment accumulation at a changing hydrological regime and periodic rise and drop in sea level is continuing. These dada are proven by observation of the interstratifying sandy and clayey sediments and periodic occurrence of the salt crystals in the sediment section.
8. The results of isotope, mineralogy, and chemistry studies of the Issyl-Kul bottom sediments show that during period of 7,600–6,000 years before the present day,the sedimentation process continued in the form of permanent discharge of cold water to the Chu River. The low concentration of the organic carbon, monohydrocalcite, δ^{18}O and low Sr/Ca values, and also little increase of $CaCO_3$ and $MgCO_3$ content on the upper border of corresponding core interval prove this conclusion. In the period from 6,000 to 4,300 years before the present day, the sedimentation process has been continued but the volume of the river runoff

has decreased, salinity roused, and the water level and sedimentation rate diminished. This conclusion is based on the facts of the increasing $\delta^{18}O$ values in the mollusk's shells, decreasing Mg^{2+} content in precipitated calcites, increasing concentration of organic matter and Sr/Ca ratio in the shells. It was found by the top core interval study that during the period between 4,300 years and present-day periodic drops and rises, limited by the Chu River bed, in the lake level have happened.

9. It was shown by the results of δ^2H, $\delta^{18}O$ measurements and ^{14}C dating that the Yaskhan Lake of fresh water in the west Turkmenistan, having less than 0.6 g/l mineralization and 17–22% of radiocarbon content (compared with modern standard), has been formed about 12,000 years ago, which proves its genetic relationship with the Aral-Sarykamysh paleo-basin.
10. It was found on the basis of isotope studies in the Amu-Darya, Syr-Darya, Chu, and Ily rivers, and also in the Syrian Desert that in the regions of Central Asia and Middle East, the formation of ground waters has had close connection with the pluvial epoch.

The following natural event and respective changes were found by the above enumerated facts:

1. The inverse character of the sedimentation rate change in the middle and southern Caspian Sea, which is accompanied by changes in the isotopic composition of carbonates, salinity of water, chemical and mineral content of sediments, appear to be the evidence of changes in hydrologic regime taking place in the Central Asian and European catchment basins in the late Holocene. This change in hydrological regime was initiated by change in the climatic conditions, which started about 23,000–25,000 years ago and continued up to 1000–12,000 years ago. It was the last long-periodic cycle of climatic changes on the Earth. As a result, the moderate humidity in the Central Asian region climate has changed by a hot and arid or semi-arid climate, and the cold semi-arid climate of the arctic desert in the Volga and Urals basin on the European territory gave way to the moderate and wet regime. The coming aridity in the Central Asian region was accompanied by a plentiful melting in the Pamir and Tian-Shan glaciers and by river runoff of large amount of water to the Aral Sea basin and through Uzboi River to the Caspian Sea. By the data of isotopes in carbonates, salinity of pore water and mineral content of bottom sediments, the aridity process in the Central Asian region has begun 12,000 years ago and later on the humidity of the territory periodically changed.
2. The keenly different rate of sedimentation, salinity of pore water and other sedimentation and water characteristics of the southern Caspian Sea, and also the bathymetric section of the Apsheron sill demonstrate that not far before now the Caspian Sea was divided by the Apsheron isthmus of about 50 km width and represented the two separate lakes, water of which has flown over the natural piece of land. The geological history of the Apsheron sill proves this conclusion. The bathymetrical structure of the sill shows that general direction of the erosive water flow moved from the south to the north, and the total erosion of the isthmus and

joining the two lakes into common sea has happened between 12,000 and 8,000 years ago.

3. The pre-Aral Sea discharged its water to the Caspian Sea since the Turanian depression was filled up to the absolute mark of +56 m from 2,300 to 8,000 years and when it was a shallow Aral-Sarykamysh Sea. All that time the excess water was discharged to the Caspian Sea through the Uzboi River. The erosion traces of the Uzboi River stormy activity in the foot of Kopet Dag Mount are observed up to nowadays.
4. Periodical change in clayey mineral content from the two deep-water Caspian depressions demonstrate effects of the eight transgression-regression cycles in the southern and middle Caspian Sea. These events happened continuously over the period of 1,500–2,000 years. More detailed analysis of the oxygen and carbon isotopic content in carbonate sediments showed evidences on the shorter periodicity of the events of hundreds and even tens of years. As it follows from the river runoff observation, the atmospheric precipitation, temperature and Caspian Sea water level over the last 150 years, there are inter-annual and 30–50 year periodicity in the hydrological regime of the sea and all Aral-Caspian basin changes. The amplitude of the sea level change reaches 1–2 m and even more. These changes have important meaning for economic and living activity of the people.

The understanding of the Aral-Caspian catchment basin's hydro-climatic changes and also the analogical facts observed in the other regions of the globe brought the authors to the idea to find the initial physical cause of the short-periodic processes governing these effects. Following the known and proved by observation of the astronomist Milankovich's approach, in which he links the climate change and the distribution of solar energy flux coming to the Earth surface, it was decided to extend his approach on the short-periodic processes. For this purpose it was necessary to formulate and solve a new problem of rotation and oscillation of the Earth in its own force field (Ferronsky and Ferronsky 2010). The results in short will be presented in the next chapter.

Chapter 20
The Nature and Mechanism of the Earth Shell Separation and Origin of Hydrosphere

The samples of the Moon rocks brought by the astronauts and space apparatus allowed in obtaining data on their physical-chemical properties and on this basis to make a number of new, in principle, cosmophysical and cosmochemical conclusions. Owing to the use of new methods of matter analysis, the information about chemical, isotope content, and absolute age of meteorite, Moon and Earth's rocks, water, and gases has substantially grown.

By the programs on deep-sea drilling in the oceans and super-deep drilling in the land the studies on structure of the ocean and continental crust were carried out, and finally, by geodetic satellites the advantageous investigations of dynamical parameters of the Earth gravitational field were realized. All these actions and obtained results and also a new approach to dynamics of the Earth enabled in coming back to discuss the oldest scientific problem of the creation of our planet and its hydrosphere.

20.1 Existing Approaches to the Problem Solution

Let us briefly review the existing hypotheses that are available for research into the nature of the solar system. At present the most common theories of formation of the Earth and the other planets of the solar system are based upon the idea of accretion of substances from a proto-planetary gas-dust cloud. There are several options on the origin of the cloud itself. Some scientists believe that the proto-planetary cloud was captured from interstellar nebula by the existing Sun during its motion through interstellar space (Schmidt 1957). Others consider it to be the product of evolution of a more massive cloud from which the Sun itself originated (Cameron and Pine 1973; Cameron 1973).

Before the 'accretion' theory was advanced, there were a lot of other ideas and theories concerning the origin of the solar system which have been discussed repeatedly in great detail (Spencer 1956); but the majority of these theories were rejected, being unable to explain a number of observed astronomical facts and aspects of celestial mechanics and cosmochemistry. However, in the light of effects of the dynamic equilibrium approach presented in this book, some of them, such as the theories of Laplace and Descartes deserve to be rehabilitated.

V. I. Ferronsky, V. A. Polyakov, *Isotopes of the Earth's Hydrosphere*,
DOI 10.1007/978-94-007-2856-1_20, © Springer Science+Business Media B.V. 2012

Table 20.1 The content of volatile components near the surface of the Earth and in fractured rocks. (From Rubey 1964)

Object	Volatiles ($\times 10^{20}$ g)					
	H_2O	Carbon in CO_2	CI	N	S	H, B, Br and others
Present atmosphere, hydrosphere, biosphere	14,600	1.5	276	39	13	1.7
Ancient sedimentary rocks	2,100	920	30	4	15	15
Total	16,700	921.5	306	43	28	16.7
Fractured igneous rocks	130	11	5	0.6	6	3.5
Excess of volatile components	16,570	910.5	301	42.4	22	13.2

According to the accretion hypothesis, the planets and satellites of the solar system were formed as a result of successive accumulation of dust particles. The subsequent growth of these accumulations into planetary and satellite bodies has been considered on the basis of the mechanism of gravitational instability and the amalgamation of small bodies with larger ones.

The majority of geological and geochemical reconstructions concerning the formation of the Earth and its shells, based on the accretion hypothesis, postulate the initial existence of a chemically homogeneous substance. It is assumed that at the initial stage of the Earth's formation the temperature of the condensing matter was low and water was retained, along with carbon, nitrogen, and other volatile constituents (Urey 1957). Subsequently, due to the gravitational heat generated during accretion and the energy produced by radioactive decay, the matter constituting the Earth underwent complete or partial melting. During the melting of this matter, the water and the volatile components degassed to the surface producing the hydrosphere and the atmosphere.

Rubey (1964) notes that the CO_2 concentration in the atmosphere and the oceans remains more or less constant over a considerable period of geologic time. In his opinion, if only 1/100 of the carbonates in sedimentary rocks were transferred to the oceans and the atmosphere, life on Earth would not be possible. From analytical computations based on the geochemical balance of the volatile substances contained in old sedimentary rocks and in the present hydrosphere, atmosphere, and biosphere on the one hand, and in disintegrated igneous rocks on the other, he obtained the very interesting data shown in Table 20.1.

As can be seen from Table 20.1, there is an enormous excess of volatile substances in the surface zone of the Earth which could not have been produced by the disintegration of igneous rock material and its subsequent transfer into sediments and solutions. Analyzing possible reasons for the excess, Rubey concludes that the present-day oceans formed gradually through the emergence of water together with other volatile substances from the deep interior of the Earth or, more specifically, from what he called hydrothermal sources. Therefore, in Rubey's opinion, the present hydrosphere is of juvenile origin.

Although satisfactory explanations exist for a number of observed facts in the framework of the cold origin hypothesis, many authors have noted certain fundamental contradictions, for example, Ringwood (1966, 1979) has drawn attention to the extremely high concentrations of iron and nickel oxides in ultramafic basic rocks and basalts and to very large amounts of CO and CO_2 which would have been released through the reduction of the iron constituting the planet's core. If the theory is correct, these gases would have had a mass half of the entire core, which is hardly possible. The mechanism of formation of the core and the other shells of the Earth also remains problematic.

Cosmochemical facts, which have been obtained in recent years by many researchers studying meteorites, the Moon, and the planets, indicate that other ideas concerning the formation of the bodies in the solar system should be employed. These new ideas have shown that the processes of condensation of the chemical elements and compounds that might have occurred in the proto-planetary substance have actually taken place. Due to these processes the bodies of the solar system could have become inhomogeneous in chemical composition and differentiated into shells. The development of these ideas is closely related to the problem of the original formation and evolution of the Earth and its shells and is therefore considered in detail below, including the effects following from our theory of the planet's motion based on its dynamical equilibrium. We consider that the main effect of the planet body formation is differentiation of the cloud's mass in density and their separation due to development of the normal and the tangential components of the force function at gravitational interaction of the nonuniform masses as discussed in the next sections.

The fact of the existence of gaseous clouds and nebulae is proved by observation. The evolutionary process inevitably leads to their density separation. The study of the isotopic composition of a substance is one of the most efficient ways of learning about the origin of that substance. We carried out an analysis of vast experimental data on the isotopic composition of hydrogen and oxygen of water on the Earth in order to solve the problem of origin of the hydrosphere by the means of its separation during formation. Analogous analysis was done on the basis of isotopic composition of the carbon and sulfur in natural objects of organic and inorganic origin. The results prove the fact of the separation process which has taken place in accordance with the chosen theory. We present our analysis in the next sections.

20.2 Separation of Hydrogen and Oxygen Isotopes in Natural Objects

At present we definitely know the physical ground for the separation of isotopes, and much experimental data concerning the isotopic composition of water in different objects has been obtained. These objects are: atmospheric moisture, water of the oceans and their sediments, inland surface and ground waters, liquid inclusions in rocks and minerals, deuterium and oxygen-18 contents in sedimentary and igneous rocks and minerals, in meteorites, and in lunar samples. The available data permit us to discuss the problem of origin of the Earth's hydrosphere as the planet's shell.

When considering the evolution of the isotopic composition of any element through what is assumed to be a wide range of temperatures, the isotopic separation can be expected to occur only within strictly defined temperature and pressure limits corresponding to the phase transition of the element to and from the gaseous, liquid, and solid state. The isotopes of a given element in the gaseous state will be distributed uniformly throughout the gas volume. If, on the other hand, an element in the gas phase reacts with other substances on reaching to an appropriate temperature for the formation of new compounds constituting a condensed phase, then the separation effect will begin to manifest itself as a depletion of the gas phase in these same isotopes. The transition of the compounds in question to each successive phase state will be accompanied by enrichment of the new phase in heavy isotopes.

Close examination of this process from the thermodynamic point of view shows that, with increasing system temperature, isotopic separation among the various compounds decreases. Under certain equilibrium conditions the system components or phase will be depleted or enriched in one or another isotope according to the isotopic exchange constant, and as the isotopic exchange constant is a function of temperature, changes in temperature must lead to change in isotopic composition of the components or phases through isotopic exchange reactions. For example, CO_2 gas, being in isotopic equilibrium with water in the liquid phase, will be enriched in ^{18}O by approximately 40% relative to the oxygen constituting H_2O molecules at 25°C.

It should be borne in mind that during the Earth's formation and subsequent evolution, isotopic separation of the elements also occurred as a result of many secondary effects associated with global and local changes in temperature due to internal and external release or absorption of energy. Global changes in the isotopic composition of water through the secondary effects associated with variation in the temperature regime are very pronounced when there is interaction between the open water basins of the Earth and the atmosphere in the form of evaporation, condensation, and precipitation. The same effects are observed locally in closed systems where underground waters interact with rocks and where individual minerals and other components with water during the formation of rocks of different origin.

The temperature range within which the isotopes of water undergo detectable fractionation during the phase transition of water is −40 to +374°C. Within 220–374°C the separation factor of hydrogen isotopes becomes less than 1. At higher temperatures, the various isotopes of water in a given gas volume will have a statistically uniform distribution. The separation of hydrogen and oxygen isotopes in other compounds during their formation and under the influence of secondary temperature effects will occur within other temperature ranges. It is known, for example, that when oxygen interacts with the silicate phase of rocks, isotopic separation occurs up to a temperature of 1,000°C (Ferronsky and Polyakov 1982). If one knows the exact temperature which induced separation relationship between two interacting components from the actual isotopic ratios then it is possible to calculate their exact

Table 20.2 Deuterium and oxygen-18 content in natural objects

Object	δ^2H, ‰ -1000 -600 -200 0 + 200	$\delta^{18}O$, ‰ -60 −20 0 +20 +60
Galaxy	+2860 - +5360	
Solar atmosphere		
Earth's atmosphere		
Molecular hydrogen		
Carbon dioxide		
Molecular oxygen		
Atmospheric precipitation		
Methane		
Surface waters		
Oxygen dissolved in water		
Rivers and lakes		
Oceans		
Underground waters and rocks		
Sedimentary carbonates and silicates		
Igneous and metamorphic rocks		
Basalts and granites		
Brines		
Hydrothermal waters		
Steam and gas from thermal waters and volcanoes		
Natural organic material		
Meteorites		
Tectites		
Chondrites		
Iron meteorites		
Water in carbonaceous chondrites		
Organic matter in carbonate chondrites		
Lunar material		
Water in rocks		
Molecular hydrogen		
Carbon dioxide		
Rock samples		

interaction temperature. This is the principle underlying oxygen paleothermometry, which is used for determining the formation temperature of sedimentary rocks, glaciers, and other terrestrial features.

Many papers containing data on the isotopic composition of water and other natural and cosmic substances have been published. The authors of this book have analyzed many of these with a view to determine the limits of variation in the concentration of the heavy isotopes deuterium and oxygen-18 in various substances. The results are presented in Table 20.2, where the isotopic values are given in units relative to the mean ocean water standard SMOW.

From Table 20.2 it can be seen that the isotopic composition of the surface waters of the continents and of the Earth's atmosphere, which is determined exclusively by the present temperature regime of the Earth's surface, varies within broad limits. As one would expect, the water most depleted in heavy isotopes is fixed in the polar regions (δ^2H up to −500‰, δ^{18}O up to −60‰ (Craig 1963)), while that most enriched is in closed basins in arid zones (δ^2H = +129‰, δ^{18}O = +31.3‰ (Fontes and Gonfiantini 1967)). It should be noted that the isotopic composition of surface waters, atmospheric moisture, and gases is controlled by secondary effects of the continuous, natural evaporation-condensation-precipitation cycle. All these processes, which occur at widely varying temperatures at the Earth's surface and in the atmosphere, cause fractionation of hydrogen and oxygen isotopes within broad limits.

Ocean water is uniform in its isotopic composition. Investigations carried out by many authors have shown that at the depth of 500 m or more ocean water has a very uniform isotopic composition with small regional variation (δ^2H from −25 to +10‰ and δ^{18}O from 0.6 to +2‰) observed in the surface layer. Despite the enormous amount of evaporation from the ocean surface, which enriches the surface layer in heavy isotopes, any appreciable change in the isotopic composition as a whole does not appear there. This is because most of the evaporated water ultimately returns to the ocean. Such a situation can obviously persist as long as the overall temperature balance on the Earth is maintained. If this balance is upset over a long time, the balance of isotopic composition in ocean water will break down. However, as has been shown by recent paleotemperature measurements of old ocean sediments and glaciers based on ^{18}O/^{16}O (Emiliani 1970, 1978), temperature variations at the ocean surface in the equatorial zone have not exceeded 5–6°C during the past 0.7 million years, and the ^{18}O/^{16}O ratios in the water have not deviated by more than 0.5‰ from the present-day value. Similar results have been obtained by determining deuterium in water contained in clay minerals of marine origin and of various ages.

The isotopic and chemical composition of the ocean water may therefore be considered to have remained virtually constant over the past 250–300 million years. This is confirmed by a number of facts obtained during paleotemperature studies based on the comparison of the oxygen isotope ratios of organogenic carbonates and shells of ancient and modern mollusks (Bowen 1966, 1991; Tays and Naydin 1973).

Taylor (1974) came to a similar conclusion after analyzing data on the hydrogen and oxygen isotope ratios in Cambrian and Precambrian siliceous charts obtained by Knauth and Epstein (1971, 1976). Moreover, he has suggested that the isotopic composition of the ocean water has remained unchanged, at least during the majority of the upper Precambrian era.

Similar results were derived while analyzing the carbon and oxygen isotope composition of Precambrian limestone and dolomites in Africa, Canada, and Europe (Schidlowski et al. 1975). It was found that the near-constancy of $\delta^{13}C \approx 0$ during geologic time corresponds to the constant ratio of the organic carbon to the total amount of carbon in sedimentary rocks ($C_o/C_t \approx 0.2$). On this basis, it has been concluded that at least 80% of the modern oxygen isotopes were formed earlier then 3×10^9 years ago.

A very important question concerns the isotopic composition of deep underground waters and also the δ^2H and δ^{18}O composition of liquid inclusions in the minerals of magmatic origin and in rocks belonging to the Earth's upper mantle. The data available concerning the studies of isotopic ratios in various types of mineralized underground waters and brines in numerous regions of the United States, Canada, Japan, and the former Soviet Union with oil and gas fields were considered in detail and showed that the relative deuterium concentration in such waters varies from +29 to −109‰ with a clear tendency to depletion in heavy isotopes relative to ocean water.

It has been shown by many researchers that the water-bearing carbonate and silicate rocks are always enriched in heavy isotopes of oxygen. Oxygen exchange occurred through the interaction of deep underground water (depleted in heavy isotopes as a result of mixing with local atmospheric precipitation in the course of their formation) with water-bearing carbonates and silicates at appropriate temperatures. This process led to some enrichment of the brines in heavy oxygen compared with local precipitation and sometimes compared with ocean water.

It may be assumed that one of the secondary temperature effects leading to the enrichment of deep underground waters in deuterium and oxygen-18 was that of the surface evaporation of water of these basins being included in inland seas and lakes.

The range of variation in the isotopic composition of the Earth's thermal waters is from −18 to −207‰ for δ^2H, and from +7.5 to −22.5‰ for δ^{18}O. The corresponding values for steam and gas from springs and volcanoes are from −40 to −520‰ for δ^2H and from +0.7 to 22.5‰ for δ^{18}O. Rubey (1964) and a number of other authors think that those places where thermal waters emerge at the Earth's surface (from crystalline rocks and in active volcanic regions) are where juvenile water emerges and mark the sources which gave rise to our present oceans. Craig (1963) has made a detailed analysis of the isotopic composition of water from hot springs in many parts of the Earth. His investigations have shown that all hot springs discharge water having an isotopic composition identical with that of the local atmospheric precipitation. The widely observed higher enrichment is explained by oxygen substitution, which takes place when these waters interact with water-bearing carbonate and silicate rocks.

Measurement of tritium concentrations in thermal waters performed by Theodorsson (1967) in Iceland, Begemann (1963) in the United States, and other investigators, have yielded values that are absolutely identical with the tritium concentrations in local surface waters and rainfalls. As tritium is an isotope formed primarily by cosmic rays and concentrations can only be increased during nuclear bomb tests, it is impossible that any tritium could have been derived from underground sources. In any case, the circulation period of waters to hot springs fed by atmospheric and surface waters ranges from several months to several years. Thus, the above-mentioned investigations suggest that hot springs do not supply juvenile water.

Let us consider the δ^2H and δ^{18}O variation in sedimentary rocks, liquid inclusions, and minerals in magmatic rocks, granites, and basalts. The limits of variations of these isotopes are shown in Table 20.2. It can be seen that for all the types of rocks in question, the isotopic ratios of hydrogen are negative relative to ocean water, whereas the oxygen values for all the rock types listed are positive relative to ocean water. The oxygen isotope ratios range from 6 to 7‰ for basalts and 7 to 12‰ for granites and igneous rocks. In younger carbonate and silicate sedimentary rocks (no older than Riphean), the relative content of heavy oxygen ranges from +25 to +38‰, whereas in the ancient (Proterozoic, Archean, and older) carbonates, clay, and siliceous shale it is much lower: from +7 to +25‰. The global trend of the heavy oxygen content to decrease over time is interpreted in different ways.

Weber (1965), who examined more than 600 samples of ancient and modern calcites and dolomites from various geographical regions of the world, reported that the gradual increase of the oxygen-18 content in younger carbonate rocks is the result of enrichment of the ocean water in heavy oxygen due to the discharges of juvenile water during the growth of the ocean.

Silverman (1951) has reported that if the primary ocean was created by the emergence of juvenile waters, its water should be enriched in oxygen-18 compared with modern ocean water. The result of the melting of rocks of the Earth's deep interior and the degassing of water in equilibrium with large masses of the mantle silicates should be enrichment in oxygen-18 by at least +7‰.

Craig (1963) has also considered that if primary oceans were formed due to the emergence of the juvenile waters, then it should exhibit a relative $\delta^{18}O$ content of about +7‰. The subsequent depletion of the ocean water in ^{18}O could have occurred as a result of the accumulation of carbonate and silicate sedimentary rocks enriched in heavy oxygen compared with the original igneous rocks. Craig has assumed that the primary calcium carbonates and silicates precipitate from the solution with relative ^{18}O contents up to +30‰. The relative content of heavy oxygen in sedimentary rocks decreases on average to +20‰ in the course of isotope exchange with fresh waters. Since the primary igneous waters had an average ^{18}O content of about +10‰ before their destruction, then during the formation of sedimentary rocks a continuous depletion of the ocean in heavy oxygen should occur.

Migdisov et al. (1974) have pointed to the importance of such processes as the volcanic activity of the Earth in the evolution of the isotopic composition of the ocean. Such activity was accompanied by increases in the weathering of the igneous rocks and the sedimentation of the weathering products. In addition, the volcanic activity is accompanied by the emergence of juvenile water and CO_2, leading to the enrichment of the ocean in ^{18}O. A similar effect is produced during the process of the deep metamorphism of sedimentary rocks accompanied with the emergence, to the ocean, of water and CO_2 enriched in heavy oxygen.

It should be pointed out that the volcanic activity, and related processes of rock sedimentation which may have taken place in certain geologic epochs, could not have markedly affected the ^{18}O change in the ocean because of the great differences between the total masses of oxygen in the ocean water and in sedimentary rocks formed during volcanic activity. Savin and Epstein (1970C) have evaluated the global effect of the sedimentary process upon changes in the isotopic composition of ocean water during the whole observed history of the Earth. They found that all the accumulated sedimentary rocks (pelagic sediments, carbonates, sandstones, and clays) could have resulted in the depletion of the hydrosphere in oxygen-18 by 3‰ and during the whole postPrecambrian period by only 0.6‰. As for deuterium, the ocean could have been enriched with it during the whole history in the course of sedimentation by less than 0.3‰, i.e., less than the errors in the measurements.

Considering the possible periodic changes in the isotopic composition of the ocean water, it should be noted that there are more effective processes than the sedimentation of rocks during volcanic activities. Those include temperature variations on the Earth over time and the associated accumulation and melting of ice in glacial and nonglacial times. According to the data of Craig (1963), only in the Pleistocene have ^{18}O variation in the ocean amounted to 1‰, and 7‰ for deuterium; in the earlier epochs these variations could have been even greater.

The contribution of juvenile water to the oceans during volcanic activity has not yet been confirmed experimentally. Some researchers (Arnason and Sigurgeirsson 1967; Friedman 1967; Kokubu et al. 1961), while studying the water isotopic composition of volcanic lavas and gases, do not consider this water to be juvenile; only Kokubu et al. (1961), analyzing liquid inclusions in samples of basalts in Japan, have concluded that they dealt with water of magmatic origin. Later, Craig showed that in terms of the hydrogen and oxygen isotope ratios this water lies on the curve for local atmospheric precipitation and, therefore, is of meteoric origin.

The deuterium content in waters of the volcanic lavas and gases varies over wide ranges. According to the data obtained by Friedman (1967), who carried out investigations during volcanic eruptions on the Hawaii Islands in 1959–1960, the deuterium content in the water of lava samples at the top of the volcano ranges from −55 to −79‰ and in samples taken from its slopes from −57 to −91‰.

Arnason and Sigurgeirsson (1967) studied the isotopic composition of steam and gases in the eruption of one of the Iceland volcanoes during 3.5 years (1964–1967). Their data are close to those obtained by Friedman and indicate an average deuterium content of −55.3‰.

Sheppard and Epstein (1970c), while studying the oldest (to 1.14 billion years) samples of rocks and minerals (peridotite, olivine, dunite, mylonite, kimberlite, etc.), which, in their opinion, are representative of rocks of the upper mantle of the lower part of the Earth's crust, found an average value of deuterium of -58 ± 18‰. On the basis of the obtained data, they concluded that the hydrogen isotope fractionation between phlogopites and water can amount to about 10‰ at 700°C. In view of this fact, the $^2H/^1H$ ratio for juvenile water should be -48 ± 20‰ relative to the standard SMOW.

Craig and Lupton (1976) studied the hydrogen isotope composition of basalts sampled from the mid-oceanic ridges of the Atlantic and Pacific Oceans. It was found that the studied samples of deep water troilite exhibit $\delta^2H = -77$‰. On the basis of analysis of the obtained data together with results of neon and helium isotope composition studies, the authors have concluded that the obtained δ^2H value may possibly reflect the juvenile hydrogen isotope composition.

A wide range in the deuterium content of terrestrial materials is found in various organic compounds (from −12 to −430‰). The corresponding range of oxygen-18 concentrations is from +16 to −5‰. This can only be attributed to great ranges of hydrogen and oxygen isotope fractionation during the oxidation-reduction processes leading to the formation and evolution of natural organic compounds (Schiegl and Vogel 1970; Ester and Hoering 1980).

It is worth noting that oil in the deposits of different origin and age has a stable deuterium concentration of about −100‰ relative to the standard. The natural photosynthetic process, despite its importance in the geochemistry of the upper terrestrial shells, could not have had a marked effect upon the evolution of hydrogen and oxygen isotopes in the hydrosphere. This is because the mass of organic substance accumulated during the Earth's history is too small in comparison with the mass of the sedimentary rocks and especially with the water mass in the hydrosphere.

With regard to the deuterium and oxygen-18 concentration in different types of meteorites, there is a wide range of deuterium variation in carbonaceous chondrites. As investigations of the isotopic composition of water and organic compounds in carbonaceous chondrites have shown (Briggs 1963), the upper limits for the deuterium content of these two compounds are very similar (+290 and +275‰), whereas lower limits are −154 and −15‰ respectively.

While studying the oxygen isotope ratios in the minerals of stone meteorites (Reuter et al. 1965; Taylor et al. 1965; Vinogradov et al. 1960), like many common features which make them similar to their terrestrial analogues, a number of peculiarities have been found. The ^{18}O content in the meteorite piroxene varies from −0.5 to +8.6‰ depending upon meteorite type. In piroxene of terrestrial igneous rocks the range is from +5.5 to +6.6‰. Olivine of carbonaceous chondrites exhibits considerable ^{18}O variations, being approximately 6‰ lower than those in olivine of other stone meteorites. If the terrestrial basalts and gabbro are enriched in ^{18}O as compared with the ultrabasic rocks, then the basaltic chondrites become depleted compared with the ultramafic chondrites.

An interesting interpretation of the data on oxygen isotope composition in different types of meteorites has been given by Clayton et al. (1976); Clayton and Mayeda (1978a, b). On the basis of the measured $^{18}O/^{16}O$ and $^{17}O/^{16}O$ isotope ratios and the construction of a diagram with $\delta^{18}O$–$\delta^{17}O$ coordinates they have distinguished six types of meteoric substances. Each of these types, characterized by its own oxygen isotope composition, cannot be obtained from any other type by means of fractionation or differentiation of substance. These groups are: (1) the terrestrial group, consisting of substances of the Earth, Moon, and enstatite ordinary chondrites; (2) the L and LL types of ordinary chondrites; (3) the H type of the ordinary chondrites; (4) the nonhydrated minerals of the C2 and C4 chondrites; (5) the hydrated minerals of the base of C2 chondrites; (6) urelites.

The authors found that the substances of the solar nebulae from which the bodies formed were inhomogeneous in the oxygen isotope composition. They pointed out that, as a first approximation, a two-component model of this substance can be assumed. The portion of the substance whose oxygen isotope composition has been formed on the basis of mass fractionation effects provided by chemical reactions and diffusion processes, can be indicated on the $\delta^{18}O$–$\delta^{17}O$ diagram by a straight line with a slope of 0.52. The other portion of the substance, like common oxygen, contains the component enriched in ^{16}O and having independent origin. Figure 20.1 indicates on the three-isotope diagram $\delta^{18}O$–$\delta^{17}O$ and the distribution of the oxygen isotopes in various types of meteorites.

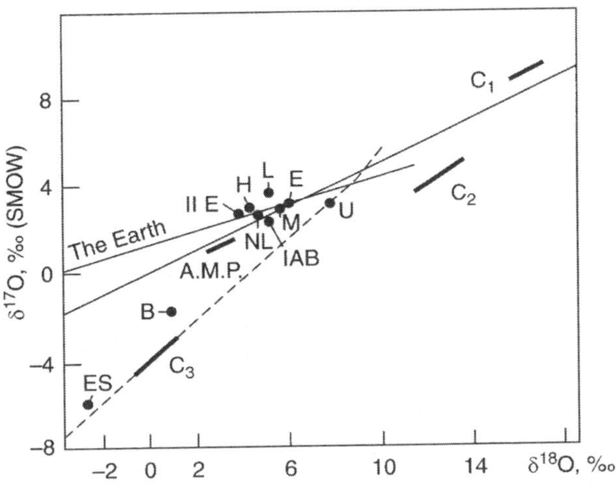

Fig. 20.1 Oxygen isotopic composition of various types of meteorites on the three-isotope diagram: carbonaceous chondrites (*C_1*, *C_2*, *C_3*); ordinary chondrites (*HL*); enstatites and autrites (*E*); eucrites, hovardites, diagenites, mezosiderites, pallasites(*AMP*); nakhelites and lafayettites (*NL*); uraelites (*U*); irons (*IAB, IIE*); Bencubbin and Wetherford (*B*); Eagle Station and Itzavisis (*ES*); Moon (*M*). (After Clayton and Mayeda 1978a, b. © IAEA, reproduced with permission of IAEA)

In their δ^2H and δ^{18}O contents, meteorites in general are not—when considered as chemical complexes—identical with any type of terrestrial rock. At the same time, the range of isotopic ratios points to their close affinity with certain types of rock. It is possible to say only that the conditions of the formation of meteorites were different from that of terrestrial rocks.

In connection with this question, there is natural interest in information on the isotopic composition of lunar material brought to Earth under the space programs of the United States and the former Soviet Union. Published data on the isotopic composition of water in lunar dust and breccia from the Sea of Tranquillity (concentration of 81–810 ppm) and of hydrogen gas (concentration of 18–66 ppm), point to extremely low deuterium concentrations in these materials (Epstein and Taylor 1970; Friedman et al. 1970). However the deuterium content in such small samples varies over wide ranges (from −158 to −870‰ for water and from −830 to −970‰ for hydrogen gas). This clearly demonstrates the nonterrestrial origin of water and hydrogen gas on the lunar surface. Moreover, the wide fractionation range for hydrogen isotopes suggests that the lunar rock samples probably have different temperature history.

The published data (Epstein and Taylor 1970; Onuma et al. 1970; Friedman et al. 1970) indicate a much narrower range of oxygen-18 concentration from +19 to +14‰ for carbon dioxide and from +7.2 to +2.8‰ for rock samples.

A very interesting oxygen isotopic composition pattern emerges from the analysis of individual minerals in lunar rocks. The isotopic composition of the oxygen in individual minerals varies appreciably; this is also an evidence of the complex temperature history which these minerals experienced during formation of the corresponding compounds. The latest data on deuterium and oxygen-18 in samples from the Sea of Storms (Epstein and Taylor 1971; Friedman et al. 1971; Onuma et al. 1970) presents another picture. While the oxygen isotopic composition lies within the same range (from +7.15 to +3.83‰), deuterium concentrations are much higher, in one of the crystalline samples a value of +283‰ was found (Friedman et al. 1971).

As to the problem of the existence of water in lunar rocks, there are conflicting opinions. Friedman et al. (1970, 1971) affirm that water is present in the lunar rock samples. Epstein and Taylor (1970, 1971, 1972) consider that Friedman and coworkers found water which was brought back by astronauts or obtained during transportation and processing of samples. Analyzing the lunar rock samples brought to the Earth by Apollo-14 and 15, Epstein and Taylor (1972) noted a rather great homogeneity of rocks on the whole in oxygen-18 (the observed changes range from +5.3 to 6.62‰). The extracted amounts of water (~10 mol/g) contain deuterium from −227 to 419‰. The $^{18}O/^{16}O$ ratio of this water is changed from −5.9 to −18.2‰. The obtained values are rather close to the ^{18}O content in atmospheric moisture in the region of Los Angeles, where analyses have been carried out. This circumstance allowed Epstein and Taylor to conclude that the studied samples could have been contaminated with local moisture. The deuterium content in the gaseous hydrogen, as earlier, was low (about molecules per million), so that the $^{2}H/^{1}H$ ratios ranged from −250 to −902‰.

When we analyze the isotopic composition of water in various terrestrial and cosmic materials the following observations are particularly striking:

1. Despite the very wide range of deuterium and oxygen-18 concentrations in various materials, the isotopic composition of ocean water remains constant over a long period of geologic time.
2. The concentration of deuterium in all terrestrial and cosmic materials tends fairly consistent toward depletion as compared with ocean water, reaching in the gas phase of molecular hydrogen values characteristic for the Sun's atmosphere (see Table 20.2). The only exceptions are carbonaceous chondrites and lunar rocks, where the water and molecular hydrogen present are sometimes enriched in deuterium. This may be due to secondary temperature phenomena which occurred during their formation and the subsequent appearance of compounds of more complex chemical composition.
3. The hypothesis that the present oceans were formed by the gradual emergence of juvenile waters from the depths of the Earth at the Earth's surface is not confirmed by evidence.
4. Underground waters are enriched in ^{18}O relative to rocks and minerals.
5. On the whole, the hydrosphere is enriched in deuterium and depleted in oxygen-18 relative to rocks and minerals.
6. Meteorites and lunar rock samples differ in hydrogen isotope composition from terrestrial materials. At the same time, the ranges of the oxygen-18 isotope variation in lunar rock samples are close to those in terrestrial rocks of similar mineralogical composition. All these facts point to a common material from which the Earth, Moon, and meteorites were formed, but this formation occurred at different temperature conditions and under different sequences of processes

accompanying the formation over time and space. It has been noted that water is contained in extremely small amounts in lunar rocks and some researchers consider its presence to be doubtful.

How do the facts established above tally with current ideas about the origin of the Earth's hydrosphere (and especially the oceans) following from the 'cold' and 'hot' creation hypotheses concerning the Earth's origin?

According to the accretion hypothesis, the hydrosphere was formed through the emergence at the Earth's surface of its light components in the course of global and local melting of the initial cold homogeneous body. If that were true, the ocean should have a relative ^{18}O content up to +7‰ corresponding to the oxygen isotope ratio in basalts. The deuterium content in water of the initial oceans should have been about −80‰, estimated by the igneous rocks (Craig 1963), about −60‰, estimated by the volcanic steam and gas (Arnason and Sigurgeirsson 1967; Friedman 1967), and -48 ± 20‰, using ancient rocks of the upper mantle (Sheppard and Epstein 1970).

As one can see, the present ocean is enriched in deuterium and depleted in oxygen-18 compared with the probable initial ocean.

Among the known processes which can result in the enrichment in deuterium of the initial ocean, the process of the dissipation of free hydrogen from the Earth's gravitational field during which protium escapes preferentially is usually invoked. In fact, within a sufficient degree of accuracy one can consider that the ratio of the escape velocities of $^1H_2$ and $^1H^2H$ molecules is proportional to $(M_{^1H^2H}/M_{^1H^1H})^{1/2} = 1.22$. This value can be taken as the fractionation factor of hydrogen isotopes analogous to the fractionation factor in the diffusion processes. In this case one can easily estimate that the enrichment of the ocean in the deuterium by 20‰ due to preferential losses of protium can be attained only by a decrease of the ocean mass by 11%. If free hydrogen is formed in the upper atmospheric layers during photodissociation of water molecules, then a reduction of the ocean by 11% (at its mass of 1.4×10^{21} kg) should result in the release into the upper atmosphere of 1.25×10^{20} kg of free oxygen. When the deuterium concentration in the juvenile water is lower than −20‰ (i.e., −48‰ according to Epstein and Taylor and −80‰ according to Craig), the amount of free oxygen in the atmosphere must increase proportionally. Taking into account the portion of oxygen which has been expended on oxidation of the igneous rocks ($\sim 2.8 \times 10^{18}$ kg; Goldschmidt 1954) and was buried together with ancient organic material ($\sim 10^{16}$ kg) then one must conclude that even in this case a large amount of free oxygen could have accumulated in the atmosphere, exceeding its modern content in the atmosphere by more than two orders of magnitude. This is not what one finds in practice, and the free oxygen has been commonly assumed to be formed due to photosynthetic processes in the biosphere.

Easy theoretical estimations show that the possible losses of protium during the hydrogen escape from the Earth's gravitational field are rather low. Therefore, there is no reason to consider that the required enrichment of the ocean in the deuterium by 50–80‰ due to the dissipation of protium has taken place. Earlier we considered the possible ranges of enrichment of the ocean in deuterium due to sedimentation of the decomposed igneous rocks. This effect can be considered as insignificant, resulting in changes of about 0.3‰. Since there are no other processes known which lead to the enrichment of the hydrosphere in deuterium, such a phenomenon is doubtful.

Let us consider now ways of depletion of the initial oceans in oxygen-18. The most effective depletion process is the sedimentation of rocks. Despite the possible errors in the estimation of initial amounts of various types of sedimentary rocks required for calculations (these values differ markedly in the works of various authors), the final effect of sedimentation of rocks estimated by Savin and Epstein (1970b), yielding 3‰ at least, has not been reduced. In this case, when the oxygen of carbon dioxide is in isotope equilibrium with rocks during its degassing, the ocean could have become only partly depleted in oxygen-18.

Moreover, we must assume that the enrichment of acid rocks in heavy oxygen has taken place due to the water of the hydrosphere since one can hardly suggest any other source of available heavy oxygen. Assuming that the average $\delta^{18}O$ value in granites is +10‰, the ocean's juvenile water should have it to +20‰. In the case of the degassing origin of the hydrosphere this last $\delta^{18}O$ value seems very improbable.

There have been attempts to resolve the above-mentioned concentrations concerning the oxygen isotopic composition with the help of the appropriate model of growth and change of the ocean isotopic composition with time, for example, it has been suggested (Chase and Perry 1972; Ahmad and Perry 1980) to use the model of recirculation through the mantle of the ocean water depleted in heavy oxygen, where it should become enriched in ^{18}O. However, such a model has difficulty in explaining the physical reasons behind such a process occurring in nature. This model also contradicts facts already mentioned, which are evidence of the very early stage of ocean formation and the constancy of its chemical and isotopic composition over long geologic time. On the other hand, the model has been suggested to account for the low δ^2H values in mantle and igneous rocks (Taylor 1974) for the modern ideas of plate tectonics and the spreading of the ocean floor. It is known that the hydroxyl-bearing minerals of the authigenic sea sediments have δ^2H values close (~ -60‰) to those of magmatic rocks. Assuming that the sea sediments are continuously contributing to the mantle through subduction zones, the process can be accepted as the mechanism of injection of light hydrogen into mantle rocks. However, in this case the same process should be accompanied by the removal of heavy oxygen from the ocean and the enrichment in heavy oxygen of the igneous rocks contained in the same sea sediments. New difficulties arise at this point.

Thus, the observed facts of the isotopic composition of water in the oceans and in underground rocks contradict the 'cold' origin hypothesis.

These same facts lead to the conclusion that juvenile water (i.e., water formed in the deep layers of the Earth and in its upper and lower mantle, which emerges at the surface) does not, and never did, exist on our planet and that the hydrosphere resulted from an atmophile process. Such a process follows from the 'hot' model of origin considered by, for example, Goldschmidt (1954).

If one assumes that during the first stage of the Earth's geochemical differentiation water remained in the gas phase and did not escape, and that there was no free oxygen, then the isotopic composition of the hydrogen and oxygen in the water falling on to the Earth's surface might have well corresponded to the concentrations now found in the ocean water. However, the 'hot' hypothesis of origin of the Earth has been subjected, as already indicated, to justified criticism. Therefore, one should look for a mechanism of the Earth's formation that can satisfy the conditions of formation of the isotopic composition of the hydrosphere, following the 'hot' hypothesis. Such a mechanism should be in agreement with the geochemical conditions which were formulated, for example, by Urey (1957) and are best fitted to the 'cold' hypothesis of the Earth's origin.

20.3 Evidence from Carbon and Sulfur Isotopes

Carbon and sulfur form a number of mobile compounds in nature (CO, CO_2, CO_3^{2-}, SO_2, SO_4^{2-}, SO_3, H_2S, etc.), which take part in the intensive cycling of the matter in the geosphere in close relationship with water of the hydrosphere. The study of the principles of the distribution of carbon and sulfur isotopes provides important data concerning the origin and sequence of evolution of the hydrosphere and the Earth itself.

Carbon Isotopes Solar carbon, according to the observational data available, is enriched in the light isotope. Some authors (Righini 1963) have considered the degree of enrichment to be rather great ($^{13}C/^{12}C \approx 10^{-5}$), others (Burnett et al. 1965) consider it to be more moderate ($^{13}C/^{12}C \approx 10^{-2}$).

Meteorites remained up to recent time the most easily available medium for studying carbon cosmic material. Comparing the $\delta^{13}C$ values obtained from iron, iron-stone and stone meteorites, it has been found that they range from −5 to −30‰ with the average value of −22‰ (relative to the PDB-1 standard). Carbon is commonly contained in insignificant amounts (hundredth %) and is present in a dispersed state. In the Canion Diablo iron meteorite, carbon has been found in the form of a graphite inclusion in troilite with $\delta^{13}C = -6.3‰$ and carbon of the carbide iron with $\delta^{13}C = -17.9‰$ (Craig 1953). The graphite of Yardimlinsky's iron meteorite has $\delta^{13}C = -5‰$ (Vinogradov et al. 1967). Generalizing the carbon isotope data in iron meteorites (Deines and Wickman 1975), it has been found that the observed depletion of taenite in heavy isotopes ($\delta^{13}C = -18.5‰$) compared with graphite ($\delta^{13}C = -5.5‰$) is common.

In view of the fact that the chondritic model of the Earth's origin is widespread, knowledge of the isotopic composition of carbon and chemical forms of its occurrence in the most widespread type of meteorites, chondrites, is of particular interest. Carbon has been found in chondrites in the form of graphite, carbonate, carbide, diamond, and organic compounds. Ordinary chondrites have a light isotopic composition of carbon with an average value of $\delta^{13}C = -24‰$. The carbon content itself amounts to only 0.1–0.01% (Hayes 1967).

Carbonaceous chondrites have been studied by Boato (1954), who reported that they can be divided into two types according to the gross content of carbon and water in them. In the first type the content of carbon ranges from 1.6 to 3.3% and $\delta^{13}C$

from −3 to −12‰, in the second type $\delta^{13}C$ varies from −13 to −18‰ with a carbon content ranging from 0.3 to 0.85%. In these meteorites the average $\delta^{13}C = -7$‰.

Thus, on the whole the following general principle is observed in meteorites: the lighter isotopic composition of carbon is observed in those meteorites where its gross content is lower. Boato has suggested that an initial undifferentiated substance of the solar system is represented by the carbonaceous chondrites of the first type. In his opinion, the observed correlation between the isotopic composition and the gross amount of carbon is likely an indicator of the preferential escape of ^{13}C in the process of depletion of the initial material in volatile elements.

Other viewpoints concerning the origin of the heavy isotopic composition of carbonaceous chondrites were proposed, for example, Galimov (1968) has reported that the enrichment of the carbon of carbonaceous chondrites in heavy isotopes can be accounted for by the excessive yield of ^{13}C in nucleosynthesis by reaction $^{16}O(n, {}^{4}He)^{13}C$ in the course of the formation of carbonaceous chondrites in a medium with a high oxygen content. Therefore, the isotopic composition of carbon in the meteorites and ^{13}C cosmogenic variations in the primeval carbon, were conditioned by the nucleosynthesis of the carbon isotopes before formation of the solar system and nuclear reactions proceeding on high-energy particles in the preplanet material during the early stage of evolution of the solar system. Later, only redistribution occurred in various geospheres of the isotope concentrations inherited from the initial course.

According to Epstein and Taylor (1972), the range of $\delta^{13}C$ variations in the lunar rock samples is wide (from −20 to −30‰). It has been found that, as well as in the case of carbonaceous chondrites, the degree by which the samples are enriched in heavy carbon is dependent on the content of carbon itself. In the lunar basalts the $\delta^{13}C$ values range between −30 and −18.6‰ and in the fine-grained soils and breccia the $\delta^{13}C$ values are greater than −3.6‰. It is of interest to note that in lunar samples no carbon with $\delta^{13}C$ within the range from −3.6 to −18.6‰ has been found. At the same time, as pointed out earlier, the average value of the carbon content in chondrites is within this range. In order to explain the observed enrichment of soil and breccia with heavy carbon a number of hypotheses were suggested concerning the effect of solar wind and meteorites upon the surface of the Moon. It is assumed that the relative content of heavy carbon in the solar wind ranges from −10 to +10‰ (Galimov 1968).

Comparing experimental data on the isotope ratios of carbon in meteorites and in the Moon it has been found that their isotope ratios are markedly different. This difference, as a rule, depends upon the gross content of carbon and its chemical form in the subject. Therefore, it follows that the determination of genetic relationships while studying the questions of the origin of cosmic bodies cannot be carried out without account being made of the isotope data, for example, since the isotopic composition of carbon in lunar rock differs from that in chondrites, the chondritic model of the Moon cannot be true.

The principles governing the carbon isotope composition in the main reservoirs of the Earth and the main mechanisms of their fractionation in nature were considered in Chap. 11 (see Table 11.1).

Thus, despite some differences in estimation of the average $\delta^{13}C$ value, obtained by different authors for the Earth's crust, all have agreed that the Earth's crust is enriched in the heavy carbon isotope compared with meteorite and solar material.

As pointed out earlier, the isotope fractionation is reduced or even absent at high-temperature processes. Assuming the chondrite model of the Earth's origin, and that carbon has appeared in its upper sphere along with other volatile components, due to degassing from its interior during geologic time, a light isotopic composition of carbon in terrestrial crust should be expected. However, in fact the opposite phenomenon is observed.

In the framework of the degassing hypothesis there is an idea that the type of material of carbonaceous chondrites, whose isotope composition is similar to that of the Earth's crust, has played an important role in the formation of the isotope composition of the Earth's crust (Galimov 1973). However, it is difficult to find the sources through which the emergence of this substance to the surface occurred.

In discussing the principles of hydrogen and oxygen isotope distribution, we have already pointed out that the existing opinion of the emergence of juvenile material through hydrotherms appears to be incorrect. Although it is impossible to pinpoint the sources of the emergence of mantle carbon at present, one can assume that they existed in the past.

The carbon isotope composition of sedimentary carbonates is known to be a sensitive indicator of the general geologic paleo-situation. If, during the Earth's history, there were periods in which a considerable emergence of the mantle carbon dioxide occurred, then it should have been reflected in the isotope composition of the sedimentary carbonates. It was concluded earlier, on the basis of data on the heavy carbon content in limestones that their isotopic composition bears no relation to geologic age (Craig 1953). Galimov (1968), generalizing the available data on the isotopic composition of sea and fresh water limestones, has come to the conclusion that carbonates were extremely enriched in the light isotope of carbon during two periods, the Carboniferous and Tertiary. He pointed out that the synchronous limestone isotope changes with the amount of organic substance on the globe during these times. Galimov suggested that the enrichment of limestones in the light isotope can be accounted for by the intensive contribution of carbon dioxide in the Hercynian and Alpine orogenesis. The enrichment of limestones in the heavy carbon isotope in subsequent epochs occurred due to the withdrawal of the light isotope during the plant's intensive photosynthetic activity and the burial in sediments of considerable amounts of the light carbon. However, this interpretation is disputable. As Schell et al. (1967) have reported, the increase in temperature of the world oceans results in a shift of equilibrium in the carbonate-calcium system in such a way that this, in turn, leads to a more active process of photosynthesis in plants. As a result, the sea and fresh water limestones become enriched in the light carbon isotopes. The decrease of temperature of the world oceans in subsequent time should result in opposite effects.

The problem of carbon origin in the upper shell of the Earth requires further investigation, but on the basis of the data available at present there are reasons to assume that neither now, nor in the past, has there been any considerable emergence of the mantle carbon to the surface through degassing.

In this connection it is of interest to note that some authors (Epstein 1969) have attempted to consider various alternatives of the possible existence of the primary form of main reservoirs of carbon on the Earth in the past. One can assume, for example, that at a certain stage in the Earth's early history, the main reservoirs of carbon contained carbonate rocks with $\delta^{13}C = -7‰$. If the carbon isotope fractionation effects between the main reservoirs have remained unchanged up to the present time, then the atmospheric carbon dioxide exhibits $\delta^{13}C = -14‰$ and the organic material $\delta^{13}C = -34‰$. On the other hand, if the major portion of the carbon was represented by the organic substance with $\delta^{13}C = -7‰$, then the atmospheric carbon dioxide had $\delta^{13}C = +13‰$ and carbon of the carbonate rocks must had $\delta^{13}C = +20‰$.

Sulfur Isotopes The available data on the sulfur isotope composition of iron, ironstone, and stone meteorites indicate that the $^{34}S/^{32}S$ ratio, characteristic of cosmic bodies, is approximately constant (Ault and Kulp 1959). Different $\delta^{34}S$ values were discovered in the Orgueil carbonaceous chondrite for individual chemical components of sulfur (for the elementary, troilite and sulfate forms). For the meteorite as a whole, the $\delta^{34}S$ value relative to the troilite standard of the iron meteorite from the Canyon Diablo appear to be zero (Monster et al. 1965).

Analyzing the sulfur isotope composition of lunar rock samples, it has been found (Kaplan and Smith 1970) that $\delta^{34}S$ in the dust ranges from +4.4 to +8.2‰. In breccias this value has been found to vary from +3.3 to +3.6‰ and in fine-grained soil from +1.2 to +1.3‰. The relatively high $\delta^{34}S$ content in the dust has been accounted for by the vaporization of the sulfur light isotope by bombardment of the Moon's surface with micrometeorites and protons of the solar wind.

As it was mentioned in Chap. 11, the main geochemical cycle of sulfur on the Earth is related to existence of the oceans. Despite the continuous process of sulfur contribution to the ocean together with river runoff, its concentration and isotope composition remains constant in the ocean due to sulfate reduction. During the cycle the relatively light continental sulfate with $\delta^{34}S = +5‰$ becomes enriched up to the oceanic average value of $\delta^{34}S = +20‰$ and the excess of the isotope is bonded in diagenetic sulfides.

The isotope composition of sulfur in oceanic sulfate is a sensitive indicator of its dynamic equilibrium in the cycle and the constancy of ocean salt composition on the whole. This equilibrium is determined by the rate of biogenic sulfate reduction and therefore by the total amount of biomass on the Earth. On the other hand, the content of sulfur in the oceans is closely related to its total salinity.

Wide studies concerning the sulfur isotope composition of evaporates of different ages have been undertaken, aimed at elucidation of the biochemical history of the oceans.

While studying sulfates in the Phanerozoic evaporates, it has been shown (Ault and Kulp 1959; Vinogradov 1980) that the sulfur isotope composition has no time-ordered changes. There occur only slight variations of the $\delta^{34}S$ values from the average value corresponding to the sulfur isotope composition of the present ocean (+20‰). The only exceptions are the Permian evaporates, which always exhibit depletion down to the value of $\delta^{34}S = +10‰$. Vinogradov (1980) has explained this phenomenon in terms of paleogeographical peculiarities in the accumulation of Permian evaporates. These peculiarities consisted, in his opinion, in an increase in the role of the inland sulfates component in the recharge of salt basins in the transition from the sulfate to the haloid accumulation of salts. On the whole, the isotopic composition of sulfur in sulfates in ancient evaporates during the whole Phanerozoic period has remained constant, and close to the modern composition of the oceans. This circumstance is evidence in favor of the constancy of the ocean's salinity, amount of the biomass, and concentration of oxygen on the Earth during the whole Phanerozoic period.

While studying the Precambrian metamorphic rocks from East Siberia, the Pamirs, and South Africa, which exhibit salt-forming features by a number of minerals (skapolite, apatite, lasurite, carbonatites), it has been found (Vinogradov 1975; Vinogradov et al. 1969) that these rocks contain relatively high concentrations of sulfide, sulfate, and native sulfur. The isotopic composition of sulfur in sulfate of the metasomatic minerals is characterized by high enrichment in the heavy isotope (from +13 to +45‰) and the identical content of $\delta^{34}S$ relative to sulfates and sulfides of the sulfur-bearing carbonate rocks. On this basis, the authors have come to a reasonable conclusion regarding the sedimentary origin of the initial sulfates. These sulfates were accumulated at the steady dynamic cycle of sulfur and served as the initial source of sulfides. In the metasamotic minerals of the Archean carbonabe rocks in Aldan (East Siberia) and rocks of the Swaziland system (South Africa) up to 3.5×10^9 year old, widely developed regional scale sulfates of sedimentary origin were found, with $\delta^{34}S = +6‰$, which are not typical for sedimentary sulfates.

On the basis of experimental data analysis, it has been found that the sources of sulfur, participating in the metasomatism and metamorphism processes, were the metamorphic sedimentary thicknesses containing sulfur in the form of sedimentary sulfide. From the existence in the section of sedimentary rocks of layers of dolomites enriched in sulfate-sulfur, it has been concluded that the process of the sedimentation of sulfates has taken place in the normal facial conditions in the course of the salinization of the sea basin. However, the sulfur cycle in the basin has not attained dynamic equilibrium and the sulfate-sulfur in a basin has not yet been enriched in the heavy isotope, but in individual sites of a basin such enrichment of the sulfate-sulfur of metasomatic minerals has occurred and the $\delta^{34}S = +20‰$ has been found there. This has led the above authors to conclude that the establishment of the dynamic equilibrium of the sulfur cycle, at a level which approximates the modern one, coincides with the age of the studied rocks, i.e., occurred about $(3\text{–}3.5) \times 10^9$ years ago. On this basis, Vinogradov (1980) has concluded that the emergence of the main mass of sulfur from the interior to the upper shell of the Earth, and also the formation of the oxygen composition of the atmosphere and the salinity of the ocean, which approximates the modern levels, ended not later than 3.5×10^9 years ago.

Let us consider now the comparative analysis of the sulfur isotope composition of the upper shell of the Earth. The balance estimations show (Grinenko and Grinenko 1974) that the major sulfur reservoir in the Earth's crust is the platform sedimentary thickness of the continents, characterized by an average value of $\delta^{34}S$ of about +4‰. In the geosynclinal areas containing up to 30% sulfur the value of $\delta^{34}S$ is close to zero. The ocean water contains 15% sulfur in the form of dissolved sulfates, enriched in the heavy isotope up to +20‰. The sulfur of the ocean sediments, amounting to about 10% of that in the sedimentary thickness, exhibits $\delta^{34}S = +7.7$‰. The average $\delta^{34}S$ value in the abyssal clays, limestone, and siliceous sediments is equal to +17‰. The ultramafic oceanic and inland rocks are characterized by an average value of $\delta^{34}S = +1.2$‰. For basic rocks this value is +2.7‰ and for acid rocks it is equal to +5.1‰. Therefore, the outer sphere of the Earth together with the ocean is enriched in ^{34}S by 5.5‰ and the terrestrial crust on the whole by 3‰ relative to meteorite material. It is of note that as one moves from the basaltic sphere toward the granite sphere and further to the sedimentary continental layers and the oceans, the amount of sulfur increases with a simultaneous enrichment in the heavy isotope. Some researches (Grinenko and Grinenko 1974) have assumed that such global processes as degassing of the mantle, crystalline differentiation, and metamorphism of rocks have the same effect. This effect manifests itself as increases in the amount of sulfur in sedimentary rocks from the Archean to the Proterozoic, Phanerozoic, and Cenozoic with simultaneous increases of sulfate-sulfur enrichment in ^{34}S.

It should be pointed out that such an approach to the observed global principle of increases in the amount of sulfur with enrichment in ^{34}S from the ultramafic rocks to the acid ones and to the oceans contradicts the evidence given above in favor of the constancy of the salt and isotopic composition of the oceans during the last $(3–3.5) \times 10^9$ years. On the other hand, the above-mentioned principle including the relationship between the gross content of the element with its isotopic composition is typical for the other elements stated: oxygen, carbon, and hydrogen. For all these elements we have observed enrichment in heavy isotopes while moving from ultramafic rocks to acid rocks and to the ocean. The only exception is the oxygen isotopic composition of the ocean. In order to explain this general principle, while considering the isotopic composition of different elements, different researchers employ assumptions which are often contradictory when compared. In addition to the above-mentioned principles of the distribution of the isotopes of the volatile elements H, O, C, S already considered, one should bear in mind one more cosmochemical principle. This principle states that the Earth's crust is, in general, enriched in the heavy isotopes of these elements relative to meteorite substance. The available experimental data on boron isotope distribution, despite being limited, shows that for isotopes of this element the above-mentioned principle also holds.

It should be pointed out that the data on the isotope distribution of noble gases (He, Ne, Ar, Kr, Xe) in the upper sphere of the Earth and in meteorite substances indicate the applicability of the above principle to this group of elements also.

The observed regularities of the distribution of volatile elements in the upper terrestrial sphere, meteorite, and lunar material, and also the results of cosmochemical and cosmophysical studies obtained during recent years suggest the importance of the processes of chemical differentiation of substance which took place at the preplanetary stage of evolution of the planets. In this case, the idea of chemical and isotope homogeneity of the forming planet material is doubtful. In this connection, let us consider in brief the cosmochemical results which have been obtained in recent years by studying meteorites, the Moon, and other planets. They have a direct bearing on the problem of the origin of the Earth and its shells including the hydrosphere and the atmosphere.

20.4 Chemical Differentiation of Proto-Planetary Substance

The fact that the average density of the material of planets decreases with increasing distance from the Sun is of importance and was noted in the early hypotheses concerning the origin of planets in the solar system. On the basis of experimental and theoretical data, it has been found that there is a sharp difference in the mass density at the transition from the planets of the terrestrial group to the major planets of the Jupiter group. The major planets were formed of low-temperature volatile elements and compounds, mainly of hydrogen, helium, methane, and ammonium. The planets of the terrestrial group, previously thought to be formed mainly of silicates and iron, exhibit decreases in iron with increasing distance from the Sun.

It has also been found that the comets moving around the Sun in eccentric orbits are constituted of light elements and compounds.

By studying different types of meteorites it has been found that they differ sharply in their chemical composition and, for some elements, difference in isotope ratios is observed.

All these facts indicate that during the formation of the planets of the solar system, processes occurred which resulted in the chemical differentiation of their material. The elucidation of the mechanism of these processes, together with that mechanism which has caused the observed principles of their motion, is one of the main problems which must be solved in the development of cosmogonic theories. The solution of this problem is closely related to the problem of the origin of the Earth and of the formation of planet shells.

Urey was one of the first researchers to point out the necessity of taking into account the observed geochemical and cosmochemical facts in developing geochemical and cosmochemical theories concerning the origin of the solar system. He formulated the boundary conditions (Urey 1957), which have formed the basis of the chemical differentiation of initial material in the course of formation of meteorites and planets of the terrestrial group and had a great effect on the course of the development of studies in cosmogony. In this respect, Urey's conclusion that the initial protoplanetary material underwent a high-temperature stage during its evolution when its temperature could have attained 2,000°C, i.e., the temperature when the initial substances were in gaseous state, was of greatest importance.

Recent studies concerning the composition and structure of meteorites have confirmed this conclusion and shown that in the course of the evolution of the protoplanetary material, processes leading to its temperature condensation took place, i.e., processes leading to the transition of the substance from gaseous to liquid and solid phases. These cosmochemical conclusions have had an effect upon theoretical studies concerning development of the general cosmogonical hypotheses, in which attempts were made to determine, from the physical viewpoint, the probability of the occurrence of a high-temperature stage in the initial cloud (Cameron 1973; Cameron and Pine 1973).

Table 20.3 Calculated temperatures of condensation of pure elements. (From Larimer 1967)

$p_\tau = 10^5$ Pa		$p_\tau = 6.6 \times 10^2$ Pa		$p_\tau = 10^5$ Pa		$p_\tau = 6.6 \times 10^2$ Pa	
Element	T(°K)	Element	T(°K)	Element	T(°K)	Element	T(°K)
Fe	1,790	Fe	1,620	Pb	655	Pb	570
V	1,760	V	1,500	Bi	620	Bi	530
Ni	1,690	Ni	1,440	Sb_2	590	Sb_2	515
Cu	1,260	Sc	1,090	Tl	540	Tl	475
Sc	1,250	Cu	1,045	Te	517	Te	460
Mn	1,195	Ge	980	Zn	503	Zn	430
Ge	1,150	Au	970	S_2	489	S_2	400
Au	1,100	Mn	920	Se	416	Se	375
Ga	1,015	Ga	880	Cd	356	Cd	318
Sn	940	Sn	806	Hg	196	Hg	181
Ag	880	Ag	780	I_2	185	I_2	169
In	765	In	670				

Let us consider now under which conditions and in what sequence the process of condensation of chemical elements and their compounds should have proceeded during evolution of the gaseous cloud, if this process took place at all.

The question of the chemical condensation of elements and their compounds from gaseous material was studied earlier by Urey (1959), Wood (1963), Ringwood (1966, 1979), Lord (1965), Anders (1972) and others. They were mainly trying to find an explanation for the observed picture of the depletion of the planets and different types of meteorites with volatile elements. On the basis of their studies it was found that, in different types of meteorites, certain principles are observed regarding the content of the high-temperature nonvolatile and low-temperature volatile fraction of the condensed material. At the same time, while explaining the chemical composition of meteorites on the basis of the obtained calculations, they met difficulties concerning the unsteadiness of the degree of depletion of different types of meteorites by different types of volatile elements.

For estimating the condensation temperatures of elements, Larimer (1967) has employed the common equations of thermodynamics of ideal gases and the data on the elements' cosmic abundances which were obtained earlier by Cameron. For the starting point of gas condensation, Latimer has assumed that when attaining corresponding thermodynamic conditions in the gaseous cloud, the gas pressure of an element of natural abundance attains the corresponding partial pressure. Estimations were carried out at equilibrium conditions of condensation on the assumption that the process proceeded in hydrogen gas and the condensing elements were in atomic or simple molecular states.

Table 20.3 shows the theoretical condensation temperatures of elements obtained by Larimer for two values of the total gas pressure p_τ in the cloud. It follows from Table 20.3 that the temperature of condensation of the elements has little effect upon the sequence of their condensation at two different gas pressures.

The calculation of the condensation temperatures of various complex chemical compounds was carried out by Larimer on the basis of the same principle of attainment, by a given compound, of a temperature corresponding to its partial pressure obtained from the gas thermodynamics equation. However, in this case it has appeared to be necessary to account for the possible distribution of each element within the different compounds. This has been done by estimating the equilibrium constants using the thermodynamic constants of the reaction products.

While considering the problem concerning condensation of different elements, compounds, and alloys, the gas pressure is of principal importance, being a function of density of the medium abundances of elements and their amounts. Larimer has considered the condensation process of the compounds and alloys to be dependent on the reaction's kinetics and the diffusion rate. In view of this, the condensation of pure elements and compounds could have occurred during quick gas cooling. During slow cooling the diffusion equilibrium in the gas-condensate system will be maintained, resulting in the formation of alloys and solid solutions.

The first estimates of the temperatures and condensation sequence of the elements and compounds, which were carried out by Larimer (1967), were later developed by Larimer and Anders (1967, 1970), and Grossman and Larimer (1974). These reestimated data, presented in Fig. 20.2, indicate that, from the cooling gas cloud at the pressure of 10 Pa, such refractory elements as Os, Re and Zr should condense first. Their formation temperature was to be higher than that of Al_2O_3, being a widespread compound, whose condensation temperature is 1,679 K. For Ti and Ca, forming $CaTiO_3$ and $Ca_2Al_2SiO_7$compounds, the condensation temperature is about 1,500 K. The rare elements U, Pu, Th, Ta and Nb form solid solutions in $CaTiO_3$. $CaMgSi_2O_6$ is formed at $T = 1{,}387$ K after about 10% of Mg and Si has been condensed. The metal iron precipitates at $T = 1{,}375$ K. The latter compound, on reacting with vapor, later forms $MgSiO_3$which absorbs the Si remaining in the gaseous phase.

The elements condensing at higher temperatures than 1,250 K form the group of the refractory compounds. At lower temperatures Cu, Ge, and Ga precipitate in the form of solid solutions in metals. Then $CaAl_2Si_2O_8$ is formed at $T = 1{,}299$ K and Na, K, and Rb, condense in the form of solid solutions. The alkali metals precipitate completely at $T = 1{,}000$ K, Ag at $T = 750$ K. The metal iron oxidizes at $T = 750$ K, troilite at $T = 700$ K being formed by the reaction between metallic iron and gaseous H_2S. Pb, Bi, In, and Tl, which are highly depleted in chondrites, condense within the temperature range 400–600 K, magnetites at $T = 405$ K, hydrated silicates at $T = 350$ K. Ar, CH_4, NH_3, H_2O, and hydrate methane condense at 200 K. The investigation of the condensational conditions of the substance over the range of pressure from 10 to 10^{-3} Pa for the high-temperature compounds, rich in Ca, Al and Ti, indicates that the sequence of their formation is preserved.

On the basis of the obtained data on the temperatures and sequences of condensation of elements and compounds, attempts were made to explain chemical fractionation in meteorites, and to reconstruct some cosmochemical events which took place in the proto-planetary cloud.

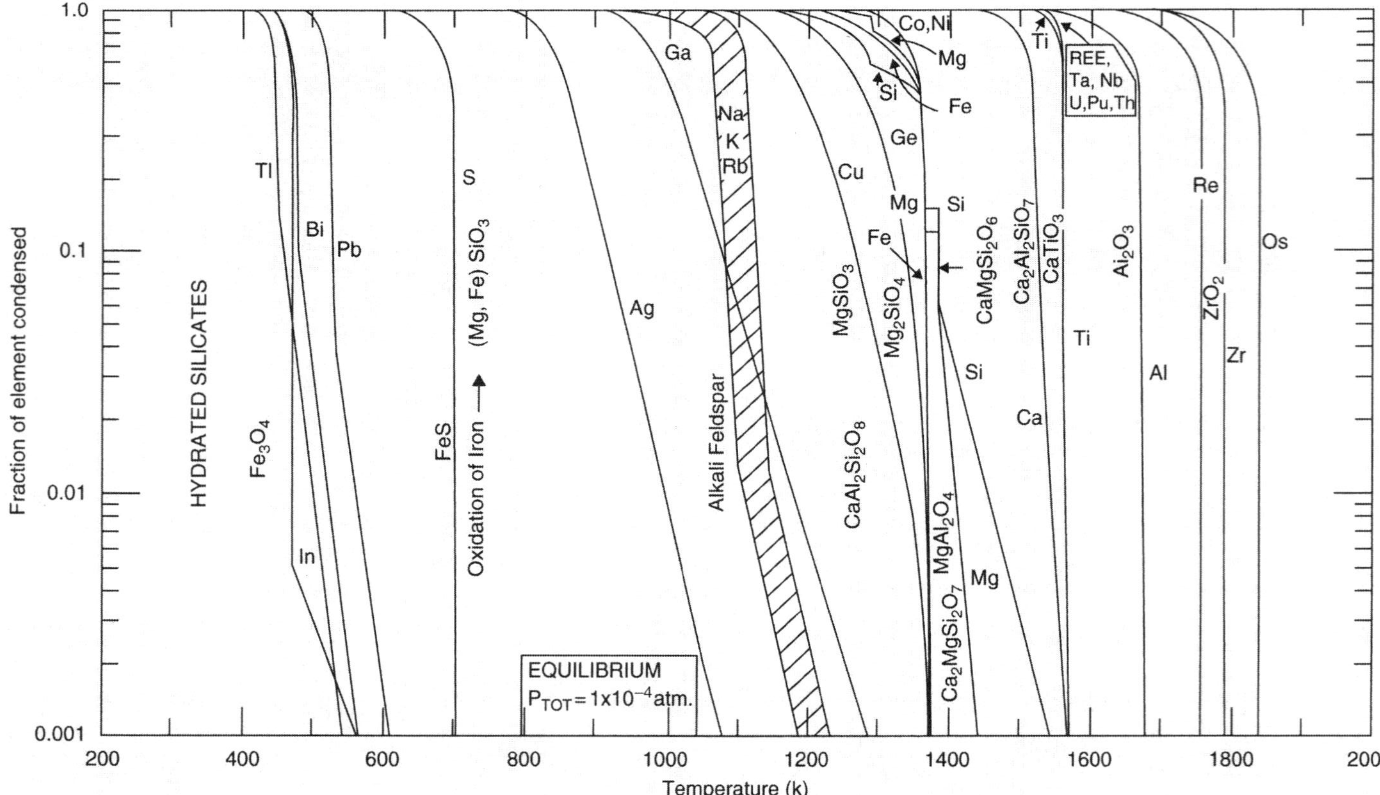

Fig. 20.2 Calculated temperatures of condensation of elements and compounds at 10 Pa. (After Grossman and Larimer 1974)

Larimer has come to a conclusion that the observed picture of fractionation of chemical elements in chondrites should be related to their fractionation in the solar nebula during the condensation of the elements and compounds. The subsequent process of heating of the meteorite bodies, being possible in his opinion, has not resulted in further fractionation except for the most volatile elements, noble gases, and mercury.

The studies of Larimer and Anders (1967), concerning the observed fractionation of the 31 most volatile elements in the carbonaceous and ordinary chondrites, led them to conclude the following. The abundance of these elements in the carbonaceous chondrites C_1, C_2, and C_3 type and in the enstatite chondrites type I gradually decrease in order of their sequences. Taking unity for the abundance of the elements in the C_1 type chondrites, their abundances in chondrites of other types are in the ratios of 1: 0.5: 0.3: 0.7. In the ordinary chondrites and enstatite type II chondrites, nine elements (Au, Cu, F, Ca, Ga, Ge, S, Sb, Se, Sn) have approximately constant depletion coefficients ~0.25 and ~0.5. The other 18 elements (Ag, Bi, Br, C, Cd, Cl, Cs, H, Hg, I, In, Kr, N, Pb, Tl, Te, Xe, Zn) have a coefficient of 0.002.

On the basis of the experimental data given above, the authors assumed that chondrites are composed of a mixture of two types of substances: (1) A low-temperature fraction representing the basis of condensed material, which holds the volatile elements and (2) A high-temperature fraction represented by individual chondrules and metallic grains, which have lost volatile elements. Further, these facts indicate the possible formation of the above-mentioned fractions directly from high-temperature nebula during cooling. They could not have been formed in any parent meteorite bodies. Conclusions were also made concerning the probable temperatures of formation of the meteorites: $T \leq 400$ K for the carbonaceous chondrites, 400–800 K for enstatite meteorites (type I), 530–680 K for ordinary and enstatite (type II) chondrites, and $T \leq 530$ K for ordinary nonequilibrium chondrites. These temperatures are considerably higher than those typical of the asteroid belt (170 K). On the basis of this hypothesis, the authors have assumed that the enstatite chondrites originate from the inner part of the asteroid belt, the ordinary chondrites from its central part, and carbonaceous chondrites from the outer belt.

Larimer, Anders, Grossman, and other researchers appear to have developed a theory which successfully explains many aspects of the chemical differentiation of substances in meteorites. However, this theory of the equilibrium condensation of chemical elements and compounds of the proto-planetary material has met with a number of difficulties, for example, in the carbonaceous chondrite Allende (fell in 1969, 2,000 kg), the high-temperature inclusions of minerals were surrounded by incompatible low-temperature minerals. The relative ^{18}O content in the three inclusions of halenite and spinel in this meteorite ranges from −9.7 to −11.5‰. This corresponds to the equilibrium temperature of their formation (approximately 800 K) for those portions of the cloud where the inclusions were formed. This temperature appears to be too low compared with that following from studying the texture and mineralogy of the inclusion. The type I enstatite chondrites must have been formed at very high pressures so that Tl remained at $T = 400$ K. The presence of graphite requires high formation temperatures and the presence of water requires low temperatures. The general conditions of the formation of carbonaceous chondrites were found to be inadequate on the whole. With regard to the formation temperatures of these meteorites they should not have been hydrated, whereas the majority of them contain a great amount of water. Besides, Clayton et al. (1977) have discovered in all the C_2, C_3 and C_4 meteorites they studied as a whole, and in the spinel, olivine, and feldspar minerals, that there is excess of ^{16}O, which has been prescribed to the presolar component, for which the mechanism of transfer to the meteorites and larger bodies including the planets remains uncertain. The fractionation picture in the iron-silicates system also remains ambiguous. The volatile elements in which all the stone meteorites are depleted, and the existence of organic materials in the form of the high-molecular hydrocarbon compounds in meteorites, create a serious problem which is unsolved as yet. The observed amounts of the noble gases should also be explained. It has been found that the content of Ar, Kr and Xe in individual types of chondrites differs by three orders of magnitude. Helium and neon were found only in the carbonaceous chondrites and in very small amounts in the ordinary chondrites, which is in disagreement with their content in the solar wind. Some promising ideas of application of the absorption effects and equilibrium dissolution of noble gases in the meteorite material during its formation have appeared recently for explanation of the observed data.

Grossman and Larimer (1974) have pointed out that in order to explain, from the viewpoint of the equilibrium of a substance and its accretion, a number of chemical, mineralogical, and textural peculiarities in stone meteorites, at least four processes are required, resulting in fractionation of the condensed material:

1. The most refractory material, rich in Ca, Al and Ti, must have transferred into the internal zone of the proto-planetary cloud. Enstatite and ordinary chondrites were preferentially depleted in this material and outer zones of the Earth and Moon were enriched in it. These formations also included the metallic portion, rich in refractory compounds.
2. Most of the metals must have been extracted from the silicate grains at $T = 1{,}000 - 700$ K. This process could only have occurred in the initial material of chondrites, which explains the observed differences in the densities of the terrestrial planets.
3. Before the accretion started, some portion of the condensed material must have been heated in order to obtain chondrites depleted in volatile elements.
4. The chemically unseparated dust fraction should have remained in equilibrium with the gaseous phase gradually being enriched in the volatile components (In, Tl, Xe, etc.). By inclusion of the chondrules in different proportions, ordinary (at $T = 450 \pm 50$ K and pressure 1 Pa) and carbonaceous (at $T = 350 \pm 50$ K and pressure 0.1 Pa) chondrites could have been formed.

As to conditions of the formation of iron meteorites, Kelly and Larimer (1977) have pointed out that under their apparent plain composition lies a complicated cosmochemical history. According to the variations of a number of elements (Au, Co, Cu, Fe, Ga, Ge, In, Mo, Ni, Os, Pd, Pt, Re, Rh, Rn) the iron meteorites are divided into 12 groups (Scott and Wasson 1976). The cosmochemical history of their metallic phase is related to the four stages of evolution:

1. Condensation and fractionation in the proto-Sun nebula and accretion of the substances.
2. Oxidation-reduction process in the nebula and parent bodies.
3. Melting and differentiation in the parent bodies.
4. Fractionation crystallization during solidification.

On the basis of the studies of distribution of the above-mentioned elements in iron meteorites on the whole, and also in their metallic, silicate, and sulfide phase, Kelly and Larimer (1977) have found that out of 12 groups of iron meteorites five satisfy the conditions of passing through all these four stages. The seven groups have unusual histories. In order to explain the observed variations of the chemical composition in some groups, high-temperature condensation ($T = 1{,}270$ K at $p_\tau = 1$ Pa) and, therefore, high-temperature accretion is required. For the other groups, there is evidence of quick metal cooling (7–200 degree/million years) and partial melting of aggregates of the parent bodies. In the third group, some features are observed which indicate the occurrence of recrystallization after partial melting.

The cosmochemical history of the metallic phase in chondrites and achondrites does not seem to be a less complicated problem.

Herndon and Suess (1976) have reported that the possibility that such minerals as TiN, Si_2N_2O and CaS, which occur in enstatite chondrites and achondrites, are formed directly from the gas nebula at the accepted conditions of equilibrium condensation is doubtful. They have noted that an extremely reducing medium is needed to obtain Ti and Fe with inclusions of the elementary silica. According to Herndon and Suess the models of formation of the above-mentioned compounds can be constructed only when assuming the pressure in the gaseous cloud to be $p_\tau \geq 10^5$ Pa.

Herndon and Suess have also thrown doubt upon the possibility of iron formation in sulfide, oxide, and pure metal forms in chondrules at equilibrium conditions. They have assumed that one way of obtaining iron in the three chemical forms occurring in ordinary chondrites is described by the nonequilibrium gas condensation model, which is being developed by a number of researchers.

Blander and Katz (1967), Blander and Abdel-Gavad (1969) have considered the problem of condensation from the gas nebula of the primary formation from which planets and meteorites could later have been formed under the nonequilibrium conditions of condensation. They suggested that the chemical composition of the cooling gas corresponded to the composition of elements in the solar system. Using data obtained earlier by Suess, Urey, and Cameron as a basis, the authors (Blander and Abdel-Gavad 1969; Blander and Katz 1967) have assumed the following sequence of ratios for the main constituents of the gas: H_2: H_2O: SiO_2: Mg: Fe as 10^{10}: 10^7: $10^{5.5}$: $10^{5.4}$: $10^{5.3}$. It was assumed that the cooling of the nebula occurred, along with other possible processes, due to energy losses by radiation as a result of which the gas cooled quickly at the boundary and more slowly in the interior of the nebula. Gas condensation was determined by its pressure and the rate of cooling. In this case as the upper limit of pressure and the lower limit of the rate of cooling in the centre of the nebula are reduced, the size of the nebula is increased. The condensation process occurred under nonequilibrium conditions and displayed a threshold character. It has been shown by the above authors that under the isobaric process of gas cooling at certain conditions, iron will precipitate first and then the silicate phase is condensed. The compounds enriched in calcium can condense earlier than silicates. It has been found that the process of material cooling at conditions of restricted equilibrium results in large variation of the chemical elements in the condensation products, which can explain the high content of oxides of iron in the silicate phase and also the observations of the different content of volatile elements in chondrites. Finally, they have studied the formation mechanism of individual chondrules in meteorites, and on the basis of the obtained experimental data have shown that they are the primeval formations of the liquid droplet condensate precipitated in the overcooled condition and subsequently solidified.

On the other hand, Arrhenius and Alfven (1977), while studying the conditions of evolution of the proto-planetary material on the basis of their own evidence, assumed that the process of its condensation could have started directly from the plasma state of the cloud, being at a temperature of 10^4 K and pressure $<10^{-2}$ Pa proceeding under disequilibrium conditions. In this case the refractory elements and compounds condense at $p_\tau = 10$ Pa and $T = 1{,}250$ K. The elements in which chondrites are normally depleted condense in the temperature range of 600–1,300 K, hydrosilicates at $T < 350$ K, CH_4, NH_3, H_2O, and hydrated methane at $T < 200$ K. Arrhenius and Alfen note that the temperature of solidification of material could have been considerably lower than the temperature of the surrounding gas. This is because the meteoric substance, according to the observation data, was not differentiated at the first stage and, therefore, it was formed under extreme disequilibrium temperature conditions with the surrounding gas.

Experimental observations and conclusions concerning the conditions of condensation and chemical fractionation of the proto-planetary material in meteorites, and also direct cosmochemical observation of the Moon and planets, are leading to a gradual revision of our ideas on the nature of the chemical differentiation of material in planets.

Analyzing the theories of possible formation of the Earth with regard to the two-component model, Larimer and Anders (1967) have considered that starting from the suggested hypothesis, the high-temperature component could have formed the planet's core at $T \geq 1{,}159$ K. The low-temperature fraction should have amounted to 10%. In view of the actual abundances of Bi, In and Tl in the Earth's core and also the higher temperature at distance of 1 A.U. from the Sun, the temperature of formation of the upper shell of the planet could have been ≤ 400 K.

Turekian and Clark (1969) and Clark et al. (1972) have suggested that the formation of the Earth occurred in a sequence determined by the process of substance cooling, which started at $T = 2{,}000$ K and pressure 10^2 Pa. According to their model the primeval Earth immediately obtained a shell structure due to a decrease of the iron-silicate ratio from the Earth's centre to the surface. In the course of the Earth's growth its gravitational energy increased, resulting in melting of the body being

Table 20.4 Accretion temperatures of the Earth, the Moon, and meteorites

Object	Heliocentric distance (A.U.)	Pressure (Pa)	Low-temperature phase content (%)	T(K)	
				Tl	$^{18}O/^{16}O$
The Earth (oceanic basalts)	1.0	10	11	458	450–470
The Earth (continental basalts)	1.0	10	11	456	450–470
The Moon	1.0	10	1.5	496	455
Chondrites type H, L, LL	2–3	1	25	420–500	445–480
Sherghotites	2–3	1	21	433	455
Nakhlites	2–3(?)	1	40	438	460
Eucrites	2–3(?)	1	0.5	432	475
Carbonaceous chondrites type C_1	3	0.1	>0.95	394	360
Carbonaceous chondrites type C_2	3	0.1	5.5	394	380

formed. In this case, the liquid iron must descend to the planet's centre, forming its core. Finally, when the Earth cooled, 20% of its substance, rich in volatile components, were added. In Turekian and Clark's model an attempt has been made to avoid the geochemical difficulties which were noted by Ringwood (1966).

Vinogradov (1971, 1975) came to the conclusion that at the high-temperature stage of the evolution of gaseous cloud physicochemical differentiation, accompanied by emergence of iron and silicate phases, had already started. These phases, during further processes of the cloud's evolution, provided the basis for formation of the cores and mantles of the future planets.

On the basis of Larimer and Anders's two-component model, Laul et al. (1973) estimated the formation temperatures of the Earth, Moon, and meteorites of different type. They took Tl as the cosmothermometer, since Tl/Rb and Tl/Cs ratios for the lunar and terrestrial rocks were found to be the same and the K/U, Rb/U, and Cs/U ratios for the same rocks approximate to constant values, where uranium was considered as a nonvolatile element. The magmatic processes are known to be inefficient in the fractionation of rocks.

In Table 20.4 the temperatures obtained by Laul et al. (1973) are shown and, for comparison, those obtained by Onuma et al. (1972) on the basis of oxygen isotope ratios for the same objects are added.

The data presented in Table 20.4 indicate, assuming that the initial assumptions are right, that the Earth, Moon, and all the above-mentioned types of meteorites were found within the narrow temperature range which could have existed in that part of the proto-planetary cloud where the formation of the considered bodies occurred. However, these data characterize the formation of the low-temperature phase. Assuming that the above-mentioned bodies contained the high-temperature phase in different proportions, this interpretation has certain difficulties with regard to the accretion hypothesis.

The latest studies of the Moon, the terrestrial group of planets, and Jupiter show that the processes of chemical differentiation of the proto-planetery substance have been more complicated in their character than those described in the existing models.

According to data obtained by the rubidium-strontium method (Wasserburg et al. 1972; Wood 1974), the age of the sampled inland rocks ranges within $(4.3–4.6) \times 10^9$ years. This interpretation of the results of the age, combined with petrological studies, leads to the conclusion that the Moon underwent a stage of melting and chemical differentiation (at least in its upper shell) at the final stage of its formation. At that time the whole of the upper layer of the Moon within the depth ≥ 100 km was in a melting state and during crystallization formed a low density core ($\rho = 2.9$ g/cm^3), mainly of plagioclase composition, within a short time interval (~200 million years). The surface melt was not formed as a result of the processes occurring in the interior of the Moon but more likely reflects the high-temperature conditions of its formation. Only some local areas of the lunar seas were filled by subsequent lava eruptions of a somewhat younger age and corresponding chemical composition. The formation of a crust with a lower density, as compared with the average density characteristic of the whole body, indicates that the chemical differentiation of the lunar substances had been completed up to the moment of the body formation, including the formation of a small size metallic core. On the basis of estimation of the energy losses and the core heat conductivity, the time of cooling of the surface layer and crystallization of the core was found to be 10^8 years. The calculations have shown that, in order to the energy released by radioactive decay to melt the lunar substance, such decay would have to continue for 2×10^9 years. According to direct measurements, the heat flux from the lunar surface at the present time has been found to be unexpectedly large, about 3×10^{-2} J/m^{2}s.

The conclusions concerning the heat history of the Moon throw doubts upon the model of the cold accretion of substance during formation of the Moon and required a reassessment of the role being played by radiogenic energy in the thermal history of the Moon, and obviously in the history of all the planets of the terrestrial group.

Cameron's model attempts to explain the early lunar heating by the release of gravitational energy during the quick accretion of substance and meets the serious cosmochemical difficulties.

Analyzing the chemical composition of the samples of lunar soils, breccias, and igneous rocks, it has been found that they are characterized by a number of cosmochemical peculiarities which distinguish them from terrestrial rocks and meteorites. The igneous lunar rocks are enriched by 10–100 times in Ca, Zn, Hf, Ta, rare earth, and other refractory elements and depleted in alkaline, halogen, and volatile metals (Bi, Tl, Hg) compared with carbonaseous chondrites of the C_1 type. The lunar rocks as a whole are enriched in high-temperature and depleted in low-temperature elements relative to their cosmic abundance. Ganapathy and Anders (1974) have given comparative data on the abundance of elements in lunar and terrestrial rocks (Table 20.5).

Table 20.5 Relative abundance of some groups of elements in terrestrial and lunar rocks

Group of elements	Temperature of condensation in the cloud (K)	Abundance relative to cosmic	
		Earth	Moon
Refractory elements (Al, Ca, Ti, Ba, Sr, U, Th, Ph, Ir)	>1,400	−1	2.7
Silicates (Mg, Si)	1,400−1,200	1[a]	1[a]
Metals (Fe, Ni, Co, Cr, Au)	1,400−1,200	~1	0.25
Volatile elements (S, Na, K, Cu, Zn, Te)	1,300−600	0.25	0.05
High volatile elements (Cl, Br, Hg, Pb, In)	<600	0.02	0.0005

[a]Given relative to the magnesium silicates

Attempts were made to explain the observed distribution of chemical elements in lunar rocks on the basis of the different models. The idea that the Moon could have been formed from high-temperature compounds, which are represented in inclusions of the carbonaseous chondrites (C_1 type), meets with difficulties in those compounds in chondrites are enriched both in high-temperature siderophylous (Re, Os, Ir) and litophylous elements. The lunar rocks are extremely depleted in the latter elements. The two-component model of the accretion of the Moon is therefore preferable. However, this suggestion meets difficulties in explaining the oxygen isotope composition of the lunar rocks where $\delta^{18}O \approx +6‰$, whereas inclusions of the carbonaseous chondrite Allende indicates $\delta^{18}O \approx -10.5‰$.

Attempts directed toward explaining the oxygen isotopic composition in the framework of the two-component model of the primeval lunar substance have therefore failed.

Other important facts which have not been explained from the viewpoint of the conditions of the Moon's substance are:

1. The Moon does not have water and obviously did not have any (at least in its upper layers). In the case of its accretion from the substance which makes up the Earth, the lunar rocks should be hydrated or have features of hydration, but these are not actually observed.
2. No signs were found of granitization of rocks, which is likely to be closely related to the absence of water.
3. The lunar minerals tend to be reduced due to a deficit of oxygen, which can also be accounted for by the absence of water.
4. Iron is also in a reduced condition and indicates a phenomenon which is usually for terrestrial silicates, also obviously accounted for by the absence of water.
5. An age paradox is observed: the lunar dust, being the product of rock destruction, appeared to be older than the rocks themselves. This phenomenon is accounted for by the presence of the 'magic' component in the dust composed of potassium, rare earths, and phosphorus and also by the effect of the solar wind.

6. There are visible signs of the formation of minerals of surface sediments from the vapor phase.

The obtained data on cosmochemistry and the thermal history of the Moon have created great difficulties in justification of the accretion, chondritic model of its formation. The other models of lunar formation based on the hypothesis of separation of its substance from that of the Earth were found to be less inconsistent.

An analysis of the high-resolution photographs of Mercury obtained during the flight of Mariner-10 allowed the following conclusions concerning this planet to be drawn (Murrey et al. 1974).

The surface of Mercury resembles the lunar seas and represents the hardened melt of large-scale fluxes of lava. According to the visible signs this melt is mainly constituted by silicates with the density of about 3 g/Cm^3. Because the mean density of the planet is equal to about 5.5 g/Cm^3, the internal layers of the planet should consist of substances heavier than silicates. The planet obviously represents a body differentiated by density and chemical composition with a massive iron core.

Mercury's surface, as with the lunar surface, is covered with craters of impact origin. It is quite justifiably assumed that the chemical differentiation of Mercury, as well as that of the Moon, was finished before the formation of the body. If it were not, a sign of the intense bombardment of its surface by meteorites would have been lost. For the same reason it has been concluded that Mercury has had neither a primary nor a secondary atmosphere, in contrast to Mars where aeolian processes had considerably changed the primary landscape of the planet.

The comparison of the surface structures of Mercury, Mars, and the Moon has led to the conclusion that the mechanisms of the formation of these bodies and their chemical differentiation were common.

The studies of Jupiter, carried out with the help of spacecraft, had provided a number of important results. The most interesting of these appeared to be the data of the high-temperature state of the planet and on the powerful flux of radiation from its surface. On the basis of measurements carried out by the cosmic spacecraft Pioner-10, the temperature of the planet's core is estimated to be about 50,000 K, the temperature of the surface layer is 2,000 K, and that of the atmosphere is 120 K. Jupiter radiates energy whose value is 2–3 times greater than that obtained from the Sun. The approximate chemical composition of the planet is estimated as the following: 82% hydrogen, 17% helium, and 1% other elements (Cotardiere 1975).

The obtained data on the temperature regime and chemical composition of Jupiter does not provide any basis for the usage of any of the existing hypotheses for an explanation of the conditions of its formation. The accretion hypothesis is the most vulnerable. It is noteworthy that Jupiter exhibits the greatest angular momentum of all the bodies of the solar system. Despite this fact, the hypotheses being developed concerning the origin of the solar system dealt mainly with planets of the terrestrial group. This is most probably the weakest point of the existing hypotheses.

The data given above on the cosmochemical studies of meteorites, planets, and the Moon, compiled in the second part of the twentieth century, leads to one general and possibly indisputable conclusion that, in both the proto-solar and proto-planetary clouds, there were common processes of chemical differentiation of the initial substances. These processes have obviously led to cosmochemical differentiation between the planets and the Sun and between the planet's shells as well. The differentiation mechanism seems to be the same. Let us apply our analytical solutions in dynamics to the above-presented cosmochemical observations.

20.5 Recent Results of Study of the Earth Gravitational Field by the Satellites

In the second part of the twentieth century, continuous study of space by artificial satellites opened a new page in space sciences. The ultimate goal of the satellite scientific program was determined to be the problem solution of the solar system origin. At the same time, in order to solve geodetic and geophysical problems, investigation of the near Earth cosmic space was initiated.

The first geodetic satellites for study of dynamical parameters of the planet about 50 years ago were launched. They gathered vast amounts of data that significantly improved our knowledge on the inner structure and dynamics of the Earth. They made it real possible in evaluation experimentally the correctness of basic physical ideas and hypotheses in geophysics, geodesy and geology, and to compare theoretical calculations with observations. Success in this direction was achieved in a short period of time.

On the basis of satellite orbit measurements, the zonal, sectorial, and tesseral harmonics of gravitational moments in expansion of the gravitational potential by a spherical function, up to ten, twenties and higher degrees were calculated. The calculations have resulted in an important discovery having far-reaching effects. The obtained results proved the long-held assumption of geophysicists that the Earth does not stay in hydrostatic equilibrium, which, in fact, is the basic principle of the theories of dynamics, figure and inner structure of the planet. The same conclusion was made about the Moon. This conclusion means that the model of the outer central force field, used for justifying hydrostatic equilibrium of the Earth and its inertial rotation, does not satisfy the observed dynamical effects of gravitational interaction of the planet's mass particles and should be revised.

Taking into account the above fundamental discovery and also the fact that the value of ratio of the potential to kinetic energy of the Earth is equal to $\sim$300, we come back to derivation of the virial theorem in classical mechanics and obtained its generalized form of the relationship between the total energy and polar moment of inertia of a gravitating body. In doing so, we obtained the equation of dynamical equilibrium of a body in its own force field where the kinetic energy in general case is represented by oscillation of the interacted mass particles and rotation of the mass. An analytical expression of the derived new form of the virial theorem is based on the energy conservation law and represents a differential equation of second order, where the variable value is kinetic energy of the body's oscillating polar moment of inertia (Ferronsky and Ferronsky 2010). Our dynamical approach is a basis for further consideration of the problem of separation of the Earth's shells and creation of the hydrosphere.

From the viewpoint of planetary dynamics, chemical differentiation of substances during separation of planets and satellites from the common solar nebula and formation of body shells is the problem of differentiation of the cloud's matter with respect to atomic and molecular weights or, generally saying, according to density. The physical basis for consideration of this problem comprises the dissipative processes in the nebula or cloud which are related to gravitational and electromagnetic interaction between the constituencies. This condition results in generation of gravitational and electromagnetic energy and its loss from the surface shell in the form of radiation. This condition follows from consideration of the structure of the potential energy of a system that is nonhomogeneous in its elementary content, which for the proto-solar system is a priori satisfied.

The problem of differentiation of substances with respect to the atomic and molecular weights of a cloud in its own force field is separation of a mass with respect to density is based on Roche's dynamics and Newton's theorem about gravitational interaction of the material point and a spherical shell. The real mechanism of separation of atoms and molecules in a diffused stage of the cloud should be the same mechanism of generation of the electromagnetic (gravitational) energy (Ferronsky and Ferronsky 2010).

The process of atomic and molecular mass separation in natural systems (in nebulae and bodies) seems to be not an episode in their history. This is the continuous process of any system connected with its evolution, the essence of which is collision and scattering of the particles (molecules, atoms, nuclei and so on) accompanied by their destruction and removal of smaller, up to elementary, particles, which form the flux that we call energy and its emission. In turn, the energy generation, i.e., the transition of the mechanical energy of a system oscillation into the energy of the electromagnetic oscillation on atomic and molecular (and nuclear for the stars) level leads to change the atomic and molecular weight and the substances themselves. Let us consider the problem of separation of the planet's shells.

20.6 The Nature and Mechanism of the Earth Shell Separation

It is well known that the Earth has a quasi-spherical shell structure. This phenomenon is fixed by the record and interpretation of the seismic longitudinal and transversal wave propagation during the earthquakes. In order to understand the physics and mechanism of the Earth mass differentiation with respect to its density we apply the Roche's tidal dynamics.

The Newton's theorem of the gravitational interaction between a material point and spherical layer states that the layer does not affect the point located inside the layer. On the contrary, the outside-located material point is affected by the spherical layer. The Roche's tidal dynamics is based on the above theorem. His approach is as follows (Ferronsky et al. 1996).

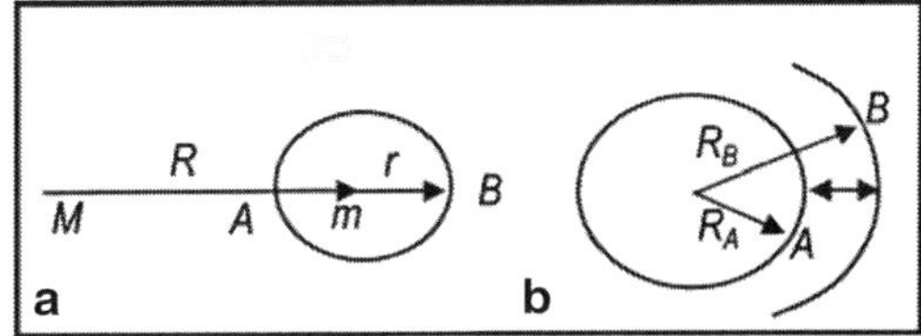

Fig. 20.3 The tidal gravitational stability of a sphere (**a**) and the sphere layer (**b**)

There are two bodies of masses M and m interacting in accordance with the Newton's law (Fig. 20.3a).

Let $M \gg m$ and $R \gg r$, where r is the radius of the body m, and R is the distance between the bodies M and m. Assuming that the mass of the body M is uniformly distributed within the sphere of radius R, we can write the accelerations of the points A and B of the body m as

$$q_A = \frac{GM}{(R-r)^2} - \frac{Gm}{r^2}, \quad q_B = \frac{GM}{(R+r)^2} + \frac{Gm}{r^2}.$$

The relative tidal acceleration of the points A and B is

$$\begin{aligned} q_{AB} &= G\left[\frac{M}{(R-r)^2} - \frac{M}{(R+r)^2} - \frac{2m}{r^2}\right] \\ &= \frac{4\pi}{3}G\left[\rho_M R^3 \frac{4Rr}{\left(R^2-r^2\right)^2} - 2\rho_m r\right] \approx \frac{8\pi}{3}Gr\left(2\rho_M - \rho_m\right). \end{aligned} \tag{20.1}$$

Here $\rho_M = M/\frac{4}{3}\pi R^3$ and $\rho_m = m/\frac{4}{3}\pi r^3$ are the mean density distributions for the spheres of radius R and r. Roche's criterion states that the body with mass m is stable against the tidal force disruption of the body M if the mean density of the body m is at least twice higher than that of the body M in the sphere with the radius R. Roche considered the problem of the interaction between two spherical bodies without any interest to their creation history and to how the forces appeared. From the viewpoint of the origin of celestial bodies and of the interpretation of dynamical effects, we are interested in the tidal stability of separate envelopes of the same body. For this purpose we can apply the Roche's tidal dynamics to study the stability of a nonuniform spherical envelope.

Let us assess the tidal stability of a spherical layer of radius R and thickness $r = R_B - R_A$(Fig. 20.3b). The layer of mass m and mean density $\rho_m = m/4\pi R_A^2 r$ is affected in point A by tidal force of the sphere of radius R_A. The mass of the sphere is M and its mean density $\rho_M = M/\frac{4}{3}\pi R_A^3$. The tidal force in point B is generated by the sphere of radius $R + r$ and mass $M + m$. Then the accelerations of the points A and B are

$$q_A = \frac{GM}{R_A^2} \quad \text{and} \quad q_B = \frac{G(M+m)}{(R_A+r)^2}.$$

The relative tidal acceleration of the points A and B is

$$q_{AB} = GM\left[\frac{1}{R_A^2} - \frac{1}{(R_A + r)^2}\right] - \frac{Gm}{(R_A + r)^2}$$
$$= \left(\frac{8}{3}\pi G\rho_M - 4\pi G\rho_m\right) r = 4\pi G r\left(\frac{2}{3}\rho_M - \rho_m\right), \ (R \gg r)\,. \qquad (20.2)$$

Eqs. (20.1) and (20.2) give the possibility to understand the nature of the Earth shell separation, which, in fact, are the Archimedes and Coriolis forces.

20.7 Physical Meaning of Archimedes' and Coriolis' Forces

The Archimedes principle states: *The apparent loss in weight of a body totally or partially immersed in a liquid is equal to the weight of the liquid displaced.* It is seen from the previous section that the principle is described by Eqs. (20.1) and (20.2) and the forces which sink down or push out the body or the shell are of a gravitational nature. In fact, in the case of $\rho_n = \rho_M$ the body immersed in a liquid (or in any other medium) is kept in the place due to equilibrium between the forces of the body's weight and the forces of the liquid reaction. In the case of $\rho_n > \rho_M$ or $\rho_n < \rho_M$ the body is sinking or floating up depending on the resultant of the above forces. Thus, the Archimedes' forces seem to have a gravity nature and are the radial component of the Earth inner force field.

It is assumed that the Coriolis' forces appeared as an effect of the body motion in the rotational system of coordinates relative to the inertial reference system. In this case rotation of the body is accepted as the inertial motion and the Coriolis, forces appear to be the inertial ones. It follows from solution of the generalized virial equation (Ferronsky and Ferronsky 2010) that the Coriolis' forces appear to be the tangential component of the Earth's inner force field, and the planet rotation is caused by the moment of those forces that are relative to the three-dimensional centre of inertia which also does not coincide with the three-dimensional gravity centre.

In accordance with Eq. (6.71) of the tidal acceleration of an outer nonuniform spherical layer at $\rho_M \neq \rho_m$ a mechanism of the gravitational density differentiation of masses is revealed. If $\rho_M < \rho_m$, then the shell immerses (is attracted) up to the level where $\rho_M = \rho_m$. At $\rho_M > \rho_m$the shell floats up to the level where $\rho_M = \rho_m$ and at $\rho_M > 2/3\rho_m$ the shell becomes a self-gravitating one. Thus, in the case when the density increases towards the sphere's center, which is the Earth's case, then each overlying stratum appears to be in the suspended state due to repulsion by the Archimedes' forces which, in fact, are a radial component of the gravitational interaction forces.

The effect of the gravitational differentiation of masses explains the nature of creation of the Earth's crust and also the ocean, geotectonic, orogenic, and seismic processes, including the earthquakes. All these phenomena appear to be a consequence of the continuous gravitational differentiation in density of the planet's masses. We assume that this effect was one of the dominating during creation of the Earth and the solar system as a whole. For instance, the mean value of the Moon density is less than 2/3 of the Earth' one, i.e., $\rho_M < 2/3\rho_m$. If one assumes, that this relation was kept during the Moon formation, then, in accordance with Eq. (20.1), this body has separated at the earliest stage of the Earth mass differentiation. Creation of the body from the separated shell should occur by means of the cyclonic eddy mechanism, which has been proposed in due time by Descartes and which was unjustly rejected. If we take into account existence of the tangential forces in the nonuniform mass, then the above mechanism seems to be realistic.

20.8 Self-Similarity Principle and Radial Component of Nonuniform Sphere

It follows from Eq. (20.1) that in the case of the uniform density distribution ($\rho_m = \rho_M$), all spherical layers of the gravitating sphere move to the centre with accelerations and velocities which are proportional to the distance from the centre. It means that such a sphere contracts without loosing its uniformity. This property of self-similarity of a dynamical system without any discrete scale is unique for a uniform body (Ferronsky et al. 1996).

A continuous system with a uniform density distribution is also ideal from the viewpoint of the Roche's criterion of stability with respect to the tidal effect. A uniform sphere is always similar in its structure in spite of the fact that it is a continuously contracting system. Here, we do not consider the Coulomb forces effect. For this case we have considered the specific proton and electron branches of the evolution of the body.

Note that in the Newton's interpretation the potential energy has a nonadditive category. It cannot be localized even in the simplest case of the interaction between two mass points. In our case of a gravitating sphere as a continuous body, for the interpretation of the additive component of the potential energy we can apply the Hooke's concept. Namely, according to Hooke there is a linear relationship between the force and the caused displacement. Therefore the displacement is in square dependence on the potential energy. Hooke's energy belongs to the additive parameters. In the considered case of a gravitating sphere, the Newton's force acting on each spherical layer is proportional to its distance from the centre. Thus, here from the physical point of view, the interpretations of Newton and Hooke are identical.

At the same time in the two approaches there is a principal difference even in the case of uniform distribution of the body density. According to Hooke the cause of displacement, relative to the system, is the action of the outer force and if the total energy is equal to the potential energy, then the equilibrium of the body is achieved. The potential energy plays here the role of elastic energy. The same uniform sphere with the Newton's forces will be contracted. All the body's elementary shells will move without change of uniformity in the density distribution. In accordance with the Newton's third law and the d'Alembert principle the attraction forces, under the action of which the shells move, should have equally and oppositely direct forces of the Hooke's elastic counteraction In the framework of the elastic gravitational interaction of shells the dynamical equilibrium of a uniform sphere is achieved in the form of its elastic oscillations with equality between the potential and kinetic energy. The uniform sphere is dynamically stable relative to the tidal forces in all of its shells during the time of the system contraction. Because the potential and kinetic energies of a sphere are equal then its total energy in the framework of the averaged virial theorem within one period of oscillation is accepted formally as equal to zero. Equality of the potential and kinetic energy of each shell means the equality of the centripetal (gravitational) and centrifugal (elastic constraint) accelerations. This guarantees the system to stay in dynamical equilibrium. On the contrary, all the spherical shells will be contracted towards the gravity centre which, in the case of the sphere, coincides with the inertia centre but does not coincide with the geometric centre of the masses. Because the gravitational forces are acting continuously, the elastic constraint forces of the body shells are reacting also continuously. The physical meaning of the self-gravitation of a continuous body consists in the permanent work which performs the energy of the interacted shell masses on one side and the energy of the elastic reaction of the same masses in the form of oscillating motion on the other side. At the dynamical equilibrium the body's equality of the potential and kinetic energy means that the shell motion should be restricted by the elastic oscillation amplitude of the system. Such an oscillation is similar to the standing wave which appears without transfer of the energy into the outer space. In this case the radial forces of the shell elastic interactions along the outer boundary sphere should have a dynamical equilibrium with the forces of the outer gravitational field. This is the condition of the system to be held in the outer force field of the mother's body. Because of this, while studying the dynamics of a conservative system its rejected outer force field should be replaced by the corresponding equilibrated forces like they do it, for instance, in the Hooke's theory of elasticity.

20.9 Charges-like Motion of Nonuniformities and Tangential Component of the Force Function

Let us now discuss the tidal motion of nonuniformities due to their interactions with the uniform body. In accordance with (20.1), the nonuniformity motion looks like the motion of electrical charges interacting on the background of a uniform sphere contraction. Spherical layers with densities exceeding those of the uniform body (positive anomalies) come together and move to the centre in elliptic trajectories. The layers with deficit of the density (negative anomalies) come together, but move from the centre on the parabolic path. Similar anomalies come together, but those with the opposite sign are dispersed with forces proportional to the layer radius. In general, the system tends to reach a uniform and equilibrium state by means of redistribution of its density up to the uniform limit. Both motions happen not relative to the empty space, but relative to the oscillating motion of the uniform sphere with a mean density. Separate consideration of motion of the uniform and nonuniform components of a heterogeneous sphere is justified by the superposition principle of the forces action which we keep here in mind. The considered motion of the nonuniformities looks like the motion of the positive and negative charges interacting on the background of the field of the uniformly dense sphere (Ferronsky et al. 1996). One can see here that in the case of gravitational interaction of mass particles of a continuous body, their motion is the consequence not only of mutual attraction, but also mutual repulsion by the same law $1/r^2$. In fact, in the case of a real natural nonuniform body it appears that the Newton's and Coulomb's laws are identical in details. Later on, while considering the Earth's bi-density differentiated masses the same picture of motion of the positive and negative anomalies will be seen.

If the sphere shells, in turn, include density nonuniformities, then by means of the Roche's dynamics it is possible to show that the picture of the nonuniformity motion does not differ from that considered above.

In physics the process of interaction of particle with different masses without redistribution of their moments is called elastic scattering. The interaction process resulting in redistribution of their moments and change in the inner state or structure is called inelastic scattering. In classical mechanics while solving the problems of motion of the uniform conservative systems (like motion of the material point in the central field or motion of the rigid body), the effects of the energy scattering do not appear. In the problem of dynamics of the self-gravitating body, where interaction of the shells with different masses and densities are considered, the elastic and inelastic scattering of the energy becomes an evident fact following from consideration of the physical meaning of the expansion of the body's energy.

Thus, we find that inelastic interaction of the nonuniformities with the uniform component of the system generates the tangential force field which is responsible for the system rotation. In the other words, in the scalar force field of the bi-density uniform body the vector component appears. In such a case, we can say that, by analogy with electromagnetic field, in the gravitational scalar potential field of the nonuniform sphere $U(R, t)$ the vector potential $A(R, t)$ appears for which $U = rotA$ and the field $U(R, t)$ will be solenoidal. In this field the conditions for vortex motion of the masses are born, where $div\,A = 0$. This vector field, which in the electrodynamics is called solenoidal, can be presented by the sum of the potential and vector fields. The fields, in addition to the energy, acquire the moments and have a discrete-wave structure. In our case the source of the wave effects appears to be the interaction between the elementary shells of the masses by means of which we can construct a continuous body with a high symmetry of the forms and properties. The source of the discrete effects can be represented by the interacted structural components of the shells, namely, atoms, molecules and their aggregates.

20.10 Differentiation of the Substances with Respect to Density and Condition for the Planet and the Satellite Separation

From the viewpoint of the planetary dynamics, chemical differentiation of the substances during separation of planets and satellites from the common solar nebula and formation of body shells is the problem of differentiation of the cloud's matter with respect to atomic and molecular weights or, generally saying, according to density. Physical basis for consideration of this problem comprises the dissipative processes in the nebula or cloud which are related to gravitational and electromagnetic interaction between the constituencies. This condition results in generation of gravitational and electromagnetic energy and its loss from the surface shell in the form of radiation. This condition follows from consideration of structure of the potential energy of a nonhomogeneous in elementary content system (Ferronsky and Ferronsky 2010).

The problem of the differentiation of the substances with respect to the atomic and molecular weights of a cloud in the own force field is based on Roche's dynamics and the Newton's theorem about gravitational interaction of the material point and a spherical shell. The real mechanism of separation of the atoms and molecules in a diffused stage of the cloud should be the same mechanism of generation of the electromagnetic (gravitational) energy.

The process of the atomic and molecular mass separation in the natural systems (in nebulae and bodies) seems to be not an episode in their history. This is the continuous process of any system connected with its evolution, essence of which is the collision and scattering of the particles (molecules, atoms, nuclei and so on) accompanied by their destruction and removal of smaller, up to elementary, particles, which form the flux what we call energy and its emission. In turn, the energy generation, i.e., the transition of the mechanical energy of a system oscillation into the energy of the electromagnetic oscillation on atomic and molecular (and nuclear for the stars) level leads to change the atomic and molecular weight and the substances themselves. Let us consider the problem of separation of the planets and satellites from the proto-solar nebula. For this we come back to the solution of equation of dynamical equilibrium of a dissipative system presented in the form (Ferronsky and Ferronsky 2010)

$$\ddot{\Phi} = -A_0[1 + q(t)] + \frac{B}{\sqrt{\Phi}}, \tag{20.3}$$

where Φ is Jacobi's function (polar moment of inertia of the system); q(t) is the time parameter continuously ascending due to dissipation of energy at 'smooth' evolution of the system in the time interval $t \in [0, \tau]$; A and B are constant values.

Solution of Eq. (20.3) was found in the form

$$\begin{aligned} \mp \arccos W \pm \arccos W_0 \mp \sqrt{1 - \frac{A_0\,[1 + q(t)]\,C}{2B^2}}\sqrt{1 - W^2} \\ - \sqrt{1 - \frac{A_0 C}{2B^2}}\sqrt{1 - W_0^2} = \sqrt{\frac{(2A_0\,[1 + q(t)])^{3/2}}{4B}}(t - t_0). \end{aligned} \tag{20.4}$$

Equations of the discriminant curves limiting the amplitude of oscillation of the Jacobi function(polar moment of inertia) are

$$\sqrt{\Phi_{1,2}} = \frac{B}{A_0\,[1 + q(t)]}\left[1 \pm \sqrt{1 - \frac{A_0\,[1 + q(t)]\,C}{2B^2}}\right], \quad t \in [0, \tau], \tag{20.5}$$

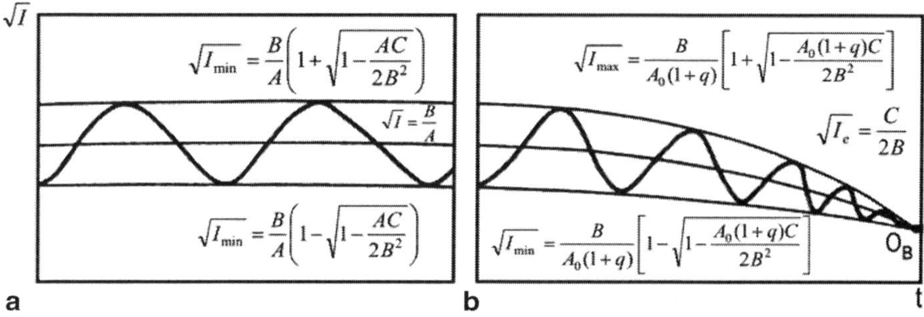

Fig. 20.4 Evolution of a dissipative system to bifurcation point

Expression for the period of oscillation T_v and amplitude of the moment of inertia of the system, obtained from Eqs. (20.4) and (20.5)

$$T_v(q) = \frac{8\pi B}{(2A_0[1+q(t)])^{3/2}}, \tag{20.6}$$

$$\Delta\sqrt{\Phi} = \frac{B}{A_0\,[1+q(t)]}\left(1 - \frac{A_0\,[1+q(t)]\,C}{2B^2}\right)^{1/2}. \tag{20.7}$$

Figure 20.4 shows changes of the polar moment of inertia due to oscillating dissipation of the system's energy. The changes are confined the discriminant curves (9.5). At the point O_b the integral curve (9.4) and the discriminant curves (9.5) tend to coincide and the amplitude of the oscillation of the moment of inertia of the system decreases up to zero. In this case the system reaches the stage close to

$$\Phi = \text{const}. \tag{20.8}$$

Expression (20.8) is the second solution of the nonlinear differential Eq. (20.3). The point O_b is physically interpreted as the bifurcation point. Here, under action of its own force field, the mass of the upper shell of the system reaches dynamical equilibrium with the inner mass and separates into a new subsystem. Equating the radicand of Eq. (20.5) to zero one finds that

$$\frac{2B^2}{A_0(1+q_b)} = C, \tag{20.9}$$

where q_b is the value of the parameter q for the bifurcation point equal to

$$q_b = \frac{2B^2}{A_0 C} - 1. \tag{20.10}$$

Then Jacobi's function (polar moment of inertia) Φ_b of the system at the bifurcation point, where the discriminant curves coincide, is

$$\Phi_b = \frac{B^2}{A_0\left(1 + \frac{2B^2}{A_0 C - 1}\right)^2} = \frac{C^2}{4B^2}. \tag{20.11}$$

While analyzing the equation of dynamical equilibrium of a celestial body by taking into account the Coulomb interactions of the charged particles shown, we found, that the relationship between the polar moment of inertia and potential energy of the Coulomb interactions, written for conservative and dissipative systems, holds. In this case solution (20.4) of Eq. (20.3) for electromagnetic interactions also holds and the system can be represented by the model of the oscillating electric dipole which generates electromagnetic oscillations (Ferronsky et al. 1982). Mechanism of that process was considered in the previous chapter from which it follows that the dissipative system cannot be electrically neutral.

The obtained solution of the equation of dynamical equilibrium of a dissipative system (20.5) represents a theoretical solution of the problem of the equilibrium or 'smooth' form of evolution of a celestial body. The condition of achievement by a dissipative system of the bifurcation point and the obtained relations for its parameters should be a qualitative solution of the problem of secondary body separation. For application to the cosmogony of the solar system, it means that all its planets were formed from the common gaseous nebula by its separation during evolution through emission of electromagnetic energy. The evolutionary process was developed by density differentiation of the gaseous substances of the nebula and by separation of the planets and satellites from the outer shell. The normal and tangential volumetric forces of the interacting mass particles there built up permanent conditions for creation of vortexes which seems to be the main mechanism in formation of new entities. The conditions for the formation of meteorites existed during the formation of both the planets and their satellites.

Now we can define the specific physical effect of separation of bodies in the solar system from the common nebula. It should be the inner energy (force function) of Coulomb interaction of the atomic, molecular and nuclei particles. It has been shown (Ferronsky et al. 1978) that for the planets in the solar system, at the moment of their formation, the magnitude of the gravitational interaction energy was equal to the electromagnetic energy. In fact, the well-known expression for the energy of the Coulomb interactions U_c, written through the Madelung energy (Kittel 1968) is:

$$U_c = n\left(e^2/R_0\right)k, \tag{20.12}$$

where $n = m/\mu$ is the quantity of interacting molecules; m is the body mass; μ is the average molecular weight; $e = 4{,}8 \times 10^{-10}$ is the charge of the electron; k is the Madelung coefficient; $R_o = (V/n)^{1/3} = R_0(\mu/\mathrm{m})^{1/3}$ is the average radius of the Coulomb interactions of the molecules; V is the volume of the body.

Equating the expressions for the potential energy of the gravitational and Coulomb interactions, one can derive a relationship between the critical mass of the planet and its average molecular weight μ:

$$\begin{gathered} \alpha^2\left(\frac{Gm^2}{R}\right) = n\frac{e^2}{R_0}k, \\ m_{\kappa p} = \frac{e^3k^{3/2}}{a^2G^{3/2}\mu^2}. \end{gathered} \tag{20.13}$$

Table 20.6 The average values of molecular weight (in a.u.m) for the planets at time of their creation

Planets	Mercury	Venus	Earth	Mars	Jupiter	Saturn	Uranus	Neptune
By Eq. (9.11)	256	66	60	183	3	6	15	14
By Eq. (9.12)	467	85	109	336	6	11	28	26

From expression (20.13) one can find the value of the average molecular weight μ for the planets of the solar system (see Table 20.6) (Ferronsky et al. 1978).

We consider relationship (20.13) to express the condition of formation of the condensing proto-planetary clouds, characterized by the corresponding average chemical compositions included in Table 20.6. On the basis of expression (9.11) we obtain an important cosmochemical dependence of the planet's mass from its average chemical composition.

$$m\mu^2 = b = const, \tag{20.14}$$

where b is a constant depending in each case on the mass density distribution and the average chemical composition of the body being formed, due to the effect of the Coulomb interactions, as follows:

$$b = \frac{e^3 k^{3/2}}{a^2 G^{3/2}}.$$

Earlier, from solution of the Chandrasekhar-Fermi equation we obtained a rigorous solution of the problem of secondary bodies separating from the nebula at its evolution and found that the constant $b = 2 \times 10^{-16}$ g^3(Ferronsky et al. 1996). The corrected average values of the molecular weight for the planets calculated by (20.14) are given also in Table 20.6.

Thus, the first condition of the planet or satellite creation from the proto-solar nebula at its evolution was equality of gravitational and Coulomb energy of interaction for the outer shell. Taking into account the mechanism of mass density differentiation of a cloud, the probable mechanism of a planetary body creation should be cyclonic vortexes. In this connection it is worth recalling the vortex hypotheses of creation of the solar system proposed by Descartes which now appears to be reasonable. The argument against this theory, related to distribution of the angular momentum between the Sun and the planets, taking into account the found oscillating (volumetric) mode, which is the moment of momentum in this book, now drops out.

The second important condition of a body separation from the common nebula is equality of the normal and tangential components of the potential energy for the outer shell. This condition determines necessity of dynamical equilibrium of a new entity as a self-gravitating body and its ability to move on the orbit in the outer parent's force field.

Accounting for the Coulomb forces one may obtain a qualitative solution of the problem of chemical differentiation of substance in the course of the formation of planets. We have already pointed out that expression (20.14) is the cosmochemical criterion for the formation of planets and satellites. Coulomb forces are more effective for the elements with greater atomic numbers. In the process of dissipation of the potential energy by radiation, a differentiation of the substance occurs on the same physical basis in accordance with boiling temperatures and the relative volatilities of the elements and compounds.

To the first approximation, while estimating the chemical diffraction of the substance of the planets and satellites, one can use the following model which is similar to the two-component model of Larimer and Anders (1967).

Assume that when the proto-solar nebula reaches the stage of the bifurcation point (see Fig. 20.4) it gives birth to proto-planetary condensation—the cloud, the substance of which consists of chemical compounds characterized by their average molecular weight μ in accordance with relationship (20.15). In this case all the proto-planetary cloud's elements and compounds, boiling temperatures of which are higher than that of the element and compound with molecular weight μ (refractory component), should remain in the gaseous phase, dissolved in the major component of the condensate. The quantitative estimation of abundances of the elements and compounds being present in the condensed and gaseous phases is determined by the abundances of these elements and compounds in the shell of the proto-solar nebula from which the formation of the proto-planet has occurred. The separation of the high-temperature and volatile components of the proto-planetary cloud and the formation of its shells, takes place due to the Coulomb forces at the stage of formation and evolution of the secondary body.

Let us use this condensational model of planet formation to explain the observed distribution of volatile elements on the Earth and also their average isotope composition in comparison with their solar abundance (Onufriev 1978). To a first approximation one can use for this purpose the thermodynamic laws of the ideal gases and write the separation factor η_{AB} of the low-temperature (volatile) component A and high-temperature (refractory) component B in the gas-condensate system in the form

$$\eta_{AB} = p_A/p_B \approx exp[(L_B - LA)/R_0 T] \tag{20.15}$$

or

$$lg\ \eta_{AB} = n(T_B - T_A),$$

where p, L and T are the vapor pressure, boiling heat of vaporization, and boiling temperature, respectively; n is the numerical factor; R_0 is the gas constant.

If the accepted model of the formation of the chemical and isotope composition of substance is true and expression (20.15) reflects, to a first approximation, the real process of such a formation, then the observed abundance of chemical elements on the Earth relative to their solar abundances can be written as

$$lg\ \eta_{AB} = lg(N^c{}_A/N^3{}_A) - lg(N^c{}_B/N^3{}_B), \tag{20.16}$$

where N^c and N^3 are the solar and terrestrial abundances of a component respectively.

In view of the above-mentioned abundances of the noble gases He, Ne, Ar, Kr, Xe were studied on the Earth and Moon relative to their solar abundances

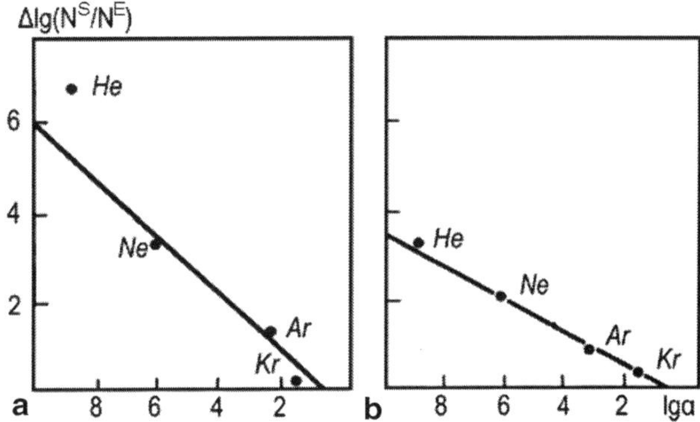

Fig. 20.5 Relationship between the value of inert gases deficit and differentiation factor for the Earth (**a**) and for the Moon (**b**)

(Onufriev 1978). Figure 20.5 indicates the dependence of the deficit of each of the elements on Earth and the Moon [$\Delta lg(N^c/N^3)$] upon the separation factor ($lg\ \eta$) defined by the boiling temperature of the elements relative to xenon. The observed linear dependence of the considered parameters indicates a correspondence between the experimental facts and the accepted model based on the ideas about the studied process. Besides the shallower slope of the line corresponding to the dependence on deficit of the element and the separation factor for the Moon as compared with the Earth, the line shows that the formation of the Moon and the distribution of the elements within the proto-Earth cloud occurred at rather high temperatures, which is also in accordance with the given theoretical and model concepts.

The probable conditions of formation of the volatile elements, some of their components, and the isotope composition of the upper shell of the Earth, were also considered (Onufriev 1978). Assuming the maximum boiling temperature of high-temperature constituents of the Earth to be equal to 3,300 K, the corresponding values of the deficit $lg(N^c/N^3)$ depended on the separation factor $lg\ \eta$ for the noble gases and a number of the refractory elements were plotted in Fig. 20.6. The relative volatility of the noble gases, which do not form chemical compounds, depends directly on boiling temperature. It follows from Fig. 20.6a that the dependence of $lg(N^c/N^3) \approx f(lg\ \eta)$ has a natural character. The group of high-temperature elements does not indicate such a dependency in the domain where the values of abundances are close to the solar ones.

The first group of elements was studied in more derail together with a number of other volatile elements and compounds at a maximum boiling temperature of 1,600 K (Fig. 20.6b). It follows from the figure that the value of the deficit of each of the considered elements decreases with a decrease of their relative volatility, which corresponds to the model considerations regarding the formation of the Earth's chemical composition concerning at least the volatile elements.

From the agreement between the corresponding thermodynamic parameters and the observed data on the deficit of a number of other volatile elements and their compounds, plotted in Fig. 20.6b, the following conclusions have been drawn (Onufriev 1978). During the formation of the Earth from the proto-solar nebula the hydrogen

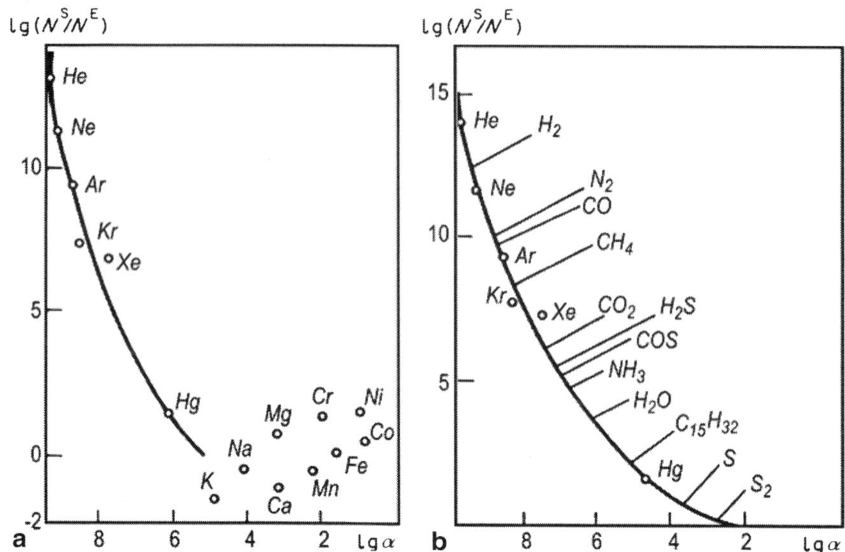

Fig. 20.6 Relationship between the value of the element and the compound deficit and differentiation coefficient for the Earth at 3,300 K (**a**) and 1,600 K (**b**)

reservoir was mainly represented by water and only a small portion of hydrogen was in the molecular form H_2. The nitrogen reservoir was represented by ammonia NH_3 and the Earth's atmosphere as a whole was in a reduced form. Searching the possible primeval carbon compounds of the Earth (CO, CH_4, CO_2, COS), it appeared that neither one of them nor all being taken in corresponding proportions can explain from the assumed viewpoint the observed deficit of carbon. The condition can only be satisfied if assuming the existence of a higher temperature compounds of the form C_nH_{2n+2} with boiling temperature of ~400 K. Assuming the initial existence of sulfur in the form of S, S_2 and so on, the existence of the lighter low-temperature compounds such as H_2S, CaS is required.

Note that investigation of the dependence of the element's deficit upon its relative volatility, expressed through the boiling temperature, together with a consideration of the different forms of its chemical compounds, provides the key for understanding the general principle of the successive increase of the element deficit on the Earth in row Hg, S, C, N, H, Xe, Kr, Ar, Ne, He.

The considered model of formation of the chemical composition of the Earth should be completely applicable for studying the conditions of formation of their isotopes.

Without any loss of generality, one can consider that an element's isotope with a greater mass number is less volatile compared with the lighter isotope. Therefore, expressions (20.15) and (20.16) are valid for estimation of the dependence of abundance on the planets of isotopes of corresponding elements upon the separation factor expressed through the boiling temperature.

It has been shown (Onufriev 1978) that the above-mentioned enrichment of the Earth's upper shell in heavy isotopes, compared with the cosmic abundance for the volatile elements, is natural from the viewpoint of our model. In view of this, there are reasons to consider that water of the hydrosphere was initially enriched in heavy oxygen to about +20‰. The hydrosphere itself should have been in the vapor phase. With decreasing temperature the water over the Earth's surface should have been converted into the liquid phase. Precipitation of large amounts of water on the Earth's surface was the trigger mechanism for magmatic processes to start. The process of granitization of rocks on the continents should have resulted in depletion of the hydrosphere in heavy oxygen and the subsequent enrichment of granite up to the present values. Thus, starting from the considered model of the formation of the Earth, with its chemical and isotopic composition, the observed enrichment of the upper shell of the Earth in heavy isotopes of the light elements can be explained uniquely.

20.11 The Third Kepler's Law as a Kinematics Basis for the Solar System Bodies Creation Problem Solution

Using the above physics, we discovered a very interesting phenomenon, which opens the way for solving the problem. It appears that the mean orbital velocities and periods of revolution of all the planets and asteroids are equal to the first cosmic velocity and corresponding period of the contracting proto-Sun, having its radius equal to the semimajor axes of each planet's orbit. The same was happened with the planets' satellites. The subsequent expansion of the space has not broken the above regularity.

In order to prove the validity of the above statement, we calculated values of the orbital periods of revolution of the planets, asteroids (small planets), and satellites obtained by applying first cosmic velocities of the proto-Sun and the proto-planets and compared the results with observations (Tables 20.7 and 20.8).

The first cosmic velocity v_1 of the proto-Sun's and proto-planetary bodies and the period of oscillation of the corresponding outer shell T_1 of the created bodies were calculated by the formulae from which, in fact, the third Kepler's law follows:

$$v_1 = \omega R = R\sqrt{\frac{Gm}{R^3}} = \sqrt{\frac{Gm}{R}}, \quad T_1 = \frac{2\pi}{\omega} = \frac{2\pi R}{v_1}, \quad \frac{(2\pi)^2}{T_1^2} = \frac{Gm}{R^3},$$

where m is the body's mass; G is the gravity constant; R is the semimajor axis; $\omega = v_1/R$ is the frequency of virial oscillation of the outer shell, which appears to be equal to the angular velocity of the orbital motion. Note that the frequency of virial oscillation of the outer weighty shell does not equal to its angular velocity because the frequency is the parameter of the force field.

For example, when the proto-Sun's radius Rextended up to the present day Earth's orbit ($m = 1.99 \times 10^{30}$ kg, $R = 1.496 \times 10^{11}$ m), then its first cosmic velocity was equal to

Table 20.7 Observable orbital periods of revolution of the planets around the Sun and calculated periods of oscillation of its corresponding outer shell

Planets	Orbital radius, $R \times 10^{11}$ (m)	Observable period of revolution T (year)	Calculating period of oscillation T_1 (year)
Mercury	0.579	0.24	0.2408
Venus	1.082	0.62	0.6153
Earth	1.496	1.0	1.00
Mars	2.28	1.88	1.8823
Vesta	3.53	3.63	3.7594
Juno	3.997	4.37	4.3733
Ceres	4.13	4.6	4.598
Themis	4.68	5.539	5.5397
Jupiter	7.784	11.86	11.8781
Saturn	14.271	29.48	29.4802
Uranus	28.708	84.01	84.1951
Neptune	44.969	164.8	164.9185
Pluto	59.466	248.09	250.8882

$$\nu_1 = \omega R = \sqrt{\frac{Gm_s}{R}} = \sqrt{\frac{6.67 \times 10^{11} \times 1.99 \times 10^{30}}{1.496 \times 10^{11}}}$$
$$= 29786.786 \text{ m/s} = 29.786786 \text{ km/s}.$$

This value corresponds to the observed mean orbital velocity of the Earth.

The period of oscillation of the interacted mass particles of the proto-Sun's outer shell ($R = 1.496 \times 10^{11}$ m, $\nu_1 = 29786.786$ m/s) was equal to

$$T_1 = \frac{2\pi R}{\nu_1} = \frac{6.28 \times 1.496 \times 10^{11}}{29786.786} = 3.1540428 \times 10^7 \text{ s} = 1 \text{ year},$$

which is equal to the observed period of the planet's orbital revolution.

When the proto-Earth's radius R extended up to the present day Moon's orbit ($m_e = 5.976 \times 10^{24}$ kg, $R = 3.844 \times 10^8$ m), then its first cosmic velocity was equal to

$$\nu_1 = \sqrt{\frac{Gm_e}{R}} = \sqrt{\frac{6.67 \times 10^{11} \times 5.976 \times 10^{24}}{3.844 \times 10^8}} = 1018.3018 \text{ m} = 1.0183918 \text{ km/s},$$

which is the present day Moon's mean orbital velocity.

The period of oscillation of the interacted mass particles of the proto-Earth's outer shell ($R = 3.844 \times 10^8$ m, $\nu_1 = 1018.3018$ m/s) was equal to

$$T_1 = \frac{2\pi R}{\nu_1} = \frac{2 \times 3.14 \times 3.844 \times 10^8}{1018.3018} = 23.706449 \times 10^5 \text{ s} = 27.438019 \text{ days},$$

which corresponds to the present day Moon's period of orbital revolution.

Table 20.8 Observable orbital periods of revolution of the satellites around the planets and calculated periods of oscillation of their corresponding outer shells

Planets	Satellites	Orbital radius R, $\times 10^6$, m	Observable period of revolution T, (year)	Calculated period of revolution T_1, (year)
Earth	Moon	384.4	27.32	27.4103
Mars	*Phobos*	9.4	0.319	0.3208
	Deimos	23.5	1.262	1.2604
Jupiter	V	181	0.498	0.4973
	Io	422	1.769	1.7706
	Europa	671	3.551	3.5508
	Ganymede	1,070	7.155	7.154
	Callisto	1,880	16.69	16.6709
	XIII	11,100	240.92	239.0960
	VII	11,750	259.14	259.5899
	XII	21,000	620.77	660.7744
	IX	23,700	758.90	745.1833
Saturn	*Janus*	151.5	0.7	0.6956
	Mimas	185.6	0.94	0.9431
	Enceladus	238.1	1.37	1.3704
	Tethys	294.7	1.89	1.8869
	Dione	377.4	2.74	2.7366
	Titan	1212.9	15.95	15.7548
	Iapetus	3560.8	79.33	79.2494
	Phoebe	12,944	548.2	549.2722
Uranus	*Cordelia*	49.751	0.3350	0.3348
	Cupid	74.8	0.618	0.6172
	Miranda	129.39	1.4135	1.4043
	Ariel	191.02	2.5204	2.5189
	Umbriel	266.3	4.1442	4.1463
	Titania	435.91	8.7058	8.6840
	Oberon	583.52	13.4632	13.4503
Neptune	*Triton*	354.8	5.877	5.8523
	Nereid	5513.4	360.14	359.8227
Pluto	*Charon*	19.571	6.387	9.5065
	Nix	48.675	24.856	37.2873
	Hydra	64.780	38.206	54.2482

The obtained results mean that all the planets and satellites were launched by first cosmic velocity of the self-gravitating proto-Sun and proto-planets after their outer shells acquired weightless. As it was said above, the process of evolutionary loss of energy by emission led to redistribution and differentiation of the body's mass density: increase it in the inner shells and decrease in the outer one by the light components dilution. In general, due to this process and contraction in the form of separation of the matter, the shell separation of a body with respect to the density was developed up to the state of weightlessness and self-gravitation of the outer shell's matter and creation of the body.

The discovered regularity of the solar system's planets and satellites creation seems to be valid for the process of separation of the proto-Sun itself and the other proto-stars from the proto-Galaxy Milky Way. If we accept for the Galaxy's known astrometric data (mass $m_g = 2.5 \times 10^{41}$ kg, distance of the Sun from the Galaxy center $R_s = 2.5 \times 10^{20}$ m), then it is not difficult to calculate that the first cosmic velocity of the proto-Galaxy, the size of which was limited by the Sun's semimajor orbital axes, is equal to 230 km/s, and the orbital period of revolution is 220×10^6 year. The values are close to those found by observation, namely: mean orbital velocity of the Sun is called as (230–250) km/s, and the orbital period of revolution $T_s = (220–250) \times 10^6$ year.

The current observed picture of the Milky Way, consisting of a bar-shaped core surrounded by a disc of gaseous matter and stars, which create two major and four or more smaller logarithmic spiral arms, prove the generally common mechanism of creation of the galactic system. This picture demonstrates that the huge mass and size of the proto-Galaxy rotating body was subjected during evolution by the polar and the equatorial oblateness. Due to redistribution of mass density and after reaching a state of their weightlessness the stars were separated from the outer shell in different surface regions. It appears that the main parts of the mass were separated in two regions (in the pericenter and the apocenter) and the left in all others. However, because the first cosmic velocity is the function of radius of the body, the regions of separation in the space formed the logarithmic spirals of the moving stars in accordance with the third Kepler's law. The same logarithmic spirals were formed also by the planets and the satellites in the solar system.

The following initial values of density ρ_i and radius R_i of the proto-Sun and proto-planets can be obtained on the basis of their dynamic equilibrium state.

The proto-solar cloud has separated from the proto-Galaxy body when its outer shell in the equatorial domain has reached the value of the first cosmic velocity. In fact, the gaseous cloud should represent chemically nonhomogeneous rotating body. As it follows from Roche's dynamics (see Eq. (20.2)), the mean density of the gaseous proto-Galaxy outer shell should be $\rho_s = 2/3\rho_g$. The condition $\rho_s = 2/3\rho_g$ is the starting point of separation and creation of the proto-Sun from the outer proto-Galaxy shell. Accepting the above described mechanism of formation of the secondary body, we can find the mean density of the proto-Galaxy at the proto-Sun separation as

$$\rho_g = \frac{m_g}{\frac{4}{3}\pi R^3} = \frac{2.5 \times 10^{41}}{\frac{4}{3} \times 3.14 \times (2.5 \times 10^{20})^3}$$
$$= 1.67 \times 10^{-21} \text{kg/m}^3 = 1.67 \times 10^{-24} \text{g/cm}^3.$$

The mean density of the separated proto-Galaxy shell is

$$\rho_s = 2/3\rho_g = 2/3 \times 1.67 \times 10^{-24} = 1.11 \times 10^{-24} g/\text{cm}^3.$$

In accordance with Eq. (20.1), the mean density and radius of the initially created proto-Sun body should be

$$\rho_s = 2\rho_g = 2 \times 1.67 \times 10^{-24} = 3.34 \times 10^{-24} g/\text{cm}^3;$$

$$R_s = \sqrt[3]{\frac{2 \times 10^{33}}{\frac{4}{3} \times 3.34 \times 10^{-24}}} = 7.5 \times 10^{18}\text{cm} = 7.5 \times 10^{16}\text{m}$$

The mean density and the radius of the initially created proto-Jupiter, proto-Earth and proto-Moon are as follows:

The proto-Jupiter: $\rho_j = 2 \times 10^{-9}$ g/cm^3, $R_j = 6.2 \times 10^{13}$ cm $= 6.2 \times 10^{11}$ m;
The proto-Earth: $\rho_e = 2.85 \times 10^{-7}$ g/cm^3, $R_e = 1.9 \times 10^{11}$ cm $= 1.9 \times 10^{9}$ m;
The proto-Moon: $\rho_m = 5 \times 10^{-4}$ g/cm^3, $R_m = 1.1 \times 10^{9}$ cm $= 1.1 \times 10^{7}$ m;

Analogous unified process was repeated for all the planets and their satellites.

Creation of the other small bodies like comets, meteors, and meteorites are also found their explanation within the considered mechanism and physics. In fact, the only condition for separation of outer body's shell is its weightlessness (its corresponding mean density relative to the body's mean density), but not a limit of some amount of mass. In this connection any volume and amount of mass could probability be separated at any time, for example, we found by calculation that the short-periodic Encke's Comet (1970 I, $T = 3.302$ years) has semimajor orbital axis $R \approx 1.5 \times 10^{11}$ m and has separated from the proto-Sun after small planet Vesta and before the Mars. The short-periodic Halley's Comet (1910 II, $T = 76.1$ years) has semimajor orbital axis $R = 2.7 \times 10^{12}$ m and has separated from the proto-Sun after Saturn and before Jupiter. The long-periodic Ikeya-Seki's Comet (1965 III, $T = 874$ years) has semimajor orbital axis $R = 1.35 \times 10^{14}$ m and has separated from the proto-Sun before the Pluto. Like the asteroid belt between Jupiter and Mars, the comet belts should definitely exist between the orbits of all the Jupiter group planets. As to the meteors and meteorites, they all should be separated from the planets by the same way. From the viewpoint of dynamical equilibrium of their orbital motion, the orbits of all the small bodies (comets, meteors, and meteorites) should have large eccentricities and dip angles of inclination to the equator of their central bodies. This is because of probable oblateness of the proto-Sun body, where its polar regions should have higher values of the first cosmic velocity. Those small bodies and meteorites, which have not reached or have later on lost dynamical equilibrium, fell down on the planet's or satellite's surface.

As it was shown in the Table 20.8, the small planets of the asteroid belt separated from the proto-Sun by the same mechanism. From the viewpoint of the orbital motion and first cosmic velocities, there are no features of their separation from a broken planet.

The above consideration takes off an old misunderstanding about the difference in the orbital planet's and the Sun's moment of momentum. The secondary body conserves the creation energy and orbital moment of momentum in accordance with the third Kepler's law . As to the direction of body's axial rotation and orbital revolution, then these parameters enter by the inner and outer force field, like in electrodynamics, in accordance with the Lenz law. As to the specific (for unit of the mass) orbital moment momentum of the planets and satellites which increases along with distance from the central body, the explanation of this gives the increasing radius from the central body.

The revealed physics and kinematics of creation and separation of the solar system bodies prove the Huygens' law of motion on semicubic parabola of his watch pendulum, which synchronously follows the Earth motion. This curve was called evolute and the curve perpendicular to the series of tangents to the evolute is called evolvent. The relationship between the evolute and evolvent represents the relationship between function and its derivative or between function and its integral. These relations exist not locally like in mathematical analysis but in integral form and geometrically visible. While plotting a series of the evolvents with fixed lengths of the pendulum a peculiarity of the same type is appeared in each point of the initial curve of evolvent. The peculiarity is the semicubic parabola of type $x^2 = y^3$ or $x = y^{3/2}$. This is just the universal law of a body motion in the nature, which is the consequence of the simple fact. The degree 3/2 is the ratio between the body's mass volume, which generates the energy and the body's surface, which emit this energy; and also, in any task of motion we always have some initial conditions, which the moving object is inherited. In the case of the Huygens' oscillating pendulum the suspension filament starts unrolling in a fixed point. In the case of a celestial body, creation of satellite starts in a fixed point of its parental body where the initial conditions are transferred by the third Kepler's law. This is just because in a fixed point, dynamical equilibrium between the generated volumetric energy (cubic degree of radius) and the irradiated from outer surface energy (square degree of radius) is broken.

First cosmic velocity was practically applied by man only in twentieth century. The nature seems to use it perpetually as the main mechanism of the Universe evolution. Our Universe seems to be a pulsating system and its basic infinitesimal particle is $\sim 10^{-36}$ g in weight, which is responsible for the system equilibrium, because of the matter evolution and energy conservation law is the process, continuing for infinitely long time.

Note that we can now derive a conclusion regarding the meteoric origin of the hydrosphere. The condensation of water to the liquid phase occurred at the final stage of our planet's formation. One can reasonably assume that the time interval between the end of condensation of the mineral part of the Earth and the beginning of water condensation could have been large in view of the difference in their boiling temperatures. This interval could have been markedly enlarged if the greenhouse effect was provided by carbon dioxide and water vapor, similar to that now observed on Venus. Thus there are reasons for considering that the Earth's hydrosphere, in the liquid phase, is a considerably younger formation than all the other shells. Further, the observed stability of chemical and isotopic composition of the oceans in time is a result of inherited thermodynamic equilibrium which it had attained in the gaseous phase.

20.12 Conclusion

Increasingly sophisticated and accurate techniques are being employed by scientists in order to try to understand natural laws. The investigation of the principles of circulation and evolution of natural waters in various shells of the Earth and in the hydrosphere as a whole, carried out on the basis of studying their isotopic composition and also isotopes of cosmogenic and radiogenic origin, is a new discipline in water science. This discipline has advanced greatly in recent years and is believed to provide many important results.

On the basis of generalization given in this book concerning the environmental isotopes in water, one can study and solve many of the principle problems in such fields as hydrology, hydrogeology, oceanography, and meteorology from the viewpoint of the unit of natural waters. This includes such fundamental problems as the origin and evolution of the hydrosphere. The studies undertaken in this direction have shown an ultimate need both for further development and widening of the theoretical concepts and accumulation of experimental data.

The generalization of numerous data on isotope distribution in natural waters and in other dissolved and suspended substances is required. In the first instance these are the noble gases (helium, argon, xenon, krypton, etc.) and also carbon, sulfur, and nitrogen. A further balance estimation of isotopes of the heavy radioactive elements for the hydrosphere as a whole is needed. These investigations will help to advance our understanding of many natural processes occurring in the Earth's hydrosphere now and also of those which took place in the past.

References

Aegerter SK, Loosli HH, Oeschger H (1967) Variation in the production of cosmogenic radionuclides. In: Radioactive dating and methods of low-level counting: proceedings of a symposium organized by the international atomic energy agency in co-operation with the joint commission on applied radioactivity (ICSU) and held in Monaco, 2–10 March 1967, IAEA, Vienna, pp 49–55

Afanasenko EA, Morkovkina IK, Romanov VV (1973) Age of groundwater in the Tas-Khayat Ridge and Selennyakh depression. Bull Moscow Univ Ser Geol 5:105–109

Afanasyev AP (1960) Water balance of Baykal lake. Proc Baykal Limnol Stn 18:85–95

Afanasyeva EA (1947) Origin and properties of the thick powerfull black earth in the Streletskaya steppe. Proc Soil Inst 227–235

Ahmad SN, Perry ES Jr (1980) Isotopic evolution of the sea. Sci Prog 66:499–511

Alder B, Oeschger H, Wasson YT (1967) Aluminium-26 in deep-sea sediments. In: Radioactive dating and methods of low-level counting: proceedings of a symposium organized by the international atomic energy agency in co-operation with the joint commission on applied radioactivity (ICSU) and held in Monaco, 2–10 March 1967, IAEA, Vienna, pp 189–195

Alekseev FA, Bondarev LG, Zverev VL, Spiridonov AI (1973) Influence of precipitation radioactivity on isotopic composition of uranium in the Issyk-Kul Lake in connection with the age determination. Geokhimiya 5:787–780

Alekseev FA, Gorbushina LV, Ovchinnikov AM, Tyminsky VG (1966) Age of water from the tashkent artesian basin. In: Alekseev FA (ed) Voprosy isotopnoi geologii, vol 3. Nedra, Moscow, pp 40–45

Alekseev FA, Gottikh RP, Saakov SA, Sokolovsky EV (1975) Radiochemical and isotopic investigations of groundwater in gas and oil-bearing areas of the USSR. Nedra, Moscow

Alekseev FA, Vetshtein VE, Malyuk GA (1974) The isotopic composition of hydrogen and oxygen in groundwater of the Amu-Darya gas and oil-bearing basin as a criterion of its genesis and dynamics. Yadern Geol 1974:62–74 (Nedra, Moscow)

Alfven H, Arrhenius G (1970) Origin and evolution of the solar system. Astrophys Space Sci 9:30–33

Allemmoz M, Olive Ph (1980) Recharge of groundwater in arid areas: case of the Dieffra plain in Tripolitania, Libyan Arab Jamahiriya. In: IAEA (ed) Arid–zone hydrology: investigations with isotope techniques: proceedings of an advisory group meeting on application of isotope techniques in arid zones hydrology. IAEA, Vienna, pp 181–192

Allison GB, Hughes MW (1974) Environmental tritium in the unsaturated zone: estimation of recharge to an unconfined aquifer. In: Isotope techniques in groundwater hydrology: proceedings of a symposium, IAEA, Vienna, 11–15 March 1974, pp 57–70

Allison GB, Turner JV, Holmes JW (1971) Estimation of groundwater inflow to small lakes. In: Isotopes in lake studies. Proceedings of an advisory group meeting, IAEA, Vienna, pp 103–113

Ambroggi RR (1966) Water under the Sahara. Sci Amer 214:21–29

V.I. Ferronsky, V. A. Polyakov, *Isotopes of the Earth's Hydrosphere*,
DOI 10.1007/978-94-007-2856-1,

Andersen LJ, Sevel T (1974) Six years environmental tritium profiles in the unsaturated and saturated zones. In: Isotope techniques in groundwater hydrology: proceedings of a symposium, IAEA, Vienna, 11–15 March 1974, vol 1, pp 3–18

Anders E (1972) Physico-chemical processes in the solar nebula, as inferred from meteorites. In: Reeves H (ed) Origin of the solar system: proc symp. Cent Nat Rech Sci, Paris, pp 179–266

Andersen LJ, Sevel T (1974) Six years' environmental tritium profiles in the unsaturated and saturated zones, Gronhoi, Denmark. In: Isotope techniques in groundwater hydrology: proceedings of a symposium, IAEA, Vienna, 11–15 March 1974, vol 1, pp 3–18

Andreev PF, Rogozina EM, Rogozin YuM (1960) Extraction of uranium from rocks by ultrasonic action. J Phys Chem 34:2429–2430

Andrusov NI (1900) The ancient shore lines along the Caspian Sea. Yearbook on geology and mineralogy of Russia 4(1–2):3–21

Arslanov KhA (1987) Radiocarbon geochemistry and geochronology. Leningrad University Press, Leningrad

Andrews JN, Kay RLF (1982) Natural production of tritium in permeable rocks. Nature 298:361–363

Appa Rao MVK (1962) The ^{3}He/(^{3}He+^{4}He) ratio in primary cosmic radiation. J Geophys Res 67:1289–1392

Arnason B (1977a) The hydrogen-water isotope thermometer applied to geothermal areas in Iceland. Geothermics 5:75–80

Arnason B (1977b) Hydrothermal systems in Iceland traced by deuterium. Geothermics 5:125–151

Arnason B, Sigurgeirsson Th (1967) Hydrogen isotopes in hydrological studies in Iceland. In: Isotopes in hydrology: proceedings of the symposium on isotopes in hydrology held by the international atomic energy agency in cooperation with the international union of geodesy and geophysics in Vienna, IAEA, Vienna, 14–18 Nov 1966, pp 35–47

Arrhenius G, Alfven H (1977) Fractionation and condensation in space. Earth Planet Sci Lett 10:253–267

Atakan YW, Roether W, Münnich KO, Matthess G (1974) The sandhausen shallow groundwater tritium experiment. In: Isotope techniques in groundwater hydrology: proceedings of a symposium, IAEA, Vienna, 11–15 March 1974, vol 1, pp 21–43

Ault WU, Kulp IL (1959) Isotopic geochemistry of sulphur. Geochim Cosmochim Acta 16:201–235

Babinets AYe, Lugova GP, Markus VI (1971) Oxigen isotopic composition of groundwater of Ukrainian Carpathians region. Dokl AN URSR 7:579–581

Babinets AYe, Vetshtein VE (1967) ^{18}O content of some different on genesis natural waters. In: Babinets AYe (ed) The problems of soil study in hydrogeology and engineering geology. Acad Sci Ukr, Kiev, pp 11–21

Bacastrow R, Keeling CD (1973) Atmospheric CO_2 in the natural carbon cycle. II. Changes from A.D. 1700 to 2007 as deduced from a geochemical model. In: Carbon and the biosphere: proceedings of the 24th Brookhaven symposium in biology, Upton, NY, 16-18 May 1972, Springfield, pp 86–135

Baertschi P (1953) Uber die relativen unter schiede im $H_2^{18}O$-gehalt naturlicher wasser. Helv Chem Acta 36:1352–1369

Baertschi P (1976) Absolute ^{18}O content of standard mean ocean water. Earth Planet Sci Lett 31:341–344

Bainbridge AE (1963) Tritium in the Northern Pacific surface water. J Geophys Res 68:3785–3789

Banis YuYu, Yodkasis VI (1981) About hydrogeological interpretation of radiocarbob data. In: Ferronsky VI (ed) Investigation of natural waters by isotope methods, Nauka, Miskva, pp 146–154

Banwell CJ (1963) Oxygen and hydrogen isotopes in New Zealand thermal areas. In: Tongiorgi E (ed) Nuclear geology on geothermal areas: Spoleto. Consiglio Nazionale delle Ricerche, Piazzale Aldo Moro, pp 95–138

Baonza E, Plata Bedmar A, Jimenez JA (1979) Estudio del comportamiento dinamico del estuario del rio Guadalquivir. In: Isotope hydrology 1978: proceedings of an international symposium on isotope hydrology, vol 2. IAEA, Vienna, pp 847–873

Baranov VI (1955) Radiometry. Izdat. AN SSSR, Moscow

Barrer RM, Denny AF (1964) Water in hydrates 1. Fractionation of hydrogen isotopes by crystallization of salt hydrates. J Chem Soc 26:4677–4684

Barry PJ, Merritt WE (1970) Perch lake evaporation study. In: IAEA (ed) Isotope hydrology: proceedings of a symposium on use of isotopes in hydrology. IAEA, Vienna, pp 139–150

Bartels OG (1972) An estimate of volcanic contributions to the atmosphere and volcanic gases and sublimates as the source of the radioisotopes ^{10}B, ^{35}S, ^{32}P and ^{22}Na. Health Phys 22:387–392

Baskov EA, Vetstein VE, Surikov SN (1973) Isotopic composition of H, O, C, Ar, and He in thermal waters and gases of the Kurilo-Kamchatka volcanous region as an indicator of their formation. Geokhimiya 2:180–189

Bath AH, Edmunds WM, Andrews JN (1979) Palaeoclimatic trends deduced from the hydrochemistry of a Triassic sandstone aquifer, United Kingdom. In: Isotope hydrology 1978: proceedings of an international symposium on isotope hydrology. IAEA, Vienna, pp 545–566

Baturin GN (1968) About geochemistry of uranium in the Baltic Sea. Geokhimiya 3:377–381

Baturin GN (1968) Relationship in forms of uranium migration in some rivers of the USSR territory. Dokl AN SSSR 178:698–701

Baturin GN, Kochenov AV (1969) Uranium migration in rivers and its residence time in waters of oceans, rivers and lakes. Geokhimiya 6:715–723

Baturin GN, Kochenov AV, Kovaleva SA (1966) Some peculiarities of uranium distribution in the Black Sea water. Dokl AN SSSR 166:698–700

Baumgartner A, Reichel E (1975) Die Weltwasserbilanz. Oldenbourg, Munich

Becker R, Clayton R (1976) Oxygen isotope study of a Precambrian banded iron-formation. Geochim Cosmochim Acta 40:1153–1165

Begemann F (1959) Neubestimmung der natürlichen irdischen Tritiumzerfallstrate und die Frage der Herkunft des natürlichen Tritium. Naturforsch 4a:334–342

Begemann F (1963) The tritium content of hot springs in some geothermal areas. In: Tongiorgi E (ed) Nuclear geology of geothermal areas: Spoleto. Consiglio Nazionale delle Ricerche, Piazzale Aldo Moro, pp 55–70

Begemann F, Libby WF (1957) Continental water balance, groundwater inventory and storage times, surface ocean, mixing rates and world-wide water circulation patterns from cosmic-ray and bomb tritium. Geochim Cosmochim Acta 12:277–296

Begemann F, Friedman I (1968) Isotopic composition of atmospheric hydrogen. J Geophys Res 73:1139–1147

Bentley HW, Phillips FM, Davis SN (1986) Chlorine-36 dating of very old groundwater 1. The Great Artesian basin, Australia. Water Resour Res 22:1991–2001

Berger R (1979) Artificial radiocarbon in the stratosphere. In: Berger R, Suess HE (eds) Radiocarbon dating. University of California Press, Berkeley, pp 309–321

Bernat M, Goldberg ED (1969) Thorium isotopes in the marine environment. Earth Planet Sci Lett 5:308–312

Berzina IG (1969) The use of the heavy radioelement fission fragments for solution of practical problems. In: The state and perspectives of nuclear geophysics in prospecting of natural resources, Nedra, Moscow, pp 298–322

Beus AA (1972) Geochemistry of the litosphere. Nedra, Moscow

Bhat SG, Krishnaswamy S, Lal D, Moore WS (1969) $^{234}Th/^{238}U$ ratios in the ocean. Earth Planet Sci Lett 5:483–491

Bhaudari W, Fruchter J, Evans J (1969) Rates of production of ^{22}Na and ^{28}Mg in the atmosphere by cosmic radiation. Earth Planet Sci Lett 7:89–92

Behrens H, Mozer H, Oerter H (1979) Models for runoff from a glaciated cachment area using measurements of environmental isotope contents. In: Isotope hydrology 1978: proceedings of an international symposium on isotope hydrology, vol 2. IAEA, Vienna, pp 829–845

Bien GS, Rakestrow P, Oldenbourg, Suess HE (1963) Investigations in marine environments using radioisotopes produced by cosmic rays. In: Berger R, Suess HE (eds) Radiocarbon dating. IAEA, Vienna, pp 159–174

Bien GS, Suess HE (1967) Transfer and exchange of ^{14}C between the atmosphere and the surface water of the Pacific Ocean. In: Radioactive dating and methods of low-level counting: proceedings of a symposium organized by the international atomic energy agency in co-operation with the joint commission on applied radioactivity (ICSU) and held in Monaco, 2–10 March 1967, IAEA, Vienna, pp 105–115

Bigeleisen J, Mayer MG (1947) Calculation of equilibrium constants for isotopic exchange reactions. J Chem Phys 15:261–267

Blanchard RL, Oakes D (1970) Relationship between uranium and radium in costal marine shells and their environment. J Geophys Res 70:2911–2921

Blander M, Abdel-Gavad M (1969) The origin of meteorites and the constrained equilibrium condensation theory. Geochim Cosmochim Acta 33:701–716

Blander M, Katz JZ (1967) Condensation of primordial dust. Geochim Cosmochim Acta 31:1025–1034

Boato G (1954) The isotopic composition of hydrogen and carbon in the carbonaceous chondrites. Geochim Cosmochim Acta 6:209–220

Boato G, Careri G, Volpi GG (1952) Hydrogen isotopes in steam wells. Nuovo Cim 9:539–540

Boella GC, Dilworth M, Panetti M, Scarsi L (1968) The atmospheric and leakage flux of neutrons produced in the atmosphere by cosmic ray interactions. Earth Planet Sci Lett 4:393–398

Bochaler P, Eberhardt P, Geiss J (1971) Tritium in lunar materials. Proc 2nd Lunar Sci Conf 2:1803–1812

Bondarenko GN (1983) Formation of isotopic composition of carbonate system in natural waters. Naukova Dumka, Kiev

Bonka H (1979) Production and emission of tritium from nuclear facilities, and the resulting problems. In: IAEA (ed) Behaviour of tritium in the environment: proceedings of the international symposium on the behaviour of tritium in the environment, vol 1. IAEA, Vienna, pp 105–122

Borevsky BV, Karasev BV, Litvak DP et al (1975) Application of radiocarbon method for justifying the relationship between aquifers in Kiev city region. In: Radioisotope methods in hydrogeology. Naukova Dumka, Kiev, pp 17–21

Borevsky BV, Polyakov VA, Subbotina LA (1981) Investigation of regularities in leaching through Neocom-Jurassic strata within the depression funnel of the Bryansk water intake well. In: Isotopes in the hydrosphere, Abstract Symp, IWP RAN, Moscow, pp 52–53

Borshchevsky YuA (1980) The nature of hydrothermal ore fluides according to oxygen and hydrogen isotopes data. Geokhimiya 11:1650–1661

Borshchevsky YuA, Khristianov VK (1965) Isotopic composition of cristalline water in salt minerals. Geokhimiya 7:844–850

Bottinga Y (1969) Calculated fractionation factors for the carbon and hydrogen isotope exchange in the system calcite–carbon dioxide–graphite–methan–hydrogen–water vapor. Geochim Cosmochim Acta 33:35–40

Bottinga Y, Craig H (1968) High temperature liquid-vapor fractionation factors for H2O–HDO–H2^{18}O. Trans Am Geophys Union 49:356–357

Bottinga Y, Javoy M (1973) Comments on oxygen isotopes geochemistry. Earth Planet Sci Lett 20:250–265

Bowen R (1966) Paleotemperature analysis. Elsevier, Amsterdam

Bowen R (1991) Isotopes and climayes. Elsevier, London

Bowen R, Williams PW (1973) Geohydrologic study of the Gort Lowland and adjacent areas of western Ireland using environmental isotopes. Water Resour Res 9:753–758

Bradley W, Stout G (1970) Vertical distribution of tritium in water vapor in the lower troposphere. Tellus 22:699–706

Brezgunov VS (1978) Regularity in distribution of hydrogen and oxygen stable isotopes distribution in natural waters during their global circulation. In: Ferronsky VI (ed) Isotopy of natural waters. Nauka, Moscow, pp 10–45

Brezgunov VS (1991) Character of atmospheric moisture exchange of the Issyk-Kul basin with the surrounding territory on the basis of the distribution of stable oxygen isotopes in natural waters. Water Resour 17:616–627

Brezgunov VS, Debolsky VK, Nechaev VV (1982) Peculiarities of formation of oxygen isotopes and salinity at marine and river waters mixing in the Barents and Kara seas. Water Resour 4:3–14

Brezgunov VS, Ferronsky VI (2005) Natural tritium as an indicator of changes in the vertical structure of Caspian Sea water mass with sea level variations. Water Resour 32:365–368

Brezgunov VS, Nechaev VV (1981) Water balance and balance of oxygen stable isotopes in the Issyk-Kul depression. In: Ferronsky VI (ed) Investigation of natural waters by isotope methods. Nauka, Moscow, pp 10–14

Brezgunov VS, Nechaev VV, Erokhin VS (1979) Study of hydrogen and oxygen stable isotope distribution during water exchange in the Issyk-Kul depression. In: Ferronsky VI (ed) Isotope studies of natural waters. Nauka, Moscow, pp 61–69

Brezgunov VS, Nechaev VV, Romanov VV, Ferronsky VI (1980) The study of genesis and dynamics of large lake basins by isotope technique. Water Resour 1:110–120

Brezgunov VS, Vlasova LS, Soifer VN (1966) Deuterium in oils and waters of oil fields. In: Alekseev FA (ed) Isotope geology problems. Nedra, Moscow, pp 24–28

Briggs MH (1963) Evidence of an extraterrestrial origin for some organic constituents of meteorites. Nature 197:1290

Broder DP, Golubev LI, Ilyasov VM (1979) Tritium distribution in the technology scheme of the New-Voronezh atomic power station. Atom Energy 47:120–122

Brodsky AI (1957) Chemistry of isotopes. Izd. AN SSSR, Moscow

Broecker WS, Gerard R, Erwing M, Heezen BC (1960) Natural radiocarbon in the Atlantic ocean. J Geophys Res 65:2903–2909

Broecker WS, Goddard J, Sarmiento JI (1976) The distribution of ^{226}Ra in Atlantic ocean. Earth Planet Sci Lett 32:220–238

Broecker WS, Kaufman A (1970) Near-surface and near-bottom radon results for the 1969 North Pacific geosecs station. J Geophys Res 75:7679–7681

Broecker WS, Kaufman A, Ku TL, Chung Y (1970) Radium-226 measurements from the 1969 North Pacific Station. J Geophys Res 75:7682–7685

Broecker WS, Li YH, Cromwell J (1967) Radium-226 and radon-222 concentration in Atlantic and Pacific oceans. Science 158:1307–1310

Broecker WS, Olson EA (1961) Lamont radiocarbon measurements VIII. Radiocarbon 21:199–216

Brown RM (1961) Hydrology of tritium in the Ottawa Valley. Geochim Cosmochim Acta 33:199–216

Brown RM (1970) Distribution of hydrogen isotopes in Canada waters. In: IAEA (ed) Isotope hydrology: proceedings of a symposium on use of isotopes in hydrology. IAEA, Vienna, pp 3–21

Brown R, Taylor CB (1974) Geohydrology of the Kaikoura-plain Marlborough, New Zealand. In: Isotope techniques in groundwater hydrology: proceedings of a symposium, IAEA, Vienna, 11–15 March 1974, vol 1, pp 169–189

Brown R, Barry PJ (1979) A review of HTO evaporation studies at Chalk River nuclear laboratories. In: Isotopes in lake studies. Proceedings of an advisory group meeting, IAEA, Vienna, pp 73–86

Brown R, Grummit WE (1956) The determination of tritium in natural waters. Canad J Chem 34:220–226

Bullard E (1978) Review of ideas of plate tectonics. In: Fischer AG, Judson S (eds) Petroleum and global tectonics. Nedra, Moscow, pp 9–20 (trans: from English)

Burchuladze AA, Gedevanishvili DD, Pagava SV, Togonidze GI (1977) Variation of radiocarbon content in the atmosphere for 1950-1975 years measured in Georgian vines. In: Low radioactivity measurements and applications. Proceedings of a symposium, High Tatras, 1975, Slovenske Pedagog, Nakland, Bratislava, pp 261–263

Burger LL (1979) Distribution and reactions of tritiated hydrogen and methane. In: IAEA (ed) Behaviour of tritium in the environment: proceedings of the international symposium on the behaviour of tritium in the environment. IAEA, Vienna, pp 47–63

Burkhard W, Fröhlich K (1970) Grundlagen hydrologischer tritium untersuchungen und Ihre anwendung bei der bestimmung der herkunft in grubenwasser einer eisenerzgrube. Bergakademie, Freiberg, H 22, 15–30

Burnett BC, Fowler WA, Hoyle F (1965) Nucleosynthesis in the early history of the solar system. Geochem Cosmochem Acta 29:1209–1241

Buttlar HV, Libby WF (1955) Natural distribution of cosmic ray produced tritium. J Inorg Nucl Chem 1:75–91

Buur CS, Thomas JM, Reines D et al (2001) Sample preparation of dissolved organic carbon in groundwater for AMS ^{14}C analysis. Radiocarbon 43:183–190

Cain WF (1979) 14^{C} in modern American trees. In: Berger R, Suess HE (eds) Radiocarbon dating. University of California Press, Berkeley, pp 495–510

Cerrai E, Lonati R, Gazzarini F, Tongeorgi E (1965) Il metodo ionio-uranio per la determinazione della'eta dei minerali vulcanici recenti. Rend Della Soc Mineralog Italia 21:109–115

Chalov PI (1975) Isotope fractionation of natural uranium. Ilim, Frunze

Chalov PI (1968) Dating by non-equilibrium uranium. Ilim, Frunze

Chalov PI (1959) Isotope ratio of $^{234}U/^{238}U$ in some secondary minerals. Geokhimiya 2:165–170

Chalov PI, Merkulova KI (1966) Relative velocity of atoms of ^{234}U and ^{238}U oxydation in some minerals. Dokl AN SSSR 167:669–671

Chalov PI, Merkulova KI, Tuzova TV (1966a) Absolute age of the Aral Sea determined by nonequilibrium uranium. Dokl AN SSSR 166:89–91

Chalov PI, Merkulova KI, Tuzova TV (1966b) Ratio of $^{234}U/^{238}U$ in water and bottom sediments of the Aral Sea and its absolute age. Geokhimiya 12:1431–1438

Chalov PI, Svetlichnaya NA, Tuzova TV (1973) Application of nonequilibrium uranium in establishing relationship between continental reservoirs in the past. Geokhimiya 6:897–902

Chalov PI, Svetlichnaya NA, Tuzova TV (1970) The results of absolute age determination of Balkhash Lake by nonequilibrium uranium. Dokl AN SSSR 195:190–192

Chalov PI, Tuzova TV, Musin YaA (1964) Isotope ratio of $^{234}U/^{238}U$ in natural waters and its application in nuclear geochronology. Geokhimiya 5:404–413

Champine WJ, Dinçer T, Woory M (1979) An evaluation of isotope composition in the groundwater of Saudi Arabia. In: IAEA (ed) Isotope hydrology, 1978: proceedings of an international symposium on isotope hydrology, vol 2. IAEA, Vienna

Chan LH (1976) Radium and barium at Geosecs stations in the Atlantic and Pacific. Earth Planet Sci Lett 32:258–267

Cherdyntsve VV (1955) Isotopic composition of radioelements in natural objects in connection with their geochronology. In: Annals of the third commission on absolute determination of geologic age, Nauka, Moscow, pp 175–233

Cherdyntsev VV (1969) Uranium-234. Atomizdat, Moscow

Cherdyntsev VV (1973) Nuclear vulcanology. Nauka, Moscow

Cherdyntsev VV, Kazachevsky IV Kuzmina EA (1965) The age of Pleistocene carbonate formations by uranium isotopes. Geokhimiya 9:1085–1092

Cherdyntsev VV, Kazachevsky IV, Kuzmina EA (1963) Isotopic composition of uranium and thorium in zone of hypergeneze. Geokhimiya 3:254–265

Cherdyntsev VV, Kazachevsky IV. Kislitsyna GI et al (1966) Nonequilibrium uranium in carbonate deposits and their age. Geokhimiya 2:1939–1946

Cherdyntsev VV, Kazachevsky IV, Kuzmina EA et al (1967) Absolute geochronology of Cenozoic deposits. Proc Acad Sci USSR, Ser Geol 1:11–20

Cherdyntsev VV, Kuptsov VM, Kuzmina EA, Zverev VL (1968) Radioisotopes and protactinium age of neovulcanic rocks of Caucasus. Geokhimiya 1:77–85

Chirkov AM (1971) ^{222}Rn content in Kamchatka's hydrothems. Dokl AN SSSR 199:202–203

Chung YC (1974a) Transient excess-radon profile in Pacific bottom water. Earth Planet Sci Lett 21:295–300

Chung YC (1974b) Radium-226 and Ra-Ba relationships in in Antarctic and Pacific waters. Earth Planet Sci Lett 23:125–135

Chung YC (1976) A deep ^{226}Ra maximum in the northeast Pacific. Earth Planet Sci Lett 32:249–257

Chung YC, Craig H, Ku TL, Goddard J, Broecker WS (1974) Radium 226 measurements from three Geosecs intercalibration stations. Earth Planet Sci Lett 23:116–124

Cambray RS, Fisher EM, Peirson DH et al (1979) Radioactive fallout in air and rain: results to end of 1978. Harwell, AERE, R-9441

Cameron AGW (1973) Accumulation processes in the primitive Solar nebulae. Icarus 18:407–450

Cameron AGW, Pine MR (1973) Numerical models of the primitive solar nebulae. Icarus 18:377–406

Cartro MC, Stute M, Schlosser P (2000) Comparison of ^{4}He ages and ^{14}C ages in sample aquifer systems: implications for groundwater flow and chronologies. Appl Geochem 15:1137–1167

Castany G, Marce A, Margat J et al (1974) Etude per les isotopes du milieu du regime des eaux souterraines dans les aquiferes de grandes dimensiones. In: Isotope techniques in groundwater hydrology: proceedings of a symposium, IAEA, Vienna, 11–15 March 1974, vol 1, pp 243–257

Champine WJ, Dinçer T, Woory M (1979) An evolution of isotope concentrations in the groundwater of Saudi Arabia. In: Isotope hydrology 1978: proceedings of an international symposium on isotope hydrology, vol 2. IAEA, Vienna, pp 443–462

Chase CG, Perry EC (1972) The oceans: growth and oxygen isotope evolution. Science 177:992–994

Chase CG, Perry EC (1973) Oceanic growth models: discussions. Science 182:602–603

Clark SP, Turekian KK, Grossman L (1972) Model for the early history of the Earth. In: The nature of the solid earth, McGraw-Hill, pp 3–18

Clayton RN (1961) Oxygen isotopic fractionation between calcium carbonate and water. J Chem Phys 34:724–726

Clayton RN, Friedman I, Graf DL et al (1966) The origin of saline formation waters: 1. Isotopic composition. J Geophys Res 71:3869–3882

Clayton RN,Mayeda T (1978a) Genetic relation between iron and stony meteorites. Earth Planet Sci Lett 42:325–327

Clayton RN Mayeda T (1978b) Multiple parent bodies of polimict brecciated meteorites. Geochim Cosmochim Acta 42:325–327

Clayton RN, O'Neil JR, Mayeda TK (1972) Oxygen isotope exchange between quartz and water. J Geophys Res 77:3057–3067

Clayton RN, Onuma N, Grossman L et al (1977) Distribution of the pre-Solar component in Allende and other carbonaceous chondrites. Earth Planet Sci Lett 34:209–224

Clayton RN, Onuma N, Mayeda TK (1976) A classification of meteorites based on oxygen isotopes. Earth Planet Sci Lett 30:10–18

Combs F, Doda RY (1979) Large-scale distribution of tritium in a commercial product. In: IAEA (ed) Behaviour of tritium in the environment: proceedings of the international symposium on the behaviour of tritium in the environment. IAEA, Vienna, pp 93–99

Compston W, Epstein SA (1958) Method for the preparation of carbon dioxide from water vapor for oxygen isotope analysis. Trans Amer Geophys Union 39:511–512

Conrad G, Fontes JCh (1970) Hydrologie du Sahara Nord—Occidental. In: IAEA (ed) Isotope hydrology: proceedings of a symposium on use of isotopes in hydrology. IAEA, Vienna, pp 405–419

Conrad G, Fontes JCh (1972) Circultions dires et periodes de recharge les nappes aquiferes du Nord-Quest Sahariens: dounees isotopiques (^{18}O, ^{13}C, ^{14}C)', C R Acad Sci 2:156–168

Cook PG, Herzeg AL (2000) Environmental tracers in subsurface hydrology. Kluwer, Dordrecht

Coplen TB, Hanshow BB (1973) Ultrafiltration by a compacted clay membrane, 1. Oxygen and hydrogen isotopic fractionation. Geochim Cosmochim Acta 37:2295–2310

Coplen TB, Schlanger SO (1973) Oxygen and carbon isotope studies of carbonate sediments from site 167, Magellan Rise, Leg 17. Initial Rep DSDP 7:505–510

Coplen TB, Winograd U, Landwerer JM, Riggs PC (1994) 500,000 year stable isotope record from Devil's Hole, Nevada. Science 263:361–365

Cortecci G (1974) Oxygen isotope ratios of sulfate ions-water pairs as a possible geothermometer. Geothermics 3:60–64

Cotardiere Ph (1975) Jupiter: un piege fluide. La Decouverte des Planetes, Paris, pp 71–79

Cotecchia V, Tasioli GS, Vaari G (1974) Isotopic measurements in research on seawater ingression in the carbonate aquifer of the Salentine peninsula, Southern Italy. In: Isotope techniques in groundwater hydrology: proceedings of a symposium, IAEA, Vienna, 11–15 March 1974, vol 1, pp 445–463

Craig H (1953) The geochemistry of stable carbon isotopes. Geochim Cosmochim Acta 3:53–92

Craig H (1957) Isotopic standards for carbon and oxygen and correction factors for mass-spectrometric analysis of carbon dioxide. Geochim Cosmochim Acta 12:133–149

Craig H (1961a) Isotopic variations in meteoric waters. Science 133:133–149

Craig H (1961b) Standard for reporting concentrations of deuterium and oxygen-18 in natural waters. Science 133:1833–1834

Craig H (1963) The isotopic geochemistry of water and carbon in geothermal areas. In: Tongiorgi E (ed) Nuclear geology of geothermal areas: Spoleto. Consiglio Nazionale delle Ricerche, Piazzale Aldo Moro, pp 17–53

Craig H (1965) The measurement of oxygen isotopic paleotemperatures. In: Tongiorgi E (ed) Stable isotopes in oceanographic studies and paleotemperatures: Spoleto. Consiglio Nazionale delle Ricerche, Piazzale Aldo Moro, pp 161–182

Craig H (1966) Isotopic composition and origin of the Red Sea and Salton Sea geothermal brines. Science 154:1544–1548

Craig H (1969) Geochemistry and origin of the Red Sea brines. In: Degens ET, Ross DA (eds) Hot brines and recent heavy metal deposits in Red Sea. Springer, New York, pp 208–242

Craig H Gordon L (1965) Deuterium and oxygen-18 variations in the ocean and the marine atmosphere. In: Tongiorgi E (ed) Stable isotopes in oceanographic studies and paleotemperatures: Spoleto. Consiglio Nazionale delle Ricerche, Piazzale Aldo Moro, pp 9–130

Craig H, Gordon L, Horibe Y (1963) Isotopic exchange effects in the evaporation of water 1. Low-temperature experimental results. J Geophys Res 68:5079–5087

Craig H, Lal D (1961) The production rate of natural tritium. Tellus 13:85–105

Craig H, Lupton JE (1976) Primordial neon, helium and hydrogen in oceanic basalts. Earth Planet Sci Lett 31:365–385

Crouset E, Hubert P, Olive Ph et al (1970) Le tritium dans les mesures d'hydrologie de surface: Determination experimentale du coefficient de russellement. J Hydrol 11:341–346

Crozaz G (1967) Datation des glaciers par le plomb-210. In: IAEA (ed) Radioactive dating and methods of low-level count. IAEA, Vienna, pp 385–392

Daly IC, Manchester AV, Gabay IJ, Sax NI (1968) Tritiated moisture in the atmosphere, surrounding a nuclear fuel reprocessing plant. Radiol Health Data and Repts 11:217–229

Damon PE (1970) Climatic versus magnetic perturbation of the atmospheric C-14 reservoir. In: Radiocarbon variations and absolute chronology, XII nobel symposium, Wiley, New York, pp 571–593

Dansgaard W (1964) Stable isotopes in precipitation. Tellus 19:435–463

Dansgaard W, Johnson SJ, Möller J et al (1969) One thousand centuries of climatic record from Camp Century on the Greenland ice sheet. Science 166:377–380

Dansgaard W, Johnson SJ, Clausen HB, Langway CC (1971) Climatic record reveald by Camp Century ice core. In: Turekian KK (ed) Late Cenozoic glacial ages. Yale University Press, Yale, pp 37–56

Dass L, Zouari K, Fage S (2005) Identifying sources of groundwater recharge in the Merguellil basin (Tunisie) using isotope methods: implication of dam reservoir water assounting. Environ Geol 49:114–123

Davis SB, Fabrika-Martin J, Wolfsberg L et al (2000) Chlorine-36 in groundwater contaminating low chloride concentrations. Groundwater 38:912–921

Deak J (1974) Use of environmental isotopes to investigate the connection between surface waters in the Nagynusad region, Hungary. In: Isotope techniques in groundwater hydrology: proceedings of a symposium, IAEA, Vienna, 11–15 March 1974, vol 1, pp 157–167

Deak J (1979) Environmental isotopes and water chemical studies for groundwater research in Hungary. In: IAEA (ed) Isotope hydrology, 1978: proceedings of an international symposium on isotope hydrology, vol 1. IAEA, Vienna, pp 221–247

Degens ET (1961) Project new walley. Eng Sci 25:20–26

Degens ET (1962) Geochemische Untersuchungen von Wasser aus der ägyptischen Sahara. Geol Rdsch Bd 52:625–637

Degens ET, Epstein S (1962) Relationship between $^{18}O/^{16}O$ ratios in coexisting carbonates, cherts and diatomits. Bull Amer Assoc Petrol Geol 46:534–542

Degens ET, Epstein S (1964) Oxygen and carbon isotopic ratios in coexisting calcites and dolomites from resent and ancient sediments. Geochim Cosmochim Acta 28:23–44

Degens ET, Hunt JM, Reuter JH et al (1964) Data on the distribution of aminoacides and oxygen isotopes in petroleum brine waters of various geologic ages. Sedimentology 3:199–225

Deines P, Wickman FE (1975) A contribution to the stable isotope geochemistry of iron meteorites. Geochim Cosmochim Acta 39:547–557

Dementyev VS, Syromiatnikov NG (1965) The residence form of thorium isotopes in groundwater. Geokhimiya 2:211–268

Dergachev VA (1975) Variations in solar activity and radiocarbon content in the atmosphere. Izv AN SSSR, Ser Phys 32:325–333

Dergachev VA (1977) Optimal model for residence time determination of exchange reservoirs. In: Low-radioactivity measurements and applications. Proceedings of a symposium, IAEA, Vienna, pp 269–277

Dergachev VA, Kocharov GE (1977) The ceqular cycles of radiocarbon variation in the Earth atmosphere. In: Low-radioactivity measurements and applications. Proceedings of a symposium, IAEA, Vienna, pp 279–286

Devis DH (1970) Geohydrologic interpretations of a volcanic island from environmental isotopes. Water Resource Res 6:652–671

Dinçer T (1968) The use of oxygen-18 and deuterium concentrations in the water balance of lakes. Water Resour Res 4:1289–1306

Dinçer T, Halevy E (1968) Leakiage from lakes and reservoirs. In: IAEA (ed) Guidebook on nuclear techniques in hydrology. IAEA, Vienna, pp 77–83

Dinçer T, Martinec J, Payne BR et al (1970) Variation of the tritium and oxygen-18 content in precipitation and snowpack in a representative basin in Czechoslovakia. In: IAEA (ed) Isotope hydrology: proceedings of a symposium on use of isotopes in hydrology. IAEA, Vienna, pp 23–41

Dinçer T, Noory M, Javed ARK et al (1974) Study of groundwater recharge and movement in shallow and deep aquifers in Saudy Arabia with stable isotopes and salinity data. In: Isotope techniques in groundwater hydrology: proceedings of a symposium, IAEA, Vienna, 11–15 March 1974, vol 1, pp 364–374

Dinçer T, Payne BR (1971) An environmental isotope study of the south-western karst region of Turkey. J Hydrol 13:233–258

Dinçer T, Payne BR, Florkowski T et al (1970) Snow melt runoff from measurements of tritium and oxygen-18. Water Resour Res 6:110–124

Dmitriev LV (1973) Geochemistry and petrology of rocks from middle oceanic ridges. Nauka, Moscow

Dobrovolsky AD, Zalogin BS (1963) Seas of the USSR. Misl, Moscow

Dockins KO, Bainbridge AE, Houtermans JC, Suess HE (1967) Tritium in the mixed layer of the North Pacific Ocean. In: Radioactive dating and methods of low-level counting: proceedings of a symposium organized by the international atomic energy agency in co-operation with the joint commission on applied radioactivity (ICSU) and held in Monaco, 2–10 March 1967, IAEA, Vienna, pp 120–160

Doe BR, Stacey JS (1974) Review of studies on lead isotopes application in the problems of ore deposite origin. In: Stable isotopes as applied to problems of ore deposits. Econ Geol 11–57

Domanitsky AP (1971) Rivers and lakes of the Soviet Union. Gidrometizdat, Leningrad

Dooley JR, TatsumotoV Jr, Rosholt JN (1964) Radioactive disequilibrium studies of roll features, Shirley Basin, Wyoming. Econ Geol 59:586–595

Douglas RG, Savin SM (1975) Oxygen and carbon isotope analysis of Tertiary and Cretaceous microfossils from Shatsky Rise and other sites in the North Pacific, log. 32. Initial Rep DSDP 32:509–521

Downing RA, Smith DB, Pearson FJ et al (1977) The age of groundwater in the Lincolnshire, England limestone and its relevance to the flow mechanism. J Hydrol 33:201–213

Drever JI, Lawrence JR, Antweller RC (1979) Gypsum and halite from the mid-Atlantic ridge, DSDP site395. Earth Planet Sci Lett 42:98–102

Drevinsky PJ, Wasson JT, Conble EC, Dimond NA (1964) ^{7}Be, ^{32}P, ^{33}P, and ^{35}S: stratospheric concentration and artificial products. J Geophys Res 69:1457–1467

Dreving VP, Kalashnikov YaA (1964) Phase rule with discussion of thermodynamics. Moscow University Press, Moscow

Drost W, Mozer H, Neumaier F et al (1972) Isotopenmetoden in der Grundwasserkunde, Inf. 61. Büro Eurisotop, Brussels, p 178

Drozdov OA, Grigorieva AS (1963) Hydrologic cycle in the atmosphere. Gidrometeoizdat, Leningrad

Dubinchuk VT (1978) Isotope sampling of groundwater the stratified by age. Water Resour 6:91–96

Dubinchuk VT (1979) What is the groundwater age? MOIP. Geol Sect 54:70–79

Dubinchuk VT, Polyakov VA (1983) Identification of hydrogeological systems by isotope idicator data. Water Resour 3:5–15

Dubinchuk VT, Polyakov VA, Kornienko ND et al (1988) Nuclear geophysical methods in hydrogeology and engeneering geology. Nedra, Moscow

Dubinchuk VT, Polyakov VA, Kuptsov VM et al (1974) Investigation of groundwater recharge by stable and radioactive isotopes and analogous modelling. In: Isotope techniques in groundwater hydrology: proceedings of a symposium, IAEA, Vienna, 11–15 March 1974, vol 1, pp 399–428

Dubinchuk VT, Yusova YeN, Seleznev VP et al (1981) Water inflow to the Arpa-Sevan tunnel studied by isotope methods. In: Isotopes in the hydrosphere. Abstract symposium, IWP Publ, pp 124–125

Dubinchuk VT, Ferronsky VI, Polyakov VA (2004) Two-year D, ^{18}O, T observations in water of the Volga River and precipitation at the Astrakhan AMS (46.15 N, 48.08 E). In: Isotope hydrology and integrated water resources management. Proceedings of the international symposium, IAEA, Vienna, 19–23 May 2003, pp 424–427

Dudey ND, Malewski RL, Rymas SL (1972) Tritium yield from fast-neutron fission of ^{235}U. Trans Am Nucl Soc 15:483

Dzens-Litovsky AI (1967) Kara-Bogaz Gol. Nedra, Leningrad

Eckelmann WR, Broecker WS, Whitlock DW, Allsup JR (1962) Implications of carbon isotopic composition of total organic carbon of some recent sediments and ancient oils. Bull Am Assoc Petrol Geol 46:699–704

Ehhalt DN (1966) Tritium and deuterium in the atmospheric hydrogen. Tellus 18:249–255

Ehhalt DN (1971) Vertical profiles and transport HTO in the troposphere. J Geophys Res 76:7351–7367

Ehhalt DN (1974) The atmospheric cycle of methane. Tellus 26:58–63

Eichler R (1965) Deuterium–Isotopengeochemie des Ground- und Oberelächenwassers. Geol Rundsch 55:144–159

Eisenbund M, Bennett D, Blanco RE et al (1979) Tritium in the environment—NCRP report no 62. In: IAEA (ed) Behaviour of tritium in the environment: proceedings of the international symposium on the behaviour of tritium in the environment. IAEA, Vienna, pp 585–587

Ekdahl CA, Keeling CD (1973) Atmospheric CO_2 in the natural carbon cycle: 1. Quantitative reduction from records at Mouna Loa Observatory and at the South Pole. In: Carbon and biosphere: 24th Brukhaven symposium Biol, Springfield, 51–85

Emiliani C (1970) Pleistocene paleotemperatures. Science 168:822–824

Emiliani C (1978) The cause of the ice ages. Earth Planet Sci Lett 37:349–352

Epstein S (1969) Distribution of carbon isotopes and their biochemical and geochemical significance. In: Proceedings of a simp on CO_2, Haverford, 5–14

Epstein S (1978) The D/H ratio of cellulose in a New Zealand Pinus radiata. A reply to the criticism of A.T. Wilson and V.J. Grinsted. Earth Planet Sci Lett 39:303–307

Epstein S, Mayeda T (1953) Variation of ^{18}O content of waters from natural sources. Geochim Cosmochim Acta 4:213–214

Epstein S, Taylor HP (1967) Variation of $^{18}O/^{16}O$ in minerals and rocks. Researches in geochemistry, vol 2. Wiley, New York, pp 29–62

Epstein S, Taylor HP (1970) $^{18}O/^{16}O$, $^{30}Si/^{28}Si$, D/H and $^{13}C/^{12}C$ studies of lunar rocks and minerals. Science 167:533–535

Epstein S, Taylor HP (1971) $^{18}O/^{16}O$, $^{30}Si/^{28}Si$, D/H and $^{13}C/^{12}C$ ratios in lunar samples. In: Proceedings of second lunar conferences 1421–1441

Epstein S, Taylor HP (1972) $^{18}O/^{16}O$, $^{30}Si/^{28}Si$, $^{13}C/^{12}C$ and D/H studies of Apollo 14 and 15 samples. Geochim Cosmochim Acta 3(2):1429–1454

Epstein S, Yapp CJ (1976) Climatic implications of the D/H ratio of hydrogen in C–H groups in tree cellulose. Earth Planet Sci Lett 30:252–261

Epstein S, Buchbaum R, Lowenstam L, Urey H (1951) Carbonate-water isotopic temperature scale. Bull GSA 62:417–426

Epstein S, Buchbaum R, Lowenstam L, Urey H (1953) Revised carbonate-water isotopic temperature scale. Bull GSA 64:1315–1325

Epstein S, Sharp RP, Gow AJ (1970) Antarctic ice sheet: stable isotope analyses of Bird Station cores and interhemispheric climatic implications. Science 168:1570–1572

Epstein S, Yapp CJ, Hall JH (1976) The determination of the D/H ratio of non-exchangeable hydrogen in cellulose extracted from aquatic and land plants. Earth Planet Sci Lett 30:241–251

Eremeev VN (1972) Oxygen isotope formation in Atlantic Ocean. Sevastopol MGI Pabl 57:172–178

Eriksson E (1963) Atmospheric tritium as a tool for the study of certain hydrologic aspects of river basins. Tellus 15:303–308

Eriksson E (1965a) An account of the major pulses of tritium and their effects in the atmosphere. Tellus 17:118–130

Eriksson E (1965b) Deuterium and oxygen-18 in precipitation and other natural waters. Tellus 17:498–512

Eriksson E (1967) Isotopes in hydrometeorology. In: Isotopes in hydrology: proceedings of the symposium on isotopes in hydrology held by the international atomic energy agency in cooperation with the international union of geodesy and geophysics in Vienna, IAEA, Vienna, 14–18 Nov 1966, pp 21–33

Eriksson E, Hoering TC (1980) Biogeochemistry of the stable hydrogen isotopes. Geochim Cosmochim Acta 44:1197–1206

Escudie AS, Blanc G, Chalie F et al (1998) Understanding present and past Caspian Sea evolution. In: IAEA (ed) Isotope techniques in the study of environmental change: proceedings of an international symposium on isotope techniques in the study of past and current environmental changes in the hydrosphere and the atmosphere. IAEA, Vienna, pp 623–631

Esikov AD (1980) Mass-specrometry analysis of natural waters. Nauka, Moscow

Esikov AD, Erokhin VE, Chernikova NS et al (1979a) Genesis of mud volcanos, south-west of Turkmenia, by hydrogen isotope content. In: Ferronsky VI (ed) Isotope studies of natural waters. Nauka, Moscow, pp 70–74

Esikov AD, Romanov VV, Erokhin VE et al (1979b) Deuterium and tritium in groundwaters of Moscow artesian basin. In: Isotope studies of natural waters. Nauka, Moscow, pp 55–60

Ester VF, Hoering TC (1980) Biogeochemistry of the stable hydrogen isotopes. Geochem Cosmochem Acta 44:1197–1206

Evans GV, Otlet RL, Downing RA et al (1979) Some problems in the interpretation of isotope measurements in United Kingdom aquifers. In: IAEA (ed) Isotope hydrology, 1978: proceedings of an international symposium on isotope hydrology, vol 2. IAEA, Vienna, pp 679–706

Evin J, Vuillaume Y (1970) Etude par le radiocarbone de la nappe captive de l'Albien du Bassin de Paris. In: IAEA (ed) Isotope hydrology: proceedings of a symposium on use of isotopes in hydrology. IAEA, Vienna, pp 315–331

Fabrica-Martin J, Davis SN, Elmore D (1987) Application of ^{129}I and ^{36}Cl in hydrology. Nucl Instr Meth Phys Res B 29:361–371

Fairbridge RW (1964) The importance of limestone and its Ca/Mg content to paleoclimanology. In: Nairn AEM (ed) Problems in paleoclimatology. Intersciences, London, pp 431–478

Fairhall AW (1971) Radiocarbon in the seas. Rep. no RLO-225-T20-3, US AEC

Fairhall AW, Buddemeir RW, Yang IA, Young YA (1969) Radiocarbon from nuclear testing and air-sea exchange of CO_2. Antarctic J 4:14–18

Fairhall AW, Young YA (1970) Radiocarbon in the environment. In: Radionuclides in the environment. Adv Chem Ser no 9, Am Chem Soc 401–418

Faltings V, Harteck P (1950) Der Tritium Gehalt der Atmosphere. Ztschr Naturforsch 8:438–439

Faure G, Powell JL (1972) Strontium isotope geology. Springer, New York

Fedorov PV (1957) Stratigraphy of quaternery deposits. Annals of GIN AN SSSR, p 10.

Fedorov YuA (1999) Stable isotopes and evolution of the hydrosphere. Istina, Moscow

Fergussen GY (1963) Upper tropospheric carbon-14 levels during spring 1961. J Geophys Res 68:3933–3941

Ferhi AM, Long A, Lerman JC (1977) Stable isotopes of oxygen in planets: a possible paleohydrometer. Hydrol Water Resour Arizona and South-West 7:191–198

Ferhugen J, Ternet F, Weiss L et al (1974) The earth: introduction to general geology, vol 2. Nauka, Moscow, pp 389–483 (trans: from English)

Ferrara GC, Gonfiantini R, Panichi G (1965) La composizione isotopica della vapore di alcuni soffioni di Larderello e della'acqua di alcune sorgenti e moffete della Toscana. Atti Soc Tosc Sci Natur 15:113–140

Ferronsky VI (1974) Origin of the Earth's hydrosphere based on isotopic composition of water. Water Resour 4:21–34

Ferronsky VI (2008) Nonaveraged virial theorem for natural systems. Investigated in Russia. http// zhurnal.ape.relarn.ru/articles/2008/066.pdf. pp 716–719

Ferronsky VI, Brezgunov VC, Vlasova LC et al (2003a) Investigation of water exchange processes in the Caspian Sea on the bases of isotopic and oceanographic data. Water Resour 30:10–22

Ferronsky VI, Polyakov VA, Brezgunov VC et al (2003b) Variations in the hydrological regime of Kara-Bogaz-Gol Golf, Lake Issyk-Kul, and the Aral Sea assessed on data of bottom sediments. Water Resour 30:252–259

Ferronsky VI, Danilin AI, Dubinchuk VT et al (1977) Radioisotope methods of investigation in ingeneering geology and hydrogeology. Atomizdat, Moscow

Ferronsky VI, Denisik SA, Ferronsky SV (1978) The solution of Jacobi's virial equation for celestial bodies. Celest Mech 18:113–140

Ferronsky VI, Denisik SA, Ferronsky SV (1996) Virial oscillations of celestial bodies: V. The structure of the potential and kinetic energies of a celestial body as a record of its creation history. Celest Mech Dynam Astron 64:167–183

Ferronsky VI, Denisik SA, Ferronsky SV (1987) Jacobi dynamics. Reidel, Dordrecht

Ferronsky VI, Dubinchuk VT, Polyakov VA et al (1975) Environmental isotopes of the hydrosphere. Nedra, Moscow

Ferronsky VI, Ferronsky SV (2010) Dynamics of the Earth. Springer, Dordrecht

Ferronsky VI, Ivanova LYu, Kikichev HG (1980) Possibilities of tritium in study of global circulation of atmospheric moisture. Water Resour 5:144–152

Ferronsky VI, Polyakov VA (1983) Isotopy of the hydrosphere. Nauka, Moscow

Ferronsky VI, Polyakov VA, Kuprin PN, Lobov AL (1999) Caspian Sea water level change (based on bottom sediment study). Water Resour 26:652–666
Ferronsky VI, Polyakov VA, Romanov VV (1984) Cosmogenic isotopes of the hydrosphere. Nauka, Moscow
Ferronsky VI, Polyakov VA, Ferronsky SV (1993) Variation of water isotope content in hydrological cycle as a tool in study mechanism of climate change. Water Resour 3:285–295
Ferronsky VI, Polyakov VA (1982) Environmental isotopes in the hydrosphere. Wiley, Chichester
Ferronsky VI, Polyakov VA, Froelich K et al (1998) Isotope studies of the Caspian Sea: climate record from bottom sediments (preliminary results). In: IAEA (ed) Isotope techniques in the study of environmental change: proceedings of an international symposium on isotope techniques in the study of past and current environmental changes in the hydrosphere and the atmosphere. IAEA, Vienna, pp 633–644
Ferronsky VI, Vlasova LC, Esikov AD et al (1982) Variation of isotopic composition of water, atmospheric precipitation, and organic matter in alluvial sediments due to climate change. Water Res 5:3–25
Fleishman DG, Kanevsky YuP, Gridchenko ZG (1975) Application of cosmogemic ^{22}Na for dating of low-mineralized surface and groundwater. In: Styro BI (ed) Cosmogenic radioactive isotopes, vol 3. Izd. IFM Lit. AN SSR, Vilnius, pp 144–152
Flohn G (1980) Basic geophysical models of glaciation. In: Gribbin J (ed) Climate change. Hydrometeoizdat, Leningrad, pp 331–356 (trans: from English)
Fireman EL (1962) Tritium in meteorites and recovered satellite material. In: Tritium in physical and biological sciences, proceedings of a symposium, vol 1, IAEA, Vienna, pp 1691–1700
Fireman EL (1967) Radioactivities in meteorites and cosmic-ray variation. Geochim Cosmochim Acta 31:1197–1206
Fireman EL, Stoenner RW (1982) Carbon and carbon-14 in lunar soil 14163. Proc 12 Lunar and Planet Sci Conf New York 12:559–565
Fluss MJ, Dudey ND (1971) Tritium and helium yields in fast fission of ^{235}U. Trans Am Nucl Soc 14:809–812
Fonselius S, Östlund HG (1959) Natural radiocarbon measurements on surface water from the North Atlantic and the Arctic Sea. Tellus 11:77–82
Fontes JCh (1981) Stable isotopes in paleowaters. In: Stable isotope hydrology: deuterium and oxygen-18 in the water cycle. IAEA, Vienna, pp 273–302
Fontes JCh (1976) Les isotopes du milieu dans les laux naturells. Le Houille Blanche 3/4:205–221
Fontes JCh, Garnie JM (1977) Determination of the initial ^{14}C activity of the total dissolved carbon, age estimation of waters in confine aquifers. In: IAEA (ed) Water-rock interactions: proceedings of a symposium. IAEA, Strasbourg, pp 1363–1376 (August 17–25 1977)
Fontes JCh, Bortolami GC, Zuppi GM (1979) Hydrologie isotopique du Massif du Mont-Blanc. In: IAEA (ed) Isotope hydrology, 1978: proceedings of an international symposium on isotope hydrology, vol 1. IAEA, pp 411–436
Fontes J Ch Garnier JM (1979) Determination of the initial ^{14}C activity of the total dissolved carbon: review of the existing models and new approach. Water Resour Res 15:399–413
Fontes JCh, Gonfiantini R (1967) Component isotopique au cours de l'evaporation de deux basins sahariens. Earth Planet Sci Lett 3:258–266
Fontes JCh, Letolle R, Maree A (1965) Some results of oxygen isotope studies of marine waters. In: Tongiorgi E (ed) Stable isotopes in oceanographic studies and paleotemperatures: Spoleto. Consiglio Nazionale delle Ricerche, Piazzale Aldo Moro, pp 131–141
Fontes JCh, Pouchen P et al (1976) Preliminary isotopic study of the Lake Asal system (Republic of Djibouti). In: Isotopes in lake studies. Proceedings of an advisory group meeting, IAEA, Vienna, pp 163–174
Fotiev SG (1978) Hydrometric pecularities of the USSR Cryogenic regions. Nedra, Moscow
Friedman I (1953) Deuterium content of natural waters and other substances. Geochim Cosmochim Acta 4:89–103
Friedman I (1965) Interstital waters from deep sea sediments. J Gephys Res 70:4066–4067
Friedman I (1967) Water and deuterium in pumice from the 1959–1960 eruption of Kölauea volcano, Hawaii. U.S. Geol Surv Prof Pap 57B:120–127

Friedman I, Hardcastle K (1973) Interstital water studies, Leg. 15. Isotopic composition of water. Initial Rep DSDP 20:901–903

Friedman I, Smith RL (1958) The deuterium content of water in some volcanic gasses. Geochim Cosmochim Acta 15:218–228

Friedman I, O'Neil JR, Gleason JD et al (1970) Water, hydrogen, deuterium, carbon, carbon-13 and oxygen-18 content of selected lunar material. Sciences 167:538–540

Friedman I, Sigurgeirsson T, Gardarsson O (1963) Deuterium in island waters. Geochim Cosmochim Acta 27:553–561

Friedman I, O'Neil JR, Gleason JD et al (1971) The carbon, hydrogen content and isotopic composition of some Apollo-12 materials. Proc 2nd Lunar Sci Conf, vol 2, pp 1707–1415

Friedman I, Redfield AC, Schoen B et al (1964) The variation of the deuterium content of natural waters in the hydrologic cucle. Rev Geophys 2:177–224

Fritz P (1983) Paleoclimatic studies using freshwater deposits and fossil ground water in Central and Northern Canada. In: Plaleoclimates and paleowaters. Proceedings of an advisory group meeting, IAEA, Vienna, p 157

Fritz P, Drimmie RJ, Render FW (1974) Stable isotope contents of a major prairie aquifer in central Manitoba, Canada. In: Isotope techniques in groundwater hydrology: proceedings of a symposium, IAEA, Vienna, 11–15 March 1974, vol 1, pp 379–396

Fritz P, Silva C, Suzuki O, Salati E (1979) Isotope hydrology in Northern Chile. In: IAEA (ed) Isotope hydrology, 1978: proceedings of an international symposium on isotope hydrology, vol 2. IAEA, Vienna, pp 525–543

Fritz P, Fontes J Ch (eds) (1980, 1986, 1989) Handbook of environmental isotope geochemistry, vol 1–3. Elsevier, Amsterdam

Fröhlich K (1990) On dating of old groundwater. Isotopenpraxis 26:557–560

Fröhlich K (ed) (2010) Enviromental radionuclides. Elsevier, Amsterdam

Fröhlich K, Jordan H, Hebert D (1977) Radioactive Umveltisotope in der Hydrologie. Grundstoffindustrie, Leipzig

Gagen M, Mc Carrol D, Hicks S (2006) The millenium project: European climate of the last millenium. Pages News 14

Galakhovskaya TV (1967) Distribution of boron, litium stroncium and bromium at evaporation of marine water. In: Valiashko MG (ed) Physical-chemical study of salts and brines. Nedra, Moscow, pp 84–107

Galimov EM (1968) Geochemistry of stable carbon isotopes. Nedra, Moscow

Galimov EM (1973) Carbon isotopes in oil and gas geology. Nedra, Moscow

Ganapathy R, Anders E (1974) Bulk compositions of the Moon and Earth, estimated from meteorites. Geochim Cosmochim Acta 5:1181–1206

Garnier J-M (1985) Retardation of dissolved radiocarbon throught a carbonated matrix. Geochem Cosmochem Acta 49:683–693

Garrels RM, Christ CL (1965) Solutions, minerals and equilibria. Harper and Row. New York

Gat JR (1970) Environmental isotope balance of lake Tiberias. In: IAEA (ed) Isotope hydrology: proceedings of a symposium on use of isotopes in hydrology. IAEA, Vienna, pp 109–127

Gat JR, Carmi I (1970) Evolution of the isotopic composition of atmospheric waters in the Mediterranean Sea area. J Geophys Res 75:3039–3078

Gat JR, Dansgaard W (1972) Stable isotope survey of the fresh water occurrence in Israel and Northern Jordan rift valley. J Hydrol 16:177–212

Gat JR, Issar A (1974) Desert isotope hydrology: water sources of the Sinai Desert. Geochim Cosmochim Acta 38:1117–1124

Gat JR, Tzur Y (1967) Modification of the isotopic composition of rainwater by processes which occur before groundwater recharge. Isotopes in hydrology: proceedings of the symposium on isotopes in hydrology held by the international atomic energy agency in cooperation with the international union of geodesy and geophysics in Vienna, IAEA, Vienna, 14–18 Nov 1966, pp 49–60

Gat JR, Gonfiantini R (eds) (1981) Stable isotope hydrology; deuterium and oxygen-18 in the water cycle. IAEA, Vienna

Gat JR, Gonfiantini R, Tongiorgi E (1968) Atmosphere-surface water interaction. In: IAEA (ed) Guidebook on nuclear techniques in hydrology. IAEA, Vienna, pp 175–184

Gelpherich F (1962) Ionite. Ionic exchange fundamentals. Inlit, Moscow (trans: from German)

Geyh MA (1970) Carbon -14 concentration of lime in soil and aspects of the carbon-14 dating of groundwater. In: IAEA (ed) Isotope hydrology: proceedings of a symposium on use of isotopes in hydrology. IAEA, Vienna, pp 215–222

Geyh MA (1972) On the determination of the initial ^{14}C content in groundwater. In: Proceedings of 8th international conference on radiocarbon dating, Lower Hutt, Wellington, pp D58–D59

Geyh MA (1974) Erfahrungen mit der ^{14}C und ^{3}H methode in der angewandten Hydrologie. Österr Wasserwirt 3/4:49–54

Geyh MA (1980) Interpretation of environmental isotopic groundwater data: arid and semi-arid zones. In: IAEA (ed) Arid–zone hydrology: investigations with isotope techniques: proceedings of an advisory group meeting on application of isotope techniques in arid zones hydrology. IAEA, Vienna, pp 31–46

Geyh MA, Jäkel D (1977) The climate of the Sahara during the late Pleistocene and Holocene on the basis of available radiocarbon data. Natur Res Devel 6:65–79

Giggenbach W (1971) Isotopic composition of waters of the Broadlands geothermal field. N Z J Sci 14:959–970

Giggenbach W, Gonfiantini R, Panichi C (1983) Geothermal systems. In: IAEA (ed) Guidebook on nuclear techniques in hydrology. IAEA, Vienna, pp 359–376

Gillespie R, Polach YA (1979) The suitability of marine shells for radiocarbon dating of Australian prehistory. In: Berger R, Suess HE (eds) Radiocarbon dating. University of California Press, Berkeley, pp 404–421

Giletti BJ, Bazan F, Kulp JL (1958) The geochemistry of tritium. Trans Am Geophys Union 39:807–818

Gischler CE (1976) Present and future trends in water resources development in Arab countries. UNESCO, New York

Glasstone S, Laider J, Eyring H (1941) Theory of rate processes. McGraw Hill, New York

Glasstone S (1962) Effects of nuclear weapons. USAEC, Washington

Glynn PD, Plummer LN (2005) Geochemistry and the understanding of groundwater systems. Hydrogeol J 13:263–287

Godnev IN (1956) Calculation of thermodynamical functions by molecular data. Gostechisdat, Moscow

Goldberg E (1963) Geochronology with Lead-210. In: IAEA (ed) Radioactive dating: proceedings of the symposium on radioactive dating. IAEA, Vienna, pp 121–131

Goldschmidt R (1954) Geochemistry. University Press, Oxford

Golubev LI, Ilyasov VM, Mechedov BN (1979) Tritium content in thermocarrier of the reactor WWER. Atomic Energy 46:79–81

Godfrey J (1962) The deuterium content of hydrous minerals from the East-Central Sierra Nevada and Yosemite National Park. Geochim Cosmochim Acta 26:1215–1245

Gonfiantini R (1965) Effetti isotopici nell'evapjrazione di acque salate. Atti Soc Tosc Sci Natur Ser A 72:550–588

Gonfiantini R (1978) Standards for stable isotope measurements in natural compounds. Nature 271:534–536

Gonfiantini R, Gratzini S, Tongiorgi E (1965) Oxygen isotopic composition of water in leaves. In: Isotopes and radiation in soil-plant nutrition studies. Proceedings of a symp, IAEA, Vienna, pp 405–410

Gonfiantini R, Dinçer T, Derekoy AM (1974) Environmental isotope hydrology in the Bodna region, Algeria. In: Isotope techniques in groundwater Hydrology: proceedings of a symposium, IAEA, Vienna, 11–15 March 1974, vol 1, pp 293–316

Gonfiantini R, Conrad C, Fontes JCh et al (1976) Etude isotopique de la nappe du Continental intercalaire et de ses relations avec les autres nappes du Sahara septentional. In: Isotope techniques in groundwater hydrology: proceedings of a symposium, IAEA, Vienna, 11–15 March 1974, vol 1, pp 227–240

Gonfiantini R, Gallo G, Payne BR et al (1976) Environmental isotope and hydrochemistry in groundwater of Gran Canaria. In: Interpretation of environmental isotope and hydrochemical data in groundwater hydrology. Proceedings of an advisory group meeting, IAEA, Vienna, pp 159–170

Gonfiantini R, Zuppi GM, Eccles DH, Ferro W (1979) Isotope investigation of lake Malavi. In: Isotopes in lake studies. Proceedings of an advisory group meeting, IAEA, Vienna, pp 195–207

Gorbushina LV, Vetshtein VE, Malyuk et al (1974) Hydrogen and oxygen isotopic content in sulphide waters of the Sochi-Adler artesian basin. Geochimiya 9:1102–1106

Gorbushina LV, Gratsiansky VG, Tyminsky VG (1968) The experience with ultrasound use for recovery of thoron and actinon from solutions. Radiokhimiya 10:495–496

Gorbushina LV, Salmenkova NA, Tyminsky VG (1967) The ages and mixture proportions of mineral waters in the Tashkent artesian basin. Izv Vissh Uch Zaved Ser Geol i Razv 2:92–95

Gorbushina LV, Tyminsky VG, Spiridonov AI (1972) On the mechanism of radiohydrogeological anomalies appearance in seismic regions and their significance in earthquake prediction. Sovetskaya Geol 1:153–156

Gorbushina LV, Tyminsky VG (1974) Radioactive and stable isotopes in geology and hydrology. Atomizdat, Moscow

Gordienko FG, Kotliakov VM (1976) On relationship between climatic and glaciating factors in formation of continental glaciar paleotemperatures. In: Kalinin GP, Klige RK (eds) Problems of paleohydrology. Nauka, Moscow, pp 282–283

Graf DL, Friedman J, Meents WF (1965) The origin of saline formation waters. II. Isotopic fractionation by shale micropore systems. U.S. State Geol Surv Circular no p 392

Graf DL, Friedman J, Meents WF (1966) The origin of saline formation waters. III. Calcium chloride waters. U.S. State Geol Surv Circular no p 397

Grashchenko SM, Nikolaev DS, Kolyadin LV et al (1960) Radium content in the Black Sea waters. Dokl AN SSSR 132:1171–1172

Gray DC, Damon PE (1970) Sunspots and radiocarbon dating in the Middle ages. In: Scientific methods in medieval archeology. University of California Press, Berkley, pp 167–182

Gratsiansky VG, Gorbushina LV, Tyminsky VG (1967) On the release of radioactive gases under ultrasound affection. Izv AN SSSR, Ser Phys Zem 10:91–94

Gribbin J (1980) Search of cyclicity. In: Gribbin J (ed) Climate change. Gidrometeoisdat, Leningrad, pp 102–121 (trans: from English)

Gribbin J, Lem GG (1980) Climate change in hystorical period. In: Gribbin J (ed) Climate change. Gidrometeoisdat, Leningrad, pp 122–140 (trans: from English)

Grinenko VA, Grinenko LN (1974) Sulphur isotope geochemistry. Nauka, Moscow

Gritchenko ZG, Flegontov BM, Fleishman DG (1975) Radioactive fallouts of ^{7}Be and ^{22}Na near Leningrad in 1960–1973. In: Styro BI (ed) Cosmogenic radioisotopes, 3. Izd. IFM AN LitSSR, Vilnius, pp 71–81

Gross VG, Tracey JT (1966) Oxygen and carbon isotopic composition of limestones and dolomites, Bikini and Eniwetok Atolls. Science 151:1082–1084

Grossman L, Larimer JW (1974) Early chemical history of the solar system. Rev Geophys Space Phys 1:17–101

Gudzenko VV, Dubinchuk VT (1987) Radium and radon isotopes in natural waters. Nauka, Moscow

Gutsalo LK (1980) The rules and factors governing changes in isotopic composition of brines during evaporation (in connection with genesis of underground brines) Geokhemiya 11:1734–1746

Gutsalo LK, Vetshtein VE, Lebedev LM (1978a) Isotopic composition of hydrogen and oxygen of hydrothermal fluids in recent metal-bearing ore deposits of Cheleken. Geokhemiya 8:1173–1179

Gutsalo LK, Vetshtein VE, Artemchuk BG, Petrichenko OI (1978b) Oxygen isotopes of micro and macro fluid inclusions from salt formations. In: 7th Symp Stab Isotop GEOHI, Moscow, pp 151–153

Hagemann FJ, Grey J, Machta L, Turkevich A (1959) Stratospheric carbon-14, carbon dioxide, tritium. Science 130:542–552

Hagemann R, Nief G, Roth T (1970) Absolute D/H ratio for SMOW. Tellus 23:172–175

Hall WE Friedman I (1963) Composition of fluid inclusions, Cave-in-Rock fluorite district, Illinois, and upper Vississippi Valey zinc-lead district. Econ Geol 58:886–911

Hall WE, Friedman I, Nash JT (1974) Fluid inclusion and light stable isotope study of the climax molibdenium deposits. Colorado Econ Geol 69:884–901

Hamilton EI (1968) Applide geochronology. Academic Press, London

Hamza MS Epstein S (1980) Oxygen isotopic fractionation between oxygen of different sites in hydroxyl-bearing silicate minerals. Geochim Cosmochim Acta 44:173–182

Harteck P (1954) The relative abundance of HT and HTO in the atmosphere. J Chem Phys 22:1746–1751

Hayes IM (1967) Organic constituents of meteorites: A review. GeochimCosmochim Acta 31:1395–1440

Hedge CE (1974) Strontium isotopes in economic geology. Econ Geol 69:823–825

Helfferich, von F (1959) Ionenaustauscher. Bd I. Grundlagen Struktur-Herstellung-Theorie, Weinheim Bergstr., Chemie

Heller D, Roedel W, Münnich KO (1977) Decreasing release of ^{85}K into the atmosphere. Naturwissenschaften 64:383

Herdon JM, Suess HE (1976) Can enstatite meteorites from a nebula of solar composition? Geochim Cosmochim Acta 40:395–399

Herdon JM, Suess HE (1977) Can the hondrites have condensed from a gas phase? Geochim Cosmochim Acta 41:233–236

Hermann A, Martinec J, Stichler W (1978) Study of snowmelt-runoff components using isotope measurement. In: Modeling of snow cover runoff. Proceedings of a symposium, Hanover (US), pp 288–296

Higashi S (1959) Estimation of microgram amount of Th in sea water. J Oceanogr Soc Jpn 15:64

Hesslein RH, Quay PD, Thomas M, Broecker WS (1979) Radionuclides in sediments. In: Isotopes in lake studies. Proceedings of an advisory group meeting, IAEA, Vienna, pp 251–254

Hitchon B, Friedman F (1969) Geochemistry and origin of formation waters in the Western Canada sedimentary basin, I. Stable isotopes of hydrogen and oxygen. Geochim Cosmochim Acta 33:1321–1349

Hitchon B, Krouse HB (1972) Hydrogeochemistry of surface waters of the Mackenzie River drainage basin, Canada, III. Stable isotopes of oxygen, carbon, and sulfur. Geochim Cosmochim Acta 36:1337–1358

Hoang CT, Servant J (1972) Le flux de radon de la mer. CR Acad Sci 274(24):1321–1349

Hoefs J (1973) Stable isotope geochemistry. Springer, New York

Hoernes S, Friedrichsen H (1979) '^{18}O/^{16}O and D/H investigations of basalts of Leg 46. In: Initial Rep DSDP, 46:253–255

Horton I, Hurry C, Johnns C (1971) Tritium loss from water exposed to the atmosphere. Environ Sci Technol 5:338–348

Houtermans J, Suess HE, Munk W (1967) Effect of industrial fuel combustion on the carbon-14 level of atmospheric CO_2. In: Radioactive dating and methods of low-level counting: proceedings of a symposium organized by the international atomic energy agency in co-operation with the joint commission on applied radioactivity (ICSU) and held in Monaco, 2–10 March 1967, IAEA, Vienna, pp 57–68

Hübner H, Richter W, Kowski P (1979a) Studies on relationship between surface water and surrounding groundwater of Lake Schwerin (GDR). In: Isotopes in lake studies. Proceedings of an advisory group meeting, IAEA, Vienna, pp 95–102

Hübner H, Kowski P, Hermichen WD et al (1979b) Regional and temporal variations of deuterium in precipitation and atmospheric moisture of Central Europe. In: IAEA (ed) Isotope hydrology, 1978: proceedings of an international symposium on isotope hydrology. IAEA, Vienna, pp 289–305

Hulston JR (1977) Isotope work applied to geothermal systems at the Institute of Nuclear Sciences, New Zealand. Geothermics 5:89–96
Hyde E, Perlman I, Seaborg G (1969) Thorium, protactimium, and uranium isotopes. Atomizdat, Moscow (trans: from English)
International Atomic Energy Agency (1969, 1970, 1971, 1973, 1975, 1979, 1983, 1986, 1990, 1994) Environmental isotope data: world survey of isotope concentrations in precipitation, no 1–10, IAEA, Vienna
International Atomic Energy Agency (1968) Guidebook on nuclear techniques in hydrology. IAEA, Vienna
International Atomic Energy Agency (1983) Guidebook on nuclear techniques in hydrology, 1983 ed. IAEA, Vienna
International Atomic Energy Agency (1979) Behaviour of tritium in the environment: proceedings of the international symposium on the behaviour of tritium in the environment. IAEA, Vienna
International Atomic Energy Agency (1973) Bulletin no 4. IAEA, Vienna, pp 10–16
International Atomic Energy Agency (1976) Interpretation of environmental isotope and geochemical data in groundwater hydrology. Proceedings of an advisory group meeting, IAEA, Vienna
International Atomic Energy Agency (1979) Isotopes in lake studies. Proceedings of an advisory group meeting, IAEA, Vienna
International Atomic Energy Agency (1980) Arid-zone hydrology: investigations with isotope techniques. In: Proceedings of an advisory group meeting on application of isotope techniques in arid zones hydrology, IAEA, Vienna, 6–9 Nov 1978
International Atomic Energy Agency (1980) New tritium scale, circular latter. IAEA, Vienna
International Atomic Energy Agency (2005) Guidebook on the use of chlorofluocarbons in hydrology. IAEA, Vienna
Imboden DM (1979) Complete mixing in Lake Tahoe, California-Nevada, traced by tritium. In: Isotopes in lake studies. Proceedings of an advisory group meeting, IAEA, Vienna, pp 209–212
Imbrie J (1985) A theoretical framework for the Pleistocene in ages. J Geol 142:417–432
Imbrie J, Imbrie KP (1979) Ice age: solving the mystery. Enslow, New York
Ingerson E, Pearson FJ (1964) Estimation of age and rate of motion of groundwater by ^{14}C method. In: Miyake Y (ed) Recent research in the field of hydrosphere, atmosphere and nuclear geochemistry. Maruzen, Tokyo, pp 263–283
Issar A, Bein A, Michaell A (1972) On the ancient water of the upper Nubian Sandstone aquifer in central Sinai and Southern Israel. J Hydrol 14:353–374
Izmaylov NA (1966) Electrochemistry of solutions, 2nd ed. Khimiya, Moscow
Jacobi CGJ (1884) Vorlesungen über dynamik. Klebsch, Berlin
James AT, Baker DR (1976) Oxygen isotope exchange between illite and water at 22°C. Geochim Cosmochim Acta 40:235–239
Javoy M (1978) '$^{18}O/^{16}O$ and D/H ratios in high temperature peridotites. Geol Surv Open-File Rep 701:202
Jordan H, Frëhlich K, Herbert D (1983) Application of isotope methods in hydrogeology. Water Resour 3:116–122
Jordan H, Fröhlich K, Hebert D (1976) Some results of T and ^{14}C investigations of groundwater in the DDR. J Hydrol 3:61–69
Joshi LV, Mahadevan TN (1967) Radiochemical determination of lead-210 concentrations in ground level air in India. Nucl Radiat Chem Pros 1:519–523
Junge CE (1963) Air chemistry and radioactivity. Academic Press, New York
Jouzel J, Pourchet M, Lorius C, Merlivat L (1979) Artificial tritium fall-out at the South Pole. In: IAEA (ed) Behaviour of tritium in the environment: proceedings of the international symposium on the behaviour of tritium in the environment. IAEA, Vienna, pp 31–45
Kamensky IL, Lobkov VA, Prasolov EM et al (1976) The gas components of the upper mantle at Kamchatka. Geokhimiya 5:682–695

Kaplan IR, Smith JW (1970) Concentration and isotopic composition of carbon and sulfur in Appolo 11 Lunar samples. Science 167:61–69

Kaplan IR, Uspenskaya TA, Zarembo YuI, Chirkov IV (1960) Thorium, its resources, chemistry and technology. Atomizdat, Moscow

Karapetyans MN (1953) Chemical thermodynamics. Geokhimisdat, Moscow

Karasev BV (1979) The effect of isotope dilution at determination of radiocarbon age of water. Annales VSEGINGEO 131:41–49

Karasev BV, Nechaev VI, Polyakov VA, Seletsky YuB (1975) Genesis of the Yaskhan lens of fresh water by radiocarbon age data. Annales VSEGINGEO 90:98–108

Karasev BV, Sokolovsky LG, Kuznetsova LA (1981a) Application of carbon isotopes for identification of break zones gassing by carbon dioxide. In: Ferronsky VI (ed) Investigation of natural waters by isotope methods. Nauka, Moscow, pp 155–157

Karasev BV, Krishtal NN, Polyakov VA (1981b) Water inflow separation at the Ayatsk bauxite excavation by the hydrogen, oxygen, and carbon isotope distribution. In: Ferronsky VI (ed) Investigation of natural water by isotope methods. Nauka, Moscow, pp 38–45

Kartsev AA, Vagin SV (1973) The role of clay minerals interlayer water in a history of groundwater formation. Izv Vissh Uch Zaved 3:64–66

Katrich IYu (1990) Tritium in natural waters after the Chernobyl atomic power accident. Meteorol Hydrol 5:92–97

Kaufman A (1969) The ^{232}Th concentration on surface ocean water. Geochim Cosmochim Acta 33:717–724

Kaufman A, Broeker W (1965) Comparison of ^{230}Th and ^{14}C ages for carbonate materials from lakes lachontan and Bonneville. J Geophys Res 70:4039–4054

Kaufman A, Trier R, Broecker WS, Feely HW (1973) Distribution of ^{228}Ra in the world ocean. J Geophys Res 78:8827–8848

Kaufman MI, Rydell HS, Osmond JK (1969) $^{234}U/^{238}U$ disequilibrium as an air to hydrologic study of the Floridian aquifer. J Hydrol 9:374–386

Kaufman S, Libby WF (1954) The natural distribution of tritium. Phys Rev 93:1337–1344

Kawabe I (1978) Calculation of oxygen isotope fractionation in quartz-water system with special reference to the bond temperature fractionation. Geochim Cosmochim Acta 42:613–621

Kazhdan AV, Boitsov VE, Zimin DF (1971) Geology and methods of the uranium ore exploration. Nedra, Moscow

Keeling CD (1973) The carbon dioxide cycle: reservoir models to depict the exchange of atmospheric carbon dioxide with oceans and land plants. In: Rasool SI, Cadle RD (eds) Chemistry of the lower atmosphere. Plenum, New York, pp 251–329

Kellogg WW (1980) Effect of human activities on global climate. Tech. Note No. 486, World Meteorological Organiztion, Geneva, p 47

Kelly WR, Larimer JW (1977) Chemical fractionation in meteorites: VIII. Iron meteorites and the cosmochemical history of the metal phase. Geochim Cosmochim Acta 41:93–111

Kharaka YK, Berry FAF, Friedman I (1973) Isotopic composition of oil-field brines from Kettleman North Dome, California, and their geologic implications. Geochim Cosmochim Acta 37:1899–1903

Khristianov VK, Korchuganov BN (1971) Radon content in the Upper Volga River waters. Geokhimiya 4:492–496

Kigoshi K (1973) Uranium 238/234 disequilibrium and age of underground water. Working paper of the IAEA panel meet, IAEA, Vienna

Kigoshi K, Yoneda K (1970) Daily variations in the tritium concentration of atmospheric moisture. J Geophys Res 75:2981–2984

Kind NV (1974) Geochronology of late antropogen by isotrope data. Nauka, Moscow

Kinman TD (1956) An attempt to detect deuterium in the solar atmosphere. Month Not Roy Astrophys Soc 116:77

Kirshenbaum I (1951) Physical property and analyses of heavy water. McGraw-Hill, New York

Kittel Ch (1968) Introduction to solid state physics, 3rd ed. Wiley, New York

Kittel Ch, Knight WD, Ruderman MN (1965) Mechanics, Berkley physics course, vol 1. McGrow Hill, New York

Klitzsch E, Sonntag C, Weistroffer K et al (1976) Grundwasser der Zentralsahara: Fossile Vorräte. Geol Rundsch 65:264–287

Knauss KG, Ku TL, Moore WS (1978) Radium and thorium isotopes in the surface waters of the east Pacific and coastal Southern California. Earth Planet Sci Lett 39:235–249

Knauth LP Epstein S (1971) Hydrogen and oxygen isotope relations in cherts and implications regarding the isotope history of the hydrosphere. Geol Soc Amer Ann Mts Abstr Program 3:624

Knauth LP, Epstein S (1975) Hydrogen and oxygen isotope ratios in silica from the JOIDES DSDP. Earth Planet Sci Lett 25:1–10

Knauth LP, Epstein S (1976) Hydrogen and oxygen isotope ratios in nodular and bedded cherts. Geochim Cosmochim Acta 40:1095–1108

Knauth LP, Lowe DR (1978) Oxygen isotope geochemistry of cherts from the onverwacht group (3.4 billion years), Transvaal, South Afrika, with implication for secular variations in isotope composition of cherts. Earth Planet Sci Lett 41:209–222

Kobayakawa HY, Horibe Y (1960) Deuterium abundance of natural waters. Geochim Cosmochim Acta 20:273–283

Kochenov AV, Baturin PN (1967) Uranium distribution in the Aral Sea sediments. Oceanology 7:623–627

Kocherov GE (1975) Isotope content of the solar corpuscular flux and the Earth's atmosphere. In: Styro BI (ed) Cosmogenic radioactive isotopes, 3. IB AN LitSSR 5–16

Koczy FF, Picciotto E, Poulaert G, Wilgain S (1957) Mesure des isotopes dy thotium dans l'ean de mer. Geochim Cosmochim Acta 11:103–129

Kokubu N, Mayeda T, Urey HC (1961) Deuterium content of minerals, rocks and liquid inclusions from rocks. Geochim Cosmochim Acta 21:247–256

Kolodny Y (1978) Participation of fresh water in chert diagenesis: evidence for oxygen isotopes and boron—a track mapping. Geol Surv Open-File Rep 701:228–229

Kolodny Y, Epstein S (1976) Stable isotope geochemistry of deep sea cherts. Geochim Cosmochim Acta 40:1195–1209

König LA (1979) Impact of the environment of tritium releases from the Karlsruhe Nuclear Research Center. In: IAEA (ed) Behaviour of tritium in the environment: proceedings of the international symposium on the behaviour of tritium in the environment. IAEA, Vienna, pp 591–610

Kotlyakov VM, Klige RK (2000) Global cycles of the Antarctic climatic change. In: Global change of natural media (climate and water regime), Nauchniy Mir, Moscow, pp 70–90

Krejči K, Zeller A (1979) Tritium pollution in the Swiss luminous compound industry. In: IAEA (ed) Behaviour of tritium in the environment: proceedings of the international symposium on the behaviour of tritium in the environment. IAEA, Vienna, pp 66–77

Kritsuk LN, Polyakov VA (2005) Isotopic and chemical content of underground ice and natural waters of Western Siberia. In: Geological study and entails protaction. A review. Geoinformcenter, Moscow

Ku TL, Lin MC (1976) ^{226}Ra distribution in the Atlantic Ocean. Earth Planet Sci Lett 32:236–248

Kulchitsky LI (1975) Role of water in argillite properties formation. Nedra, Moscow

Kuprin PN, Ferronsky VI, Popovchak VP (2003) Caspian Sea bottom sediment content as an indicator of water regime change. Water Resour 30:154–172

Kuptsov VM (1989) Methods of chronology of quaternery deposits in oceans and seas. Nauka, Moscow

Kuptsov VM (1986) Absolute chronology of ocean and sea bottom sediments. Nauka, Moscow

Kuptsov VM, Cherdyntsev VV (1969) Uranium and thorium decay products in the USSR active volcanism. Geokhimiya 6:643–658

Kuptsov VM, Cherdyntsev VV (1968) Radon and thoron in fumarole gases. Dokl AN SSSR 2:436–438

Kuptsov VM, Cherdyntsev VV, Zverev VL (1969) Radioisotopes in waters of Caucasus new volcanic region. Geokhimiya 9:1153–1155

Kuptsov VM, Cherdyntsev VV, Kuzmina EA, Sulerzhitsky LD (1969) Ionic age and conditions of effusion rock formation. Geokhimiya 7:829–834

Kuroda Y, Susuoki T, Matsuo S (1974) A preliminary study of D/H ratios of hornblends from granites and ultramafic rocks. Geochim J 8:135–139

Kusakabe M, Wada H, Matsuo S (1970) Oxygen and hydrogen isotope ratios of monthly collected waters from Nasudake volcanic area, Japan. J Geophys Res 75:5941–5951

Kuznetsov YuV (1962) On the forms of ionium and thorium in the oceans. Geokhimiya 2:177–184

Kuznetsov YuV, Elizarova AN, Frenklikh MS (1966) Study of sedimentation in oceanic waters by ^{231}P and ^{230}Th isotopes. Radiochemistry 8:459–468

Kuznetsov YuV, Elizarova AN, Frenklikh MS (1966) Protactinium and thorium content in oceanic waters. Radiochemistry 8:455–458

Kuznetsov YuV, Legin VK, Lisitsin AP et al (1967) Radioactivity of oceanic suspension, 2. Uranium in oceanic suspension. Radiochemistry 9:498–499

Labeyrie L Jr (1974) New approach to surface seawater paleotemperatures using ^{18}O/^{16}O ratios in silica diatom frustules. Nature 248:40–42

Lal D (1963) Study of long and short-term geophysical processes using natural radioacivity. In: IAEA (ed) Radioactive dating: proceedings of the symposium on radioactive dating. IAEA, Vienna, pp 149–157

Lal D, Peters B (1962) Cosmic-ray-produced isotopes and their application to problems in geophysics. Program Elem Part Cosm Ray Phys 6:1–74

Lal D, Peters B (1967) Cosmic-ray-produced radioactivity on the Earth. Encycl Phys 46:551–612

Lal D, Nijampurkar VN, Rama S (1970) Silicon-32 hydrology. In: IAEA (ed) Isotope hydrology: proceedings of a symposium on use of isotopes in hydrology. IAEA, Vienna, pp 847–863

Lal D, Rajan RS, Venkatavaradan VS (1967) Nuclear effects of solar and "galactic" cosmic-ray particles in near-surface regions of meteorites. Geochim Cosmochim Acta 31:1859–1869

Lal D, Venkatavaradan VS (1967) Activation of cosmic dust by cosmic-ray particles. Earth Planet Sci Lett 3:293–310

Lancet MS, Anders E (1970) Carbon isotopes fractionation in the Fisher-Tropch synthesis and in meteorites. Science 170:980–982

Larimer LW (1967) Chemical fractionation in meteorites. 1. Condensation of the eltments. Geochim Cosmochim Acta 31:1215–1238

Larimer LW, Anders E (1967) Chemical fractionation in meteorites, II. Abundance patterns and their interpretation. Geochim Cosmochim Acta 31:1234–1270

Larimer LW, Anders E (1970) Chemical fractionation in meteorites, III. Major element fractionation in chondrites. Geochim Cosmochim Acta 34:367–387

Larionov VV (1963) Nuclear geology and geophysics. Gostoptechizdat, Moscow

Laul JC, Ganapathy R, Anders E et al (1973) Chemical fractionation in meteorites. 6. Accretion temperatures of H-, LL-, and E-chondrites, from abundance of volatile trace elements. Geochim Cosmochim Acta 37:329–357

Lawrence JR (1973) Interstitial water studies: Stable oxygen and carbon isotope variations in water, carbonates and silicates from the Venezuela Basin (site 149) and the Avec Rise (site 148) Leg. 15. Initial Rep DSDP 17:377–406

Lawrence JR, DreverJI, Anderson TF, Bruecker HK (1967) Importance of alteration of volcanic material in the sediments of DSDP site 323: chemistry. ^{18}O/^{16}O and ^{87}Sr/^{86}Sr'. Geochim Cosmochim Acta 43:573–588

Lawrence JR, Gieskes JM, Broecker WS (1975) Oxygen isotopes and carbon composition of DSDP pore water and the alteration of layer II basalts. Earth Planet Sci Lett 27:1–10

Lawrence JR, Taylor HP (1971) Deuterium and oxygen-correlation clay minerals and hydroxides in Quaternary soils compared to meteoric waters. Geochim Cosmochim Acta 35:993–1003

Lawrence JR, Taylor HP (1972) Hydrogen and oxygen isotope systematics in weathering profiles. Geochim Cosmochim Acta 36:1377–1393

Lazarenko EK (1958) Certain problems of clay minerals and argillites study. In: Lazarenko EK (ed) The study and use of clays. Lvov University Press, L'vov, pp 34–41

Lehmann BE, Fabrica-Martin JT, Davis SN et al (2003) A comparison of grounwater dating with Kr-81, Cl-36, and He-4 in four wells of the Great Artesian Basin, Australia. Earth Planet Sci Lett 211:237–250

Lehmann BE, Davis SN, Love A et al (1993) Atmospheric and subsurface sources of stable and radioactive nuclides used for groundwater dating. Water Resour Res 29:2027–2040

Lemeile E, Letolle R, Meliere F, Olive P (1983) Isotope and other physico-chemical parameters of paleocarbonates. In: Paleoclimates and paleowaters. Proceedings of an advisory group meeting, IAEA, Vienna, p 135

Lepin VS, Solodyankina BN, Bankovskaya YeV et al (1975) Application of strontium isotopes in some hydrogeological problems. In: Modern studies of the Earth's crust, East-Sib. Pub., Irkutsk, pp 175–178

Lerman JC (1972) Carbon-14 dating: origin and correlation of isotope fractionation errors in terrestrial living matter. In: Proceedings of 8th international conference on radiocarbon dating, Lower Hutt, Wellington, pp 612–624

Letolle R (1980) Nitrogen-15 in the natural environment. In: Fritz P, Fontes JCh (eds) Handbook of environmental isotope geochemistry. Elsevier, Amsterdam, pp 407–433

Letolle R, Olive P (1983) Isotopes as pollution tracers. In: IAEA (ed) Guidebook on nuclear techniques in hydrology. IAEA, Vienna, pp 411–420

Levin M, Gat JR, Issar A (1980) Precipitation, flood and groundwaters of the Negev Highlands: an isotopic study of desert hydrology. In: IAEA (ed) Arid-zone hydrology: investigations with isotope techniques: proceedings of an advisory group meeting on application of isotope techniques in arid zones hydrology. IAEA, Vienna, pp 3–22

Li YH, Ku TL, Mathieu GG, Wolgemuth K (1973) Barium in the Antarctic Ocean and implications regarding the marine geochemistry of Ba and ^{226}Ra. Earth Planet Sci Lett 19:352–358

Li YH, Mathieu GG, Biscye P, Simpson HJ (1977) The flux of ^{226}Ra from estuarine and continental shell sediments. Earth Planet Sci Lett 37:237–241

Libby WF (1963) Moratorium tritium geophysics. J Geophys Res 68:4485–4494

Libby WF (1955) Radiocarbon dating. Chicago University Press, Chicago

Libby WF (1967) History of radiocarbon dating. In: Radioactive dating and methods of low-level counting: proceedings of a symposium organized by the international atomic energy agency in co-operation with the joint commission on applied radioactivity (ICSU) and held in Monaco, 2–10 March 1967, IAEA, Vienna, pp 3–25

Libby WF (1973) History of tritium. In: Moghissi A, Carter M (eds) Tritium, Messenger graphics, Phoenix, pp 3–11

Libby WF, Palmer CE (1960) Stratospheric mixing from radiation fallout. J Geophys Res 65:3307–3317

Libby LM, Pandolfi LJ (1979) Isotopic tree thermometers: anticorrelation with radiocarbon. In: Berger R, Suess HE (eds) Radiocarbon dating. University of California Press, Berkeley, pp 661–669

Lingenfelter RE, Ramaty R (1970) Astrophysical and geophysical variations in ^{14}C production. In: Radiocarbon variations and absolute chronology, XII nobel symposium, Wiley, New York, pp 513–537

Lloyd RM (1966) Oxygen isotope enrichment of sea water by evaporation. Geochim Cosmochim Acta 30:801–814

Lloyd RM (1968) Oxygen isotope behaviour in the sulphate-water system. J Geophys Res 73:6099–6110

Locante J (1971) Tritium in pressurized water reactor. Trans Am Nucl Soc 14:161–162

Long A Lerman JC (1978) Oxygen-18 in tree rings as a paleohydrometer. In: 7th Symp Stab Isot Geochim, Tez Dokl, Moscow, pp 322–323

Longinelli A (1968) Oxygen isotope composition of sulfate ions in water from thermal springs. Earth Planet Sci Lett 4:206–210

Longinelli A, Nuti S (1965) Oxygen isotope composition of phosphate and carbonate from living and fossil marine organisms. In: Tongiorgi E (ed) Stable isotopes in oceanographic studies and paleotemperatures: Spoleto. Consiglio Nazionale delle Ricerche, Piazzale Aldo Moro, pp 183–197

Loosli HH, Oeschger H (1979) Argon-39, carbon-14 and kripton-85 measurements in groundwater samples. In: IAEA (ed) Isotope hydrology, 1978: proceedings of an international symposium on isotope hydrology, vol 2. IAEA, Vienna, pp 931–947

Loosli HH, Oeschger H (1980) Use of ^{39}Ar and ^{14}C for groundwater dating. Radiocarbon 22:863–870

Lopes JS, Pinte RE, Almendra ME, Machado JA (1977) Variation of ^{14}C activity in portuguese wines from 1940 to 1974. In: Low-radioactivity measurements and applications. Proceedings of the symposium, High Tatras, IAEA, Vienna, pp 265–268

Lord HC III (1965) Molecular equilibrium and condensation in solar nebula and cool stellar atmospheres. Icarus 4:279–288

Lowenstam HA, Epstein S (1954) Paleotemperatures of the post-Albian Cretaceous as determined by the oxygen isotope method. J Geol 62:207–248

Lowenstam HA, Epstein S (1959) Cretaceous paleotemperatures as determined by the oxygen isotope method, their relations to and nature of rudistid reefs. In: 20th International geological congress, Cretaceous, Mexico, pp 65–76

Lugovaya IP, Snezhko AM, Proskurenko LA (1978) 13C/12C and 18O/16O ratio in Krivoy Rog Precambrian graphite-carbonates as indicator of their organogenity. In: 7th Symp Stab Isotop Geochi, Moscow, pp 135–137

Lujanas VYu (1975) On the rate of cosmogenic radionuclides production. In: Cosmogen Radioact Isotop 3:17–25 (Vilnius)

Lujanas VYu (1979) Cosmogenic radionuclides in the atmosphere. Mokslas, Vilnius

Lujanas VYu, Styro BI, Zinkevichus PK, Shopauskas KK (1975) Regimen of cosmogenic ^{22}Na and ^{7}Đ'Đµ in the atmosphere. Cosmogen Radioact Isotop 3:99–107 (Vilnius)

L'vovich MI (1969) Water resources in future. Prosveshchenie, Moscow

Maass I (1962) Contributions to the isotope geology of the elements hydrogen, carbon and oxygen. Isotopentechnik 111–116

Machta L (1969) Evaporation rates based on tritium measurements for hurricane Betcy. Tellus 21:407–408

Maev EG, Karpichev YuA (1999) Radiocarbon dating of the Aral Sea bottom sediments and sea level variation. Water Resour 26:212–220

Magaritz M, Taylor HP Jr (1976a) Oxygen, hydrogen and carbon isotope studies of the Franciscon formation, cost rages, California. Geochim Cosmochim Acta 40:215–234

Magaritz M, Taylor HP Jr (1976b) Isotopic evidence for meteoric-hydrothermal alteration of plutonic igneous rocks in the Yakutat Bay Skagway areas, Alaska. Earth Planet Sci Lett 30:179–190

Maloszewski P, Zuber A (1996) Lanped parameter models for the interpretation of environmental tracer data. In: Manual on the mathematical models in isotope hydrology, Tecdoc no 910, IAEA, Vienna, pp 9–58

Mamaev OI (1970) Marine turbulence. Moscow University Press, Moscow

Martell EA (1963) On the inventory of artificial tritium and its occurrence in atmospheric methane. J Geophys Res 68:3759–3769

Martin ID, Hackett IP (1974) Tritium in atmospheric hydrogen. Tellus 26:603–607

Martinec J, Siegenthaler U, Oeschger H et al (1974) A new insights into the run-off mechanism by environmental isotopes. In: Isotope techniques in groundwater hydrology: proceedings of a symposium, IAEA, Vienna, 11–15 March 1974, vol 1, pp 129–142

Martinec J, Kumar MB (1980) Thermal migration of formation waters in salt domes. EOS Trans Am Geophys Union 61:1177–1178

Martma TA, Pirrus RO, Punning YaMK (1985) Isotopic profiles of lacustrine lime as an indicator on climatic changes in the Holocene. In: Lacustrine carbonates of nonchernozem zone of the USSR, Politechn Inst Perm, p 160

Mason B (1966) Principles of geochemistry, 3rd ed. Wiley, New York

Mason AS, Öslund HG (1979) Atmospheric HT and HTO: V. Distribution and large-scale circulation. In: IAEA (ed) Behaviour of tritium in the environment: proceedings of the international symposium on the behaviour of tritium in the environment. IAEA, Vienna, pp 3–15

Matsubaya D, Sakai H (1973) Oxygen and hydrogen isotopic studies on the water of crystallization of gypsum from the Kuroko type mineralisation. Geochim Jpn 3:153–165

Matsuo S, Friedman I, Smith CI (1972) Studies of quaternary saline lakes. 1. hydrogen isotope fractionation in saline minerals. Geochim Cosmochim Acta 36:427–435

Matsuo S, Kuroda Y, Suzuoki T et al (1978) Mantle water based on the hydrogen isotope ratios of hydrous silicates in the mantel. Geol Surv Open-file Rep 701:278–280

Matthews A, Kolodny Y (1978) Oxygen isotope fractionation in decarbonation metamorphysm: mottled Zone Event. Earth Planet Sci Lett 39:179–192

Matthess G, Münnich KO, Sonntag C (1976) Practical problems of groundwater model ages for groundwater protection studies. In: Interpretation of environmental isotope and hydrochemical data in groundwater hydrology. Proceedings of an advisory group meeting, IAEA, Vienna, pp 185–194

Mazor E (1976) The Ram lake crater: a not on the revival of a 2000-year old groundwater traicing experiment. In: Interpretation of environmental isotope data in groundwater hydrology. Proceedings of an advisory group meeting, IAEA, Vienna, pp 179–181

Mazor E, Verhagen BT, Negreanu E (1974a) Hot springs of the igneous terrain of Swasiland. In: Isotope techniques in groundwater hydrology: proceedings of a symposium, IAEA, Vienna, 11–15 March 1974, vol 2, pp 29–45

Mazor E, Bosch A (1992) Helium as semi-quantitative indicator for groundwater dating in the range 10^4–10^8 years. In: Isotopes in noble gases as tracers in environmental studies. Proc Consult Meet., IAEA, Vienna, pp 163–178

Mazor E, Verhagen BT, Sellschop JPF et al (1974b) Kalahari groundwaters: their hydrogen, carbon and oxygen isotopes. In: Isotope techniques in groundwater hydrology: proceedings of a symposium, IAEA, Vienna, 11–15 March 1974, vol 1, pp 203–223

Mayer B (1997) Potential and limitations of using sulfur isotope abundance rations as an indicator natural and anthropogenic induced environmental change. In: IAEA (ed) Isotope techniques in the study of environmental change: proceedings of an international symposium on isotope techniques in the study of past and current environmental changes in the hydrosphere and the atmosphere. IAEA, Vienna, pp 432–435

Meniaylov IA, Vetshtein VE, Nikitina LP, Artemchuk VG (1981) D/H and $^{18}O/^{16}O$ ratios in magmatic water and gas of the Tolbachik great fracture eruption, Kamchatka. Reports of the USSR Acad Sci 273:258–272

McKenzie WF, Truesdell AH (1977) Geothermal reservoir temperatures estimated from the oxygen isotope compositions of dissolved fate and water from hot springs and shallow drill-holes. Geotermics 5:51–61

Meison B (1971) Geochemistry fundamentals. Nedra, Moscow

Merlivat L (1970) D'etude quantitative de bilans de lacs à l'aide des concentrations en deuterium et oxygen-18 dans lean. In: IAEA (ed) Isotope hydrology: proceedings of a symposium on use of isotopes in hydrology. IAEA, Vienna, pp 89–107

Merlivat L, Vuillanme Y (1970) Caracterisation de l'intrusion marine dans la nappe de la crau, a l'aide du deuterium. Bull Bur Rech Geol Et Miniers. Sec, 3, Paris, pp 81–85

Michel RL (1976) Tritium inventories of the world oceans and their implications. Nature 2263:103–106

Michel RL, Suess HE (1978) Tritium in the Caspian Sea. Earth Planet Sci Lett 39:309–312

Michel RL, Suess HE (1975) Bomb tritium in the Pacific Ocean. J GeophysRes 40:4139–4152

Migdisov AM, Dontsova EI, Kuznetsova LV, Ronov AV (1974) Possible mechanism for oxygen isotopic composition evolution in the upper Earth's shells. In: Proceedings of the 1st internatinal geochemistry congress, 4, Sediment rocks, Nauka, Moscow, pp 173–184
Miller AR, Densmore CD, Degens ET et al (1966) Hot briens and recent iron deposits in deeps of the Red Sea. Geochim Cosmochim Acta 30:341–359
Miskel JA (1973) Production of tritium by nuclear weapons. In: Moghissi A, Carter M (eds) Tritium. Messenger graphics, Phoenix, pp 79–85
Miyake Y (1965) Elements of Geochemistry, Maruzen, Japan
Miyake Y (1969) Fundamentals of geochemistry. Nedra, Moscow (trans: from English)
Mizutani Y (1972) Isotopic composition and undeground temperature of the Otake geothermal water, Kyushu, Japan. Geochim J 6:67–73
Mizutani Y, Hamasuna T (1972) Origin of the Shimogama brine, Izu. Volcan Soc Jap Bull 17:123–134
Mizutani Y, Rafter TA (1969) Oxygen isotope composition of sulphates, 3. oxygen isotopic fractionation in the bisulfate ion-water system. N Z J Sci 12:54–59
Möller P, Wagner K (1965) Dating soil layers by ^{10}B. In: IAEA (ed) Radioactive dating and methods of low-level counting: proceedings of a symposium. IAEA, Vienna, pp 177–187
Monster L, Anders E, Thode HC (1965) ^{34}S/^{32}S ratios for the different forms of sulphur in Orgueil meteorite and their mode of formation. Geochim Cosmochim Acta 29:773–779
Moody DJ (1978) Geography and geology of gigantic petroleum fields. In: Fisher AG, Judson S (eds) Petroleum and global tectonics. Nedra, Moscow, pp 112–160 (trans: from English)
Mook WG (1970) Stable carbon and oxygen isotopes in natural waters in the Netherlands. In: IAEA (ed) Isotope hydrology: proceedings of a symposium on use of isotopes in hydrology. IAEA, Vienna, pp 163–189
Mook WG (1972) On the reconstraction of the initial ^{14}C content of groundwater from the chemical and isotopic composition. In: Proceedings of the 8th international conference on radiocarbon dating, Lower Hutt, Wellington, pp 343–352
Mook WG (1976) The dissolution-exchange model for dating groundwater with ^{14}C. In: Interpretation of environmental and hydrochemical data in groundwater hydrology. In: Proceedings of an advisory group meeting, IAEA, Vienna, pp 213–226
Mook WG (1977) The radiocarbon time scale. In: Low-radioact Meas Appl, Proceedings of the symposium, High Tatras, Bratislava, pp 193–298
Mook WG (1980) Carbon-14 in hydrogeological studies. In: Fritz P, Fontes JCh (eds) Handbook of environmental isotope geochemistry, vol 1. Elsevier, Amsterdam, pp 49–74
Mook WD (2005) Introduction to isotope hydrology: stable and radioactive isotopes of hydrogen, carbon and oxygen. Taylor and Francis, The Netherlands
Moore JG (1970) Water content of balalt erupted on the ocean floor. Contrib Miner Petrol 28:272–279
Moore WS (1969) Measurement of ^{228}Ra and 228Th in the sea water. J Geophys Res 74:694–704
Mopper K, Garlick GD (1968) Oxygen isotope fractionations between biogenic silica and ocean water. Trans Am Geophys Union 49:336
Morgenstern U (2000) Silicon-32. In: Cock P, Herczeg A (eds) Environmental tracers in subsurface hydrology. Kluwer, Boston, pp 498–502
Morkovkina IK (1978) Tritium application in study of groundwater recharge. In: Ferronsky VI (ed) Isotopy of natural waters. Nauka, Moscow, pp 165–179
Morkovkina IK (1979) Tritium use in hydrogeological studies. In: Ferronsky VI (ed) Isotope investigation of natural waters. Nauka, Moscow, pp 75–84
Mörner NA (1971) The position of the ocean level during the interstadial at about 30,000 B.P. A discussion from a climatic, glaciologic point of view. Canad J Earth Sci 8:8
Moser H (1980) Trends in isotope hydrology. In: Isotope Hydrol Proc Symp Bogota, IAEA, Vienna, pp 3–11
Moser H, Stichler W, Zötl J (1972) Altergliederung von tiefliegendenartesischen Wasser. Naturwissenschaften 59:122–123

Münnich KO (1957) Messungen des ^{14}C-Gehaltes von hartem Grundwasser. Naturwissenschaften 44:32–34

Münnich KO (1968) Isotopen datierung von Grundwasser. Naturwissenschaften 55:158–163

Münnich KO, Roether W (1963) A comparison of carbon-14 and tritium ages of groundwater. In: IAEA (ed) Radioisotopes in hydrology: proceedings of a symposium. IAEA, Vienna, pp 397–404

Münnich KO, Roether W (1967) Transfer of bomb ^{14}C and tritium from the atmosphere to the ocean on the basis of tritium and ^{14}C profiles. In: Radioactive dating and methods of low-level counting: proceedings of a symposium organized by the international atomic energy agency in co-operation with the joint commission on applied radioactivity (ICSU) and held in Monaco, 2–10 March 1967, IAEA, Vienna, pp 93–104

Münnich KO, Vogel JC (1962) Untersuhungen an pluviulen Wassern der Ost-Sahara. Geol Rdsch 52:611–624

Münnich KO, Vogel JC (1963) Investigation of meridional transport in the troposphere by means of carbon-14 measurement. Radiocarbon dating. Proceedings of a symposium, IAEA, Vienna, pp 189–197

Münnich KO, Roether W, Thilo L (1967) Dating of groundwater with tritium and ^{14}C. In: Isotopes in hydrology: proceedings of the symposium on isotopes in hydrology held by the international atomic energy agency in cooperation with the international union of geodesy and geophysics in Vienna, IAEA, Vienna, 14–18 Nov 1966, pp 305–319

Murrey BC., Belton JS, Danielson GE et al (1974) Mariner 10 pictures of Mercury: first results. Science 184:459–461

Naydenov BM, Polivyannyi EYa, Bogolepov VG (1978) Status of knowledge and reason for study of argon isotopes variation in mineral gas and fluid inclusions. Geokhimiya 12:1877–1882

Nikanorov AM, Vetshtein VE, Tarasov MG (1976) The origin and formation of groundwaters in near Caucasus Mesozoic sediments by means of hydrogen and oxygen isotopes. In: VI Symposium on application of stable isotopes in geochemistry, Moscow, pp 62–63

Nikanorov AM, Yakubovsky AV, Shalaev LN et al (1980) On isotope and chemical anomaly of fresh water in oil fields. In: 8th Vses Symp on Stable Isotop Geochim, Moscow, pp 224–226

Nikolaev SD (1995) Isotopic paleogeography of inner continental seas. Izd VNIRO, Moscow

Nikolaev SD, Korn OP, Lazarev LF et al (1960) Uranium content in the Black Sea waters. Dokl AN SSSR 6:1411–1412

Nikolaev SD., Lazarev KF, Grashchenko SM (1961) Thorium isotopes content in the Asov Sea waters. Dokl AN SSSR 138:674–676

Nikolaev SD., Lazarev KF, Korn OP, Drozhin VM (1966) Geochemical balance of radioactive elements in the Black Sea and Asov Sea basins: 1. uranium balance. Radiochemiya 11:688–698

Nir A (1964) On the interpretation of tritium age measurements of groundwater. J Geophys Res 69:423–431

Nozaki Y, Thompson J, Turekian KK (1976) The distribution of in the surface waters of the Pacific Ocean. Earth Planet Sci Lett 32:304–312

Nozaki Y, Tsunogai S (1976) ^{226}Ra, ^{210}Pb and ^{210}Po distribution in the western north Pacific. Earth Planet Sci Lett 32:313–321

Northrop DA, Clayton RN (1966) Oxygen isotope fractionation in system containing dolomite. J Geol 74:174–196

Nydal R (1967) On the transfer of radiocarbon in nature. In: Radioactive dating and methods of low-level counting: proceedings of a symposium organized by the international atomic energy agency in co-operation with the joint commission on applied radioactivity (ICSU) and held in Monaco, 2–10 March 1967, IAEA, Vienna, pp 119–128

Nydal R, Lövseth K, Gulliksen S (1979) A survey of radiocarbon variation in nature since the Test Ban Treaty. In: Berger R, Suess HE (eds) Radiocarbon dating. University of California Press, Berkely, pp 313–323

O'Brien K (1979) Secular variation in the prodLustion of cosmogenic isotopes in the Earth's atmosphere. J Geophys Res 84:2589–2595

Oeschger H, Siegenthaller U (1979) Prognosis for expected CO_2 increase to fossil fuel combustion. In: Berger R, Suess HE (eds) Radiocarbon dating. University of California Press, Berkely, pp 633–642

Oeschger H, Gugelmann A, Loosli H et al (1974) ^{39}Ar dating of groundwater. In: Isotope techniques in groundwater hydrology: proceedings of a symposium, IAEA, Vienna, 11–15 March 1974, vol 2, pp 179–189

Ohmoto H, Rye RO (1974) Oxygen and hydrogen isotope composition of fluid inclusions in the Kuroko deposits, Japan. Econ Geol 69:947–953

Olsson IU (1979) The radiocarbon contents of various reservoirs. In: Berger R, Suess HE (eds) Radiocarbon dating. University of California Press, Berkely, pp 613–618

Olshtinsky SP, Gudzenko VV, Bondarenko GM, Kostyuchenko NA (1981) Isotope and geochemical studies of groundwaters in the North Crimea. In: Ferronsky VI (ed) Natural waters study by isotope methods. Nauka, Moscow, pp 222–225

Onufriev VG (1978) Principles of volatile elements in the Earth's upper shell and formation of their isotopic composition. In: Ferronsky VI (ed) Isotopy of natural waters. Nauka, Moscow, pp 90–118

Onufriev VG, Karpychev YuA, Pavlov VA (1975) The cycle of radiocarbons as a geochemical basis for natural waters radiocarbon research. In: Ferronsky VI (ed) Isotopy of natural waters. Nauka, Moscow, pp 193–212

Onufriev VG, Karpychev YuA, Pavlov VA et al (1975) Origin and recharge of the Yaskhan lens of fresh water, Turkmenia, by 14C data. In: Babinets AYe (ed) Radioisotope methods in hydrogeological studies. Naukova Dymka, Kiev, pp 68–69

O'Neil JR (1979) Stable isotope geochemistry of rocks and minerals. In: Jäger E, Hunziker JC (eds) Lectures in isotope geology. Springer, New York, pp 235–263

O'Neil JR. Adomi EH, Epstein S (1975) Revised value for the ^{18}O fractionation between CO_2 and water at 25°C. United States Geol Surv J Res 3:623–624

O'Neil JR, Kharaka JK (1976) Hydrogen and oxygen isotope exchange reactions between clay minerals and water. Geochim Cosmochim Acta 40:214–245

Onuma N, Clayton RN, Mayeda TK (1970) Oxygen isotope fractionation between minerals and an estimate of the temperature of formation. Science 167:536–537

Onuma N, Clayton RN, Mayeda TK (1972) Oxygen isotope cosmothermometer. Geochim Cosmochim Acta 36:169–188

Osmond JK, Rydell HS, Kaufman MI (1968) Uranium disequilibrium in groundwater: an isotope delution approach in hydrologic investigations. Science 162:997–999

Östlund HG (1967) Hurricane tritium: I. Preliminary results on Hild 1964 and Betsy 1965. In: Isotope techniques in hydrologic cycle, IAEA, Vienna, pp 58–60

Östlund HG (1970) Hurricane tritium: III. evaporation of sea water in hurricane faith 1966. J Geophys Res 75:2203–2209

Östlund HG (1970) Tritium in the atmosphere and ocean. In: Moghissi A, Carter M (eds) Tritium. Messenger graphics, Phoenix, pp 382–301

Östlund HG (1973) Tritium in the atmosphere and ocean. In: Moghissi A, Carter M (eds) Tritium. Messenger graphics, Phonex, pp 382–391

Östlund HG, Berry E (1970) Modification of atmospheric tritium and water vapour by Lake Tahoe. Tellus 22:463–468

Östlund HG, Fine RA (1979) Oceanic distribution and transport of tritium. In: IAEA (ed) Behaviour of tritium in the environment: proceedings of the international symposium on the behaviour of tritium in the environment. IAEA, Vienna, pp 303–312

Östlund HG, Mason AS (1974) Atmospheric HT and HTO: 1. experimental procedures and tropospheric data 1968–1972. Tellus 26:91–102

Östlund HG, Rinyel MO, Rooth C (1969) Tritium in the equatorial Atlantic current system. J Geophys Res 74:4535–4540

Panichi C, Celati R, Noto P et al (1974) Oxygen and hydrogen isotope studies of the Larderello (Italy) geothermal system. In: Isotope techniques in groundwater hydrology: proceedings of a symposium, IAEA, Vienna, 11–15 March 1974, vol 2, pp 3–28

Panichi C, Ferrara GC, Gonfiantini R (1977) Isotope geothermometry in the Larderello geothermal field. Geothermics 5:81–88

Panichi C, Nuti S, Noto P (1979) Use of isotopic geothermometers in the Larderello geothermal field. In: IAEA (ed) Isotope hydrology, 1978: proceedings of an international symposium on isotope hydrology, vol 2. IAEA, Vienna, pp 613–629

Parck R, Epstein S (1960) Carbon isotope fractionation during photosynthesis. Geochim Cosmochim Acta 21:110–126

Pavlov VA., Petrukhin VA, Romanov VV (1981) The study of hydogeological and hydrochemical regime of the Mazhaysk reservoir by nuclear-physical methods. In: Ferronsky VI (ed) Natural waters study by isotopic methods. Nauka, Moscow, pp 114–128

Payne BR (1970) Water balance of lake Chala and its relation to groundwater from tritium and stable isotope data. J Hydrol 11:47–58

Payne BR, Yurtsever Y (1974) Environmental isotopes as a hydrogeological tool in Nicaragua. In: Isotope techniques in groundwater hydrology: proceedings of a symposium, IAEA, Vienna, 11–15 March 1974, vol 1, pp 193–201

Payne BR, Schotterer P (1979) Importance of on filtration from the Chimbo River in Equador to groundwater: a study based on environmental isotope variation. In: IAEA (ed) Isotope hydrology, 1978: proceedings of an international symposium on isotope hydrology, vol 1. IAEA, Vienna, pp 145–156

Pearson FJ (1965) Use of $^{13}C/^{12}C$ ratios to correct radiocarbon ages of materials initially diluted by limestone. In: Proceedings of the 6th international conference, USAEC, Pullman, Washington pp 357–366

Pearson DH, Cambray RS, Spiser GS (1966) Lead-210 and polonium-210 in the atmosphere. Tellus 18:427–433

Pearson FJ, Fisher DM, Plummer LN (1978) Correction of groundwater chemistry and carbon isotopic composition for effect of CO_2 outgasing. Geochim Cosmochim Acta 42:1799–1807

Pearson FJ, Hanshaw BB (1970) Sources of dissolved carbonate spacies in groundwater and their effects of carbon-14 dating. In: IAEA (ed) Isotope hydrology: proceedings of a symposium on use of isotopes in hydrology. IAEA, Vienna, pp 271–286

Pearson FJ, Swarzenki WV (1974) ^{14}C Evidence for origin of arid region groundwater, Northeastern province, Kenia. In: Isotope techniques in ground water hydrology: proceedings of a symposium, IAEA, Vienna, 11–15 March 1974, vol 1, pp 95–108

Pearson FJ, White DE (1967) Carbon-14 ages and flow rates of water in Carriso Sand Atascosa County, Texas. Water Resour Res 3:251–261

Pelmegov SV, Munaev Ye, Bondarenko GN (1978) Isotopic and geochemical studies of groundwater from a boundary of an artesian basin. Sov Geol 4:119–125

Perry EG (1967) The oxygen isotope chemistry of ancient cherts. Earth Planet Sci Lett 3:62–66

Perry EG, Tan FS (1972) Significance of oxygen and carbon isotope variations in early Precambrien cherts and carbonate rocks of Southern Africa. Bull Geol Soc Amer 83:647–664

Perry EA, Giskes JM, Lawrence JR (1976) Mg, Ca and $^{18}O/^{16}O$ exchange in the sediment-pore water system, Hole 149, DSDP. Geochim Cosmochim Acta 40:413–423

Petrov VP (1975) Stories about white clay. Nedra, Moscow

Phillips FM (2000) Chlorine. In: Cock P, Herzeg A (eds) Environmental tracers in subsurface hydrology. Kluwer, Boston, pp 299–348

Pichon XLe, Francheteau XJ, Bonnin J (1973) Plate tectonics. Elsevier, Amsterdam

Pinneker EV (1974) Role of isotope investigation at groundwater resources exploration in Eastern Siberia. In: Pinneker EV (ed) Groundwaters of Irkutsk region. Nedra, Leningrad, pp 14–31

Pinneker EV (1975) Formation of modern hydrotherms in the dead volcanic regions (in the light of isotopic data). In: Pinneker EV (ed) Geothermal process in the regions of active structural magmatism. Nauka, Moscow, pp 38–43

Pinneker EV, Vetshtein VE, Dzyuba AA et al (1973) Oxygen-18 content in Siberian Platform brines. In: Pinneker EV (ed) Outlines on hydrogeology of Siberia. Nauka, Novosibirsk, pp 86–92

Pinneker EV, Romanov VV, Dzyuba AA (1978) The peculiarities of tritium distribution in the near-Baykal Lake natural waters. In: Pinneker EV (ed) Regional hydrogeology and engineering geology of the Eastern Siberia. Nauka, Novosibirsk, pp 3–7

Plesset M, Latter A (1960) Transient effects in the distribution of carbon-14 in nature. Proc Nat Acad Sci 46:232–241

Plummer LN (2005) Dating of young groundwater. In: Aggarwal P, Gat J, Froehlich K (eds) Isotopes in the water cycle. Springer, Dordrechht, pp 193–218

Plummer LN, Busenberg E (2000) Chlorofluorocarbons. In: Cock P, Herzeg A (eds) Environmental tracers in subsurface hydrology. Kluwer, Boston, pp 441–478

Plummer LN, Busenberg E, Riggs FC (2000) In-situ grows of calcite at Devil Hole, Nevada: comparison of field and laboratory rates to a 500,000 year record of near-equilibrium calcite growth. Aquatic Geochim 6:254–274

Plyusnin GS, Lomonosov IS, Posokhov BF (1978a) The problem of nowaday hydrothermaes genesis using $^{7}Li/^{6}Li$ and $^{87/86}Sr$ ratios. Dokl AN SSSR 243:481–484

Plyusnin GS, Lomonosov IS, Posokhov BF, Sandimirova GP (1978b) $^{7}Li/^{6}Li$ and $^{87/86}Sr$ ratios as a groundwater genesis criterion. In: 7th Vses Simp Stab Isotop Geokhim, Moscow, pp 113–114

Poldervaart A (1955) Geochemistry of the Earth's crust. Geol Soc Am Pap 62:119–144

Polivanova AI (1970) On the isotopic composition of underground brines as their origin feature. Geokhimiya 7:829–837

Polyakov VA (1981) Radiocarbon dating of groundwaters. Izd VIEMS, Moscow

Polyakov VA, Golubkova EV (2007) Protection study of groundwaters by isotopic and hydrochemical data. Prospect Prot Min Resour 5:48–52

Polyakov VA, Karasev BV (1979) Correction method of groundwater radiocarbon dating. In: Juodkazis VI (ed) Use of ^{14}C dating in geology. LithNIGRI, Visnius, pp 25–28

Polyakov VA, Kolesnikova LN (1978) Some regional specifics in formation of isotopic content of precipitation. In: 7th Vses Simp Stab Isotop Geokhim, Moscow, pp 148–149

Polyakov VA, Seletsky YuB (1972) Experimental study of deuterium behaviour at marine water evaporation. In: Galimov EM (ed) Stable isotope geochemistry, abstracts. Geochemistry Institute, Moscow, pp 104–105

Polyakov VA, Seletsky YuB (1978) Radiocarbon and tritium study of groundwater dynamics in the Assel-Kliasinan aquifer at Sudogda River region. Geokhimiya 8:1230–1238

Polyakov VA, Seletsky YuB (1979) Natural factors affecting the prcision of groundwater radiocarbon dating the methods of their account. In: Ferronsky VI (ed) Isotope studies of natural waters. Nauka, Moscow, pp 122–130

Polyakov VA, Sokolovsky LG (2005) Genesis and dynamics of the Caucasus mineral waters by isotope data. Reviews in hydrogeology and engineering geology. Geoinform, Moscow

Polyakov VA, Tkachenko AE (1980) Use of stable isotopes of carbon, hydrogen and oxygen in study of formation of the Bakharden mineral waters, Turkmenia. In: Reports to 8th Vses Simp Stab Isotop Geokhim, Moscow, pp 344–345

Polyakov VA, Seletsky YuB, Yakubovsky AV et al (1974) Deuterium in the Naftusya' mineral water. Annals VSEGINGEO 59:80–87

Polyakov VA, Borevsky BV, Plugina TA (1978) The study of groundwater overflow through a relative watertight layer based on paleotemperature variation of stable isotopes. In: 7th Vses Symp Stab Isotop Geokhim, Moscow, pp 149–150

Polyakov VA, Ezhova MP, Seletsky YuB et al (1979) Isotope studies of thermal and mineral waters in the Caucasus region. In: Ferronsky VI (ed) Nuclear and isotope methods of natural waters study, abstracts. IWP AN SSSR, Moscow, pp 12–13

Polyakov VA, Kryshtal NN, Tkachenko AE et al (1981) Use of stable isotopes and tritium in hydrogeological mapping of 1:200,000 scale in West Siberian depression. In: Isotopes in the hydrosphere, abstract, IWP AN SSSR, Moscow, pp 122–124

Polyakov VA, Seletsky YuB, Dubinchuk VT (2005) Mineral water formation in the light of isotopic and geochemical studies. In: Borevsky BV (ed) A.M. Ovchinnicov's ideas on hydrogeology development. GYDEC, Moscow, pp 50–56

Polyakov VA, Dubinchuk VT, Golubkova EV et al (2008) Isotopic studies on the Tomsk polygon. Geol Explor Prot 11:47–52

Polynov VV (1953) Geologic role of living organisms. Voprosy Geografii 3:45–64

Ponomarev GM, Antipov-Karataev IN (1947) Steppe type soils formed on eruptive rocks. Annals Soil Inst 25:6780–6790 (Moscow)

Pugin VA, Khitarov NI (1978) Experimental petrology of abissal magmatism. Nauka, Moscow

Quay PD, Broecker WS, Hesslein RH et al (1979) Whole lake tritium spikes to measure horizontal and vertical mixing rates. In: Isotopes in lake studies. Proceedings of an advisory group meeting, IAEA, Vienna, pp 175–193.

Rabinovich IB (1968) Isotopic effects. In: Physical and chemical properties of solutions. Nauka, Moscow

Rachinsky VV (1964) Introduction to general theory of sorption dynamics and chromatography. Nauka, Moscow

Olsson I (ed) (1970) Radiocarbon variation and absolute chronology. Proceedings of the 12th nobel symposium, Stockholm

Radnell CL, Aitken MJ, Otlet RL (1979) In situ ^{14}C production in wood. In: Berger R, Suess HE (eds) Radiocarbon dating. University of California Press, Berkely, pp 643–650

Rafter TA, Mizutani Y (1967) Oxygen isotope composition of sulfates: 2. Preliminary results of oxygen isotope variation in sulphates and relationship to their environment and to their ^{34}S values. N Z J Sci 10:815–840

Rafter TA, O'Brien BJ (1972) C-14 measurements in the atmosphere and in the South Pacific Ocean. In: Proceedings of the 8th international conference on radiocarbon dating, Lower Hutt, Wellington, p 241

Ralf EK (1972) A cyclic solution for the relationship between magnetic and atmospheric C-14 changes. In: Proceedings of the 8th international conference on radiocarbon dating, Lower Hutt, Wellington, 76–84

Ralf EK, Klein J (1979) Composite computer plots of ^{14}C dates for treering-dated Bristlecone Pine and Sequoies. In: Berger R, Suess HE (eds) Radiocarbon dating. University of California Press, Berkely, pp 545–553

Rama SNI (1968) Chronology. In: IAEA (ed) Guide book on nuclear techniques in hydrology. IAEA, Vienna, pp 167–174

Rank L, Stroll E (1979) Test for the applicability of combines nuclear and geochemical methods in relation to the water balance of Lake Neusiedler, Austria. In: Isotopes in lake studies. Proceedings of an advisory group meeting IAEA, Vienna, pp 121–130

Rankama K (1963) Progress in isotope geology. Interscience, New York

Rauert W, Stichler W (1974) Groundwater investigations with environmental isotopes. In: Isotope techniques in groundwater hydrology: proceedings of a symposium, IAEA, Vienna, 11–15 March 1974, vol 1, pp 429–431

Ravoire I, Lorius C, Robert J, Roth E (1970) Tritium content in a firn core from Antarctica. J Geophys Res 75:2331–2336

Reardon EJ, Mozeto AA, Fritz P (1980) Recharge in northern clime calcareous sandy soils: soil water chemical and carbon-14 evolution. Geochim Cosmochim Acta 44:1723–1735

Redfield AC, Friedman I (1964) Factors affecting the distribution of deuterium in the ocean. Proc Symp Mar Geochim 149–168

Redlich O (1935) Eine allgemeine Beziehung zwischen den Molekuln (nebst Bemerkung über die Berechnung harmonischer Kraftkonstanten). Ztschr Phys Chem Bd 28:371–382

Reuter JH, Epstein S, Taylor HP (1965) $^{18}O/^{16}O$ ratios of some chondritic meteorites and terrestrial ultramafic rocks. Geochim Cosmochim Acta 29:481–488

Richet P, Bottinga Y, Javoy MA (1977) A review on hydrogen, carbon, nitrogen, oxygen, sulfur and chlorine isotope fractionation among gaseous molecules. Ann Rev Earth Planet Sci 5:65–110

Righini G (1963) Gli isotopi nell'atmosphera solara. Ric Sci Rev 3:145
Ringwood AE (1966) Chemical evolution of the terrestrial planets. Geochim Cosmochim Acta 30:41–104
Ringwood AE (1979) Origin of the Earth and Moon. Springer, New York
Robinson BW (1974) The origin of mineralization at the Tui Mine, Te Aroha, New Zealand, in the light of stable isotope studies. Econ Geol 69:910–925
Rodnon P (1976) Essai d'interpretation des variations climatiques au Sahara depuis 40,000 ans. Rev Geogr Phys et Geol Dyn 18:251–270
Rodnon P, Williams MAJ (1977) Late quaternary climatic changes in Australia and North-Africa: a preliminary interpretation. Palaeogeogr Paleoclimatol Palaeoecol 21:285–302
Rodriguez CO (1979) Identification del origin del aqua Subterranea en problemas de la Cordillera Andina. In: IAEA (ed) Isotope hydrology, 1978: proceedings of an international symposium on isotope hydrology, vol 1. IAEA, Vienna, pp 113–124
Roether W, Münnich KO, Östlund HE (1970) Tritium profile at the North Pacific (1969) Geosecs intercalibration station. J. Geophys Res 75:7672–7675
Roether W, Münnich KO, Ribbat B, Sarmiento JL (1980) A transatlantic ^{14}C section near 40°N of F/S meteor. Ergebnisse A 21:57–69
Romanov VV (1978) Regularities in tritium distribution for natural waters. In: Ferronsky VI (ed) Isotopy of natural waters. Nauka, Moscow, pp 46–89
Romanov VV (1982) Tritium use in study of marine and river water mixing. Water Res 5:22–26
Romanov VV, Zverev VL, Demin NV et al (1981) Tritium in the Pleshcheevo Lake waters and groundwaters of the Pereyaslavl intake well. In: Ferronsky VI (ed) Natural waters study by isotope methods. Nauka, Moscow, pp 79–84
Romanov VV, Kikichev HG (1979) Tritium in atmospheric hydrogen. In: Ferronsky VI (ed) Isotope studies of natural waters. Nauka, Moscow, pp 85–92
Romanov VV, Kikichev HG (1981) Tritium study of marine and continental water vapor portion in the near-surface air layer of the Black Sea north-west area. In: Ferronsky VI (ed) Natural waters study by isotope methods. Nauka, Moscow, pp 111–114
Romanov VV, Salnova LV, Seryegina LA (1979) Tritium use in studying dynamics of the Baykal Lake waters. In: Ferronsky VI (ed) Isotope studies of natural waters. Nauka, Moscow, pp 46–54
Romanov VV, Soyfer VN (1968) Application of tritium in hydrogeological investigations. In: Alekseev FA (ed) Geochemical and nuclear methods in oil and gas exploration. Nauka, Moscow, pp 289–304
Romanov VV, Ferronsky VI, Vakulovsky SM et al (1983) Tritium content in the USSR surface waters over 1979–1980. Water Res 3:109–121
Ronov AB, Yaroshevsky AA (1967) Chemical structure of the Earth's crust. Geokhimiya 11:1285–1309
Rona D, Akers LK, Noakes JE, Supernew I (1965) Geochronology in the Gulf of Mexico. Prog Oceanogr 3:289–295
Rooth CG, Östlund HE (1972) Penetration of tritium into the Atlantic thermocline. Deep-Sea Res 19:481–492
Rosholt JN, Harshman EN, Shields WR, Garner EL (1964) Isotopic fractionation of uranium related to roll features in sandstone, Shirley basin, Wyoming. Econom Geol 59:570–585
Rowland FS (1959) Ratio of HT/HTO in the atmosphere. J Chem Phys 30:1098–1099
Rozanski K, Florkowski T (1979) Kripton-85 dating of groundwater. In: IAEA (ed) Isotope hydrology, 1978: proceedings of an international symposium on isotope hydrology, vol 2. IAEA, Vienna, pp 949–961
Rozkowski A, Przewlocki K (1974) Application of stable environmental isotopes in mine hydrology taking Polish coal basin as an example. In: Isotope techniques in groundwater hydrology: proceedings of a symposium, IAEA, Vienna, 11–15 March 1974, vol 1, pp 481–501
Rubey WW (1964) Geologic history of sea water. In: Brancuzio Cl, Cameron AGW (eds) The origin and evolution of atmospheres and oceans. Wiley, New York pp 1–73

Rye RO (1966) The carbon, hydrogen and oxygen isotopic composition of the hydrothermal fluids responsible for the lead-zinc deposits at Providencia, Zacatecas, Mexico. Econ Geol 61:1399–1427

Rye RO., Hall WE, Ohmoto H (1974) Carbon, hydrogen, and sulfur isotope study of the Darwin lead-silver-zinc deposits, Northern California. Econ Geol 69:468–481

Rye RO, O'Neil JR (1968) The ^{18}O content of water in primary fluid inclusions from Providencia, north-central Mexico. Econ Geol 63:232–238

Sackett WM (1963) Geochemistry of ocean water. Trans Am Eophys Union 44:483–485

Sackett WM (1960) Protactinium-231 content of ocean water and sediments. Science 132:1761–1762

Sackett WM, Mo R, Sapaldin RF, Exnet ME (1973) A revalution of the marine geochemistry of uranium. In: IAEA (ed) Radioactive contamination of the marine environment: proceedings of a symposium. IAEA, Vienna pp 757–769

Sakai H (1977) Sulphate-water isotope thermometry applied to geothermal systems. Geotermics 5:67–74

Sakai H, Matsubaya O (1974) Isotope geochemistry of the thermal waters of Japan and its bearing on the Kuroko ore solutions. Econ Geol 69:674–991

Sakai H, Matsubaya O (1977) Stable isotope studies of Japanese geothermal systems. Geotermics 5:97–123

Sakai H, Tsutsumi M (1978) D/H fractionation factors between serpentine and water at 100°C to 500°C and 2000 bar water pressure, and the D/H ratios of natural serpentines. Earth Planet Sci Lett 40:231–242

Salati E, Leal JM, Campos MM (1974) Environmental isotopes used in a hydrological study of Northeastern Brazil. In: Isotope techniques in groundwater hydrology: proceedings of a symposium, IAEA, Vienna, 11–15 March 1974, vol 1, pp 259–282

Salati E, Matsui E, Leal JM et al (1980) Utilization of natural isotopes in the study of salination of the water in the Pejeu River valley, Northeast Brazil. In: IAEA (ed) Arid-zone hydrology: investigations with isotope techniques: proceedings of an advisory group meeting on application of isotope techniques in arid zones hydrology. IAEA, Vienna, pp 133–151

Salem O, Visser JH, Dray M, Gonfiantini R (1980) Groundwater flowpattern in the West Libyan Arab Jamahiriya. In: IAEA (ed) Arid-zone hydrology: investigations with isotope techniques: proceedings of an advisory group meeting on application of isotope techniques in arid zones hydrology. IAEA, Vienna, pp 165–180

Salvamoser J (1981) Vergleich der Alterbestimmung von Grundwasser mit Hilfe des Tritium und Kripton-85 gehalts. Naturwissenschaften 68:(6)328–329

Samoilov OYa (1957) Structure of aqueous electrolyte solutions and ion hydration. USSR Acad Sci Pub. House, Moscow

Sandimirova GP, Plyusnin GS, Pampura VD (1978) Strontium isotopic composition of thermal waters in water-bearing rocks of the Pauzhetskaya volcanic-tectonic depression. In: 7th Symp Stab Isot Geochim GEOHI, Moscow, pp 110–112

Sapozhnikov Yu A, Aliev ZA, Kalmikov SN (2006) Environmental radioactivity. Binom, Moscow

Sarmiento JL, Feely HW (1976) The relationship between vertical eddy diffusion and buoyancy gradient in the deep sea. Earth Planet Sci Lett 32:357–370

Savin SM (1977) The history of the Earth's surface temperature during the past 100 million years. Ann Rev Earth Planet Sci 5:319–355

Savin SM, Epstein S (1970a) The oxygen and hydrogen isotope geochemistry of clay minerals. Geochim Cosmochim Acta 34:25–42

Savin SM, Epstein S (1970b) The oxygen and hydrogen isotope geochemistry of ocean sediments and shales. Geochim Cosmochim Acta 34:43–63

Savin SM, Epstein S (1970c) The oxygen isotopic composition of coarse grained sedimentary rocks and minerals. Geochim Cosmochim Acta 34:323–329

Schell WR (1970) Investigation and comparison of radiogenic argon, tritium and C-14 in atmospheric reservoir. In: Radiocarbon variations and absolute chronology, XII nobel symposium, Wiley, New York, pp 447–466

Schell WR, Sauzay G (1970) Global sampling and analysis of tritium and stable isotopes. In: Report to panel on procedures for establishing limits for radionuclides in the sea. IAEA, Vienna, pp 1970

Schell WR, Sauzay G, Payne B (1970) Tritium injections and concentration distribution in the atmosphere. J Geophys Res 75:2251–2266

Schell WR, Fairhall DW, Harp GD (1967) An analytic model of carbon-14 distribution in the atmosphere. In: IAEA (ed) Radioactive dating and low-level counting. IAEA, Vienna, pp 79–92

Scheppard SMF, Epstein S (1970) D/H and 18 O/^{16}O Rtios of minerals of possible mantle of lower crustal origin. Earth Planet Sci Lett 9:232–239

Schidlowski M (1988) A 3.8 million-year isotopic record of life from carbon in sedimentory rocks. Nature 333:313–318

Schidlowski M, Eichmann R, Lunge CR (1975) Precambrian sedimentary carbonates: carbon and oxygen isotope geochemistry and implications for the terrestrial oxygen budget. Precambrian Res 2:1–60

Schidlowski M, Appel PWU, Eichmann R et al (1979) Carbon isotope geochemistry of the 3.7Â·10^9 year-old Isua sediments, West Greenland: implications for the Archaean carbon and oxygen cycles. Geochim Cosmochim Acta 43:189–199

Schiegl WE, Vogel LC (1970) Deuterium content of organic matter. Earth Planet Sci Lett 7:307–313

Schlosser P, Shapiro SD, Stute M (2000) Tritium/^{3}He measurements in young groundwater: progress in applications to complex hydrological systems. In: Tracers and modelling in hydrogeolog: proceedings of international conference, IASH Liege, pp 481–486

Schlosser P, Stute M, Sonntag C, Münnich KO (1988) Tritium/^{3}He dating of shallow groundwater. Earth Planet Sci Lett 89:353–368

Schotterer U, Wildberger A, Siegenthaler U et al (1979) Isotope study in the Alpine karst region of Rawil, Switzerland. In: IAEA (ed) Isotope hydrology, 1978: proceedings of an international symposium on isotope hydrology, vol 1. IAEA, Vienna, pp 351–366.

Schotterer U, Oldfield F, Froehlich K (1996) Global network for isotopes in precipitation. Peges, Bern

Schmidt U (1974) Molecular hydrogen in the atmosphere. Tellus 26:78–90

Schmidt OYu (1957) Four lectures on the earth origin theory. Izd AN SSSR, Moscow

Scholz TG, Ehhalt DH, Heidt LE, Martell EA (1970) Water vapour, molecular hydrogen, methane and tritium concentrations near the stratopause. J Geophys Res 75:3049–3054

Scott ERD, Wasson JT (1976) Chemical classification of iron meteorites–VIII: I, IIE, IIIF, and 97 other irons. Geochim Cosmochim Acta 40:103–115

Sehgal BR, Remport HH (1971) Tritium production in fast reactors, containing B_4C. Trans Am Nucl Soc 14:779–780

Seletsky Yu B (1977) Deuterium and oxygen-18 in study of active exchange groundwaters. Izd VIEMS, Moscow

Seletsky Yu B, Nechaev VI, Polyakov VA (1979) Radiocarbon as an indicator of groundwater recharge and discharge location. In: Ferronsky VI (ed) Isotope studies of natural waters. Nauka, Moscow, pp 111–121

Seletsky YuB, Polyakov VA, Yakubovsky AV, Isaev NV (1973) Deuterium and oxygen-18 in Groundwaters. Nedra, Moscow

Seletsky YuB, Polyakov VA, Yakubovsky AV, Isaev NV (1974) Preliminary results of deuterium content in certain types of North Caucasus groundwaters. Annals VSEGINGEO 59:70–79

Sentürk F, Bursali S, Omay Y et al (1970) Isotope techniques applied to groundwater movement in the Kenya plain. In: IAEA (ed) Isotope hydrology: proceedings of a symposium on use of isotopes in hydrology. IAEA, Vienna, pp 153–161

Sergeev EM, Ilyinskaya GG, Rekshinskaya LG (1963) On the distribution of clay minerals for their engineering geological study. Vestnic MGU, Ser Geolvol 4, Mosikva, pp 3–9

Serdiukova AS, Kapitonov YuT (1969) Radon isotopes and their short-living products in the nature. Atomisdat, Moscow

Shackleton MJ, Opdyke ND (1973) Oxygen isotope and paleomagnetic stratigraphy of equatorial Pacific core v. 28—v. 238: oxygen isotope temperatures and ice volumes on a 10^9 years and 10^6 year scale. Quatern Res 3:39–55

Shannou LV, Cherry RD, Orren MJ (1970) Polonium-210 and lead-210 in the marine environment cycles. Geochim Cosmochim Acta 34:701–711

Sheppard SMF, Taylor HP (1974) Hydrogen and oxygen isotope evidence for the origin of waters in the Boulder batholith and the Butte ore deposits, Montana. Econ Geol 69:926–946

Sheppard SMF, Epstein S (1970) D/H and $^{18}O/^{16}O$ ratios of minerals of possible mantle or lower crustal origin. Earth Planet Sci Lett 9:232–239

Sheppard SMF, Nielsen RL, Taylor HP (1969) Oxygen and hydrogen isotope ratios in minerals from porphyry copper deposits. Econ Geol 64:755–777

Sheppard SMF, Nielsen RL, Taylor HP (1971) Oxygen and hydrogen isotope ratios of clay minerals from porphyry copper deposits. Econ Geol 66:515–542

Shiro Y, Sakai N (1972) Calculation of the reduced partition function ratios of α, β-quartz and calcite. Bull Chem Soc Jap 45:2355–2359

Shukolyukov YuA (1980) Use of isotope methods in solution of the hydrothermal ore formation problem. Geokhimiya 12:1763–1779

Shukolyukov YuA (1970) Uranium nuclear fission in nature. Atomizdat, Moscow

Shukolyukov YuA, Komarov AN (1966) Possibilities of paleothermometry by uranium fission tracks. Izv AN SSSR, Ser Geol 9:137–141

Silverman S (1951) The isotope geology of oxygen. Geochim Cosmochim Acta 2:26–42

Singer SF (1958) The primary cosmic radiation and its time variations. Prog Cosm Ray Phys 4:205–335

Smirnov SI (1971) Origin of groundwater salinity in sedimentary basins. Nedra, Moscow

Smith DB, Downing RA, Monkhouse RA et al (1976) The age of groundwater in the Chalk of the London Basin. Water Resour Res 12:392–404

Smith RM (1966) Computations of the fluxes of tritium and water vapour over North America for July 1962. Ann Progr Rep Contract AT30-1, IAEA, Vienna

Sobotovich EV, Bondarenko GN, Vetshtein VE et al (1977) Isotope and geochemical estimates of a degree of surface and groundwater interconnection. Naukova Dumka, Kiev

Sofer Z (1978) Isotopic composition of hydratation water in gypsum. Geochim Cosmochim Acta 42:1141–1149

Sofer Z, Gat JR (1972) Activities and concentrations of oxygen-18 in concentrated aqueous salt solutions: analytical and geophysical implications. Earth Planet Sci Lett 15:232–238

Sofer Z, Gat JR (1975) Activities and concentration of oxygen-18 in concentrated aqueous salt solutions: analytical and geophysical implications. Earth Planet Sci Lett 26:179–186

Sokolov IYu (1974) Tables and nomograms for calculation of natural waters chemical analyses results. Nedra, Moscow

Sokolovsky LG, Polyakov VA. (2002) Isotopic and chemical content of snow and ice cover in the regions with technogenic contamination of atmosphere. In: Demetieva IA (ed) Geological studies and land protection: a review. Geoinformcenter, Moscow

Sokolovsky LG, Polyakov VA, Golubkova EV (2007a) Light isotopes of waters of the Asdov-Kuban artesian basin: conditions of formation and balneological significance. Prospect Prot Min Resour 5:44–47

Sokolovsky LG, Polyakov VA, Sokolova AV et al (2007b) Isotope-hydrochemical studies of ground and surface water in the West Siberian artesian basin and Ural complicated hydrogeological folded zone. Prospect Prot Min Resour 7:65–71

Solyankin EV (1963) On the Black Sea water balance. Okeanologiya 3:125–130

Somayajulu BLK, Craig H (1976) Particulate and soluble ^{210}Pb activities in the deep-sea. Earth Planet Sci Lett 32:269–276

Sonntag C, Klitzsch E, El-Shazly EM et al (1978) Paläoklimatische information in Isotopengehalt C-14 datierter Sahara wasser: Kontinentaleffekt in D und O-18. Geol Rdsch 67:413–324

Sonntag C, Klitzsch E, Löhnert EP et al (1979) Paleoclimatic information from deuterium and oxygen-18 in carbon-14-dated North Saharian groundwater. In: IAEA (ed) Isotope hydrology, 1978: proceedings of an international symposium on isotope hydrology, vol 2. IAEA, Vienna, pp 569–580

Sonntag C, Thoma G, Münnich KO et al (1980) Environmental isotopes in North Africa groundwaters and the Dahna sand dune study, Saudi Arabia. In: IAEA (ed) Arid-zone hydrology: investigations with isotope techniques: proceedings of an advisory group meeting on application of isotope techniques in arid zones hydrology. IAEA, Vienna, pp 77–84

Sorokhtin OG (2002) Green house effect: myth and reality. Inf Analit Vest, Rus Center, vol 1, Moscow, pp 27–28

Soyfer VN, Brezgunov VS, Vlasova LS (1967) Role of hydrogen stable isotopes in study of geological processes. Geokhimiya 5:599–606

Soyfer VN, Brezgunov VS, Verbolov VI et al (1970) Isotope application in the Baykal Lake water exchange study. Trudi Limn Inst So AN SSSR 14:146–153

Spencer JH (1956) The origin of the solar system. In: Ahrens LH, Urey HC, Rankama K, Runcorn SK, Press F (eds) Physics and chemistry of the Earth, vol 1. Pergamon Press, pp 1–16

Spiridonov AI, Tyminsky VG (1971) On $^{234}U/^{238}U$ ratio variation in groundwaters. Izv AN SSSR, Ser Phys Zem 3:91–93

Srdoč D, Siepčevič A, Obelič B et al (1980) Isotope investigations as a tool for regional hydrogeological studies in the Libyan Arab Jamahiria. In: IAEA (ed) Arid-zone hydrology: investigations with isotope techniques: proceedings of an advisory group meeting on application of isotope techniques in arid zones hydrology. IAEA, Vienna, pp 153–164

Stankevich EF (1968) On the hypothesis of underground evaporation. Sov Geol 5:90–96

Starik IE (1961) Nuclear geochronology. Izd AN SSSR, Moscow-Leningrad

Starik IE, Kolyadin LB (1957) On the conditions of uranium existance in oceanic water. Geokhimiya 3:204–213

Starik IE, Kuznetsov YuV, Nikolaev DS et al (1959) Radioactive element distribution in the Black Sea sediments. Dokl AH SSSR 129:1142–1145

Starik IE, Lazarev KF, Nikolaev DS et al (1959) Thorium isotope concentration in the Black Sea waters. Dokl AH SSSR 129:919–921

Starik IE, Melikova OS (1957) Emanation ability of minerals. Trudy Radievogo Inst 5:184–202

Stahl W, Aust H, Daounas A (1974) Origin of artesian and thermal waters determined by oxygen, hydrogen and carbon isotope analyses of water samples from the Sperkhios Valley, Greece. In: Isotope techniques in groundwater hydrology: proceedings of a symposium, IAEA, Vienna, 11–15 March 1974, vol 1, pp 317–337

Stenhause MJ (1979) Further application of bomb ^{14}C as a biological tracer. In: Berger R, Suess HE (eds) Radiocarbon dating. University of California Press, Berkely, pp 342–354

Stenhause MJ, Baxter MS (1979) The uptake of bomb ^{14}C in humans. In: Berger R, Suess HE (eds) Radiocarbon dating. University of California Press, Berkely, pp 324–341

Sternberg RS, Damon PE (1979) Sensitivity of radiocarbon fluctuations and inventory to geomagnetic and reservoir parameters. In: Berger R, Suess HE (eds) Radiocarbon dating. University of California Press, Berkely, p 691

Stewart GL (1965) Experinces using tritium in scientific hydrology. In: Radiocarbon and tritium dating, Proc 6th Intern Conf, USAEC, Washington pp 645–658

Stewart MK, Hulston JR (1975) Stable isotope ratios of volcanic steam from White Island, New Zealand. Bull Vulcanol 39:28–46

Stiel G, Haendel D, Runge A et al (1979) Isotopenverhältnisse und hydrogeologischen Praxis sowie in der Umwelt. Zeits Ang Geol 25:9–14

Stuiver M (1980) ^{14}C distribution in the Atlantic Ocean. J Geophys Res 85:2711–2718

Stuiver M (1965) Carbon-14 content of 18th and 19th century wood: variations correlated with sunshot activity. Science 149:533–535

Stuiver M, Suess HE (1966) On the relations between radiocarbon dates and true sample ages. Radiocarbon 8:534–540

Stuiver M, Quay PD (1981) Atmospheric ^{14}C change resulting from fossil fuel CO_2 release and cosmic ray flux variability, Earth Planet. Sci Lett 53:349–362

Styro BI (1968) Self-Purification of the Atmosphere from Radioactive Contamination. Gidrometeoizdat, Leningrad

Styro BI, Shpirkauskayte IK, Kuptsov VM (1970) ^{238}U, ^{232}Th and ^{239}Pu altitude distribution in atmospheric precipitation. At Energ 29:135–136

Suess HE (1969) Tritium geophysics as an international reswarch project. Science 163:1405–1410

Suess HE (1970) The three causes of the secular C-14 fluctuations, their amplitudes and time constants. In: Radiocarbon variations and absolute chronology, XII nobel symposium, Wiley, New York pp 595–605

Suess HE (1979) A calibration table for conventional radiocarbon dates. In: Berger R, Suess HE (eds) Radiocarbon dating. University of California Press, Berkely, pp 777–784

Sulerzhitsky LD, Forova VS (1966) Radiocarbon in woods from the modern volcanous areas. Dokl AN UzbSSR 6:1421–1423

Sultanov BI (1961) Deep condensed waters of gas-condensates and their formation conditions. Dokl AH AzSSR 17:1165–1166

Sultankhodzhaev AN, Tyminsky GV, Taneev RN (1970) Non-equilibrium uranium in groundwaters of the Tashkent artesian basin. Uzb Geol J 3:75–77

Suzuoki T, Epstein S (1976) Hydrogen isotope fractionations between OH-bearing minerals and water. Geochim Cosmochim Acta 40:1229–1240

Syromyatnikov NG (1961) Uranium, thorium and radium isotope migration and interpretation of the radioactive anomalies. Izd ANKazSSR, Alma-Ata

Tamers MA (1960) C-14 dating with liquid scintillation techniques. Science 132:668–669

Tamers MA (1965) Routine carbon-14 dating using liquid scintillation techniques. In: Radiocarbon and tritium dating. Proceedings of the 6th international conference, USAEC, Pulman, pp 53–59

Tamers MA (1967) Radiocarbon ages of groundwater in arid and unconfined aquifer. In: Stout GE (ed) Isotope techniques in the hydrological cycle. AGU Monograph 11, AGU, Washington, pp 143–152

Tamers MA (1979) Radiocarbon transmutation mechanism for spontaneous somatic cecular mutations. In: Berger R, Suess HE (eds) Radiocarbon dating. University of California Press, Berkely, pp 335–364

Tamers MA, Scharpenseel HW (1970) Sequential sampling of radiocarbon in groundwater. In: IAEA (ed) Isotope hydrology: proceedings of a symposium on use of isotopes in hydrology. IAEA, Vienna, pp 241–256

Tatevsky VM (1951) The calculation of equilibrated reactions for isotopic exchange. J Phys Chim 58:261–273

Taube H (1954) Use of oxygen isotope effects in the study of hydration of ions. J Phys Chem 58:523–530

Tauber H (1979) ^{14}C activity of arctic marine mammlas. In: Berger R, Suess HE (eds) Radiocarbon dating. University of California Press, Berkely, pp 447–452

Taylor HP (1974) The application of oxygen and hydrogen isotope studies to problem of hydrothermal alteration and ore deposition. Econ Geol 69:213–298

Taylor CR (1968) A comparison of tritium and strontium-90 in fallout in the southern hemisphere. Tellus 20:559–576

Taylor JR, Pefers FE (1972) Tritium transport in LMFBR's. Trans Am Nucl Soc 15:431–432

Taylor HP (1978) Oxygen and hydrogen isotope studies of plutonic granitic rocks. Earth Planet Sci Lett 38:177–210

Taylor HP, Silver LT (1978) Oxygen isotope relationships in plutonic igneous rocks of the Peninsular ranges batholith, southern Baja, California. Geol Surv Open-File Rep 701:423–426

Taylor HP, Duke MB, Silver LT et al (1965) Oxygen isotope studies in stone meteorites. Geochim Cosmochim Acta 29:489–512

Tays RV, Naydin DP (1973) Paleothermometry and oxygen isotopic composition of organogenic carbonates. Nauka, Moscow

Thatcher LL, Payne BR (1965) The distribution of tritium in precipitation over continents and its significance to groundwater dating. In: Radiocarbon and tritium dating. Proceedings of the 6th international conference, USAEC, Washington, 604–629

Theodorsson P (1967) Natural tritium in groundwater studies. In: Isotopes in hydrology: proceedings of the symposium on isotopes in hydrology held by the international atomic energy agency in cooperation with the international union of geodesy and geophysics in Vienna, IAEA, Vienna, 14–18 Nov 1966, pp 371–380

Thilo L, Münnich KO (1970) Reliability of carbon-14 dating of groundwater: effect of carbon exchange. IAEA (ed) Isotope hydrology: proceedings of a symposium on use of isotopes in hydrology. IAEA, Vienna, pp 259–269

Thod HG, Monster D (1964) Distribution of sulphur isotopes in ancient seas based on evaporates study in sediment deposites. Chemistry of the Earth's crust, vol 2. Nauka, Moscow, pp 589–599

Thomson J, Turekian KK (1976) ^{210}Po and ^{210}Pb distribution in oceaqn water profiles from the Eastern South Pacific. Earth Planet Sci Lett 32:297–303

Thurber DL (1962) Anomalous $^{234}U/^{238}U$ in nature. J Geophys Res 67:4518–4520

Thurber DL (1963) Natural variations in the ratio $^{234}U/^{238}U$. In: IAEA (ed) Radioactive dating: proceedings of the symposium on radioactive dating. IAEA,Vienna, pp 113–120

Thurber DL (1965) The concentration of some natural radioelements in the waters of the Great Basin. Bull Volcanol 28:195–201

Titaeva NA (1966) On the possibilities of orogenic sediments absolute age determination by ionium method. Geokhimiya 10:1183–1192

Titaeva NA, Filonov VA, Ovchenkov VYa et al (1973) Uranium and thorium isotopes behaviour in crystalline rocks-surface water system for cold humid climate conditions. Geokhimiya 10:1522–1528

Tkachuk VG, Vetshtein VE, Malyuk GA, Altshuler PG (1975) Hydrogen and oxygen isotopes of the Pripyat depression brines and possibilities of their use in oil and gas exploration. Geokhimiya 7:999–1006

Tokarev AV, Shcherbakov AV (1956) Radiohydrogeology. Gosgeoltekhizdat, Moscow

Tokarev I, Zubkov AA, Rumynin VG et al (2005) Origin of high $^{234}U/ ^{238}U$ ratio in post-permafrost aquifers. In: Merkel BJ, Hasche-Berger A (eds) Uranium in the environment, mining impact and consequences, Springer, New York, pp 854–863

Tolstikhin IN, Kamensky IL (1969) Determination of of groundwater age by the T–^{3}He method. Geochem Int 6:810–811

Trifonov DN, Krivomazov AN, Lisnevsky YuI (1974) A story of periodicity and radioactivity (comments on chronology of the most significant events). Atomizdat, Moscow

Tsunogai Sh, Nozaki Y (1971) Lead-210 and plutonium-210 in the surface water of the Pacific. Geochim J 5:165–173

Turekian KK, Clark SP Jr (1969) Ingomogeneous accumulation of the Earth from the primitive solar nebula. Earth Planet Sci Lett 6:346–348

Tyminsky BG, Sultankhodzhaev AN, Rozanov IM (1966) Paleohydrological estimations for the waters of the Tashkent artesian basin, Uzbekistan. Geol J 3:64–68

Urey HC (1947) Thermodynamic properties of isotopic substance. J Chem Soc 6:562–581

Urey HC (1957) Boundary conditions for theories of the origin of the solar system. Phys Chem Earth 2:46–76

Urey HC (1959) The planets, 2nd ed. Yale University Press, New Haven

Urey HC, Lowenstam HA, Epstein S et al (1951) Measurement of paleotemperatures and temperatures of the Upper Cretaceous England, Denmark, and South-Eastern United States. Bull Geol Soc Amer 62:399–416

Ustinov VI, Strizhov VA, Grinenko VA, Zairi NM (1978) Informativity of isotopic composition of silica in the Earth objects. In: 8th Vses Symp Stab Isotop Geochim GEOHI, 108–110

Vakulovsky SM, Vorontsov AI, Katrich IYu et al (1978) Tritium in atmospheric precipitation, rivers, and surrounding the USSR seas. At Energ 44:432–435

Van der Straaten CM, Mook WG (1983) Stable isotopic composition of precipitation and climatic variability. In: Paleoclimates and paleowaters. Proceedings of an advisory group meeting, IAEA, Vienna, p 53

Van Hook WA (1968) Condensed phase isotope effects. Isotopenpraxis 5:161–169

Varshavsky YaM, Veisberg SE (1955) On the equilibrium distribution of deuterium at the isotopic exchange reaction of hydrogen. Dokl AN SSSR 100:97–100

Vasil'chuk YuK, Kotlyakov VM (2000) Principles of isotopic geocriology and glaciology. Moscow University Press, Moscow, p. 616

Veizer J, Hoefs J (1976) The nature of $^{18}O/^{16}O$ and $^{13}C/^{12}C$ secular trends in sedimentary rock. Geochim Cosmochim Acta 40:1387–1395

Verhagen BTh, Smith PE, McGregore I et al (1979) Tritium profiles in Kalachari sands as a measure of rain-water recharge. In: Isotope hydrology 1978: proceedings of an international symposium on isotope hydrology, vol 2. IAEA, Vienna, pp 733–749

Verhoogen J, Turner EJ, Weiss LS et al (1970) The Earth: an introduction to physical geology, vol 2. Holt, Rinehart and Winston, New York

Vernadsky VI, Vinigradov AP, Teys RV (1941) Determination of isotopic composition in water from metamorphic rocks and minerals. Dokl AN SSSR 31:574–577

Vetshtein VE (1967) ^{18}O content of some genetically different natural waters. In: Problemy Gidrogeologii I Inzhenernogo Gruntovedeniya. Naukova Dumka, Kiev, pp 11–21

Vetshtein VE, Baskov EA, Klimov GI et al (1971) New data on oxygen-18 content in volcanic thermal and mineral waters from Kurili Islands, Kamchatka, and Baykal region. Sov Geol 9:98–108

Vetshtein VE, Gutsalo LK, Malyuk GA, Miroshnichenko AG (1973) On the origin of formation waters in the Dnepr-Donetsk gas and oil-bearing sedimentary basin by oxygen and hydrogen isotopic composition. Geokhimiya 3:327–338

Vetshtein VE, Malyuk GA, Lapshin FV (1972) Oxygen and hydrogen isotopic composition of mineral waters in Ukrainian Carpathy as their genesis criterion. Dop AN URSR 12:1062–1066

Vetshtein VE, Malyuk GA, Rusanov VP (1974) ^{18}O distribution in waters of the Central Arctic basin. Oceanologiya 14:642–648

Vinograd IJ, Friedman I (1972) Deuterium as a tracer of regional groundwater flow, Southern Great basin, Nevada and California. Bull Geol Soc Amer 83:3691–3706

Vinogradov AP (ed) (1963) The main features of uranium geochemistry. Izd AN SSSR, Moscow

Vinogradov AP, Dontsova EI, Chupakhin MC (1960) Isotope ratios of oxygen in meteorites and igneous rocks. Geochim Cosmochim Acta 18:278–293

Vinogradov AP (1967a) Introduction to ocean geochemistry. Nauka, Moscow

Vinogradov AP (1967b) Origin of the oceans. Izv AN SSSR, Ser Geol 4:3–9

Vinogradov AP (1971) High temperature protoplanetary processes. Geokhimiya 11:1238–1296

Vinogradov AP (1975) The Moon differentiation. In: Vinogradov AP (ed) Cosmochemistry of the Moon and planets. Nauka, Moscow, pp 5–28

Vinogradov AP, Kropotov OI, Vdovikin GP, Grinenko VA (1967) Isotopic composition of different phases in carbonaceous meteorites. Geokhimiya 3:267–273

Vinogradov VI (1980) Role of the sedementation cycle in geochemistry of sulphur isotopes. Nauka, Moscow

Vinogradov VI, Ivanov IB, Litsarev MA et al (1969) On the age of Earth's oxygenous atmosphere. Dokl AN SSSR 188:1144–1147

Vlasova LS, Ageev LS, Esikov AD et al (1979) The origin of hydrothermal solutions by oxygen, carbon, and hydrogen isotopes data. In: Ferronsky VI (ed) Isotope study of natural waters. Nauka, Moscow, pp 93–110

Vlasova LS, Brezgunov VS (1978) The distribution of hydrogen and oxygen isotopic composition in natural brines by model calculations. In: Ferronsky VI (ed) Isotope study of natural waters. Nauka, Moscow, pp 119–139

Vlasova LS, Ferronsky VI (2008) Water transport over West Europe and its correlation climate variations based on data of isotopic composition in precipitation. Water Resour 35:525–545

Vlasova LS, Lebedev NA, Romanov VV (1978) The use of natural tritium for water exchange study in Aragats massive friable rocks. In: Ferronsky VI (ed) Isotopy of natural waters. Nauka, Moscow, pp 180–192

Vogel JC (1967) Investigation of groundwater flow with radiocarbon. In: Isotopes in hydrology: proceedings of the symposium on isotopes in hydrology held by the international atomic energy agency in cooperation with the international union of geodesy and geophysics in Vienna, IAEA, Vienna, 14–18 Nov 1966, pp 355–369

Vogel JC (1970) Carbon-14 dating of groundwater. In: IAEA (ed) Isotope hydrology: proceedings of a symposium on use of isotopes in hydrology. IAEA, Vienna, pp 225–237

Vogel JC, Ehhalt D (1963) The use of the carbon isotopes in groundwater studies. In: IAEA (ed) Radioisotopes in hydrology: proceedings of a symposium. IAEA, Vienna, pp 383–395

Voytkevich GV (1961) Radiogeology problems. Gosgeoltekhizdat, Moscow

Voytov GI, Gureev EV, Erokhin BK et al (1976) Hydrogen isotopic composition of thermal waters from South Belozersk iron ore deposits. Dokl AN SSSR 231:1226–1229

Wallick WR (1976) Isotopic and chemical consideration in radiocarbon dating of groundwater within the semi-arid Tucson basin, Arizona. In: Interpretation of environmental isotope and hydrochemical data in groundwater hydrology. Proceedings of an advisory group meeting, IAEA, Vienna, pp 195–211

Wasserburg CJ, Nurner G, Tera F et al (1972) Comparison of Rb–Sr, K–Ar and U–Th–Pb ages: lunar chronology and evolution. In: Watkins C (ed) Lunar science III. Lunar Science Institute, Moffett Field, pp 788–790

Webber W (1967) The spectrum and charge composition of the primary cosmic radiation. Encycl Phys 46:181

Weber JN (1965) Extension of the carbonate paleotermometer to premesozoic rocks. In: Tongiorgi E (ed) Stable isotopes in oceanographic studies and paleotemperatures: Spoleto. Consiglio Nazionale delle Ricerche, Piazzale Aldo Moro, pp 227–309

Weiss W (1979) Tritium and helium-3 studies in Lake Constance. In: Isotopes in lake studies. Proceedings of an advisory group meeting, IAEA, Vienna, pp 227–231

Weiss W, Roether W (1975) Der Tritium abfluss des Rheins 1961–1973. Dt. Gewasser Kd Mitt Bd 19:1–10

Weiss W, Bullacher J, Roether W (1979) Evidence of pulsed discharge of tritium from nuclear energy installations in Central European precipitation. In: IAEA (ed) Behaviour of tritium in the environment: proceedings of the international symposium on the behaviour of tritium in the environment. IAEA, Vienna, pp 17–30

Wendland WM, Donley DL (1971) Radiocarbon calendar age relationship. Earth Plane Sci Lett 11:135–139

Wendt I, Stahl W, Geyh M, Fauth F (1967) Model experiments for ^{14}C water-age determinations. In: Isotopes in hydrology: proceedings of the symposium on isotopes in hydrology held by the international atomic energy agency in cooperation with the international union of geodesy and geophysics in Vienna. IAEA, Vienna, pp 321–336 (14–18 Nov 1966)

White DE (1974) Diverse origins of hydrothermal ore fluides. Econ Geol 69:954–973

White DE (1965) Saline waters of sedimentary rocks. In: Fluids in subsurface environments. Am Assoc Petrol Geol 4:343–366

White DE, Craig H, Begemann F (1963a) Sammary of the geology and isotope geochemistry of Steamboat Springs, Nevada. In: Tongiorgi E (ed) Nuclear geology on geothermal areas: Spoleto. Consiglio Nazionale delle Ricerche, Piazzale Aldo Moro, pp 9–16

White DE, Anderson ET, Grubls DK (1963b) Geothermal brine well mile-deep drill hole may tap ore-bearing magmatic water and rocks undergoing metamorphism. Science 139:919–992

Wigley TML (1976) Effect of mineral precipitation of isotopic composition and ^{14}C dating of groundwater. Nature 263:219–221

Wigley TML, Plummer LN, Pearson FJ (1978) Mass transfer and carbon isotope evolution in natural water systems. Geochim Cosmochim Acta 42:1117–1139

Wilson AT, Grinsted MJ (1975) Paleotemperatures from tree rings and D/H ratio of cellulose as biochemical thermometer. Nature 257:287–288

Wilson AT, Grinsted MJ (1977) The D/H ratio of cellulose as biochemical thermometer (a comment on "Climatic implication of D/H ratio of hydrogen in C–H groups in tree cellulose" by S. Epstein and C.J. Yapp). Earth planet Sci Lett 36:246–248

Winograd IJ, Farlekas GM (1974) Problems in ^{14}C dating of water from aquifers of deltic origin: an example from the New Jersey costal plain. In: Isotope techniques in groundwater hydrology: proceedings of a symposium, IAEA, Vienna, 11–15 March 1974, vol 2, pp 69–91

Winograd IJ, Riggs AC, Coplen TB (1998) The relative contributions of summer and cold-season precipitation to groundwater recharge, Spring Mountaine, Nevada, USA. J Hydrol 6:77–93

Wong WW, Sackett WM (1978) Fractionation of stable carbon isotopes be by marine phytoplankton. Geochim Cosmochim Acta 42:1809–1815

Wood YA (1963) On the origin of chondrules and chondrites', Icarus, 2, 152–180

Wood JA (1974) Origin of the Earth's Moon, preprint series no 139. Center for Astrophysics, Cambridge (Mass.)

Yakubovsky AV, Isaev NV, Polyakov VA, Tereshchenko VA (1978) On the formation of low-mineralized groundwater with high level of deuterium and oxygen-18 content. In: 7th Vses Symp Stab Isotop Geokhim, GEOHI, Moscow, pp 202–203

Yang A, Fairhall AW (1972) Variations of natural radiocarbon during the last 2,000 years and geophysical mechanism for producing them. In: Proceedings of the 8th international conference on radiocarbon dating, Lower Hutt, Wellington, pp A44–A54

Yapp CJ, Epstein S (1977) Climatic implication of D/H ratio of meteoric waters over North America (9,500–22,000 B.P.) as inferred from ancient wood cellulose C–H hydrogen. Earth Planet Sci Lett 34:333–350

Yeh HW, Epstein S (1980) D/H ratios and late-stage dehydration of shales during burial. Geochim Cosmochim Acta 44:341–352

Yeh HW, Epstein S (1978) Hydrogen isotope exchange between clay minerals and sea water. Geochim Cosmochim Acta 42:140–143

Yeh HW, Savin SM (1976) The extent of oxygen isotope exchange between clay minerals and the sea water. Geochim Cosmochim Acta 40:743–748

Yezhova MP, Polyakov VA, Tkachenko AE et al (1996) Palaeowaters of North Estonia and their influence on changes in the resourses and the quality of fresh groundwaters of large costal water supplies. Geologiya 19:37–40

Young H (1965) Chemical content and radioactivity of the atmosphere. Mir, Moscow (trans: from English)

Young JA, Wogman NA, Thomas CW, Perkins R (1970) Short lived cosmic ray produced radionuclides as tracers of atmospheric processes. In: Gould RF (ed) Radionuclides in the environment. Advances in chemistry, series no 93. American Chemical Society, Washington, pp 506–521

Yurtsever Y, Gat JR (1981) Stable isotopes in atmospheric waters. In: Gat JR, Gonfiantini R (eds) Stable isotope hydrology. IAEA, Vienna, pp 103–142

Yurtsever Y, Payne BR (1979) Application of environmental isotopes to groundwater investigations in Qatar. In: IAEA (ed) Isotope hydrology 1978: proceedings of an international symposium on isotope hydrology, vol 2. IAEA, Vienna, pp 465–490

Zavelsky FS (1968) One more clarification to radiocarbon method. Dokl AN SSSR 180:1189–1192

Zaitsev IK (1967) Hydrochemical and hydrothermal zoning of the artesian basins of the USSR in connection with underground evaporation hypothesis criticism. In: 5th Meeting of Siberia and Far East, Irkutsk-Tyumen, pp 39–40

Zlobina VL, Kovalevsky VS, Morkovkina IK et al (1980) On the use of helium and tritium mapping for groundwater recharge study. Water Res 1:166–170

Zimmerman U (1979) Determination by stable isotopes of underground inflow and outflow and evaporation of young artificial groundwater lakes. In: Isotopes in lake studies. Proceedings of an advisory group meeting, IAEA, Vienna, pp 87–94.

Zimmerman U, Ehhalt D, Münnich KO (1967) Soilwater movement and evapotranspiration: changes in the isotopic composition of the water. In: Isotopes in hydrology: proceedings of the symposium on isotopes in hydrology held by the international atomic energy agency in cooperation with the international union of geodesy and geophysics in Vienna, IAEA, Vienna, 14–18 Nov 1966, pp 567–584

Zuber A (1994) On calibration and validation of mathematical models for the interpretation of environmental tracer data in aquifer. In: Mathematical models and their application to isotope studies in groundwater hydrology, IAEA, Vienna, pp 11–41

Zuber A, Grabczak J, Kolonko M (1979) Enironmental and artificial tracers for investigating leakages into salt mines. In: IAEA (ed) Isotope hydrology: proceedings of an international symposium on isotope hydrology, vol 1. IAEA, Vienna, pp 45–62

Zuber A, Michalczyk Z, Maloszewsky P (2001) Great tritium ages explain the occurrence of good-quality groundwater in a phreatic aquifer of an urban area, Lublin, Poland. Hydrogeol J 9:451–460

Index

V.I. Ferronsky, V. A. Polyakov, *Isotopes of the Earth's Hydrosphere*,
DOI 10.1007/978-94-007-2856-1, © Springer Science+Business Media B.V. 2012